10th International Conference on Clustering Aspects of Nuclear Structure and Dynamics

(CLUSTER'12)

Journal of Physics: Conference Series Volume 436

Debrecen, Hungary
24-28 September 2012

ISBN: 978-1-62748-313-1
ISSN: 1742-6588

Printed from e-media with permission by:

Curran Associates, Inc.
57 Morehouse Lane
Red Hook, NY 12571

Some format issues inherent in the e-media version may also appear in this print version.

Printed by Curran Associates, Inc. (2013)

Institute of Physics
Dirac House, Temple Back
Bristol BS1 6BE UK

Phone: 44 1 17 929 7481
Fax: 44 1 17 920 0979

techtracking@iop.org

Additional copies of this publication are available from:

Curran Associates, Inc.
57 Morehouse Lane
Red Hook, NY 12571 USA
Phone: 845-758-0400
Fax: 845-758-2634
Email: curran@proceedings.com
Web: www.proceedings.com

Introduction

Alpha clustering

Cluster reactions

Few-body and *ab initio* approaches

Microscopic approach to cluster structure and reactions

Algebraic cluster models

Structure and reactions of exotic nuclei

Nuclear molecules

Elongated states, dacay and fission; superheavy elements

Clustering in nuclear matter

Nuclear reactions for astrophysics

High-energy and multifragmentation reactions

Hypernuclei and quark clusters

Application of cluster theory and clustered nuclei

Closing

List of participants

Journal of Physics: Conference Series **436** (2013) 011001 doi:10.1088/1742-6596/436/1/011001

10th International Conference on Clustering Aspects of Nuclear Structure and Dynamics (CLUSTER'12)

Preface

As Editors of this Proceedings volume and organizers as well as participants of the Conference, let us sum up a few facts about the Conference and let us add some subjective notes.

The conference was held at Köolcsey Centre, a prestigious site of scientific and business meetings and cultural events in the city of Debrecen, Hungary, 24–28 September 2012. The city was chosen as the home of the Institute of Nuclear Research of the Hungarian Academy of Sciences, the home institution of the organizers. The Institute was visited by some of the participants. The organizing team was complemented by some belonging to the University of Debrecen, which made it possible to hold the conference dinner in the marvellous covered 'ceremonial court' of the University. The participants were taken for an excursion and wine tasting to a wine cellar at Tokaj, centre of, historically, the most significant wine producing area in Hungary. By coincidence, the closing day coincided with the day of what is called the Researchers' Night in the European Union. That night, (or rather, that evening) every year there is a public lecture in the Institute of Nuclear Research for a general audience, mainly secondary-school pupils. The public lecture scheduled on this occasion was held by one of the participants of the conference who represented the world outside Hungary and yet was able to deliver a talk in Hungarian. He is Professor Kálmán Varga, Vanderbilt University. The title of his talk was Simulation of Nanosystems http://kutatokejszakaja.hu/2012/esemenynaptar/esemeny.php?id=112&menu_id=4".

There were 115 registered participants, representing 22 countries of Europe, Africa, Asia, Central, South and North America.

We had 44 plenary talks and 47 talks presented in parallel sessions. In the Programme list the talks are arranged following the session structure of the Conference. In this list the invited speakers are marked by asterisks. All the talks in the plenary sessions were invited talks, but we had some invited talks even in the parallel sessions. Written versions of 86 talks have been submitted. Quite a number of the speakers have modified the title of their talk in the written version. To keep the correspondence between the Programme list and the written papers, we have accordingly changed the titles in the Programme list as well. The papers are arranged according to their subjects, without regard to whether they were delivered in a plenary or in a parallel session. There was a talk classified wrongly; this is now classified correctly. In the Programme list and in the list of the papers submitted the names are those of the speakers.

You can read in the Opening Address that, by ruling of the organizing institution, no members of the Debrecen Institute of Nuclear Research were selected as invited speakers and no contributions were accepted from them. We doubted the wisdom of this rigour ourselves, and I think the criticism we received for it is justifiable.

The success of a conference depends primarily on the speakers. In retrospect, we can say that this conference was extremely successful, and that is owing to the great many wonderful talks delivered. This reflects very well on the present status of the field as well as on the work of the International Advisory Board, which proposed the list of excellent speakers. The Japanese dominance characteristic of all cluster conferences has only been complained about by some of the Japanese who came to Europe primarily to learn things that they could not learn at home.

We would like to express our gratitude to our sponsors:

- the Hungarian Academy of Sciences
- the Paks Nuclear Power Ltd
- HMP Logic Ltd
- International Workshop for Theoretical Physics(Budapest)

We were also supported indirectly by a JSPS–MTA bilateral cooperation project, which made it possible for five Japanese colleagues to participate. It just remains to be announced that the next, number 11 in this series of conferences, will be held in Naples in 2016.

Zs Dombrádi
G G Kiss
A T Kruppa
G Lévai
R G Lovas

Conference photograph.

Journal of Physics: Conference Series **436** (2013) 011002 doi:10.1088/1742-6596/436/1/011002

List of committees

International Advisory Board

Yasuhisa Abe, Osaka
Juha Äystö, Jyväskylä
Zoran Basrak, Zagreb
Daniel Baye, Brussels
R. Russell Betts, Chicago
Christian Beck, Strasbourg
Roman Čaplar, Zagreb
József Cseh, Debrecen
Massimo Di Toro, Catania
Martin Freer, Birmingham
Mamoru Fujiwara, Osaka
Walter Greiner, Frankfurt-am-Main
Florent Haas, Strasbourg
Muhsin N Harakeh, Groningen
Hisashi Horiuchi, Osaka
Björn Jonson, Gothenburg
Taka Kajino, Tokyo
Masayasu Kamimura, Wako
Rezső G. Lovas, Debrecen
Witold Nazarewicz, Oak Ridge
Yuri Ts. Oganessian, Dubna
Wolfram von Oertzen, Berlin
Nigel Orr, Caen
Katsuhiko Sato, Tokyo
Peter Schuck, Orsay

Local Organizing Committee

Chairman: Rezső G. Lovas
Secretary: András T. Kruppa

Margit Csatlós
József Cseh
Judit Darai
Zsolt Dombrádi
János Farkas
Zsolt Fülöp
György Gyürky
Zoltán Halász
István Hornyák
Attila Krasznahorkay
Géza Lévai
Zoltán Máté
Dorottya Sohler
János Tímár
Zoltán Trócsányi
Zsolt Vajta
Tamás Vertse

Host institutions

MTA Institute of Nuclear Research
University of Debrecen

Journal of Physics: Conference Series **436** (2013) 011002 doi:10.1088/1742-6596/436/1/011002

Sponsors

Hungarian Academy of Sciences
Paks Nuclear Power Ltd.
HMP Logic Ltd.
International Workshop for Theoretical Physics (Budapest)

Journal of Physics: Conference Series **436** (2013) 011003 doi:10.1088/1742-6596/436/1/011003

Conference programme[1]

MONDAY, 24th September, 2012

9.00	Registration	
10.00	R G Lovas	Opening address
Session 1		
chaired by	**R G Lovas**	**Introduction**
10.15	*H Horiuchi:	Recent developments in nuclear cluster theory
10.50	*M Freer:	Challenges to the field of nuclear clustering
11.25	Coffee break	
Session 2		
chaired by	**Y. Kanada-En'yo**	**Alpha clustering**
11.40	*Y Funaki:	Gas-like alpha cluster states and condensates in nuclei
12.05	*M Dufour:	Nonmicroscopic and microscopic descriptions of condensate states in the ^{12}C and ^{16}O nuclei
12.35	*M Itoh:	Nature of 10 MeV state in ^{12}C
13.00	Lunch break	
Session 3		
chaired by	**P Descouvemont**	**Cluster reactions**
15.00	*C Beck:	Clustering effects in reactions induced by light nuclei
15.35	*A Lépine-Szily:	Cluster structures observed in ^{40}Ca from $^{12}C+^{28}Si$ scattering
16.00	*D G Jenkins:	Electromagnetic transitions as a probe of clustering in nuclei
16.25	*C Wheldon:	Absolute partial decay branching-ratios in ^{16}O
16.50	Coffee break	
Session 4		
chaired by	**P Schuck**	**Few-body and *ab initio* approaches**
17.20	*J P Draayer:	Symmetry-adapted *ab initio* no-core shell model calculations for ^{12}C
17.55	*K Arai:	Tensor force manifestations in *ab initio* study of the $^2H(d,\gamma)^4He$, $^2H(d,p)^3H$ and $^2H(d,n)^3He$ reactions
18.25	*Th Neff:	Microscopic calculations for the structure and reactions of light nuclei
19.00	Welcome drink	

[1] Asterisks mark invited speakers.

Journal of Physics: Conference Series **436** (2013) 011003 doi:10.1088/1742-6596/436/1/011003

TUESDAY, 25th September, 2012

Session 5		
chaired by	**J P Draayer**	**Microscopic approach to cluster structure and reactions**
9.00	*K Katō:	Multi-cluster problems: resonances, scattering and condensed states
9.30	*M Ito:	Studies of light neutron-excess systems from bound states to continuum
9.55	*H Masui:	Cluster-shell competition in systems with a few alpha particles and valence neutrons
		Algebraic cluster models
10.20	*P O Hess:	The geometric interpretation of the semimicroscopic algebraic cluster model and the role of the Pauli principle
10.45	Coffee break	
Session 6		
chaired by	**Th Neff**	**Structure and reactions of exotic nuclei**
11.15	*Y Kanada-En'yo:	Cluster correlation in light nuclei
11.50	*P Descouvemont:	Core excitations in exotic nuclei
12.20	*M Kimura:	Clustering aspects in highly excited states and in neutron-rich nuclei
12.50	Lunch break	

THREE PARALLEL SECTIONS

Session 7A		
chaired by	**A Lépine-Szily**	**Alpha clustering**
15.00	*T Yamada:	Cluster structure and isoscalar monopole excitation in light nuclei
15.25	*T Kawabata:	Search for alpha inelastic condensed state in ^{24}Mg
15.50	*H Akimune:	Alpha cluster structure in ^{56}Ni
16.15	Y Fukuoka:	Large scale calculations for cluster structure of light nuclei with Skyrme interaction
16.35	Coffee break	
Session 8A		
chaired by	**M Kimura**	
17.05	A A Cowley:	Alpha-Cluster structure in the ground state of ^{40}Ca displayed in a (p,pα) knockout reaction
17.25	M Norrby:	Alpha-particle resonances and clusters in ^{32}S, ^{34}S, ^{36}Ar and ^{40}Ca
		Structure and reactions of exotic nuclei
17.45	G V Rogachev:	Clustering in $A = 10$ nuclei
18.05	F Kobayashi:	Two novel cluster states in ^{10}Be
18.25	T Suhara:	Cluster structures of excited states in ^{11}B
18.45	Dinner time	

Journal of Physics: Conference Series **436** (2013) 011003 doi:10.1088/1742-6596/436/1/011003

Session 7B		
chaired by	**M Ito**	**Cluster Reactions**
15.00	J Kasagi:	Anomalous behavior of the d+d reaction in liquid indium bombarded by low-energy deuterons: In-d subpico-size molecule?
15.20	E C Simpson:	Spatial correlations probed in direct two-nucleon removal reactions
15.40	H Shimoyama:	Di-neutron correlation in monopole pair transfer modes near the neutron-separation energy in Sn, Ni, Ca, Zr isotopes
16.00	L Prepolec:	$^{13}C+^{4}He$ resonant elastic scattering on thick gas target
16.20	I Lombardo:	Resonant elastic scattering of α particles on ^{9}Be and cluster states in ^{13}C
16.40	M S Hussein:	Complete fusion of weakly bound cluster-type nuclei near barrier energies
17.00	Coffee break	
Session 8B		
chaired by	**K Katō**	**Microscopic approach to cluster structure and reactions**
17.30	T Myo:	Role of the tensor force in light nuclei with the tensor-optimized shell model
17.50	J Dohet-Eraly:	Microscopic description of $\alpha + \alpha$ bremsstrahlung from a realistic nucleon-nucleon interaction
18.10	W Horiuchi:	Cluster structure in light nuclei studied with few-body models
18.30	N Kurihara:	Study of the isospin mixing in ^{6}Li
18.50	Dinner time	

Session 7C		
chaired by	**J Cseh**	**Few-body and *ab initio* approaches**
15.00	*S Aoyama:	Excited states of ^{4}He
		Algebraic cluster models
15.25	Yu M Tchuvil'sky:	Multicluster solutions of the Elliott SU(3) nucleonic Hamiltonian and alpha-cluster level density
15.45	T Yoshida:	Algebraic approach for α-cluster structure in ^{12}C and ^{16}O
16.05	A I Georgieva	Comparing the group-theoretical approaches to the description of dipole and quadrupole collectivity in nuclei
16.35	Coffee break	

Session 8C		
chaired by	**M Milin**	**Structure and reactions of exotic nuclei**
17.05	M Uroić:	States in ^{10}Be populated in $^{6,7}Li(^{7}Li,^{3,4}He)^{10}Be^{*}$ reactions at 30 and 52 MeV
17.25	E Pinilla:	Breakup of ^{11}Li in a three-cluster model
17.45	Y Kikuchi:	Mechanisms of Coulomb breakup reactions of ^{6}He and ^{11}Li
18.05	N Kobayashi:	Breakup reactions of neutron drip-line nuclei near $N = 20$, 28
18.25	Y L Ye:	^{8}He cluster structure studied by recoil proton tagged knockout reaction
18.45	Dinner time	

Journal of Physics: Conference Series **436** (2013) 011003 doi:10.1088/1742-6596/436/1/011003

WEDNESDAY, 26th September, 2012

Session 9		
chaired by	**Y Abe**	**Nuclear molecules**
9.00	*E Uegaki:	Mystery of resonances in heavier systems
9.30	*M Milin:	Cluster and molecular structures observed in reactions of light nuclei
10.00	*S Courtin:	Electromagnetic decay of molecular states
10.30	Coffee break	
Session 10		
chaired by	**M Dufour**	**Elongated states, decay and fission; superheavy elements**
11.00	*Y Taniguchi:	Cluster correlations in largely deformed states
11.30	*D N Poenaru:	Heavy-particle radioactivity
11.55	*Ch E Düllmann:	Search for superheavy elements at GSI
		Alpha clustering
12.25	*M Gai:	The structure of the Hoyle state via a measurement of the 'Hoyle Rotational Band' in ^{12}C
		Structure and reactions of exotic nuclei
12.50	*P Capel:	Mechanisms of direct reactions with halo nuclei
13.15	Lunch break	

THREE PARALLEL SECTIONS

Session 11A		
chaired by	**D Poenaru**	**Elongated states, decay and fission; superheavy elements**
15.00	*J Giovinazzo:	Two-proton radioactivity: 10 years of experimental progresses
15.25	*N Schunck	Microscopic description of induced fission
15.50	*N V Antonenko:	Cluster aspects of binary fission
16.15	*G Adamian:	Formation of strongly deformed states in entrance channels
16.40	Coffee break	
Session 12A		
chaired by	**T Vertse**	
17.10	*R Id Betan:	Complex-energy shell model description of alpha decay
17.35	A S Zubov:	Population and properties of superdeformed bands in $A \sim 150$ region
17.55	N Eren:	Charge distribution in disintegration of uranium isotopes
18.15	Repose	

Journal of Physics: Conference Series **436** (2013) 011003 doi:10.1088/1742-6596/436/1/011003

Session 11B		
chaired by	**M Freer**	**Nuclear molecules**
15.00	Y Iwata:	Power series expansion method for determining collective quantities
15.20	F R Xu:	Alpha-decay quantum-tunnelling calculations based on a folded Woods-Saxon potential
15.40	E Uegaki:	Wobbling excitation of triaxial nuclear molecule ^{28}Si–^{28}Si
16.00	E Vardaci:	Clustering effects in ^{48}Cr composite nucleus produced via the reaction $^{24}Mg+^{24}Mg$ at the excitation energy of 60 MeV
16.20	Coffee break	
Session 12B		
chaired by	**A Tumino**	**Clustering in nuclear matter**
16.50	G Röpke:	Quantum statistical calculation of cluster abundances in hot dense matter
17.10	*A Tohsaki:	A new step of microscopic cluster model
		Nuclear reactions for astrophysics
17.35	V Tokić:	Study of ^{24}Mg resonances relevant to carbon burning nucleosynthesis
17.55	L Lamia:	Electron screening effects in (p,α) reactions induced on boron isotopes studied via the Trojan Horse Method
18.15	Repose	

Session 11C		
chaired by	**Ad R Raduţa**	**High-energy and multifragmentation reactions**
15.00	*I Gnesi:	Collective and thermal signatures in π^4He interaction in the Δ excitation energy region
15.25	R Oğul:	Nuclear symmetry energy in charge and isotopic distributions
15.45	Coffee break	
Session 12C		
chaired by	**Yu M Tchuvil'sky**	
16.35	D Gruyer:	Pseudo-critical clusterization in nuclear multifragmentation
		Hypernuclei and quark clusters
16.55	M Isaka:	Structure of Be hyper isotopes
17.15	Repose	
19.30	Conference dinner	

THURSDAY, 27th September, 2012

Session 13		
chaired by	**G Röpke**	**Clustering in nuclear matter**
9.00	*P Schuck:	Alpha-particle condensation in nuclear systems: present status and perspectives
9.25	*Q N Usmani:	Nuclear matter and surface clustering
9.50	*V de la Mota:	Dynamical description of exotic structures at sub-nuclear densities
10.20	*A Ono:	Nuclear matter properties in fragmentation reactions
10.45	Coffee break	
Session 14		
chaired by	**Zs Fülöp**	**Nuclear reactions for astrophysics**
11.15	*S Kubono:	Role of clusters in nuclear astrophysics with Cluster Nucleosynthesis Diagram (CND)
11.50	*O S Kirsebom:	^{12}C and the triple-α reaction rate
12.15	*A Tumino:	Low-energy d+d fusion via the Trojan Horse Method
		High-energy and multifragmentation reactions
12.40	*Ad R Raduţa:	Nuclear (multi-)fragmentation
13.10	Lunch break	
14.30	Excursion & wine tasting	

FRIDAY, 28th September, 2012

Session 15		
chaired by	**D Baye**	**Hypernuclei and quark clusters**
9.00	*E Hiyama:	Four- and five-body cluster structure of hypernuclei
9.35	*H Tamura:	Gamma spectroscopy in hypernuclei
10.00	*K Wisniewski:	Clustering of strangeness in heavy-ion reactions
10.25	*Y Fujiwara:	Few-nucleon systems via non-Local baryon-baryon interaction
10.55	Coffee break	
Session 16		
chaired by	**F Beck**	**Application of cluster theory and clustered nuclei**
11.25	*K Varga:	Interaction of electromagnetic fields and atomic clusters
11.55	*A J T Jull:	Some interesting and exotic applications of carbon-14 dating by accelerator mass spectrometry
		Closing
12.25	*Y Suzuki:	Summary talk – from a theoretical point of view
12.50	*W N Catford:	Conference summary: an experimental perspective
13.15	The end	
15.15	(optional) Visit to the Institute of Nuclear Research	
18.00	Researchers' Night	

Opening address

R G Lovas[1] **on behalf of the organizers**

[1] Institute of Nuclear Research, Hungarian Academy of Sciences
Debrecen, PO Box 51, H–4001, Hungary

E-mail: rgl@atomki.hu

Ladies and Gentlemen,

hearty welcome to the 10th conference on nuclear clustering.

Looking at the audience, I see that nuclear cluster physics is as popular as before. I don't mean that 'cluster' is a vogue-word though it is at least among innovation-ridden Eurocrats. What I mean is that we are as many here as expected. In fact, we are many more at such a conference when it is held in Japan, but this is the usual number in Europe.

This conference could have taken place a year earlier, but, in the meantime, a series of workshops on the state of the art in nuclear cluster physics was started, and it was important to avoid interference with them. The first one was held in Strasbourg in 2008, and the second in Brussels in 2010. This obviously shows that our field is flourishing. That is especially conspicuous regarding the crises that have come round in recent years:

(1) the world economic recession spead from America since 2008;

(2) the earthquake and tidal wave, in Japan, that demolished a Fukushima nuclear power station, which may cause a long-term effect on the funding of nuclear science as well;

(3) the current European financial crisis;

(4) finally, in this country we have a feeling that the outside world has turned upside down. The matter is that nowadays the country is portrayed as if it had turned upside down just in a period when it is straightening up its own matters, and, in spite of a bumpy start, it is turning from abnormal to normal. I was afraid that the bad news about Hungary flooding the world press may have deterred some prospective participants.

Considering all these forebodings, we are happy that you have come here, and we can start the 10th conference of our discipline.

You must have noticed that there is no poster session. We pay for this bold decision by having more parallel sessions than is usual. Receiving the contributions, we found that there was just one contributor who wanted to show a poster, while the oral presentations enjoyed overwhelming popularity. That is why we decided so.

Another unusual feature of this conference is that the programme is not flooded by local contributions. In fact, there is none. That is so because the local organizers have decided not to accept any local contributions. Our great scientific feats should appear in the talks of others. We hoped that the invitees know the works of the Debrecen groups, and we are curious to see whether that is so. We apologize to the Debrecen colleagues if this hope proves to be false.

Let me thank the International Advisory Board for their enthusiastic and in fact perplexingly great number of recommendations. I apologize for not being able to attract all recommended speakers here. Let us hope that the Board will bring a wise decision on the site of the next conference as well.

You can guess that in these critical years this country is not very well off. Well, we still have a nuclear power plant in Paks, which works very well, and still takes care of initiatives like ours, and they, together with a small enterprise working in energy industry, have given us some non-negligible support. Our problem was that the state support scheme of international conferences in Hungary has ceased to exist. In view of this, I approached the Academy of Sciences, and as a result, a fund was separated within the budget of the Academy to support such conferences. I am grateful to Professor József Pálinkás, President of the Academy for this step. That is how it was possible to reduce the conference fees for the young participants recommended by their elders in due course.

We promised that the proceedings will appear in a reputable journal. Well, there is an adequate fund put aside for that purpose, and we are discussing the publication with Journal of Physics Conference Series. Short papers of all talks can be included in this volume, which is foreseen to be almost 500 pages long. Details are to be found on a leaflet in the conference bag.

Let me finally tell you a few words about our logo. I noticed a similarity between birds, especially swallows, perching on electric wires and the nuclear levels populated by nucleons long ago. Thus a proton perching on a s.p. level could be depicted as a swallow facing us. Since it is obviusly directed upwards, this can be considered to have spin-up, and a spin-down proton should be depicted as a swallow facing us upside down. The neutrons are similar but they express their neutrality by turning their backs to us. In an α-particle each one of them should be represented by one sample. The most popular clustered nucleus, ^{12}C, is to be displayed like a triangle of three α-particles. The usual symbol of ^{12}C is at the same time a start of the acronymic of the conference, $'^{12}$Cluster. To express that this conference takes place in Hungary, the oscillator valleys holding the wires are to be coloured as the Hungarian tricolor. Thus as a symbol of a cluster conference held in Hungary in 2012, one can use our logo:

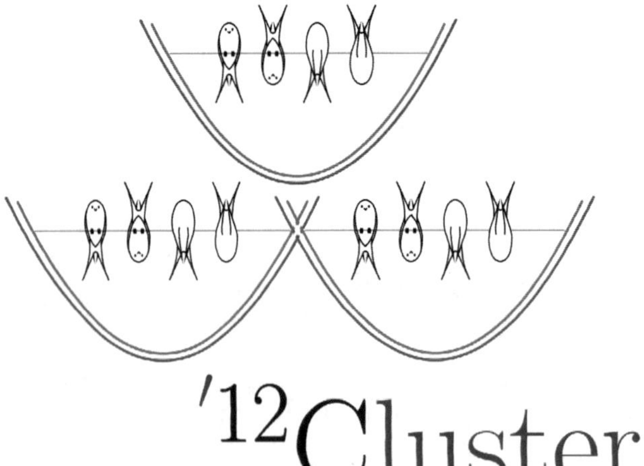

That is all I wanted to say. Have a nice conference and enjoy your stay in Debrecen.

Journal of Physics: Conference Series **436** (2013) 012002 doi:10.1088/1742-6596/436/1/012002

Challenges to the field of nuclear clustering

M Freer

School of Physics and Astronomy, University of Birmingham, Birmingham, B15 2TT, UK.

Email: M.Freer@bham.ac.uk

Abstract. This contribution reviews some of the challenges that nuclear cluster science has taken on focussing on those surrounding nuclear molecules for example in ^{24}Mg. A view is taken that the most exciting challenges link to the structure of light nuclei, where they may be used to inform *ab inito* theories which attempt to utilise realistic nuclear forces. In some cases the limit of these theories is the nucleus ^{12}C. The experimental progress in understanding this crucial nucleus is examined.

1. Introduction

The subject of clustering has, for many years, sat at the edge of nuclear physics – a curiosity rather than the main stream. There are many reasons for this, but perhaps the most obvious is that the subject has often promised significant discoveries or new ways of understanding nuclear phenomena and subsequently not delivered.

Between the 1960s and 1990s a colossal effort was devoted to the study of molecular structures populated in resonant scattering [1]. These measurements populated the intermediate short-lived systems at excitation energies typically in excess of 20 MeV, far from the ground-states, where the number degrees of freedom are high and the correspondingly complexity is large. Given this, it is surprising that simple-modes are resistant to damping into the reservoir of the compound system. The best known of these are the ^{12}C+^{12}C resonances where close to the barrier 100 keV wide structures indicate the formation of an intermediate ^{24}Mg complex which survives for a period which significantly longer than the collision time. But as indicated in Fig. 1, such resonances are not only limited to this single system but are widespread. Such a dramatic phenomenon has attracted many, just as the Sirens in Greek mythology, but a definitive understanding remains to be grasped. In some ways the field has struggled to extract itself from this grasp. Some aspects are clear; the resonances are only observed when the underlying absorption is low (the number of open reaction channels is limited [2]), the energy-spin systematics follow a rotational behavior (but with a number of resonances of the same spin), often the resonances are associated with very large partial widths for the entrance channel [1]. This has been interpreted as the formation of a cluster-cluster intermediate complex associated with shape-isomeric minima in the potential energy of the composite system [3]. The resonances are then rotational states formed in the

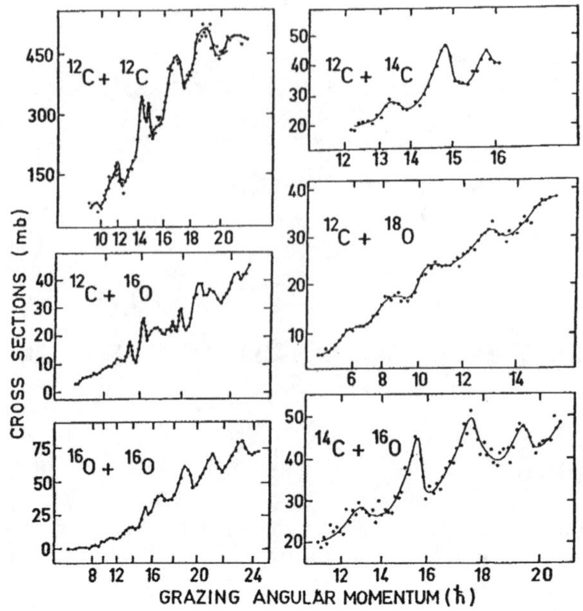

Figure 1. Resonances observed in the scattering of light nuclear systems from Ref. [2].

second minimum with the fragmentation arising from the coupling with states of the same spin in the first. The demonstration that such a picture is accurate lies at the limits of experimental possibilities – for example the measurement of the electromagnetic transition rates, $B(E2)$s, between the rotational states. The one valiant attempt to detect the gamma-rays from transitions between the states produced, realistically, only produced an upper limit [4]. An alternative approach to characterize the states is through their radiative decay to the ground-state; so-called radiative capture. This is a technique pioneered by Nathan and Sandorfi using NaI detectors [5], but has more recently been pursued using high resolution Ge detectors (and reported in the present conference [6]). But again an unambiguous understanding has not emerged. Complexity still dominates.

In recent times, there has been a refocusing of the efforts of the cluster community on the outstanding questions in lighter systems. Here as indicated in Fig. 2, the clustering degree of freedom moves closer to the ground state and the competing degrees of freedom diminish. Indeed here it has been possible to observe the electromagnetic decay of the 4^+ to 2^+ state in the nucleus ^{8}Be – confirming its deformation, consistent with a 2α cluster structure [7]. Similarly, the quality of information available for the states in ^{20}Ne point very strongly to its ^{16}O+α structure [8]. The attraction of the study of light nuclear systems is that they are also the dominion of the ab initio type calculation. Such models utilize realistic nucleon-nucleon forces in a bid to replicate nuclear properties. These include the Greens Function Model Carlo (GFMC) [9], No Core Shell Model (NCSM) [10] and Effective Field Theories (EFT) [11]. In the context of clustering the images of the density of the ^{8}Be nucleus calculated in the framework of the GFMC approach are particularly stimulating [9].

Structural changes in nuclei are often driven by the valence particles, interactions between these particles and the core then can induce collective changes in shape or deformation. In light systems the valence nucleons dominate and correlations are extremely important. Understanding the correlations, both spatial and momentum, that drive the formation of various nuclear structures is strongly entwined with understanding the nature of the strong

Journal of Physics: Conference Series **436** (2013) 012002 doi:10.1088/1742-6596/436/1/012002

interaction at play within the nucleus. It is here that perhaps the understanding of nuclear correlations or clusters has its greatest impact, as understanding the nature of the strong interaction in nuclei is currently one of the most important frontiers for nuclear science.

It has not gone unnoticed (and also reported in the present conference [12]), that one of the other great conquests for nuclear science, the determination of the abundance of the elements, is strongly linked to the nature of threshold states. This is depicted in both the Ikeda and modified-Ikeda diagrams, which indicate that clusterisation occurs close to the threshold for the formation of a nucleus from its cluster constituents. Correspondingly, structural

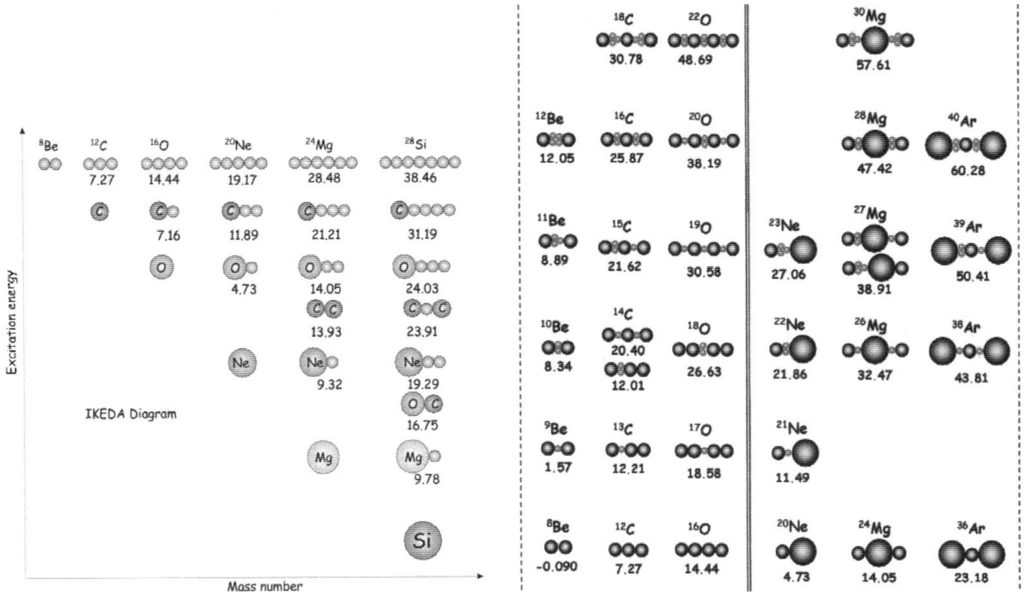

Figure 2. Threshold diagrams – revealing potential excitation energies at which clustering should appear. *Left* The Ikeda diagram [13] which reveals the cluster thresholds for alpha-particle like nuclei. *Right* Modified Ikeda diagram [14] for systems in which covalent neutrons may be exchanged between the cluster cores.

clusterisation is intimately entwined with the formation of the elements and even life itself.

Remarkably, these two aspects, nucleosynthesis and the nature of the strong interaction, intersect in ^{12}C and the Hoyle-state. Within the field, understanding the nature of this state is the foremost challenge and is perhaps the reason why so many contributions to the conference focused on this element. The structure of this state has variously been described in terms of a Bose condensate, a cluster state and an alpha-particle gas. It is not clear yet which picture is appropriate and even if the various descriptions are non-degenerate. Similarly, the spectroscopy of ^{12}C above the alpha-decay threshold remains to be determined definitively.

The challenges in determining the structure of light nuclei both from a theoretical and experimental perspective require a new approach; one which is coherent rather than disjointed and has a genuine strategy in terms of its approach.

For the remainder of this contribution, I will focus on the nature of ^{12}C as an illustration of the advances that have been made in this single nucleus and the questions that remain.

2. Structure of ^{12}C

The ^{12}C nucleus has, potentially, a symmetry which is related to a three-alpha structure. The ground-state has an oblate character which may be understood in terms of the arrangement of the three alpha-particles in a plane. The collective excitations of this structure would result in the population of 0^+, 2^+ and 4^+ ….. states associated with rotations around an axis which lies within the plane of alpha-particles and 3^-, 4^-, 5^-, ….. corresponding to a rotation perpendicular to the plane, corresponding to K=0 and K=3 bands, respectively. Within the ground-state the alpha-particle structure is expected to be strongly suppressed, though the symmetry remains. This suppression would coincide with the fact that the 3alpha decay threshold lies some 7 MeV higher in energy and hence the cluster structure should not be strongly developed. Indeed, quasi free proton scattering $^{12}C(p,p\ \alpha)$, reported in the present conference [15], indicates that the alpha-particle spectroscopic factor is close to that which would be predicted by the shell model. The 3^- state on the other hand lies at 9.6 MeV which is above the alpha-decay threshold and hence could in principle display a more developed 3alpha cluster structure. An analysis of the width of the 3^- resonance in the high resolution inelastic scattering data shown in Fig. 3, would indicate a width of 40(2) keV. This would, for a channel radius of $1.35(8^{1/3}+4^{1/3})$ fm, suggest a width which is 25% of the Wigner limit.

The complexity of the Greens Function Monte Carlo (GFMC) approach prohibits its extension currently to nuclei with A>12, and ^{12}C is currently the focus of the calculations with some preliminary results for the 7.65 MeV, 0^+, Hoyle-state just being reported. The Hoyle-state was proposed by Sir Fred Hoyle some 50 years ago [16]. This state is the gateway through which carbon is synthesized via what is known as the triple-alpha process. Without this state there would be no organic life. Calculations of the properties of the Hoyle-state using the GFMC approach are still on-going. However, calculations using the

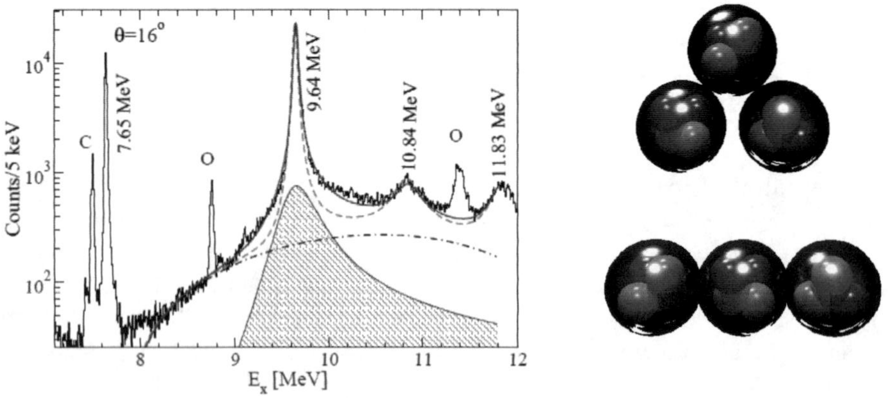

Figure 3. Left-hand-side. ^{12}C excitation energy spectrum showing the broad 2^+ state required to reproduce the shape of the experimental spectrum, from Ref. 8. Right-hand-side. Possible arrangements of the alpha-clusters in either a triangle (top) or chain (bottom). The experimental data indicate that the triangular structure is preferred.

Antisymmetrized Molecular Dynamics (AMD) [17] and its cousin the FMD approach

Journal of Physics: Conference Series **436** (2013) 012002 doi:10.1088/1742-6596/436/1/012002

(Fermionic Nuclear Dynamics) [18] reveal a highly clustered state. Its' very dilute nature, together with the 3alpha cluster structure has been interpreted in terms of an alpha-condensate by Schuck and co-workers [19], this would be a new bosonic state of nuclear matter. Very recent calculations using the framework of chiral effective field theory have been able to closely reproduce the energy of the Hoyle-state [11]. Although they offer little detailed structural information, they do suggest a larger charge radius than the ground-state, as observed experimentally. As with the ^{8}Be ground-state, the width of the 7.65 MeV state indicates a significant cluster structure and, in agreement with theoretical predictions, a large charge radius (see Ref. [17]) indicating a volume which is 3-4 times that of the ^{12}C ground state.

Revealing the collective excitations of such a state could have significant implications for understanding its structure. In the alpha-condensate model this would correspond to a 2^+ state which is associated with the conversion of one of the bosons from s-wave to d-wave. In a more traditional cluster model it would correspond to a collective rotation and the moment of inertia would then reveal the details of the cluster arrangement.

The closest 2^+ state which has been tabulated with these characteristics is at 11.16 MeV [20]. This state was observed in the ^{11}B(^{3}He,d)^{12}C reaction, but has not been observed in measurements subsequently.

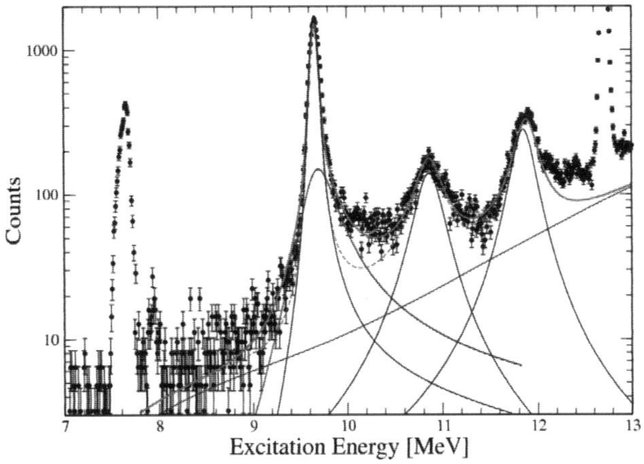

Figure 4. ^{12}C excitation energy spectrum from the ^{11}B(^{3}He,d) reaction. The measurement clearly demonstrates the absence of the 11.16 MeV state in ^{12}C [9].

A recent re-measurement of this reaction using the K600 spectrometer at iThemba demonstrates that the earlier observation of a state at 11.16 MeV was almost certainly an experimental artifact and no such state exists [21]. The existence of the 2^+ excitation is then an open question.

Recent studies of the ^{12}C(α,α') [22] and ^{12}C(*p*,*p*') [23,24] (see Fig. 3) reactions indicate the presence of a 2^+ state close to 9.6-9.7 MeV with a width of 0.5 to 1 MeV. The state is only weakly populated in these reactions, presumably due to its underlying cluster structure, and is broad. Consequently, its distinction from other broad-states and dominant collective excitations (e.g. the 9.6 MeV, 3^-) makes its unambiguous identification challenging. Further evidence for such an excitation comes from measurements of the ^{12}C(γ,3α) reaction performed at the HIGS facility, TUNL [25]. Here a measurable cross section for this process was observed in the region of 9-10 MeV which cannot be attributed to known states in this region. Furthermore, the angular distributions of the alpha-particles are consistent with an L=2

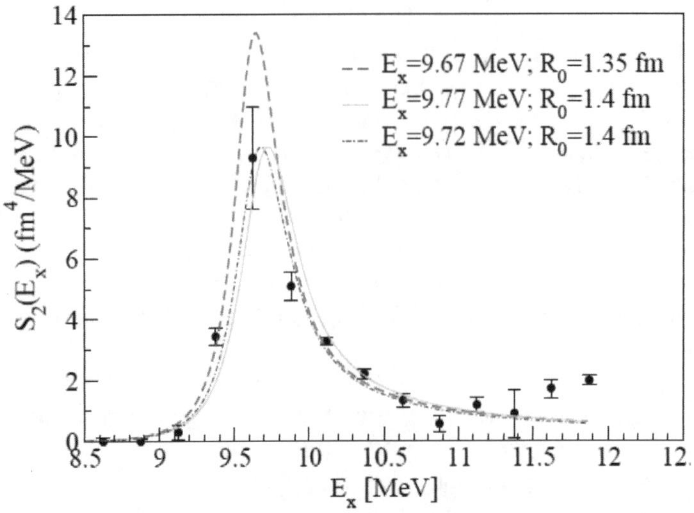

Figure 5. The 2^+ strength function extracted from $^{12}C(\alpha,\alpha')$ inelastic scattering measurements. The various curves correspond to R-matrix line shapes and show that there is a consistent agreement between the 2^+ line shape found in the excitation energy spectra and extracted from the *L*-dependence of the angular distributions. Ref. [26]

pattern, indicating a dominant 2^+ component. The alpha-particle inelastic scattering measurements also included a multipole decomposition analysis of the angular distributions [22] the amplitudes of the *L*=2 component is shown in Fig. 5. The variation of this amplitude with energy matches well that deduced from an R-matrix type analysis of the component found in the excitation energy spectra [26].

Based on a rather simple description of this state in terms of three alpha-particles with radii given by the experimental charge radius, it is possible to use the 2 MeV separation between the Hoyle-state and the proposed 2^+ excitation to draw some conclusions as to the arrangements of the clusters (see Fig. 3). This would indicate that rather than a linear arrangement of the three clusters that a more appropriate description would be a loose arrangement of the alpha-particles in something approaching a triangular.

A natural extension of such a conclusion is that there should also be a collective 4^+ state. Using the simple $j(j+1)$ scaling, a 4^+ excitation close to $E_x(^{12}C)$ = 14 MeV would be expected. We have performed recent measurements of the two reactions $^9Be(\alpha,3\alpha)n$ and $^{12}C(\alpha,3\alpha)^4He$ [27]. In these measurements three alpha-particles were detected in an array of four silicon strip detectors (shown in Fig 6). The analysis required that two of the three alpha-particles came from the decay of the ground-state of ^8Be. For the decay of ^{12}C to $^8Be+\alpha$ this ensures that the decay process can proceed through only natural parity states (i.e. 0^+,1^-, 2^+ ....). This restricts the complexity of the excitation energy spectrum. The measurements for both the ^9Be and ^{12}C targets reveal the known 3^-, 1^- and 4^+ states at 9.64, 10.84 and 14.08 MeV, respectively. However, there is an additional component to the spectrum close to 13.3 MeV with a width estimated to be 1.7 MeV (Fig. 6). It is believed that this is not a contaminant and is observed with similar properties in all spectra. Angular correlation measurements made using the ^{12}C target are not definitive, but indicate a 4^+ assignment.

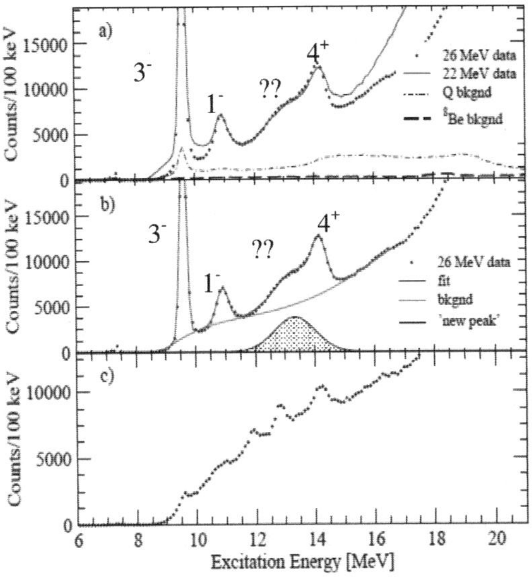

Figure 6. Carbon-12 excitation energy spectra. a) The blue line shows the measurement at 22 MeV. The spectrum corresponding to measurements at 26 MeV is shown by the dots. b) Fit to the 26 MeV data is given by the blue solid line. The polynomial background (red-line) and line-shape for the new peak (shaded area) is shown. c) Excitation energy spectrum for events not proceeding via the decay to the ^{8}Be ground-state. The proposed new state is indicated by "??". From [27].

Such measurements are challenging and it is difficult to be absolutely certain that the feature does not correspond to an experimental artifact. Nevertheless, the appearance of a 4^{+} state close to the energy predicted based on the extension of the Hoyle-state and proposed 2^{+} excitation. It is clear that further measurements are required to confirm, or otherwise, this new state in ^{12}C. Moreover, analysis of beta-decay measurements populating alpha-decay states in ^{12}C indicate that there may be broad 2^{+} strength in ^{12}C but at somewhat higher energies than suggested in the proton and alpha inelastic scattering measurements [20]. This contradiction between the two types of measurements remains to be reconciled and hence a firm conclusion regarding the excitation structure of ^{12}C above the alpha-decay threshold remains to be established.

3. Summary

The characterization of clustering is extremely challenging, but there is a broad spectrum of evidence for this behavior. Providing a detailed characterization of the cluster states has been made for a number of systems. However, even for nuclei as simple as ^{12}C there remain unanswered questions. We have shown here that we are moving towards a more detailed understanding, but have not finally arrived. This will require precise experimental measurements coupled with state-of-the-art theory.

References

[1] Erb K A and Bromley D A 1985 in *Treatise on Heavy--Ion Science*, edited by D.A. Bromley (Plenum, New York, 1985), Vol. 3.
[2] Haas F and Abe Y 1981 *Phys. Rev. Lett.* **46** 1667
[3] Leander G and Larsson S E 1975, *Nucl. Phys.* **A239** 93
[4] Haas F, *et al.* 1997 *Il Nuovo Cimento A* **110** 989
[5] Nathan A M, Sandorfi A M and Bowles T J, 1981, *Phys. Rev.* C **24** 932
[6] Jenkins D, contribution to the present proceedings

[7] Datar V M, *et al.* 2004 *Phys. Rev. Lett.* **94** 122502
[8] Freer M 2007 *Reports on Progress in Physics* **70** 2149
[9] Wiringa R B, Pieper S C, Carlson J and Pandharipand V R 2000 *Phys. Rev. C* **62** 014001
[10] Navrátil P, Vary J P, and Barrett B R 2000 *Phys. Rev. Lett.* **84** 5728
[11] Epelbaum E, Krebs H, Lee D and Meißner U-G 2011 *Phys. Rev. Lett.* **106** 192501
[12] Kubono S, contribution to the present proceedings
[13] Ikeda K *et al.* 1968, *Suppl. Prog. Phys.* (Japan) Extra Numbers 464.
[14] von Oertzen W, Freer M and Kanada En'yo Y 2006 *Phys. Rep.* **432** 43
[15] Cowley A, *et al.* contribution to the present proceedings
[16] Hoyle F 1954 *The Astrophysical Journal, Supplement Series* Vol. 1 12
[17] Kanada En'yo Y and Horiuchi H 2001 *Prog. Theo. Phys. Suppl.* **142** 205
[18] Chernykh M, *et al.* 2007 *Phys. Rev. Lett.* **98** 032501
[19] Tohsaki A, *et al.* 2001 *Phys. Rev. Lett.* **87** 192501
[20] Reynolds G M, Rundquist D E, Poichar M 1971 *Phys. Rev.* C **3** 442
[21] Smit F D, *et al.* 2012 *Phys. Rev. C* **86**, 037301
[22] Itoh M, *et al.* 2004 *Nucl. Phys.* A **738** 268 and *Phys. Rev.* C **84**, 054308
[23] Freer M, *et al.* 2009 *Phys. Rev.* C **80** 041303
[24] Zimmerman W R, Destefano N E, Freer M, Gai M, Smit F D 2011 *Phys.Rev.* C 84 027304
[25] Gai M for the UConn-Yale-Duke-Weizmann-PTB-UCL Collaboration 2011 *Acta Phys. Pol.* B **42** 0775 and contribution to the current proceedings.
[26] Freer M, *et al.* 2012 *Phys. Rev.* C **86**, 034320
[27] Freer M, *et al.*, 2011 *Phys. Rev.* C **83** 034314

Recent developments in nuclear cluster theory

Hisashi Horiuchi

Research Center for Nuclear Physics, Osaka University, Ibaraki, Osaka 567-0047, Japan
International Institute for Advanced Studies, Kizugawa, Kyoto 619-0225, Japan

E-mail: horiuchi@rcnp.osaka-u.ac.jp

Abstract. The major novelties that have emerged in nuclear cluster theory since the last, 9th, cluster conference is reviewed. Among the marked developments made in many traditional and novel subjects of nuclear cluster physics, four subjects are chosen in this introductory talk. They are 1. neutron-rich nuclei, 2. cluster gas, 3. coexistence of cluster and mean-field dynamics, and 4. *ab initio* approaches. A representative example of the present status of the field is that it is studied in the framework of *ab initio* nuclear theory as well. We discuss a recent *ab initio* calculation of the Hoyle state of ^{12}C, which supports its cluster-gas-like structure.

1. Introduction

Complying with the request of the organizers of the present conference, the purpose of this talk is to review the major novelties in nuclear cluster theory since the last, 9th, cluster conference at Stratford-upon-Avon. Actually, marked progress has been made in many traditional and novel subjects of nuclear cluster theory. I have chosen four subjects among them in this talk. The substantial progress in these subjects shows the importance of cluster dynamics as a basic dynamics of the nuclear many-body system. The topics to be covered are 1. neutron-rich nuclei, 2. cluster gas, 3. coexistence of cluster and mean-field dynamics, and 4. *ab initio* approaches. The discussions are largely based on a recent publication [1]. A representative result of present-day nuclear cluster physics is the application of *ab initio* nuclear theory to the study of the Hoyle state of ^{12}C. We discuss that a recent calculation [2] confirms the cluster-gas-like structure of the Hoyle state.

2. Neutron-rich nuclei: novel types of clustering

2.1. Cluster thresholds and neutron thresholds

In most cases, cluster structure in neutron-rich nuclei consists of two parts: a clustered core and valence neutrons around the core. Therefore the cluster dynamics in neutron-rich nuclei consists of cluster dynamics in the core part and neutron dynamics around the core. The cluster dynamics in the core part is the same as that in stable nuclei. The neutron dynamics around a clustered core is a new ingredient, which does not exist in stable nuclei. Because of this new feature, when the ground state and/or low-lying states have molecular-orbit-type cluster structure, these cluster states do not obey Ikeda's threshold rule proposed for stable nuclei. This is a very distinct character of clustering in neutron-rich nuclei. Since neutron-rich nuclei have weakly-bound valence neutrons, neutron thresholds are located low in excitation energy, but, on the contrary, cluster thresholds are located high in excitation energy. This fact means [1] that the ground state and/or low-lying states having molecular-orbit-type cluster

Journal of Physics: Conference Series **436** (2013) 012003 doi:10.1088/1742-6596/436/1/012003

structure are deeply bound from the cluster thresholds although they are weakly bound from the neutron thresholds. Fig. 1 shows this situation in ^{12}Be, whose ground state has, as a dominant component, molecular-orbit-type clustering with a 2α core.

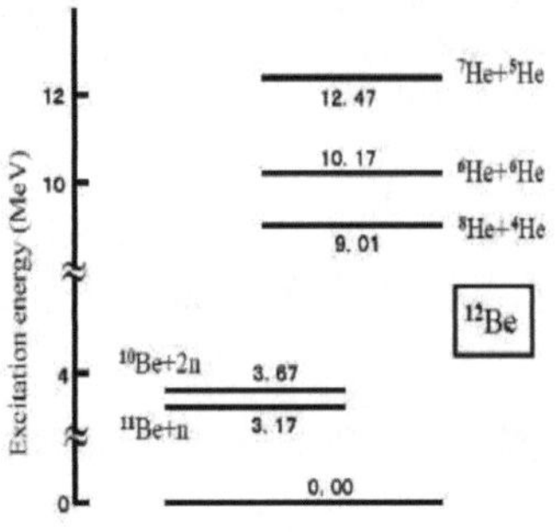

Figure 1. Cluster threshold energies (^{8}He + ^{4}He, ^{6}He + ^{6}He, ^{7}He + ^{5}He) are high, while neutron threshold energies (^{11}Be + n, ^{12}Be + $2n$) are low in ^{12}Be, whose ground state has a dominant cluster structure. The figure is taken from Ref. [1].

2.2. Clustering in the core and neutron dynamics around the core depend on each other

Cluster dynamics in the core and neutron dynamics around a clustered core do of course interact with each other. A very characteristic example is seen when the so-called molecular σ-orbit is occupied by valence neutron(s). The σ-orbit is the intruder orbit which comes down from the higher shell when the clustering of the core part is increased [1]. In the case of Be isotopes, the σ-orbit around the 2α core corresponds to the $(n_x, n_y, n_z) = (0, 0, 2)$ orbit of the anisotropic harmonic oscillatior model with axial symmetry. In the case of F isotopes, the σ-orbit around the ^{16}O + α core corresponds to the $(n_x, n_y, n_z) = (0, 0, 3)$ orbit of the anisotropic harmonic oscillatior model with axial symmetry. When the σ-orbit is occupied by valence neutron(s), the clustering of the core part is forced to become larger because the enhancement of the core clustering lowers the single-particle enegy of the σ-orbit. For example, many authors have shown that the ground-state configurations of ^{11}Be and ^{12}Be are dominantly of $(\pi_{3/2})^2\sigma$ and $(\pi_{3/2})^2(\sigma)^2$, respectively around the 2α core [1]. Fig. 2 shows predictions [3] of antisymmetrized molecular dynamics (AMD) for three types of excited cluster bands having ^{15}N + α core in F isotopes. All three bands have one proton hole in the $0p$-shell. The bands labelled by σ^2 (first band) and $\sigma\pi$ (second band) have two neutrons and one neutron in the valence σ-orbit, respectively. A very interesting prediction of this figure is that the excited bands of the first and second types come down rapidly to low-excitation-energy region as the neutron number increases, as low as around 2 MeV in ^{27}F and ^{29}F.

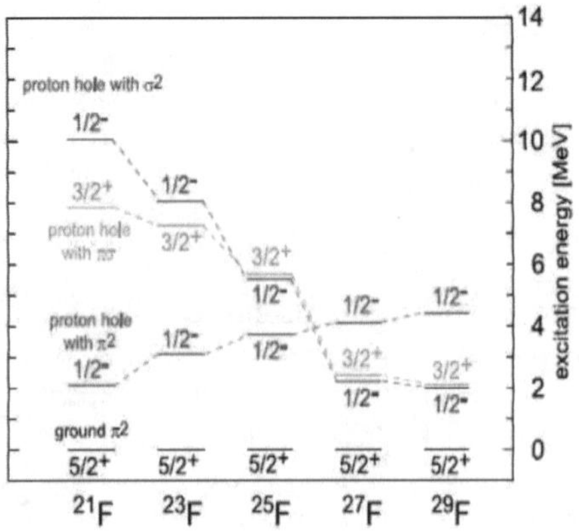

Figure 2. Excitation energies of band-head states of three types of excited cluster bands in F isotopes predicted by AMD [3]. All three bands have a ^{15}N + α core due to one proton hole in the $0p$-shell. The bands labeled by σ^2 and $\sigma\pi$ have two neutrons and one neutron in the valence σ-orbit, respectively.

2.3. Atomic orbits and mixture of molecular and atomic orbits
Recent investigations of the cluster structure in neutron-rich nuclei show the formation of cluster states with atomic-orbital configurations of neutrons at somewhat higher excitation energies. In this configuration, neutrons move not over the whole system but around individual clusters constituting the core. A neutron-rich clustered nucleus having valence neutrons in atomic orbits has a cluster structure similar to that of a stable nucleus with no neutrons moving over the whole system. Namely, the structure of such a necleus is composed of neutron-rich clusters. Representative examples are ^{6}He + ^{6}He and ^{8}He + ^{4}He structures in ^{12}Be. These structures have been studied in Refs. [4, 5, 6, 7, 3, 1]. In Ref. [8] it is reported that the AMD study of ^{22}Ne gives (1) a mean-field-type structure for the ground-state band, (2) $K^\pi = 0_2^+, 0_1^-$, and 1^- bands with molecular-orbit structure around the ^{16}O + α core, and (3) $K^\pi = 0_3^+$ and 0_2^- bands with atomic-orbit structure of ^{18}O + α type. The excitation energy of a cluster state with atomic-orbit structure is rather close to the breakup threshold energy into the constituent neutron-rich clusters.

The existence of molecular-orbital clustering and atomic-orbital clustering exhibits the richness of valence-neutron dynamics, and in Ref. [7] clustering with a mixture of molecular and atomic orbits is also discussed.

2.4. Dineutron
The dineutron is unbound in free space but can be bound inside a nucleus by the Borromean effect. At present ^{11}Li is the only system where dineutron structure has been confirmed experimentally [9]. In ^{11}Li, the dineutron structure (^{11}Li = ^{9}Li + dineutron) is intimately related to the two-neutron halo structure and the breaking of N=8 magicity. A very attractive explanation was given in Ref. [10] where, in addition to the pairing-force effect of valence neutrons, the tensor-force effect in the ^{9}Li core was newly introduced to give a simultaneous explanation of the above three problems, namely the dineutron formation, the two-neutron halo, and the breaking of N=8 magicity. It is discussed that the Pauli blocking in pairing and tensor-force correlations cause that the gap between $0p$ and $1s$ orbits is small (magicity breaking), from which it follows that the $1s$ orbit enhances the halo character, and the parity mixing with approximately equal weights of the opposite parities due to the approximate degeneracy of 0p and 1s orbits induces spatial clustering. The Coulomb breakup cross section was reproduced well, with a large matter radius and a large weight of the $(1s)^2$ component.

3. Cluster-gas states: novel nuclear structure and non-localized clustering

3.1. 4α condensate-like state in ^{16}O
The Hoyle state in ^{12}C was studied in detail in 1970's by 3α RGM and GCM which succeeded in reproducing almost all experimental data. These RGM and GCM studies predicted a large radius of this state. In 2001 Ref. [11] proposed the interpretation of the Hoyle state as a 3α condensate-like state by using the THSR wave function $\mathcal{A}[\prod_{j=1}^{3}\exp(-(2/B^2)\mathbf{X}_j^2)\phi(\alpha_j)]$. In 2003 it was discovered [12] that the 3α RGM/GCM wave functions of the Hoyle state are almost equivalent to a single THSR wave function:

$$|\langle\Phi(\text{single THSR})|\Psi(3\alpha\text{RGM/GCM})\rangle|^2 \approx 100\% \quad \text{for the Hoyle state.} \tag{1}$$

Studies for the 4α condensate-like state in ^{16}O were reported in Refs. [13, 14]. In ^{16}O, six 0^+ states are observed up to the 6th 0^+ state at 15.1 MeV. The 4α OCM calculation of Ref. [13] could reproduce the energies and decay widths of six 0^+ states up to the 6th 0^+ state. This OCM study assigns the observed 0_6^+ state the 4α condensate-like structure, because the reduced-α-width amplitude $y(r)$ of the calculated 0_6^+ state has a large amplitude only in the ^{12}C(0_2^+) + α channel, whereas the amplitudes in other channels are much suppressed. Since $y(r)$ in the

Hoyle-state ($^{12}C(0_2^+)$) channel has a very long tail stretching out to about 20 fm and the Hoyle state has a 3α condensate-like structure, one can certainly say that the calculated 0_6^+ state has a character of 4α-condensate-like structure.

The 4α threshold is about 7 MeV higher than the $^{12}C + \alpha$ threshold. Thus one may think that the width of a 4α-condensate-like state expected to be around the 4α threshold must have a very large decay width. But contrary to this conjecture, the calculated 0_6^+ state has a rather small width of 136 keV. The reason of this small width is simple and twofold. The first reason is the small overlap of the exotic structure of a 4α-condensate-like state with $^{12}C + \alpha$ channels except the Hoyle-state channel $^{12}C(0_2^+) + \alpha$. The overlap is small because the ^{12}C states other than the Hoyle state are compact. The second reason is the very small decay Q-value of the decay to the Hoyle-state channel, which makes the Coulomb-barrier penetration very small although the overlap of this channel with the 4α-condensate-like state is rather large. The width of 136 keV of the calculated 0_6^+ state is in a good accord with the observed value, 166 keV. The above arguments are expected to be general and applicable to heavier $n\alpha$-condensate-like states as well.

Recently Ref. [15] reported that the fine structure of the isoscalar monopole strength function in the low energy region up to about 16 MeV in ^{16}O is rather satisfactorily reproduced within the 4α OCM. It is reported that the observed fine structure can be attributed to the 0_3^+, 0_4^+, 0_5^+, and 0_6^+ states.

3.2. Cluster-gas-like states in various nuclei

Cluster-gas-like states other than ^{12}C and ^{16}O have been studied in various nuclei by many authors. As to ^{11}B (^{11}C), Refs. [16, 17, 18, 19] discussed possible gas-like structure of $2\alpha + t$ ($2\alpha + {}^3He$) for the third $3/2^-$ state at 8.11 MeV (8.56 MeV). Refs. [20, 21, 27, 23, 24, 25] discussed candidates for gas-like states of $3\alpha + p$ ($3\alpha + n$) type in ^{13}C (^{13}N). The possibility of $n\alpha$-condensate-like states with $n > 4$ was discussed in Ref. [29]. In Refs. [26, 27, 28], the possiblity of α-gas-like states around core nuclei like ^{16}O and ^{40}Ca was discussed.

An important category of the nuclear states strongly related to the cluster-gas-like states is the states that contain cluster(s) having cluster-gas-like structure themselves. Representative examples are the molecular states of $^{12}C + {}^{12}C$ in which one or two ^{12}C cluster(s) is(are) in the Hoyle state (0_2^+). The states of this type were studied both experimentally and theoretically (see Refs. [1, 30]). The 'Hoyle + α' state is discussed by Funaki in this conference.

3.3. Non-localized clustering

The THSR wave function was introduced originally in order to describe cluster-gas-like states. However it was found already in 2003 [12] that the 3α RGM/GCM wave functions for the ground state are equivalent by 93% to a single THSR wave functions:

$$|\langle\Phi(\text{single THSR})|\Psi(3\alpha\text{RGM/GCM})\rangle|^2 \approx 93\% \quad \text{for the ground state.} \tag{2}$$

Recently it was found [31] that the $^{16}O + \alpha$ RGM/GCM wave functions for both the ground-band states and the $K^\pi = 0_1^-$ band states are almost equivalent to single THSR wave functions:

$$|\langle\Phi(\text{single THSR})|\Psi(^{16}\text{O} + \alpha\ \text{RGM/GCM})\rangle|^2 \approx 100\% \tag{3}$$

$$\text{for ground and } K^\pi = 0_1^- \text{ band states.} \tag{4}$$

The THSR wave function does not appear to represent localized clustering, but the intercluster Pauli repulsion creates non-zero separation between the clusters. We may say that the dynamics prefers non-localized clustering but the Pauli principle makes the system look like consisting of localized clusters. Non-localized clustering in ordinary cluster states looks to be intimately related to cluster-gas-like structure.

4. Coexistence of cluster and mean-field dynamics

It was pointed out in Ref. [17] that the observed $E0$ excitation strength to cluster states is in general large, with its magnitude comparable to the single-particle strength. For example in ^{16}O we have large observed $E0$ values for three excited states, 0_2^+, 0_3^+, and 0_5^+ (3.6, 4.0, and 3.3 fm^2, respectively). In mean-field theories there is no simple way to explain the three observed strong $E0$ transitions of similar magnitudes. On the other hand, cluster model calculations can easily reproduce them. The reason is explained [33] by the duality property of the ground state, which means that the ground state wave function has both mean-field-type and clustering characters. This duality is well-known as the Bayman-Bohr theorem [32].

The dual character of the ground state means that the ground state has both the mean-field-type degree of freedom and the clustering degree of freedom. When the mean-field-type degree of freedom is excited, mean-field-type excited states are created, but when the clustering degree of freedom is excited, clustered excited states are formed. This argument is explained in detail in Refs. [34, 35, 1, 36].

5. Ab initio approaches

5.1. *Ground and the first-excited* 0^+ *states of* ^{4}*He*

In an *ab initio* study of ^{4}He [37] the calculated ground and first-excited 0^+ states are compared. A very important result is that the probabilities of S, P and D components of the 0_2^+ state are very different from those of the 0_1^+ state but are almost the same as those in ^{3}H (and ^{3}He). For example $P_D = 14.08\%$ in 0_1^+ while it is 8.74% in 0_2^+, which is close to 8.70% in ^{3}H. This result and the fact that the 0_2^+ state is located near the $3N + N$ threshold suggest strongly that the 0_2^+ state has $3N + N$ cluster structure. In fact the reduced-width amplitude of the 0_2^+ state in the $3N + N$ channel is stretched in the outer spatial region while that of the 0_1^+ state is mostly in the inner spatial region. What is remarkable here is that a clustering effect is inferred not only from the density distribution but also from the existence of a tensor-force effect.

5.2. *Ab initio cluster model calculation with microscopic R-matrix method*

A very good example which shows that the effective-force approach cannot be a substitute of the *ab initio* approach is given by the ^{2}H$(d,\gamma)^4$He reaction in the astrophysical low-energy region. Fig. 3 shows an *ab initio* cluster model calculation with the microscopic R-matrix method for ^{2}H$(d,\gamma)^4$He [38]. In the effective-force approach, only relative D wave is existent and therefore in the astrophysical low-energy region the reaction rate becomes very small due to the centrifugal barrier. However, in the *ab initio* approach, relative S wave can also exist due to the tensor force, and that allows the reproduction of the data.

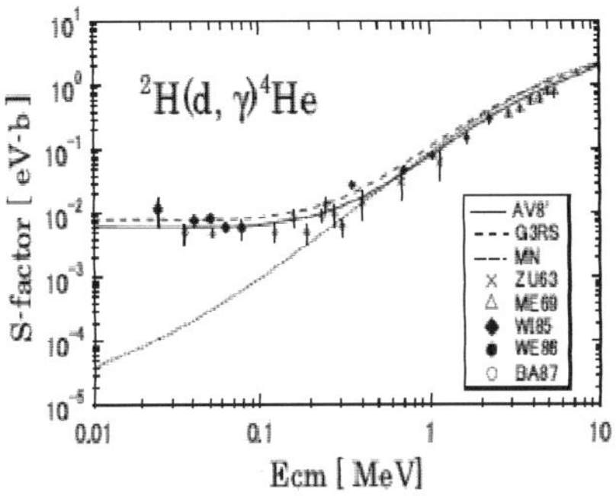

Figure 3. Astrophysical S-factor of the ^{2}H$(d,\gamma)^4$He reaction. Calculated results with the AV8' force (solid), with the realistic force G3RS (dashed) and the effective force MN (dotted) are compared with experiments.

5.3. Ab initio calculation of the Hoyle state

As we mentioned in the introduction, one of the representative features of present-day nuclear cluster physics is that the study of the Hoyle state of ^{12}C is an important and challenging frontier topic of the *ab initio* approach to nuclear physics.

The calculations in the no-core shell model (e.g., Ref. [39]) have difficulties in obtaining sufficiently low excitation energies of the Hoyle state and other cluster states although mean-field-type states are well reproduced. The Hoyle state was also studied recently by lattice simulation with effective field theory in Ref. [40]. Here we show the recent result by the quantum Monte Carlo approach [2]. In Fig. 4 we compare the density distributions $r^2\rho(r)$ of the Hoyle state (0_2^+) and the ground state (0_1^+) of ^{12}C as result from the quantum Monte Carlo calculation [2] (left figure) and the 3α THSR wave function [41] (right figure). Although the quantum Monte Carlo calculation has not fully converged for the Hoyle state, we clearly see a good correspondence of the density distributions between the quantum Monte Carlo approach and the 3α THSR wave function. Thus we see that the quantum Monte Carlo results are consistent with the gas-like structure of the Hoyle state.

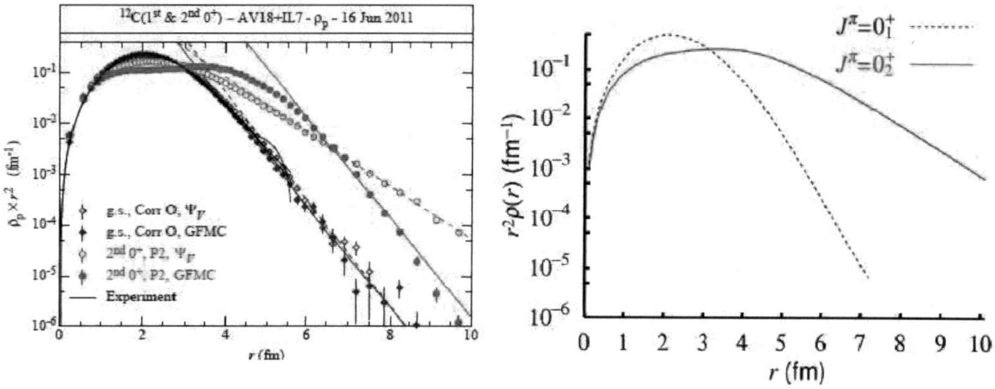

Figure 4. Comparison of the density distributions $r^2\rho(r)$ of the Hoyle state (0_2^+) and the ground state (0_1^+) of ^{12}C. Quantum Monte Carlo calculation (left figure) and 3α THSR wave function (right figure).

6. Summary

Marked progress has been made in many traditional and novel subjects of nuclear cluster physics. Four subjects were discussed in this talk. The developments in these subjects show the importance of clustering among the basic modes of motion of nuclear many-body systems.

References

[1] Horiuchi H, Ikeda K and Kato K 2012 *Prog. Theor. Phys. Suppl.* **192** 1
[2] Wiringa R B 2012 *Wilhelm und Else Heraeus-Seminar on Nuclear Ground-State Properties of the Lightest Nuclei: Status and Perspectives* p 501
[3] Kanada-En'yo Y and Kimura M 2010 *Lecture Notes in Physics* **818** 129
[4] Freer M *et al* 2001 *Phys. Rev.* C **63** 034301
[5] Descouvemont P and Baye D 2001 *Phys. Lett.* B **505** 71
[6] Kanada-En'yo Y and Horiuchi H 2003 *Phys.Rev.* C **68** 014319
[7] Ito M, Itagaki N, Sakurai H and Ikeda K 2008 *Phys. Rev. Lett.* **100** 182502

[8] Kimura M 2007 *Phys. Rev.* C **75** 034312
[9] Nakamura T *et al* 2006 *Phys. Rev. Lett.* **96** 252502
[10] Ikeda K, Myo T, Kato K and Toki H 2010 *Lecture Notes in Physics* **818** 165
[11] Tohsaki A, Horiuchi H, Schuck P and Röpke G 2001 *Phys. Rev. Lett.* **87** 192501
[12] Funaki Y, Tohsaki A, Horiuchi H, Schuck P and Röpke G 2003 *Phys. Rev.* C **67** 051306(R)
[13] Funaki Y, Yamada T, Horiuchi H, Röpke G, Schuck P and Tohsaki A 2008 *Phys. Rev. Lett.* **101** 082502
[14] Funaki Y, Yamada T, Tohsaki A, Horiuchi H, Röpke G and Schuck P 2010 *Phys. Rev.* C **82** 024312
[15] Yamada T, Funaki Y, Myo T, Horiuchi H, Ikeda K, Röpke G, Schuck P and Tohsaki A 2012 *Phys. Rev.* C **85** 034315
[16] Fujita Y *et al* 2004 *Phys. Rev.* C **70** 011306(R); Kawabata T *et al* 2004 *Phys. Rev.* C **70** 034318
[17] Kawabata T *et al* 2007 *Phys. Lett.* B **646** 7
[18] Kanada-En'yo Y 2007 *Phys. Rev.* C **75** 024302
[19] Yamada T and Funaki Y 2010 *Phys. Rev.* C **82** 064315
[20] Sasamoto Y *et al* 2006 *Mod. Phys. Lett.* A **82** 2393
[21] Fujimura H *et al* 2004 *Phys. Rev.* C **69** 064327
[22] von Oertzen W, Freer M and Kanada-En'yo Y 2006 *Phys. Rep.* **432** 43
[23] Milin M and von Oertzen W 2002 *Eur. Phys. J.* A **14** 295
[24] Yamada T and Funaki Y 2008 *Int. J. Mod. Phys.* E **17** 2101
[25] Tohsaki A 2008 *Int. J. Mod. Phys.* E **17** 2106
[26] Itagaki N *et al* 2007 *Phys. Rev.* C **75** 037303
[27] von Oertzen W 2006 *Eur. Phys. J.* A **29** 133
[28] Kokalova Tz *et al* 2005 *Eur. Phys. J.* A **23** 19
[29] Yamada T and Schuck P 2004 *Phys. Rev.* C **69** 024309
[30] Ito M, Hirabayashi Y and Sakuragi Y 2002 *Phys. Rev.* C **66** 034307
[31] Zhou B *et al* 2012 *Phys. Rev.* C **86** 014301
[32] Bayman B F and Bohr A 1958/59 *Nucl. Phys.* **9** 596
[33] Yamada T, Funaki Y, Horiuchi H, Ikeda K and Tohsaki A 2008 *Prog. Theor. Phys.* **120** 1139
[34] Horiuchi H 2009 *Int. Journ. Mod. Phys.* A **24** 2205
[35] Horiuchi H 2010 *Lecture Notes in Physics* **818** 57
[36] Horiuchi H 2012 *Rom. Journ. Phys.* **57** 227
[37] Hiyama E, Gibson B F, Kamimura M 2004 *Phys. Rev.* C **70** 031001(R)
[38] Arai K, Aoyama S, Suzuki Y, Descouvemont P and Baye D 2011 *Phys. Rev. Lett.* **107** 132502
[39] Roth R, Langhammer J, Calci A, Binder S and Navrátil P 2011 *Phys. Rev. Lett.* **107** 072501
[40] Epelbaum E, Krebs H, Lee D and Meisner U 2011 *Phys. Rev. Lett.* **106** 192501
[41] Funaki Y, Tohsaki A, Horiuchi H, Schuck P and Röpke G 2005 *Eur. Phys. J.* A **24** 321

Journal of Physics: Conference Series **436** (2013) 012004 doi:10.1088/1742-6596/436/1/012004

Gas-like alpha-cluster states and condensates in nuclei

Y Funaki, T Yamada[a], H Horiuchi[b], G Röpke[c], P Schuck[d] and A Tohsaki[b]

RIKEN, Nishina Center, Wako 351-0198, Japan
[a] Laboratory of Physics, Kanto Gakuin University, Yokohama 236-8501, Japan
[b] Research Center for Nuclear Physics (RCNP), Osaka University, Osaka 567-0047, Japan
[c] Institut für Physik, Universität Rostock, D-18051 Rostock, Germany
[d] Institut de Physique Nucléaire, 91406 Orsay Cedex, France

E-mail: funaki@riken.jp

Abstract. Gas-like α-cluster states are investigated in ^{16}O with the use of 4α OCM (Orthogonality Condition Model). The 0_6^+ state in ^{16}O is shown to have a character of a 4α condensate, an analogue to the Hoyle state in ^{12}C. The decay properties of this state and the α+Hoyle-like rotatinal states are also investigated.

1. Introduction

It is well established that α clustering plays a very important role in the structure of lighter nuclei [1, 2]. The importance of α-cluster formation has also been discussed in infinite nuclear matter, where α-particle type condensation is expected at low density [3, 4]. Although in nuclei there are only a few bosonic constituents, they may give rise to clear condensation characteristics, as is well known from nuclear pairing [5]. Concerning α-particle condensation, the Hoyle state, i.e. the 0_2^+ state in ^{12}C, has clearly been established. Several papers of the past [6, 7, 8, 9] and also more recent ones [10, 11, 12, 13, 14, 15] have by now established beyond doubt that the Hoyle state has only about one third of saturation density and can be described to good approximation as a product state of three α particles, condensed into the lowest mean field $0S$-orbit [16, 17]. This shall be the definition of an α-particle condensed state in finite nuclei, clearly reflecting the situation found in infinite matter in Ref. [3]. The emergence of this novel aspect of the Hoyle state naturally leads us to speculations about 4α-particle condensation in ^{16}O.

In the present contribution, the recent progress on the investigation of 4α-particle condensate states in ^{16}O is shown.

2. Four-alpha condensation in ^{16}O

It is well documented [18] that in ^{16}O the 0_2^+ state at 6.06 MeV contains an α particle orbiting in a $0S$-wave around the ground state of ^{12}C. However, there are many more possibilities. The α particle can be in higher nodal S-wave, or can orbit in a $0D$-wave around the 2_1^+ state of ^{12}C and couple to a 0^+ state. It can also orbit in an odd parity wave around the 1^- and 3^- states in ^{12}C etc. The ^{12}C core can also be in the Hoyle state which then leads us to the 4α gas state

Table 1. The rms radii R and monopole transition matrix elements to the ground state $M(\mathrm{E0})$ in units of fm and fm^2, respectively; $R_{\mathrm{exp.}}$ and $M(\mathrm{E0})_{\mathrm{exp.}}$ are the corresponding experimental data.

	R		M(E0)		$R_{\mathrm{exp.}}$	$M(\mathrm{E0})_{\mathrm{exp.}}$
	SW	MHN	SW	MHN		
0_1^+	2.7	2.7			2.71 ± 0.02	
0_2^+	3.0	3.0	4.1	3.9		3.55 ± 0.21
0_3^+	2.9	3.1	2.6	2.4		4.03 ± 0.09
0_4^+	4.0	4.0	3.0	2.4		no data
0_5^+	3.1	3.1	3.0	2.6		3.3 ± 0.7
0_6^+	5.0	5.6	0.5	1.0		no data

in ^{16}O, which is the state of our interest. Experimentally, there are six 0^+ states in ^{16}O up to around the 4α-disintegration threshold at 14.4 MeV: the ground state 0_1^+, 0_2^+ at 6.06 MeV, 0_3^+ at 12.05 MeV, 0_4^+ at 13.6 MeV [19], 0_5^+ at 14.01 MeV, and 0_6^+ at 15.1 MeV.

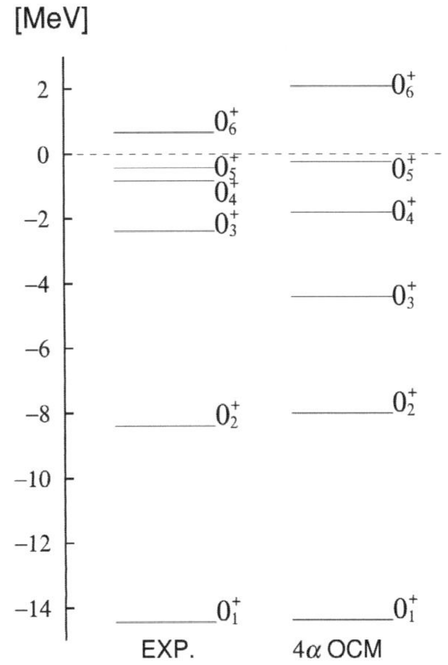

Figure 1. Comparison between energy spectra of experiments [20] and of 4α OCM calculations [21]. The 0_4^+ state in experiment is given in Ref. [19].

We recently performed a quite complete OCM calculation for ^{16}O, including many of the cluster configurations [21]. We were able to reproduce the full spectrum of the observed 0^+ states. As shown in Fig. 1, we tentatively make a one-to-one correspondence of our states with the six lowest 0^+ states of the experimental spectrum. In view of the complexity of the situation, the agreement can be considered as very satisfactory.

In Table 1 we show the calculated rms radii and monopole matrix elements to the ground state, together with the corresponding experimental values. The $M(\mathrm{E0})$ values for the 0_2^+ and 0_5^+ states are consistent with the corresponding experimental values. The consistency for the

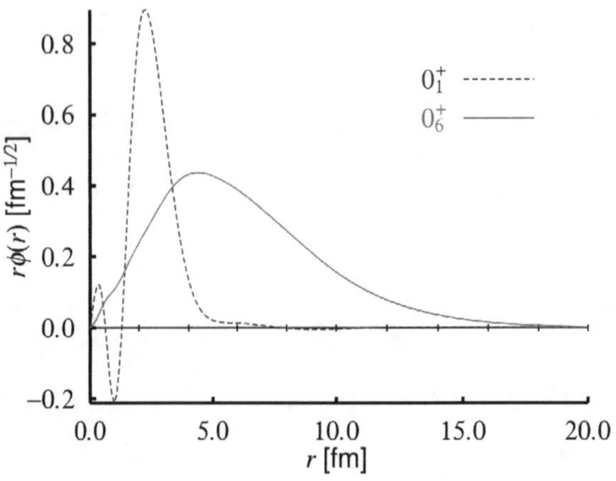

Figure 2. Single-α orbits with $L = 0$ belonging to the largest occupation number, for the ground and 0_6^+ states. The radial part of the orbits are shown.

0_3^+ state is within a factor of two. The structures of the 0_2^+ and 0_3^+ states are well established as having α+^{12}C(0_1^+) and α+^{12}C(2_1^+) cluster structures, respectively [18, 22, 23, 24]. These structures of the 0_2^+ and 0_3^+ states are confirmed in the present calculation. We also mention that the ground state is described as having a shell-model configuration consistent with the present framework, the calculated rms value agreeing with the observed one (2.71 fm). On the contrary, the structures of the observed 0_4^+, 0_5^+ and 0_6^+ states have, in the past, not clearly been understood. As shown in Fig. 1, the present calculation succeeds in reproducing the 0_4^+, 0_5^+ and 0_6^+ states, together with the 0_1^+, 0_2^+ and 0_3^+ states. This puts us in a favorable position to discuss the 4α-condensate state, expected to exist around the 4α threshold. In Table 1, the largest rms value of about 5 fm is found for the 0_6^+ state. Compared with the relatively smaller rms radii of the 0_4^+ and 0_5^+ states, this large size indicates that the 0_6^+ state is composed of a weakly interacting gas of α particles of the condensate type.

The analysis of the diagonalisation of the α-particle density matrix $\rho(\boldsymbol{r}, \boldsymbol{r}')$ (as was done in [16, 17, 25, 26]) showed that the newly discovered 0^+ state at 13.6 MeV [19], as well as the well known 0^+ state at 14.01 MeV, corresponding to our states at 12.6 MeV and 14.1 MeV, respectively, have very little condensate occupancy of the 0S-orbit (about 20%). On the other hand, the sixth 0^+ state, whose calculated energy is 16.5 MeV, to be identified with the experimental state at 15.1 MeV, has 61% of the α particles being in the 0S-orbit. The corresponding single-α 0S orbit is shown in Fig. 2. It has a strong spatially extended behaviour, and in the interor no node (0S), showing almost no effect of the Pauli principle exchanging nucleons belonging to different α particles. This indicates that the α particles are condensed into the very dilute 0S single-α orbit (see also Ref. [27]). Thus, the 0_6^+ state clearly has a 4α-condensate character. We should note that the orbit is very similar to the single-α orbit of the Hoyle state [17]. In Fig. 2 we also show the single-α orbit for the ground state. It has maximum amplitude at around 3 fm and oscillations in the interior with two-nodal (2S) behaviour, due to the Pauli principle and reflecting the shell-model configuration. The 0_6^+ state also has a very large radius of 5.6 fm, though this value may be somewhat overestimated because of the too high energy of the 0_6^+ state.

The condensate nature for the 0_6^+ state can also be seen clearly by the following analysis. We

Journal of Physics: Conference Series **436** (2013) 012004 doi:10.1088/1742-6596/436/1/012004

calculate the reduced-width amplitude, which is defined as follows:

$$\mathcal{Y}(r) = \sqrt{\frac{4!}{3!1!}} \Big\langle \Big[\frac{\delta(r'-r)}{r'^2} Y_L(\hat{\boldsymbol{r}}') \Phi_L(^{12}\mathrm{C}) \Big]_0 \Big| \Psi(0_6^+) \Big\rangle. \quad (1)$$

Here, $\Phi_L(^{12}\mathrm{C})$ is the wave function of ^{12}C, given by the 3α OCM calculation [17], and r is the relative distance between the centre-of-mass of ^{12}C and the α particle. From this quantity we can see how large is the component in a certain α+^{12}C channel which is contained in the 0_6^+ state. The amplitudes for the 0_6^+ state are shown in Fig. 3. It only has a large amplitude in the α+^{12}C(0_2^+) channel, whereas the amplitudes in other channels are much suppressed. The amplitude in the Hoyle-state channel has no oscillations and its long tail stretches out to $\sim$ 20 fm.

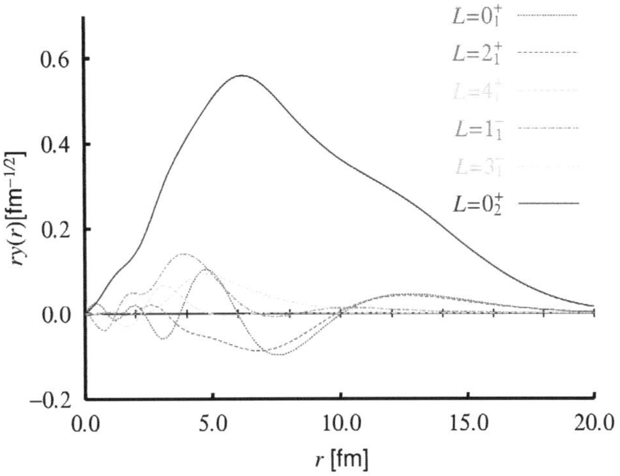

Figure 3. The reduced-width amplitudes $r\mathcal{Y}(r)$ for the 0_6^+ state, with $\mathcal{Y}(r)$ defined by Eq. (1); L denotes the orbital angular momentum of the remaining α-particle coupled to ^{12}C.

3. Decay properties

The α-particle condensate states occur near the α-particle disintegration threshold, which rapidly grows in energy with mass, and thus the level density in which such a condensate state is embedded raises enormously. For example the α-disintegration threshold in ^{12}C is at $E_x = 7.27$ MeV and in ^{16}O it is already at $E_x = 14.4$ MeV. Under ordinary circumstances this could mean that the α-particle condensate state in ^{16}O has a very short life-time. However, on the one hand, it is an experimental fact that the supposed ^{16}O 'Hoyle'-state at $E_x = 15.1$ MeV has a startlingly small width, 160 keV, for such a high excitation energy, and, on the other hand, it is easily understandable that such an exotic configuration as four α particles moving almost independently within the common Coulomb barrier, has great difficulties to decay into states, lower in energy, which all have very different configurations.

We make a quantitative estimate of the decay width of the $E_x = 15.1$ MeV state. Based on the R-matrix theory [28], the decay width Γ_L can be given by the following fomulae:

$$\Gamma_L = 2P_L(a)\gamma_L^2(a), \qquad P_L(a) = \frac{ka}{F_L^2(ka) + G_L^2(ka)},$$

$$\gamma_L^2(a) = \theta_L^2(a)\gamma_W^2(a), \ \gamma_W^2(a) = \frac{3\hbar^2}{2\mu a^2}, \tag{2}$$

where k, a and μ are the wave number of the relative motion, the channel radius, and the reduced mass, respectively, and F_L, G_L, and $P_L(a)$ are the regular and irregular Coulomb wave functions and the corresponding penetration factor, respectively. The reduced width of $\theta_L^2(a)$ is related to Eq. (1) like $\theta_L^2(a) = (a^3/3)\mathcal{Y}_L^2(a)$. In Table 2, we show the partial α-decay widths of the 0_6^+ state Γ_L decaying into the $\alpha+^{12}\mathrm{C}(0_1^+)$, $\alpha+^{12}\mathrm{C}(2_1^+)$ and $\alpha+^{12}\mathrm{C}(0_2^+)$ channels, the total α-decay width, which is obtained as a sum of the partial widths, and the reduced widths $\theta_L^2(a)$. The wave numbers k were calculated from the experimental decay energies. Thus the excitation energy of the 0_6^+ state was assumed to be 15.1 MeV, the one of the observed 0_6^+ state.

Table 2. Partial α widths in the 0_6^+ state of ^{16}O decaying into possible channels and the total width. The reduced widths defined as $\theta_L^2(a) = (a^3/3)\mathcal{Y}_L^2(a)$ at a certain channel radius a are also shown.

	$^{12}\mathrm{C}(0_1^+)+\alpha$ ($a = 8.0$ fm)	$^{12}\mathrm{C}(2_1^+)+\alpha$ ($a = 7.4$ fm)	$^{12}\mathrm{C}(0_2^+)+\alpha$ ($a = 8.0$ fm)	Total
Γ_L (keV)	104	32	8×10^{-7}	136
$\theta_L^2(a)$	0.024	0.016	0.6	

The very small total α-decay width of 136 keV obtained is in good agreement with the corresponding experimental value of 160 keV, and indicates that this state is unusually long-lived. This fact can be explained in terms of the present analysis as follows: Since this state has a very exotic structure composed of gas-like four α particles, the overlap between this state and the $\alpha+^{12}\mathrm{C}(0_1^+)$ or $\alpha+^{12}\mathrm{C}(2_1^+)$ wave functions with a certain channel radius becomes very small, as this is, indeed, indicated by the small $\theta_L^2(a)$ values, 0.024 and 0.016, respectively, which implies small $\gamma_L^2(a)$ values. These largely suppress the decay widths expressed by Eq. (2) in spite of the large values of the penetration factors caused by the large decay energies 7.9 MeV and 3.5 MeV into the $\alpha+^{12}\mathrm{C}(0_1^+)$ and $\alpha+^{12}\mathrm{C}(2_1^+)$ channels, respectively. On the other hand, the decay into $\alpha+^{12}\mathrm{C}(0_2^+)$ is also suppressed due to the very small penetration caused by the very small decay energy, 0.28 MeV, into this channel, even though the corresponding reduced width takes a large value, $\theta_L^2(a) = 0.6$, which is natural since the 0_2^+ state of ^{12}C has a gas-like three-α-particle structure. It is very likely that the above mechanism holds generally for the α-gas states in heavier $n\alpha$ systems, and therefore such states can also be expected to exist in heavier systems as relatively long-lived resonances.

4. Alpha + Hoyle rotatinal states

We further investigated the non-zero-spin excited states in ^{16}O, which are analogous to the Hoyle state, i.e., are members of the same family as the 0_6^+ state. We found that some states characteristically have large components of $\alpha+$ Hoyle state, where the α-particle orbits around the Hoyle state in S, P, D, F, and G waves for $J^\pi = 0^+, 1^-, 2^+, 3^-$, and 4^+, respectively. In Fig. 4, we show the spectra as a function of $J(J+1)$. One can suppose most naturally that these form rotational bands with $\alpha+$ Hoyle-state structure, with $K^\pi = 0^+$ for the positive parity states and $K^\pi = 0^-$ for the negative parity states. This was also very recently discussed in Ref. [29] and provides another possibility of interpreting the old [30] and rather recent [31] experimental data, which have been discussed for a long time in terms of a 4α-linear-chain structure.

It is interesting to see that the 0_6^+ state at 16.5 MeV strays off the $J(J+1)$ rotational line and gets more strongly bound. This indicates that it has a rather different structure from the

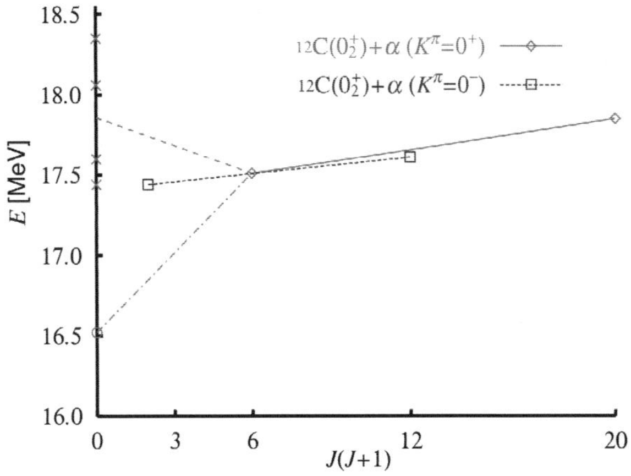

Figure 4. Rotational band of the α+Hoyle state. The dashed line starts from the average excitation energy, 17.9 MeV, of the four 0^+ states over the 0_6^+ state at 16.5 MeV.

other rotatinal members. One can say that when the α particle rotates around the Hoyle state in an S-wave, it seems to drop into the S-orbit which is already occupied by the three α particles in the Hoyle state, leading to the energy gain for the state. This might be interpreted as a transition from the local condensate to the complete condensate (see also Ref. [29]). On the other hand, we found four 0^+ states above the 0_6^+ state, which all have relatively large overlaps with an α+Hoyle configuration. They can be the real bandhead of the α+ Hoyle rotational band, instead of the 0_6^+ state. However, the average excitation energy, 17.9 MeV, is 0.5 MeV higher than the $J(J+1)$ rotational line. Due to the appearance of the 0_6^+ state, the real bandhead seems to be pushed up energetically and fragmented, since the α+ Hoyle state component must already be rather exausted in the 0_6^+ state.

5. Conclusion

The theoretical investigation of the Hoyle state in ^{12}C has established beyond doubt that it is a dilute gas-like state of three α particles, held together only by the Coulomb barrier, where the three bosons are condensed into the $0S$-orbital. There is no objective reason, why in ^{16}O, ^{20}Ne,$\cdots$ there should not exist similar 'Hoyle-like' states. In this contribution, we have given the results of a complete OCM calculation which reproduces the first six 0^+ states of ^{16}O. In that calculation the 0_6^+ state at 16.5 MeV, to be identified with the experimental 0^+ state at 15.1 MeV, shows characteristics typical for the Hoyle state. We argued that the 0_6^+ state has a very unusual structure and thus couples only very weakly to states at lower energies, leading to the very small width of 160 keV. Going on with this reasoning, it is not at all excluded that α-gas states in even heavier self-conjugate nuclei will have unusually long life-times, given their excitation energy high up in the continuum. We further discussed the rotational band of the α+ Hoyle state. The interplay between such rotational states and the 4α-codensate state is very interesting to investigate in much more detail. Further experimental information is also very much required to confirm this novel interpretation of these states.

References

[1] Wildermuth K and Tang Y C 1977 *A Unified Theory of the Nucleus* (Braunschweig: Vieweg)
[2] Ikeda K *et al* 1980 *Prog. Theor. Phys. Suppl.* **68** 1

[3] Röpke G *et al* 1998 *Phys. Rev. Lett.* **80** 3177
[4] Sogo T, Lazauskas R, Röpke G and Schuck P 2009 *Phys. Rev.* C **79** 051301(R)
[5] Ring P and Schuck P 1980 *The Nuclear Many-Body Problem* (Berlin: Springer-Verlag)
[6] Horiuchi H 1974 *Prog. Theor. Phys.* **51** 1266; 1975 *Prog. Theor. Phys.* **53** 447
[7] Kamimura M 1981 *Nucl. Phys.* A **351** 456
[8] Uegaki E *et al* 1977 *Prog. Theor. Phys.* **57** 1262; 1978 *Prog. Theor. Phys.* **59** 1031; 1979 *Prog. Theor. Phys.* **62** 1621
[9] Descouvemont P *et al* 1987 *Phys. Rev.* C **36** 54
[10] Tohsaki A *et al* 2001 *Phys. Rev. Lett.* **87** 192501
[11] Funaki Y *et al* 2003 *Phys. Rev.* C **67** 051306(R)
[12] Funaki Y *et al* 2005 *Eur. Phys. J.* A **24** 321
[13] Funaki Y *et al* 2006 *Eur. Phys. J.* A **28** 259
[14] Chernykh M *et al* 2007 *Phys. Rev. Lett.* **98** 032501; 2010 *Phys. Rev. Lett.* **105** 022501
[15] Funaki Y *et al* 2009 *Phys. Rev.* C **80** 064326
[16] Matsumura H *et al* 2004 *Nucl. Phys.* A **739** 238
[17] Yamada T *et al* 2005 *Eur. Phys. J.* A **26** 185
[18] Horiuchi H and Ikeda K 1968 *Prog. Theor. Phys.* **40** 277
[19] Wakasa T *et al* 2007 *Phys Lett* B **653** 173
[20] Ajzenberg-Selove F 1986 *Nucl. Phys.* A **460** 1
[21] Funaki Y *et al* 2008 *Phys. Rev. Lett.* **101** 082502
[22] Suzuki Y 1976 *Prog. Theor. Phys.* **55**, 1751; *Prog. Theor. Phys.* 1976 **56** 111
[23] Libert-Heinemann M *et al* 1980 *Nucl. Phys.* A **339** 429
[24] Fukatsu K *et al* 1992 *Prog. Theor. Phys.* **87** 151
[25] Suzuki Y *et al* 2002 *Phys. Rev.* C **65** 064318
[26] Yamada T *et al* 2008 *Phys. Rev.* A **78** 035603
[27] Funaki Y *et al* 2008 *Phys. Rev.* C **77** 064312
[28] Lane A M and Thomas R G 1958 *Rev. Mod. Phys.* **30** 257
[29] Ohkubo S and Hirabayashi Y 2010 *Phys. Lett.* B **684** 127
[30] Chevallier P *et al* 1961 *Phys. Rev.* **180** 827
[31] Freer M *et al* 1995 *Phys. Rev.* C **51** 1682

Journal of Physics: Conference Series **436** (2013) 012005 doi:10.1088/1742-6596/436/1/012005

Nonmicroscopic and microscopic descriptions of condensate states in the ^{12}C and ^{16}O nuclei

Marianne Dufour[1], **Rimantas Lazauskas**[1] **and Pierre Descouvemont**[2]

[1] IPHC Bat27, IN2P3-CNRS/Université Louis Pasteur BP28, F-67037 Strasbourg Cedex 2 - France

[2] Physique Nucléaire Théorique et Physique Mathématique, C.P. 229, Université Libre de Bruxelles (ULB), B 1050 Brussels, Belgium

Abstract. ^{12}C and ^{16}O nuclei are investigated within nonmicroscopic and microscopic theoretical frameworks, respectively. For the ^{12}C nucleus viewed as a 3α system, 3-body Faddeev equations are solved in configuration space. Positions of the resonant states are obtained through the complex scaling method. We show that the nonlocal potential developed by Z. Papp and S. Moszkowski appears to be well-adapted to study 3α system. In particular, we show evidence for ^{12}C states of positive-parity which share common features with the well-known 0_2^+ Hoyle state, currently interpreted as a condensate state. For the ^{16}O nucleus, a ^{12}C+α multicluster generator coordinate method is used to solve the 16-nucleon problem. The ^{16}O nucleus is described by four α cluster. We comment the difficulty to interpret broad resonant states. The phase-shift analysis of the ^{12}C$(0_2^+) + \alpha$ channel reveals the existence of two 0^+ states located above the 4α threshold which can be interpreted as condensate states.

1. Introduction

Owing to the compactness and strong binding of the α particles, the idea that they tend to keep their own identity inside the nuclei is at the origin of cluster models. According to the description of the α as an elementary or as a composite particle, nonmicroscopic and microscopic theoretical frameworks have been developed. The first applications were devoted to the study of so-called α-particle nuclei, that is $^{4n}_{2n}X$ type nuclei, with nucleon number $A = 4n$. In nonmicroscopic approaches, α being boson, nuclear states are described by $n\alpha$ symmetrized wave functions. Phenomenological potentials between the α's have therefore to be considered. In microscopic approaches, the nucleus is described at the nucleon scale. It is the analysis of the antisymmetrized wave functions described in adapted basis which allows to show evidence for clustering effects.

The transition from descriptions in terms of A fermions toward those in terms of n bosons is far from being obvious. A current challenge in nuclear structure is to show evidence for states in which α clustering is expected to be strongly dominant such as in so-called condensate states [1, 2]. A typical example is the 0_2^+ in the ^{12}C nucleus, known as the Hoyle state [1]. Condensate states being usually located near the relevant $n\alpha$ breakup threshold, a proper treatment of many-body resonances is necessary in most cases. This subject remains however a difficult theoretical topic.

In this publication, we present recent calculations performed within either a nonmicroscopic theoretical framework [3] or a microscopic one [4]. The nonmicroscopic calculation based on the Faddeev equations [5] is used to study the ^{12}C viewed as a 3α system (section 1). The chosen $\alpha\alpha$ interaction is the nonlocal potential developed by Z. Papp and S. Moszkowski [6]. The Complex Scaling Method (CSM) is used to analyze the resonances [7, 8]. The microscopic calculation is devoted to the ^{16}O nucleus (section 2). A multicluster $^{12}C + \alpha$ model is used to solve the 16-nucleon problem. The total wave functions of ^{16}O are fully antisymmetrized. They are obtained by the Generator Coordinate Method (GCM) combined with the Microscopic R Matrix method (MRM) [9]. This method allows an exact treatment of the asymptotic behavior of wave functions. We show the importance to calculate phase shifts in each channel of reaction to interpret broad resonances. In both approaches, we focuss here on the existence of states showing a clear $n\alpha$ structure interpreted here as the signature of α condensate states.

2. Nonmicroscopic calculations of 3α bosonic states in ^{12}C

One of the most important challenges in nonmicroscopic calculations is to replace the forces acting between nucleons by phenomenological potentials acting between the α-particles. These forces must be able to simulate not only the internal structure of the composite cluster but also, and most importantly, the Pauli principle between the single nucleons. In such a context, several local and nonlocal $\alpha\alpha$ potentials have been developed [3]. Among the well known local interactions, let us cite the Ali and Bodmer potential developed in the 1960's [10] and the B. Buck *et al.* potential [11]. However non-microscopic calculations based on these interactions fail to describe bound states of the ^{12}C nucleus and thus reveal the necessity to introduce phenomenological multi-body forces involving additional parametrization [12].

In this paper, we report on calculations performed with the nonlocal Papp and Moszkowski potential [6], which is of fish-bone type [13]. This potential represents a very accurate parametrization of the $\alpha\alpha$ system. It takes into consideration not only fully Pauli forbidden states of relative motion between the clusters but also partially Pauli forbidden states and thus may be considered as an extension of the Orthogonality Condition Model (OCM) [14]. Based on single parameter set for each partial wave this potential reproduces together the $\ell = 0, \ell = 2, \ell = 4$ $\alpha\alpha$ phase shifts up to 20 MeV, the $E_{2\alpha} = 91.6$ keV resonance in the ^{8}Be nucleus as well as the ground-state binding-energy of the ^{12}C.

For the ^{12}C nucleus viewed as a 3α system, 3-body Faddeev equations [5] are solved in configuration space. Resonant state energies and widths are determined by applying the CSM [7]. Complex scaling operation rotates the continuous spectra clockwise into the complex energy plane by an angle 2θ (where θ is the complex scaling angle parameter used in the calculation), whereas positions of bound or resonant states remain unaffected. If $\theta > \arctan(-Im(E_{res})/Re(E_{res}))$ complex scaled wave-functions of the resonant states become exponentially convergent and thus may be described using bound-state like techniques. In Figure 1 we present spectra obtained for 0^+ ^{12}C nucleus. Here, huge three-body matrices are not diagonalised. We are searching for the eigenvalues with the smallest $|Im(E_{res})/Re(E_{res})|$ value by using iterative methods and thus only small part of the eigenvalues are presented in this figure. We can identify two types of the continuum states which set the branches making 2θ angle with the real axis: one branch is associated with three-alpha particle continuum starting from the origin, the other branches are based on the ^{8}Be+α continuum and start from the respective positions of ^{8}Be states [8]. In Figure 1 the branch of the three-alpha particle continuum is difficult to dissociate from the ^{8}Be$(0^+) + \alpha$, ^{8}Be$(2^+) + \alpha$ branches. Since both ^{8}Be(0^+) and ^{8}Be(2^+) resonances are situated very close to the three-alpha particle continuum branch. On the contrary, the ^{8}Be$(0^+) + \alpha$ branch is clearly visible. In addition to these continuum states we identify one bound and three resonant states of the ^{12}C nucleus.

In Figure 2, we summarize ^{12}C positive parity spectrum. A very interesting result related to our study is the existence of a set of states $\{0_2^+, 2_2^+, 0_3^+, 0_4^+, 4_2^+\}$ which shares common feature with the well reproduced 0_2^+ Hoyle state. Indeed, they all show similarities in the wave functions and can be shown to belong to same rotational bands. We refer to the reference [3] for detailed

Journal of Physics: Conference Series **436** (2013) 012005 doi:10.1088/1742-6596/436/1/012005

discussions. Let us just mention that for the Hoyle state we get a rms radius of 3.98 fm compatible with the analysis of electron inelastic scattering data [15]. Furthemore, the 2_2^+ state at 4.42 MeV with a width of 0.66 MeV is interpreted as the 2^+ member of the Hoyle-state band. Its energy is however overestimated by ≈ 2 MeV as compared to the measure of Freer *et al.* [16]. We also assign the 4_2^+ state to a possible 4^+ member of the Hoyle-state band.

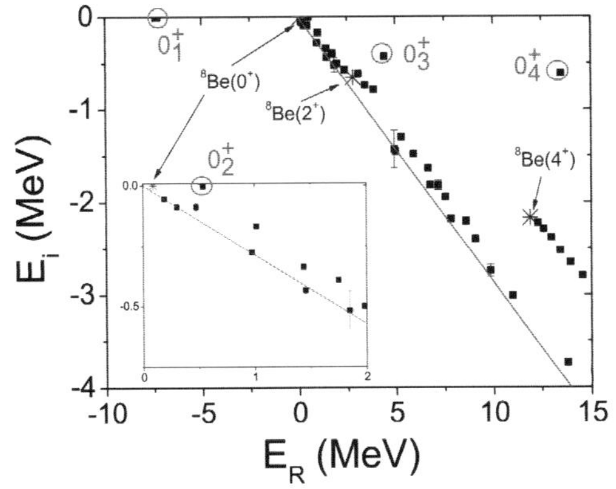

Figure 1. The 0^+ ^{12}C states obtained using the CSM. The bound and resonant states of ^{12}C are identified with circles. States representing discretized continuum form the bands making 2θ angle with the real energy axis. In this calculation complex scaling angle $\theta = 8°$ has been used.

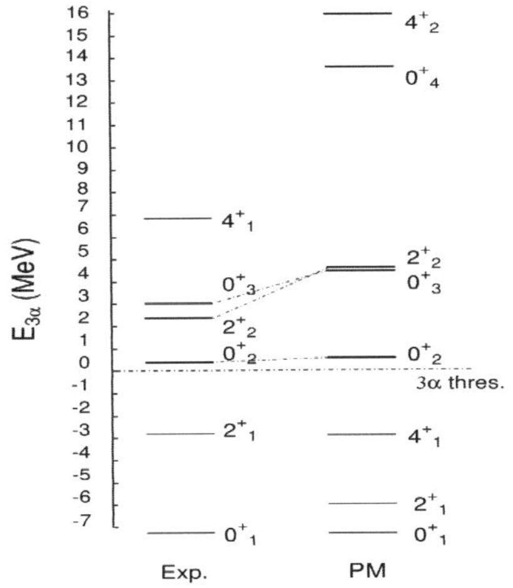

Figure 2. Ground-State band and positive-parity condensate states (bold) obtained with the nonlocal Papp and Moszkowski potential (PM). Corresponding experimental data (Exp.) are taken from Ref. [16] for the 2_2^+ and from Ref. [17] for the other data.

3. Microscopic calculations of 0^+ condensate states in ^{16}O

3.1. The multicluster model

In the present investigation, a multicluster generator coordinate method is applied to a microscopic study of the ^{16}O nucleus [4]. Details on the related theoretical framework and on the multicluster description can be found in Ref. [9, 18] and in references therein. We restrict ourselves to the particularities of the four-α ^{12}C+α model used here.

A schematic representation of the four-α cluster model is given in Figure 3. The ^{12}C is described by 3 α clusters located at the apexes of an isosceles triangle. The fourth cluster

Journal of Physics: Conference Series **436** (2013) 012005 doi:10.1088/1742-6596/436/1/012005

is devoted to the α particle. The side R_S and the angle θ of the isosceles triangle are the generator coordinates associated with the ^{12}C nucleus. The generator coordinate R is defined as the distance between the center of mass of the triangle and the fourth cluster. It controls the relative motion. This model generalizes a previous approach where the ^{12}C nucleus was described only in terms of equilateral triangles [19].

Let $\phi_\alpha(S)$ be a Slater determinant for the α particle, involving $0s$ harmonic-oscillator wave functions with parameter $b = 1.36$ fm, and centered at S, we then consider four α particles located at S_i, where i labels the ith apex of the multicluster model. S_i is a function of the generator coodinates. In these conditions, a basic GCM ^{16}O basis function can be written as follow

$$\phi(R_S, \theta, R) = \mathcal{A}\,\phi_{cm}^{-1}\,\phi(S1)\phi(S2)\phi(S3)\phi(S4), \tag{1}$$

where $\mathcal{A}$ is the antisymmetrization operator. Here, the center of mass function $\phi_{c.m.}$ is removed. The total wave functions of the ^{16}O nucleus are then expressed as a sum of ^{12}C+α multichannel functions. More precisely, in partial wave $JM\pi$, they can be written as

$$\Psi^{JM\pi} = \sum_{c,\ell I} \int f^{J\pi}_{c\ell I}(R)\ \Phi^{J\pi}_{c\ell I}(R)\, dR, \tag{2}$$

where the $\Phi^{J\pi}_{c\ell I}(R)$ functions are obtained from the $\phi(R_S, \theta, R)$ functions by usual techniques of projection. Here, label c represents the various ^{12}C+α channels. I is the ^{12}C spin, ℓ, the relative angular momentum, and $J = I + \ell$, the ^{16}O spin. In our calculations, two I^π values are considered: 0^+ and 2^+. The ^{16}O parity is then given by $\pi = (-1)^\ell$. The generator functions $f^{J\pi}_{c\ell I}(R_S, \theta, R)$ are determined by solving the MRM which restores the correct Coulomb behavior of the wave functions. This step is crucial to treat properly resonating states.

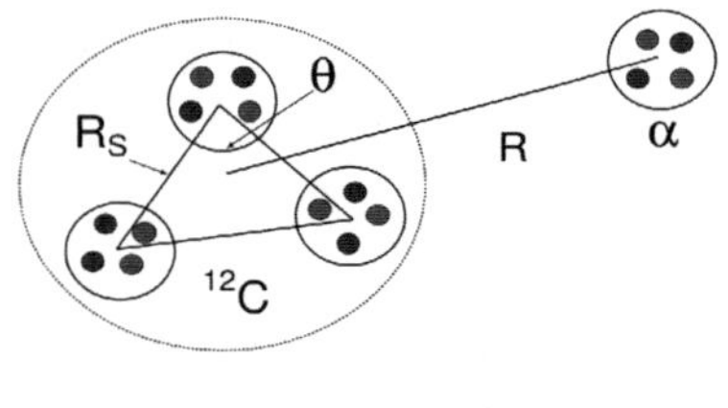

Figure 3. Schematic representation of the GCM basis wave functions used in the four-alpha multicluster model.

Our calculations have been performed with the Volkov V2 [20] and the Minnesota [21] nucleon-nucleon interactions. Both contain one adjustable parameter which is tuned to reproduce the 0_2^+ state located at $E_{c.m.} = -1.11$ MeV below the ^{12}C+α threshold. Indeed, previous studies show that this state can be well reproduced by cluster models [22]. Value of the parameter is $m = 0.63726$ for V2 and $u = 0.82225$ for Minnesota, respectively.

3.2. The $^{12}C(0_2^+)$ +α channel

In practice, Eq. (2) becomes a finite sum over the generator coordinates. R takes values from $R = 2.0$ fm to $R = 10.4$ fm in steps of 1.2 fm. The generator coordinates associated with the isosceles triangles are chosen in order to optimize the ^{12}C(0_2^+) - ^{12}C(0_1^+) energy gap. This point is crucial. Indeed, the interpretation of ^{16}O resonances in terms of condensate of α particles is linked to the importance of the ^{12}C(0_2^+) +α channel in the corresponding wave functions. A good description of this channel is then necessary. In our calculation, seven different triangles have been considered. The corresponding values of R_S and θ are gathered in Table 1. They

	R_S (fm)	θ (degree)
T1	2.1	28.1
T2	3.0	100.0
T3	4.0	30.0
T4	5.0	45.0
T5	6.0	30.0
T6	4.0	110.0
T7	1.0	60.0

Table 1. Generator coordinates of the isosceles triangles considered in the multicluster calculations.

have been chosen step by step in order to minimize the ground state and the gap energy. We can notice that most of the triangles are isosceles and that the chain configuration (θ = 180 degree) is not selected.

The evolution of the $^{12}C(0_2^+)$ - $^{12}C(0_1^+)$ energy gap as a function of the number of triangles introduced in the GCM variational basis is given in Figure 4. We can notice the clear improvement obtained when the number of triangles increases. For seven triangles, the result becomes close to experiment for both interactions.

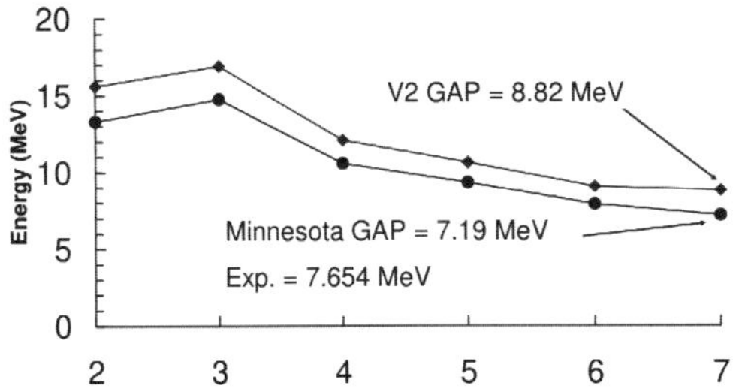

Figure 4. Evolution of the $^{12}C(0_2^+)$ - $^{12}C(0_1^+)$ energy gap as a function of the number of triangles introduced in the GCM calculations.

3.3. Many-body resonance analysis

In this sub-section, we comment the difficulties to interpret broad resonances. In Table 2, we show typical calculations performed with the V2 interaction for the 0^+ states. Results obtained with a bound-state approximation are given in the first column. The MRM output involving the correction of the asymptotic behavior, is given in the second column for the energies and the third for the widths, respectively. As expected, we get similar results for the two bound states. Then we get resonance energies which appear to be close to the bound-state values. The corresponding widths calculated with usual techniques based on Breit-Wigner approximation [23] are however (very) broad. The interpretation of such states as physical ones is then questionable. The way to disentangle between physical and non physical states is to compute the phase shifts in each channel of reaction. Results obtained for 0^+ partial waves obtained with V2 are given in Figure 5. The stability between six and seven triangle calculations is also tested. We find that only three resonances can be considered as physical. They are indicated with a X in Table 2. Furthermore, the analysis of reduced α widths θ_c^2 shows that the resonances at $E_{c.m.}$ = 9.742 MeV and at $E_{c.m.}$ = 11.498 MeV have dominant components in the $^{12}C(0_2^+)$ $+\alpha$ channel. This result supports an interpretation in terms of α condensates, in agreement with calculations based on a 4α THSR wave-function by Y. Funaki *et al.* [24] Corresponding values are gathered in Table 3. This study illustrates the importance to compute phase shifts. It also shows that bound-state approximations are generally not well appropriate to describe resonances.

	$E^{BSA}_{c.m.}$ (MeV)	$E^{MRM}_{c.m.}$ (MeV)	Γ (MeV)	
1	-14.045	-14.045		
2	-1.112	-1.112		
3	2.257	2.200	0.46	X
4	3.686	3.818	1.86	
5	5.973	6.369	1.96	
6	6.635	6.859	3.71	
7	9.060	9.724	0.37	X
8	9.705	10.137	4.33	
9	11.083	11.498	1.45	X
10	11.676	13.897	6.76	
11	13.810	13.997	0.92	
12	14.226	15.701	4.68	

Table 2. ^{16}O 0^+ states obtained with a bound-state approximation (BSA) and with the MRM. The corresponding total width is given in the third column.

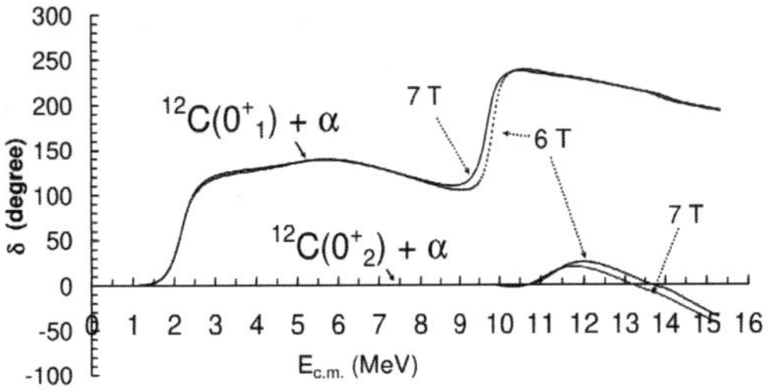

Figure 5. Phase shift analysis of the 0^+ resonances performed for 6 and 7 isosceles triangles.

$0^+ E_{c.m.}$	I, ℓ Channel	θ_c^2
9.72 MeV		
	$I = 0_1^+\ \ell = 0$	6.4 $\times 10^{-2}$
	$I = 0_2^+\ \ell = 0$	6.8 $\times 10^{-2}$
	$I = 2_1^+\ \ell = 0$	2.4 $\times 10^{-2}$
11.49 MeV		
	$I = 0_1^+\ \ell = 0$	1.5 $\times 10^{-2}$
	$I = 0_2^+\ \ell = 0$	7.7 $\times 10^{-1}$
	$I = 2_1^+\ \ell = 2$	4.1 $\times 10^{-2}$

Table 3. Three dominant reduced α widths (in %) for the two resonances interpreted as α condensate states obtained with V2.

3.4. The ^{16}O spectrum

In Figure 6, we show the ^{16}O 0^+ spectra obtained with our multicluster approach. The 0_1^+ binding energy is overestimated by both the V2 and the Minnesota interactions and are not reproduced on the picture. This problem is well known in cluster model [22]. The calculated 0_3^+ resonance at $E_{c.m.} = 2.2$ MeV (with V2) with a width of $\Gamma = 0.46$ MeV cannot be directly assigned. The calculated 0_4^+ and 0_5^+ resonances previously discussed as possible condensate states are located above the 4α threshold. Experiment reveals the presence of a 0^+ state located at $E_{c.m.} = 7.935$ MeV with a width of $\Gamma = 0.166$ MeV [25] which could correspond to one of them. Furthemore, if we compare with the band measured by Chevallier *et al.* [27], we find that the extrapolated 0^+ state at $E_{c.m.} = 9.64$ MeV is very close to the lowest condensate energy obtained with both interactions.

Journal of Physics: Conference Series **436** (2013) 012005 doi:10.1088/1742-6596/436/1/012005

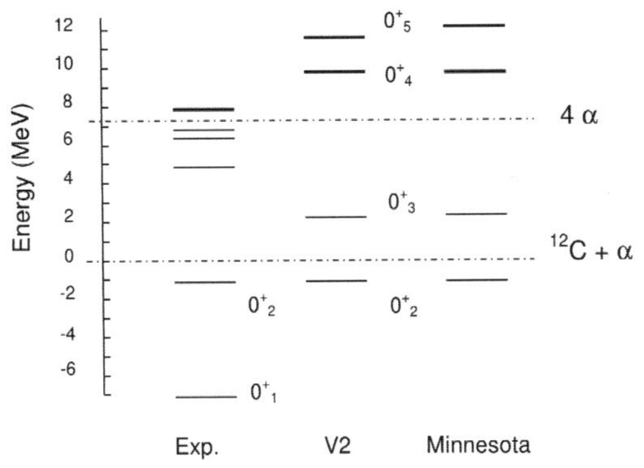

Figure 6. ^{16}O 0^+ energy spectra obtained with the four-cluster calculations for V2 and Minnesota interactions. Condensates are indicated in bold. Experimental data are taken from Ref. [25] and from Ref. [26] for the 0_4^+.

4. Conclusion

In this paper, we have seen that both nonmicroscopic and microscopic approaches are able to show the existence of condensate states.

References

[1] Hoyle F 1954 *Astrophys. J. Suppl.* **1** 121
[2] Tohsaki A, Horiuchi H, Schuck P and Röpke G 2001 *Phys. Rev. Lett.* **87** 192501
[3] Lazauskas R and Dufour M 2011 *Phys. Rev.* C **84**(6) 064318
[4] Dufour M and Descouvemont P in preparation
[5] Faddeev L D 1961 *Sov. Phys. JETP* **12** 1014–1019
[6] Papp Z and Moszkowski S 2008 *Mod. Phys. Lett.* B **22** 2201
[7] Nuttal J and Cohen H L 1969 *Phys. Rev.* **188** 1542–1544
[8] Aoyama S, Myo T, Katō K and Ikeda K 2006 *Prog. Theor. Phys.* **116** 1
[9] Descouvemont P and Dufour M *Clusters in Nuclei, Vol.2* (Springer, Lectures Notes in Physics, 848)
[10] Ali S and Bodmer A R 1966 *Nucl. Phys.* **80** 99
[11] Buck B, Friedrich H and Wheatley C 1977 *Nucl. Phys. A* **275** 246
[12] Funaki Y, Yamada T, Horiuchi H, Röpke G, Schuck P and Tohsaki A 2008 *Phys. Rev. Lett.* **101** 082502
[13] Schmid E W 1980 *Z. Phys.* A **297** 105
[14] Saito S 1977 *Prog. Theor. Phys. Suppl.* **62** 11
[15] Funaki Y, Tohsaki A, Horiuchi H, Schuck P and Röpke G 2006 *Eur. J. Phys.* A **28** 259
[16] Freer M *et al.* 2009 *Phys. Rev.* C **80** 041303
[17] Ajzenberg-Selove F 1990 *Nucl. Phys.* A **506** 1
[18] Dufour M, Descouvemont P and Baye D 1994 *Phys. Rev.* C **50** 795
[19] Descouvemont P 1991 *Phys. Rev.* C **44** 306
[20] Volkov A B 1965 *Nucl. Phys.* **74** 33
[21] Thompson D R, LeMere M and Tang Y C 1977 *Nucl. Phys.* A **286** 53
[22] Dufour M and Descouvemont P 2008 *Phys. Rev.* C **78** 015808
[23] Descouvemont P and Vincke M 1990 *Phys. Rev.* A **42** 3835
[24] Funaki Y, Yamada T, Tohsaki A, Horiuchi H, Röpke G and Schuck P 2010 *Phys. Rev.* C **82**(2) 024312
[25] Ajzenberg-Selove F 1982 *Nucl. Phys.* A **375** 1
[26] Wakasa T *et al.* 2007 *Physics Letters* B **653** 173
[27] Chevallier P and Scheibling F 1967 *Phys. Rev.* **160** 827

Journal of Physics: Conference Series **436** (2013) 012006 doi:10.1088/1742-6596/436/1/012006

Nature of 10 MeV state in ^{12}C

Masatoshi Itoh[1], Hidetoshi Akimune[2], Mamoru Fujiwara[3], Umesh Garg[4], Takahiro Kawabata[6], Keigo Kawase[5], Tetsuya Murakami[6], Kosuke Nakanishi[5], Yohei Nakatsugawa[5], Harutaka Sakaguchi[5], Satoru Terashima[7], Makoto Uchida[8], Yuusuke Yasuda[5], Masaru Yosoi[5] and Juzo Zenihiro[7]

[1]Cyclotron and Radioisotope Center, Tohoku University, Sendai, Japan
[2]Department of Physics, Konan University, Kobe, Japan
[3]Faculty of Education, Gifu University, Gifu, Japan
[4]Department of Physics, Notre Dam University, USA
[5]Research Center for Nuclear Physics (RCNP), Osaka University, Osaka, Japan
[6]Department of Physics, Kyoto University, Kyoto, Japan
[7]RIKEN, Saitama, Japan
[8]Department of Physics, Tokyo Institute of Technology, Tokyo, Japan

E-mail: itoh@cyric.tohoku.ac.jp

Abstract. We performed a coincidence measurement of decay-α particles by using inelastic scattering of 386 MeV α particles on ^{12}C. We obtained further evidence for the second 2^+ state at 9.84 MeV, which we found in previous work, from angular correlation functions for decay-α particles. The coincidence energy spectra with the α-decay to the ground state and the first excited 2^+ state in ^{8}Be from the 10 MeV states in ^{12}C were obtained. They indicated the broad 0^+ state at 10 MeV consists of two components. The lower 0^+ state do α decay to the ground state of ^{8}Be only. The higher 0^+ state has a distinct peak at 10.8 MeV with a width of 0.4 MeV in the coincidence spectrum for the first excited state of ^{8}Be channel. They have consistent properties predicted for a higher nodal state of the Hoyle state and a linear-like 3α state, respectively.

1. Introduction

The 0_2^+ state at E_x = 7.65 MeV in ^{12}C known as Hoyle state plays important role of a nucleosynthesis in the universe and is considered to have a typical 3α cluster structure. In an early stage of the study on the structure of Hoyle state, Morinaga proposed it had a linear 3α chain structure [1]. However, this configuration of the 3α clusters was unstable energetically. According to the microscopic α-cluster models such as 3α-GCM [2] and 3α-RGM [3], this state is considered to have a 3α gas-like structure in which 3α clusters are weakly coupled. About ten years ago, Tohsaki *et al* showed this 3α gas-like structure was similar to the Bose-Einstein condensate of the 3α clusters by using the condensate model wave function [4]. In this new concept, the 3α clusters entered into the lowest S orbit.

These α cluster models also predicted the 2^+ state at around 10 MeV. In the previous experiment, we have found the 2_2^+ state at 9.84 MeV with the width of 1.0 MeV by inelastic α scattering [5]. The 2^+ state at 9.6 MeV was also found in the (p,p′) experiment [6]. From the combined analysis of the (α,α') and (p,p') reactions, the 2^+ states at 9.84 and 9.6 MeV are confirmed to be identical [7]. Recently, M. Gai *et al* performed the ^{12}C(γ,3α) experiment at High Intensity Gamma-Ray Source (HIGS) facility in the Duke University [8]. They also measured the 2^+ strength at around 10 MeV.

In this way, the 2_2^+ state at 10 MeV was established. This result supports the 3α gas-like structure of the Hoyle state. The next problem is the structure of the broad 0^+ state at 10 MeV. This broad 0^+ state is also considered to have a 3α cluster structure. Kurokawa and Katō claimed the broad 0^+ state at 10 MeV was a higher nodal state of the Hoyle state by using the Complex Scaling Method [9]. Kanada-En'yo predicted the 0^+ state of the linear-like 3α structure appeared at around 10 MeV in the framework of AMD [10]. In our previous study, we have obtained the 0^+ strength distribution at around 10 MeV. It was almost consistent with the excitation energy and the width of the broad 0^+ state, if it consisted of a single 0^+ state. However, our result indicated it consists of two 0^+ components at 9.04 MeV and 10.56 MeV, as shown in Ref. [5] Fig.8(a).

In this report, we present results of the coincidence measurement of the ^{12}C(α,$\alpha'\overline{\alpha}$) reaction. We obtained angular correlation functions (ACFs) for the α_0 decay to the ground state of ^{8}Be and further evidence for the existence of the 2_2^+ state at 9.84 MeV. We also have found a new peak at 10.8 MeV in the coincidence energy spectrum for the α_1 decay to the first excited 2^+ state of ^{8}Be at 0°. This is additional evidence for consisting of two 0^+ components in the broad 0^+ state at 10 MeV. By using the coincidence energy spectrum, natures of the 10 MeV state including this new peak are discussed.

2. Experiment

The experiment was performed at the ring cyclotron facility of Research Center for Nuclear Physics (RCNP), Osaka University. The ^{4}He^{++} beam was accelerated up to 386 MeV by using both of the AVF cyclotron and the ring cyclotron, and bombarded to the self-supporting natural Carbon foil target with a thickness of 100 μg/cm^2. The size of the beam spot was less than a diameter of 1 mm. The target was rotated by 10 degrees in order to optimize the energy loss of decay-α particles for the detectors configuration.

The scattered α particles were momentum-analyzed in the Grand Raiden spectrometer and detected in the focal-plane detectors system comprised of two multi-wire drift chambers and two plastic scintillators, which provided both the horizontal and vertical trajectories. The measurement angles (θ_{GR}) were 0° and 4° in the laboratory frame, which correspond to those of the cross section maximum for the 0^+ and 2^+ states, respectively. The angles inside the opening angle and momentum of scattered particles were determined by the ray-tracing method.

Decay-α particles were detected by the silicon detectors array, which was composed of fourteen silicon detectors of the surface-barrier type and installed into the scattering chamber at backward angles from 115° to 255° by the 10° step. A 2 mm thick brass collimator with a circular opening of 22 mm diameter was mounted at 250 mm from the target center in front of each silicon detector to define the solid angle. The defining solid angle was about 6 msr, which correspond to the opening angle of 5°. In front of the collimator, a pair of small magnets was attached in order to reduce the background noise due to electrons ejected out of the target. Thickness of silicon detectors were 300 μm (four silicon detectors) and 500 μm (the others), which were enough to measure energies of decay-α particles from the 10 MeV state in ^{12}C. The identification of decay particles was done by the time of flight method, although the excited states at around 10 MeV in ^{12}C decay essentially only by α particles except very small γ-decay.

Figure 1 shows the two-dimensional coincidence energy spectrum. The locus corresponding to the decay channel of ^{12}C* $\rightarrow$ ^{8}Be + α_0 is clearly observed.

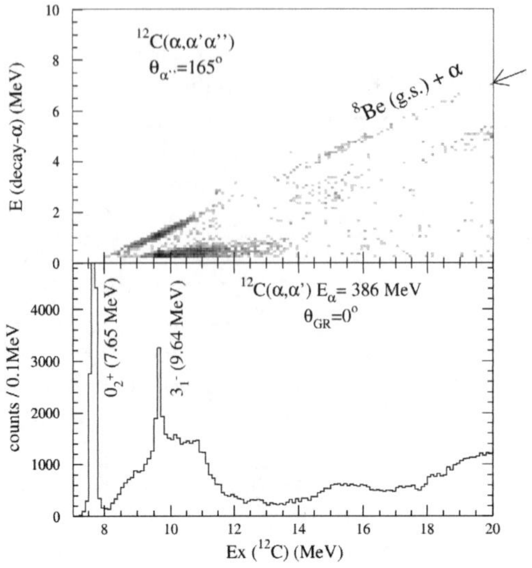

Figure 1. Upper panel shows a two-dimensional scattered plot for coincidence events induced by the ^{12}C(α,α') reaction at 0°. Lower panel is a singles spectrum of the ^{12}C(α,α') reaction at 0°.

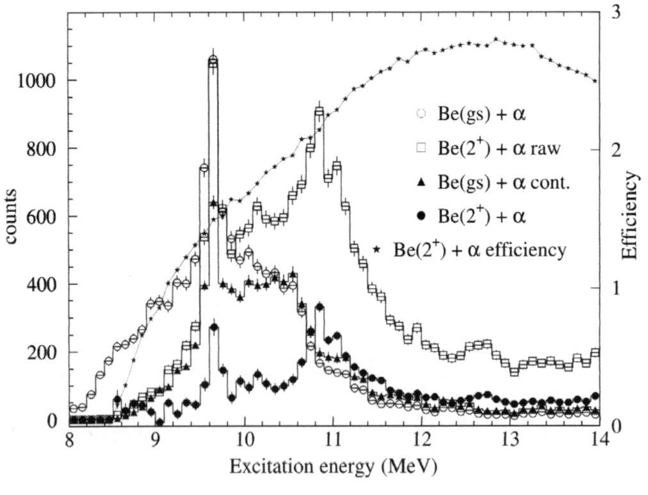

Figure 2. Comparison of energy spectra for the ^{12}C(α,α')^{12}C*[α + X] reaction at 0° in coincidence with decay-α particles for ^{8}Be(0_1^+) and ^{8}Be(2_1^+) channels.

3. Analysis and Results

The coincidence energy spectra with the ground state (g.s.) and the 2_1^+ state of ^{8}Be decay channels were obtained. The excitation energy of ^{12}C* was determined by the missing mass method of inelastically scattered α particles. The excitation energy of ^{8}Be was obtained from the kinematics by assuming the recoil ^{12}C* decay to ^{8}Be and α particles. In Fig. 2, open circles and open squares show coincidence energy spectra at $\theta_{GR} = 0°$ for the ^{8}Be(g.s.) and the ^{8}Be(2_1^+) decay channels, respectively. Decay-α particles for the excitation energy of ^{8}Be from 0.5 to 5 MeV were identified to be those from the 2_1^+ state at 3.0 MeV with the width of 1.5 MeV. However, that of the ^{8}Be(2_1^+) channel contains decay-α particles from ^{8}Be(g.s.) and ^{8}Be(2_1^+). The contribution from the decay of ^{8}Be(g.s.), which is shown by closed triangles, was estimated from the coincidence energy spectrum with the ^{8}Be(g.s.) channel and subtracted from that with the ^{8}Be(2_1^+) channel. The contribution from the decay of ^{8}Be(2_1^+) was estimated by the Monte-Carlo simulation, which is displayed as the detection efficiency (closed stars) for the ^{8}Be(2_1^+) decay channel. The real coincidence energy spectrum with the ^{8}Be(2_1^+) channel, which is shown by closed circles in Fig. 2, was obtained by multiplying this efficiency to the raw spectrum. A distinct peak at about 10.8 MeV with the width of 0.4 MeV still remains in the coincidence energy spectrum for the ^{8}Be(2_1^+) channel.

The ACF for decay-α particles from the intermediate state to the final 0^+ state is proportional to $[P_0^L(\cos\theta)]^2$ in the PWBA calculation. L is the angular momentum of the intermediate state. θ is the angle of the decay-α particle with respect to the recoil particle. This is a good approximation in the 0° measurement, since there are no m-substate populations except m = 0 at $\theta_{GR} = 0°$. If resonances overlapped in the same excitation energy, the ACF is expressed as a coherent summation of the square of Legendre functions. Therefore, the ACF in an energy interval is expressed as,

$$W(\theta) \propto \Big(\sum_L a_L P_L(\cos\theta)\Big)^2 + \sum_L \Big(b_L P_L(\cos\theta)\Big)^2. \quad (1)$$

a$_L$ and b$_L$ are amplitudes of states for angular momentum L in coherent and incoherent parts, respectively. Figure 3 shows the ACFs for decay-α particles of the ^{12}C* → ^{8}Be(g.s.) + α channel at $\theta_{GR} = 0°$. Dashed-lines show incoherent calculations in the PWBA. On the other hand, solid lines show calculations taken into account for overlapping of resonances. Since the intrinsic width of the 3_1^- state at 9.64 MeV is much smaller than the broad 0^+ state and the second 2^+ state, the ACF was almost same as the incoherent summation of the 0^+ and 3^- states. In Fig. 3(b), 3(d), and 3(c), ACFs summed 0^+ and 2^+ states coherently reproduce experimental data better than incoherent summed ACFs. Figure 3(a) and 3(f) show the ACF for only the 0^+ state.

Figure 4 shows ACFs at $\theta_{GR} = 4°$. The 0^+ contributions are almost disappeared at this angle due to the angular distribution of inelastic α scattering. Therefore, we calculated ACFs by assuming resonance was not overlapped with each other. ACFs were generally written as,

$$W(\theta) \propto \sum_\nu G_\nu C(LL\nu;00)W(j_ij_iLL;\nu j_f)P_\nu(\cos\theta). \quad (2)$$

$C(LL\nu;00)$ and $W(j_ij_iLL;\nu j_f)$ are a Clebsch-Gordan coefficient and a Racah coefficient, respectively. G_ν, which is a statistical tensor, was calculated by ECIS95 [11]. The ACFs calculated in the DWBA were shifted by 10° in order to reproduce experimental data. This effect was reported in the measurement of decay-α particles from giant resonances in ^{24}Mg and ^{40}Ca on inelastic α scattering at 120 MeV [12]. Figure 4(a) shows the ACF for the excitation energy range from 9.55 MeV to 9.75 MeV which corresponds to the 3_1^- state at 9.64 MeV. Since the calculated ACF (solid line) reproduces experimental data well, a contribution from the 2^+ state may be small in this energy region. ACFs calculated for the 2^+ state were in agreement with experimental data in energy regions from 9.8 to 10.2 MeV and from 10.2 to 10.6 MeV except at around 100°, as shown in Fig. 4(b) and 4(c). On the other hand, the ACF in the region from 10.6 to 11.0 MeV was failed to describe experimental data. It was not a simple ACF for the 1_1^- state at 10.84 MeV.

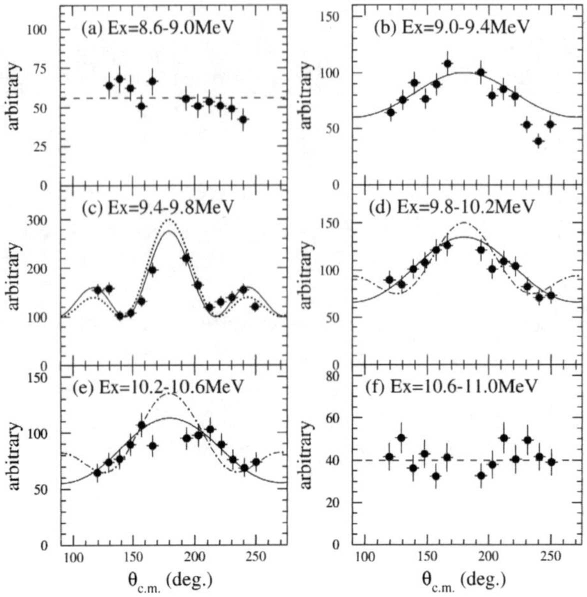

Figure 3. Angular correlation functions for decay-α particles to the ^{8}Be(g.s.) + α channel in the ^{12}C(α,α')^{12}C* reaction at $\theta_{GR} = 0°$. Vertical axis means counts normalized by solid angles in the c.m. frame.

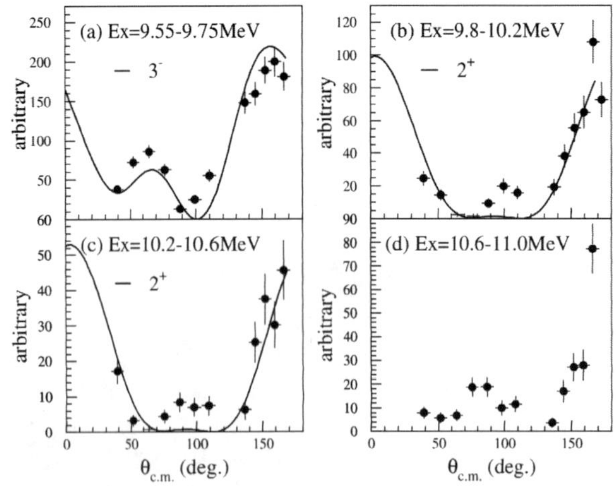

Figure 4. Angular correlation functions for decay-α particles to the ^{8}Be(g.s.) + α channel in the ^{12}C(α,α')^{12}C* reaction at $\theta_{GR} = 4°$.

4. Discussion

Figure 5 shows the comparison of energy spectra in the singles measurement of our previous experiment [5] (thick solid lines) and in the coincidence measurement of the present work

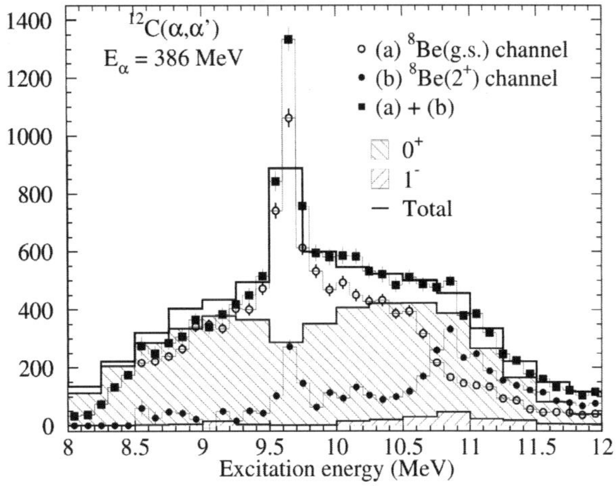

Figure 5. Comparison between the result of the multipole decomposition analysis and the present one at $\theta_{GR} = 0°$.

(symbols). The energy spectrum estimated by summing both decay channels is displayed with closed squares. The ACF for the azimuthal angle was assumed to be uniform. The fact that the summed energy spectrum is in good agreement with the singles spectrum in ^{12}C indicates coincidence energy spectra for the ^{8}Be(g.s.) and ^{8}Be(2_1^+) decay channels were reasonably obtained. The ratio of energy spectra between these two decay channels is directly related to the branching ratio of the state, although several states are contained in energy spectra. However, in the energy spectrum at 0°, main parts are 0^+ states, as shown with the right hatched area in Fig. 5. Although the peak energy at 10.8 MeV in coincidence spectrum for the ^{8}Be(2_1^+) channel appears at the almost same energy as the 10.84 MeV 1^- state, the 1^- contribution to this peak is very small, as shown with the left hatched area in Fig. 5.

Consequently, following results are obtained. The lower 0^+ component do α-decay only to the ^{8}Be(g.s.) channel. This property is consistent with the higher nodal 0^+ state of Hoyle state predicted by Kurokawa and Katō [9]. On the other hand, the higher 0^+ component has a large branching ratio to the ^{8}Be(2_1^+) channel and is expected to have a linear-like 3α structure, which is dominated by the ^{8}Be(2_1^+) + α component according to the AMD calculation [10].

Figure 6 shows the energies of positive-parity resonances around 10 MeV in relation to this topic. Recently, Freer *et al* found the 4^+ state at 13.3 MeV with a width of 1.7 MeV in ^{12}C [13]. If the 2_2^+ state at 9.84 MeV has a strong coupling to this 4^+ state as a member of a rotational band of the ^{8}Be + α structure, the band head 0^+ state appears at around 8.5 MeV, as shown in Fig. 6. For that reason, the band head 0^+ state of the ^{8}Be + α structure is considered to fragment into two parts of the Hoyle state at 7.65 MeV and the lower 0^+ state at 9.04 MeV.

5. Summary

The decay properties from the 10 MeV states in ^{12}C have been investigated by the coincidence measurement of the ^{12}C(α,$\alpha'\overline{\alpha}$) reaction. Angular correlation functions for decay-α particles from the 10 MeV states in ^{12}C were measured and compared with calculations. The results were

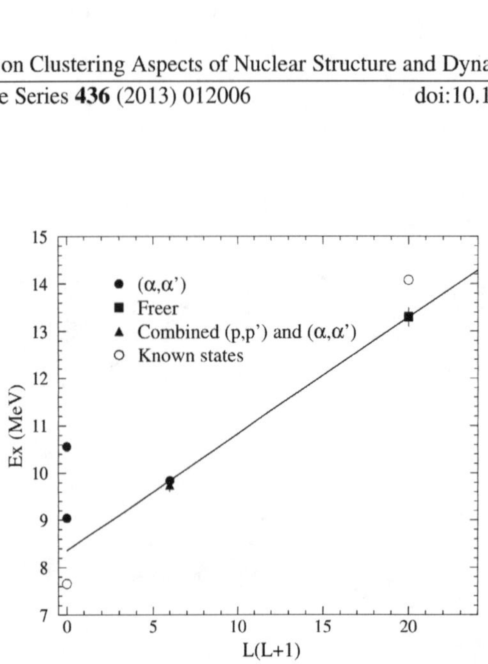

Figure 6. The energies of positive-parity resonances around 10 MeV on an $L(L+1)$ scale. The line was drawn for the eye-guide.

consistent with the previous work on the singles measurement of the ^{12}C(α,α') reaction. The existence of the 2_2^+ state at 9.84 MeV in ^{12}C has been confirmed.

The coincidence energy spectra on ^{12}C for decay channels of ^{8}Be(g.s.) and ^{8}Be(2_1^+) were obtained. In the coincidence spectrum for the ^{8}Be(2_1^+) channel, a distinct peak at 10.8 MeV with a width of 0.4 MeV was observed. It indicates that the broad 0^+ state at 10 MeV in ^{12}C can consist of two components. The lower 0^+ state is considered to have the ^{8}Be(g.s.) + α structure and to be the higher nodal state of the Hoyle state. On the other hand, the higher 0^+ state can be a linear-like 3α structure, which has a large ^{8}Be(2_1^+) + α component.

If the 2_2^+ state at 9.84 MeV and recently found 4^+ state at 13.3 MeV have a rotational structure, the band head 0^+ state can be fragmented into two parts. One is the Hoyle state and another is the lower 0^+ state at 9.04 MeV.

Acknowledgments

Authors acknowledge to the RCNP cyclotron staffs for providing very stable and high quality beam during the experiment. This work was supported in part by JSPS KAKENHI Grant Number 24740139.

References

[1] Morinaga H 1956 *Phys. Rev.* **101** 254
[2] Uegaki E *et al.* 1977 *Prog. Theor. Phys.* **57** 1262
[3] Kamimura M 1981 *Nucl. Phys.* A **351** 456
[4] Tohsaki A, Horiuchi H, Schuck P and Röpke G 2001 *Phys. Rev. Lett.* **87** 192501
[5] Itoh M *et al* 2011 *Phys. Rev.* C **84** 054308
[6] Freer M *et al* 2009 *Phys. Rev.* C **80** 041303(R)
[7] Freer M *et al* 2012 *Phys. Rev.* C **86** 034320
[8] Gai M *et al* 2011 *Act. Phys. Pol.* B **42** 775
[9] Kurokawa C and Kato K 2007 *Nucl. Phys.* A **792** 87
[10] Kanada-Enyo Y 2007 *Prog. Theor. Phys.* **117** 655
[11] Raynal J 1995 computer code ECIS95 NEA0850-14
[12] Zwarts F, Drentje A G, Harakeh M N and Van der Wounde A *Nucl. Phys.* A **439**
[13] Freer M *et al* 2011 *Phys. Rev.* C **83** 034314

Journal of Physics: Conference Series **436** (2013) 012007 doi:10.1088/1742-6596/436/1/012007

The structure of the Hoyle state via a measurement of the 'Hoyle Rotational Band' in ^{12}C

M Gai[1,2] for the UConn-Yale-TUNL-Weizmann-PTB-UCL collaboration

[1] LNS at Avery Point, University of Connecticut, Groton, CT 06340-6097, USA
[2] WNSL Room 100, Yale University, New Haven, CT 06520-8124, USA

E-mail: moshe.gai@yale.edu

Abstract. We have measured the ^{12}C$(\gamma, 3\alpha)$ reaction with an Optical Time Projection Chamber (O-TPC) detector operating with the CO_2(80%) + N_2(20%) gas mixture and gamma-ray beams from the HIγS facility of the TUNL at Duke. We measured complete angular distributions (between 9.1 - 10.7 MeV) from which we determine the cross section yield curve and E1-E2 relative phases leading to an unambiguous identification of the second 2^+ state in ^{12}C at 10.03(11) MeV. The observed spectrum of ^{12}C below 12 MeV including the 2_2^+ observed in this work resembles the rotation-vibration spectrum predicted for a triangular shape oblate spinning top in which the Hoyle state is the first vibrational breathing mode of the triangular three alpha particle system. The predicted rotation-vibration spectrum of a triangular shape oblate spinning top (with a D_{3h} symmetry) allows us to compare the moment of inertia of the predicted Hoyle rotational band to the ground state rotational band and in this way extract the model radius parameter of the Hoyle state of 3.22(8) fm ($\sqrt{2}$ time the r.m.s. radius of the ground state) which is close to the radius extracted from ^{12}C(x,x') data.

1. Introduction

The second 0^+ state at 7.654 MeV in ^{12}C that was first predicted (in 1953) by Hoyle [1] plays a central role in the study of clustering in nuclei. Ever since Brink suggested that the Hoyle state is a very extended object with the structure of three alpha-particles arranged in a linear chain [2] many theoretical models were developed to describe the Hoyle state and the structure of ^{12}C. One of the issues of great current interest is whether the Hoyle state is an extended object with an rms radius considerably larger than the ground state of ^{12}C. Indeed clusterization is thought to occur in nuclear matter with low density (e.g. one third of the central density of nuclei). As such for an extended Hoyle state clusterization can occur throughout the nucleus and not as typically found on the surface of nuclei. In this paper we deduce the rms radius of the Hoyle state from the moment of inertia of the rotational band built on top of the Hoyle state and we conclude that the rms radius of the Hoyle state is considerably (35%) larger than the ground state.

A number of models have been proposed to describe the structure of ^{12}C including a group theoretical U(7) model with a D_{sh} symmetry [3], a microscopic Fermionic Molecular Dynamic (FMD) model [4] together with 'BEC-like' cluster model [5], ab-initio No Core Shell Model (NCSM) [6], ab-initio No Core Simplectic Model (NCSpM) [7], and ab-initio Lattice Effective Field Theory (L-EFT) [8, 9]. It is shown by the ab-initio NCSpM calculations that the Hoyle state can only be described by up to 20 $\hbar\omega$ shell model space [7], suggesting that it is an extended alpha-clustering configuration.

Current models differ on the shape of the Hoyle state. In the U(7) model and the FMD model the Hoyle state is predicted to be an oblate triangular three alpha-particle configuration. In the frame of the L-EFT theory [9] the Hoyle state is primarily of the bent-arm chain (or obtuse triangular) shape. Both ab-initio NCSpM calculations [7] and L-EFT calculations [9] predict the Hoyle state to be of deformed prolate shape. In addition while the FMD model predicts an rms radius of the Hoyle state (3.48 fm) that is $\sqrt{2}$ larger than the ground state (2.39 fm) [4], calculations in the frame of the ab-initio L-EFT theory [9] predicts an rms radius (2.4 fm) equal (within the predicted error bar) to the rms radius of the ground state of ^{12}C and the ab-initio NCSpM calculations [7] predict an rms radius (2.93 fm) that is only 22% larger than the ground state of ^{12}C.

The three alpha-particle structure of ^{12}C naturally leads to models that utilize triangular geometry [3, 4, 8, 9]. Such triangular systems are ubiquitous in physics including the X_3 molecular system [10], and the three quark system [11, 12] and their spectrum resemble the one predicted by the oblate spinning top with a D_{3h} symmetry [10, 11]. The rotation-vibration spectrum predicted for the triangular three alpha-particle oblate spinning top appears to be a very useful phenomenological framework to discuss the (rotation-vibration) spectrum of the three alpha-particle system of ^{12}C [3]. It is phenomenological schematic model that serves as a useful guide in which to discuss the essential degrees of freedom of the three alpha-particle system. In nuclei unlike molecules the energy scale of the vibrational and rotational states is similar leading to large mixing among rotation-vibration states [3].

For example in the U(7) model the Hoyle state is predicted to be the first vibrational breathing mode of the (oblate) deformed state on top of which a rotational band is predicted which is very similar to the ground state rotational band with the same triangular geometry. In contrast for example the FMD model [4] and the 'BEC-like' cluster models [5] do not predict the 2_2^+ to be a member of a rotational band. The EFT calculations [9] on the other hand predict a Hoyle rotational band [9]. The 1^- state at 10.84 MeV is predicted by the U(7) model to be the bending mode of the three alpha-particle system with a rotational band that includes the 1^- and almost degenerate 2^+ and 2^- states.

2. Measurement of the ^{12}C($\gamma, 3\alpha$) Reaction

The current measurement of the ^{12}C($\gamma, 3\alpha$) reaction was performed at the HIγS facility that produces an intense, nearly monoenergetic gamma-ray beam by Compton backscattering photons from a free-electron laser [13]. Beam energy spreads of 300 - 350 keV were measured at beam energies between 9.1 and 10.7 MeV, with on-target intensities of $\approx 2 \times 10^8$ γ/sec. The beam intensity was measured by detecting neutrons from the $d(\gamma, n)p$ reaction using a D_2O target, cross-calibrated against a large NaI(Tl) detector. The energy profile of the beam was measured using a large HPGe detector, and the spectra were unfolded using a Monte Carlo technique [15, 16]. The alignment of the detector with respect to the beam was achieved using a gamma camera and lead absorbers placed in the front and back of the detector [17] as discussed in [14].

An optical-readout time projection chamber (O-TPC) operating at 100 torr with the gas mixture of CO_2(80%) + N_2(20%) described in [14], was used. The O-TPC consists of a time projection chamber with an opto-electronic chain to record the image of each event. The O-TPC was used as target material as well as to detect the outgoing alpha-particles from the ^{12}C($\gamma, 3\alpha$) reaction.

The events recorded in the O-TPC include protons from the ^{14}N(γ, p) reaction, alpha particles from the ^{16,18}O(γ, α) and the ^{12}C($\gamma, 3\alpha$) reactions and cosmic rays. Nearly all (98%) of the ^{12}C dissociation events proceed via the ^{12}C(γ, α_0)^{8}Be and the subsequent (immediate) decay of the

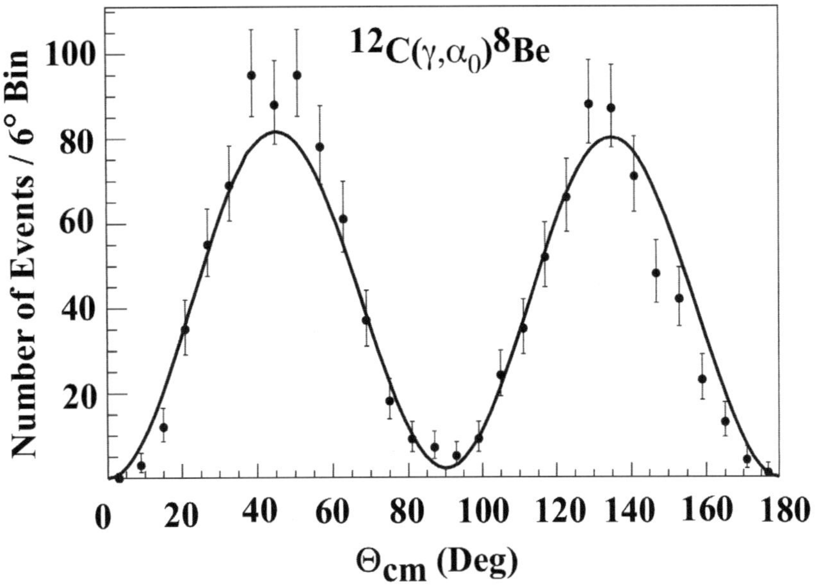

Figure 1. Angular distribution of the $^{12}C(\gamma, \alpha_0)^8Be$ reaction measured at $E_\gamma = 9.6$ MeV.

ground sate of ^8Be by two co-linear alpha-particles. As discussed in [14] all events were easily distinguished from the $^{12}C(\gamma, 3\alpha)$ events except for the $^{16}O(\gamma, \alpha)$ events. The recorded energy and track of the ^{12}C and ^{16}O dissociation events are very similar and the events are separated using the line shape analysis described in [14]. Each measured time projection signal was fit using the calculated line shapes of the $^{12}C(\gamma, \alpha_0)^8Be$ and $^{16}O(\gamma, \alpha)^{12}C$ events as discussed in [14]. The goodness-of-fit parameters, χ^2_C and χ^2_O of the predicted line shapes of $^{12}C(\gamma, \alpha_0)^8Be$ and $^{16}O(\gamma, \alpha)^{12}C$ respectively, were used to classify the event. Discrimination between $^{12}C(\gamma, \alpha_0)^8Be$ and $^{16}O(\gamma, \alpha)^{12}C$ events is carried out using the fitted line shape of the time projection signals.

Complete angular distribution of $^{12}C(\gamma, \alpha_0)^8Be$ events were measured at seven energies between 9.1 and 10.7 MeV. The events recorded in the O-TPC (in three dimension) are transformed to the (θ, ϕ) coordinate system as discussed in [14]. The angular distributions were fit in terms of $E1$ and $E2$ amplitudes and their relative phase ϕ_{12} [18]. Since angular information was available for each $^{12}C(\gamma, 3\alpha)$ event individually, unbinned maximum likelihood fits were used to avoid losing information through binning. This also reduced the number of fitting parameters, as the fitting function is normalized to unity. An angular distribution at a gamma-ray beam energy of 9.6 MeV is shown in Fig. 1. The solid curve is the fitted angular distribution calculated for $\frac{\sigma(E2)}{\sigma} = 96.75\%$ and $\phi_{12} = 80.3°$. For almost all beam energies the angular distributions were dominated by the $E2$ component.

The total cross section yield curve was deduced from the angular distribution measured at seven different gamma-ray beam energies between 9.1 and 10.7 MeV, as shown in Fig. 2 with error bars that include both statistical and systematic uncertainties. The systematic uncertainties associated with each measured cross section are dominated by a 5% uncertainty in the gamma-ray beam intensity. In Fig. 2a we show the separated E1 and E2 cross section components measured at these energies together with the calculated Breit-Wigner resonances with energy-dependent level shifts and widths [19], convoluted with the measured gamma-ray

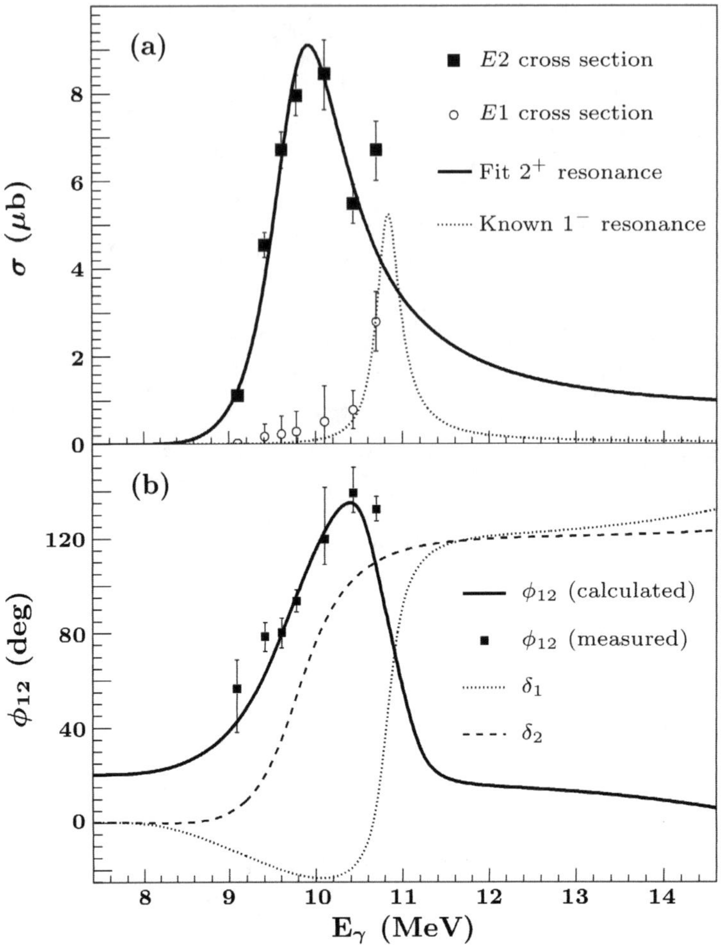

Figure 2. (a) The measured (separated) E1 and E2 cross sections of the $^{12}C(\gamma, \alpha_0)^8Be$ reaction together with fits to two energy-dependent Breit Wigner functions, as discussed in the text. (b) The measured E1-E2 relative phase angle (ϕ_{12}) together with the predicted phase angle, as discussed in the text.

beam energy distribution. Coulomb wave functions were calculated using the continued-fraction expansion technique [20] with $R_0 = 1.4\ fm$. Each Breit-Wigner term includes three free parameters: the partial widths Γ_α, Γ_γ and the resonance energy.

The E2 cross section data allow us to unambiguously identify a 2^+ resonance at 10.03(11) MeV with a total width of 800(130) keV and gamma-decay width of 60(10) meV; $B(E2: 2_2^+ \rightarrow 0^+) = 0.73(13)\ e^2fm^4$ or 0.45 W.u. The measured $B(E2: 2_2^+ \rightarrow 0^+)$ is close to prediction of the FMD model (0.46 e^2fm^4) [21], and smaller than predicted in the L-EFT calculations [2(1) e^2fm^4] [9]. Note that the slight difference between the maximum of the calculated cross section (at 9.8 MeV) and the resonance energy (at 10.03 MeV) is due to the energy-dependent widths used in the fit. The highest energy data point at 10.7 MeV is not consistent with this single resonance hence in order to estimate the error in the measured resonance energy we also analyzed our data including another 2^+ that was previously suggested at 11.1 MeV leading to total error of the resonance energy of 110 keV.

3. Conclusion

The measured 2_2^+ state at 10.03(11) MeV reported in this work lies 2.38(11) MeV above the 0^+ Hoyle state which is approximately half the excitation energy of the 2_1^+ state of ^{12}C. Since the U(7) model predicts the ground state rotational band and the Hoyle bands to arise from the same geometrical shape we conclude that in this model the radius parameter of the Hoyle state is approximately $\sqrt{2}$ larger than the r.m.s. radius of the ground state. This conclusion is in fact consistent with (but slightly larger than) the r.m.s. radius of the Hoyle state [2.89(4) fm] measured using ^{12}C(x,x') data [23]. Indeed all thus far extracted radius parameter of the Hoyle state including using the moment of inertia as discussed here, or from ^{12}C(e,e') data [4], or from ^{12}C(x,x') data [23], are all model dependent. Our inability to measure elastic scattering off (the very short lived) Hoyle state will continue to inhibit a measurement of the r.m.s. radius of the Hoyle state.

[1] Hoyle F, Dunbar D N F, Wenzel W A and Whaling W 1953 *Phys. Rev.* **92** 1095c in Minutes of the New Mexico Meeting Held at Alberquerque September 2-5.
[2] Brink D M 1966 *Proc. Intern. School of Physics: 'Enrico Fermi'* course 36 ed Block C (Academic Press)
[3] Bijker R and Iachello F 2002 *Ann. Phys.* **298** 334
[4] Chernykh M, Feldmeier, Neff T, von Neumann-Cosel P and Richter A 2007 *Phys. Rev. Lett.* **98** 032501
[5] Funaki Y, Horiuchi H, von Oertzen W, Röpke G, Schuck P, Tohsaki A and Yamada 2009 *Phys. Rev.* C **80** 64326 and references therein.
[6] Roth R, Langhammer J, Calci A, Binder S and Navratil P 2011 *Phys. Rev. Lett.* **107** 072501
[7] Dreyfuss A C, Launey K D, Dytrych T, Draayer J P and Bahri C 2012 to be published, arXiv:1212.2255v1.
[8] Epelbaum E, Krebs H, Lee D and Meissner Ulf-G 2011 *Phys. Rev. Lett.* **106** 192501
[9] Epelbaum E, Krebs H, Lahde T, Lee D and Meissner Ulf-G 2012 *Phys. Rev. Lett.* **109** 252501
[10] Bijker R, Dieperink A E L and Leviatan A 1995 *Phys. Rev.* A **52** 2786
[11] Bijker R, Iachello F and Leviatan A 1994 *Ann. Phys.* **236** 69
[12] Capstick S and Isgur N 1986 *Phys. Rev.* D **34** 2809
[13] Weller H R, Ahmed M W, Gao H, Tornow W, Wu Y K, Gai M and Miskimen R 2009 *Prog. in Part. and Nucl. Phys.* **62** 257
[14] Gai M, Ahmed M W, Stave S C, Zimmerman W R, Breskin A, Bromberger B, Chechik R, Dangendorf V, Delbar Th, France III R H, Henshaw S S, Kading T J, Martel P P, McDonald J E R, Seo P-N, Tittelmeier K, Weller H R and Young A H 2010 *JINST* **5** 12004
[15] Sun Cand Wu Y K 2011 *Phys. Rev. ST Accel. Beams* **14** 044701
[16] Sun C, Wu Y K, Rusev G and Tonchev A 2009 *Nucl. Inst. Meth. Phys. Res.* A **605** 312
[17] Sun C 2010 Ph.D. thesis Duke University
[18] Dyer P and Barnes C 1974 *Nucl. Phys.* A **233** 495
[19] Lane A M and Thomas R G 1958 *Rev. Mod. Phys.* **30** 257
[20] Barnett A R, Feng D H, Steed J W and Goldfarb L J B 1974 *Comput. Phys. Commun.* **8** 377
[21] Neff T 2012 private communication
[22] Ajzenberg-Selove F 1990 *Nucl. Phys.* A **506** 1
[23] Danilov A N, Belyaeva T L, Demyanova A S, Goncharov S A and Ogloblin A A 2009 *Phys. Rev.* C **80** 054603

Cluster structure and isoscalar monopole excitation in light nuclei

T Yamada

Laboratory of Physics, Kanto Gakuin University, Yokohama 236-8501, Japan

E-mail: yamada@kanto-gakuin.ac.jp

Abstract. Isoscalar monopole excitation to cluster states in light nuclei is in general strong as to be comparable with the single particle strength and shares about 20 % of the sum rule value. In the present paper, the isoscalar monopole strength function in ^{16}O is discussed up to $E_x \lesssim 40$ MeV as a typical example. We found that 1) two different types of monopole excitations exist in ^{16}O; one is the monopole excitation of cluster states which is dominant in the lower energy part, and the other is the monopole excitation of the mean-field type such as one-particle one-hole ($1p1h$) which is attributed mainly to the higher energy part, 2) this character of the monopole excitations originates from the fact that the ground state of ^{16}O with the dominant doubly closed shell structure has a duality of the mean-field-type as well as alpha-clustering character, and 3) the monopole strength is much enhanced by the α-type ground state correlation.

1. Introduction

Isoscalar monopole excitation is related to a density fluctuation of nucleus. The typical example is the isoscalar giant monopole resonance (ISGMR) observed as a single bump in medium and heavy nuclei, which exhausts almost 100 % of the energy weighted sum rule (EWSR) value. It is interesting to study what happens for the ISGMR in lighter nuclei. When the nuclear masses decrease from medium nuclei to light nuclei, the isoscalar monopole strengths are in general fragmented. Table 1 shows the experimental monopole strengths of the low-lying states in ^{16}O, ^{12}C, ^{11}B, and ^{4}He [1, 2]. The strengths of the clusters states are rather strong they are comparable with the single particle strength (~ 4 fm^2) and shares about 20 % of the EWSR value [3]. In this paper, we concentrate on discussing the case of ^{16}O.

The histogram in Fig. 1 shows the experimental isoscalar monopole strength function in ^{16}O obtained by the inelastic α scattering [11] (the data below $E_x \sim 10$ MeV is absent due to the experimental condition). The monopole strengths split over wide energy region, and one can see discrete peaks in low energy ($E_x \lesssim 16$ MeV) and gross three-bump structure in higher energy region ($16 \lesssim E_x \lesssim 40$ MeV). The experimental data is compared with the RRPA calculation [12], the result of which is shown by the solid line in Fig. 1. In order to match their calculation to the experimental centroid, the calculated strength function was shifted down in energy by 4.2 MeV and furthermore they normalized it by multiplying the RRPA curve by a factor of 0.25 [11], although their calculation failed to reproduce the 0^+ states found in the low energy region. In the non-relativistic RPA (SRPA) calculations for ^{16}O [13] a significant discrepancy is also revealed as compared with the experimental data, in particular, in the low energy region, although the gross structures at the higher energy region in the RPA calculations are in rather good agreement with the data.

Table 1. Experimental monopole strengths ($M^{\rm exp}$(E0) [fm^2]) in ^{16}O, ^{12}C, ^{11}B, and ^{4}He [1, 2]. In ^{11}B, the isoscalar monopole transition rates (B(E0, IS) [fm^4]) are presented with "a)". $P^{\rm e.w.}$ stands for the percentage of the energy weight strength to the isoscalar monopole EWSR value. The assignment of the structure of each state together with the calculated monopole strength ($M^{\rm cal}$(E0)) is referred from Refs. [4, 5, 6, 7, 8, 9]. h. o. denotes the higher nodal (see text).

	J_i	J_f	$E_x(J_f)$ [MeV]	$M^{\rm exp}$(E0)	$P^{\rm e.w.}$	Structure of J_f	$M^{\rm cal}$(E0)
^{16}O	0_1^+	0_2^+	6.05	3.55 ± 0.21	3.5%	α+^{12}C(0_1^+)	3.9 [5] (3.88 [10])
		0_3^+	12.1	4.03 ± 0.09	8.9%	α+^{12}C(2_1^+)	2.4 [5] (3.50 [10])
		0_4^+	13.6	no data	—	h. n. α+^{12}C(0_1^+)	2.4 [5] (— — — —)
		0_5^+	14.1	3.3 ± 0.7	6.9%	α+^{12}C(1_1^-)	2.6 [5] (— — — —)
		0_6^+	15.1	no data	—	4α-gas	1.0 [5] (— — — —)
^{12}C	0_1^+	0_2^+	7.65	5.4 ± 0.2	16%	3α-gas	6.7 [6]
^{11}B	$3/2_1^-$	$3/2_2^-$	5.02	$< 9^{\rm a)}$	< 1%	shell-model-like	$0.6^{\rm a)}$ [8]
		$3/2_3^-$	8.56	$96 \pm 12^{\rm a)}$	12%	$\alpha + \alpha + t$	$92^{\rm a)}$ [8]
^{4}He	0_1^+	0_2^+	20.2	1.10 ± 0.16	11%	$3N - N$	1.38 [9]

Recently the structure study of ^{16}O has made a great advance up to $E_x \simeq 16$ MeV around the 4α disintegration threshold. The six lowest 0^+ states of ^{16}O, up to $E_x \simeq 16$ MeV, including the ground state, have for first time been reproduced very well with the 4α orthogonality condition model (OCM) [4]. The six 0^+ states have the following characteristic structures [4]: 1) the ground state (0_1^+) has dominantly a doubly-closed-shell structure, 2) the 0_2^+ state at $E_x = 6.05$ MeV and the 0_3^+ state at $E_x = 12.05$ MeV have mainly α+^{12}C structures where the α-particle orbits around the ^{12}C(0_1^+) core in an S-wave and around the ^{12}C(2_1^+) core in a D-wave, respectively, 3) the 0_4^+ ($E_x = 13.6$ MeV) and 0_5^+ ($E_x = 14.1$ MeV) states mainly have α+^{12}C(0_1^+) structure with higher nodal behavior and α+^{12}C(1^-) structure, respectively, and 4) the 0_6^+ state at 15.1 MeV is a strong candidate of the 4α condensate, $(0S)_\alpha^4$, with the probability of 61 %. The results of 1) $\sim$ 3) are consistent with those obtained by the α+^{12}C(0_1^+,2_1^+,4_1^+) OCM calculation [10]. The 4α OCM reasonably reproduces the experimental monopole strengths within a factor of 1.5 (see Table 1) as well as the decay widths [4, 5]. The purpose of this paper is to study whether the 4α OCM can reproduce the experimental monopole strength function in the low energy region ($E_x \lesssim 16$ MeV) in ^{16}O, a region which is difficult to be treated in the mean-field theory, and to show the excitation mechanism of the cluster states by the monopole transition.

2. Calculated monopole strength function of ^{16}O with 4α OCM

Fig. 2 [5] shows the calculated isoscalar monopole strength function of ^{16}O with the 4α OCM, where we use the calculated monopole matrix elements and the calculated decay widths for the six 0^+ states up to $E_x \simeq 16$ MeV obtained by the 4α OCM calculation, also the experimental excitation energies for the six 0^+ states are employed. We can see a rather good correspondence with the experimental data. The fine structures in the calculated strength function, i.e. one peak at $E_x = 12.1$ MeV (corresponding to the 0_3^+ state), one shoulder-like peak at $E_x = 13.8$ MeV (0_4^+), two peaks at $E_x = 14.1$ MeV (0_5^+) and 15.1 MeV (0_6^+), are well reproduced. On the other hand, it has been recently suggested that some 0^+ states with the dominant α+^{12}C(Hoyle:0_2^+) configuration [14] can contribute to the energy region of the first bump in Fig. 1 ($E_x \sim 18$ MeV) and their contributions are estimated to be less than several % of the EWSR value.

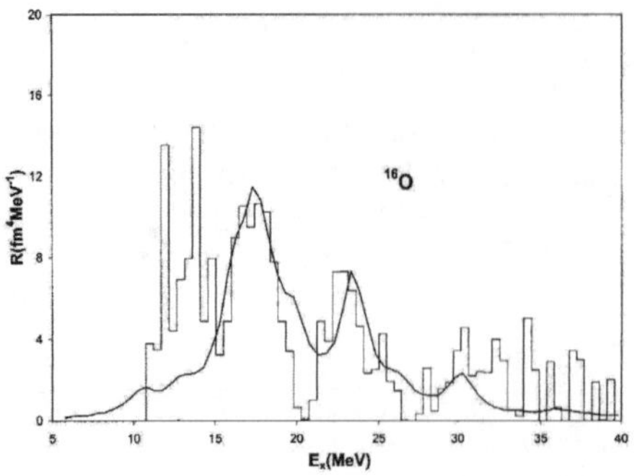

Figure 1. Experimental isoscalar monopole strength function of ^{16}O [11] is shown by the histogram. The real line is the calculated result by the relativistic RPA calculation [12].

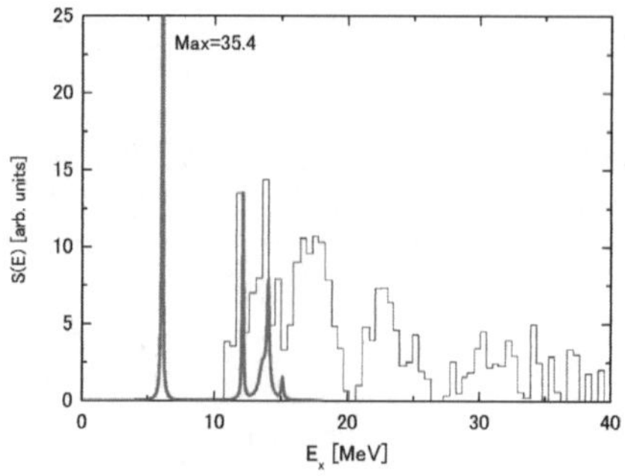

Figure 2. (Color online) Calculated isoscalar monopole strength functions of ^{16}O with the 4α OCM (bold line) [5] and experimental data (thin line: see Fig. 1 [11]).

3. Mechanism of monopole excitation of cluster states from the ground state

It is instructive to discuss the mechanism of why the five α cluster states (0_2^+, 0_3^+, 0_4^+, 0_5^+, and 0_6^+) of ^{16}O are excited relatively strongly from the shell-model-like ground state by the isoscalar monopole transition.

The wave function of the ^{16}O ground state has the dominant $(0s)^4(0p)^{12}$ configuration, corresponding to the SU(3) $(\lambda,\mu)=(0,0)$ wave function. This character is in common in the mean-field calculations and cluster-model calculations, and is also obtained by the no-core shell-model calculation [16]. According to the Bayman-Bohr theorem [15], this doubly closed shell-model wave function is mathematically equivalent to a single cluster-model wave function of α+^{12}C as well as 4α with the total harmonic oscillator (h.o.) quanta $Q=12$ [3, 5],

$$\begin{aligned}
\det|(0s)^4(0p)^{12}| &= N_0 \times \mathcal{A}\left\{\left[u_{40}(\boldsymbol{\xi}_3,3\nu)\phi_{L=0}(^{12}\mathrm{C})\right]_{J=0}\phi(\alpha)\right\}\phi_{\rm cm}(\boldsymbol{R}_{\rm cm}), \quad (1)\\
&= N_2 \times \mathcal{A}\left\{\left[u_{42}(\boldsymbol{\xi}_3,3\nu)\phi_{L=2}(^{12}\mathrm{C})\right]_{J=0}\phi(\alpha)\right\}\phi_{\rm cm}(\boldsymbol{R}_{\rm cm}), \quad (2)\\
&= \hat{N}_0 \times \mathcal{A}\Big\{\,[u_{40}(\boldsymbol{\xi}_3,3\nu)\,[u_{40}(\boldsymbol{\xi}_2,8\nu/3)u_{40}(\boldsymbol{\xi}_1,2\nu)]_{L=0}]_{J=0}\\
&\qquad \times\phi(\alpha_1)\phi(\alpha_2)\phi(\alpha_3)\phi(\alpha_4)\Big\}\phi_{\rm cm}(\boldsymbol{R}_{\rm cm}), \quad (3)
\end{aligned}$$

where $\phi_{\rm cm}$ denotes the wave function of the c.o.m. motion of ^{16}O, and $\boldsymbol{\xi}_k$ $(k=1\sim 3)$ stand for the Jacobi coordinates between the clusters. The functions $\phi(\alpha)$ and $\phi_L(^{12}\mathrm{C})$ represent, respectively, the internal wave function of the α cluster and that of ^{12}C with the angular momentum of L. $N_L(\hat{N}_0)$ is the normalization factor. The relative wave function between the α and ^{12}C clusters in Eqs. (1) and (2) is described by the h.o. wave function $u_{QL}(\boldsymbol{\xi},\beta)$ with $Q=4$. Eqs. (1) and (2) mean that the doubly closed shell-model wave function has an α+^{12}C cluster degree of freedom. In addition, Eq. (3) demonstrates that the doubly closed shell-model wave function also possesses a 4α cluster degree of freedom. Thus the ground state of ^{16}O has the mean-field of degree of freedom as well as the cluster degree of freedom. We call this the dual nature.

On the other hand, the isoscalar monopole operator of ^{16}O can be decomposed into the internal part (acting on the constituent clusters) and relative part (with respect to the

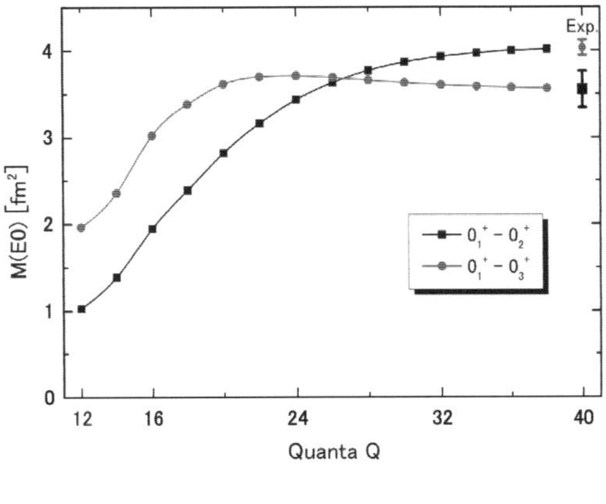

Figure 3. (Color online) Dependence of the monopole strength $M_Q(\mathrm{E0})$ on the model space of the ^{16}O ground state wave function characterized by the harmonic oscillator quanta Q with the α+^{12}C(0_1^+,2_1^+,4_1^+) OCM [3]. The square and circle points correspond to $M_Q(\mathrm{E0}; 0_1^+ \to 0_2^+)$ and $M_Q(\mathrm{E0}; 0_1^+ \to 0_3^+)$, respectively, and the experimental data are also shown there.

clusters) [3, 5]:

$$\mathcal{O}^{\mathrm{IS}}_{\mathrm{E0}}(^{16}\mathrm{O}) = \sum_{i=1}^{16} (\boldsymbol{r}_i - \boldsymbol{R}_{\mathrm{cm}})^2 = \mathcal{O}^{\mathrm{IS}}_{\mathrm{E0}}(\alpha) + \mathcal{O}^{\mathrm{IS}}_{\mathrm{E0}}(^{12}\mathrm{C}) + 3\boldsymbol{\xi}_3^2 \tag{4}$$

$$= \sum_{k=1}^{4} \mathcal{O}^{\mathrm{IS}}_{\mathrm{E0}}(\alpha_k) + \sum_{k=1}^{4} 4(\boldsymbol{R}_{\alpha_k} - \boldsymbol{R}_{\mathrm{cm}})^2, \tag{5}$$

where $\boldsymbol{R}_k$ denotes the c.o.m. coordinate of the k-th α cluster in the 4α system. ¿From Eqs. (1), (2), and (4), we can understand the reasons why the 0_2^+ and 0_3^+ states of ^{16}O with the $\alpha + {}^{12}$C cluster structure are excited by the monopole transition as follows: 1) the ground state of ^{16}O is of the SU(3) $(\lambda, \mu) = (0, 0)$ nature with the α-clustering degree of freedom, 2) the relative part of the monopole operator referring to the α+^{12}C relative motion, $3\boldsymbol{\xi}_3{}^2$, in Eq. (4) can activate the α-cluster degree of freedom, and 3) the $\alpha + {}^{12}$C states are excited by the monopole operator.

Here, we discuss the important role of the α-cluster-type ground-state correlation significantly enhancing the monopole strength. This is studied with the α+^{12}C(0_1^+,2_1^+,4_1^+) OCM [3], which reproduces the monopole strengths of the 0_2^+ and 0_3^+ states in ^{16}O [10] (see Table 1). The component of the SU(3) $(\lambda, \mu) = (0, 0)$ wave function with the total h.o. oscillator $Q = 12$, i.e. $(0s)^4(0p)^{12}$, in the ground state of ^{16}O, is about 90 % in the α+^{12}C OCM calculation as well as the 4α OCM [3, 4, 5, 10]. The model space of the α+^{12}C OCM is characterized by the total harmonic oscillator quanta Q of ^{16}O. The remaining component of about 10 % in the ground-state wave function corresponds to the α-type ground-state correlation with $Q > 12$. When the pure SU(3) $(\lambda, \mu) = (0, 0)$ wave function is adopted as the ^{16}O ground-state wave function, the monopole strengths to the 0_2^+ and 0_3^+ states are about half or one third smaller than the experimental data, although the calculated values reproduce the order of magnitude of the experimental values (see Fig. 3). When the model space of the ground state increases from the lowest value of $Q = 12$, the calculated monopole strengths to the 0_2^+ and 0_3^+ states are gradually increasing and reproduce the experimental values at $Q \simeq 30$ within a factor of 1.13, as shown in Fig. 3. This means that the component with $Q > 12$ in the ground-state wave function (corresponding to the α-type ground-state correlation) gives a coherent contribution to enhancement of the monopole strengths. From these results, we can learn that the α-cluster-type ground-state correlation plays the important role in reproducing the monopole strengths.

The reason why the 0_6^+ state of ^{16}O with the 4α-gas-like character is excited by the monopole transition can be also understood from the property of the ground state of ^{16}O. The doubly closed shell-model wave function is mathematically equivalent to the single 4α cluster wave function with $Q = 12$ in Eq. (3). This equation means that the ground state of ^{16}O inherently has a 4α-cluster degree of freedom. The relative part (or second term) of the monopole operator in Eq. (5) can excite the relative motion among the 4α particles. In other words, the monopole operator has an ability to populate democratically 4α particles by $2\hbar\omega$ with respect to the c.o.m. coordinate of ^{16}O. The resultant state, thus, has some amount of the overlap with the 4α-gas-like state or $\alpha + {}^{12}\mathrm{C}(0_2^+)$, i.e. 0_6^+, with the 4α-condensate-like structure [4]. It is noted that the mechanism of the 4α-gas-like state being populated by the monopole transition is similar to that of the Hoyle state (0_2^+) with the 3α-gas-like structure, excited by the monopole transition, although the ground state of ^{12}C has a shell-model-like compact structure with the main configuration of SU(3) $(\lambda, \mu) = (0, 4)$ [3].

As for the 0_5^+ state, its main configuration is $\alpha+{}^{12}\mathrm{C}(1_1^-)$. According to the Bayman-Bohr theorem [15], the SU(3) $(0,0)$ state of ^{16}O has no component of the $\alpha+{}^{12}\mathrm{C}(1_1^-)$ channel. However, the monopole strength to the 0_5^+ state is as large as 3 fm^2 [5]. This is the reason that the 0_5^+ state has small but important components of the $\alpha+{}^{12}\mathrm{C}(0_1^+, 2_1^+, 0_2^+)$ configurations. Since these three configurations can be excited from the ground state of ^{16}O by the monopole operator as discussed above, their respective contributions are coherently added to provide the relatively large monopole strength to the 0_5^+ state. Concerning the 0_4^+ state, the situation is similar to the case of the 0_5^+ state. The 0_4^+ state has also small but non negligible components of the $\alpha+{}^{12}\mathrm{C}(0_1^+, 2_1^+, 0_2^+)$ channels, which contribute to the monopole strength for the 0_4^+ state.

4. Summary

The isoscalar monopole excitation is useful to search for cluster states in light nuclei. Two different types of the monopole excitations and the duality of the ground state discussed in this paper seem to exist in general in light nuclei [3, 5]. It is desirable that systematically study the isoscalar monopole excitations in $4n$ and neutron-rich nuclei as well as light hypernuclei may be studied.

Acknowledgments

This contribution is in collaboration with Y. Funaki, T. Myo, H. Horiuchi, K. Ikeda, G. Röpke, P. Schuck, and A. Tohsaki.

References

[1] Ajzenberg-Selove F 1986 *Nucl. Phys.* A **460** 1
[2] Kawabata T *et al* 2007 *Phys. Lett.* B **646** 6
[3] Yamada T *et al* 2008 *Prog. Thor. Phys.* **120** 1139
[4] Funaki Y *et al* 2008 *Phys. Rev. Lett.* **101** 082502
[5] Yamada T *et al* 2012 *Phys. Rev.* C **85** 034315
[6] Kamimura M 1981 *Nucl. Phys.* A **351** 456
[7] Funaki Y *et al* 2003 *Phys. Rev.* C **67**, 051306(R)
[8] Yamada T *et al* 2010 *Phys. Rev.* C **82** 064315
[9] Hiyama E *et al* 2004 *Phys. Rev.* C **70** 031001(R)
[10] Suzuki Y 1976 *Prog. Theor. Phys.* **55** 1751; Suzuki Y 1976 *Prog. Theor. Phys.* **56** 111
[11] Lui Y W, Clark H L and Youngblood D H 2001 *Phys. Rev.* C **64** 064308
[12] Ma Z *et al* 1997 *Phys. Rev.* C **55** 2385
[13] Gambacurta D *et al* 2010 *Phys. Rev.* C **81** 054312
[14] Ohkubo S *et al* 2010 *Phys. Lett.* B **684** 127; Funaki Y *et al* 2012 private communication

Journal of Physics: Conference Series **436** (2013) 012008 doi:10.1088/1742-6596/436/1/012008

[15] Bayman B F and Bohr A 1958/59 *Nucl. Phys.* **9** 596
[16] Dytrych T *et al* 2007 *Phys. Rev. Lett.* **98** 162503

Journal of Physics: Conference Series **436** (2013) 012009 doi:10.1088/1742-6596/436/1/012009

Search for alpha inelastic condensed state in ^{24}Mg

T Kawabata[1], T Adachi[2], M Fujiwara[2], K Hatanaka[2], Y Ishiguro[1], M Itoh[3], Y Maeda[4], H Matsubara[5], H Miyasako[4], Y Nozawa[1], T Saito[4], S Sakaguchi[6], Y Sasamoto[5], Y Shimizu[6], T Takahashi[3], A Tamii[2], S Terashima[6], H Tokieda[5], N Tomida[1], T Uesaka[6], M Uchida[7], Y Yasuda[2], N Yokota[1], H P Yoshida[3] and J Zenihiro[6]

[1]Department of Physics, Kyoto University
[2]Research Center for Nuclear Physics, Osaka University
[3]Cyclotron and Radioisotope Center, Tohoku University
[4]Faculty of Engineering, University of Miyazaki
[5]Center for Nuclear Study, University of Tokyo
[6]RIKEN (The Institute for Physical and Chemical Research)
[7]Department of Physics, Tokyo Institute of Technology

E-mail: kawabata@scphys.kyoto-u.ac.jp

Abstract. The alpha inelastic scattering from ^{24}Mg was measured to obtain the isoscalar natural-parity excitation strengths and to search for α-condensed states. Multipole decomposition analysis was performed for the measured cross sections, and the strength distributions for the ΔL = 0–3 transitions were successfully obtained. The decay particles from the excited states in ^{24}Mg were also measured in concidence with the inelastically scattered alpha particles. Candidates were found for α-condensed states around the ^{16}O core.

1. Introduction

Alpha particle clustering is an important concept in nuclear physics. On the basis of the Ikeda diagram [1], the α cluster structures are expected to emerge near the α-decay threshold energies. For instance, it has been suggested that the 7.65-MeV 0_2^+ state in ^{12}C, which is located at an excitation energy higher than the 3α-decay threshold by 0.39 MeV, has a spatially well-developed 3α-cluster structure.

This 0_2^+ state is described by introducing a novel concept into nuclear structure. This state has a dilute-gas-like structure, in which three α clusters are weakly interacting and are condensed into the lowest s-orbit [2]. The next natural question addressed is whether such α-condensed states exist in heavier self-conjugate $4n$ nuclei.

Such α-condensed states are theoretically predicted up to $n = 10$ [3]. The energy of the $n\alpha$-condensed state relative to the $n\alpha$-decay threshold increases with n due to the short-range nature of the attractive force between α clusters and the long-rage nature of the Coulomb repulsion. Finally, the $n\alpha$ condensed state becomes unstable beyond $n = 10$.

A recent theoretical work proposed a new conformation of the α-condensed state: α clusters may be condensed into the lowest s-orbit around a core nucleus [4]. The attractive potential for α clusters provided by the core nucleus stabilizes the α-condensed state around the core. Thus, such α-condensed states around core nuclei are expected to appear at excitation energies lower than the corresponding cluster-decay threshold energies.

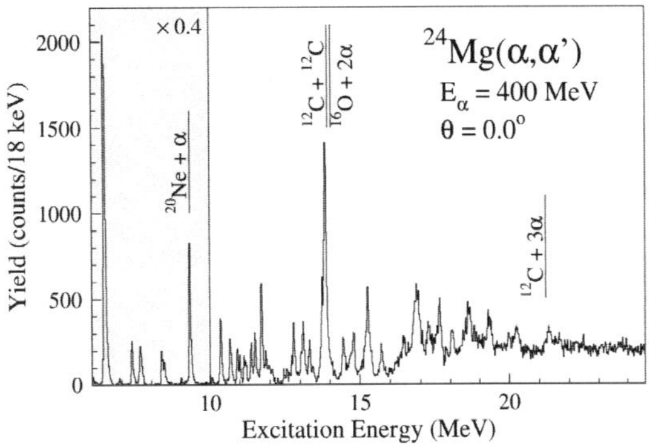

Figure 1. Typical spectrum for the ^{24}Mg(α,α') reaction measured at 0°. The excitation spectrum below E_x = 10 MeV is downscaled by a factor of 0.4.

From the experimental point of view, the 3α- and 4α-condensed states in ^{12}C and ^{16}O were extensively studied in the last decade [5, 6, 7, 8, 9]. However, the other nuclei heavier than ^{16}O are still unexplored due to experimental difficulties. Since the fully condensed states in ^{20}Ne and ^{24}Mg are expected to appear about 5 MeV above the $n\alpha$-decay threshold, these states are obscured by continuum states in the highly excited region. Although the α-condensed states around the core nuclei should appear at relatively low excitation energies, the level densities near these condensed states are still high and the experimental identifications are not easy.

In our previous works [10, 11], it was demonstrated that, in light nuclei, spatially well-developed cluster states whose spins and parities are the same as in the ground state, are strongly excited by isoscalar monopole transitions. The 0_2^+ state in ^{12}C, a typical 3α cluster state, is strongly excited with an isoscalar monopole strength of 121$\pm$9 fm^4, which is about 3 times larger than the single particle estimate. Besides the 0_2^+ state in ^{12}C, the $3/2_3^-$ state in ^{11}B with a well-developed $2\alpha + t$ structure is also strongly excited by an isoscalar monopole transition. These facts are theoretically explained on the basis of the Bayman-Bohr theorem and the ground-state correlation [12]. Since the ground-state wave functions in light nuclei inherently possess the clustering degrees of freedom, it is naturally expected that spatially developed cluster states are excited by perturbing the relative motion between the clusters with the monopole operator r^2. Therefore, the isoscalar monopole strengths are key observables in searching for the α-condensed states.

The alpha inelastic scattering at intermediate energies and at forward angles is one of the most useful probes to measure the isoscalar monopole strengths because its reaction mechanism is simple. It has a selectivity for isoscalar natural-parity transitions, and there is a good linear relation between the reaction cross sections and the relevant nuclear transition matrix elements. The alpha inelastic scattering from ^{24}Mg was previously measured at Texas A&M University [13, 14]. However, these measurements were devoted to studying giant resonances in ^{24}Mg, and the excitation-energy resolution was not good enough to separate the α-condensed states from the other states.

In the present work, we performed high-resolution measurement of alpha inelastic scattering from ^{24}Mg and obtained the isoscalar natural-parity excitation strengths to search for the α-condensed states in ^{24}Mg. Since the α-condensed states are expected to have large branching ratios to the α-decay channels, the particles emitted from the decaying excited states were also measured in coincidence with the inelastically scattered alpha particles.

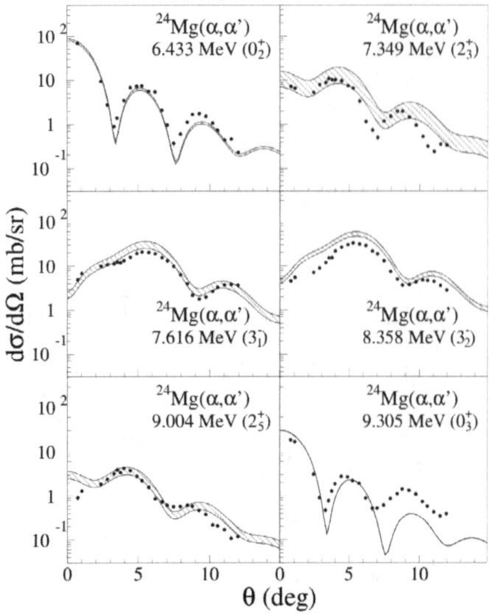

Figure 2. Cross sections for the prominent low-lying states in ^{24}Mg compared with the DWBA calculation. The hatched bands show uncertainties of the calculated cross sections due to the errors of the electromagnetic transition strengths.

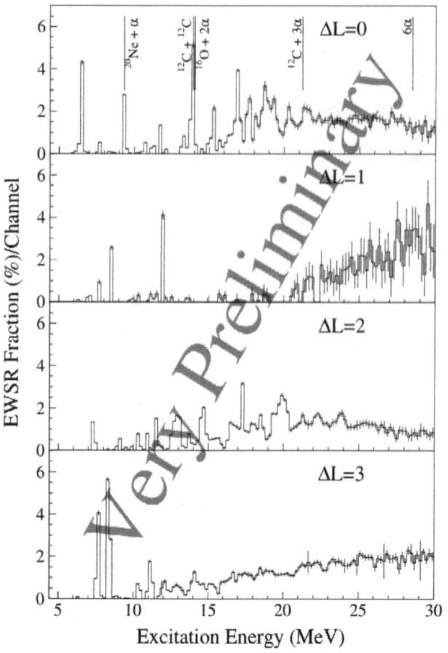

Figure 3. Strength distributions for ΔL = 0–3 transitions obtained from the multipole decomposition analysis for the ^{24}Mg(α,α') reaction.

2. Experiment

The experiment was performed at the Research Center for Nuclear Physics, Osaka University, using a 400-MeV alpha beam. The halo-free alpha beam extracted from the ring cyclotron was transported to the ^{24}Mg target. The scattered alpha particles were momentum-analysed by the high-resolution spectrometer Grand Raiden [15]. The focal-plane detector system of Grand Raiden consisting of two multi-wire drift chambers and plastic scintillation detectors allowed the reconstruction of the scattering angle at the target via ray-tracing techniques. A typical spectrum for the ^{24}Mg(α,α') reaction measured at 0° is shown in Fig. 1. An energy resolution of 80 keV full width at half maximum was achieved.

Six silicon counter telescopes were installed at a distance of 200 mm from the target in the scattering chamber to measure decay particles from excited states. Each telescope consists of 4 silicon detectors with a sensitive area of 50 mm × 50 mm. The thickness of the first silicon detector, nearest to the target, was 65 μm, while those of the other silicon detectors were 525 μm. The sensitive area of the first silicon detector was segmented into four strips, and the dimension of each strip was 12.5 mm in width and 50 mm in height.

3. Result and Discussion

The measured cross sections for the ^{24}Mg(α,α') reaction exciting the prominent low-lying states are compared with a distorted-wave Born-approximation (DWBA) calculation in Fig. 2. The transition potentials used in the DWBA calculation were obtained by folding the macroscopic form factors with the phenomenological αN interaction $V_{\alpha N}$ given by $V_{\alpha N}(r) = -V\exp(-r^2/\alpha_V) - iW\exp(-r^2/\alpha_W)$. The interaction strengths and range parameters of $V = 13.1$ MeV, $W = 8.8$ MeV, and $\alpha_V = \alpha_W = 5.0$ fm^2 were determined so as to reproduce the cross section for the α+^{24}Mg elastic scattering.

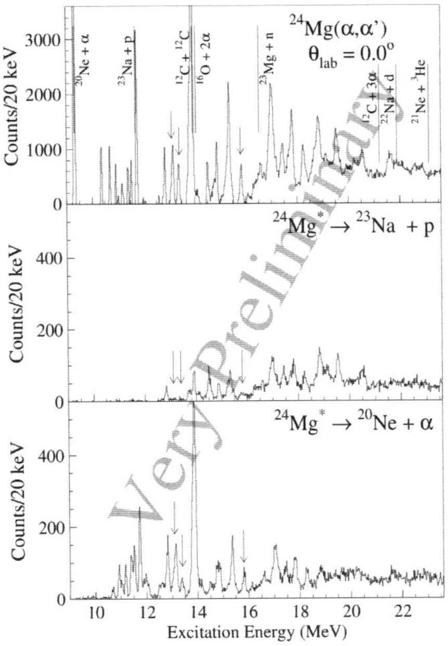

Figure 4. Excitation energy spectra in the ^{24}Mg(α,α') reaction at E_x = 9–23.6 MeV taken from the singles measurement (top) and from the coincidence measurements with decay protons (middle) and α particles (bottom). The vertical arrows show the 0^+ states at 13.1, 13.4, and 15.8 MeV.

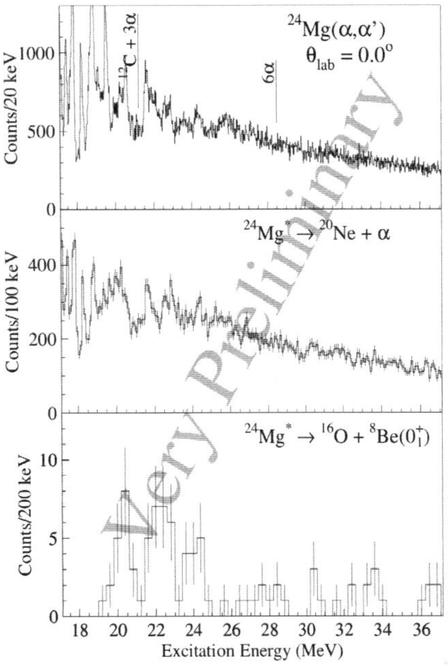

Figure 5. Excitation energy spectra in the ^{24}Mg(α,α') reaction at E_x = 17.1–37.1 MeV taken from the singles measurement (top) and from the coincidence measurements with decay α particles (middle) and ^{8}Be (bottom).

The normalization factors for the macroscopic form factors were determined from the known electromagnetic transition strengths except for the 0_3^+ state at E_x = 9.305 MeV. The hatched bands in Fig. 2 show the uncertainties of the calculated cross sections due to the errors in the electromagnetic transition strengths. The normalization factor for the 0_3^+ state was determined by fitting to the present data. As seen in Fig. 2, the DWBA calculation describes the measured cross sections reasonably well. It should be noted that all the parameters in the present DWBA calculation were determined from previous experimental data and no parameters are tuned to reproduce the present result except for the normalization factor for the 0_3^+ state.

Since the angular distribution of the cross section for each multipole transition depends on its transferred angular momentum, it is possible to decompose the cross section into multipole components by fitting the measured angular distribution. In the fitting procedure, the multipole contributions up to ΔL = 15 were taken into account. The strength distributions for the ΔL = 0–3 transitions obtained from the multipole decomposition analysis are shown in Fig. 3. All the prominent low-lying states shown in Fig. 2 are successfully decomposed into the correct multipole components. This proves the validity of the present multipole decomposition analysis.

It is remarkable that fine structures are observed in the $\Delta L = 0$ strength distribution and some of them are located near the $n\alpha$-decay threshold energies. Although no prominent structure is found near the 6α-decay threshold, a narrow peak at E_x = 13.8 MeV and a broad bump at E_x = 21.5 MeV are very close to the ^{16}O + 2α and ^{12}C + 3α threshold energies, respectively. These states are considered to be candidates of the α-condensed states around the core nuclei ^{16}O and ^{12}C predicted in Ref. [4]. However, it is difficult to conclude that these states are indeed the α-condensed states around the core nuclei because the alpha inelastic scattering excites not only α-cluster states but also single-particle states such as giant monopole resonances (GMR). These candidates, thus, might be attributed to the GMR instead of the α-condensed states, and further information is needed to distinguish between the two possibilities.

In order to clarify the microscopic structure of the excited states, we detected the decay particles from the excited states in coincidence with the inelastically scattered α particles. If the excited states above the particle-decay thresholds have spatially well-developed α-cluster structures, these states should dominantly decay into α-emission channels rather than into proton- or neutron-emission channels. On the other hand, single-particle states such as GMR should prefer proton- or neutron-emission channels to α-emission channels. The decay branching ratio of the α-emission channels to the proton-emission channels, thus, provide complementary information to identify the α-cluster states.

The excitation energy spectrum in the ^{24}Mg(α, α') reaction from the singles measurement at 0° is compared with those from the coincidence measurements with decay protons and α particles in Fig. 4. It is remarkable that the strong 0^+ state at 13.9 MeV near the ^{16}O + 2α and ^{12}C $+^{12}$ C decay thresholds has a large decay width to the α-emission channel. This state is considered to be the most probable candidate of the 2α-condensed state around the ^{16}O core predicted in Ref. [4]. The three 0^+ states at E_x = 13.1, 13.4, and 15.8 MeV also have large decay widths to the α-emission channel and their proton-decay widths are quite small. This result suggests these three states should also have well-developed alpha-cluster structures.

The excitation energy spectrum in the highly excited region is shown in Fig. 5. The top, middle, and bottom figures present the spectra taken from the singles measurement and the coincidence measurements with the α particles and ^{8}Be. Unfortunately, no significant structures are observed near the 6α decay threshold where the 6α-condensed state is expected. It is still interesting to observe bump structures at 20.5, 22.2, and 24.3 MeV in the coincidence spectrum with ^{8}Be. These states should be attributed to multi-alpha-cluster structures.

The analysis of the decay is still in progress. The results will be presented soon elsewhere.

Acknowledgments

The authors acknowledge the effort of the RCNP cyclotron crew for providing a stable and clean beam. This research was supported in part by the Nogami memorial fund for young nuclear physicists, the Grant-in-Aids for Scientific Research No. 21740171 and for the Global COE Program "The Next Generation of Physics, Spun from Universality and Emergence" from the Japan Ministry of Education, Sports, Culture, Science, and Technology.

References

[1] Ikeda K, Takigawa N and Horiuchi H 1968 *Prog. Theor. Phys. Suppl.* **Extra Number** 464–475
[2] Tohsaki A, Horiuchi H, Schuck P and Röpke G 2001 *Phys. Rev. Lett.* **87** 192501
[3] Yamada T and Schuck P 2004 *Phys. Rev.* C **69** 024309
[4] Itagaki N, Kimura M, Kurokawa C, Ito M and von Oertzen W 2007 *Phys. Rev.* C **75** 037303
[5] Itoh M *et al* 2004 *Nucl. Phys.* A **738** 268–272
[6] Chernykh M, Feldmeier H, Neff T, von Neumann-Cosel P and Richter A 2007 *Phys. Rev. Lett.* **98** 032501
[7] Freer M *et al* 2009 *Phys. Rev.* C **80** 041303
[8] Belyaeva T L *et al* 2010 *Phys. Rev.* C **82** 054618
[9] Wakasa T *et al* 2007 *Phys. Lett.* B 173–177
[10] Kawabata T *et al* 2007 *Phys. Lett.* B **646** 6–11
[11] Kawabata T *et al* 2004 *Phys. Rev.* C **70** 034318
[12] Yamada T, Funaki Y, Horiuchi H, Ikeda K and Tohsaki A 2008 *Prog. Theor. Phys.* **120** 1139–1167
[13] Youngblood D H, Lui Y W, Chen X F and Clark H L 2009 *Phys. Rev.* C **80** 064318
[14] Youngblood D H, Lui Y W and Clark H L 1999 *Phys. Rev.* C **60** 014304
[15] Fujiwara M *et al* 1999 *Nucl. Instrum. Methods Phys. Res.* A **422** 484–488

Journal of Physics: Conference Series **436** (2013) 012010 doi:10.1088/1742-6596/436/1/012010

Alpha cluster structure in ^{56}Ni

H Akimune[1], M Fujiwara[2], J Gibelin[3], M N Harakeh[4], L Achouri[3], S Bagchi[4], B Bastin[5], K Boretzky[12], H Bouzomita[5], L Caceres[5], S Damoy[5], F Delaunay[3], B. Fernández-Domínguez[3,14], M Caamano[8], U Garg[9], G F Grinyer[5], N Kalantar-Nayestanaki[4], O Kamalou[5], E Khan[6], A Krasznahorkay[10], G Lhoutellier[3], S Lukyanov[13], K Mazurek[11], M Najafi[4], J Pancin[5], Y Penionzkhevich[13], L Perrot[6], R Raabe[7], C E Rigollet[4], T Roger[5], S Sambi[7], H Savajols[5], M Senoville[3], C Stodel[5], L Suen[5], J C Thomas[5], J van de Walle[4] and M Vandebrouck[6]

[1] Department of Physics, Konan University, Kobe 658-8501, Japan
[2] RCNP, Osaka University, Osaka, Japan
[3] LPC Caen, ENSICAEN, Université de Caen, CNRS/IN2P3, Caen, France
[4] KVI, University of Groningen, Groningen, The Netherlands
[5] GANIL, Caen, France
[6] IPN Orsay, Paris, France
[7] Katholieke Universiteit, Leuven, Belgium
[8] University Santiago de Compostela, Spain
[9] University of Notre-Dame, USA
[10] Institute of Nuclear Research, (ATOMKI), Hungary
[11] Instytut Fizyki Jadrowej, Poland
[12] GSI, Darmstadt, Germany
[13] Joint Institute for Nuclear Research, Dubna, Russia
[14] Universidade de Santiago de Compostela, Santiago de Compostela, Spain

E-mail: akimune@konan-u.ac.jp

Abstract. The inelastic α–scattering experiment on ^{56}Ni in inverse kinematics was performed at an incident energy of 50 MeV/u at GANIL. A very high multiplicity for α–particle emission was observed with our phase-space limited experimental set-up. The maximum observed multiplicity, which cannot be explained by means of the statistical decay model, amounted to seven. The ideal classical gas model at kT = 3 MeV fairly well reproduced the experimental momentum distribution and multiplicity of alpha particles. This result strongly suggests that an alpha-gas state in ^{56}Ni may be excited via inelastic alpha scattering.

1. Introduction

Alpha cluster states were predicted in the 1960s by Ikeda et al.[1]. About 30 years later, several theorists started to discuss the possibility of α–particle condensation in low-density nuclear matter [2], as various exotic features were expected because this might correspond to a nuclear phenomenon analogous to the Bose-Einstein condensation for a finite number of dilute bosonic atoms. They predicted that the condensation could occur in the low-density region, and that the condensate states in finite nuclei may exist in excited states with a dilute density composed of weakly interacting gas of α–particles [3, 4]. However, experimentally, almost no information about such states has been obtained except for light nuclei such as ^{8}Be, ^{12}C [5] and ^{16}O.

2. Experiment

We performed the ^{4}He(^{56}Ni,^{4}He) inelastic scattering experiment at an incident energy of 50 MeV/u at GANIL. Originally, the aim of the experiment has been to search for the compression modes in ^{56}Ni which is almost the heaviest $N=Z$ nucleus to be obtained as secondary beam at the present RI beam facility. A recoiled ^{4}He particle as a target was measured using the active target system MAYA [6]. The MAYA detector is optimized to measure nuclear reactions with RI beams in reverse kinematics. It consists of an active gas target and an ancillary SSD+CsI detector wall located at forward angles.

3. Analysis

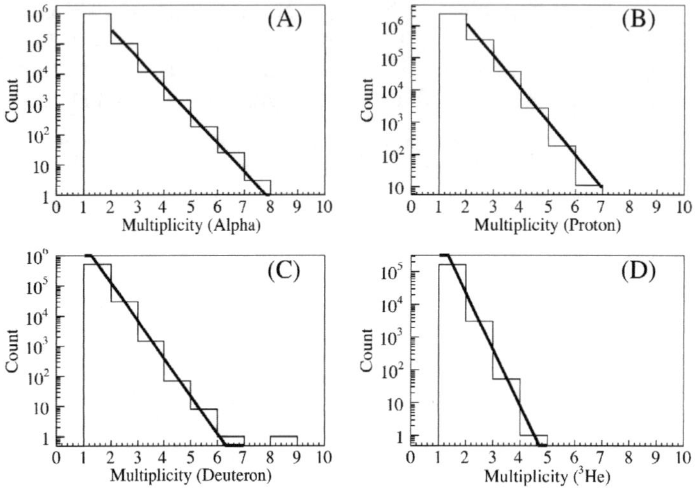

Figure 1. Multiplicity distributions of ^{4}He (A), proton (B), deuteron (C), and ^{3}He (D). The histograms and solid lines correspond to the experimental data and the results of fitting.

An analysis of the data was performed for the detector wall. In this analysis, we found that many particles are emitted at forward angles with high multiplicities. In Fig. 1, multiplicity distributions of ^{4}He, proton, deuteron, and ^{3}He are shown. As shown in Fig. 1 the maximum observed multiplicity for alpha particles is seven. It means that half of the mass of ^{56}Ni is observed to be carried away by means of emitting alpha particles. The multiplicity distributions are well reproduced by an exponential function $exp(-\lambda m)$, where m is the multiplicity. The decay constant λ is determined to be 2.17 for alpha particles. This λ_{4He} value is the smallest among the decay constants for p, d, and ^{3}He, here λ_p= 2.37, λ_d = 2.88, and λ_{3He} = 3.98. The λ constant is considered to be the sum of the multiplicity dependence of the cross-section for the ^{56}Ni(^{4}He,m-α) reaction and the experimental acceptance for particles. The experimental acceptance for particles is roughly estimated to be around 15%, and to be the major part of λ. Consequently, the multiplicity dependence of the cross section is relatively small. This experimental result is surprising, because it is natural to expect that the charged particle decay from highly excited states is strongly suppressed by the Coulomb barrier. Usually, in medium-heavy and heavy nuclei neutron decay is the predominant process for damping of highly excited states at the excitation energy above the particle decay threshold. The experimental result obtained in the present

Journal of Physics: Conference Series **436** (2013) 012010 doi:10.1088/1742-6596/436/1/012010

experiment shows strong evidence for multiple-particle decay, and may suggest a possible existence of a condensed state.

3.1. Statistical decay model

In order to eliminate the possibility for statistical decay from compound states which have small overlap with an alpha cluster state, we performed statistical decay model calculation based on the Hauser-Feshbach formalism using the code CASCADE [7]. In the calculation, spin, isospin, and parity of the excited states in daughter nuclei are used as the input parameters. Individual level parameters for the low-excitation energy region were taken from the experimental data. Global parameters for the level-density formula were used to describe the unknown levels at high excitation energies. All the decay channels: γ, proton-, deuteron-, and alpha- decays are taken into account.

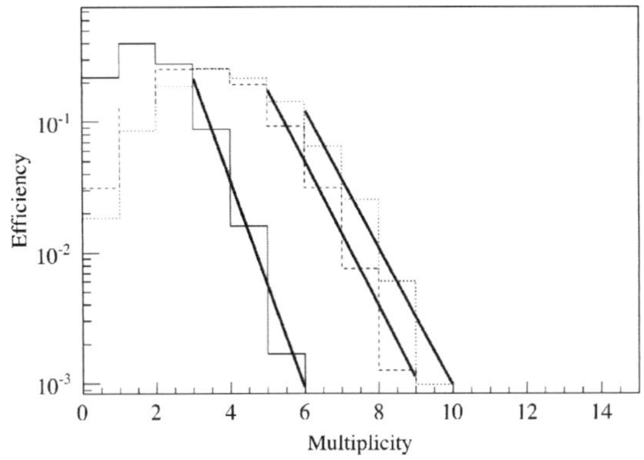

Figure 2. Detection efficiency for a fourteen alpha event at $kT = 1$ MeV (solid histogram), 3 MeV (dashed histogram), 5 MeV(dotted histogram). The solid lines are the results of fitting with an exponential function *exp(-λm)*.

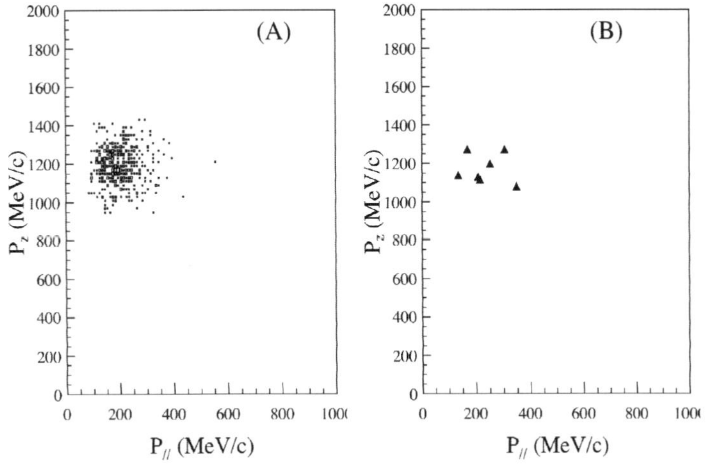

Figure 3. Results of simulations based on the classical ideal gas model. (A) Momentum distributions of alpha particles from events with multiplicity larger than 5. (B) Momentum distribution for one event with multiplicity seven.

The spin and parity for the excited states in ^{56}Ni were set to bs J^{π}=0^{+}, assuming that a strong monopole resonance was strongly excited at high excitation energy of 150 MeV. The maximum angular momentum for decay products was set to 40$\hbar$. According to the result of the statistical decay model calculation, the probability for 7-alpha decay is less than 0.01%. Therefore, the statistical decay model cannot explain the experimental result. One of the possible explanations is that these many-alpha particle emissions may come from an exotic excited state such as a diluted α–gas state.

3.2. Ideal gas model

In order to understand the events with high multiplicity of alpha particles, we performed a simulation based on the simple statistical model. In this model, we assumed that the excited state in ^{56}Ni consists of seventeen independent alpha clusters, and that the distribution of the kinetic energy for each alpha particle follows the Boltzmann distribution for the classical ideal gas at temperature *T*. In the simulation, the detection probability was estimated event by event for fourteen alpha particles. The kinetic energy and the direction of momentum in the centre of mass system of ^{56}Ni were randomly given by the Boltzmann distribution. The SSD+CsI detector wall of MAYA is segmented into 10×8 ΔE-E cells. Because each cell can identify only one charged particle in one event, we omitted in the present analysis the cell signals with a multiple-hit of charged particles. Since the thickness of the ^{4}He gas target is about 40 cm and it is located in front of the detector wall, the scattering angles depend on the vertex point of the reaction. In order to take into account this effect, the vertex points are determined in the active gas target, when the polar angles and azimuthal angles of the recoil alpha particles are reconstructed.

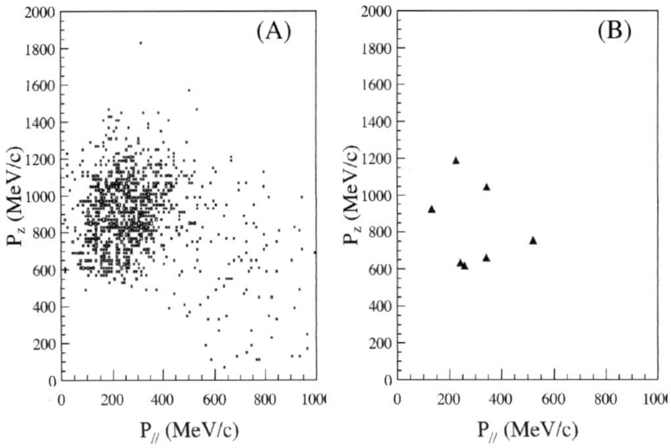

Figure 4. (A) The experimental momentum distributions of alpha particles from events with multiplicity larger than 5. (B) The momentum distribution for one event with the multiplicity seven.

In Fig. 2, the results of simulation for the detection efficiency of alpha particles from ^{56}Ni at kT= 1, 3, and 5 MeV are shown, where k is the Boltzmann constant. The temperature of kT = 3 MeV corresponds to the excitation energy of 150 MeV in ^{56}Ni. In Fig. 2, we see that the efficiency for detecting many alpha particles is low. Furthermore, the efficiency is especially low at lower temperature. Because at the lower temperature, alpha particles moves with small kinetic energies in the c.m. system of ^{56}Ni, and are therefore boosted to very small angles in the lab system, only a few alpha particles can be identified. At kT = 1, 3, and 5 MeV, the detection efficiency for an event with seven alpha particles detected out of fourteen in total is <0.1%, 0.8% and 1.8% respectively. Taking this into account, the cross-section for excitation of the alpha gas state via inelastic alpha scattering is estimated to be 1~10 mb at 50 MeV/u. This uncertainty mainly comes from the ambiguity in the excitation energy. Fig. 3 shows the two-dimensional scatter plot of the momentum distribution of alpha particles obtained by simulation at kT = 3 MeV. The left panel shows the superposition of all the events with a multiplicity larger than five. The right panel shows an example of the momentum distributions for one event of multiplicity seven.

Fig. 4 shows the two-dimensional scatter plot of the momentum distributions of alpha particles obtained in the experiment. The left panel shows the superposition of all events with multiplicity larger than five. The right panel shows the momentum distribution for an event with multiplicity seven. In Fig. 4(A), alpha particles with $P_{//}$ higher than 500 MeV/c are interpreted as recoiled target alpha particles, which are not shown in Fig. 3(A) obtained by the simulation. Although the momentum distribution of alpha particles is slightly wider compared with that of the simulation, the characteristics of the momentum distribution are well reproduced by the simple ideal gas model.

4. Summary and conclusion

We performed the ^{4}He(^{56}Ni,^{4}He) inelastic scattering experiment at an incident energy of 50 MeV/u. The inverse kinematics method was utilized to measure the reaction on the unstable ^{56}Ni nucleus. Multi-alpha decay events were observed. The maximum multiplicity observed for alpha particle was seven. This result is striking, because such high multiplicity cannot be explained by using a simple statistical-decay-model calculation. In order to understand this situation, we performed a simulation based on the ideal alpha-gas model. In this model, the momentum distribution follows the Boltzmann distribution. The experimental data are fairly well reproduced by taking into account the acceptance of the detector for multi-alpha particles events. This result strongly suggests the existence of an alpha gas state at high excitation energies in ^{56}Ni.

References

[1] Ikeda S, Takigawa N, and Horiuchi H 1968 *Prog. Theor. Phys. Suppl.Extra Number*, 464
[2] Tohsaki A, Horiuchi H, Schuck P and Ropke G 2001 *Phys. Rev. Lett.* **87** 192501
[3] Yamada T and Schuck P 2004 *Phys. Rev.* C **69** 024309
[4] Itagaki N et al. 2008 *Phys. Rev.* C **77** 037301
[5] Itoh M et al. 2011 *Phys. Rev.* C **84** 054308
[6] Demonchy C E et al. 2007 *Nucl. Inst. Meth. Phys. Res.* A **583** 341
[7] Pühlhofer F, 1977 *Nucl. Phys.* A **280**, 267, Harakeh M N, Computer code CASCADE, extended version

Journal of Physics: Conference Series **436** (2013) 012011 doi:10.1088/1742-6596/436/1/012011

Alpha-cluster structure in the ground state of ^{40}Ca displayed in a (p,pα) knockout reaction

A A Cowley[1,2]

[1] Department of Physics, Stellenbosch University, Private Bag X1, Matieland 7602, South Africa

[2] iThemba Laboratory for Accelerator Based Sciences, P O Box 722, Somerset West 7129, South Africa

E-mail: aac@sun.ac.za

Abstract. The analyzing power is very sensitive to details of the reaction mechanism of $(p,p\alpha)$ knockout reactions in the incident energy range of approximately 100 MeV and higher. Whereas distorted wave impulse approximation calculations in the past proved to give an excellent reproduction of analyzing power angular distributions for quasifree $(p,p\alpha)$ reactions on light targets such as ^{6}Li, ^{9}Be and ^{12}C, the situation for ^{40}Ca was not as simple. It is now shown that the theory also offers good agreement with the experimental distribution of the heaviest target nucleus if care is taken to use proper distorted waves which treat α–^{36}Ar properly as a system for which α–elastic scattering is anomalous. Thus it is shown that ^{40}Ca reveals its ground state α–cluster structure in an unambiguous way similar to the light target nuclei.

1. Introduction

Whereas clustering in specific excited states of atomic nuclei is a clearly-identified phenomenon, the presence of α–cluster configurations in the ground-states of nuclides is more controversial. This scepticism is directly due to the success of shell-model descriptions of nuclei. However, it should be recognized that by exploiting single particle shell-model wave functions, it is possible to project ground-state α–like cluster structures out of those configurations. Experimentally it is consequently of interest to determine whether knockout and pickup reactions support this refined theoretical expectation.

Of course, the $(p,p\alpha)$ reaction at energies of 100 to 300 MeV, not surprisingly, provides direct evidence of ground state clustering in very light target nuclei, such as ^{6}Li and ^{9}Be. However, for even such a light-mass target as ^{12}C a very clear signature in terms of a $(p,p\alpha)$ analyzing power energy and angular distribution was only recently demonstrated [1, 2]. On the other hand, the analyzing power, which is very sensitive to details of the reaction mechanism of knockout, recently failed to confirm the expected response from clustering in the ground state of ^{40}Ca [3].

2. New results for ^{40}Ca

We have further investigated the ^{40}Ca$(p,p\alpha)^{36}$Ar(g.s.) reaction at an incident energy of 100 MeV. The results are shown in Figs. 1 and 2.

It is found that, in contrast with the cross section, the analyzing power depends very strongly on the optical potential employed in a distorted-wave impulse approximation (DWIA) [4, 5]

Journal of Physics: Conference Series **436** (2013) 012011 doi:10.1088/1742-6596/436/1/012011

description of the outgoing α channel of the reaction. More specifically, this observable requires an optical potential for the α–particle which generates a wave function that describes the α–^{36}Ar interaction accurately. This system manifests an anomalous large angle effect [6], which is not present for adjacent target masses. This is shown in Fig. 1 where a DWIA calculation of the analyzing power with an α–particle optical potential from the work of Reidemeister *et al.* [6] is compared with results employing a standard global optical potential. The Reidemeister potential is of a Woods-Saxon squared shape, which is needed to describe elastic scattering of α–particles, and we use a parameter set appropriate to the projectile energy corresponding to that required by the kinematics of the $(p,p\alpha)$ knockout reaction.

The result with an α–particle optical potential, which gives a reasonable description of elastic scattering in a global mass range, but not specifically for ^{36}Ar, is clearly unacceptable, as may be seen from Fig. 1.

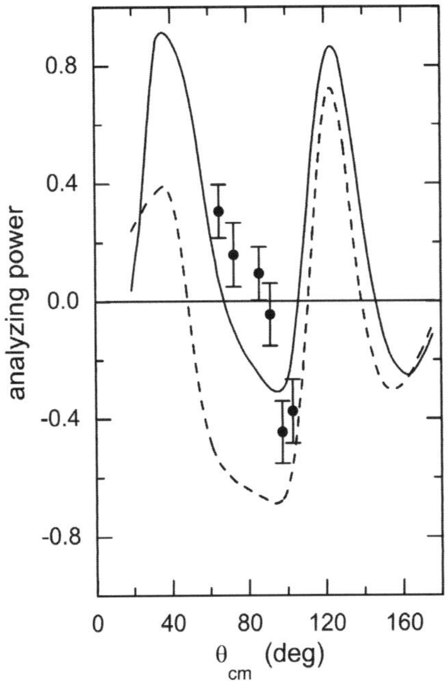

Figure 1. Analyzing power angular distribution for the ^{40}Ca$(p,p\alpha)^{36}$Ar reaction at an incident energy of 100 MeV. Results are shown as a function of the centre-of-mass scattering angle θ_{cm} of the p–α two-body system under quasifree kinematic conditions. Experimental data are from Ref. [3]. The solid line corresponds to a DWIA calculation with an α–particle potential set [6] which fits anomalous large angle elastic scattering from the ^{36}Ar residual nucleus. The dashed line is a calculation with a standard representative potential from Carey *et al.* [7].

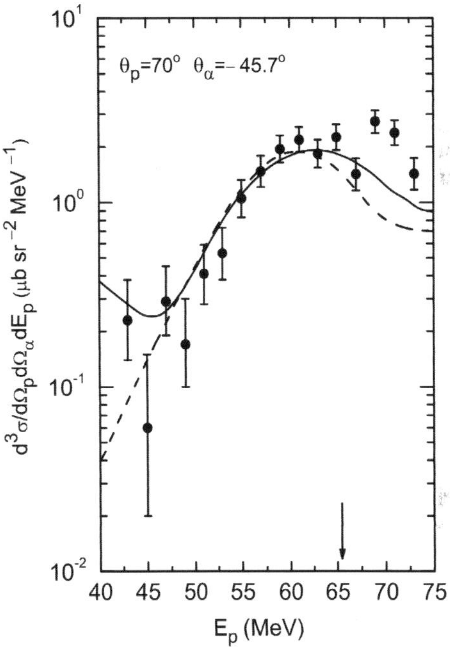

Figure 2. Energy sharing cross section distribution for the reaction ^{40}Ca$(p,p\alpha)^{36}$Ar at an incident energy of 100 MeV. Results for the quasifree angle pair indicated in the figure are shown as a function of the energy of the observed scattered proton. Experimental data from Ref. [7]. The solid curve is from a DWIA calculation with the α–particle optical potentials of Ref. [6]. The dashed line is calculated with a standard potential set from Carey *et al.* [7] which fits elastic scattering from the general target-mass range of the residual nucleus.

Journal of Physics: Conference Series **436** (2013) 012011 doi:10.1088/1742-6596/436/1/012011

The cross section distribution, on the other hand is not sensitive to this detail (See Fig. 2), and α–potentials extracted for generic target masses are equally acceptable for that case, as was also found in earlier investigations [7]. Differences are observed at the extreme ends of the cross section distribution where the quality of experimental data is poor (low proton energy) or where sequential α–particle decay [8] may contribute at higher proton energies. Those two regions are far removed from the zero recoil kinematic condition (shown as an arrow in Fig. 2) at which the data in Fig. 1 were measured for that specific angle pair.

Clearly the use of standard Woods-Saxon type potential shapes for the α–particle in the work of Neveling *et al.* [3] is the direct cause of the difficulty encountered in that study. However, because cross section distributions did not suffer from this deficiency, it was not realised that those potential shapes are inappropriate for analyzing power, with its increased sensitivity to correct distorted waves in the DWIA.

Such an effect is not encountered for the proton channels, for which elastic scattering is generally accurately reproduced by a number of global optical model parameter sets. As long as the different optical parameter sets all give reasonable fits to elastic scattering in the relevant mass and incident energy range, DWIA results are roughly similar for all sets.

The increased sensitivity of the analyzing power distribution relative to cross section data in the knockout reaction ^{40}Ca(p,$p\alpha$)^{36}Ar, together with the agreement between the DWIA calculation and the experimental data when an appropriate α–particle optical potential is used in the DWIA, removes any doubt as to the ground state clustering of the target nucleus.

3. Summary and conclusion

We have investigated the DWIA description of the knockout reaction ^{40}Ca(p,$p\alpha$)^{36}Ar(g.s.) at an incident energy of 100 MeV. It is shown that previous difficulties in the theoretical reproduction of the analyzing power angular distribution, which is expected to be directly related to knockout of a preformed cluster in the taget nucleus, is sensitive to the use of correct distorted waves for the outgoing α–particle in the calculations.

The observed sensitivity of the analyzing power to the specific distorting potentials used is very different from the behaviour experienced for lighter target masses. In the case of mass 12 and lighter, results do not differ appreciably for different sets of equivalent optical potentials.

The present analysis clearly supports a cluster interpretation of the reaction. Thus ^{40}Ca displays a degree of α–clustering in the ground state similar to lighter target nuclei such as ^{12}C.

Further experimental and theoretical work are desirable. An investigation of the trend as one moves away from stable to rare nuclear species may be especially useful.

Acknowledgement

This work was performed with funding from the South African National Research Foundation (NRF). The financial support is gratefully acknowledged.

References

[1] Mabiala J, Cowley A A, Förtsch S V, Buthelezi E Z, Neveling R, Smit F D, Steyn G F and Van Zyl J J 2009 *Phys. Rev.* C **79** 054612
[2] Cowley A A, Mabiala J, Buthelezi E Z, Förtsch S V, Neveling R, Smit F D, Steyn G F and Van Zyl J J 2009 *EPL* **85** 22001
[3] Neveling R, Cowley A A, Buthelezi Z, Förtsch S V, Fujita H, Hillhouse G C, Lawrie J J, Steyn G F, Smit F D, Wyngaardt S M, Botha N T, Mudau L and Ntshangase S S 2008 *Phys. Rev.* C **77** 037601
[4] Chant N S and Roos P G 1977 *Phys. Rev.* C **15**, 57–68
[5] Chant N S and Roos P G 1983 *Phys. Rev.* C **27** 1060–1072
[6] Reidemeister G, Ohkubo S and Michel F 1990 *Phys. Rev.* C **41** 63–74
[7] Carey T A, Roos P G, Chant N S, Nadasen A and Chen H L 1984 *Phys. Rev.* C **29** 1273–1288
[8] Roos P G, Chant N S, Cowley A A, Goldberg D A, Holmgren H D and Woody R 1977 *Phys. Rev.* C **15** 69–83

Journal of Physics: Conference Series **436** (2013) 012012 doi:10.1088/1742-6596/436/1/012012

Large scale calculations for cluster structure of light nuclei with Skyrme interaction

Y Fukuoka[1], **Y Funaki**[2], **T Nakatsukasa**[2,3] **and K Yabana**[1,3,2]

[1]Graduate School of Pure and Applied Sciences, University of Tsukuba, Tsukuba 305-8571, Japan

[2]RIKEN Nishina Center, Wako 351-0198, Japan

[3]Center for Computational Sciences, University of Tsukuba, Tsukuba 305-8571, Japan

E-mail: fukuoka@nucl.ph.tsukuba.ac.jp

Abstract. We present a computational approach to describe structure of light nuclei including cluster states. Apart from the use of an empirical nuclear force, Skyrme interaction, our scheme does not utilize any a priori knowledge on the structure of nuclei. In our framework, we first generate a number of Slater determinants in a stochastic way. We then make projections of parity and angular momentum, and perform configuration mixing calculation. We show results for ^{12}C and ^{16}O nuclei. Our calculation provides a reasonable description for the ground state rotational band, Hoyle state, and low-lying negative parity states of ^{12}C. We may also describe the 0_2^+ rotational band of ^{16}O, although excitation energies are slightly overestimated.

1. Introduction

In light nuclei, a variety of cluster structures appear in excited states. For a theoretical description of these states, microscopic cluster models such as resonating group method and generator coordinate method have been developed [1]. In these methods, existence of nuclear clusters are assumed from the beginning. To understand mechanisms of a formation of clusters, theoretical studies which do not assume existence of clusters are required. The anti-symmetrized molecular dynamics is one of the successful approaches for that purpose [2]. In its simplest version, however, the model wave function is rather simple, a Slater determinant composed of Gaussian wave packets.

In last two decades, ab-initio calculations starting from a realistic nucleon-nucleon force have been developed. They have been successful for description of the ground and low-lying excited states. However, at a moment, they are not able to describe typical clustering states, such as the second 0^+ states in ^{12}C and ^{16}O [3, 4].

In this paper, we report an application of the configuration mixing approach which we have recently developed [5] for light nuclei, ^{12}C and ^{16}O. Superposing a number of Slater determinants which include both shell-model like and cluster configurations, we can obtain a converged excitation spectrum of these nuclei for a given Skyrme Hamiltonian. We will show that our approach is capable of describe both ground state and excited states with cluster structures simultaneously.

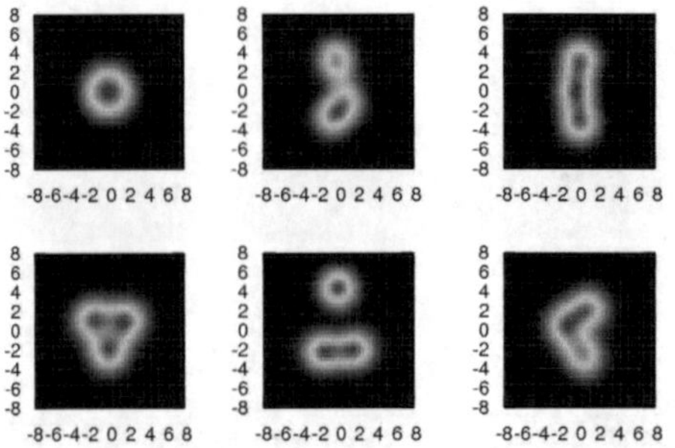

Figure 1. Density distribution of several Slater determinants which are prepared by the imaginary-time method and are used for a configuration mixing calculation of ^{12}C.

2. Formulation

We use a method developed by us [5]. The method is composed of the following three steps. First we prepare a number of Slater determinants with different correlation structures employing the imaginary-time method. The imaginary-time method is usually used to calculate self-consistent Hartree-Fock solutions. We use this method for different purpose, generating a number of Slater determinants with different correlation structures. In practice, we prepare an initial orbitals of a Slater determinant in a stochastic way. We then perform the imaginary time evolution. We pick up and store a number of Slater determinants in the course of iterations before reaching the self-consistent solution. In Fig. 1, we show density distributions of several Slater determinants which are generated in the above mentioned procedure. One can see that both shell-model-like and cluster-like configurations are included. We typically generate 50 Slater determinants. We then make projections of parity and three-dimensional angular momentum for the Slater determinants. Finally we achieve a configuration mixing calculation using the projected states. We use SLy4 parameter in our calculation. A real-space Cartesian grid representation is used with a grid spacing of 0.8 fm. Grid points inside a sphere of 8 fm are used.

3. Results

3.1. ^{12}C

We first show results for ^{12}C. We show the energy spectrum and the transition strength in Fig. 2. We also show density distributions of Slater determinants which are main components of 0_2^+ state. The binding energy is reasonably reproduced, 95.3 MeV after configuration mixing. This consists of the Hartree-Fock energy of the ground state, 90.6 MeV, and the correlation energy, 4.7 MeV. The binding energy in our calculation is slightly larger than the measured value, 92.1 MeV. As seen from the figure, our calculation accurately reproduces the excitation energies of the ground rotational band. The transition strengths are also reproduced reasonably. The 0_2^+ state, so-called the Hoyle state, appears somewhat high compared with the measurement. We show density distributions of two Slater determinants which are the main components of this state in the configuration mixing calculation. They show a well-developed 3-α cluster structure. We find it necessary to superpose a number of Slater determinants to describe the state, consistent with a picture of alpha-condensation [6].

3.2. ^{16}O

We show results for ^{16}O in Fig. 3. The main component of the ground state is the Hartree-Fock solution which accounts for more than 80% in the configuration mixing calculation. Our calculation describes second 0^+ rotational band, 0_2^+, 2_1^+ and 4_2^+. The band head energy of the

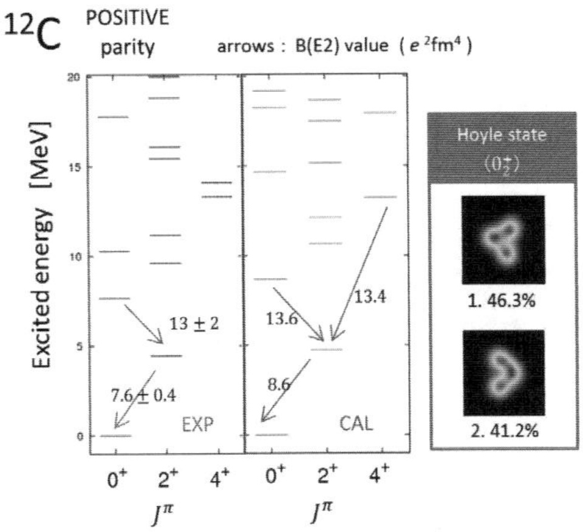

Figure 2. Energy spectrum for ^{12}C. Density distributions of Slater determinants which are main components of 0_2^+ state is shown as well. The experimental data are taken from Refs. [7].

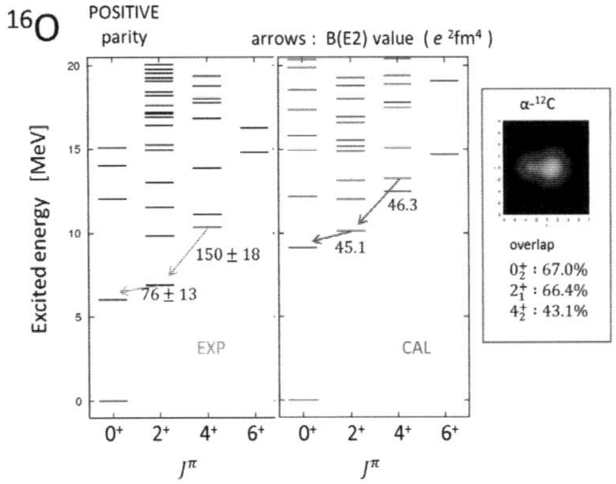

Figure 3. Energy spectrum for ^{16}O. Density distribution of a Slater determinant which is the main component of 0_2^+ band is also shown. The experimental data are taken from Refs.[8].

0_2^+ is 9.5 MeV, higher than the measured energy, 6.1 MeV. The B(E2) transition strengths of the 0_2^+ band is underestimated in our calculation. We show the density distribution of a Slater determinant which is a major component of the 0_2^+ band. It shows a cluster structure of α-^{12}C.

4. Conclusion

We have presented configuration mixing calculations using Skyrme interaction for ^{12}C and ^{16}O nuclei superposing a number of Slater determinants which include cluster configurations as well as shell-model-like configurations. For ^{12}C, our calculation reproduces the ground state rotational band and the 0_2^+ Hoyle state simultaneously. For ^{16}O, our calculation reproduces the 0_2^+ rotational band having α+^{12}C cluster state, though excitation energies are slightly too high.

5. Acknowledgments

The numerical calculations were performed on T2K-Tsukuba System at Center for Computational Sciences in University of Tsukuba and SR16000 at YITP in Kyoto University.

References

[1] Ikeda K, Horiuchi H and Saito S 1980 *Prog. Theor. Phys. Suppl.* **68** 1; Fujiwara Y *et al* 1980 *Prog. Theor. Phys. Suppl.* **68** 29

[2] Horiuchi H 1991 *Nucl. Phys.* A **522** 257; Kanada-En'yo Y and Horiuchi H 2001 *Prog. Theor. Phys. Suppl.* **142** 205

[3] Wiringa R B, Pieper S C, Carlson J and Pandharipande V R 2000 *Phys. Rev.* C **62** 014001

[4] Navrátil P, Vary J P and Barrett B R 2000 *Phys. Rev.* C **62** 054311

[5] Shinohara S, Ohta H, Nakatsukasa T and Yabana K 2006 *Phys. Rev.* C **74** 054315

[6] Funaki Y *et al* 2003 *Phys. Rev.* C **67** 051306

[7] Ajzenberg-Selove F 1990 *Nucl. Phys.* A **506** 1; Zimmerman W R *et al* 2011 *Phys. Rev.* C **84** 027304; Itoh M *et al* 2011 *Phys. Rev.* C **84** 054308; Freer M *et al* 2012 *Phys. Rev.* C **86** 034320; Freer M *et al* 2011 *Phys. Rev.* C **83** 034314

[8] Tilley D R *et al* 1993 *Nucl. Phys.* A **564** 1

Journal of Physics: Conference Series **436** (2013) 012013 doi:10.1088/1742-6596/436/1/012013

Alpha-particle resonances and clusters in ^{32}S, ^{34}S, ^{36}Ar and ^{40}Ca

Markus Norrby

Åbo Akademi University, Teacher Training School, Skolhusgatan 31, 65100 Vasa, Finland

E-mail: mnorrby@abo.fi

Abstract. Recently the nuclei ^{32}S, ^{34}S and ^{40}Ca have been examined in detail from an alpha-cluster point of view, adding to previous knowledge on cluster structures in medium light nuclei. But as often in science, new results give rise to more new questions than old ones they answer. In the text below some of these open questions and suggestions for future work are stated.

1. Introduction

In the last decade or so there has been a clear shift in the nuclear physics community. Many smaller laboratories have been closed down and the focus has shifted to a few large-scale facilities around the world. This concentration of experimental resources has opened many new and very interesting fields of exploration through exotic beams and reactions, and highly advanced detector set-ups.

At the same time, as the community has invested heavily in these specialized facilities, it has become increasingly difficult for smaller research groups to compete for beam-time for experiments examining less exotic nuclei. Still there is clearly a lot of work to be done among stable nuclei and nuclei close to stability. The medium-light nuclei (roughly between oxygen and titanium in the nuclide chart) are in a region of special interest. They represent the transition from light nuclei, where various *ab-initio* methods can make increasingly precise theoretical predictions, and the heavy nuclei where collective models can be used with relatively good precision. In the gap between these very different models it can be expected that descriptions based on cluster models are important.

2. Open questions and suggestions for future work

The Åbo Akademi University nuclear structure group has been working with alpha-cluster structures in medium light nuclei for more than 25 years [1], beginning as a study of the very old problem of Anomalous Large-Angle Scattering (ALAS) [2]. A lot of results have been published over the years, see e.g. [3, 4, 5] and the references therein. As the focus of the department has shifted to other areas of interest, this endeavor in nuclear structure physics is now coming to an end at Åbo Akademi University. But there are still many mysteries and open questions to be explored among these nuclei. In the following some suggestions are made where the work could be continued.

2.1. A systematic examination of alpha-particle resonances among medium light nuclei

The last few years have shown a great increase in interest for resonance scattering, especially in inverse kinematics. The Tick-Target Inverse Kinematics technique (TTIK) [6] has made it possible to measure excitation functions in large energy intervals in a relatively short time. Still the analysis of such data, usually based on R-matrix theory [7], is very time consuming. It has been shown, though, that the combination of TTIK with a simplified R-matrix analysis can be a time efficient and relatively reliable method for mapping large numbers of resonances [3, 8].

Only a few of the alpha-conjugate nuclei in the region have been examined thoroughly (^{32}S, ^{36}Ar and ^{40}Ca) in this way and it would clearly be of interest to carry out similar work on ^{20}Ne, ^{24}Mg, ^{28}Si and ^{44}Ti. The measurements also need to be extended at least to the neutron rich side as the report on ^{34}S [4] shows many interesting features. Especially the nuclei ^{40}Ar and ^{44}Ca should be easily accessible with stable beams and have an interesting conjugate plus four neutrons configuration possibility.

2.2. Verification of the nature of the observed resonances

The large amount of strong and narrow resonances found in these nuclei cannot yet be explained by theory. While it can be shown that the probability is low that the observed resonances are mainly caused by statistical effects (either so called Ericsson fluctuations or Porter-Thomas fluctuations), their contributions are not insignificant and the interplay with real quantum mechanical states is not well understood [9]. More experiments with high resolution and detailed analysis similar to [10] are needed to independently verify the resonances and their fine structure. New theory needs to be developed, perhaps along the promising lines of [11, 12].

2.3. Verification of and explanation for the proposed rotational model

The spin distributions of the observed resonances have been interpreted as an indication of a quantum mechanical rotation with the rotational levels split into many close-lying states with the same spin (see figure 1). If the proposed model is valid, the mechanisms behind the splitting need to be explained. The extracted moments of inertia point to a rotation involving a few alpha-particles orbiting an inert core. This strange type of dynamical structure needs to be verified experimentally and explained theoretically.

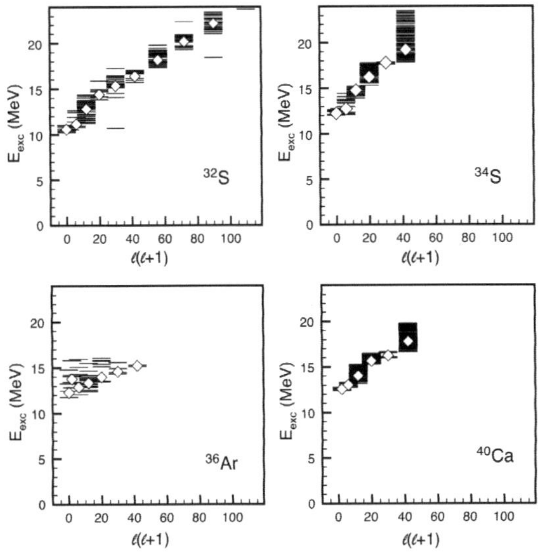

Figure 1. Comparison of resonances in ^{32}S, ^{34}S, ^{36}Ar and ^{40}Ca. The diamonds represent average energies of resonances with the same spin, weighted with the reduced widths of the resonances [5].

Closely connected to the rotational interpretation are the large reduced widths that are observed. Due to experimental limitations the values given in e.g. [3, 4, 5] have a large uncertainty and the measurements need to be improved in order to verify the cluster structure in these nuclei.

2.4. Explanation of the general structure of the observed spectra

Many of the excitation functions for medium light alpha-conjugate nuclei show interesting average features beside the resonances. For example, the spectrum of the nucleus ^{32}S shows a number of bumps followed by sharp decreases in cross section [13, 14]. At very high excitation energy the elastic cross section disappears completely. These features seem to coincide in energy with various alpha particle break-up thresholds, but due to the very high angular momenta involved such a process is difficult to imagine. It would, however, be an interesting experimental challenge to do detailed multichannel particle spectroscopy at these thresholds in order to understand their structure.

3. Summary

In the text above some challenges for both experimentalists and theorists have been expressed. Many more potentially interesting routes of exploration exist for resonances and clusters in medium light nuclei, like the connection to nuclear astrophysics and the importance of cluster resonances for nucleosynthesis in massive stars. Hopefully many of these questions can be settled through new experiments and new developments in theory in the near future. And hopefully experiments of this relatively straightforward type will not drown in the concentrated efforts to push the limits of rare beams and exotic reactions.

Acknowledgments

The generous support from Högskolestiftelsen i Österbotten is noted with gratitude.

References

[1] Manngard P, Brenner M, Alam M, Reichstein I and Malik F 1989 *Nucl. Phys.* A **504** 130–142
[2] Corelli J C, Bleuler E and Tendam D J 1959 *Phys. Rev.* **116** 1184–1193
[3] Lönnroth T *et al* 2010 *Eur. Phys. J.* A **46** 5–16
[4] Norrby M *et al* 2011 *Eur. Phys. J.* A **47** 73
[5] Norrby M *et al* 2011 *Eur. Phys. J.* A **47** 1–7
[6] Goldberg V Z and Pakhomov A E 1993 *Phys. At. Nucl.* **56** 1167
[7] Lane A M and Thomas R G 1958 *Rev. Mod. Phys.* **30** 257–353
[8] Goldberg V Z, Rogachev G V, Brenner M, Källman K M, Lönnroth T, Rozhkov M V, Torilov S, Trzaska W H and Wolski R 2000 *Phys. At. Nucl.* **63** 1518–1526
[9] Brenner M and Gridnev K A 2010 *Cluster Structure of Atomic Nuclei* (Kerala: Research Signpost) p 43
[10] Brenner M, Fletcher N R, Liendo J A, Belov S E, Caussyn D D, Nieto T K and Myers S H 2000 *Acta Phys. Hung. N.S.* **11** 221
[11] Gridnev K A, Torilov S, Kartavenko V G and Greiner W 2007 *Int. J. Mod. Phys.* E **16** 1059–1063
[12] Gnilozub I A, Kurgalin S D and Tchuvil'sky Y M 2011 *AIP Conf. Proc.* **1377** 441–443
[13] Brenner M, Lattuada M, Gulino M, Khlebnikov S V, Chengbo L, Prete G, Trzaska W H, Zadro M and Belov S E 2006 *Phys. Scr.* **74** 692–696
[14] Norrby M, Brenner M, Lönnroth T and Kallman K M 2008 *Int. J. Mod. Phys.* E **17** 2019

Journal of Physics: Conference Series **436** (2013) 012014 doi:10.1088/1742-6596/436/1/012014

Clustering effects in reactions induced by light nuclei

C Beck

Département de Recherches Subatomiques, Institut Pluridisciplinaire Hubert Curien, IN_2P_3-CNRS and Université de Strasbourg - 23, rue du Loess BP 28, F-67037 Strasbourg Cedex 2, France

E-mail: `christian.beck@iphc.cnrs.fr`

Abstract. Since the pioneering discovery of $^{12}C+^{12}C$ molecular resonances half a century ago, a great deal of research work has been undertaken in the α-clustering study. Our knowledge on the physics of nuclear molecules has increased considerably and nuclear clustering remained one of the most fruitful domains of nuclear physics, facing some of the greatest challenges and opportunities in the years ahead. In this work, the occurrence of "exotic" shapes in light N=Z α-like nuclei is investigated. Various approaches of the superdeformed and hyperdeformed bands associated with quasimolecular resonant structures are presented. Clustering aspects are also briefly discussed for light nuclei with neutron excess trough very recent results on neutron-rich Oxygen isotopes.

1. Introduction

The search for resonant structures in the excitation functions for various combinations of light α-cluster (N=Z) nuclei in the energy regime from the Coulomb barrier up to regions with excitation energies of E_x=20−50 MeV remains a subject of contemporary debate [1, 2]. These resonances have been interpreted in terms of nuclear molecules [1]. The question of how quasimolecular resonances may reflect continuous transitions from scattering states in the ion-ion potential to true cluster states in the compound systems is still unresolved [1, 2]. In many cases, these resonant structures have been associated with strongly-deformed shapes and with alpha-clustering phenomena [3, 4], predicted from the Nilsson-Strutinsky approach, the cranked α-cluster model [3], or other mean-field calculations [4, 5]. In light α-like nuclei clustering is observed as a general phenomenon at high excitation energy close to the α-decay thresholds [3, 6]. This exotic behavior has been perfectly illustrated by the famous "Ikeda"-diagram for N=Z nuclei in 1968 [7], which has been recently extended by von Oertzen [8] for neutron-rich nuclei, as shown in the left panel of Fig.1. Clustering is a general phenomenon [9] not only observed in light neutron-rich nuclei [10], but also in halo nuclei such as ^{11}Li [11] or ^{14}Be, for instance [12]. The problem of cluster formation has also been treated extensively for very heavy systems by R.G. Gupta [5] and by Zagrebaev and W. Greiner [13] where giant molecules and the collinear ternary fission may exist [14].

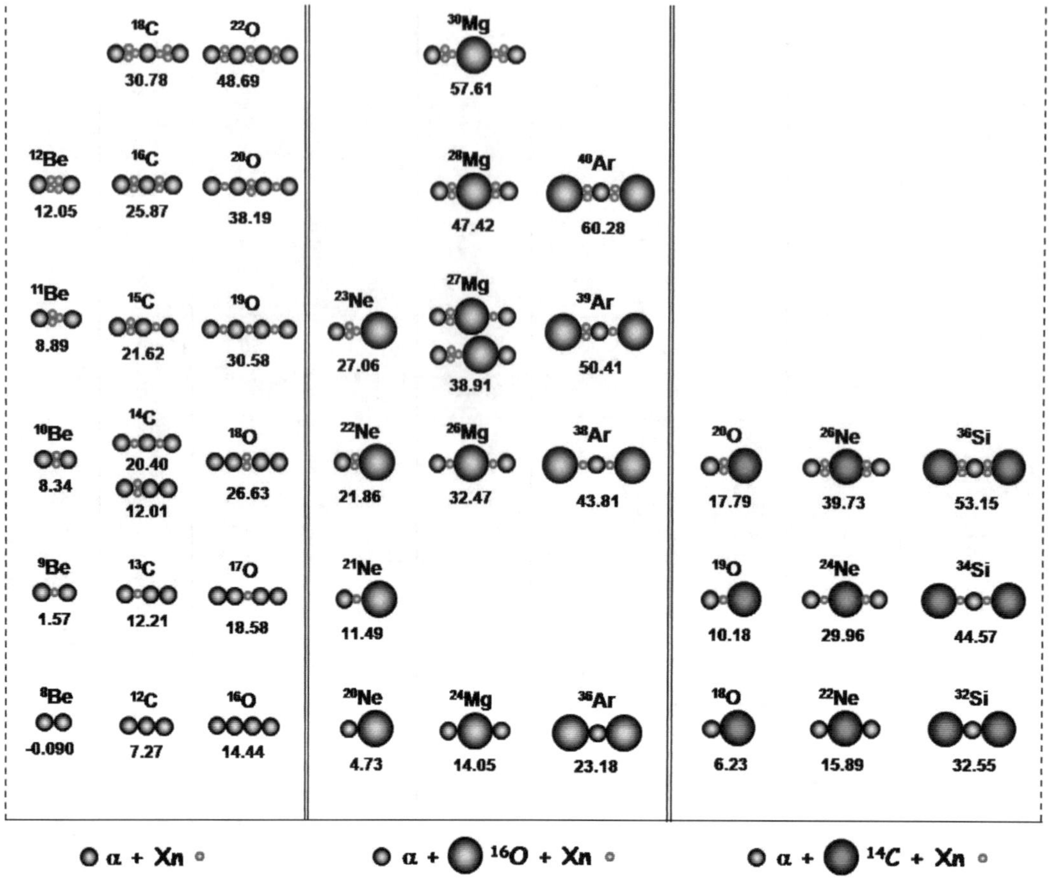

Figure 1. Schematic illustration of the structures of molecular shape isomers in light neutron-rich isotopes of nuclei consisting of α-particles, ^{16}O- and ^{14}C-clusters plus some covalently bound neutrons (Xn means X neutrons) [9]. The so called "Extended Ikeda-Diagram" [8] with α-particles (left panel) and ^{16}O-cores (middle panel) can be generalized to ^{14}C-cluster cores (right panel). The lowest line of each configuration corresponds to parts of the original Ikeda diagram [7]. However, because of its deformation, the ^{12}C nucleus is not included, as it was earlier [7]. Threshold energies (in MeV) are given for the relevant decompositions.

2. Alpha clustering, deformations and alpha condensates

The real link between superdeformation (SD), nuclear molecules and alpha clustering [4, 15, 16, 17] is of particular interest, since nuclear shapes with major-to-minor axis ratios of 2:1 have the typical ellipsoidal elongation for light nuclei (with quadrupole deformation parameter $\beta_2 \approx 0.6$). Furthermore, the structure of possible octupole-unstable 3:1 nuclear shapes (hyperdeformation (HD) with $\beta_2 \approx 1.0$) has also been discussed for actinide nuclei [17] in terms of clustering phenomena. Typical examples for possible relationship between quasimolecular bands and extremely deformed (SD/HD) shapes have been widely discussed in the literature for $A = 20-60$ α-conjugate N=Z nuclei, such as ^{28}Si [18, 19, 20, 21, 22], ^{32}S [4, 23, 24, 25, 19], ^{36}Ar [17, 26, 27, 28, 29], ^{40}Ca [30, 31, 32, 33], ^{44}Ti [4, 34, 35], ^{48}Cr [36] and ^{56}Ni [37, 38, 39, 40].

In fact, highly deformed shapes and SD rotational bands have been discovered in several light α-conjugate (N=Z) nuclei, such as ^{36}Ar and ^{40}Ca by using γ-ray spectroscopy techniques [27, 30]. In particular, the extremely deformed rotational bands in ^{36}Ar [27] (shown as crosses in Fig. 2) might be comparable in shape to the quasimolecular bands observed in both ^{12}C+^{24}Mg [28, 41, 42, 43] (shown as open triangles in Fig. 2) and ^{16}O+^{20}Ne [44, 45] (shown as full rectangles) reactions. Ternary clusterizations are also predicted theoretically, but were not found experimentally in ^{36}Ar so far [29]. On the other hand, ternary fission of ^{56}Ni – related to its hyperdeformed shapes – was identified from out-of-plane angular correlations measured in the ^{32}S+^{24}Mg reaction with the Binary Reaction Spectrometer (BRS) at the VIVITRON Tandem facility of the IPHC, Strasbourg [46]. This finding [46] is not limited to light N=Z compound nuclei, true ternary fission [13, 14, 48] can also occur for very heavy [14, 48] and superheavy [47] nuclei.

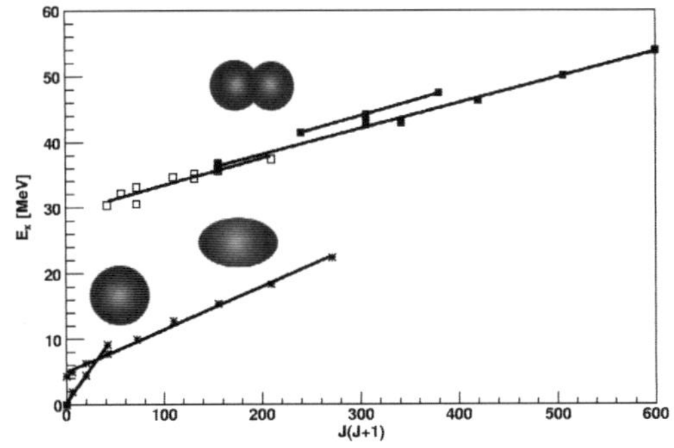

Figure 2. Rotational bands and deformed shapes in ^{36}Ar. Excitation energies of the ground state (spherical shape) and SD (ellipsoidal shape) bands [27], respectively, and the energies of HD (dinuclear shape) band from the quasimolecular resonances observed in the ^{12}C+^{24}Mg (open rectangles) [28, 41, 42, 43] and ^{16}O+^{20}Ne (full rectangles) [44, 45] reactions are plotted as a function of J(J+1). This figure has been adapted from Refs. [26, 28].

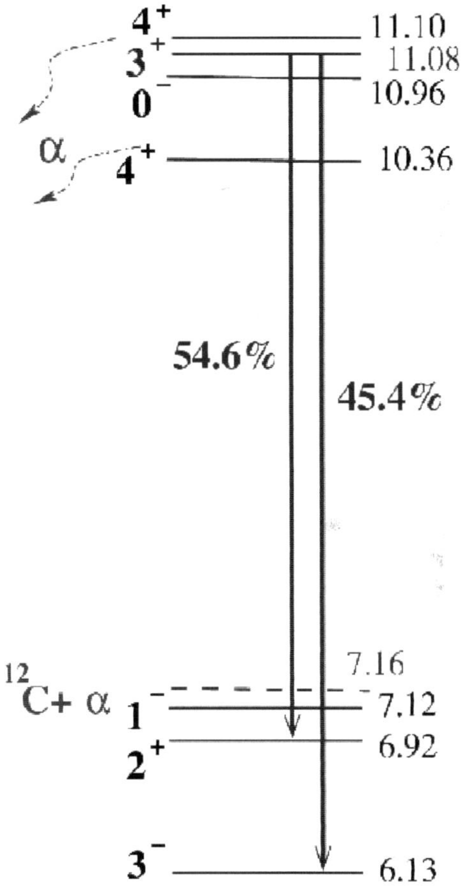

Figure 3. New partial (high-energy) level scheme of ^{16}O corresponding to γ-ray transitions observed in the ^{12}C(^{24}Mg,^{20}Ne)^{16}O* α-transfer reactions. This figure has been adapted from Ref. [29].

There is a renewed interest in the spectroscopy of the ^{16}O nucleus at high excitation energy [29]. Exclusive data were collected on ^{16}O in the inverse kinematics reaction ^{24}Mg+^{12}C studied at $E_{lab}(^{24}Mg)$ = 130 MeV with the BRS in coincidence with the EUROBALL IV installed at the VIVITRON facility [29]. From the α-transfer reactions (both direct transfer and deep-inelastic orbiting collisions [49]), new information has been deduced on branching ratios of the decay of the 3^+ state of ^{16}O at 11.085 MeV $\pm$ 3 keV. The high-energy level scheme of ^{16}O shown in Fig. 3 indicates that this state does not α-decay because of its non-natural parity (in contrast to the two neighbouring 4^+ states at 10.36 MeV and 11.10 MeV), but it γ decays to the 2^+ state at 6.92 MeV (54.6 $\pm$ 2 %) and to the 3^- state at 6.13 MeV (45.4%). By considering all the four possible transition types of the decay of the 3^+ state (*i.e.* E1 and M2 for the $3^+ \rightarrow 3^-$ transition and, M1 and E2 for the $3^+ \rightarrow 2^+$ transition), our calculations yield the conclusion that $\Gamma_{3^+} < 0.23$ eV, a value fifty times lower than known previously, which is an important result for the well studied ^{16}O nucleus [29]. Clustering effects in the light neutron-rich oxygen isotopes ^{17,18,19,20}O will be discussed in Section 3.

In the study of the Bose-Einstein Condensation (BEC) the α-particle states in light N=Z nuclei [50, 51, 52], are of great importance. At present, the search for an experimental signature of BEC in ^{16}O is of highest priority. A state with the structure of the "Hoyle" state [53] in ^{12}C coupled to an α particle is predicted in ^{16}O at about 15.1 MeV (the 0_6^+ state), the energy of which is $\approx$ 700 keV above the 4α-particle breakup threshold [54]. However, any state in ^{16}O equivalent to the so-called "Hoyle" state [53] in ^{12}C is most certainly going to decay by particle emission with very small, probably un-measurable, γ-decay branches, thus, very efficient particle-detection techniques will have to be used in the near future to search for them. BEC states are expected to decay by alpha emission to the "Hoyle" state and could be found among the resonances in α-particle inelastic scattering on ^{12}C decaying to that state (an early attempt to excite these states by α inelastic scattering was presented in Ref. [55]), or could be observed in an α-particle transfer channel leading to the ^{8}Be–^{8}Be final state. Another possibility might be to perform Coulomb excitation measurements with intense ^{16}O beams at intermediate energies.

3. Clustering in light neutron-rich nuclei

As discussed previously, clustering is a general phenomenon observed also in nuclei with extra neutrons as it is presented in an extended "Ikeda"-diagram [7] proposed by von Oertzen [8] (see the left panel of Fig. 1). With additional neutrons, specific molecular structures appear with binding effects based on covalent molecular neutron orbitals. In these diagrams α-clusters and ^{16}O-clusters (as shown by the middle panel of the diagram of Fig. 1) are the main ingredients. Actually, the ^{14}C nucleus may play similar role in clusterization as the ^{16}O one since it has similar properties as a cluster: i) it has closed neutron p-shells, ii) first excited states are well above E^* = 6 MeV, and iii) it has high binding energies for α-particles.

A general picture of clustering and molecular configurations in light nuclei can be drawn from a detailed investigation of the light oxygen isotopes with A $\geq$ 17. Here we will only present recent results on the even-even oxygen isotopes: ^{18}O [56] and ^{20}O [57]. But very striking cluster states have also been found in odd-even oxygen isotopes such as: ^{17}O [58] and ^{19}O [59].

Fig. 4 gives an overview of all bands in ^{18}O as a plot of excitation energies as a function of J(J+1) together with their respective moments of inertia. In the assignment of the bands both the dependence of excitation energies on J(J+1) and the dependence of measured cross sections on 2J+1 [56] were considered. Slope parameters obtained in a linear fit to the excitation energies [56] indicate the moment of inertia of the rotational bands given in Fig. 4. The intrinsic structure of the cluster bands is reflection asymmetric, the parity projection gives an energy splitting between the partner bands. The assignment of the experimental molecular bands are supported by either generator-coordinate-method [60] or Antisymmetrized Molecular Dynamics (AMD) calculations [61].

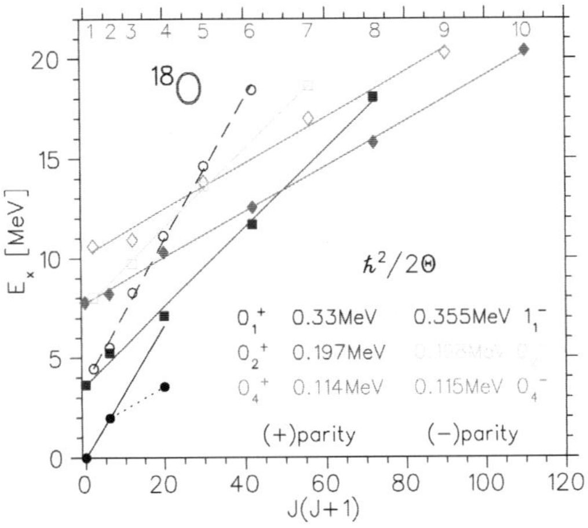

Figure 4. Overview of six rotational band structures observed in ^{18}O. Excitation energy systematics for the members of the rotational bands forming inversion doublets with K=0 are plotted as a function of J(J+1). The curves are drawn to guide the eye for the slopes. The indicated slope parameters contain information on the moments of inertia. Square symbols correspond to cluster bands, whereas diamonds symbols correspond to molecular bands. This figure is adapted from [56].

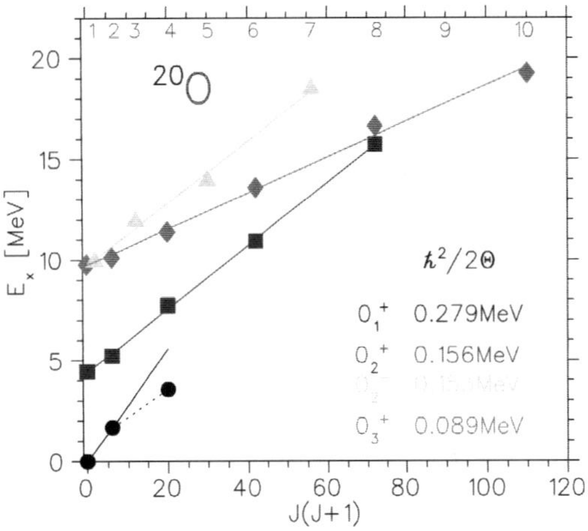

Figure 5. Overview of four rotational band structures observed in ^{20}O. Excitation energy systematics for the members of the rotational bands forming inversion doublets with K=0 are plotted as a function of J(J+1). The curves are drawn to guide the eye for the slopes. The indicated slope parameters contain information on the moments of inertia. Square and triangle symbols correspond to cluster bands, whereas diamonds symbols correspond to molecular bands.This figure is adapted from [57]

We can compare the bands of ^{20}O [57] shown in Fig. 5 with those of ^{18}O displayed in Fig. 4. The first doublet (K=$0_2^{\pm}$) has a slightly larger moment of inertia (smaller slope parameter) in ^{20}O, which is consistent with its interpretation as ^{14}C–^{6}He or ^{16}C–^{4}He molecular structures (they start well below the thresholds of 16.8 MeV and 12.32 MeV, respectively). The second band, for which the negative parity partner is yet to be determined, has a slope parameter slightly smaller than in ^{18}O. This is consistent with the study of the bands in ^{20}O by Furutachi et al [61], which clearly establishes parity inversion doublets predicted by AMD calculations for the ^{14}C–^{6}He cluster and ^{14}C-2n-α molecular structures. The corresponding moments of inertia given in Fig. 4 and Fig. 5 are strongly suggesting large deformations for the cluster structures. We may conclude that the reduction of the moments of inertia of the lowest bands of ^{18,20}O is consistent with the assumption that the strongly bound ^{14}C nucleus having equivalent properties to ^{16}O, has a similar role as ^{16}O in relevant, less neutron rich nuclei. Therefore, the Ikeda-diagram [7] and the "extended Ikeda-diagram" consisting of ^{16}O cluster cores with covalently bound neutrons [8] must be further extended to include also the ^{14}C cluster cores as illustrated in Fig. 1.

4. Summary, conclusions and outlook

The connection of α-clustering, quasimolecular resonances, orbiting phenomena and extreme deformations (SD, HD, ...) has been discussed in this work by using γ-ray spectroscopy of coincident binary fragments from either inelastic excitations and direct transfers (with small energy damping and spin transfer) or from orbiting (fully damped) processes [49] in the ^{24}Mg+^{12}C reaction. From a careful analysis of the ^{16}O+^{20}Ne α-transfer exit-channel (strongly populated by orbiting) new information has been deduced on branching ratios of the decay of the 3^+ state of ^{16}O at 11.089 MeV. This result is encouraging for a complete γ-ray spectroscopy of the ^{16}O nucleus at high excitation energy. In addition, we have presented new results on neutron-rich oxygen isotopes displaying very well defined molecular bands in agreement with AMD predictions. Consequently, the extended Ikeda diagram has been further extended for light neutron-rich nuclei by inclusion of the ^{14}C cluster, similarly to the ^{16}O one. Of particular interest is the quest for the 4α states of ^{16}O near the ^{8}Be+^{8}Be and ^{12}C+α decay thresholds, which correspond to the so-called "Hoyle" state. The search for extremely elongated configurations (HD) in rapidly rotating medium-mass nuclei, which has been pursued by γ-ray spectroscopy measurements, will have to be performed in conjunction with charged-particle techniques in the near future since such states are most certainly going to decay by particle emission (see [46, 62, 63]).

Acknowledgments

We thank the VIVITRON staff and the EUROBALL group of Strasbourg for their excellent support. C.B. would like to acknowledge Christian Caron (Springer) for initiating the new series of three volumes of *Lecture Notes in Physics* dedicated to "Clusters in Nuclei".

References

[1] Greiner W, Park Y Jae, and Scheid W 1995 *Nuclear Molecules*, (Singapore: World Scientific)
[2] Beck C *et al* 1994 *Phys. Rev.* C **49** 2618
[3] Freer M 2007 *Rep. Prog. Phys.* **70** 2149
[4] Horiuchi H 2010 Clusters in Nuclei - Vol.1, ed. Beck C, *Lecture Notes in Physics* **818** 57; Horiuchi H, this conference.
[5] Gupta R K 2010 Clusters in Nuclei - Vol.1, ed. Beck C, *Lecture Notes in Physics* **818** 232
[6] von Oertzen W, Freer M, and Kanada-En'yo Y 2006 *Phys. Rep.* **432** 43
[7] Horiuchi H and Ikeda K 1968 *Prog. Theor. Phys.* **40** 277
[8] von Oerzten W 2001 *Eur. Phys. J.* A **11** 403
[9] von Oerzten W and Milin M, 2013 Clusters in Nuclei - Vol.3, ed. Beck C, *Lecture Notes in Physics* (in press); Milin M, this conference
[10] Kanada-En'yo Y and Kimura M 2010 Clusters in Nuclei - Vol.1, ed. Beck C, *Lecture Notes in Physics* **818** 129
[11] Ikeda K *et al* 2010 Clusters in Nuclei - Vol.1, ed. C. Beck, *Lecture Notes in Physics* **818** 165
[12] Nakamura N and Kondo Y 2012 Clusters in Nuclei - Vol.2, ed. Beck C, *Lecture Notes in Physics* **848** 67
[13] Zagrebaev V and Greiner W 2010 Clusters in Nuclei - Vol.1, ed. Beck C, *Lecture Notes in Physics* **818** 267
[14] Pyatkov Y and Kamanin D 2013 Clusters in Nuclei - Vol.3, ed. Beck C, *Lecture Notes in Physics* (in press)
[15] Beck C 2004 *Nucl. Phys.* A **738** 24
[16] Beck C 2004 *Int. J. Mod. Phys.* **E13** 9
[17] Cseh J *et al* 2009 *Phys. Rev.* C **80** 034320
[18] Taniguchi Y *et al* 2009 *Phys. Rev.* C **80**, 044316
[19] Ichikawa T, Kanada-En'yo Y, and Möeller P, 2011 *Phys. Rev.* C **83** 054319
[20] Jenkins D G *et al* 2012 *Phys. Rev.* C **86** 064308
[21] Jenkins D G 2013 Clusters in Nuclei - Vol.3, ed. Beck C, *Lecture Notes in Physics* (in press); Jenskins D G this conference.
[22] Darai J, Cseh J and Jenkins D G 2012 *Phys. Rev.* C **86** 064309
[23] Kimura M and Horiuchi H 2004 *Phys. Rev.* C **69** 051304
[24] Lönnroth T *et al* 2010 *Eur. Phys. J.* A **46** 5
[25] Pandit D *et al* 2010 *Phys. Rev.* C **81** 061302

[26] Beck C *et al* 2008 *AIP Conf. Proc.* **1098** 207
[27] Svensson C E, *et al* 2000 *Phys. Rev. Lett.* **85** 2693
[28] Sciani W *et al* 2009 *Phys. Rev.* C **80** 034319
[29] Beck C *et al* 2009 *Phys. Rev.* C **80** 034604
[30] Ideguchi E *et al* 2001 *Phys. Rev. Lett.* **87** 222501
[31] Rousseau M *et al* 2002 *Phys. Rev.* C **66** 034612
[32] Taniguchi Y *et al* 2007 *Phys. Rev.* C **76** 044317
[33] Norrby M *et al* (unpublished); Norrby M, private communication and Norrby M, this conference.
[34] O'Leary C D *et al* 2000 *Phys. Rev.* C **61** 064314
[35] Fukada M *et al* 2009 *Phys. Rev.* C **80** 064613
[36] Salsac M D *et al* 2008 *Nucl. Phys.* A **801** 1
[37] Nouicer M *et al* 1999 *Phys. Rev.* C **60** 041303(R)
[38] Rudolph D *et al* 1999 *Phys. Rev. Lett.* **82** 3763
[39] Beck C *et al* 2001 *Phys. Rev.* C **63** 014607
[40] Bhattacharya C *et al* 2002 *Phys. Rev.* C **65** 014611
[41] Cindro N *et al* 1979 *Phys. Lett.* **84B** 55
[42] Mermaz M C *et al* 1984 *Nucl. Phys.* A **429** 351
[43] Pocanic D and Cindro N 1985 *Nucl. Phys.* A **433** 531
[44] Schimizu J *et al* 1982 *Phys. Lett.* **112B** 323
[45] Gai M *et al* 1984 *Phys. Rev.* C **30** 925
[46] von Oertzen W *et al* 2008 *Eur. Phys. J.* A **36** 279
[47] Zagrebaev V I *et al* 2010 *Phys. Rev.* C **81** 044608
[48] Pyatkov Y F *et al* 2010 *Eur. Phys. J.* A **45** 29
[49] Sanders S J, Szanto de Toledo A and Beck C 1999 *Phys. Rep.* **311** 487
[50] Tohsaki A, *et al* 2001 *Phys. Rev. Lett.* **87** 192501
[51] von Oertzen W 2010 Clusters in Nuclei - Vol.1, ed. Beck C, *Lecture Notes in Physics* **818** 102
[52] Yamada T *et al* 2012 Clusters in Nuclei - Vol.2, ed. Beck C, *Lecture Notes in Physics* **848** 229
[53] Hoyle F 1954 *Astrophys. J. Suppl. Ser.* **1** 121
[54] Funaki Y *et al* 2008 *Phys. Rev. Lett.* **101** 082502
[55] Itoh M *et al* 2004 *Nucl. Phys. A* **738** 268
[56] von Oertzen W *et al* 2010 *Eur. Phys. J.* A **43** 17
[57] von Oertzen W *et al* 2009 *AIP Conf. Proc.* **1165** 19
[58] Milin M *et al* 2009 *Eur. Phys. J.* A **41** 335
[59] Bohlen H G *et al* 2011 *Eur. Phys. J.* A **47** 44
[60] Descouvemont P and Baye D 1985 *Phys. Rev.* C **31** 2274
[61] Furutachi N *et al* 2008 *Prog. Theor. Phys.* **119** 403
[62] Wheldon C *et al* 2008 *Nucl. Phys.* A **811** 276
[63] Papka P and Beck C 2012 Clusters in Nuclei - Vol.2 ed. Beck C, *Lecture Notes in Physics* **848** 299

Journal of Physics: Conference Series **436** (2013) 012015 doi:10.1088/1742-6596/436/1/012015

Cluster structures observed in ^{40}Ca from ^{12}C+^{28}Si scattering

E A Benjamim[1], A Lépine-Szily[1], J M Oliveira Jr[1,3], A C C Villari[1,4], R Lichtenthäler[1], L C Chamon[1], V Guimarães[1], W Sciani[2], R Y R Kuramoto[1,5], J A Alcantara[1], A Barioni[1,6], O Camargo Jr[1], R Z Denke[1], P N de Faria[1], G. F. Lima[1,7], D R Mendes Jr[1,8], K C C Pires[1,9]

[1] Instituto de Fisica, Universidade de São Paulo, CP 66318, 05389-970, São Paulo, Brasil

[2] Universidade Presbiteriana Mackenzie, Rua da Consolacão, 930 - São Paulo, Brasil.

E-mail: alinka@if.usp.br

Abstract. We discuss evidences of strong cluster structures observed in nuclear reactions even at quite high excitation energies. The oscillatory angular distributions of the elastic scattering of the ^{12}C+^{28}Si system measured at 13 energies close to the Coulomb barrier were reproduced introducing in the scattering S-matrix additional Regge-pole form terms. These poles were interpreted as describing doorway states with ^{12}C+^{28}Si cluster structure in the composite nucleus ^{40}Ca. The excitation energies of these states indicate that they can be part of a rotational band in ^{40}Ca, together with similar states observed in the elastic scattering and α-transfer reactions of the ^{16}O+^{24}Mg system. We propose that the presently observed, largely deformed molecular band corresponds to the hyper-deformed band in ^{40}Ca, which has been found in a new Nilsson-model-calculation, which predicts preference of the ^{12}C + ^{28}Si and ^{16}O + ^{24}Mg molecular structures, in accordance with our present results.

1. Introduction

Atomic nuclei often display characteristics which can be related to clusterization. Due to the very large binding energy and compact structure of ^{4}He, the alpha-clusterization is the best known example. Reaction studies have also revealed interesting aspects about the presence of clusters [1, 2] in nuclei. Elastic, inelastic scattering experiments and α-transfer reactions realized

[3] Present address, Universidade de Sorocaba - UNISO Cidade Universitária - Rod. Raposo Tavares, KM 92.5 CEP-18023-000, Sorocaba, SP, Brasil

[4] Present address, Pantechnik S.A. 13 rue de la Résistance 14400 Bayeux France

[5] Present address, Instituto de Pesquisas Energéticas e Nucleares, Av. Prof. Lineu Prestes 2242, São Paulo, Brasil

[6] Present address, Universidade Federal da Bahia, Rua Barão de Jeremoabo, Ondina, Salvador, Brasil

[7] Present address,Universidade Paulista - UNIP, (Campus Alphaville) Av. Yojiro Takaoka, 3500 - Santana de Parnaiba/SP, Brasil

[8] Present address, Instituto de Fisica, Universidade Federal Fluminense, Campus da Praia Vermelha, Boa Viagem, Niterói, RJ, Brasil,

[9] Present address, Universidade Tecnológica Federal do Paraná, Av. Alberto Carazzai, 1640. Cornélio Procópio, Paraná, Brasil

with systems of light n-α nuclei, have shown anomalies called ALAS (Anomalous Large Angle Scattering) with strong oscillations in the angular distributions correlated with large peaks in the excitation functions [3]. These anomalies got mainly two interpretations: a non-resonant i.e. potential scattering description, where the angular momentum associated with the oscillations is simply the grazing angular momentum in the entrance channel. The other interpretation describes these states as molecular resonances with cluster structure in the composite system.

In previous papers [1, 2] we could explain fifteen strongly oscillating elastic scattering angular distributions of the $^{12}C+^{24}Mg$ system by adding energy-dependent resonance terms to the 'background' scattering matrix obtained from the 'São Paulo potential' (SPP) [4, 5]. These resonances belong to a rotational band, with a moment of inertia close to that of a hyperdeformed band, provided by α-cluster [6] and Nilsson [2] calculations. Similar quasi-molecular states observed previously in the $^{20}Ne+^{16}O$ reaction [7–9] fall on the same rotational band, giving support to the interpretation of ^{36}Ar composite system resonances.

In this work we measured complete angular distributions of the elastic scattering of the $^{12}C+^{28}Si$ system at 13 energies, between $E_{c.m.}$=12.4 to 22.4 MeV, all energies close to the Coulomb barrier [10–12]. The $^{12}C+^{28}Si$ system has been extensively studied over the past three decades for a number of reasons. It shows a strong orbiting behavior [13, 14] and demonstrates strong ALAS phenomena [15–17]. Superdeformed (SD) bands have also been discovered in the $N = Z$ nuclei ^{36}Ar and ^{40}Ca [18].

2. Experimental method

The measurements were performed at the Pelletron laboratory of the University of São Paulo, using ^{28}Si and ^{12}C beams on, respectively, ^{12}C and ^{28}Si targets. The data were obtained at center of mass energies close and above the Coulomb barrier ($V_{CB} \sim 13.0$ MeV), namely at $E_{c.m.}$= 12.4, 13.3, 14.0, 14.7, 16.1, 16.8, 17.5, 18.9, 19.6, 20.3, 20.9, 21.9 and 22.4 MeV (see fig. 1). The simultaneous measurement of the scattered projectile (^{28}Si or ^{12}C) and the recoiling target nucleus (^{12}C or ^{28}Si) allowed to cover an angular range from $\theta_{c.m.}$=27° to 164°. A strong oscillatory behavior was observed in the angular distributions of $^{12}C+^{28}Si$ system at energies above $E_{c.m.}$ = 16.1 MeV and they are presented in figure 1.

3. Data analysis and results

The starting point in our calculations was a double-folding, deep optical potential, also called São Paulo Potential (SPP) [4, 5], where the Pauli non-locality results in energy dependence of the potential depth. The imaginary part of the potential has the same form factor as the real part, and the normalizations of the real and imaginary parts are, respectively, N_r and N_i. In SPP, Fermi-Dirac type density distributions are used with diffuseness a = 0.56 fm and the normalization are N_r=1.0 and N_i = 0.78. These are values obtained through the systematic study of many systems [19] with SPP. The angular distributions calculated with this strongly absorbing, deep potential present no oscillations. For the energies below and near the Coulomb barrier, at $E_{c.m.}$= 12.4, 13.3, 14.0 and 14.7 MeV no oscillations were observed in the angular distributions and they could be well fitted by the SPP. The optical model analysis clearly does not reproduce the strong oscillations at energies above $E_{c.m.}$ = 16.1 MeV and we have added Regge poles [20] to the optical model S-matrix S_l^0, as shown below:

$$S_l = S_l^o \left(1 + \frac{iDe^{2i\Phi}}{(l-l_0) - \frac{i\Delta_l}{2}}\right), \qquad (1)$$

where S_l^0 is the optical model 'background' S-matrix, and D, Φ and Δ_l are the amplitude, phase and width, respectively, of a pole centered at l_0. The second term in eq. (1) is related to a resonance centered in l_0 in the angular momentum space. For a complete analysis we also used angular distributions found in the literature [21, 22].

The data were fitted using the Monte-Carlo method, varying D, Φ, Δ_l and l_0. The angular distribution of $E_{c.m.} = 16.1$ MeV could be well reproduced with a single pole with $l_0 = 8\hbar$ and width $\Delta_l = 1.47\ \hbar$. All the other angular distributions needed 2 poles with different values of l_0 in order to obtain a good fit. As a matter of fact, for most of cases Δ_l was very small and the only term affected in the S-matrix was S_{l_0}, the term at the resonance l_0-value.

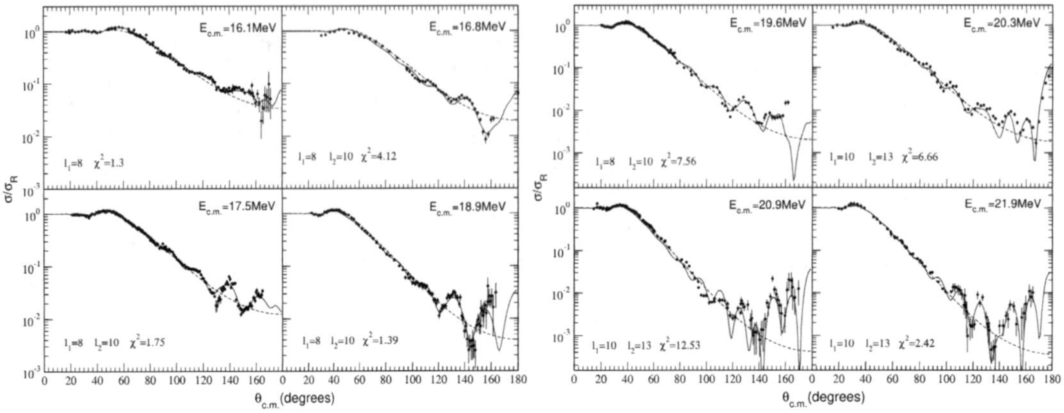

Figure 1. The elastic angular distributions for the $^{12}C+^{28}Si$ system. The dashed and solid lines are fits without and with the introduction of Regge poles in the S-matrix form, respectively.

Figure 1 shows the fits obtained for the $^{12}C+^{28}Si$ system using the São Paulo Potential without poles (dashed line) and introducing the Regge poles (solid line). In figure 1 the l_0 values of the poles are indicated for every angular distribution. Thus poles with $l_0 = 8, 10$ and $13\hbar$ are needed to reproduce the angular distributions between energies E_{cm}= 16.1 and 21.9 MeV. For energies $E \geq 22.4$ MeV [22], there are no measurements at backward angles ($\theta_{c.m.} \geq$ 100 degrees) and the fits become worse. The determination of l_0 at these energies is ambiguous.

By dividing the absolute value of the S_{l_0} by the absolute value of the background $S^0_{l_0}$ (eq. (1)) $|S_l|/|S^0_l|$ for each $l = l_0$ and for each energy we can determine the effect of the poles individually. In this way, deviations from unity of the $|S_l|/|S^0_l|$ will indicate the effect of a pole centered in l_0. The center of mass energy of the $^{12}C+^{28}Si$ system was transformed into excitation energy in ^{40}Ca by summing $Q = 13.35$ MeV and $|S_l|/|S^0_l|$ was plotted as a function of the excitation energy (see figure 2(a)). The existence of peaks is suggested on this figure, which could correspond to resonances with cluster structure of $^{12}C+^{28}Si$, with $J^{\pi} = 8^+$, 10^+, 13^- and $15^-\hbar$, respectively, at E_{cm} = 19.0, 20.6, 22.0 and 23.6 MeV and excitation energies of 32.4, 34.0, 35.3 and 37.0 MeV in the compound nucleus ^{40}Ca. The J= $15\hbar$ and $18\hbar$ states are represented in figure 2(e). Figure 2 compares $|S_l|/|S^0_l|$ with backward angle cross sections of the elastic scattering excitation functions of the $^{12}C+^{28}Si$ [15, 16] and the $^{16}O+^{24}Mg$ [23, 25] systems found in the literature. We indicate by vertical dashed lines the approximative energy position of our peaks in $|S_l|/|S^0_l|$ for each $l_0 = J$, and our resonance energies are in agreement with the structures observed in the excitations functions. For $J = 18\hbar$ we adopted the energy found by Ost [16].

Ost *et al* [16, 17] have fitted backward angle angular distributions of the $^{12}C+^{28}Si$ elastic scattering using Legendre polynomials of J = 13, 15 and 20(19)$\hbar$ at E_{cm}= 21.5, 23.2 and 33.6 MeV, and J= $18\hbar$ at 26.0 and 30.2 MeV, which correspond to $E^*(^{40}Ca)$=34.9, 36.6, 39.4, 43.6 and 47.0 MeV of excitation energy in ^{40}Ca. Similar measurements (excitation functions and angular distributions on the peaks) have been performed for the elastic scattering of the

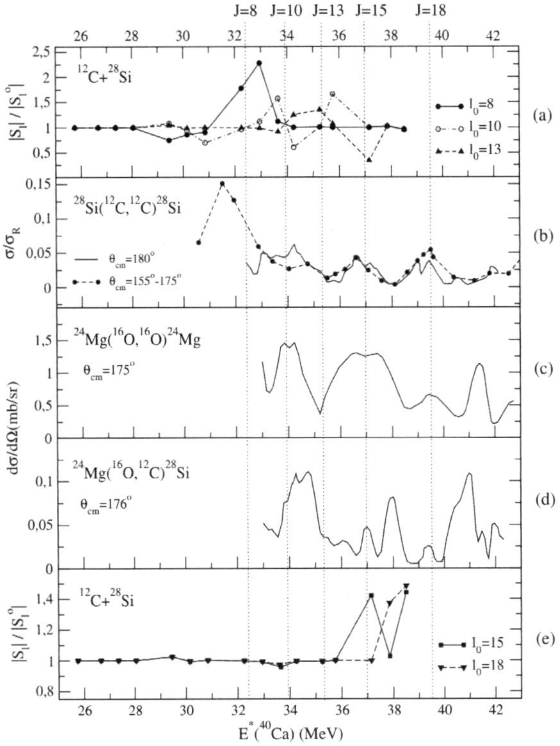

Figure 2. (a) and (e): $|S_l|/|S_l^0|$ for the ^{12}C+^{28}Si system, as a function of the excitation energy in ^{40}Ca. (b), (c) and (d): Excitation functions for ^{12}C+^{28}Si [15, 16] and ^{16}O+^{24}Mg [23–25].

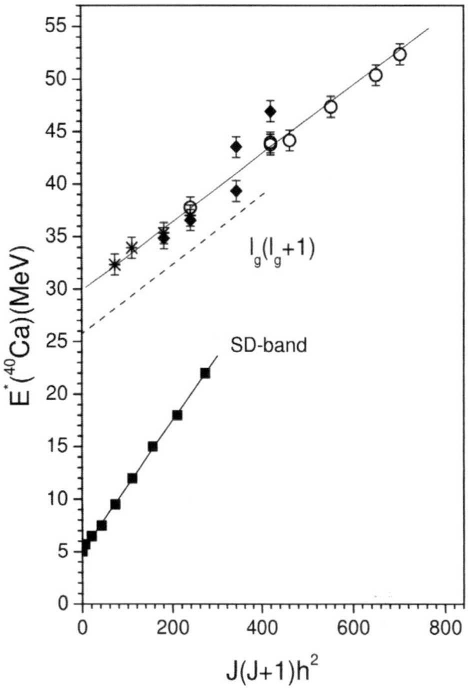

Figure 3. Resonances obtained in this work (crosses), in [16, 17] (full diamonds) and in the ^{16}O+^{24}Mg system [23–25] (circles), as a function of $J(J+1)$.

^{16}O+^{24}Mg system, and for the α-transfer reaction ^{24}Mg(^{16}O,^{12}C)^{28}Si at forward and backward angles, respectively. The ^{24}Mg(^{16}O,^{12}C)^{28}Si forward angular distributions [23, 24] indicated molecular states with J=20, 21, 23 and 25$\hbar$ at $E_{c.m.}$=27.6, 28.2, 31.2 and 34.2 MeV. The backward angular distributions of [23, 25] yielded J=15, 20, 21, 26$\hbar$ at $E_{c.m.}$ = 21.6, 27.8, 28.0 and 36.2 MeV. Transforming the $E_{c.m.}$ into excitation energy in the compound nucleus ^{40}Ca, through the addition of Q=16.176 MeV, the following results were obtained for the ^{16}O+^{24}Mg entrance channel: molecular states with J = 15, 20, 21, 23, 25 and 26$\hbar$ at excitation energies of 37.8, 44.0(43.8), 44.2, 47.0(47.4), 50.4 and 52.4 MeV.

We interpret the five poles added to the S-matrix in our analysis, as well as the other states observed in the ^{12}C+^{28}Si and ^{16}O+^{24}Mg systems as corresponding to molecular resonances with di-nuclear cluster structure in the ^{40}Ca compound nucleus. In figure 3 we plot their excitation energies against the $J(J+1)$ values and we observe a rotational band with a clear linear dependence,

$$E_J^* - E_0 = E_{rot} = \frac{J(J+1)\hbar^2}{2\mathcal{I}}, \tag{2}$$

where E_J^* is the excitation energy of a state with spin J in the ^{40}Ca nucleus, E_0 = 29.8(7) MeV is the band head energy and $\mathcal{I}$ = 15.3(9)$\hbar^2$/MeV = 6.0(4) $\times$ 10^5 MeV fm^2/c^2 is the moment of inertia. The superdeformed (SD) band, observed in ^{40}Ca [18] and the grazing angular momenta

l_g for $^{12}C+^{28}Si$ are also included in fig. 3. The moment of inertia of the nucleus in the SD band is $\mathcal{I} = 8.07\hbar^2$/MeV=3.1 $\times$ 10^5 MeV fm^2/c^2. Thus the molecular rotational band has a much larger deformation than the SD band, indicating a possible hyperdeformation in ^{40}Ca. The grazing angular momenta l_g were calculated using SPP and they fall on a parallel line below the molecular band, showing that they have similar moments of inertia. However, as $J \leq l_g$, the molecular states are more compact or internal than the grazing situation. The facts that $J \leq l_g$ and that the molecular resonances of $^{12}C+^{28}Si$ and $^{16}O+^{24}Mg$ lie on the same rotational band indicate that there is a strong cluster structure mixing in the composite system ^{40}Ca.

4. Theoretical interpretation

The hyperdeformed state and other shape isomers of the $N = Z$ nuclei can be determined theoretically from different models. Here we refer to a symmetry-adapted method, which provides not only the shape isomers, but also their possible clusterizations [26]. It consists of two parts. First the shape isomers are obtained from a Nilsson-calculation by requiring the self-consistency of the quadrupole shape [2, 27, 28]. This procedure is based on the application of the quasi-dynamical U(3) symmetry, which coincides with the real U(3) symmetry in the simple cases when this latter one is valid too [29]. Due to to this fact the U(3) selection rule for the determination of the allowed clusterizations can be extended to the quasi-dynamical symmetry [30–32]. Therefore, in the second step, the allowed clusterizations can be obtained both for the binary [33, 34] and for the ternary configurations [35, 36]. Work is in progress to carry out this kind of investigations systematically for the shape isomers of the ^{40}Ca nucleus; the preliminary results indicate a hyperdeformed state, which is similar to the suggestion of the experimental band, and has the preferred cluster configurations of $^{12}C+^{28}Si$ and $^{16}C+^{24}Mg$.

References

[1] Sciani W, Otani Y, Lépine-Szily A, Benjamim E A, Chamon L C, Lichtenthäler Filho R, Darai J and Cseh J 2009 *Phys. Rev.* C **80** 034319
[2] Cseh J, Darai J, Sciani W, Otani Y, Lépine-Szily A, Benjamim E A, Chamon L C and Lichtenthäler Filho R 2009 *Phys. Rev.* C **80** 034320
[3] Braun-Munzinger P and Barrette J 1982 *Phys. Rep.* **87** 209
[4] Chamon L C, Carlson B V, Gasques L R, Pereira D, De Conti C,Alvarez M A G, Hussein M S, Candido Ribeiro M, Rossi Jr E S and Silva C P 2002 *Phys. Rev.* C **66** 014610
[5] Alvarez M A G, Chamon M C, Hussein M S, Pereira D, Gasques L R, Rossi E S and Silva C P 2003 *Nucl. Phys.* A **723** 93
[6] Rae W D M and Merchant A 1992 *Phys. Lett.* B **279** 207
[7] Schimizu J *et al* 1982 *Phys. Lett.* B **112** 323
[8] Gai M, Berkowitz G M, Braun-Munzinger P, Jachcinski C M, Ordonez C E, Renner T R, Uhlhorn C D 1984 *Phys. Rev. C* **30** 925
[9] Miao Y, Zurmuhle R W, Barrow S P, Wimer N G, Murgatroyd J T, Lee C, Liu Z 1996 *Phys. Rev.* C **53** 815
[10] Benjamim E A, PhD thesis, University of São Paulo
[11] Oliveira Jr J M, Masters thesis, University of São Paulo
[12] Villari A C C, Masters thesis, University of São Paulo
[13] Shapira D, Novotny R, Chan Y D, Erb K A, Ford J L C, Peng J C, Moses J D 1982 *Phys. Lett.* B **114** 111
[14] Ray A, Shapira D, Halbert L M, Delcampo J G, Kim H J, Sullivan J P, Shivakumar B, Mitchell J 1991 *Phys. Rev.* C **43** 1789
[15] Barrette J, Levine M J, Braun-Munzinger P, Berkowitz G M, Gai M, Harris J W, Jachcinski C M 1978 *Phys. Rev. Lett.* **40** 445
[16] Ost R, Clover M R, Devries R M, Fulton B R, Gove H E, Rust N J 1979 *Phys. Rev.* C **19** 740
[17] Clover M R, Devries R M, Ost R, Rust N J A, Cherry R N, Gove H E 1978 *Phys. Rev. Lett.* **40** 1008
[18] Ideguchi E *et al* 2001 *Phys. Rev. Lett.* **87** 2693
[19] Chamon L C, Pereira D, Rossi E S, Silva C P, Dias H, Losano L, Ceneviva C A P 1996 *Nucl. Phys.* A **597** 253
Chamon L C, Pereira D, Hussein M S 1998 *Phys. Rev.* C **58** 576
[20] Lichtenthäler R and Gomes L C 1994 *Phys. Rev.* C **50**, 3163

Journal of Physics: Conference Series **436** (2013) 012015 doi:10.1088/1742-6596/436/1/012015

[21] Ost R *et al* unpublished, private communication
[22] Cheng C M, Maher J V, Oelert W, Snyder F D 1977 *Phys. Lett.* B **71** 304
[23] Paul M, Sanders S J, Cseh J, Geesaman D F, Henning W, Kovar D G, Olmer C, Schiffer J P 1978 *Phys. Rev. Lett.* **40** 1310
[24] Sanders S J, Paul M, Cseh J, Geesaman D F, Henning W, Kovar D G, Kozub R, Olmer C, Schiffer J P 1980 *Phys. Rev.* C **21** 1810
[25] Lee S M, Adloff J C, Chevallier P, Disdier D, Rauch V, Scheibling F 1979 *Phys. Rev. Lett.* **42** 429
[26] Darai J *et al* 2012 J. Cseh, G. Adamian, N. Antonenko, *Proc. Int. Conf. Nuclear Structure*, Dubna, in press
[27] Darai J, Cseh J, Antonenko N V, Royer G, Algora A, Hess P O, Jolos R V and Scheid W 2011 *Phys. Rev.* C **84** 024302
[28] Darai J, Cseh J and Jenkins D G 2012 *Phys. Rev.* C **86** 064309
[29] Hess P O, Algora A, Hunyadi M and Cseh J 2002 *Eur. Phys. J.* A **15** 449
[30] Cseh J and Scheid W 1992 *J. Phys.* G **18** 1419
[31] Cseh J, Darai J, Algora A, Yépez-Martinez H and Hess P O 2006 *Rev. Mex. Fis. S.* **52** 1
[32] Cseh J and Darai J 2009 *AIP Conf. Proc.* **1098** (Fusion08) 225
[33] Cseh J, Algora A, Darai J and Hess P O 2004 *Phys. Rev.* C **70** 034311
[34] Cseh J, Darai J, Algora A, Yépez-Martinez H and Hess P O 2008 *Rev. Mex. Fis. S.* **54** 30
[35] Algora A, Cseh J, Darai J and Hess P O 2006 *Phys. Lett.* B **639** 451
[36] Cseh J, Darai J, Algora A, Antonenko N V and Adamian G G 2011 *Eur. Phys. J. Web of Conf.* **17** 16001

Journal of Physics: Conference Series **436** (2013) 012016 doi:10.1088/1742-6596/436/1/012016

Electromagnetic transitions as a probe of clustering in nuclei

David Jenkins

Department of Physics, University of York, York, YO10 5DD, UK

E-mail: david.jenkins@york.ac.uk

Abstract. Clustering in nuclei is traditionally explored through reaction studies but observation of electromagnetic transitions can be of high value in establishing, for example, that highly-excited states with candidate cluster structure do indeed form rotational sequences. A topical example is given of the identification of a candidate super deformed band in ^{28}Si where super deformation in this nucleus has been described as originating from ^{24}Mg+α clustering.

1. Introduction

Alpha-clustering is often suggested in various alpha-conjugate light nuclei on the basis of the location of excited states which appear to form rotational sequences from their energy spacing. Confirmation of such assignments and a deeper understanding of the clustering phenomenon can only come, however, from observation of electromagnetic transitions connecting these states, or indeed electromagnetic transitions between these cluster states and excited states of standard shell model character. Unfortunately, there are very few cases where this has been done in practice. The best example, perhaps, where clustering is observed even in the ground state is the case of ^{8}Be. Datar *et al* carried out a "brute-force" determination of the $4^+ \rightarrow 2^+$ transition strength in ^{8}Be in an experiment at the Tata Institute for Fundamental Research in Mumbai [1]. The measurement comprised a coincidence between a detected gamma ray in an array of BGO detectors with alpha particles from the break-up of the 2^+ state in ^{8}Be. The B(E2) value obtained is 25(8) e^{2}fm^4. This value is consistent with the predictions of both *ab initio* and cluster model calculations. Recently, Freer *et al* [2] reported the possible existence of a 2^+ state at 9.6(1) MeV in ^{12}C with a width of 600(100) keV. It is argued that this state corresponds to the first member of the rotational band built on the Hoyle state. Locating this state was extremely challenging as it sits underneath an extremely broad 0^+ state at 10.3 MeV. The observation of an E2 transition connecting this 2^+ state to the 0^+ "Hoyle" state and measuring its transition strength would be sensational as it would provide extremely important information regarding the nature of the "Hoyle" state. It would, however, be extremely difficult to realise as the gamma width might be expected to be of the order of 10^{-5} of the width of the state.

In this paper, we describe a recent example where the observation of electromagnetic transitions was studied as a means of probing clustering behaviour in nuclei, specifically, in the context of super deformed bands in light nuclei.

2. Candidate superdeformed band in ^{28}Si

Superdeformed (SD) states in nuclei were first reported in rare-earth isotopes like ^{152}Dy [3], and were later found to exist in several mass regions, including those with A$\sim$150, A$\sim$130, and A$\sim$190 [4, 5]. The identification of these weakly-populated, highly-excited structures came about through a step-change in technology with the advent of highly-segmented, high-resolution gamma-ray detector arrays. These same techniques led to the discovery around ten years ago, of SD bands in the light, alpha-conjugate nuclei, ^{36}Ar [6] and ^{40}Ca [7]. These form fascinating examples of superdeformation since complementary descriptions can be found in terms of particle-hole excitations in the shell model [8, 9], and α-clustering configurations within various cluster models [10, 11, 12]. Key theoretical questions center on whether the clustering is a real feature of the system, or whether it simply corresponds to the appearances but is not a true physical description. In addition, a major question is how such clustered configurations evolve into deformed ones. It is particularly important to locate SD bands in lighter, alpha-conjugate nuclei such as ^{32}S and ^{28}Si for which long-standing theoretical predictions exist and which continue to attract the interest of theory. Recent examples of theory initiatives in this area include AMD calculations for ^{28}Si [13] and ^{32}S [14], and macroscopic-microscopic calculations for both nuclei [15]. In all cases, it is predicted that the SD bands in ^{28}Si and ^{32}S should lie at high excitation energy; *i.e.*, with bandheads around 10 MeV. This has two consequences in terms of the challenge in identifying such states experimentally: firstly, phase space favours high-energy, out-of-band transitions compared to low-energy, in-band ones despite the strong collective character of the latter. Secondly, the bandhead lies on or above the particle-decay threshold meaning that there is competition with particle emission.

Recently, Taniguchi *et al* [13] carried out an extensive study of collective structures in ^{28}Si using the AMD model. They explore clustering degrees of freedom of the type: ^{24}Mg+α and ^{12}C+^{16}O. These studies reveal a rich diversity of rotational behaviors. An SD band is identified in the AMD calculations [13] with a strong ^{24}Mg+α configuration as well as some ^{12}C+^{16}O component. Such a cluster configuration for the SD minimum is supported by recent macroscopic-microscopic potential-energy surface calculations for ^{28}Si [15] as well as by Nilsson model calculations. The AMD calculations [13] suggest that the SD band should have a moment of inertia $\mathcal{J}^{(1)} \approx 6\ \hbar^2/MeV$, related to the large associated deformation, $\beta_2 \approx 0.8$. It is difficult to identify experimental counterparts for the predicted SD states. Taniguchi *et al* [13] compare their predictions for the SD band in ^{28}Si with the properties of a so-called "excited prolate" band identified in the early 1980s by Kubono *et al* [16] using the ^{12}C(^{20}Ne,α)^{28}Si reaction. The experimental assignment of this "excited prolate" band rests on peaks in a charged-particle spectrum, and many of the associated states do not have well-established spin/parity. As shown by Taniguchi *et al* [13], the states identified by Kubono *et al* [16] do not form a smooth sequence characteristic of a rotational band even when making plausible allowance for mixing, and the suggested moments of inertia are higher than the calculated values. Moreover, γ-ray transitions between these states are not observed, and, consequently, transition strengths are unknown. Without the observation of in-band transitions, assigning candidate rotational bands is difficult and potentially ambiguous, although such an approach has been a common procedure in the past for "cluster" bands in light nuclei.

The fact that both recent AMD [13] and other [15, 17] calculations suggest that the SD band in ^{28}Si should have a strong ^{24}Mg+α component, raises the question as to whether the ^{24}Mg(α,γ) radiative capture reaction might prove to be a favoured one to selectively populate SD states in ^{28}Si. Such a possibility was not considered by Taniguchi *et al* [13], but a detailed review of the literature suggests, in fact, that plausible candidates for SD states may already exist. In a series of articles, Brenneisen *et al* [18] collate data from studies they carried out with the ^{27}Al(p,γ)^{28}Si and ^{24}Mg(α,γ)^{28}Si reactions. Of the large number of states identified in this systematic study, a number stand out as having unusual characteristics. In particular, a 6^+ state at 12.86 MeV is identified which is populated in the (α,γ), but not in the (p,γ) reaction. This 12.86-MeV level has decay branches to a number of states including a 4^+ state at 10.945 MeV, via a 1.921-MeV transition. The observation of a relatively intense, low-energy E2 transition, in competition with high-energy γ rays, immediately suggests that it must have a large transition strength. Brenneisen *et al* [18] infer that $(2I+1)\Gamma_\gamma > 0.37$ eV for the 12.86-MeV state which means that the transition to the 10.945-MeV level has an associated B(E2) value exceeding 25 Wu [18]. The unusual character of the 10.94-MeV and 12.86-MeV states becomes clear in conjunction with other work such as the ^{12}C(^{20}Ne,α)^{28}Si reaction studied by Kubono *et al* [16]. In particular, the 10.94-MeV state is the most strongly populated level below 12 MeV (see Figure 1 (a) of Ref. [16]), and it is populated with more than ten times the cross-section of the 4^+ levels in the prolate and oblate ground-state bands. A ^{24}Mg(^{6}Li,d) reaction by Tanabe *et al* [19] also shows a remarkably similar spectra of states with selective population. Again, the 10.94-MeV level is the most strongly populated one below 12 MeV, exceeding the cross-section to the other 4^+ levels by a similar factor. These observations taken together would suggest that the 10.94-MeV state has a dominant ^{24}Mg+α configuration. Indeed, it is interesting to consider this in the light of studies of the ^{32}S(^{12}C,α) reaction by Middleton *et al* [20], where the 0^+ state attributed to the 4p-4h configuration is excited ten times more strongly than the level associated with the 0p-0h configuration. The 8p-8h level is excited 1.5 times more strongly than the 4p-4h one. Indeed, the state most strongly excited in this reaction is at 7.98 MeV in ^{40}Ca which has later been shown to correspond to the 6^+ member of the SD band based on the 8p-8h configuration [7].

A state at 12.8 MeV is strongly excited in both the ^{12}C(^{20}Ne,α)^{28}Si [16] and ^{24}Mg(^{6}Li,d) reactions [19]. Analysis of angular correlations in the ^{12}C(^{20}Ne,α)^{28}Si reaction provides a firm assignment of 6^+ to a 12.8-MeV state [16]. This level is also shown to have a direct proton branch to the $5/2^+$ ground state of ^{27}Al [16], implying L = 4 decay and, hence, there must be an associated $g_{9/2}$ component. This is estimated by Kubono *et al* [21] as corresponding to a spectroscopic factor for the $g_{9/2}$ component of S = 0.3. This result is reinforced by a parallel ^{24}Mg(α,t) study by Kubono *et al* [16, 21] which also indicated a sizable $g_{9/2}$ component in the 12.82-MeV state. In this scenario, a consistent picture emerges where the candidate intruder states discussed by Brenneisen *et al* [18] appear with unusual selectivity in the ^{12}C(^{20}Ne,α)^{28}Si reaction, and with the suggestion of strong deformation, in the case of the 12.86-MeV state.

3. Gammasphere study

Confirmatory information on the presence of candidate super deformed states in ^{28}Si has recently been obtained from analysis of a data-set related to a γ-ray spectroscopy study where ^{28}Si was one of the main channels [22]. The original objective of the experiment was, in fact, the study of mirror symmetry in ^{31}S and ^{31}P, for which results were published some years ago [23]. Excited states in ^{28}Si were populated via the ^{12}C(^{20}Ne,α) reaction using a 32-MeV ^{20}Ne beam from the ATLAS accelerator at Argonne National Laboratory. A self-supporting ^{12}C target of 90 μg/cm^2 was bombarded with a 40 pnA ^{20}Ne beam for a period of two days. The resulting γ decays were detected by Gammasphere, an array of 100 Compton-suppressed germanium detectors [24]. The array was operated in stand-alone mode with a trigger condition of two or more coincident γ rays. Since evaporated alpha particles were not detected, γ rays associated with ^{28}Si were strongly Doppler-broadened, but the use of high-fold coincidence data still permitted a level scheme for ^{28}Si to be produced from the analysis of a γ-γ matrix and a γ-γ-γ cube. The analysis confirms the location and decay branching of the candidate states in the intruder band identified by Brenneisen *et al* [18] (see figure 1).

4. Conclusions and future work

In conclusion, observation of electromagnetic transitions is of high value in probing clustering in nuclei. A concrete example is given of the identification of candidate super deformed states

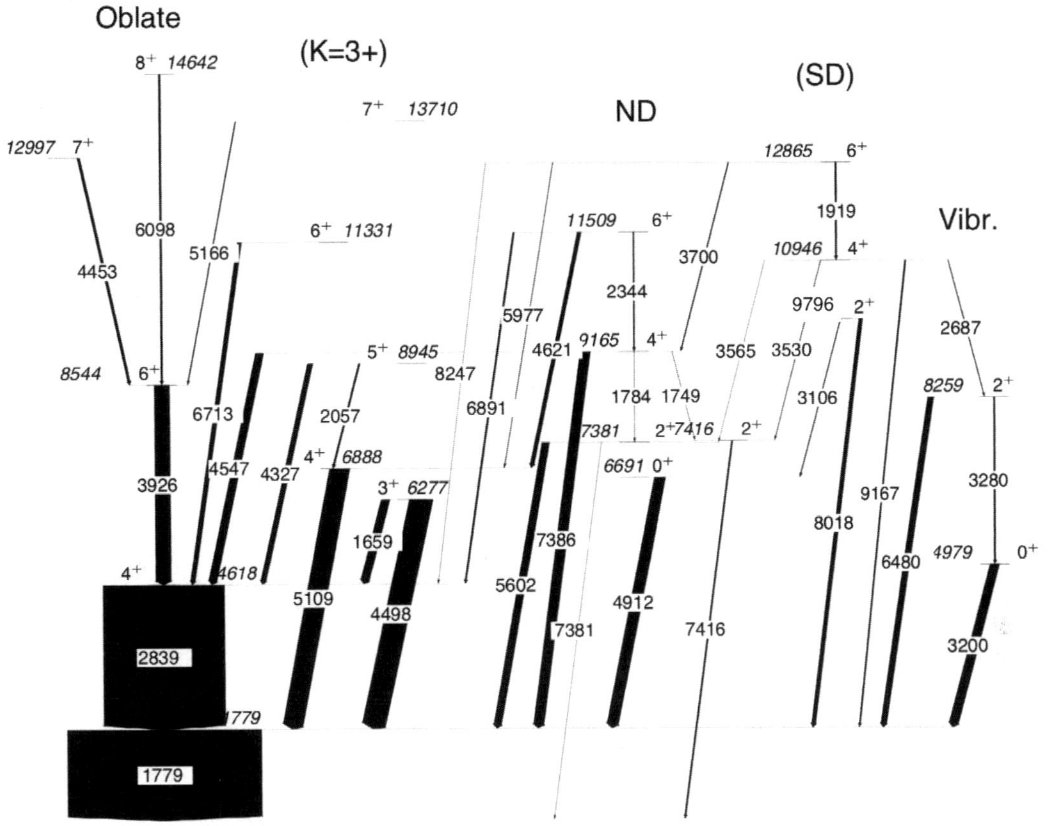

Figure 1. Subset of positive-parity levels in ^{28}Si derived from the analysis of the Gammasphere dataset . Excited states and transition energies are labeled with their energy in keV, while the width of the arrows corresponds to the relative intensity of the observed transitions. The different structures are labeled according to previous assignments as oblate, prolate (ND), vibrational and with different K values.

in ^{28}Si which find various theoretical descriptions including alpha cluster models. Important questions remain outstanding, however. It would be highly desirable to obtain, for example, a precise value for the B(E2) strength of the $6^+ \rightarrow 4^+$ transition rather than a lower limit as at present. It would also be important to locate additional states in such a band and indeed, complete a study of high spin states in ^{28}Si. An experiment is approved at iThemba laboratory and will take place in May 2013. In this experiment, the ^{12}C(^{20}Ne,α)^{28}Si reaction originally employed by Kubono *et al* [16] will be used to produce high spin states in ^{28}Si with selectivity for the candidate super deformed band. The K600 spectrometer at iThemba will be used to detect the reaction alpha particles at zero degrees as a means of selecting the ^{28}Si states of interest. The particle decay of highly-excited states will be examined with a silicon detector system around the target position, while a very large sodium iodide detector and germanium detectors will be used to detect the gamma-ray decay of these states. It is hoped that this experiment may shed fresh light on high spin behaviour and alpha-clustering in ^{28}Si.

Journal of Physics: Conference Series **436** (2013) 012016 doi:10.1088/1742-6596/436/1/012016

References

[1] Datar V M *et al* 2005 *Phys. Rev. Lett.* **94** 122502
[2] Freer M *et al* 2009 *Phys. Rev.* C **80** 041303
[3] Nyakó B M *et al* 1984 *Phys. Rev. Lett.* **52** 507
[4] Nolan P J and Twin P J 1988 *Ann. Rev. Nucl. Part. Sci.* **38** 533
[5] Janssens R V F and Khoo T L 1991 *Ann. Rev. Nucl. Part. Sci.* **41** 321
[6] Svensson C E *et al* 2000 *Phys. Rev. Lett.* **85** 2693
[7] Ideguchi E *et al* 2001 *Phys. Rev. Lett.* **87** 222501
[8] Poves A, Caurier E, Nowacki F, Zuker A 2004 *Eur. Phys. J.* A **20** 119
[9] Caurier E, Menendez J, Nowacki F, Poves A 2007 *Phys. Rev.* C **75** 054317
[10] Taniguchi Y, Kimura M, Kanada-En'yo Y, Horiuchi H 2007 *Phys. Rev.* C **76** 044317
[11] Sakuda T and Okhubo H 2004 *Nucl. Phys.* A **744**, 77
[12] Cseh J, Algora A, Darai J, Hess P O 2004 *Phys. Rev.* C **70** 034311
[13] Taniguchi Y, Kanada-En'yo Y and Kimura M 2009 *Phys. Rev.* C **80** 044316
[14] Kimura M and Horiuchi H 2004 *Phys. Rev.* C **69** 051304
[15] Ichikawa T, Kanada-En'yo Y and Möller P 2011 *Phys. Rev.* C **83** 054319
[16] Kubono S *et al* 1986 *Nucl. Phys.* A **457** 461
[17] Darai J, Cseh J and Jenkins D G 2012 *Phys. Rev.* C **86** 064309
[18] Brenneisen J *et al* 1995 *Z. Phys.* A **352** 149; ibid. p.279; ibid. p.403.
[19] Tanabe T *et al* 1983 *Nucl. Phys.* A **399** 241
[20] Middleton R, Garrett J D and Fortune H T 1972 *Phys. Lett.* B **39** 339
[21] Kubono S *et al* 1986 *Phys. Rev.* C **33** 1524
[22] Jenkins D G *et al* 2012 *Phys. Rev.* C **86** 064308
[23] Jenkins D G *et al* 2005 *Phys. Rev.* C **72**, 031303
[24] Lee I Y 1990 *Nucl. Phys.* A **520** 641c

Journal of Physics: Conference Series **436** (2013) 012017 doi:10.1088/1742-6596/436/1/012017

Absolute partial decay branching-ratios in ^{16}O

C Wheldon[1], N I Ashwood[1], M Barr[1], N Curtis[1], M Freer[1], Tz Kokalova[1], J D Malcolm[1], S J Spencer[1], V A Ziman[1], Th Faestermann[2], R Krücken[2], H-F Wirth[3], R Hertenberger[3], R Lutter[3] and A Bergmaier[4]

[1]School of Physics and Astronomy, University of Birmingham, Edgbaston, Birmingham B15 2TT, UK.
[2]Physik Department, Technische Universität München, D-85748 Garching, Germany
[3]Fakultät für Physik, Ludwig-Maximilians-Universität München, D-85748 Garching, Germany
[4]Institut für Angewandte Physik und Messtechnik LRT2, Fakultät für Luft- und Raumfahrttechnik, Werner-Heisenberg-Weg 39, D-85577 Neubiberg, Germany

E-mail: `c.wheldon@bham.ac.uk`

Abstract. The α-transfer reaction $^{12}_{6}$C($^{6}_{3}$Li, d)$^{16}_{8}$O* has been performed at a ^{6}Li bombarding energy of 42 MeV to populate excited states in ^{13}C and ^{16}O. Absolute branching ratios have been unambiguously determined for states in the excitation energy range 13.85 to 15.87 MeV and reduced widths are extracted.

1. Introduction

Clustering is a well-established phenomenon in certain light nuclei [1], the most ubiquitous form of which is based on systems with at least one α-particle sub-unit. Precise measurements can be made to elucidate the underlying structure. One of the most useful properties in this regard is the reduced width for a particular state and decay branch. This enables the characterisation of the partial decay widths with the effect of barrier penetrabilities removed, allowing the degree of clustering to be quantified. In order to extract reduced widths a precise determination of the total width and the decay branching ratios is needed. This paper reports such measurements for excited states in ^{16}O.

2. Experimental Method

Beams of 42 MeV ^{6}Li^{3+} ions were provided by the 14 MV tandem accelerator of the Maier Leibnitz Laboratory of the Technical and Ludwig-Maximillian Universities, Munich. A self-supporting target of $^{nat.}$C (100 μg/cm^2) was bombarded at the centre of the target chamber and the deuteron ejectiles analysed by the Q3D high-resolution spectrograph [3] yielding focal-plane position (recoil excitation energy), energy-loss and energy [4, 5]. The operation of the spectrograph is such that ejectiles of a particular energy are focused to the same point on the focal plane, independent of angle. The recoils or the recoil break-up products were detected in a 2×2 array of 50×50 mm^2 double-sided silicon-strip detectors covering a large angular range: 8.4° to 84.5° and −35.4° to +36.4° in the horizontal and vertical planes respectively. The master trigger condition required one good event at the Q3D focal plane after which a 5 μs time window was opened for reading all ADC events.

Energy calibration of the focal-plane detector was done using known resonances in ^{16}O while a triple-α source (^{239}Pu, ^{241}Am and ^{244}Cm) provided the energy calibration for the silicon detectors. More details for this experiment can be found in Ref. [6]. The Q3D settings used for the experiment are as follows:

Q3D at 21.5°; Excitation energy settings: 6.23 MeV (5.20 to 7.95 MeV),
14.6 MeV (13.85 to 15.87 MeV).

3. Results

The analysis was performed by identifying the recoiling products, which for the main, 14.6 MeV setting was dominated by break-up particles. The method employed for this was Catania plots. This can be understood by examining the energy balance for the final-state particles. Briefly, for a reaction between a beam, b, and target, t, $m_b + m_t \rightarrow m_{eject.} + m_{rec.}$. If the recoiling nucleus subsequently undergoes the break-up, $m_{rec.} \rightarrow m_1 + m_2$, an overall three-body Q-value can be defined by combining the kinetic energies such that

$$Q_3 = E_d + E_1 + E_2 - E_b, \tag{1}$$

where E_d is the deuteron (ejectile) kinetic energy. The data recorded in the current work were dominated by multiplicity 1 events in the silicon array, corresponding to detection of only one of the two break-up particles. The assumption is made that particle 1 is detected and its energy is corrected for losses in the target ($E_{1corr.}$). Via conservation of momentum the total momentum of the undetected particle 2 is

$$p_2(tot)^2 = \sum_{i=x,y,z} (p_{b\,i} - p_{1\,i} - p_{d\,i})^2. \tag{2}$$

Substituting into Eqn. 1 gives

$$E_b - E_d - E_{1corr.} = \frac{p_2(tot)^2}{2}\frac{1}{m_2} - Q_3. \tag{3}$$

Plotting the square of the 'missing' momentum $p_2(tot)^2/2$ on the abscissa and the 'missing' energy ($= E_b - E_d - E_{1corr.}$) on the ordinate yields the mass, m_2, of the undetected particle as the inverse gradient of the loci, and the intercept on the ordinate is minus the three-body Q-value. All that is required is an assumption of the mass of particle 1 to calculate its initial momentum ($p_1(tot) = \sqrt{2E_{1corr.}m_1}$). In the current analysis, for each pair of break-up particles, Catania plots were constructed assuming that the heavier particle was detected as particle 1. Figure 3a shows the Catania plot assuming the ^{16}O $\rightarrow p+^{15}$N break-up channel. In the adjacent panel (Fig. 3b) Monte-Carlo simulations [7, 8] have been performed including the proton, $\alpha+^{12}$C(g.s.) and $\alpha+^{12}$C(2_1^+) break-up channels. Five loci can be clearly seen, four corresponding to the detection of the α and carbon nuclei for each channel respectively, and one representing detection of ^{15}N nuclei. The final locus would correspond to detection of the proton, which has a large spatial distribution and, therefore, is only very weakly present in Fig. 3. The particle identification is confirmed with a line of gradient 1 ($=1/m_p$) and intercept 6.439 MeV ($=-Q_3$) on Fig. 3b.

The Monte-Carlo simulations have been used to extract the geometrical efficiency for each gate/break-up channel. For the proton channel, $\epsilon_{N(g.s.)}$=0.41±0.01 and for the α channels $\epsilon_{C(g.s.)}$=0.40±0.01 and $\epsilon_{C(2_1^+)}$=0.37±0.01. Finally, the efficiency-corrected ratio of [counts detected at the focal plane in coincidence with a particle event in the silicon array] to [all events at the focal plane] yields the decay branching ratios,

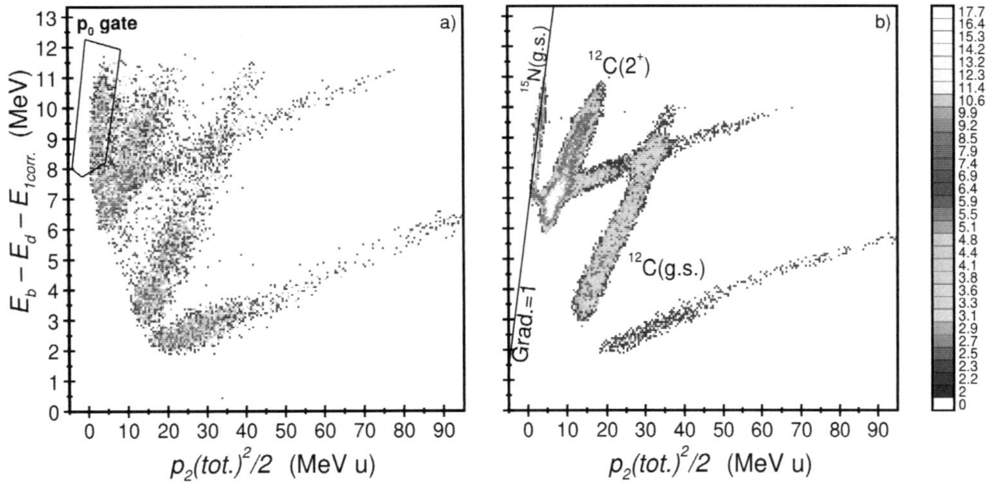

Figure 1. Catania plots constructed assuming the detected particle was ^{15}N. a) Data showing the proton, p_0 gate to the ^{15}N ground-state. b) Simulated events corresponding to $2{\times}10^4$ p+^{15}N events and $1{\times}10^5$ events for each of the α+^{12}C(g.s.) and α+^{12}C(2_1^+) channels. The verical scale has been normalised to match that of the data in panel a). For both panels a threshold of >1 has been used when plotting the data. See text for details.

$$\frac{\Gamma_{p0}}{\Gamma_{tot.}} = \frac{I(^{15}N(g.s.))}{I(tot)\epsilon_{N(g.s.)}}, \quad \frac{\Gamma_{\alpha 0}}{\Gamma_{tot.}} = \frac{I(^{12}C(g.s.))}{I(tot)\epsilon_{C(g.s.)}} \quad \text{and} \quad \frac{\Gamma_{\alpha 1}}{\Gamma_{tot.}} = \frac{I(^{12}C(2_1^+))}{I(tot)\epsilon_{C(2_1^+)}}. \tag{4}$$

The results obtained from this analysis are summarised in Table 1. Due to the conservation of angular momentum, $I(^{16}\mathrm{O}^*) - l_\alpha = I(^{12}\mathrm{C}^*)$, only natural parity states can undergo α-decay to the ^{12}C ground state. The 14.6 MeV $I^\pi = 5^-$ level is the only state in this region to decay exclusively to the ^{12}C ground state. The only significant proton decay branch is observed for the 14.911 MeV level with a branching ratio obtained in the current work of $\Gamma_{p0}/\Gamma_{tot.}$=0.22±0.12. This is in agreement with the published value of $\Gamma_{p0}/\Gamma_{tot.}$=0.36±0.06 [9].

4. Discussion

In the preceding section precise measurements have been reported for all of the absolute partial decay branching ratios on a state-by-state basis. These can be used together with the total widths to extract the reduced widths, γ_i^2, with the effect of the Coulomb barrier (*i.e.* the barrier penetrabilities, P_i) removed;

$$\gamma_i^2 = \frac{\Gamma_i}{2P_i}. \tag{5}$$

Expressing the penetrabilities in terms of the Coulomb wavefunctions, $P_i = k_i r_i/[F_l^2(k_i r_i) + G_l^2(k_i r_i)]$ for the ith decay channel, with wavenumber, k_i, and radius, $r_i = 1.4(A_1^{1/3} + A_2^{1/3})$ fm, dependent on the masses, 1 and 2, of the break-up fragments leads to

$$\gamma_i^2 = \Gamma_{tot.}\frac{\Gamma_i}{\Gamma_{tot.}}\frac{[F_l^2(k_i r_i) + G_l^2(k_i r_i)]}{2k_i r_i}. \tag{6}$$

Table 1. Branching ratios, barrier penetrabilities and reduced widths for states observed in ^{16}O for each decay channel in turn. Where an energy level is not given, no decay strength has been observed in that channel.

E_{level} (MeV)	$\Gamma_{tot.}$ [a] (keV)	I^π	l	Branching ratios	Barrier penetrabilities [b]	Reduced widths [c]
$p0 + {}^{15}N$(g.s.)			l_{p0}	$\Gamma_{p0}/\Gamma_{tot.}$	$P_{p0}/(k_{p0}r_{p0})$	θ^2_{p0}
14.911	84(37)	2^+	1	0.22(12)	0.402	0.005(3)
$\alpha0 + {}^{12}C$(g.s)			$l_{\alpha0}$	$\Gamma_{\alpha0}/\Gamma_{tot.}$	$P_{\alpha0}/(k_{\alpha0}r_{\alpha0})$	$\theta^2_{\alpha0}$
14.566	446(27)	5^-	5	1.14(8)	0.195	0.33(3)
14.808	71(9)	6^+	6	0.46(6)	0.0819	0.049(9)
15.790	122(15)	3^+	4	<0.3	0.457	$<9.350\times10^{-3}$
$\alpha1 + {}^{12}C(2^+_1)$			$l_{\alpha1}$	$\Gamma_{\alpha1}/\Gamma_{tot.}$	$P_{\alpha1}/(k_{\alpha1}r_{\alpha1})$	$\theta^2_{\alpha1}$
13.983	36(13)	2^-	1	0.87(11)	0.152	0.046(18)
14.297	27(19)	$4^{(-)}$	3	1.04(15)	0.0353	0.17(12)
14.396	22(18)	5^+	4	0.92(10)	0.00841	0.5(4)
14.808	71(9)	6^+	4	0.59(4)	0.0190	0.42(6)
14.911	84(37)	2^+	0	~0.2	0.426	~0.007
15.790	122(15)	3^+	2	0.88(18)	0.358	0.050(12)

[a] The experimental resolution of 60(4) keV has been removed from the widths measured in the current work.
[b] The quantity $P_i/(k_ir_i) = 1/[F^2_l(k_ir_i) + G^2_l(k_ir_i)]$.
[c] Calculated using excitation energies and $\Gamma_{tot.}$ measured in the current work; columns 1 and 2 respectively. See text for details.

Note that the orbital angular momentum carried by the decay particle is denoted by l. Taking the ratio of the reduced width to to the Wigner limit [10, 11] yields

$$\theta^2_i = \frac{\gamma^2_i}{\gamma^2_{Wi}} = \Gamma_i \frac{2\mu_i r^2_i \left[F^2_l(k_ir_i) + G^2_l(k_ir_i)\right]}{6\hbar^2 k_i r_i}, \tag{7}$$

[12], where μ_i is the reduced mass of the break-up particles. These reduced widths, θ^2_i, are tabulated (see Table 1) for each state and decay channel observed in the current work.

Examining the final column of Table 1 the majority of states exhibit relatively low reduced widths. The two exceptions are the 14.6 and 14.8 MeV level for the $\alpha0$ and $\alpha1$ channels respectively. The 14.6 MeV state is a strong candidate for the $I^\pi = 5^-$ member of the cluster band build on the 9.585 MeV 1^- bandhead. The current result demonstrates a significant degree of α clustering (33%) which is backed up by its exclusive decay to the ^{12}C ground state – implying a large radius compared to other states in this excitation energy range. The increased radius for the resonance makes it easier for the α particle to accommodate the orbital angular momentum. The large total width is also consistent with that of other members of the cluster band in ^{16}O [6].

For the 14.8 MeV level, the situation is less clear. With almost equal decay branches to both the ground state and first excited state of ^{12}C, it is only the latter that shows a large α-reduced width (42%), suggesting a configuration with smaller radius than the above mentioned 14.6

MeV state. The 14.8 MeV state is strongly populated in α-transfer reactions, but has yet to be assigned to a rotational band and more work is needed to unravel its associated structure.

5. Summary

A high resolution measurement has been made of the 14-16 MeV excitation region of ^{16}O using the Munich Q3D magnetic spectrograph coupled to the Birmingham large angular coverage silicon array. The break-up products for all decay channels have been unambiguously identified and the absolute branching ratios determined. States in this region are observed to decay via protons to the ^{15}N ground state and via α particles to the ^{12}C ground and first excited states. Reduced widths have been extracted, confirming the nature of the $I^{\pi} = 5^{-}$ level as an $\alpha+^{12}$C cluster-band member.

Acknowledgments

It is a pleasure to thank the accelerator operators of Maier-Leibnitz Laboratory for providing a stable ^{6}Li beam. We are grateful for the financial support of the UK Science and Technology Facilities Council (STFC) and the DFG Cluster of Excellence 153, 'Origin and Structure of the Universe'. CW acknowledges receipt of an STFC Advanced Fellowship. TzK acknowledges receipt of an STFC-funded Daphne Jackson Fellowship.

References

[1] Freer M 2007 *Rep. Prog. Ph.* **70** 2149
[2] von Oertzen W, Freer M and Kanada-En'yo Y 2006 *Phys. Rep.* **432** 43
[3] Löffler M, Scheerer H J and Vonach H, 1973 *Nucl. Instr. and Meth.* **111** 1
[4] Wirth H-F, Angerer H, von Egidy T, Eisermann Y, Graw G and Hertenberger R 2000 *Beschleunigerlaboratorium München, Annual Report* p71
[5] Wirth H-F 2001 Ph.D Thesis, Technical University, München, see *http://tumb1.biblio.tu-muenchen.de /publ/diss/ph/2001/wirth.html*
[6] Wheldon C *et al* 2011 *Phys. Rev.* C **83** 064324 and references therein
[7] Curtis N *et al* 1995 *Phys. Rev.* C **51** 1554
[8] Curtis N *et al* 1996 *Phys. Rev.* C **53** 1804
[9] Tilley D R, Weller H R and Cheves C M 1993 *Nucl. Phys.* A **565** 1
[10] Nilson R, Jentschke W K, Briggs G R, Kerman R O and Snyder J N 1958 *Phys. Rev.* **109** 850
[11] Teichmann T and Wigner E P 1952 *Phys. Rev.* **87**, 123
[12] Sanders S J, Martz L M and Parker P D 1979 *Phys. Rev.* C **20** 1743

Journal of Physics: Conference Series **436** (2013) 012018 doi:10.1088/1742-6596/436/1/012018

Spatial correlations probed in direct two-nucleon removal reactions

E C Simpson

Department of Physics, Faculty of Engineering and Physical Sciences, University of Surrey, Guildford, Surrey, GU2 7XH U.K.

E-mail: `e.simpson@surrey.ac.uk`

Abstract. Direct two-nucleon removal reactions provide an effective technique for studying both stable and highly exotic nuclides. Grazing collisions with targets of beryllium or carbon sample the projectile structure near its surface, with the geomertic selectivity of the reaction favouring two-nucleon configurations exhibiting strong spatial correlations. We discuss the correlations relevant to two-nucleon removal reactions and consider knockout from ^{12}C, which offers the opportunity to benchmark theoretical calculations and test ab-initio structure theories.

1. Correlations in two-nucleon removal reactions

Fast two-nucleon removal reactions are a key spectroscopic tool in exotic nuclei. Two well-bound nucleons are suddenly removed from an intermediate energy (secondary) beam in a collision with a target of beryllium or carbon. The technique, relying only on measurement of properties of the (heavy) projectile core or *reaction residue*, provides an efficient route to highly asymmetric isotopes [1]. The requirement that the projectile core interacts (at most) elastically with the target constrains the reaction to grazing collisions, probing a cylindrical volume near the projectile surface determined by the core- and nucleon-target interactions. The cross sections are calculated for transitions between the $(A+2)$-body projectile ground state i and A-body residue final states f, sampling the two-nucleon overlap function defined as

$$\Psi_i^f(1,2) = \sum_{I\mu T\alpha} C_\alpha^{IT}\ (I\mu J_f M_f|J_i M_i)\ (T\tau T_f\tau_f|T_i\tau_i)\ \overline{[\psi_{\beta_1}(1)\otimes\psi_{\beta_2}(2)]}_{I\mu}^{T\tau}. \qquad (1)$$

The signed two-nucleon amplitudes (TNA) C_α^{IT} describes the parentage of a particular residue state f coupled to the two-nucleon configuration $\alpha \equiv (\beta_1, \beta_2)$ in the projectile ground state i. The square bracketed term in the antisymmetrized two-nucleon wave function coupled to total angular momentum I and isospin T. Under the assumption that the residue acts as a spectator and that its state is not coupled by the eikonal reaction dynamics, the reaction samples the square modulus of $\Psi_i^f(1,2)$ projected onto the impact parameter plane. The volume sampled is determined by the core and valence nucleon S-matrices, typically calculated using density folding models [2]. Configurations α (or coherent mixtures thereof) that exhibit spatial correlations are thus favoured due to the *geometric* selectivity of the reaction. Further details of the approximations made may be found in e.g., Refs. [3, 4].

To first order the two-nucleon removal cross sections are determined by (a) the absolute magnitudes of the parentage amplitudes C_α^{IT} and (b) the total angular momentum I. I is

Table 1. Stripping cross sections for ^{28}Mg$(-2p)$ assuming a single configuration $\pi[0d_{5/2}]^2$ with $C_\alpha^{IT} = 1$. Model inputs (wave functions, S-matrices) are taken from previous work [3].

	Uncorrelated	$I=0$	$I=2$	$I=4$
σ_{str} (mb)	0.27	0.39	0.27	0.21

critical in determining the spatial proximity of the two-nucleons and this sensitivity is evident in the impact parameter plane two-nucleon position probability distribution (see e.g., Figs 1 and 2 of Ref. [5]). For $I = 0$, the probability for finding the core-nucleon vectors $\vec{r}_1$ and $\vec{r}_2$ with a small angular separation ω, where $\cos\omega = \hat{r}_1 \cdot \hat{r}_2$, is strongly enhanced. This has a direct impact on two-nucleon removal cross sections. Consider the example of ^{28}Mg$(-2p)$, assuming a simplified $\pi[0d_{5/2}]^2$ structure with $C_\alpha^{IT} = 1$. The calculated cross sections for $I = 0$, 2 and 4 are shown in Table 1, in addition to the case where the two protons are assumed to be entirely uncorrelated save for being bound to the same core (see Ref. [1]). The $I = 0$ cross section is $\sim 40\%$ larger than the uncorrelated case, whereas for $I = 4$ there is a $\sim 20\%$ suppression. These basic spatial correlations are a crucial factor in calculating final-state exclusive cross sections and branching ratios. With full sets of shell-model TNA, coherent mixing of configurations α within a single major shell will further modify the spatial correlations in the overlap function [3, 5]. Beyond this, mixing of differing parity single-nucleon states, absent from most conventional shell-model model spaces, introduces interference effects that can further enhance spatial correlations [4, 6, 7].

2. Two-nucleon removal from ^{12}C

We now discuss these aspects in the context of ^{12}C$(-2N)$ where both conventional shell-model and *ab-initio* structure descriptions are available, allowing a first assessment of major shell mixing on two-nucleon removal. The existing 2.1 GeV/nucleon carbon on carbon data [8], though final-state inclusive, show a significant enhancement for $(T = 0, 1)$ np-removal (^{10}B residue) over $(T = 1)$ nn-removal (^{10}C residue). This enhancement is of particular interest as a potential signal and measure of strong np correlations at the nuclear surface - the spatially selective nature of the reaction providing the leverage to test this. A first study [9] using conventional $0p$-shell truncated-basis shell-model TNA found reasonable agreement with experiment for nn-removal, but significantly underestimated the np-removal cross section.

Here we compare the earlier shell model results ($0p$-shell, WBP interaction) [9] to recent No-Core Shell Model (NCSM) calculations in model spaces allowing excitations up to $N_{max} = 6$ [10]. Two chiral EFT NN+3N interaction denoted NCSM1 and NCSM2 were used, softened using the similarity renormalisation group (SRG) technique. The NCSM1 NN+3N Hamiltonian used a 3N cutoff of 400 MeV with parameters fit to the ^{3}H lifetime and the ^{4}He binding energy, whereas for NCSM2 the 3N cutoff was 500 MeV and the parameters were fit to the lifetime and binding energy of ^{3}H. Consistent with the NCSM calculations, harmonic oscillators were used for the valence nucleon radial wave functions with $\hbar\omega = 16$ MeV throughout. Further details of interactions and density-folding model S-matrices may be found in Refs. [9, 10].

Four $T = 0$ states (3_1^+, 1_1^+, 1_2^+, 2_1^+) in ^{10}B and two $T = 1$ states (0_1^+,2_1^+) common to both ^{10}B and ^{10}C are predicted to be populated. The ^{10}B final-state inclusive cross sections σ_{-np} are 18.36, 25.00 and 26.01 mb for the WBP, NCSM1 and NCSM2 interactions respectively, compared to the experimental cross section 35.10±3.40 mb. The respective cross sections for ^{10}C are 5.12, 6.22 and 7.28 mb, compared to 4.11±0.22 mb. Evidently the NCSM amplitudes lead to enhanced cross sections, but the relative enhancement of the np channel above the nn channel is small and the significant underestimation of the np-channel measurement remains.

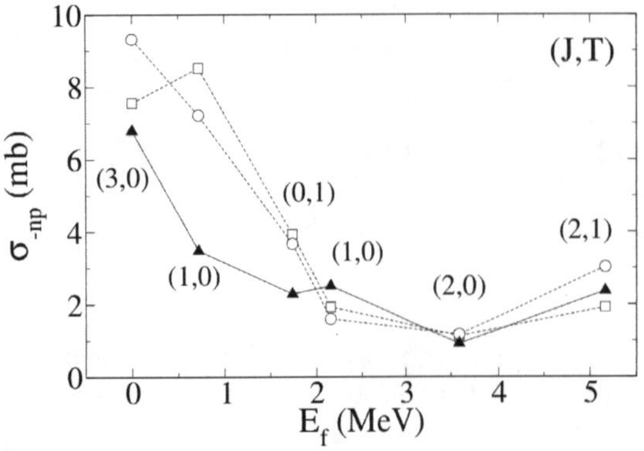

Figure 1. Two-nucleon removal cross sections for ^{12}C($-np$) calculated using the WBP (black triangles), NCSM1 (open squares) and NCSM2 (open circles) interactions.

The final-state exclusive cross sections are shown in Figure 1. The three structural approaches predict measurably different patterns of final-state-exclusive cross sections. The large enhancements found for the lowest three states in the NCSM calculations are partly due to a redistribution of $0p$-shell strength, but the (much) larger model space is also important. The mixing of different major shells enhances the probability for finding the two nucleons in close proximity, leading to larger cross sections. For example, the cross section for the first 1^+ state in the full NCSM1 calculations is 8.52 mb; truncating these TNA to the $0p$-shell gives 6.09 mb (see also Figure 7 of Ref. [10]). The associated residue momentum distributions have widths characteristic of each state, though are largely independent of the interaction.

3. Summary

Two-nucleon removal reactions are sensitive to the spatial correlations of the nucleons removed, primarily determined by their total angular momentum I. Coherent mixing of two-nucleon configurations within and between major shells can also lead to strong enhancements of spatial correlations and, consequently, cross sections. For ^{12}C($-$2N), the extended basis NCSM calculations give larger cross sections for the low-lying states, in part due to mixing between major shells. However, the existing inclusive data suggest that the $T = 0$ strength is still insufficient. New final-state-exclusive measurements would offer a means to interrogate these shell-model inputs and the direct reaction mechanism predictions in detail.

Acknowledgments

This work was supported by the UK Science and Technology Facilities Council under Grant ST/J000051/1. Ongoing collaboration with Jeff Tostevin, Petr Navrátil and Robert Roth is gratefully acknowledged.

References

[1] Bazin D *et al.* 2003 *Phys. Rev. Lett.* **91** 012501
[2] Al-Khalili J S and Tostevin J A 1996 *Phys. Rev. Lett.* **76** 3903
[3] Tostevin J A and Brown B A 2006 *Phys. Rev.* C **74** 064604
[4] Simpson E C and Tostevin J A 2010 *Phys. Rev.* C **82** 044616
[5] Simpson E C and Tostevin J A 2012 *50 Years of Nuclear BCS* (Singapore: World Scientific)
[6] Catara F, Insolia A, Maglione E and Vitturi A 1984 *Phys. Rev.* C **29** 1091;
[7] Pinkston W 1984 *Phys. Rev.* C **29** 1123
[8] Lindstrom P J, Greiner D E, Heckman H H, Cork B, and Bieser F S 1975 LBL Report No. LBL-3650 (unpublished)
[9] Simpson E C and Tostevin J A 2011 *Phys. Rev.* C **83** 014605
[10] Simpson E C, Navrátil P, Roth R and Tostevin J A 2012 *Phys. Rev.* C **86** 054609

Di-neutron correlation in monopole pair transfer modes near the neutron-separation energy in Sn, Ni, Ca, Zr isotopes

H Shimoyama[1] **and M Matsuo**[1,2]

[1] Graduate School of Science and Technology, Niigata University, Niigata 950-2181, Japan
[2] Department of Physics, Faculty of Science, Niigata University, Niigata 950-2181, Japan

E-mail: `shimoyama@nt.sc.niigata-u.ac.jp`

Abstract. We discuss the monopole pair transfer by means of the Hartree-Fock-Bogoliubov method and the continuum quasiparticle random phase approximation. We investigate the low-lying pair-vibrational state and the pair-rotational ground state having small separation energy in neutron-rich Sn isotopes and Ni, Zr and Ca isotopes. We show that the transfer amplitudes of the 0^+ states are characterized by the coherent superposition of two-quasiparticle configurations up to high orbital angular momenta $l \sim 12$. It suggests the transfer motion of a di-neutron.

1. Introduction

The neutron-rich nuclei have anomalous features, for instance, the low density distributions outside the nuclear surface such as neutron-halo and -skin, which are not seen in stable nuclei. In such a situation the di-neutron correlation is expected [1-4] because the pairing in low-density neutron matter is predicted to be stronger [5, 6]. If the neutron pairing becomes strong around the surface we may expect a large probability for a neutron pair to be added or removed from a nucleus [7-9].

In this work, we discuss the relation between the di-neutron correlation and the characteristic 0^+ states, i.e. the ground state and the pair-addition vibrational state in neutron-rich nuclei. To clarify the di-neutron behavior we analyze the microscopic structure of the pair-transfer transition densities.

2. Two-neutron transfer to the weakly bound 0^+ states

We have studied two-neutron transfer modes populating low-lying 0^+ states and the ground states in even-even Sn, Zr, Ni and Ca isotopes. We use Skyrme-Hartree-Fock-Bogoliubov mean-field model and the continuum quasiparticle random phase approximation (continuum QRPA) theory [4]. The two-neutron transfer strength connecting the ground states is enhanced in the very neutron-rich regions (see Fig. 1(a)). In the case of the neutron-rich Sn isotopes, the strength of the ground state transfer is significantly enhanced above A=140 [8, 9]. Also the pair-vibrational 0_2^+ states with anomalously large pair-additional strength appear in the Sn isotopes with A=132-140 [8].

We first analyze microscopic structure of these low-lying pair vibrational states in $^{132-140}$Sn. Within the QRPA theory, one can characterize an excitation mode in terms of the forward-

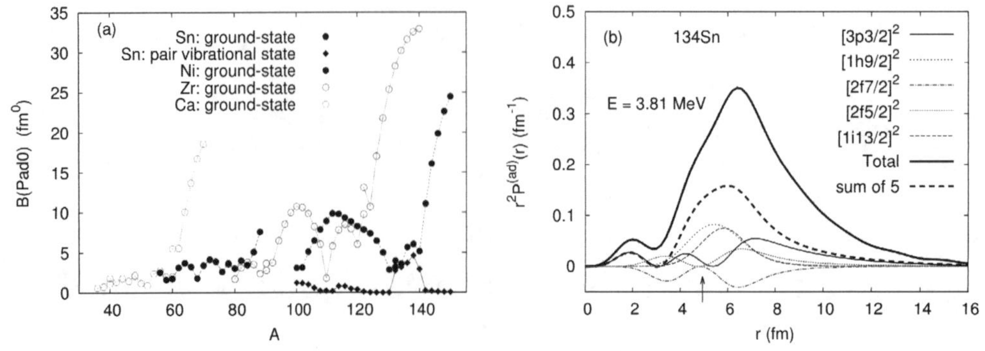

Figure 1. (a) The two-neutron transfer strengths populating the ground states for Sn(A=100-150), Zr(A=80-140), Ni(A=56-88) and Ca(A=36-70) isotopes, plotted with circles, and the pair-addition strengths of the low-lying pair vibrational state, plotted with diamonds. Adopted Skyrme parameter is SLy4 for Sn and Ni isotopes, and SkI4 for Zr and Ca isotopes. (b) The pair-addition transition density $\rho^{tr}_{\alpha L}(r)$ of the low-lying pair vibration in ^{134}Sn and its dominant five components $\rho^{tr}_{\alpha L,ii'}(r)$. The thick dashed curve is the sum of the five components.

and backward-going amplitudes X^{ν}_{ii} and $Y^{\nu}_{ii'}$ associated with the creation and the annihilation of two-quasiparticle configurations $[ii']_L = (nlj)(n'l'j')$. Similarly, the transition density $\rho^{tr}_{\alpha L}(r)$ for the pair transfer operator can be decomposed as $\rho^{tr}_{\alpha L}(r) = \sum_{ii'} \rho^{tr}_{\alpha L,ii'}(r)$ in terms of the two-quasiparticle configurations. The component of the transition density $\rho^{tr}_{\alpha L,ii'}(r)$ is expressed using X^{ν}_{ii} and $Y^{\nu}_{ii'}$ as

$$\rho^{tr}_{\alpha L,ii'}(r) = (-)^{l'} X^{\nu}_{ii'} \frac{\langle l'j'||Y_L||lj\rangle^*}{\sqrt{2L+1}} \langle 0|\hat{\rho}_\alpha(r)|ii'\rangle + (-)^{l} Y^{\nu}_{ii'} \frac{\langle l'j'||Y_L||lj\rangle^*}{\sqrt{2L+1}} \langle ii'|\hat{\rho}_\alpha(r)|0\rangle. \tag{1}$$

Fig. 1(b) shows the transition density of the anomalous 0^+ states in ^{134}Sn. Decomposed transition densities $\rho^{tr}_{\alpha L,ii'}(r)$ are also plotted for dominant five two-qasiparticle configurations $[3p_{3/2}]^2$, $[1h_{9/2}]^2$, $[2f_{7/2}]^2$, $[2f_{5/2}]^2$ and $[1i_{13/2}]^2$, which have the largest probabilities $|X^{\nu}_{ii'}|^2 - |Y^{\nu}_{ii'}|^2$= 0.565, 0.136, 0.087, 0.078 and 0.058, respectively. The probability of $[3p_{3/2}]^2$ is significantly larger than the other four, but amplitudes of the five transition densities are comparable as seen in Fig. 1(b). It suggests that the 0^+_2 state has a certain degree of collectivity.

It is seen in Fig. 1(b) that the $[3p_{3/2}]^2$ component has an extended tail reaching r ~15 fm. Its amplitude in external part is small comparing with the total transition density. The superposition of the five transition densities can reproduce only approximately half of the total one although the sum of the probabilities of the five components is about 91%. Other configurations play an important role to construct the pair vibrational state in ^{134}Sn. These characters are also seen in the pair vibrational state in A=132-140 isotopes.

To investigate the behavior of all the two-quasiparticle configurations, we plot in Fig. 2(a) the transition density decomposed with respect to the orbital angular momentum of the quasiparticle states contributing to the phonon amplitude X^{ν}_{ii} and $Y^{\nu}_{ii'}$. The largest occupied orbital angular momentum is l_{occ}=5 ($1h_{11/2}$) in the ground state of ^{134}Sn.

Although the amplitude (X^{ν}_{ii}) of each high-l component is very small, the superposition of these small transition densities reproduces the total transition density in Fig. 2(a). The accumulating coherent high-l contributions in the pair-vibrational state suggests the di-neutron correlation discussed in the same way in Ref.[3, 4]. The high-l two-quasiparticle configurations $l > l_{occ}$ are seen outside the nuclear surface in Fig. 2(a).

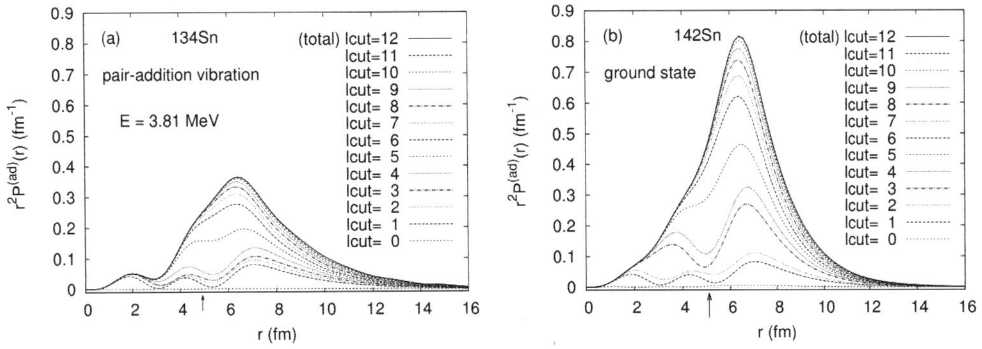

Figure 2. (a) Decomposition of the pair-addition transition density of the low-lying pair vibrational state in ^{134}Sn with respect to the orbital angular momentum cutoff l_{cut}=0, 1, 2, $\cdots$, 12. (b) The same as (a) but for the ground state transfer in ^{142}Sn.

Fig. 2(b) shows the transition density for the transfer to the ground state in ^{142}Sn. We also found that the weakly bound $3p_{3/2}$ orbit and unbound orbits in continuum contribute to the ground state of $N > 90$ Sn isotopes. Comparing Figs. 2(a) with 2(b) we find that the pair vibrational state in ^{134}Sn and the ground state transfer in ^{142}Sn have similar characters: the extended tail and the accumulating high-l contribution.

We have studied also the Ni, Zr and Ca isotopes in order to explore whether the di-neutron correlation appears systematically. The strengths of the ground state transfer are significantly enhanced in $N > 40$ for Ca, $N > 56$ for Ni, and $N \geq 82$ for Zr isotopes. These nuclei have the extended pair density $\tilde{\rho}(r)$ and the di-neutron behavior like Fig. 2(b).

We have also examined the di-neutron character in the low-lying 0^+ states in neutron-rich Ca, Ni and Zr isotopes. The low-lying pair vibrational state appearers only in Ni isotopes with A=78-84. The pair vibrational states in $^{78-84}$Ni and the ground state transfer in $^{86-88}$Ni have the similar di-neutron character.

3. Conclusion

We find that two-quasiparticle configurations involving continuum states up to around $l \sim 12$ accumulate coherently to form the total transition density of the pair-vibrational 0_2^+ state in $^{132-140}$Sn. It suggests that the pair vibrational state is characterized by the motion of a di-neutron. The large ground state transfer in very neutron-rich isotopes is also characterized by the di-neutron behavior.

[1] Pillet N, Sandulescu N and Schuck P 2007 *Phys. Rev.* C **76** 024310
[2] Hagino K, Sagawa H, Carbonell J and Schuck P 2007 *Phys. Rev. Lett.* **99** 022506
[3] Matsuo M, Mizuyama K and Serizawa Y 2005 *Phys. Rev.* C **71** 064326
[4] Serizawa Y and Matsuo M 2009 *Prog. Theor. Phys.* **121** 97
[5] Matsuo M 2006 *Phys. Rev.* C **73** 044309
[6] Margueron J, Sagawa H and Hagino K 2008 *Phys. Rev.* C **77** 054309
[7] Khan E, Grasso M and Margueron J 2009 *Phys. Rev.* C **80** 044328
[8] Shimoyama H and Matsuo M 2011 *Phys. Rev.* C **84** 044317
[9] Grasso M, Lacroix D and Vitturi A 2012 *Phys. Rev.* C **85** 034317

Journal of Physics: Conference Series **436** (2013) 012020 doi:10.1088/1742-6596/436/1/012020

^{13}C+^{4}He resonant elastic scattering on a thick gas target

L Prepolec[1], M Freer[2], N I Ashwood[2], N Curtis[2], A Di Pietro[3], P Figuera[3], M Fisichella[3], L Grassi[1], D Jelavić Malenica[1], Tz Kolalova[2], T Mijatović[1], M Milin[4], V Scuderi[3], N Skukan[1], N Soić[1], S Szilner[1], V Tokić[1], D Torresi[3,5] and C Wheldon[2]

[1] Ruđer Bošković Institute, Bijenička 54, HR-10000 Zagreb, Croatia

[2] School of Physics and Astronomy, University of Birmingham, Birmingham B15 2TT, United Kingdom

[3] INFN, Laboratori Nazionali del Sud, Via S. Sofia 44, Catania, Italy

[4] Department of Physics, Faculty of Science, University of Zagreb, Bijenička 32, HR-10000 Zagreb, Croatia

E-mail: `lovro@lnr.irb.hr`

Abstract. Resonant elastic scattering of ^{13}C on a thick ^{4}He target has been measured. Reaction products from ^{13}C(^{4}He, ^{4}He) were detected at several beam energies and gas pressures. Raw yield was efficiency corrected using Monte Carlo simulations and an averaged yield spectrum was produced. The observed peaks correspond well to the previously measured elastic scattering data.

1. Introduction

The cluster structure of light nuclei is a topic of great interest. Such structure has been well established in the excited states of many nuclei [1], among them in ^{16}O and ^{18}O, but not yet in ^{17}O. Cluster state candidates in ^{17}O have been observed by studying ^{13}C+^{4}He coincidences from the ^{13}C+^{9}Be $\rightarrow$ ^{13}C+^{4}He+^{4}He+n reaction [2] (Fig. 1). The authors proposed spin and parity assignments for known states which could be members of a rotational band analogous to the known one in ^{16}O (Fig. 2).

The aim of experiment was to investigate ^{17}O in the excitation energy range between 7.5 and 13.5 MeV. Using the resonant elastic scattering on a thick gas target method we measured the excitation function. An R-matrix analysis will be used to obtain J^{π} assignments.

2. Experimental setup

The experimental apparatus (Fig. 3) consists of a single large-area double-sided silicon strip detector of 50 $\times$ 50 mm size. The active area of the detector was divided into sixteen strips, both at front and back face, which were mutually perpendicular. The detector was placed at 0°, 42 cm far from the foil.

[5] Present address: Dipartimento di Fisica "Galileo Galilei", Università di Padova, via Marzolo 8, I-35131 Padova, Italy

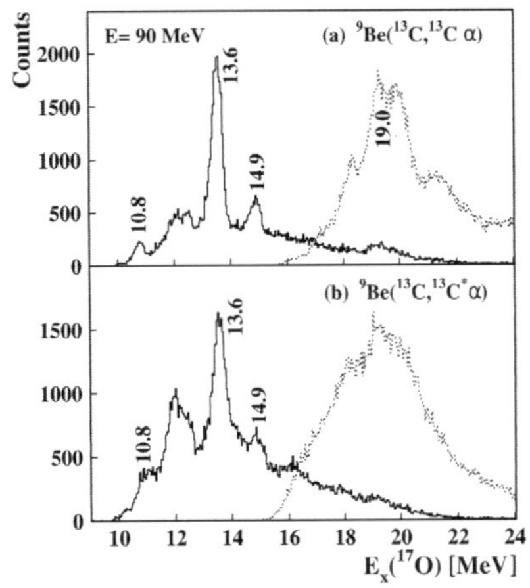

Figure 1. Figure adopted from [2]. ^{17}O excitation energy spectrum obtained from the ^{13}C+^{4}He with ^{17}O in its (a) ground state, and (b) excited state(s) at $E_x = 3.7$ MeV.

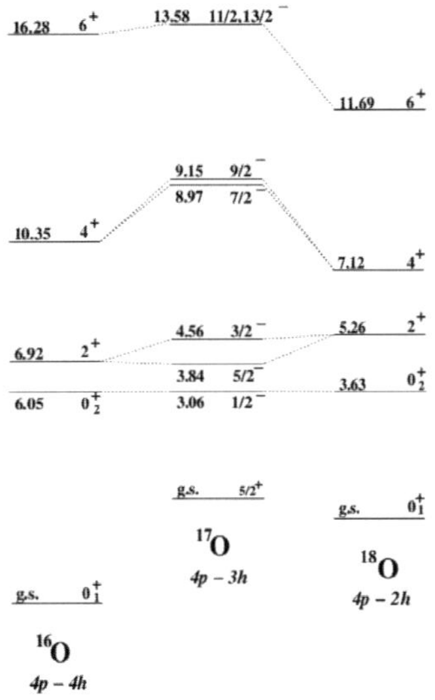

Figure 2. Figure adopted from [2]. Plot of the 4p-nh (where $n = 2, 3, 4$ energy levels for $^{16-18}$O.

The entire volume of the scattering chamber was filled with a pure helium gas, separated from the vacuum by a 2 μm thick Havar membrane.

The gas pressures and beam energies for every run were adjusted in order to stop the ^{13}C beam approximately 5 cm from the detector. Beam energies of 20.0, 25.0, 30.0, 33.0 and 35.0 MeV were used with ^{4}He gas pressures of 313, 461, 591, 699 and 790 mbar, respectively.

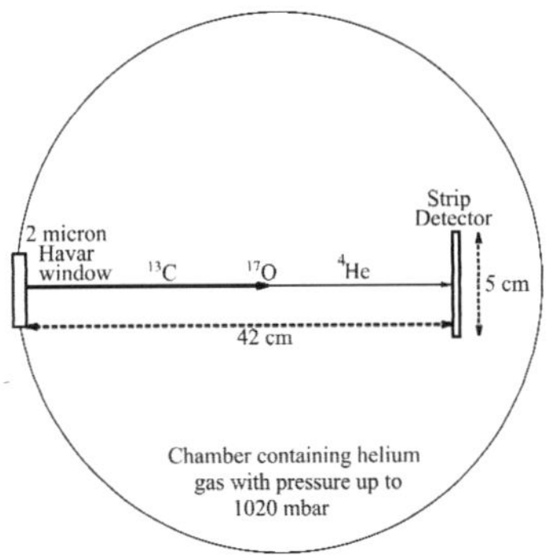

Figure 3. Sketch of the experimental setup.

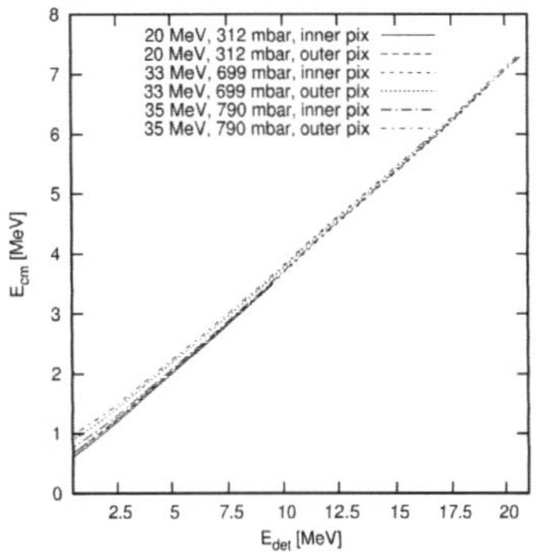

Figure 4. Plots used to calculate centre-of-mass energy from the detected energy.

In this way for every incident beam energy we covered continuous range of centre-of-mass energies in the gas. Scattering and reactions occurred along the entire beam path in the gas. Due to a smaller stopping power of ^{4}He than ^{13}C ions in the gas, we were able to detect the outgoing ^{4}He ions for all the depths of the interaction in a gas.

3. Data analysis

The data analysis was performed event-by-event, assuming that all detected events were ^{4}He ions from the elastic scattering. The analysis is similar to one described in [3].

The measured energy was related to the centre-of-mass energy using kinematics and energy loss calculations [4]. In order to take into account appropriate kinematics and energy loss for the events originating close to the detector (low centre-of-mass energy), the detector was divided into 28 rings of pixels (Fig. 4). For each ring the calculated points were interpolated with polynomials. Interpolations were used to calculate raw yield versus centre-of-mass energy.

The next step in our analysis was to correct the data for the detector efficiency. The detection efficiency depends on: *i)* the solid angle of the detector, *ii)* the energy loss of the scattered ^{4}He ions and *iii)* the centre-of-mass decay angular distribution, assumed to be isotropic.

A Monte Carlo simulation, taking into account the three aforementioned contributions, was used to calculate the efficiency correction factors.

Efficiency corrected yields from all runs are displayed in Fig. 5. All runs give consistent yield without significant contributions from inelastic processes.

All spectra in the Fig. 5 were averaged and the efficiency corrected averaged yield was obtained (Fig. 6).

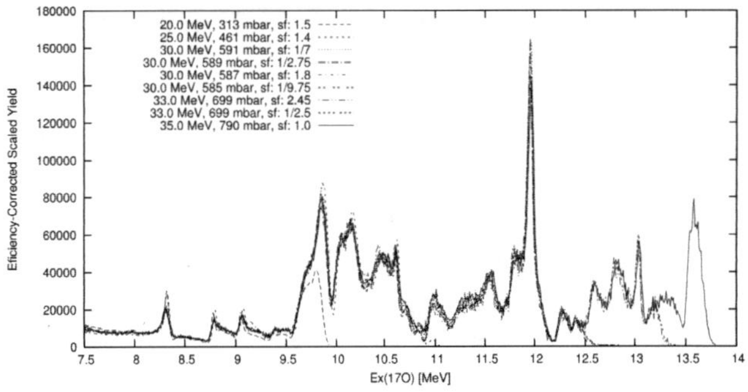

Figure 5. Monte-Carlo efficiency-corrected yields for all runs (sf is the scaling factor).

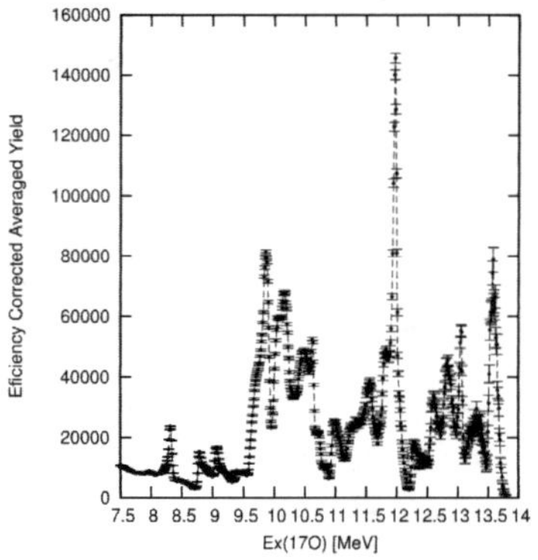

Figure 6. Averaged Monte-Carlo efficiency-corrected yield.

The averaged yield corresponds well to other elastic scattering data measured at similar angles [5].

In our experiment we had no means for the beam current measurement. In order to convert the Monte Carlo-corrected averaged yield to the cross-section, we will try to fit the low-energy part to the elastic scattering cross section.

The R-matrix analysis, which is in progress, has to contain many levels and, apart from the elastic channel, at least three other channels [5]. The starting point for our fit are the energy level parameters published in [5].

References

[1] von Oertzen W, Freer M and Kanada-En'yo Y 2006 *Phys. Rep.* **432** 43
[2] Milin M *et al* 2009 *Eur. Phys. J.* A **41** 335-339
[3] Freer M *et al* 2011 *Phys. Rev.* C **84** 034317
[4] DEDX code, unpublished
[5] Heil M *et al* 2008 *Phys. Rev.* C **78** 025803

Journal of Physics: Conference Series **436** (2013) 012021 doi:10.1088/1742-6596/436/1/012021

Resonant elastic scattering of α particles on ^{9}Be and cluster states in ^{13}C

I Lombardo[1], L Campajola[1], E Rosato[1], G Spadaccini[1] and M Vigilante[1]

[1]Dipartimento di Scienze Fisiche, Università di Napoli Federico II, and INFN-Sezione di Napoli, Via Cinthia, 80126, Napoli, Italy

E-mail: ivlombardo@na.infn.it

Abstract. We discuss the preliminary results obtained by studying the α+^{9}Be elastic scattering at bombarding energies from 3.5 to 10 MeV. Experimental spectra show the presence of elastic and inelastic scattering phenomena. Elastic scattering excitation functions have been obtained at various angles; their shape is in good agreement with data at backward angles reported in the literature.

1. Introduction

An important topic in Nuclear Physics concerns the investigation on the possible presence of α cluster structure in light nuclei [1]. This aspect plays a fundamental role in the description of the structure of light self-conjugated nuclei (such as ^{8}Be, ^{12}C, ^{16}O, ^{20}Ne), for which the study on the existence of α cluster states with large deformation has been the subject of many investigations since long time ago. Recently, accurate studies on ^{12}C spectroscopy pointed out the possible existence of a rotational band built on the (very-intriguing) Hoyle state [2]. Furthermore, it has been suggested that α cluster effects can also arise in neutron rich isotopes of beryllium, carbon, oxygen and neon [3]. For example, a clear molecular structure of ^{9}Be, that is characterized by very large deformation [4], has been evidenced by investigating rotational states characterizing the level scheme of this nucleus [5, 6].

Other interesting physics cases are constituted by ^{13}C and other, more neutron rich, carbon isotopes. As observed in [7], the spectroscopy of the states in ^{13}C around and above the α emission threshold shows some recurrences that could suggest the existence of rotational bands associated to large-deformation molecular-like configurations. Unfortunately, many ambiguities in J^{π} assignments do exist for ^{13}C high-lying states (especially in the excitation energy region between 12 and 18 MeV, containing some members of the suggested rotational bands). In order to clarify this interesting question, an improved spectroscopic evaluation of high-lying ^{13}C states is needed. To populate excited states that are characterized by large α structure, the most suited nuclear reactions seem to be the α-transfer (as the ^{9}Be(^{6}Li,d)^{13}C, [8, 9]) and the Resonant Elastic Scattering ^{9}Be(α,α)^{9}Be. In the last case, experimental excitation functions of ^{9}Be(α,α)^{9}Be scattering, obtained both in direct kinematics and with the Thick Target Inverse Kinematic (TTIK) method, have been reported [10, 11]. Data have been analyzed by means of Blatt-Biedenharn or R-matrix formalisms, thus contributing to improve the spectroscopy of ^{13}C states. In order to enlarge the experimental investigation on this topic, we performed a new

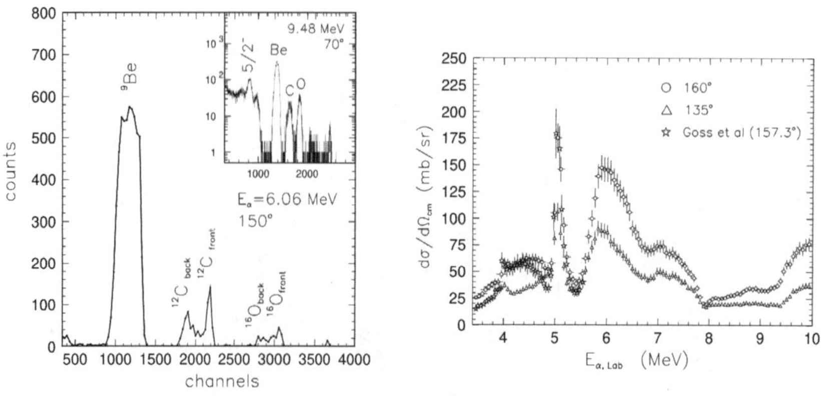

Figure 1. (Left panel) Experimental spectra observed in the α+^{9}Be scattering at 150° and 70° (inset), at 6.06 and 9.48 MeV bombarding energies, respectively. (Right panel) Preliminary elastic scattering excitation functions observed at 160° and 135°, together with data reported by Goss et al at 157.3° [10].

resonant elastic scattering experiment α+^{9}Be in direct kinematics, in the bombarding energy range from 3.5 to 10 MeV. Excitation functions have been obtained at various polar angles. The preliminary results of this experiment will be discussed in the following sections.

2. Experimental Setup

The experiment has been performed at the TTT-3 Tandem facility of the University of Naples Federico II [12]. After magnetic analysis, ^{4}He^{++} beams of energies ranging from 3.5 up to 10 MeV bombarded a self-supporting ^{9}Be target (0.66 μm thick). The beam energy was varied at 60 keV steps in order to obtain suitable excitation functions. The beam energy spread was of the order of 1 per thousand. Beam current was of the order of 10 enA, and was measured by a Faraday cup followed by a current integrator. Ejectiles have been detected at various angles (principally in the backward hemisphere) by means of 10 collimated silicon detectors, their resolution being 0.7-1.0% for the 5.48 MeV peak of a standard α source. Data have been stored in memory for subsequent analysis, and dead-time corrections have been applied to calculate absolute differential cross sections.

3. Preliminary results

An example of experimental spectra obtained at 70° and 150° is reported in Figure 1. The strongest peak in both spectra is due to the elastic scattering of α particles by the ^{9}Be layer. The two peaks at higher energies are due to the elastic scattering of α particles by carbon and oxygen contaminants, that are present both in the back and front face of the target. Due to the combination of kinematics and energy loss effects, the scattering by the front and back contaminants becomes explicitly visible at backward scattering angles, as we can see in the 150° spectrum, presented in Figure 1. The nature of the carbon and oxygen contaminants is confirmed by studying the trends of excitation functions obtained from peaks corresponding to the scattering on the front contaminants; they are very similar to those reported in the literature for the α+^{12}C and α+^{16}O elastic scattering. The low energy products observed at 70° are due to inelastic scattering of α particles on ^{9}Be and ^{12}C. In particular, the peak associated to the ^{9}Be(α,α_2) scattering is clearly recognized.

Starting from the experimental spectra, it is possible to obtain the excitation functions for the α+^{9}Be elastic scattering process at various scattering angles. As an example, in Figure 1 (right panel) we report the 160° and 135° excitation functions, together with the results of Goss et al ([10], 157.3°, direct kinematics). As we can observe, the shapes of the excitation functions at around 160° are in good agreement. In particular, a strong peak is observed at 5.05 MeV and a large bump at around 5.88 MeV. The first state corresponds to a member of the $K = \frac{3}{2}^-$ proposed rotational band [7] at ^{13}C excitation energy of 14.13 MeV. Assignment of spin-parity for this level is quite ambiguous [10, 11, 13]. In the systematics [13], it is reported as a $\frac{3}{2}^+$ state, while in Ref. [10], the existence of a doublet of states at 14.11 and 14.16 MeV in ^{13}C is claimed, with $J^\pi = \frac{5}{2}^-$ and $\frac{7}{2}^+$ assignments respectively. In the more recent paper by [11], the trend of the 180° elastic scattering excitation function in correspondence of the 5 MeV bump is nicely reproduced by R-matrix calculations performed by assuming a single state at 14.13 MeV excitation energy, having $J^\pi = \frac{5}{2}^-$. The large bump at around 5.88 MeV would correspond to a state with excitation energy around 14.72 MeV. This state has not been reported in the systematics [13] but has been observed in ^{9}Be(α,α)^{9}Be [11] and ^{9}Be(^{6}Li,d)^{13}C reactions [8]. We plan to perform R-matrix fit of our experimental excitation functions, obtained at various angles (from 70 to 160° in the lab. frame), in order to extract refined spectroscopic evaluations of high-lying excited states in ^{13}C that could be of interest to unveil the (possible) cluster structure in such non self-conjugated nucleus.

4. Conclusions

In this paper we report preliminary results obtained by studying the scattering of α particles on ^{9}Be. The experiment has been performed by using a ^{4}He beam with energies from 3.5 to 10 MeV, and by varying the energy at 60 keV energy step. Experimental spectra, taken at various polar angles, show clearly the presence of elastic and inelastic scattering events. Excitation functions of the elastic scattering process α+^{9}Be have been obtained at various angles. Their shape is in good agreement with data at backward angles reported in the literature and obtained both by using the direct kinematics method and the TTIK method. Many anomalies are clearly visible in the excitation functions. An analysis performed in term of the R-matrix model would allow to improve the spectroscopy of the ^{13}C compound nucleus in an excitation energy region where the possible existence of rotational bands, due to the molecular structure of this nucleus, has been suggested.

Acknowledgments

We acknowledge A. Brondi, E. Perillo and D. Pierroutsakou for their support. We are indebted to C. Marchetta and E. Costa for providing us ^{9}Be targets.

References

[1] von Oertzen W, Freer M and Kanada En'yo Y 2006 *Phys. Rep.* **432** 43
[2] Itoh M *et al* 2011 *Phys. Rev.* C **84** 054308
[3] von Oertzen W 1996 *Zeit. Phys.* A **354** 37
[4] Vinciguerra D and Stovall D 1969 *Nucl. Phys.* A **132** 410
[5] Harakeh M N *et al* 1980 *Nucl. Phys.* A **344** 15
[6] Soic N *et al* 1998 *Eur. Lett.* **41** 489
[7] Milin M and von Oertzen W 2002 *Eur. Phys. Jour.* A **14** 295
[8] Rodrigues M R D *et al* 2011 *AIP Conf. Proc* **1351** 125
[9] Wheldon C *et al* 2012 *Phys. Rev.* C **86** 044328
[10] Goss J D *et al* 1973 *Phys. Rev.* C **7** 1837
[11] Freer M *et al* 2011 *Phys. Rev.* C **84** 034317
[12] Campajola L *et al* 1987 *Nucl. Instrum. Methods* B **29** 129
[13] Ajzenberg-Selove F 1991 *Nucl. Phys.* A **523** 1

Complete fusion of weakly bound cluster-type nuclei near barrier energies

M S Hussein[1], P R S Gomes[2], J Lubian[2], R Linares[2] and L F Canto[3]

[1] Instituto de Estudos Avançados and Instituto de Física, Universidade de São Paulo, CP 66318, cep 05314-970, São Paulo, S.P., Brazil
[2] Instituto de Fisica, Universidade Federal Fluminense, Av. Litoranea s/n, Niteri, R.J. 24210-340, Brazil
[3] Instituto de Fisica, Universidade Federal do Rio de Janeiro, CP 68528, Rio de Janeiro, Brazil

E-mail: hussein@if.usp.br

Abstract. We consider the influence of breakup channels on the complete fusion of weakly bound cluster-type systems in terms of dynamic polarization potentials. It is argued that the enhancement of the cross section at sub-barrier energies may be consistent with recent experimental observations that nucleon transfer, often leading to breakup, is dominant compared to direct breakup. The main trends of the experimental complete fusion cross sections are analysed in the framework of the Dynamic Polarization Potential approach. The qualitative conclusions are supported by CDCC calculations including a sequential breakup channel, the one-neutron stripping of ^{7}Li followed by the breakup of ^{6}Li.

The fusion and breakup of weakly bound, cluster-type, nuclei, both stable and radioactive, have been a subject of great interest in the last years [1, 2, 3]. Several systems have been studied, both theoretically and experimentally, including stable weakly bound projectiles (^{6}Li, ^{7}Li and ^{9}Be) and radioactive halo-type projectiles, like ^{6,8}He, ^{7,11}Be, ^{8}B and ^{17}F, on targets with masses ranging from ^{7}Li to ^{238}U. The basic question is whether the coupling to the breakup process enhances or hinders the fusion cross section. First, it should be stated what is considered as fusion cross section. Is it the complete fusion of the projectile with the target (CF) or the total fusion (TF), defined as the sum of the complete fusion and the incomplete fusion (ICF), the latter being the fusion with a fragment of the projectile after breakup on the target? Most of the fusion data reported in the literature are for TF, since it is very difficult to separate experimentally CF from ICF. One may compare a particular result to a well-defined reference calculation to find an enhancement or suppression. Also, different breakup effects may occur, like static and dynamic effects. The first one is related to different barrier characteristics, when compared with those for similar tightly bound systems, and the latter is related to the coupling to breakup channels, which feed continuum states, and to other direct reactions. Therefore, if one compares data with theoretical predictions, the choice of the bare interaction potential plays a major role, and contradictory conclusions can be drawn with the same data set, depending on the potential used [4].

The accepted picture of the fusion and breakup of a cluster-type nucleus on a certain target, is based on the following decomposition of the different processes involved: (1) the complete capture of the whole projectile by the target (complete prompt fusion), (2) the breakup of the projectile followed by the sequential capture of both fragments (complete sequential fusion), (3) the breakup of the projectile followed by the capture of one of the fragments, while the other fragment flies on (incomplete fusion), and (4) the breakup of the projectile with no capture of any of the fragments (non-capture breakup). The above processes compete and affect each other. In particular, the study of the complete fusion, defined as the sum of (1) and (2), and the total fusion, defined as the sum of (1), (2) and (3), are in general very much influenced by the coupling to the non-capture breakup process (also called 'elastic' breakup). In this contribution we discuss the above processes and supply evidence of the occurrence of yet another process, which we may call (5) transfer followed by breakup. This has been confirmed recently by the Canberra group [7], and should be considered as another process to reckon with in discussing fusion of weakly bound stable or radioactive nuclei.

In the analysis of experimental data, the usual procedure has been to define the background or benchmark cross section, which describes the tunnelling of the given system with due reference given to its general overall geometrical features, as well as its charge and mass. This is accomplished through what has been known as the Universal Fusion Function (UFF), introduced and discussed extensivly in [5, 6]. This function hinges on an appropriate rescaling of the Wong formula for fusion [8], given by

$$\sigma_F = \frac{R_B^2 \hbar\omega}{2E} \ln\left[1 + \exp\left(\frac{2\pi}{\hbar\omega}(E - V_B)\right)\right].$$

The Wong cross section is useful in describing the fusion of structureless nuclei, in the vicinity of the Coulomb barrier, and fails at energies well below the barrier owing to the parabolic (symmetrical) barrier used in its derivation. The actual Coulomb barrier is quite asymmetrical. The limits of the Wong formulae are well known. At above-barrier energies it reduces to the geometrical formula,

$$\sigma_F = \pi R_B^2 \left[1 - \frac{E}{V_B}\right],$$

and, as expected, no reference to the curvature of the barrier is maintained. At below-barrier energies, the Wong formula reduces to the usual exponential tunnelling form,

$$\sigma_F = \frac{R_B^2 \hbar\omega}{2E} \exp\left(\frac{2\pi}{\hbar\omega}(E - V_B)\right).$$

Again we remind the reader that the above formula is not appropriate since the barrier is not an inverted parabola. An appropriate use of the correct classical action should be employed at these sub-barrier energies [1].

The rescaling of the Wong formula is done by defining the variable $x = (E - V_B)/(\hbar\omega)$ and defining the UFF through

$$F_0(x) = \frac{2E}{R_B^2 \hbar\omega}\sigma_F = \ln[1 + \exp(2\pi x)].$$

Note that $F_0(x) \to 2\pi x$ even for small positive x. The function $F_0(x)$ is universal in the sense that no specific reference to the system is made. The experimental $F(x)$ is defined [6] as $F(x) = F_{exp}(x)\frac{F_0(x)}{F_{CC}(x)}$, with $F_{CC}(x)$ being the CC fusion function which includes couplings to all important bound channels. The deviations of $F(x)$ from the UFF ($F_0(x)$) is traced to the breakup and transfer couplings not included in the coupled-channels (CDCC) calculation, $F_{CC}(x)$. A large body of data have been analysed using the above picture [6]. In the following we consider the particular data for collisions of ^{6,7}Li projectiles incident upon a ^{209}Bi target, which have been measured with high precision [10].

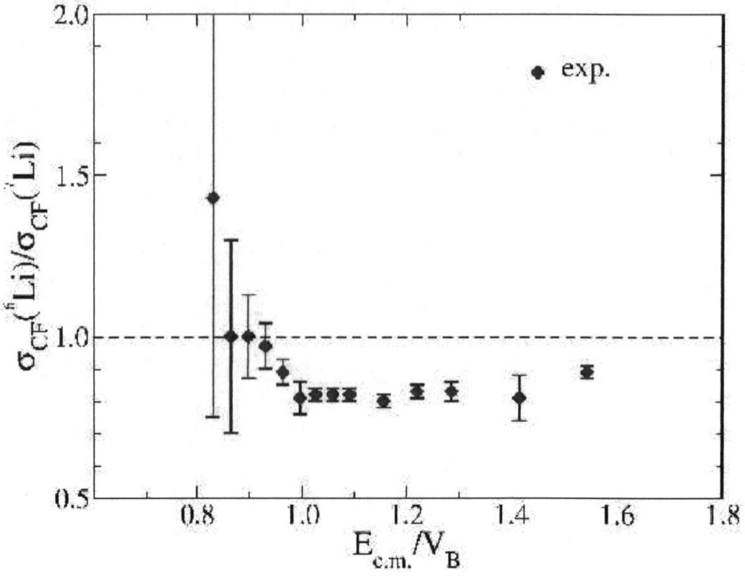

Figure 1. Ratio between the complete fusion cross section of the ^{6}Li + ^{209}Bi system and of the ^{7}Li + ^{209}Bi system, as a function of the centre-of-mass energy, divided by the fusion barrier, obtained from the measured fusion barrier distributions [10]. Data from Ref. [10].

Figure 1 shows the ratio between the complete fusion cross sections of the ^{6}Li + ^{209}Bi and of the ^{7}Li + ^{209}Bi reaction, as a function of the centre-of-mass energy divided by the fusion barrier energies, obtained from the measured fusion barrier distributions [10]. Above the barrier, the

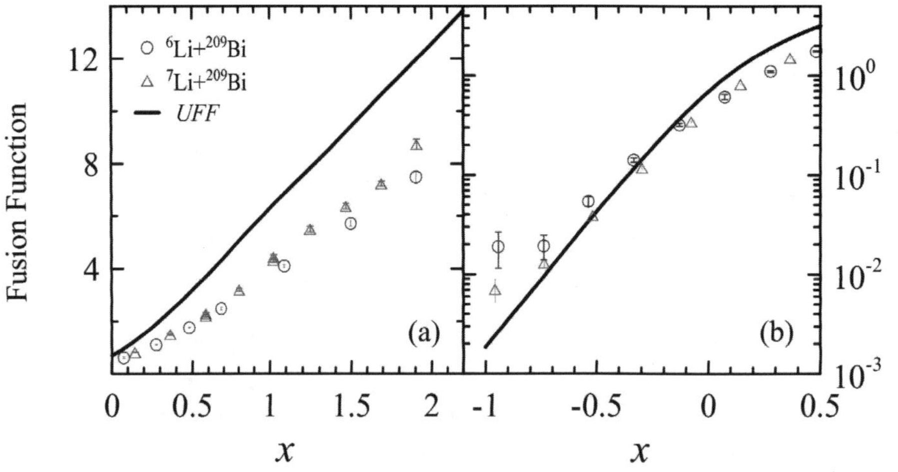

Figure 2. Renormalized fusion functions (see text) for complete fusion plotted against $x = (E - V_B)/\hbar\omega$ for the two systems. The reduced data F are constructed from [10] and the full curves are the universal fusion functions (UFF), F_0, given in the text.

stronger the couplings that lead to prompt breakup, the larger is the suppression. Below the barrier, the couplings lower the barrier height. Because of the exponential dependence of the tunnelling probabilities on the barrier energy, this outweighs the linear reduction in cross-section due to prompt breakup. The behaviour seen by plotting the ratio of the cross sections for the two reactions (Fig. 1) is consistent with this picture.

Figure 2 shows the renormalized experimental complete fusion function, $F(x)$, for these two systems. These $F(x)$'s are obtained by using the São Paulo potential [11] for $F_{CC}(x)$. One can observe that the renormalized experimental complete fusion functions are below the UFF (full curve), F_0, at energies above the barrier. The main features of the data are summarized as follows: (i) the CF cross sections are suppressed by about 30% at energies above the barrier; (ii) the CF cross sections at sub-barrier energies are enhanced; (iii) the above two effects are more pronounced for ^{6}Li than for ^{7}Li. The behaviour below the barrier can be traced to the afore-mentioned process (5), transfer followed by breakup. This can be seen in Fig. 3, where the real part of the dynamic polarization potential (DPP) is calculated for this process, and it shows a significant increase in attraction, resulting in a lower barrier.

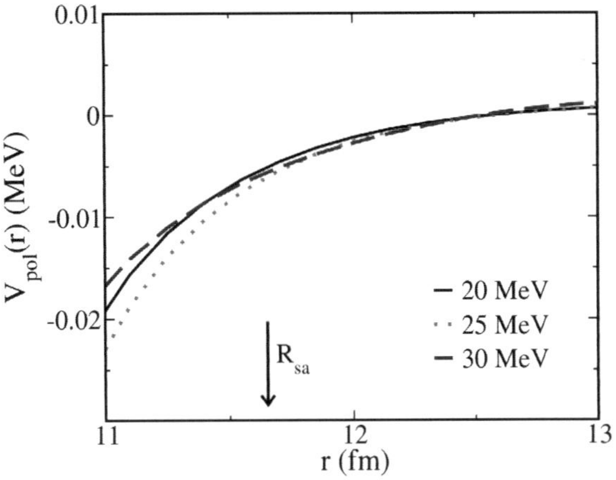

Figure 3. Real part of the DPP around the strong absorption radius for ^{7}Li + ^{144}Sm at laboratory energies above (30 MeV), below (20 MeV) and close to the barrier (25 MeV) for breakup of ^{6}Li into alpha + d following the one-neutron stripping of ^{7}Li.

Thus, the inclusion of the neutron pickup (transfer) followed by breakup can explain the below-barrier enhancement seen in the fusion of ^{6}Li on a ^{209}Bi target. The content of this contribution is a summary of a recent publication [12].

References

[1] Canto L F, Gomes P R S, Donangelo R, M. S. Hussein 2006 *Phys. Rep.* **424** 1
[2] Liang J F, Signorini C 2005 *Int. J. Mod. Phys.* E **14** 1121
[3] Keeley N, Raabe R, Alamanos N, Sida J L 2007 *Prog. Part. Nucl. Sci.* **59** 579
[4] Gomes P R S *et al* 2011 *Phys. Lett.* B **695** 320
[5] Gomes P R S, Lubian J, Padron I, Anjos R M 2005 *Phys. Rev.* C **71** 017601
[6] Canto L F *et al* 2009 *J. Phys.* G **36** 015109; 2009 *Nucl. Phys* A **821** 51
[7] Luong D H *et al* 2011 *Phys. Lett.* B **695** 105 *Eur. Phys. J. Web of Conf.* **17** 03002
[8] Wong C Y 1973 *Phys. Rev. Lett.* **31** 766
[9] Gomes P R S, Lubian J, Canto L F 2009 *Phys. Rev.* C **79** 027606
[10] Dasgupta M *et al* 2002 *Phys. Rev.* C **66** 041602(R) (2002); 2004 *Phys. Rev.* C **70** 024606
[11] Chamon L C *et al* 2002 *Phys. Rev.* C **66** 014610; 1997 *Phys. Rev. Lett.* **79** 5218
[12] P. R. S. Gomes *et al* 2013 *J. Phys. G: Nucl. Part. Phys.* **39** 115103

Symmetry-adapted *ab initio* no-core shell model calculations for ^{12}C

K D Launey[1], A C Dreyfuss[2], T Dytrych[1], J P Draayer[1], D Langr[3], P Maris[4], J P Vary[4], and C Bahri[5]

[1]Department of Physics and Astronomy, Louisiana State University,
Baton Rouge, Louisiana 70803, USA
[2] Keene State College, Keene, New Hampshire 03435, USA
[3] Department of Computer Systems, Czech Technical University in Prague,
Prague, Czech Republic
[4] Department of Physics and Astronomy, Iowa State University, Ames, Iowa 50011, USA
[5] Department of Physics, University of Notre Dame, Notre Dame, Indiana 46556, USA

E-mail: draayer@lsu.edu

Abstract. Symmetry-adapted no-core shell-model calculations reveal dominant symmetry patterns in the structure of light nuclei, independent of whether the system Hamiltonian is phenomenological in nature or derived from realistic interactions. We show results of large-scale nuclear structure computations based on the *ab initio* symmetry-adapted no-core shell model that use only a fraction of the model space. In addition, the symmetry patterns unveiled in these results are employed to explore ultra-large model spaces for ^{12}C. The outcome suggests a possible path forward for realizing collective theories that target correlated highly-deformed and alpha-cluster structures in terms of microscopic degrees of freedom that build forward from the nucleon-nucleon interaction itself.

1. Introduction

Complex nuclear systems often display striking simplicities. Experimental evidence supports the fact that low-lying nuclear states in light and medium-mass nuclei favor a dominance of low spin and high deformation, which has been also recently demonstrated through our first-principle studies [1, 2]. This, in turn, points to a remarkable new insight into the symmetry patterns observed in atomic nuclei, namely, understanding the mechanism on how such simple structures emerge from the fundamental level of the underlying quark/gluon physics.

The symmetry-guided approach we have developed utilizes symmetry to reduce the dimensionality of the model space through a very structured winnowing of the basis states to physically relevant subspaces. We achieve significantly enhanced reach beyond that of the current no-core shell model (NCSM) by taming the scale explosion problem through the development of a practical, *ab initio* symmetry-adapted no-core shell model (SA-NCSM) framework. This framework exploits our knowledge of dominant symmetries, first of the interaction itself and then of those that emerge as a result of the many-body dynamics, and utilizes the power of the current and unfolding next-generation of high performance computing (HPC) facilities. This combination opens up new regions of the periodic chart, the '*sd*-shell', to investigation with *ab initio* methods. Our approach enables the inclusion of higher-order correlations as required, for example, for understanding excited 0^+ states – e.g., the elusive Hoyle state – in the lighter nuclei, and sets the stage for advancing the applicability of the current *ab initio* methods to systems beyond (e.g., '*fp*-shell' nuclei, and even beyond to 'upper *fp*-shell' and 'lower *gds*-shell' as well as rare-earth and actinide species where deformation plays an even more dominant role) that are necessary, for example, for a better understanding of astrophysical processes among higher-mass nuclei.

The theory is applied to a study of the structure of ^{12}C, and in particular, to shape deformations and cluster substructures that are well pronounced in the low-lying states of this nucleus. Comparisons with results from conventional no-core shell model analyses illustrate the efficacy of the symmetry-adapted framework. In addition, a simple algebraic interaction, which reduces to the Elliott model in its single-shell limit, is shown to reproduce all features of the elusive Hoyle state in ^{12}C. While this requires that the no-core underpinning of the model be extended beyond the current reach of the standard no-core shell model to excitations as high as eighteen major oscillator shells, only the most deformed configurations in the extended region are required. This further affirms the importance of exploiting algebraic methods in exploring special correlations in many-particle and multi-shell environments.

2. Symmetry patterns revealed by the *ab initio* SA-NCSM model

The *ab initio* symmetry-adapted no-core shell model (SA-NCSM) [2] adopts the first-principle concept. The conventional NCSM basis spaces are constructed using harmonic oscillator (HO) single-particle states and are characterized by the $\hbar\Omega$ oscillator strength as well as by the cutoff in total oscillator quanta, $N_{\max}$, above the lowest energy configuration for a given nucleus. The basis states of the SA-NCSM for a given $N_{\max}$ are constructed in the proton-neutron formalism using also HO single-particle states and are labeled by the SU(3)$\supset$SO(3) subgroup chain quantum numbers $(\lambda\,\mu)\kappa L$, together with proton, neutron, and total intrinsic spins S_p, S_n, and S. The orbital angular momentum L is coupled with S to the total orbital momentum J and its projection M_J. Each basis state in this scheme is labeled schematically as $|\vec{\gamma}\,(\lambda\,\mu)\kappa L; (S_pS_n)S; JM_J\rangle$. The label κ distinguishes multiple occurrences of the same L value in the parent irrep $(\lambda\,\mu)$, and $\vec{\gamma}$ distinguishes among configurations carrying the same $(\lambda\,\mu)$ and $(S_pS_n)S$ labels.

The organization of the model space allows the full space to be down-selected to the physically relevant subspace [1, 3]. The significance of the SU(3) group for a microscopic description of the nuclear dynamics can be seen from the fact that it is the symmetry group of the Elliott model [4, 5, 6], and a subgroup of the $\mathrm{Sp}(3,\mathbb{R})$, the underpinning symmetry of the symplectic model [7].

Ab initio SA-NCSM results for *p*-shell nuclei reveal a dominance of shapes of large deformation (typically large $|\lambda\ -\mu|$) in the $0\hbar\Omega$ subspace. For example, the *ab initio* SA-NCSM results with the bare JISP16 realistic interaction [8] for the 0^+ ground state (*g.st.*), first 2^+ and first 4^+ states of ^{12}C reveal the dominance of the $0\hbar\Omega$ component with the foremost contribution coming from the leading $(0\,4)$ $S=0$ irrep [9]. In addition, important SU(3) configurations are organized into structures with $\mathrm{Sp}(3,\mathbb{R})$ symplectic symmetry, that is, the $(0\,4)$ symplectic irrep gives rise to $(0\,2)$ and $(2\,4)$ configurations in the $2\hbar\Omega$ subspace, and those configurations indeed realize the major components of the wavefunction in this subspace. This further confirms the significance of the symplectic symmetry to nuclear dynamics and points to the fact that the relevant model space can be systematically determined by down-selecting to important configurations (low-spin and high-deformation) in the high $N_{\max}$ spaces. This is illustrated in table 1, where results for the binding energy of ^{12}C and the point-particle matter rms radius of the ground state calculated in the symmetry-selected space of $N_{\max}=8[6]$ are compared to the corresponding counterparts calculated in the complete $N_{\max}=8$ space [10]. Here, the SA-NCSM basis space of $N_{\max}=8[6]$ (shown in fig. 1, filled circles) includes all the configurations up through $6\hbar\Omega$ (complete space,

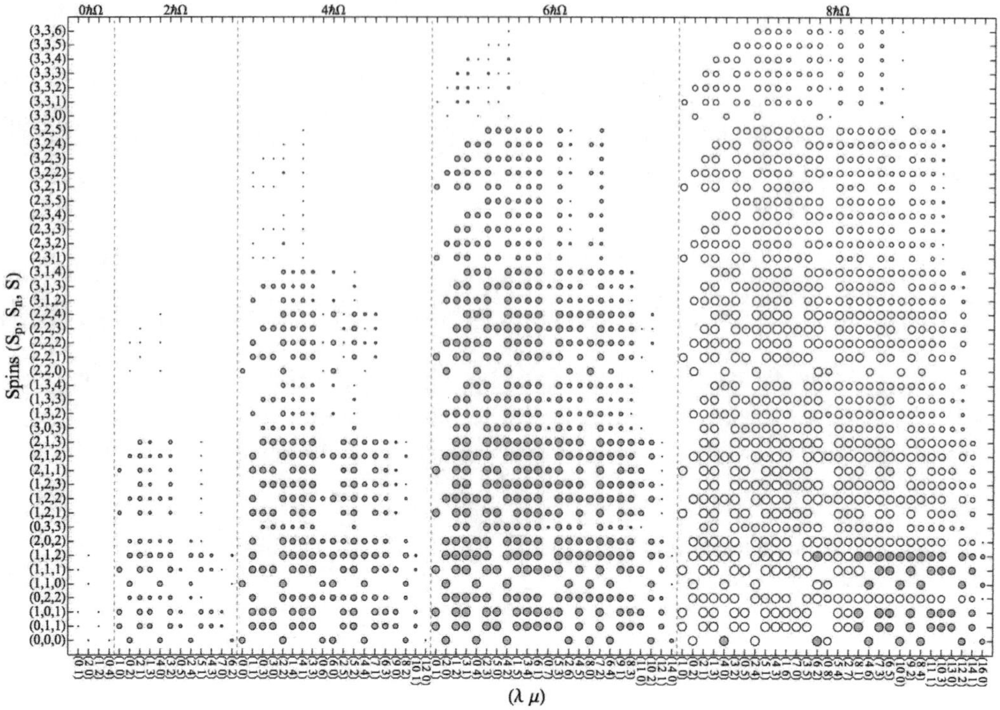

Figure 1. Complete $J = 0$ model space for ^{12}C and $N_{\max} = 8$ given in terms of spin values (vertical axis) and shapes (horizontal axis). Each circle represents basis states carrying the same (S_p, S_n, S) and $(\lambda\,\mu)$ quantum numbers within an $\hbar\Omega$-subspace, with the radius being proportional to $\log_{10}$ of the number of such states. Filled circles indicate the $N_{\max} = 8[6]$ restricted model space used in the present study for SA-NCSM calculations of 0^+ states in ^{12}C.

labeled as '[6]') and a restricted subspace thereof up to $N_{\max} = 8$.

The results clearly show that using only a 0.47% fraction of the complete $N_{\max} = 8$ model space, the $N_{\max} = 8[6]$ SA-NCSM outcome reproduces 98%-99% of the observables obtained in the complete $N_{\max} = 8$ space. This also holds for other observables, such as electric quadrupole moments and $B(E2)$ transition strengths [10]. While this simple example allows one to demonstrate the efficacy of the SA-NCSM as compared to the NCSM complexity, the significance of the SA-NCSM ability to drastically reduce the space dimensionality is critical for higher $N_{\max}$ spaces that are inaccessible to the NCSM.

3. Alpha-clustering phenomena described by the NCSpM model

The no-core symplectic shell model (NCSpM) is a fully microscopic no-core shell model that uses a symplectic $\mathrm{Sp}(3,\mathbb{R})$ basis and $\mathrm{Sp}(3,\mathbb{R})$-preserving interactions. The NCSpM employed within a full model space up through a given $N_{\max}$ coincides with the NCSM for the same $N_{\max}$ cutoff. However, in the case of the NCSpM, the symplectic irreps divide the space into 'vertical slices' that are comprised of basis states of a definite deformation $(\lambda\,\mu)$. Hence, the model space can be reduced to only a few important configurations that are chosen among all possible $\mathrm{Sp}(3,\mathbb{R})$ irreps within the $N_{\max}$ model space.

The NCSpM, while selecting the most relevant symplectic configurations, is employed to provide shell model calculations beyond current NCSM limits, namely, up through $N_{\max} = 20$

Table 1. ^{12}C binding energy and point-particle matter rms radius of the ground state for SA-NCSM reduced $N_{\rm max} = 8[6]$ and complete $N_{\rm max} = 8$ model spaces. The observables are calculated using bare JISP16 for $\hbar\Omega$=25 MeV and compared to experiment ('Expt.'). The fraction of the spaces used in the calculations as compared to the complete $N_{\rm max}$ space is also specified.

	$N_{\rm max} = 8[6]$	$N_{\rm max} = 8$	Expt.
Model space (%)	0.47% (J=0)	100%	
BE (MeV)	88.822	90.869	92.162
matter rms r (fm)	1.920	1.937	2.43(2)

for ^{12}C, the model spaces we found sufficient for the convergence of results [11]. We employ a very simple Hamiltonian with an effective interaction derived from the long-range expansion of the two-body central nuclear force,

$$H_{\rm eff} = H_0 - \frac{\chi}{2}\frac{1}{\gamma}\left(e^{\gamma(Q.Q-\langle Q.Q\rangle_n)} - 1\right), \tag{1}$$

which includes the spherical HO potential (which together with the kinetic energy yields the HO Hamiltonian, H_0) and the $Q.Q$ quadrupole-quadrupole interaction not restricted to a single shell. For the latter term, the average contribution, $\langle Q.Q\rangle_n$, of $Q.Q$ within a subspace of n HO excitations is removed [12]. Here, $\langle Q.Q\rangle_n$ is the trace of $Q.Q$ divided by the space dimension for a fixed n. Hence, the large monopole contribution of the $Q.Q$ interaction is removed, which, in turn, helps eliminate the spurious renormalization of the zero-point energy, while retaining the $Q.Q$-driven behavior of the wavefunctions.

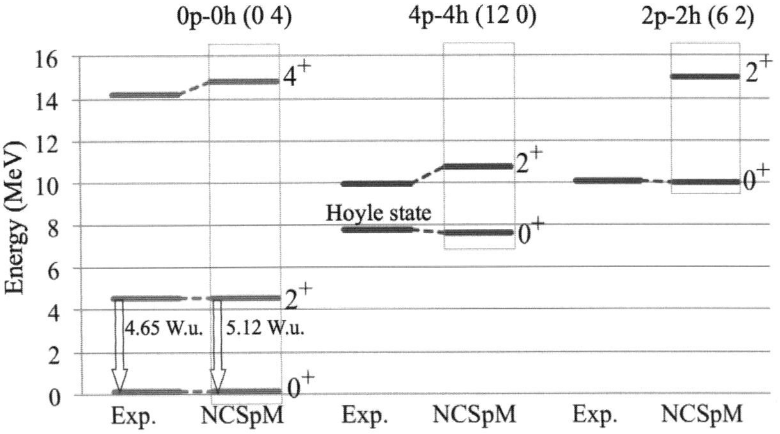

Figure 2. ^{12}C energy spectrum calculated by the $N_{\rm max} = 20$ NCSpM and compared to experiment. The NCSpM calculations are performed using eq. (1) with $\gamma = -1.71 \times 10^{-4}$ and three Sp(3, $\mathbb{R}$) irreps, $0\hbar\Omega$ 0p-0h (0 4) and $4\hbar\Omega$ 4p-4h (12 0), as well as $2\hbar\Omega$ 2p-2h (6 2). Experimental energies for all states, except the latest results for the 10-MeV 2^+ and 0_3^+ [13], are from [14].

This Hamiltonian in its zeroth-order approximation (for parameter $\gamma \to 0$) and for a single valence shell goes back to the established Elliott model [4, 6]. We take the coupling constant

χ to be proportional to $\hbar\Omega$ and, to leading order, to decrease with the total number of HO excitations, as shown by Rowe [15] based on self-consistent arguments. It is important to note that such a choice for χ renders the eigenstates $\hbar\Omega$-independent.

As the interaction and the model space are carefully selected to reflect the most relevant physics, the outcome reveals a quite remarkable agreement with experiment. The low-lying energy spectrum and eigenstates for ^{12}C were calculated using the NCSpM with H of eq. (1) for $\hbar\Omega = 18$ MeV given by the empirical estimate $\approx 41/A^{1/3} = 17.9$ MeV. The results are shown for $N_{\rm max} = 20$, which we found sufficient to yield convergence. This $N_{\rm max}$ model space is further reduced by selecting the most relevant symplectic irreps, namely, the spin-zero ($S = 0$) $0\hbar\Omega$ 0p-0h (0 4), $2\hbar\Omega$ 2p-2h (6 2), and $4\hbar\Omega$ 4p-4h (12 0) symplectic bandheads together with all multiples thereof up through $N_{\rm max} = 20$ of total dimensionality of 4.5×10^3. In comparison to the experimental energy spectrum (fig. 2), the outcome reveals that the lowest 0^+, 2^+, and 4^+ states of the $0\hbar\Omega$ 0p-0h (0 4) symplectic irrep calculated for $\gamma = -1.71 \times 10^{-4}$ closely reproduce the *g.st.* rotational band, while the calculated lowest 0^+ states of the $4\hbar\Omega$ 4p-4h (12 0) and the $2\hbar\Omega$ 2p-2h (6 2) slices are found to lie close to the Hoyle state and the 10-MeV 0^+ resonance (third 0^+ state), respectively. We note that the NCSpM energies given in fig. 2 are rescaled by an overall factor of ~ 2. This factor is determined by fixing the lowest 2^+ by its experimental value. This, however, has no implications on the underlying physics, as an overall factor for H does not change its properties and eigenstates, together with associated observables. Indeed, the model successfully reproduces other observables for ^{12}C that are informative of the state structure, such as matter rms radii, electric quadrupole moments and $B(E2)$ transition strengths. These quantities are shown in fig. 3, where the values obtained for $\gamma = -1.71 \times 10^{-4}$ lie remarkably close to the experimental data.

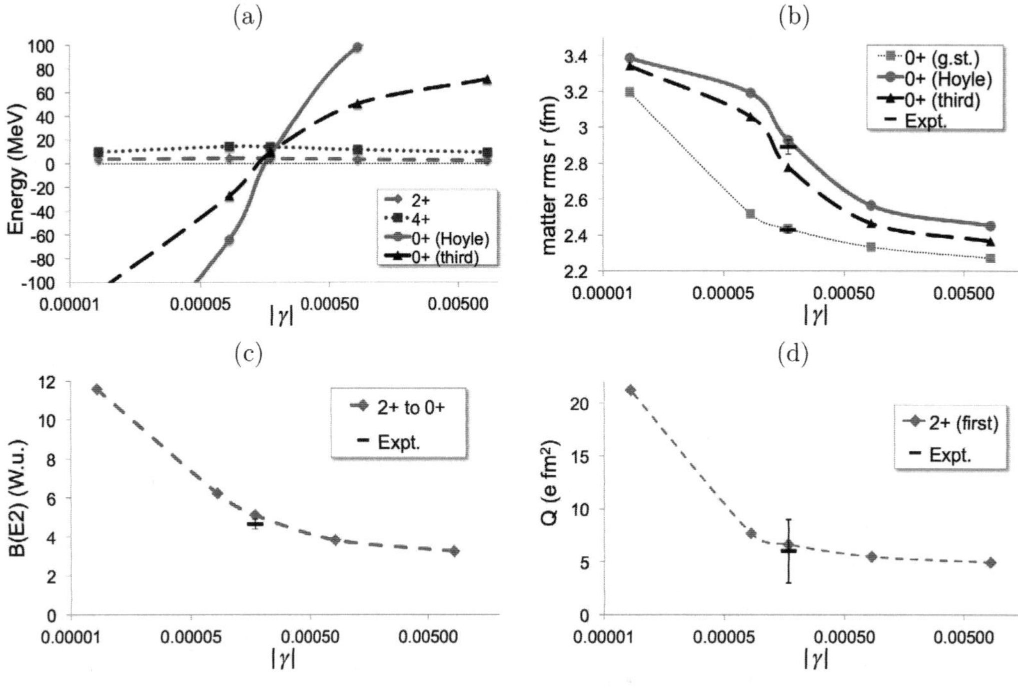

Figure 3. Dependence of the ^{12}C NCSpM energy spectrum on the γ model parameter for $N_{\rm max} = 20$. Available experimental values are shown for (b), (c), and (d) and compared to the NCSpM results with $\gamma = -1.71 \times 10^{-4}$.

Furthermore, fig. 3 reveals that the additional degree of freedom associated with the γ model parameter is in fact substantially limited by the lowest 0^+ states (with only a small effect on the *g.st.* rotational band). Clearly, while the model includes an adjustable parameter, γ only controls the decrease rate of the $Q.Q$ interaction with increasing n; the entire many-body apparatus is fully microscopic and no adjustments are possible. Hence, given the dramatic variation with γ of the energy and the relative position of the three lowest 0^+ states (including the *g.st.*), there is only a small window of reasonable γ values where observables are also found in agreement with experiment.

Table 2. Probability distribution for ^{12}C ($\geq 0.1\%$) over the n total excitations and the SU(3) $(\lambda\,\mu)$ of the lowest 0^+, 2^+, and 4^+ states as calculated by the NCSpM and by the *ab initio* SA-NCSM using the bare JISP16 for $\hbar\Omega = 20$ MeV and an $N_{\max} = 8$ $S_{p,n} = 0$ space.

	0^+			2^+			4^+	
$n(\lambda\,\mu)$	Probability (%)		$n(\lambda\,\mu)$	Probability (%)		$n(\lambda\,\mu)$	Probability (%)	
	SA-NCSM	NCSpM		SA-NCSM	NCSpM		SA-NCSM	NCSpM
0 (0 4)	69.12	64.49	0 (0 4)	68.17	64.82	0 (0 4)	66.11	65.02
0 (2 0)	0.13		2 (2 4)	8.40	17.71	2 (2 4)	9.26	17.84
2 (2 4)	8.35	17.85	2 (0 2)	4.58	0.68	2 (1 3)	8.60	0.3
2 (0 2)	7.07	0.95	2 (1 3)	3.41		2 (6 2)	0.76	
2 (6 2)	1.82		2 (6 2)	1.48		2 (4 3)	0.16	
4 (4 4)	6.34	9.44	4 (4 4)	6.25	9.24	4 (4 4)	6.08	8.98
4 (2 2)	1.56	0.98	4 (2 2)	0.91	0.76	4 (3 3)	1.86	0.11
4 (0 0)	0.55		4 (1 1)	0.84		4 (2 2)	1.35	0.18
4 (8 2)	0.54		4 (3 3)	0.61		4 (1 4)	0.35	1.19
4 (0 6)	0.23	0.67	4 (8 2)	0.45		4 (0 6)	0.31	0.37
4 (6 0)	0.16		4 (0 6)	0.25	0.58	4 (8 2)	0.30	
6 (6 4)	1.50	3.22	4 (1 4)	0.16	0.32	6 (6 4)	1.39	3.22
6 (4 2)	0.63	0.32	4 (5 2)	0.11		6 (5 3)	0.49	
6 (2 6)	0.31	0.11	6 (6 4)	1.45	3.18	6 (2 6)	0.33	
6 (10 2)	0.20		6 (4 2)	0.42	0.18	6 (3 4)	0.29	0.37
6 (2 0)	0.18		6 (2 6)	0.31		6 (4 2)	0.26	
8 (8 4)	0.51	1.1	6 (5 3)	0.18		6 (10 2)	0.11	
8 (6 2)	0.20	0.12	6 (10 2)	0.16		6 (3 1)	0.11	
8 (4 6)	0.14		6 (3 1)	0.14		8 (8 4)	0.47	1.08
10 (10 4)	—	0.32	6 (3 4)	0.12		8 (7 3)	0.21	
			8 (8 4)	0.50	1.09	8 (4 6)	0.13	
			8 (4 6)	0.13		8 (5 4)		0.16
			8 (6 2)	0.13		10 (10 4)	—	0.28
			10 (10 4)	—	0.25			

In addition, a close similarity is observed when the probability distributions for the *g.st.* rotational band are compared to *ab initio* results when only configurations of zero proton and neutron spins ($S_{p,n} = 0$) are selected (table 2). In particular, NCSpM eigenstates, which are $\hbar\Omega$-independent, are compared to SA-NCSM calculations with the bare JISP16 realistic interaction for $\hbar\Omega = 20$ MeV (around the minimum of the calculated binding energy for ^{12}C) and a $N_{\max} = 8$ model space. This space appears to be sufficient to yield convergent results for the *g.st.* rotational band for both models. The close agreement points to the fact that the schematic interaction used in NCSpM has effectively captured most of the underlying physics of the realistic interaction important to the low-energy nuclear dynamics.

Furthermore, in accordance with the mapping [16] between the shell-model $(\lambda\,\mu)$ SU(3) labels and the shape variables of the Bohr-Mottelson collective model [17], the symplectic basis states bring forward important information about the nuclear shapes and deformation. The simplest cases, $(0\,0)$, $(\lambda\,0)$, and $(0\,\mu)$, describe a spherical, prolate, and oblate shape, respectively, while a general nuclear state is typically a superposition of several hundred various triaxial shapes. Thus, while a preponderance of the $(0\,4)$ shape (of $\sim 65\%$ probability) is observed for the *g.st.* rotational band (table 2), indicating a pronounced oblate shape of the *g.st.*, the $(12\,0)$ bandhead of the Hoyle-state rotational band includes shapes of even larger deformations (more prolate) with the largest contribution ($\sim 30\%$) of $(16\,0)$.

4. Conclusion

We showed that employing the symmetry-adapted no-core shell-model with SU(3) the underpinning symmetry is effective in providing an efficient description of low-lying eigenstates of ^{12}C, which also holds for various other *p*-shell nuclei. The symmetry patterns revealed in the *ab initio* calculations were, in turn, employed to reach ultra-large model spaces that are currently inaccessible. We carried forward a no-core shell-model study with a schematic many-nucleon interaction to further unveil the underlying physics behind various phenomena important to the low-energy nuclear dynamics of ^{12}C. We showed, for the first time, how both collective and cluster-like structures emerge out of a shell-model framework, which can extend to and take into account essential high-lying configurations.

Acknowledgments

We thank all members of the NSF PetaApps supported collaboration for useful discussions. This work was supported by the U.S. NSF (OCI-0904874 & OCI-0904782), the U.S. DOE (DE-SC0005248, DE-FG02-95ER-40934, DE-SC0008485 & DE-FG02-87ER40371), and SURA. ACD acknowledges support by the U.S. NSF (grant 1004822) through the REU Site in Dept. of Physics & Astronomy at LSU. We acknowledge LSU/LONI for providing HPC resources. A portion of the computational resources were provided by the NERSC Center, which is supported by the Office of Science of the U.S. DOE under Contract DE-AC02-05CH11231.

References

[1] Dytrych T, Sviratcheva K D, Bahri C, Draayer J P and Vary J P 2007 *Phys. Rev. Lett.* **98** 162503
[2] Dytrych T *et al.* 2013 *to be submitted to Phys. Rev. Lett.*
[3] Dytrych T, Sviratcheva K D, Draayer J P, Bahri C and Vary J P 2008 *J. Phys. G: Nucl. Part. Phys.* **35** 123101
[4] Elliott J P 1958 *Proc. Roy. Soc. A* **245** 128
[5] Elliott J P 1958 *Proc. Roy. Soc. A* **245** 562
[6] Elliott J P and Harvey M 1962 *Proc. Roy. Soc. A* **272** 557
[7] Rosensteel G and Rowe D J 1977 *Phys. Rev. Lett.* **38** 10
[8] Shirokov A M, Vary J P, Mazur A I and Weber T A 2007 *Phys. Lett.* **B 644** 33
[9] Draayer J P, Dytrych T, Launey K D and Langr D 2012 *Prog. Part. Nucl. Phys.* **67** 515
[10] Dytrych T, Maris P, Launey K D, Draayer J P, Vary J P, Caprio M A, Langr D, Çatalyürek Ü V, Saule E and Sosonkina M 2013 *to be submitted*
[11] Dreyfuss A C, Launey K D, Dytrych T, Draayer J P and Bahri C 2012 *Preprint* arXiv:1212.2255
[12] Rosensteel G and Draayer J P 1985 *Nucl. Phys. A* **436** 445
[13] Itoh M *et al.* 2011 *Phys. Rev. C* **84** 054308
[14] Ajzenberg-Selove F 1990 *Nucl. Phys. A* **506** 1
[15] Rowe D J 1967 *Phys. Rev.* **162** 866
[16] Castaños O, Draayer J P and Leschber Y 1988 *Z. Phys. A* **329** 33
[17] Bohr A and Mottelson B R 1969 vol 1 (New York: Benjamin)

Tensor force manifestations in *ab initio* study of the ^{2}H$(d,\gamma)^{4}$He, ^{2}H$(d,p)^{3}$H and ^{2}H$(d,n)^{3}$He reactions

K Arai[1], S Aoyama[2], Y Suzuki[3], P Descouvemont[4] and D Baye[4,5]

[1] Division of General Education, Nagaoka National College of Technology, 888 Nishikatakai, Nagaoka, Niigata, 940-8532, Japan
[2] Center for Academic Information Service, Niigata University Niigata 950-2181, Japan
[3] Physics Department, Niigata University, Niigata 950-2181, Japan
[4] Physique Nucléaire Théorique et Physique Mathématique, C.P.229, Université Libre de Bruxelles, B 1050 Brussels, Belgium
[5] Physique Quantique, CP165/82, Université Libre de Bruxelles, B-1050 Brussels, Belgium

E-mail: arai@nagaoka-ct.ac.jp

Abstract. The ^{2}H$(d,\gamma)^{4}$He capture reaction and the ^{2}H$(d,p)^{3}$H and ^{2}H$(d,n)^{3}$He transfer reactions at very low energies are studied in an extended microscopic cluster model, in which the ^{3}H, ^{3}He, and d clusters are given by the precise few-body wave functions with a realistic nucleon-nucleon force and the cluster relative motions are solved with the same realistic nucleon-nucleon force. Our results show that the tensor force in realistic interactions plays an essential and indispensable role to reproduce the very low energy astrophysical S factor of these reactions.

1. Introduction

The microscopic cluster model(RGM) [1, 2] is a successful model to study the nuclear structure and reactions between light nuclei. This model usually employs simple cluster wave functions, S-wave wave function for the s-shell clusters, and accordingly employs the effective N-N interaction such as the Minnesota (MN) potential [3]. In this effective potential, the effect of the tensor term and the short-range repulsion are presumed to be renormalized in the central force of the effective interaction.

Recently, we have developed a microscopic cluster model in which the cluster wave functions are given by precise few-body wave functions including higher partial waves up to the D-wave. The cluster wave functions and the cluster relative motion are solved with the same realistic NN interaction.

In the present paper, we calculate the low energy cross sections of the ^{2}H$(d,\gamma)^{2}$H, ^{2}H$(d,p)^{3}$H, and ^{2}H$(d,n)^{3}$He reactions. We employ the extended microscopic cluster model and the results with the AV8$'$ and G3RS potentials are compared with those with the MN potential in order to clarify the role of the tensor force at very low energies. Along with the radiative capture reaction, we calculate the transfer reactions, ^{2}H$(d,p)^{3}$H and ^{2}H$(d,n)^{3}$He, to show the role of the tensor force in the zero energy S factor of these two reactions.

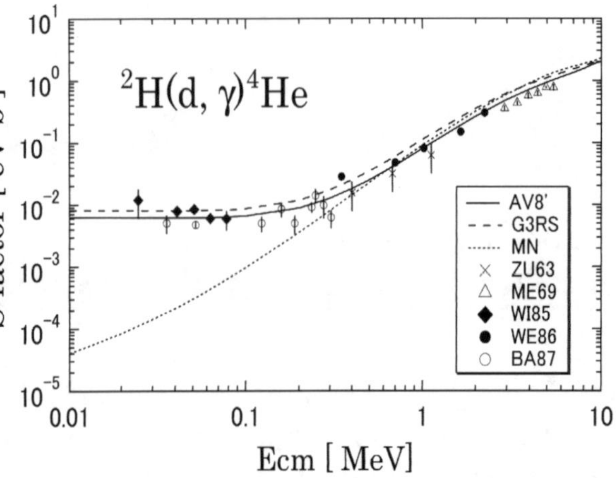

Figure 1. Astrophysical S-factor of the ^{2}H(d,γ)^{4}He reaction [4]. Results calculated with the realistic (AV8′, G3RS) and effective (MN) potentials are compared to experiment [12].

2. Model

Capture and transfer reactions at low energies are calculated in the microscopic cluster model [4] with the multi-channel configurations of the ^{3}H($1/2^+$)+p, ^{3}He($1/2^+$)+n, $d(1^+)+d(1^+)$, $pn(0^+)+pn(0^+)$, and $pn(0^+)+pn(0^+)$. In this model, all cluster wave functions are given by precise three- or two-body wave functions with the AV8′ [5] and G3RS [6] realistic interactions. In these cluster wave functions the higher partial wave up to the D-wave are taken into account. Since it is very essential to reproduce correctly the two-body thresholds of d+d, t+p, and h+n in these reaction calculations, a phenomenological three-body force [7] is added to the present Hamiltonian. The cluster wave functions are selected by the stochastic variational method [8, 9] in order to make the four-body calculation feasible. Number of the basis set is 30 for the t and h clusters and 8 for the d cluster. For the 0^+ state of pn, pp, and nn clusters, the same S-wave basis set with the d cluster are employed and these wave functions are given by the bound state approximation. These channels are included in order to take into account the distortion effect. The pseudo-excited states are also taken into account for all of the clusters.

With above the precise cluster wave functions, the cluster relative wave functions are solved with the same realistic N-N interaction. The total wave function satisfies the Pauli principle exactly and is free from the spurious center-of-mass motion. Total angular momentum and the parity is correctly projected on the total wave function. The scattering problem is solved by microscopic R-matrix method(MRM) [10]. In the MRM, the cluster relative wave function is expanded with the Gaussian basis set within the channel radius and then it is connected the exact Coulomb function at the channel radius and so we get the S-matrix. The cross section of the capture and transfer reaction are given in Ref. [11]. Cross sections with the realistic interactions are compared with those with the MN effective potential, in which the tensor force is simulated by a renormalized central force, in order to make clear the role of the tensor force in these reactions.

3. Results

Figure 1 shows our astrophysical S factor for the ^{2}H$(d,\gamma)^4$He capture reaction, and a comparison with the experimental data. We consider only the dominant $E2$ transition, that is, the transition from the 2^+ continuum state to the 0^+ ground state of ^{4}He. Our results with the AV8′ and

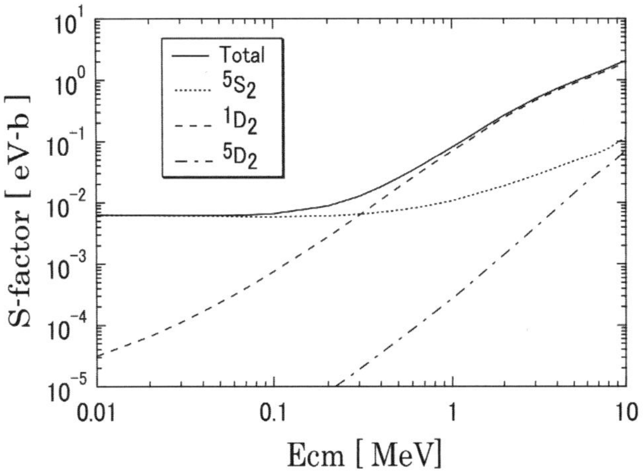

Figure 2. Astrophysical S-factors of the ^{2}H(d,γ)^{4}He reaction with the AV8$'$ potential, and contributions of the three incoming dd channels, $^5S_2$, $^1D_2$, and $^5D_2$ [4].

G3RS potentials reproduce the experimental data very well, especially the flat behaviour below 0.3 MeV whereas the MN potential fails to reproduce the data below 0.3 MeV. In order to make clear the role of the tensor force, we decomposed the S factor with the AV8$'$ potential into the three different entrance d+d channels, $^5S_2$, $^1D_2$, and $^5D_2$, as shown in figure 2. This shows that the S factor below the 0.3 MeV is dominated by the $^5S_2$ entrance channel which gives the flat behaviour at low energies, while the $^1D_2$ channel plays a dominant role above 0.3 MeV. Through the $E2$ transition, the $^5S_2$ entrance channel transits into the D-wave component of the 0^+ ground state where the tensor force of the N-N interaction is indispensable to produce this D-wave component. However, since the MN potential gives a pure S-wave component (L=0, S=0) in the 0^+ ground state, the $E2$ transition from the $^5S_2$ entrance channel is forbidden. Therefore the S factor below the 0.3 MeV with the MN potential comes from the the $^1D_2$ entrance channel. As a result, the MN potential fails to reproduce the data below 0.3 MeV.

Figure 3 shows our astrophysical S factors of the ^{2}H$(d,p)^3$H and ^{2}H$(d,n)^3$He transfer reactions, in comparison with experimental data. We have taken into account up to the $2^\pm$ states in the present calculation. The results with the AV8$'$ and G3RS potentials reproduce the experimental data very well whereas the MN potential underestimates the S factor. In figure 4(a), the S factor of the ^{2}H$(d,p)^3$H with the AV8$'$ potential is decomposed into the different spin/parity states J^π. Below 0.1 MeV, the most important contribution to the S factor comes from the 2^+ state. In figure 4(b), this 2^+ partial S factor is further decomposed into the three dominant transitions, d+d $^5S_2 \rightarrow t$+p $^1D_2$, d+d $^5S_2 \rightarrow t$+p $^3D_2$, and d+d $^1D_2 \rightarrow t$+p $^1D_2$. This figure shows that the $^5S_2 \rightarrow {}^3D_2$ transition gives dominant contribution below 0.1 MeV where the tensor force of the N-N interaction is indispensable for this transition. Therefore, this transition is forbidden in the MN potential and the 2^+ contribution comes mainly from the $^1D_2 \rightarrow {}^1D_2$ transition. This is the reason why the MN potential fails to reproduce the data in this transfer reaction.

4. Conclusion

The radiative capture reaction, ^{2}H$(d,\gamma)^4$He and the transfer reactions, ^{2}H$(d,p)^3$H and ^{2}H$(d,n)^3$He, which are of astrophysical interest, are calculated in the extended microscopic cluster model with a realistic nucleon-nucleon interaction in order to make a clear the role of the

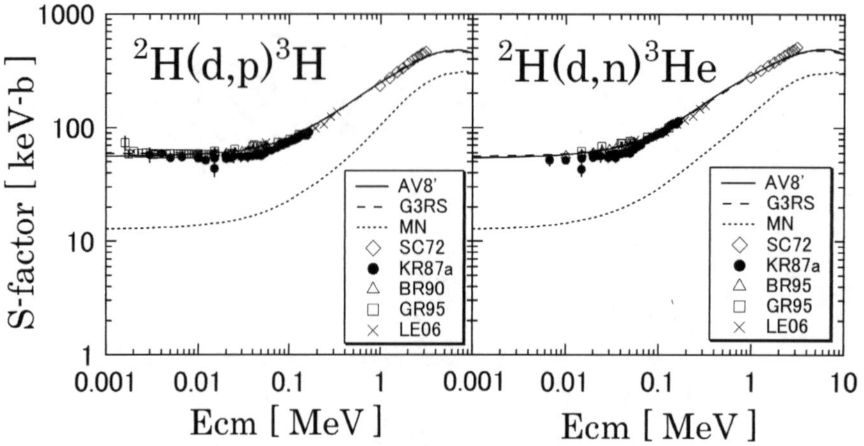

Figure 3. Astrophysical S-factors of the ^{2}H(d,p)^{3}H and ^{2}H(d,n)^{3}He reactions [4]. Results calculated with the realistic (AV8$'$, G3RS) and effective (MN) potentials are compared to experiment [12, 13].

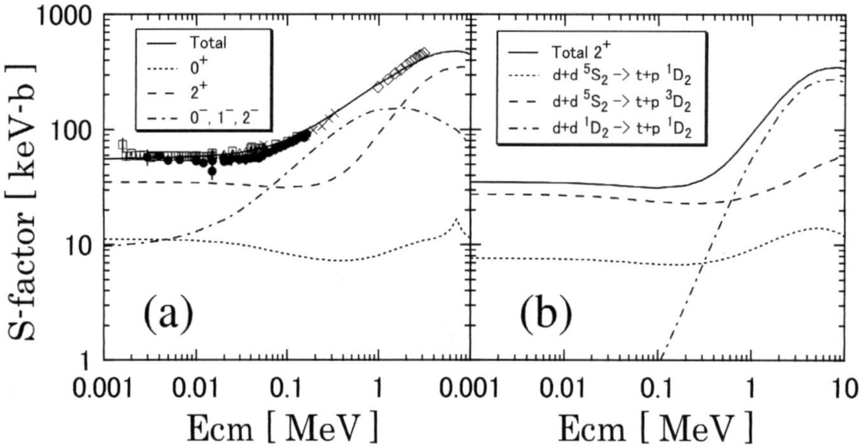

Figure 4. (a) Contributions of the different J^π states to the astrophysical S-factor of the ^{2}H(d,p)^{3}H reaction [4]. (b) Decomposition of the 2^+ contribution according to the incoming dd and outgoing tp channels. The AV8$'$ potential is used.

tensor force in these reactions. The AV8$'$ and G3RS potentials can reproduce the astrophysical S factor of these reactions at very low energies, while the MN potential fails to reproduce the experimental data. In the capture reaction, the $E2$ transition from the 2^+ d+d S wave continuum state to the D wave component of the 0^+ ground state is dominant. And the transition from the d+d S wave continuum state to the t+p(^{3}He+n) D wave channel in the 2^+ state is essential in the transfer reactions. The tensor force is indispensable for these transitions but the transitions are forbidden in the calculation with the effective potential since the tensor force is not explicitly included in the MN potential. Our results show that the tensor force plays an important role in the astrophysical reactions involving four nucleons.

Reference

[1] Wildermuth K and Tang Y C 1977 *A Unified Theory of the Nucleus* (Braunschweig: Vieweg)
[2] Langanke K 1994 *Advances in Nuclear Physics*, Vol. **21** ed Negele J W and Vogt E (New York: Plenum) p 85
[3] Thompson D R, LeMere M and Tang Y C 1977 *Nucl. Phys.* A **286** 53
[4] Arai K, Aoyama S, Suzuki Y, Descouvemont P and Baye D 2011 *Phys. Rev. Lett.* **107** 132502
[5] Pudliner B S, Pandharipande V R, Carlson J, Pieper S C and Wiringa R B 1997 *Phys. Rev.* C **56** 1720
[6] Tamagaki R 1968 *Prog. Theor. Phys.* **39** 91
[7] Hiyama E, Gibson B F and Kamimura M, 2004 *Phys. Rev.* C **70** 031001
[8] Kukulin V I and Krasnopol'sky V M 1977 *J. Phys.* G **3** 795
[9] Varga K, Suzuki Y and Lovas R G 1994 *Nucl. Phys.* A **571** 447; Varga K and Suzuki Y 1995 *Phys. Rev.* C **52** 2885
[10] Baye D, Heenen P -H, and Libert-Heinemann M 1977 Nucl. Phys. A **291** 230; Descouvemont P and Baye D 2010 *Rep. Prog. Phys.* **73** 036301
[11] Arai K, Baye D and Descouvemont P 2002 *Nucl. Phys.* A **699** 963
[12] Angulo C *et al* 1999 *Nucl. Phys.* A **656** 3
[13] Leonard D S, Karwowski H J, Brune C R, Fisher B M and Ludwig E J 2006 *Phys. Rev.* C **73** 045801

Journal of Physics: Conference Series **436** (2013) 012025 doi:10.1088/1742-6596/436/1/012025

Excited states of ^{4}He

Shigeyoshi Aoyama
Center for Academic Information Service,
Niigata University, Niigata 950-2181, Japan

E-mail: aoyama@cc.niigata-u.ac.jp

Abstract. The study of the ^{4}He nucleus is important because it is the most basic sub-unit (cluster) in nuclei. We have investigated the structures and the reaction mechanisms in ^{4}He by using the correlated Gaussian basis function with the global vector representation. In order to treat the boundary condition for the *ab-initio* calculation of the four nucleons, we employ the Microscopic R-matrix Method (MRM) and the Complex Scaling Method (CSM) . Elastic-scattering phase shifts for four-nucleon systems are studied in an *ab-initio* type cluster model with MRM in order to clarify the role of the tensor force and to investigate cluster distortions in low energy d+d and t+p scattering. For $^1S_0$, the calculated phase shifts show that the t+p and h+n channels are strongly coupled to the d+d channel for the case of the realistic interaction.

1. Introduction

"*Ab-initio*"-type studies have been carried out for the ^{4}He nucleus by many groups [1]. ^{4}He is an interesting nucleus because it is a most basic sub-unit (cluster) in nuclei and has a double closed s-shell nature. However, for excited states of ^{4}He, theoretical analyses have been limited mainly because the treatment of unbound states are very difficult. One of the useful method for unbound states is the Microscopic R-matrix Method (MRM) [2] which is combined with the microscopic cluster model. The microscopic cluster model is one of the successful models to study the structure and reactions of light nuclei [3]. Since the conventional cluster model does not directly treat the D-wave component, the strong attraction of the nucleon-nucleon interaction due to the tensor term is assumed to be renormalized into the central term of the effective interaction. However, it is well known that the ground states of the typical s-shell clusters d, t, h and ^{4}He have non-negligible admixtures of D-wave component due to the tensor interaction. Therefore, we need an extension of the cluster model to an "*ab-initio*"-type one. Recently, we have studied the importance of the tensor force in the d+d scattering [4] and the $d(d,\gamma)^4$He radiative capture [5] with the extended cluster model combined with the MRM. The calculated results for the transfer and capture reactions will be also shown by Arai in the proceedings. Another useful method to treat unbound states is the Complex Scaling Method (CSM) [6, 7]. In the MRM, the "*ab-initio*" solution in the internal region is connected to the two body asymptotic solution (Coulomb and Whittaker functions) at a channel radius. Three and four body continuums are approximately and reasonably treated through the rearrangement in the MRM. The CSM can directly treat three-body and four-body continuums in the excited region of ^{4}He. However, the computational time with the CSM is very slower than that with the MRM. Thus, the comparative analyses with both method are important. In the present paper, we present the latest studies for excited states of ^{4}He with the MRM and the CSM.

2. Method

We explain the present method in brief. The wave function of the four nucleon system in the channel representation is described,

$$\Psi_{\alpha}^{JM\pi} = \mathcal{A}\Big[\Big[[\psi_{L_a}^{(\text{space})}\psi_{S_a}^{(\text{spin})}]_{I_a}[\psi_{L_b}^{(\text{space})}\psi_{S_b}^{(\text{spin})}]_{I_b}\Big]_I \chi_{\alpha}(\boldsymbol{\rho}_{\alpha})\Big]_{JM} \psi_{T_aM_{T_a}}^{(\text{isospin})}\psi_{T_bM_{T_b}}^{(\text{isospin})}, \qquad (1)$$

where $\psi_{L_k}^{(\text{space})}$, $\psi_{S_k}^{(\text{spin})}$ and $\psi_{T_kM_{T_k}}^{(\text{isospin})}$ denote the space, spin and isospin parts of the cluster wave function. χ_{α} is a relative wave function which is labeled by a channel α. The matrix elements can easily calculated by the so-called f-function in the global vector representation[8, 4]. In order to treat unbound states, we use with the MRM and the CSM. Both method have been applied many systems, and the detail of the methods was written in [2, 6].

The four nucleon Hamiltonian is described as

$$H = \sum_{i=1}^{4} T_i - T_{c.m} + \sum_{i<j}^{4} V_{ij} + \sum_{i<j<k}^{4} V_{ijk}, \qquad (2)$$

where T_i is a kinetic energy of each nucleon, $T_{c.m}$ is a kinetic energy of the center of mass motion, V_{ij} is a two nucleon interaction and V_{ijk} is a three nucleon interaction. We want to compare the phase shifts obtained with two Hamiltonians that differ in the type of two nucleon interactions. One is a realistic interaction called the AV8$'$ potential [9] that includes central, tensor and spin-orbit components. We also add an effective three-nucleon force (TNF) in order to reproduce reasonably the binding energies of t, h and ^{4}He [10], which makes reasonable thresholds. Another is an effective central interaction called the Minnesota (MN) potential [11], which reproduce reasonably the binding energies of d, t, h and ^{4}He, though it has central terms alone (with an exchange parameter $u = 1$). The Coulomb potential is included for both potentials.

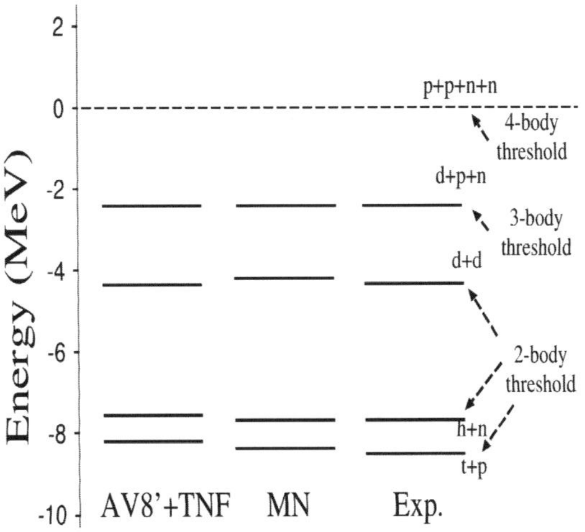

Figure 1. The calculated two-body and three-body threshold energies for AV8'+TNF, MN and the experimental threshold energy. The dashed line is the p+p+n+n four-body threshold.

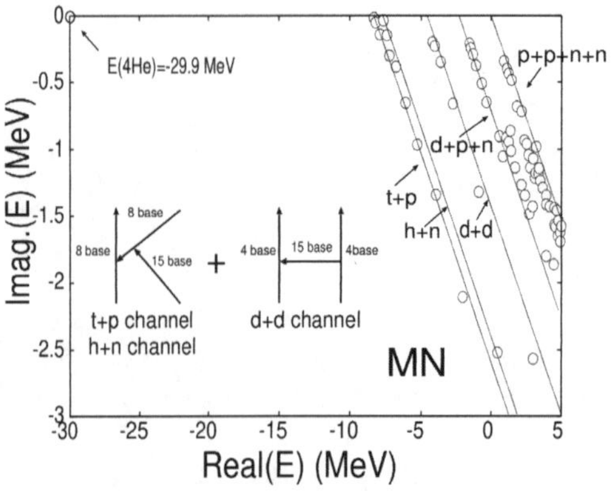

Figure 2. The complex scaled eigenvalue distribution for four nucleons system of ^{4}He (0^+) with the Minnesota potential.

3. Results

Figure 1 displays two-body and three-body decay thresholds above 20 MeV excited energy region. The lowest three thresholds are corresponding to two-body decay physical channels; t+p, h+n, d+d. The next lowest threshold are corresponding to a three-body decay physical channel; d+p+n. In the present calculation, we reproduce the experimental threshold position both for the AV8'+TNF and MN potentials as seen in Figure 1.

If we directly diagonalize the hamiltonian matrix with a bound state approximation, there are many continuum states whose origin are two-body, three-body and four-body continuum states. By using the complex scaling method, we can easily distinguish these continuum states because the complex eigenvalues corresponding continuum states are the corresponding so-called "2θ-lines" [6]. Furthermore, we can get the bound state solutions on the negative real axis as usual, and also the resonant solution in the wedged region whose real part and imaginarily part are corresponding to the resonance energy and the decay width ($\Gamma/2$), respectively. In Figure 2, we can see continuum lines of t+p, h+n, d+d, d+p+n, p+p+n+n, whose starting points are threshold energies. Furthermore, we get the bound ground state solution of ^{4}He just on the negative real axis.

Figure 3 displays the $^1S_0$ elastic-scattering phase shift (d+d) obtained with the AV8$'$+TNF potential. The dotted line is the phase shift calculated with $\{d(1^+)+d(1^+)\}$ channel (deuteron spin $I_d = 1^+$), and the dashed line is the phase shift with $\{d(1^+)+d(1^+)\}\&\{d(0^+)+d(0^+)\}$ channels ($I_d = 1^+, 0^+$). The phase shifts calculated by including further excited deuterons are also plotted by the dash-dotted and solid lines that correspond to the channels ($I_d \leq 2^+$) and ($I_d \leq 3^+$), respectively. A naive picture that the $^1S_0$ elastic-scattering phase shift might (d+d) be well described in channel $d(1^+) + d(1^+)$ alone completely breaks down in the case of the AV8$'$+TNF potential.

Because the deuteron has a virtual state $\bar{d}$ with 0^+, it is reasonable that the inclusion of the $\{d(0^+)+d(0^+)\}$ channel gives rise to a considerable attractive effect of several tens of degrees on the phase shift, as shown by the dashed line of Figure 3. However, the phase shift exhibits no converging behavior even when the higher spin states such as $d(2^+)$ and $d(3^+)$ are taken into account in the calculation. One may conclude that the deuteron is strongly distorted even in the low energy $^1S_0$ elastic scattering (d+d). But this should be a consequence that there exist two observed 0^+ states below the d+d threshold. Obviously the d+d scattering wave function is

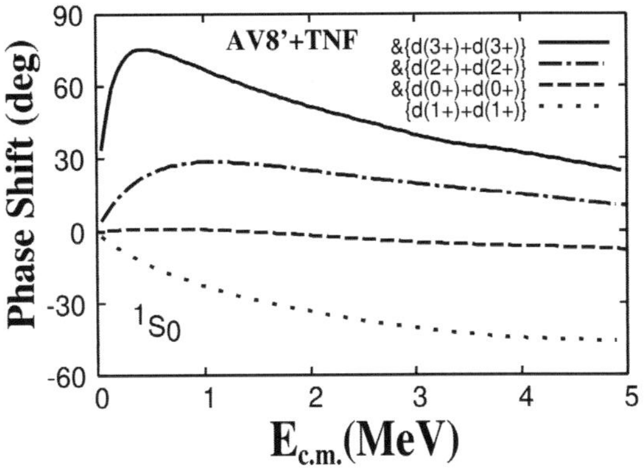

Figure 3. $^1S_0$ elastic-scattering phase shift calculated with the AV8′+TNF potential. The phase shifts are all obtained within the d+d channels. The set of included channels is successively increased from $\{d(1^+)+d(1^+)\}$ (dotted) to $\{d(1^+)+d(1^+)\}\&\{d(0^+)+d(0^+)\}$(dashed), $\{d(1^+)+d(1^+)\}\&\{d(0^+)+d(0^+)\}\&\{d(2^+)+d(2^+)\}$(dash-dotted), and $\{d(1^+)+d(1^+)\}\&\{d(0^+)+d(0^+)\}\&\{d(2^+)+d(2^+)\}\&\{d(3^+)+d(3^+)\}$(solid).

subject to the structure of those states in the internal region and the deuteron can not remain a simple deuteron ground state of 1^+. Especially, the second 0^+ state of ^{4}He lying about 4 MeV below the d+d threshold, which is known to have a $3N$+N cluster structure [12, 10]. This 3N+N state together with the ground state influence to the d+d scattering state.

Figure 4 displays the $^1S_0$ elastic-scattering phase shifts (d+d) obtained with the AV8′+TNF potential (left) and the MN potential (right). The $\{2N+2N\}$ &$\{3N+N\}$ calculation (solid line in Figure 4) means that major $2N$+$2N$ and $3N$+N channels are coupled. The R-matrix

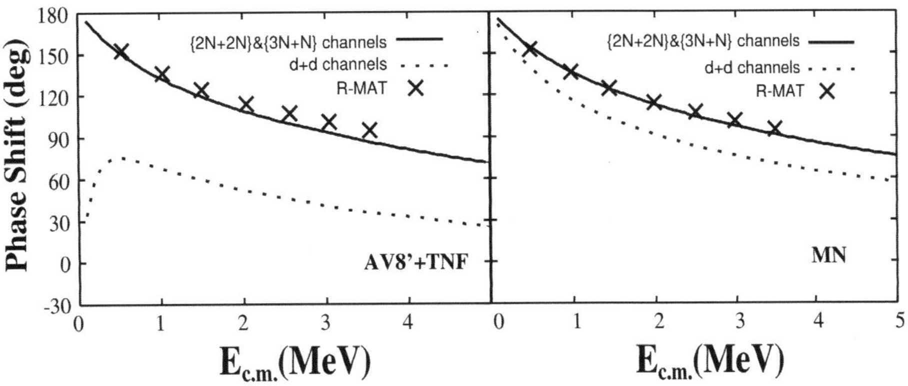

Figure 4. $^1S_0$ d+d elastic-scattering phase shift calculated with the AV8′+TNF (left) and MN (right) potentials. The solid line is a FULL calculation, while the dotted line is a d+d channel calculation. Crosses correspond to the R-matrix analysis of Ref. [13].

analysis (crosses) [13] is reproduced well by both the AV8′+TNF and MN potentials with the $\{2N+2N\}$ &$\{3N+N\}$ calculation. Compared to the uncoupled phase shift (dotted line), one clearly sees that the $3N+N$ channel produces a very large effect on the $d+d$ elastic phase shift, especially in the case of the AV8′+TNF potential. The slow convergence seen in Figure 3 is thus attributed to the neglect of the $3N+N$ channel, indicating that a proper account of the $^1S_0$ $d+d$ elastic phase shift at low energy can be possible only when the coupled channels $\{d(1^+)+d(1^+)\}$ $+\{d(0^+)+d(0^+)\}+$ $\{t(1/2^+)+p(1/2^+)\}$ $+\{h(1/2^+)+n(1/2^+)\}$ are considered. Thus, the slow convergence in Figure 3 suggests that the $2N+2N$ partition is not an economical way to include the effects of the $3N+N$ channel. In the case of the MN potential (right panel in Figue 4), the situation is very different from the AV8′+TNF case. The channel coupling effect is rather modest, and the size of the $^1S_0$ elastic phase shift($d+d$) is already accounted for mostly in the $d+d$ channel calculation.

4. Summary

We have investigated the influence of clusters' distortion in the low-energy $^1S_0$ elastic scattering phase shift using a microscopic cluster model with the global vector method. We showed that the tensor interaction changes the phase shifts very much by comparing a realistic interaction (AV8') and an effective interaction (MN). For the realistic interaction, the calculated $^1S_0$ phase shift shows that the $t+p$ and $h+n$ channels strongly couple with the $d+d$ channel because of the tensor interaction. On the contrary, the coupling of these $3N+N$ channels plays a relatively minor role for the case of the effective interaction because of the absence of tensor term. In the future work, we want to analyze the details of the excited structure of ^{4}He applying the CSM.

Acknowledgments

The main part of the present work presents research results of Bilateral Joint Research Projects of the JSPS (Japan) and the FNRS (Belgium) collaborated with K. Arai, Y. Suzki, P. Descouvemont and D. Baye. And also, this work is supported by a Grant-in-Aid for Scientific Research (No.24540262). The author thank A.T. Kruppa and R.G.Lovas for fruitful discussions for the application of the CSM to the four nucleons system, which is supported by the short visiting program and Bilateral Joint Research Projects between the JSPS and the HAS(Hungary). A part of computational calculations was carried out in T2K-Tsukuba and SR16000(Yukawa Institute).

References

[1] Kamada H, Nogga A, Glöckle W, Hiyama E, Kamimura M *et al* 2001 *Phys. Rev.* C **64**, 044001
[2] Baye D, Heenen P.H and Libert-Heinemann M 1977 *Nucl.Phys.* A **291**, 230
[3] Wildermuth K and Tang Y.C 1977 *A unified theory of the nucleus* (Braunschweig: Vieweg)
[4] Aoyama S, Arai K, Suzuki Y, Descouvemont P and Baye D 2012 *Few-Body Syst.* **52** 97
[5] Arai K, Aoyama S, Suzuki Y, Descouvemont P and Baye D 2011 *Phys.Rev.Lett.* **107** 132502
[6] Aoyama S, Myo T, Katō K and Ikeda K 2006 *Prog.Theor.Phys.* **116** 1(a review of CSM)
[7] Aguilar J and Combes J M 1971 *Commun. Math. Phys.* **22**, 269
Balslev E and Combes J M 1971 *Commun. Math. Phys.* **22** 280
[8] Suzuki Y and Varga K 1998 *Stochastic variational approach to quantum-mechanical few-body problems* (*Lecture notes in physics* Vol. 54) (Berlin Heidelberg New York: Springer)
[9] Pudliner B S, Pandharipande V R, Carlson J, Pieper S C and Wiringa R B 1997 *Phys Rev* C **56** 1720
[10] Hiyama E, Gibson B.F and Kamimura M 2003 *Phys. Rev.* C **70** 031001
[11] Thompson D R, LeMere M and Tang Y C 1977 *Nucl Phys* A **286** 53
[12] Horiuchi W and Suzuki Y 2008 *Phys Rev* C **78** 034305
[13] Hofmann H M and Hale G M 2008 *Phys Rev* C **77** 044002

Journal of Physics: Conference Series **436** (2013) 012026 doi:10.1088/1742-6596/436/1/012026

Multi-cluster problems: resonances, scattering and condensed states

K Katō[1], T Myo[2], Y Kikuchi[3] and T Yoshida[4]

[1] Nuclear Reaction Data Centre, Faculty of Science, Hokkaido University, Sapporo 060-0810, Japan
[2] General Education, Faculty of Engineering, Osaka Institute of Technology, Osaka, Osaka 535-8585, Japan
[3] Research Center for Nuclear Physics (RCNP), Osaka University, Ibaraki, Osaka 567-0047, Japan
[4] Division of Physics, Graduate School of Science, University of Tokyo, 7-3-1 Hongo, 133-0033 Tokyo, Japan

E-mail: kato@nucl.sci.hokudai.ac.jp

Abstract. This talk is mainly concerned with many-body resonances in nuclear physics. We extensively discuss the structure and reactions of multi-cluster systems using the complex scaling method (CSM). We expound three interesting problems in recent studies of multi-cluster systems. First, we discuss four- and five-body resonances in A=7 (^{7}He, ^{7}B) and A=8 (^{8}He, ^{8}C) systems, respectively. The observed states are well explained and many additional states are predicted. Second, the Coulomb breakup reactions of two-neutron halo nuclei and the α-d scattering are investigated using a three-body model with CSM. Finally, we discuss the α-condensate-like states and their symplectic excitation properties in three- and four-α models for ^{12}C and ^{16}O, respectively.

1. Introduction

Cluster states in stable nuclei are usually located near the threshold of the relevant cluster breakup. Thus the physics of clustering has always necessitated the proper treatment of resonances and sometimes even the description of relevant reaction processes. This necessity is more enhanced for cluster states of neutron-rich nuclei, because even when the states of neutron-rich nuclei are located deeply below the cluster breakup thresholds, they are often near the neutron dissociation threshold. For the investigation of resonances over a wide region of the complex energy plane, the complex scaling method (CSM) [1, 2] has been used as a very powerful method in the description of many-body resonances.

A great advantage of CSM is that we can obtain resonance poles in the complex energy plane in the same way as a standard bound state owing to the L^2 property of the resonance wave functions. In this way many-body systems can be treated as straightforward extensions of two-body calculations. We show the applications of CSM to the ^{4}He+XN model [3, 4, 5, 6, 7], where N and X indicate a nucleon (neutron or proton) and their number ($X = 1,\ 2,\ 3,\ 4$).

CSM transforms the scattering states on the positive energy axis into two separate groups: resonant states and residual continuum states on the rotated cuts. This is also an important advantage of CSM. We explain the extended completeness relation (ECR) [8], consisting of bound, resonant and rotated continuum states. For many-body systems, the rotated continuum states are further separated into sets of states lying on various rotated continua starting from different eigen-energies of subsystems. We apply CSM to the Green's function, and extensively use the complex scaled Green's function in calculations of scattering and reaction cross sections. In this paper, the Coulomb breakup cross section of two-neutron halo systems [9] and $\alpha + d$ scattering phase shifts [11] are discussed.

Furthermore, studies of the α-condensate-like states of ^{12}C and ^{16}O as many-body resonances are also presented. For these many-α systems, the description of various excited states, by symplectic basis truncations, is shown.

2. Many-body Resonant States

2.1. Complex Scaling Method (CSM)

The wave function of a resonant state with a complex energy diverges exponentially in an asymptotic distance region as shown in Fig. 1. Then the norm and matrix elements cannot be defined in the usual manner. The difficulties of resonant states due to singularity at infinity have been overcome by regularization procedures, such as CSM [1, 2]. The complex scaling consists in the following transformation of the Jacobi coordinates and their conjugate momenta:

$$U(\theta): \quad \boldsymbol{r}_i \to \boldsymbol{r}_i e^{i\theta} \quad \text{and} \quad \boldsymbol{p}_i \to \boldsymbol{p}_i e^{-i\theta}, \qquad (i = 1, \cdots, f), \tag{1}$$

where θ is a scaling angle common ($0 \le \theta < \theta_{max}$) for all coordinates of the f degrees of freedom. The maximum value $\theta_{\max}$ is determined so as to keep the analyticity of the transformed function; for instance $\theta_{\max} = \pi/4$ for a Gaussian potential.

CSM gives us a very simple way to construct resonant states by using the L^2 basis functions. Therefore, we can easily apply CSM to many-body systems. Diagonalizing the complex-scaled Hamiltonian $H(\theta) = U(\theta)HU(\theta)^{-1}$, we obtain resonant solutions with complex energies $E_R - i\Gamma/2$, which are stationary against variations of the θ value. An advantage of CSM is that matrix elements of a physical operator $\hat{O}$ can be calculated as

$$\langle O \rangle = \langle \tilde{\psi}_R^\theta | U(\theta)\hat{O}U(\theta)^{-1} | \psi_R^\theta \rangle, \tag{2}$$

where $\tilde{\psi}_R^\theta$ and ψ_R^θ are conjugate pairs in the bi-orthogonal basis. The matrix elements are generally complex, although the imaginary parts are small when the width of the resonant state is small. Then we can see a correspondence between the real part of the matrix element and the observed value. However, when the width is larger than the resonance energy, the matrix element does not correspond directly to an observable. We need to develop a new method to relate the complex energy resonant states to the physical states. In section 4, we show that

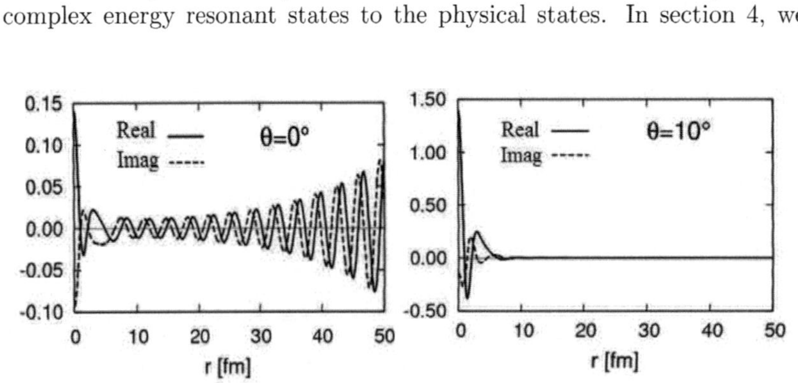

Figure 1. Wave functions of a resonant state (a) before and (b) after complex scaling. Solid and broken lines are the real and imaginary parts of the wave function, respectively.

the observed cross section can be calculated by complex resonance solutions and non-resonant continuum states in CSM.

2.2. Many-body decaying resonant states

As examples of many-body decaying resonances, we present our recent studies [3, 4, 5, 6, 7] on the neutron-rich He isotopes (A=5–8) and their proton-rich mirror nuclei. Those systems are described in the models of the α-core plus nucleons (^{4}He+XN) as shown in Fig. 2. The Hamiltonian is given by

$$H = \sum_{i=1}^{X+1} t_i - T_{cm} + \sum_{i=1}^{X} V_i^{\alpha N} + \sum_{i<j}^{X} V_{ij}^{NN}$$
$$= \sum_{i=1}^{X} \left[\frac{\boldsymbol{p}_i^2}{2\mu} + V_i^{\alpha N} \right] + \sum_{i<j}^{X} \left[\frac{\boldsymbol{p}_i \cdot \boldsymbol{p}_j}{4M} + V_{ij}^{NN} \right]. \quad (3)$$

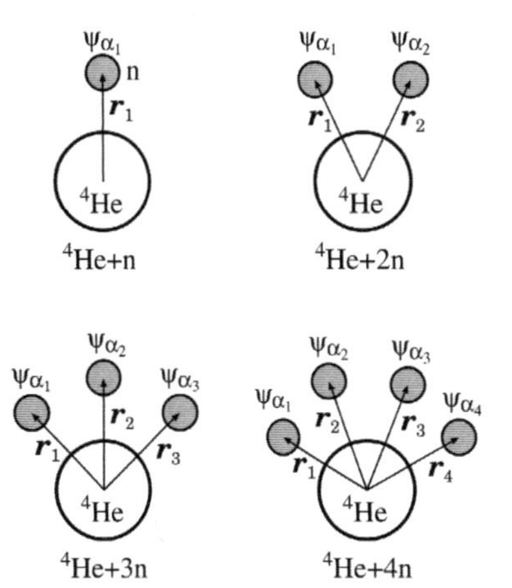

Figure 2. The ^{4}He+Xn models for He isotopes.

The second line contains the form used in the cluster orbital shell model (COSM) [12]. For the details of the interactions $V_i^{\alpha N}$ and V_{ij}^{NN}, see references [3, 4, 5, 6, 7]. The mirror nuclei are described by replacing valence neutrons with protons.

The results are presented in Fig. 3. The energy is measured from the threshold of separation into ^{4}He and nucleons. For A=5 (^{5}He, ^{5}Li), A=6 (^{6}He, ^{6}Be), A=7 (^{7}He, ^{7}B) and A=8 (^{8}He, ^{8}C) systems, two-, three-, four- and five-body decay resonances are described, respectively, under the correct boundary conditions. The observed bound and resonant states are well reproduced, and many theoretical predictions are made. Comparing spectra of the He isotopes and their mirror nuclei, we see good mirror symmetry. In ^{7}He, we found a broad $1/2^-$ resonance at the low excitation energy of E_x = 1.05 MeV, while the experimental uncertainty is still large. Recently, there is a report [13] discussing the order of energy levels, which fully agrees with our result. The energy of the ^{7}B ground state is obtained to be E_r = 3.35 MeV, which agrees with the recently observed value, 3.38 MeV [14]. For A=8 nuclei, only 0^+ states have been calculated

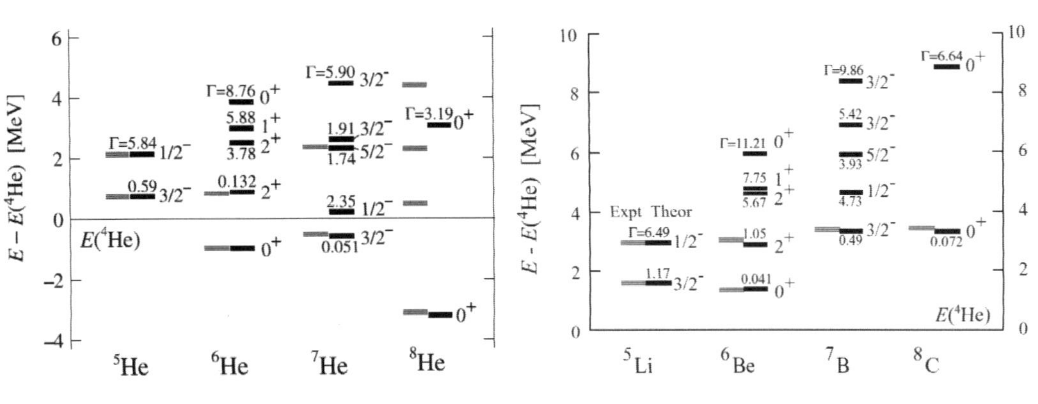

Figure 3. Calculated (black bar) and observed (gray bar) spectra of He isotopes and their mirror nuclei.

because a huge number of basis states are required by states with higher spins. The 0_2^+ state of ^{8}He is predicted to be a five-body resonance dominated by the $(p_{3/2})^2(p_{1/2})^2$ configuration, with an excitation energy of 6.3 MeV and a width of 3.2 MeV.

3. Extended completeness relation and complex scaled Green's function

The result of the complex scaling [Eq. (1)] in the momentum and energy planes is illustrated in Fig. 4. A semicircle in the upper half momentum plane is inclined clockwise by an angle θ, and then some resonance poles in the fourth quadrant are enclosed within the inclined semicircle. Integrating the Green's function along the inclined semicircle [15], we obtain an extended completeness relation (ECR) [8]

$$\mathbf{1} \;=\; \sum_{B}^{n_B} |\psi_B^\theta\rangle\langle\tilde{\psi}_B^\theta| + \sum_{R}^{n_R(\theta)} |\psi_R^\theta\rangle\langle\tilde{\psi}_R^\theta| + \int_{L_\theta} dk |\psi_{k^\theta}^\theta\rangle\langle\tilde{\psi}_{\tilde{k}^\theta}^\theta|, \tag{4}$$

where ψ_B^θ, ψ_R^θ and $\psi_{k^\theta}^\theta$ are bound, resonant and continuum solutions of the complex-scaled Schröndinger equation, respectively. The momentum $\tilde{k}$ in the bra-states is $\tilde{k}_{B,R} = -k_{B,R}^*$ for bound (B) and resonant (R) states, and $\tilde{k}^\theta = k^{\theta *}$ for continuum states. The continuum states $\psi_{k^\theta}^\theta$ are solutions of the complex-scaled Schröndinger equation for momenta k^θ along the branch cut L_θ rotated from the original position of the real momentum axis, and describe the broad resonance contributions from poles not included within the semicircle.

The above discussion is for a single channel case, but we can postulate a similar ECR for many-body systems as well:

$$\mathbf{1} = \sum_{B}^{n_B} |\psi_B^\theta\rangle\langle\tilde{\psi}_B^\theta| + \sum_{R}^{n_R(\theta)} |\psi_R^\theta\rangle\langle\tilde{\psi}_R^\theta| + \sum_{cb} \int_{L_\theta(cb)} dk_{cb} |\psi_{k_{cb}}^\theta\rangle\langle\tilde{\psi}_{k_{cb}}^\theta| + \sum_{cr} \int_{L_\theta(cr)} dk_{cr} |\psi_{k_{cr}}^\theta\rangle\langle\tilde{\psi}_{k_{cr}}^\theta|, \tag{5}$$

where the suffices *cb* and *cr* represent different channels consisting of bound subsystems and resonant subsystems, respectively.

We take a Green's function satisfying the outgoing boundary condition:

$$\mathcal{G}^{(+)}(E;\, \vec{r},\, \vec{r}\,') = \left\langle \vec{r} \left| \frac{1}{E - H + i\epsilon} \right| \vec{r}\,' \right\rangle. \tag{6}$$

Its complex-scaled version is given by

$$\mathcal{G}^{\theta}(E;\, \vec{r},\, \vec{r}\,') = \left\langle \vec{r} \left| \frac{1}{E - H(\theta)} \right| \vec{r}\,' \right\rangle. \tag{7}$$

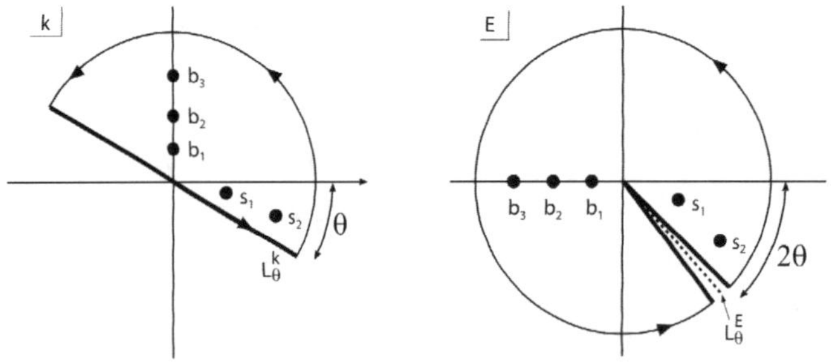

Figure 4. The rotated Cauchy integral contours in the momentum and energy planes in CSM.

Here we used a superscript 'θ' instead of '(+)' and dropped $i\epsilon$ from the operator on the right-hand side, because $H(\theta)$ automatically satisfies the outgoing boundary condition due to CSM without any additional prescription.

4. Nuclear reactions to many-body final states

Here we discuss nuclear reactions leading to many-body final states, which satisfy outgoing boundary conditions. Such a many-body final state can be easily described by the complex scaling method. As examples we show two successful results for the Coulomb breakup reaction of a two-neutron halo nucleus and the α+deuteron scattering with the deuteron breakup and rearrangement processes taken into account.

4.1. Coulomb breakup reactions of two-neutron halo nuclei

The Coulomb breakup experiment is a unique method to investigate the exotic properties of neutron halo nuclei, which are weakly bound systems. Many studies have been done on the two-neutron halo nuclei ^{6}He and ^{11}Li, but a deep understanding of the relation between the exotic structure and characteristic reaction mechanism has not been obtained yet. Since the final states of the two-neutron halo breakup are three-body continuum states, it is very appropriate to apply CSM to the Coulomb breakup reactions.

The breakup cross section of the Coulomb response is assumed to be determined by $E1$ transitions and is described as

$$\frac{d\sigma}{dE} = \frac{16\pi^3}{9\hbar c} N_{E1}(E_\gamma) \frac{dB(E1)}{dE}, \tag{8}$$

where $N_{E1}(E_\gamma)$ is the virtual photon number and

$$\begin{aligned} \frac{dB(E1)}{dE} &= \frac{1}{2J_{gr}+1} \sum_\nu \langle\tilde{\psi}_{gr}|\hat{E}1|\psi_\nu\rangle\langle\tilde{\psi}_\nu|\hat{E}1|\psi_i\rangle\delta(E-E_\nu) \\ &= -\frac{1}{\pi}\mathrm{Im}\left[\frac{1}{2J_{gr}+1}\sum_\nu \frac{\langle\tilde{\psi}^\theta_{gr}|\hat{E}1(\theta)|\psi^\theta_\nu\rangle\langle\tilde{\psi}^\theta_\nu|\hat{E}1(\theta)|\psi^\theta_i\rangle}{E-E^\theta_\nu}\right]. \end{aligned} \tag{9}$$

The last equation is calculated with CSM, where ψ^θ_ν are solutions belonging to the eigenenergy E^θ_ν and satisfy the outgoing boundary condition. It should be noted that the result is independent of the parameter θ.

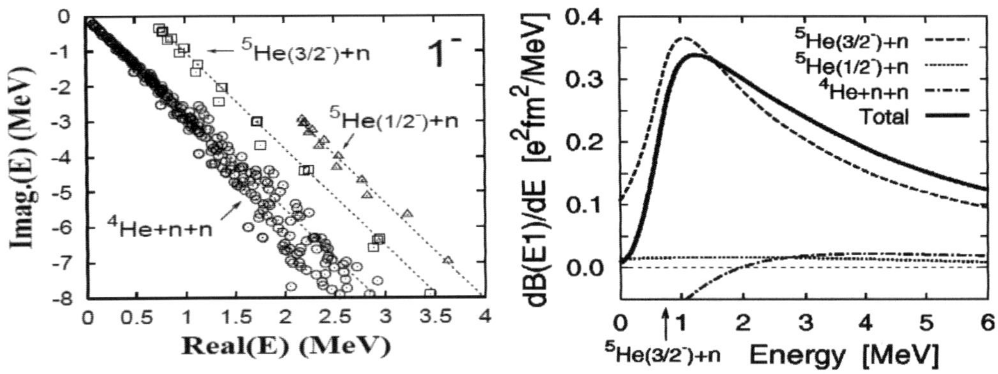

Figure 5. The 1^- energy eigenvalue distribution (left-hand panel) in the complex-scaled ^{4}He+n + n model and the Coulomb breakup reaction cross section (right-hand panel) $E1$ transition strength distributions. For details, see text.

In Fig. 5, we show the result of the calculated Coulomb breakup cross section for the two-neutron halo nucleus ^{6}He using a ^{4}He + n + n model [9]. In the left-hand panel, the eigenenergy distribution of 1^- states obtained with $\theta = 30°$ is shown, where squares and triangles indicate the two-body continuum states of ^{5}He($3/2^-$) + n and ^{5}He($1/2^-$) + n, respectively. Circles indicate the three-body continuum states of ^{4}He + n + n. In the right-hand panel, we show $E1$ transition strength distributions, where dashed, dotted and dash-dotted lines are contributions from the ^{5}He($3/2^-$)+n, ^{5}He($1/2^-$) + n and ^{4}He + n + n continuum states, respectively. The thick solid line indicates the total strength distribution. The position of the arrow stands for the two-body threshold energy of the ^{5}He($3/2^-$) + n channel. We can see that the contribution from the continuum states of ^{5}He($3/2^-$) + n is dominant in the cross section, and the total strength well reproduces the observed Coulomb dissociation cross sections [10] of ^{6}He as shown in Fig. 6.

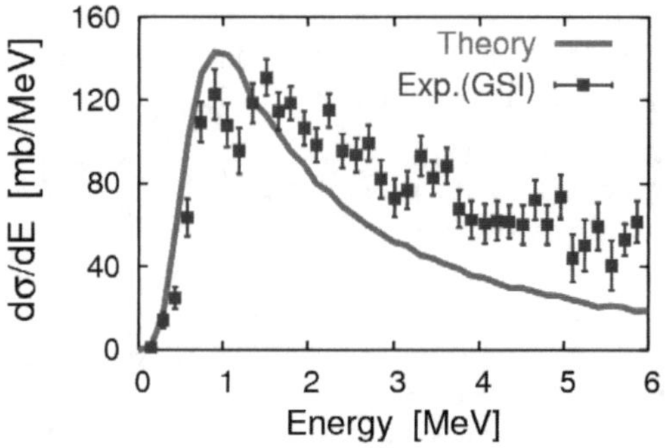

Figure 6. Coulomb dissociation cross sections of ^{6}He into the ^{4}He + n + n system. Experimental data are taken from Ref. [10]

Similar calculations for ^{11}Li are presented by Kikuchi at this conference.

4.2. Beyond the continuum discretized coupled channels

The method of continuum discretized coupled channels (CDCC) has been employed in studies of nuclear reactions in which the projectile or the target nucleus is broken up [18]. Although the CDCC method is successful in two-body breakup problems, it is difficult to apply it to three-body breakup reactions of two-neutron halo nuclei because three-body scattering states are needed explicitly in order to construct the smoothing factors for the discretized cross sections. This problem has been solved recently by applying CSM to the calculation of the smoothing factors [19].

Another interesting problem in the CDCC method is its extension to various reactions including a rearrangement process. This problem has also been discussed by using CSM recently, and applied to the $\alpha + d$ scattering [11]. Using ECR and the complex-scaled Green's function, the $\alpha + d$ scattering state is expressed as the solution of a Lippmann-Schwinger equation

$$\langle \Psi^{(-)} | \quad = \quad \langle \Phi_0 | + \sum\!\!\!\!\!\!\!\int_\nu \langle \Phi_0 | \hat{V} U^{-1}(\theta) | \psi_\nu^\theta \rangle \frac{1}{E - E_\nu^\theta} \langle \tilde{\psi}_\nu^\theta | U(\theta), \tag{10}$$

where Φ_0 is a solution of the asymptotic Schrödinger equation $H_0 \Phi_0 = E \Phi_0$. The total Hamiltonian is given by $H = H_0 + \hat{V}$, where $\hat{V}$ is an interaction. When we employ an $\alpha + p + n$ model with the asymptotic Hamiltonian belonging to the $\alpha + d$ channel, the second term of Eq. (10) describes the deuteron breakup with intermediate rearrangements of p or n.

In Fig. 7, we show results of phase shifts for the $\alpha + d$ elastic scattering with $J^\pi = 3^+$, 2^+ and 1^+. The solid lines are the calculations including all terms in Eq. (10). They reproduce the observed phase shifts well. The dashed lines show the results with only the deuteron breakup included. The interaction is obviously not attractive enough, but the overall trends are very similar to the solid lines. Furthermore, we claculate the phase shifts including the elastic channel alone described by the first term in Eq. (10). They are presented by the dotted lines, which belonging to different spin values (seen at the bottom) almost coincide, and show no attractive behaviour.

From the results in Fig. 7, it is found that the deuteron breakup has a significant effect on the production of the resonances of ^{6}Li in the $\alpha + d$ scattering, and the coupling to the rearrangement channels of ^{5}He+p and ^{5}Li+n has a sizable effect on the determination of the resonance positions.

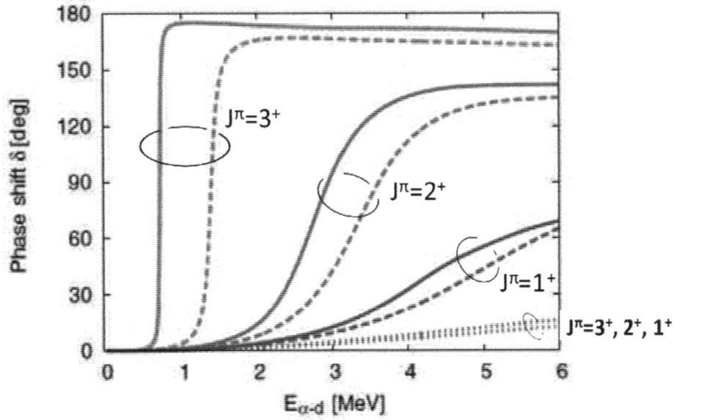

Figure 7. Effects of deuteron breakup and rearrangement on the phase-shift results.

5. Alpha-condensate-like states in ^{12}C and ^{16}O

The α-condensate-like states in ^{12}C and ^{16}O are typical examples of multi-cluster resonances in highly excited energy regions. Kurokawa [20, 21, 22] calculated many three-α resonances in the 3α orthogonality condition model (OCM). The three-α system is a Borromean system, which has no bound states in any of its subsystems. In ^{12}C, except for the ground 0^+ and an excited 2^+ states, all excited states are resonant states above the three-α threshold. The structures of the observed levels of ^{12}C have been thoroughly investigated and many resonant states are predicted as shown in Fig. 8[22]. The most interesting prediction is the 0_3^+ state [21, 22] above the Hoyle (0_2^+) state, which has recently been observed by Itoh *et al* [24, 25]. It has turned out that this broad resonant state is like an excited Hoyle state.

The structure of the Hoyle state has been discussed in connection with α-condensate-like states [27, 17]. This 0_2^+ state has been shown to have a large monopole strength of transition from the ground state. This property is understood from the viewpoint of the cluster symplectic excitation. Yoshida has recently investigated the symplectic structure and monopole strength in ^{12}C [28]. His recent multi-alpha cluster model calculations for ^{12}C and ^{16}O are presented in his talk at this conference.

6. Summary

In this talk, we discussed three problems of multi-cluster systems investigated recently. Since cluster states are usually located near the relevant breakup threshold, proper treatment of resonances and the description of the relevant reaction processes are required. To facilitate the discussion, we briefly explained the powerful method of complex scaling. As for the first problem, we discussed many-body resonances in He isotopes and their mirror nuclei using the ^{4}He+XN model. The completeness relation for the bound and scattering states were extended so as to separate the scattering states into resonant states and rotated continuum states by applying the complex scaling. The second problem to be dealt with was the three-body Coulomb breakup cross section and the extension of the continuum discretized coupled channels method. This was treated by using the extended completeness relation and the complex-scaled Green's

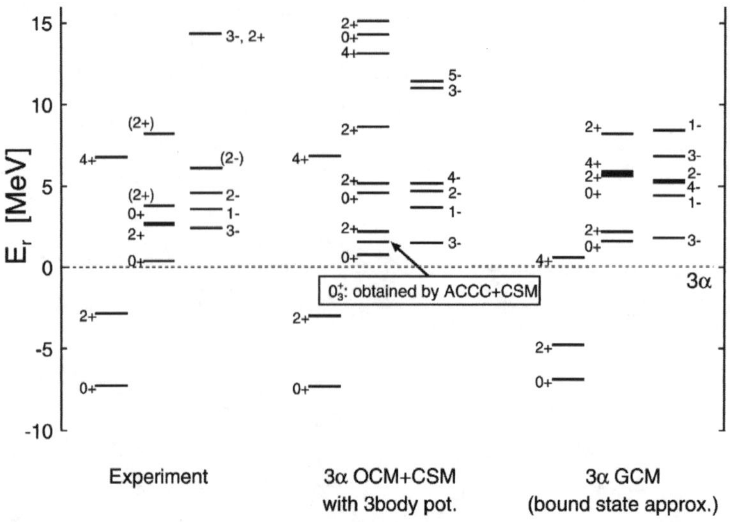

Figure 8. Energy levels for the low-lying states of ^{12}C in comparison with the experiment and other 3α theoretical calculations of the 3α GCM [23]. Experimental values of the 2_2^+ and 0_3^+ states are given in [24, 25], and other experimental data are taken from [26].

function. The third group of problems discussed was that of the α-condensate-like states and the symplectic excitations in the three- and four-α models for ^{12}C and ^{16}O, respectively.

References

[1] Aguilar J and Combes J M 1971 *Commun. Math. Phys.* **22** 269
[2] Balslev E and Combes J M 1971 *Commun. Math. Phys.* **22** 280
[3] Myo T, Katō K and Ikeda K 2007 *Phys. Rev.* C **76** 054309
[4] Myo T, Ando R and Katō K 2009 *Phys. Rev.* C **85** 034338
[5] Myo T, Ando R and Katō K 2010 *Phys. Lett.* B **691** 150
[6] Myo T, Kikuchi Y and Katō K 2011 *Phys. Rev.* C **84** 064306
[7] Myo T, Kikuchi Y and Katō K 2012 *Phys. Rev.* C **85** 034338
[8] Myo T, Ohnishi A and Katō K 1998 *Prog. Theor. Phys.* **99** 801
[9] Myo T, Katō K, Aoyama S and Ikeda K 2001 *Phys. Rev.* C **63** 054313
[10] Aumann T *et al* 1999 *PHys. Rev.* C **59** 1252
[11] Kikuchi Y *et al* 2011 *Phys. Rev.* C **84** 064610
[12] Suzuki Y and Ikeda K 1988 *Phys. Rev.* C **38** 310
[13] Wuosmaa A H *et al* 2008 *Phys. Rev.* C **78** 041302
[14] Charity R J *et al* 2011 *Phys. Rev.* C **84** 014320
[15] Newton R G 1960 *J. Math. Phys.* **1** 1
[16] Aoyama S, Myo T, Katō K and Ikeda K 2006 *Prog. Theor. Phys.* **116** 1
[17] Horiuchi H, Ikeda K and Katō K 2012 *Prog. Theor. Phys. Suppl.* **192** 1
[18] Kamimura M *et al* 1986 *Prog. Theor. Phys. Suppl.* **89** 1
[19] Matsumoto T, Katō K and Yahiro M 2010 *Phys. Rev.* C **82** 051602(R)
[20] Kurokawa C and Katō K 2004 *Nucl. Phys.* A **738** 455458
[21] Kurokawa C and Katō K 2005 *Phys. Rev.* C **71** 021301
[22] Kurokawa C and Katō K 2007 *Nucl. Phys.* A **792** 87
[23] Uegaki E, Okabe S, Abe Y and Tanaka H 1979 *Prog. Theor. Phys.* **62** 1621
[24] Itoh M *et al.* 2004 *Nucl. Phys.* A **738** 268
[25] Itoh M *et al.* 2011 *Phys. Rev.* C **84** 054308
[26] Ajzenberg-Selove F 1988 *Nucl. Phys.* A **490** 1626
[27] Tohsaki A, Horiuchi H, Schuck P and Röpke G 2001 *Phys. Rev. Lett.* **87** 192501
[28] Yoshida T, Itagaki N and Katō K 2011 *Phys. Rev.* C **83** 024301

Journal of Physics: Conference Series **436** (2013) 012027 doi:10.1088/1742-6596/436/1/012027

Studies of light neutron-excess systems from bound states to continuum

Makoto Ito[1,2] **and Hideaki Otsu**[2]

[1] Department of Pure and Applied physics, Kansai University, 3-3-35 Yamate-cho, Suita, Osaka 564-8680, Japan

[2] RIKEN Nishina Center for Accelerator-base Science, RIKEN, Wako, 351-0198, Japan

E-mail: itomk@kansai-u.ac.jp

Abstract. The generalized two-centre cluster model (GTCM), which can handle various single-particle configurations in general two-centre systems, is applied to the light neutron-rich system, ^{12}Be= α+α+4N. We discuss the change of the neutrons' configuration around two α cores as the excitation energy varies. We show that the excess neutrons form various chemical-bonding-like configurations around the two α cores in the unbound region above the α-decay threshold. The possibility of α-cluster formation in a heavier system with neutron excess, ^{28}Ne, is also discussed.

1. Introduction

In the last two decades, developments of experiments with secondary radioactive-ion beams have extensively advanced the studies of light neutron-rich nuclei. In particular, much effort has been devoted to the investigation of molecular structures in Be isotopes. The Be isotopes can be considered as typical examples of two-centre superdeformed systems, which are built on the α+α rotor of ^{8}Be. Theoretically, molecular orbitals (MO), in particular, the π^- and σ^+ orbitals, associated with the covalent bonding in atomic molecules, have been successful in explaining the low-lying states of these isotopes [1].

The MO model can describe many kinds of characteristic properties of these isotopes, but it is essentially limited to the analysis of low-lying bound states, and the theory of the highly excited states above the particle-decay threshold is still an open area. In contrast to the situation of theoretical studies, recent experiments on ^{12}Be revealed the existence of many resonant states [2, 3, 4], which strongly decay into ^{6}He$_{g.s.}$+^{6}He$_{g.s.}$ and α+^{8}He$_{g.s.}$. Similar resonances, decaying to He isotopes, have also been observed in other Be isotopes, such as ^{10}Be=α+^{6}He and ^{14}Be=^{6}He+^{8}He.

In the present study, we investigate the structural changes appearing in an unbound region of ^{12}Be. In order to investigate the continuum states above the particle-decay threshold, the intrinsic structures and their coupling to the scattering states should be treated in a unified manner. For this purpose, we introduce the generalized two-centre cluster model (GTCM) [6]. In this model, the covalent MO configurations can be smoothly connected to the atomic or ionic configurations, in which valence neutrons are localized around one of the α clusters. Furthermore, it becomes possible to describe both the formation of the covalent MO structures and reaction processes. In this report, we investigate the chemical-bonding-like structures based on the two α cores in ^{12}Be, which appear in the continuum energy region. As a natural extension of these considerations, we also discuss the formation of an α cluster in a heavier system that has neutron excess, ^{28}Ne. In the study of ^{28}Ne, we assume a simple α+^{24}O cluster model and estimate the branching ratio of the α and neutron emission. The possibility of the observation of the α+^{24}O structure is discussed.

2. Theoretical framework

The detailed explanation of GTCM has already been published in Ref. [6], and here we only show its formulation briefly. In GTCM, the total wave function of ^{12}Be is given by a superposition of the basis states $\{\Phi_{\mathbf{m}}^{J^\pi K}(S)\}$, where

$$\Phi_{\mathbf{m}}^{J^\pi K}(S) \;=\; \hat{P}_K^{J^\pi}\mathcal{A}\left\{\psi_L(\alpha)\psi_R(\alpha)\prod_{j=1}^{4}\varphi_j(m_j)\right\}_S . \tag{1}$$

The α-cluster wave function $\psi_n(\alpha)$ (n=L, R) is chosen to be the $(0s)^4$ configuration of the harmonic oscillator (HO) centred on the left (L) or right (R) side, with the relative distance parameter S [5]. The single-particle wave functions for the four valence neutrons localized around one of the α clusters is given by a $0p$ atomic orbital (AO) $\varphi(p_k, i, \tau)$, where the subscript k of p_k may be x, y or z, the symbol i denotes L or R, and τ ($=\uparrow$ or $\downarrow$) stands for the spin state. The four symbols $\mathbf{m}$=(m_1, m_2, m_3, m_4) comprise the labels (p_k, i, τ) of the four AO's. The intrinsic basis functions with full antisymmetrization expressed by $\mathcal{A}$ are projected to the eigenstate of the total spin J, its intrinsic angular-momentum projection K and the total parity π by the projection operator $\hat{P}_K^{J^\pi}$. In the present study, we discuss the level structure of the J^π=0^+ states. Thus, only the case of J=K=0 is considered. We include all possible AO basis elements which have axial symmetric (K=0) configurations.

The total wave function is finally given by taking a superposition over S and $\mathbf{m}$

$$\Psi^{0_\nu^+} \;=\; \int dS \sum_{\mathbf{m}} C_{\mathbf{m}}^{\nu}(S)\,\Phi_{\mathbf{m}}^{0^+}(S) \;\equiv\; \Psi^{\nu}. \tag{2}$$

The coefficients for the 0_ν^+ state (νth 0^+), $C_{\mathbf{m}}^{\nu}(S)$, are determined by solving an eigenvalue problem. As for the nucleon-nucleon (NN) interaction, we use the Volkov No. 2 and the G3RS force for the central and spin-orbit part, respectively. The parameters of the NN interactions and the size of the HO are optimized so as to reproduce the ^{x}He+^{y}He threshold energies as much as possible [6].

3. Results

3.1. Results for ^{12}Be

We identified two bound and four resonant 0^+ states below and above the α+^{8}He$_{g.s.}$ threshold, respectively. The ground state has a well developed MO configuration of $(\pi_{3/2}^-)^2(\sigma_{1/2}^+)^2$. Its characteristic degrees of freedom are the intercluster relative motion and the single-particle motions of the four valence neutrons. By analysing the properties of the wave functions of the excited states, we find that all the excited states can be characterized in terms of the excitations in the characteristic modes of motion of the ground state. We summarize the identified levels and their intrinsic properties in Fig. 1.

(1) Molecular orbital (MO) excitation mode. The 0_2^+ state has the MO configuration, $(\pi_{3/2}^-)^2(\pi_{1/2}^-)^2$, while the 0_5^+ state has a hybrid structure of MO and AO configurations. Namely, two of the valence neutrons are localized around individual α's as ^{5}He+^{5}He, which is an AO

Journal of Physics: Conference Series **436** (2013) 012027 doi:10.1088/1742-6596/436/1/012027

structure, and the remaining two neutrons occupy the σ^+ orbital and rotate around both clusters, which is of MO character. This state can be generated by the excitation of two $\pi^-_{3/2}$ MO's into AO's around the two α cores.

(2) Cluster excitation mode. The 0_3^+ and 0_6^+ states have large components of the ionic structures of α+^{8}He$_{g.s.}$ and ^{5}He$_{g.s.}$+^{7}He$_{g.s.}$ type, respectively. These two states are the excitation modes of the α–α relative motion from the 0_1^+ and 0_2^+ states, respectively.

(3) Double excitation mode. The 0_4^+ state has the atomic configuration, ^{6}He$_{g.s.}$+^{6}He$_{g.s.}$. It involves a simultaneous excitation of the relative motion of the two α cores and the MO motions of the excess neutrons.

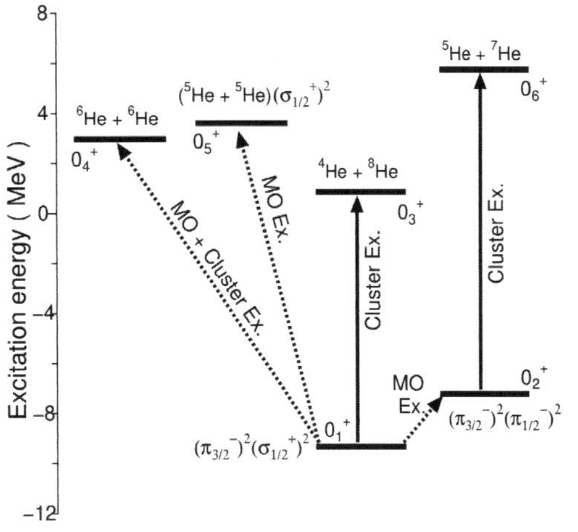

Figure 1. Energy spectra classified by the excitation mode (J^π=0^+). See the text for details. The origin is the threshold energy of the α emission.

The results in Fig. 1 show that the cluster structures can change from level to level, and various cluster configurations coexist within a small energy interval of about 1 $\sim$ 2 MeV in ^{12}Be. All the excited states contain a large fraction of the (isoscalar) monopole excitation of the ground (ν=1) state [7] expressed by the overlap

$$M(IS) = \left\langle \Psi^\nu \left| \sum_{i=1}^{12} r_i^2 \right| \Psi^1 \right\rangle \equiv \left\langle 0_\nu^+ \left| \sum_{i=1}^{12} r_i^2 \right| 0_1^+ \right\rangle . \tag{3}$$

Here, r_i denotes the radial coordinate for the ith nucleon measured from the centre of mass of ^{12}Be. To assess the magnitude of the monopole strength, it should be compared to the single-particle (S.P.) strength, which is defined as the matrix element of a one-node excitation in a simple HO shell model [8]. All the strengths have been found to be comparable with or 2 $\sim$ 3.5 times larger than the S.P. strength. Among the various 0^+ states shown in Fig. 1, strong enhancement occurs for the transition to the 0_3^+ state (α+^{8}He$_{g.s.}$), which is a direct cluster excitation mode from the ground state. Thus the monopole strength can be used as a probe of a direct cluster excitation from the ground state.

The finding of the monopole strength to be comparable with the S.P. strength is consistent with the analysis in Ref. [8], where the enhancement of the low-lying monopole strength by cluster formation is discussed. It should be noted that in the present system all the monopole strength corresponding to various cluster structures appears at $E_x \leq 20$ MeV. That is in marked contrast with a naive mean-field picture, in which a monopole excitation involves a $2\hbar\omega$ ($\sim$35 MeV) jump. This observation may be generalized. In a system with a considerable neutron excess, there are almost degenerate monopole states at much lower energies than expected in a naive shell-model picture, and the transition strengths to these states are comparable with the S.P. strength.

3.2. Discussion of ^{28}Ne

We also discuss α-cluster formation in the heavier system ^{28}Ne. This nucleus has almost the same N/Z ratio as ^{12}Be and hence the formation of the ionic α+^{24}O state is analogous to the α+^{8}He structure appearing in ^{12}Be. According to the prediction of the threshold rule [9], α-cluster formation is expected at excitation energies comparable with the dissociation energy into α+^{24}O ($E_x \sim 10$ MeV). In such a heavier system, however, there are two effects which will disturb α-cluster formation. First, the level density of ^{28}Ne is larger by about $2 \sim 4$ orders of the magnitude than that of ^{12}Be. Second, the one-neutron emission channel opens at a lower excitation energy than the α threshold ($E_x \sim 4$ MeV). These two factors strongly suggest that an α-cluster state couples to many compound nucleus states, which finally emit a neutron. Therefore, we should consider the competition of α decay and decays into the compound states from the α-cluster states.

In the present study, we discuss the decay widths for α and neutron emission and their branching ratios. The escape width of the α decay from the α cluster state, $\Gamma^{\uparrow}_{\alpha}$, can be estimated from the α+^{24}O cluster wave function. Here we applied the Orthogonalized Condition Model (OCM) to α+^{24}O. In this calculation, a simple configuration of $(\lambda, \mu) = (0, 2)$ in the SU(3) representation was assumed for the intrinsic structure of ^{24}O, and we took into account the excitation of ^{24}O to the 2^+ state in a coupled-channels framework. We employed a global α potential for the interaction between α and ^{24}O, and its derivative is used for the form factor of the coupling potential. We found that the α-cluster state appears around the corresponding threshold. From the amplitude of the wave function at the surface and from the penetration factor, we calculated $\Gamma^{\uparrow}_{\alpha}$.

As for the width of the decay of the cluster state (CL) of excitation energy E_x into the compound state (CN), $\Gamma^{\downarrow}_{CN}$, and for the neutron-emission width from CN, $\Gamma^{\downarrow\uparrow}_{n}$, we employ a simple perturbation formula

$$\begin{aligned} \Gamma^{\downarrow}_{CN} &= 2\pi < CN|V|CL >^2_{av} \rho_{28}(E_x) \\ \Gamma^{\downarrow\uparrow}_{n} &= \Gamma^{\downarrow}_{CN} P_n, \qquad P_n = \int\!\!\int dE_{27}\, d\epsilon_n \rho_{27}(E_{27}) \rho_n(\epsilon_n)/\rho_{28}(E_x)\,, \end{aligned} \tag{4}$$

where $\rho_A(E)$ is the level density of nucleus A at excitation energy E. The factor P_n represents the one-neutron emission probability, which is estimated from the level densities of ^{27,28}Ne ($\rho_{27,28}$) and of the free neutron (ρ_n). In Eq. (4), the condition of energy conservation $E_x = E_{27} + \epsilon_n + E^{th}_{27+n}$ is observed. [E_{27}, ϵ_n and E^{th}_{27+n} denote the internal energy of ^{27}Ne, the emitted neutron's kinetic energy and the ^{27}Ne+n threshold, respectively.] In this estimate, we use the density given in Ref. [10] for ^{27,28}Ne, while a simple level density calculated from a plane wave is applied to the neutron. $< CN|V|CL >_{av}$ represents the average coupling matrix element of the compound state (CN) and the cluster state (CL). Since there is no theoretical analysis for this average matrix element, we assess its order of the magnitude from the experimental spectra of the α inelastic scattering on ^{12}Be [4] and ^{24}Mg [11]. The estimate for the matrix element is in the range of $10^{-4} \sim 10^{-3}$ MeV^2.

Table 1. The decay width and branching ratio of the α and neutron emission in the α+^{24}O system. The energy and Γ_{tot} are written in units of MeV. See text for details.

E_x	J^π	Γ_{tot}	$P(\alpha)$	$P(n)$
13.4	0^+	1.45	0.69	0.10
15.9	2^+	0.86	0.35	0.22

In Table 1, the estimated branching ratios are shown. The total decay width is $\Gamma_{tot} = \Gamma_{\alpha}^{\uparrow} + \Gamma_{CN}^{\downarrow}$, and the α and neutron branching ratios are defined as $P(\alpha) = \Gamma_{\alpha}^{\uparrow}/\Gamma_{tot}$ and $P(n) = \Gamma_{n}^{\downarrow\uparrow}/\Gamma_{tot}$. The total decay width of the J^{π}=0^{+} state is about 1.5 MeV, and the dominant decay mode is the α decay, with a branching ratio of ~ 0.7. The ratio is suppressed in the J^{π}=2^{+} state, but is still substantial (~ 0.4), while the one-neutron emission probability is smaler ($P(n) \sim 0.2$).

These estimations suggest that α-decay is dominant even if the one-neutron emission channel opens at low excitation energy. Since in the heavier neutron-rich systems, the one-neutron emission channel always opens at lower energy than the α-decay threshold, the α and neutron decays generally compete. Therefore, in the study of α-cluster states, coincidence measurements for these two particles are essential. The new facility SAMURAI at RIKEN is appropriate for such measurements, and the identification of α-cluster states in ^{28}Ne will be a breakthrough in the study of α-cluster formation under the condition of strong neutron emission.

4. Summary and discussion

In summary, we have studied the exotic structure of the excited states of ^{12}Be as unbound systems by applying GTCM. Due to the excitation of the cluster and single-particle degrees of freedom, various structures appear at different excitation energies. We found that, above the particle decay threshold, there are several excited states within a small energy range. Each excited state can be characterized by a particular combination of 'atomic' and 'molecular' excitations of the excess neutrons. All the energy levels which appear close to one another have monopole strengths comparable with or larger than the single-particle strength. As a consequence, in light neutron-rich systems, strong monopole transitions can be observed in the low-energy region. This feature is completely consistent with recent observations [4].

We have also extended our study to the heavier system of ^{28}Ne and discussed the formation of the ionic α+^{24}O configuration. In such systems, there is a possibility that the formation of α-cluster resonances may be strongly disturbed by the large level density and the one-neutron emission opening at a much lower energy than the α emission. However, according to the present estimate of the branching ratios, an α+^{24}O state generated around the α threshold is expected to have a large α-decay width. This is because the estimated coupling matrix element between the α-cluster state and compound states is small, and the large enhancement of the level density is compensated for by this suppression of the coupling. In experimental studies of cluster structures in the heavier systems with neutron excess, coincidence measurements for α-particle and neutron emissions are strongly desirable. Such measurements are possible with the new facility SAMURAI at RIKEN.

References

[1] Itagaki N and Okabe S 2000 *Phys. Rev.* C **61** 044306
[2] Korsheninnikov A A *et al* 1995 *Phys. Lett.* B **343** 53
[3] Freer M *et al* 2001 *Phys. Rev.* C **63** 034301
[4] Saito A *et al* 2010 *Mod. Phys. Lett.* A **25** 1858
[5] Horiuchi H 1977 *Prog. Theor. Phys. Suppl.* **62** 90
[6] Ito M 2012 *Phys. Rev.* C **85** 044308 and references therein
[7] Ito M 2011 *Phys. Rev.* C **83** 044319
[8] Yamada T *et al* 2008 *Prog. Theor. Phys.* **120** 1139
[9] Ikeda K *et al* 1980 *Prog. Theor. Phys. Suppl.* **68** 1 and references therein
[10] Dilg W, Schantl W, Vonach H and Uhl M 1973 *Nucl. Phys.* A **217** 269
[11] Kawabata T 2012 private communication

Cluster-shell competition in systems with a few alpha particles and valence neutrons

H Masui[1] **and N Itagaki**[2]

[1] Information Processing Center, Kitami Institute of Technology, Kitami, Japan, 090-8507
[2] Yukawa Institute for Theoretical Physics, Kyoto University, Kitashirakawa Oiwake-Cho, Kyoto, Japan, 606-8502

E-mail: hgmasui@mail.kitami-it.ac.jp

Abstract. We construct nuclear wave functions from alpha clusters and some additional valence nucleons, and allow the inter-cluster distance to change and one alpha to dissolve from the $(0s)^4$ structure as a result of the nuclear interaction. The change of the inter-cluster distance and the dissolution of the alpha particle can be interpreted as resulting from the competition of the 'shell model' and 'cluster model'. We demonstrate this competition through a few parameters.

1. Introduction: cluster-shell competition

The structure of nuclei has a variety due to the particular forms in which nuclei with different proton and neutron numbers accommodate themselves to the nature of the nucleon-nucleon interaction. The structural variety consists in particular forms of the motion, localization, density etc. of the nucleons. Two of the typical structures are the mean-field and the cluster structure. The standard framework for describing the mean-field-like nature of nuclei is provided by the shell model. The shell model reproduces the nuclear magic numbers by treating valence nucleons as particles moving in the potential induced by a core, with the spin-orbit interaction between the core and valence nucleons playing an important role. The clustering aspect is embodied in the cluster model, in which the nucleons form groups called 'clusters' due to the strong correlation induced by the nucleon-nucleon interaction, and each cluster is spatially localized. These two aspects are realized not only in the two approaches of theoretical description but also in two pictures for the interpretation of nuclear phenomena.

We proposed a simplified parameterization for describing the structural change of nuclei from the shell model to the cluster model and *vice versa*. In this parameterization, the key parameters are the distance of the alpha particles inside the nucleus and a degree of the dissolution of the alpha particle(s). We call this parameterization the 'cluster-shell competition' [1, 2, 3, 4]. The dissolution of the alpha particle is caused by an explicit inclusion of the spin-orbit interaction. We also take into account the tensor correlation in the cluster model in a similar manner [2].

In this work, we briefly introduce the concept of the cluster-shell competition and show its applications to carbon isotopes. We investigate the role of valence neutrons on the structure of ^{12}C, which is assumed to be a 3α structure including one dissolved α particle [3]. Then we extend the approach to other carbon isotopes.

2. Formalism for the inclusion of the spin-orbit interaction

We briefly introduce the framework of the cluster-shell competition as a simplified modelling of the spin-orbit interaction (SMSO) [1]. We use an effective nucleon-nucleon interaction, which includes central, spin-orbit and Coulomb terms. We choose the Volkov No. 2 interaction [5] for the central part of $\hat{v}_{ij}$,

$$V(r) = (W - MP^\sigma P^\tau + BP^\sigma - HP^\tau)\,[V_1 e^{-\rho_1 r^2} + V_2 e^{-\rho_2 r^2}]\,, \tag{1}$$

where the exchange parameters are taken to be $W = 1 - M$, $M = 0.61$, and $B = H = 0.125$. For the spin-orbit term, we use the G3RS interaction [6],

$$V_{ls}(r) = V_0^{ls}[e^{-d_1 r^2} + e^{-d_2 r^2}]P(^3O)\,\boldsymbol{L}\cdot\boldsymbol{S}\,, \tag{2}$$

where $d_1 = 5.0$ fm^{-2}, $d_2 = 2.778$ fm^{-2}, $V_0 = 2000$ MeV, and $P(^3O)$ is a projection operator onto the triplet odd state. The parameters of the interaction have been determined from $\alpha + n$ and $\alpha + \alpha$ scattering phase shifts and the binding energy of the deuteron [7].

The total wave function is fully antisymmetrized and the spatial part of the ith nucleon is described by a Gaussian wave packet centered at $\vec{z}_i/\sqrt{\nu}$. Here ν is the width parameter of the wave packet. Its value is $\nu = 1/2b^2$, $b = 1.46$ fm. For the calculation of the carbon isotopes, we assume that the ^{12}C core is described as three alpha particles forming an equilateral triangle. A length parameter, R_1, is introduced for the distance between two alpha particles in the triangle. The limit of $R_1 \to 0$ results in the wave function of the lowest state of the SU(3) shell-model.

To introduce the concept of the SMSO, we loosen up one of the alpha clusters by replacing each Gaussian centre $\vec{z}_i/\sqrt{\nu}$ by an expression containing a 'dissolution parameter' Λ:

$$\vec{z}_i/\sqrt{\nu} = \vec{R} + i\Lambda(\vec{e}^{\,i}_{spin} \times \vec{R})\,, \tag{3}$$

where $\vec{e}^{\,i}_{spin}$ is the unit vector along the spin-direction and the vector $\vec{R}$ is real. The $\Lambda = 0$ limit corresponds to the $(0s)^4$ configuration in the α particle, while the $\Lambda = 1$ case corresponds to four nucleons in one of the alpha clusters being elevated to the $p_{3/2}$-orbit. In this case, the spin-orbit interaction acts attractively for the $j_> = l + 1/2$ component of the wave function [1].

In order to describe the motion of the valence neutrons around the ^{12}C core, we use a stochastic variational procedure [8, 9, 10].

3. Results

3.1. Application of SMSO to systems with several alpha particles

First, we briefly review the application of SMSO to systems with several alpha particles. This is based on the previous works [1, 2, 3, 4].

As a first step for incorporating the spin-orbit interaction within the SMSO framework, we studied ^{9}Be $(2\alpha+n)$, ^{10}Be $(2\alpha+2n)$, ^{12}C (3α) and ^{20}Ne (4α) [1]. We simplified the cluster-shell competition by using just two parameters; the alpha-alpha distance and a single dissolution parameter of one alpha cluster. By optimizing these two parameters, we found that their optimum values depend on the nuclei, and, as a general tendency, the alpha-alpha distance fixed to a smaller value favours a larger dissolution parameter and *vice versa*. Furthermore, as the dissolution parameter becomes larger, the expectation value of the kinetic energy becomes large due to the increase of the higher orbital components. On the other hand, the spin-orbit interaction causes a larger gain in binding energy for a larger value of the dissolution parameter.

Next, we used our SMSO formalism to investigate the efficiency of taking into account the tensor interaction [2]. We introduced a similar parameterization to allow for the effect of the tensor force on the wave function. Unlike for the SMSO, for the tensor case we used a real number for the geometrical shift of one spin-up proton and one spin-down neutron in opposite directions. This shift enhances the $S = 1$, $T = 0$ component, in other words, a deuteron-like component. We call this parameterization SMT (simple method to include the tensor force). By combining SMSO and SMT, we calculated the energy of some $N\alpha$ systems, ^{4}He, ^{8}Be, and ^{12}C. Since the tensor interaction in this calculation is not the bare interaction, the energy gain from the tensor part is not as large as that which can be obtained from the tensor force in an *ab initio* calculation with a realistic nucleon-nucleon interaction.

A comparison of the results of SMSO-SMT calculations for ^{8}Be and ^{12}C reveals a competition between the spin-orbit and tensor forces. For ^{8}Be, the optimum alpha-alpha distance is large and the dissolution is small. The energy gain from the tensor force is larger than that obtained in the ^{4}He case. On the other hand, for ^{12}C, the optimum value of the alpha-alpha distance is not so large as in ^{8}Be, and the dissolution parameter becomes large so as to allow the spin-orbit interaction to have a larger effect. The energy gained from the tensor interaction becomes small in this case.

The cause of such a difference between ^{8}Be and ^{12}C can be attributed to the following mechanism. A small alpha-alpha distance like that in ^{12}C hinders the particle-hole-like excitations, which would be involved in the dissolution of the alpha particle, due to the Pauli principle. A large alpha-alpha distance, in contrast, does not suppress the particle-hole excitations. Hence, in the ^{8}Be case, the energy gain of the tensor interaction becomes large.

3.2. Application to the carbon isotopes

The SMSO formalism was applied to investigating the carbon isotopes, especially ^{14}C and ^{16}C [3]. For the nucleons outside the ^{12}C core, we assume that they are described by randomly located wave packets distributed according to the weight function $w(r) = \exp(-r/R_2)$, where r is the radial coordinate of the wave packet. Here, the range parameter of the distribution R_2 was taken to be 1.5 fm. The optimum values of R_1 and Λ are specific to each nucleus. When the valence neutrons are added to ^{12}C, the optimum value of R_1 becomes smaller, and simultaneously the optimum Λ becomes larger. Therefore, it can be concluded that the 3α structure of ^{12}C is dissolved by adding valence neutrons. The results are shown in Table 1. From the values of R_1 and Λ it is seen that in ^{16}C the structure of the ^{12}C core is completely different from that of the 'free' ^{12}C.

Table 1. Calculated $B(E2; 2_1^+ \to 0_1^+)$ values of the carbon isotopes. The $B(E2)_n$ values for ^{14}C and ^{16}C are calculated by exchanging the nucleon charges as $e_p = 0$ and $e_n = e$. The units of R_1 and $B(E2)$ are fm and e^2fm^4, respectively.

Isotopes	(R_1, Λ)	$B(E2)$		$B(E2)_n$
		Calc.	Exp.	
^{12}C (3α)	(1.5, 0.4)	3.51	8.2 ± 0.1 [12]	3.51
^{14}C ($3\alpha + 2n$)	(0.5, 0.8)	2.99	3.74 ± 0.5 [12]	0.03
^{16}C ($3\alpha + 4n$)	(0.5, 0.8)	0.01	$2.6 \pm 0.2 \pm 0.7$ [13]	4.11

We sought clustering in the excited states of ^{16}C. To this end, we mixed wave functions with different R_1 and Λ values. The two terms to be mixed, (a) $R_1 = 0.5$ fm, $\Lambda = 0.8$ and (b) $R_1 = 2.5$ fm, $\Lambda = 0.0$, are shell-model-like and cluster-model-like components, respectively. First, we performed the calculation with the basis set (a). When we reached energy convergence, we appended the basis with set (b). We found that the energy of the 0_3^+ state comes down significantly, whereas the energies of the ground and second 0^+ states do not change drastically. Therefore, it is justifiable to say that the 3α cluster structure with a geometric shape appears in the $E_x = 10$ MeV region of ^{16}C, in analogy with the crystallization of the clusters in ^{14}C [11].

The observed $B(E2, 2_1^+ \rightarrow 0_1^+)$ value, $2.6 \pm 0.2 \pm 0.7$ $e^2\text{fm}^4$ [13], is smaller than that of ^{12}C, although the two nuclei have the same proton number. Here, we calculated $B(E2; 2_1^+ \rightarrow 0_1^+)$ with the optimum values of R_1 and Λ. The $B(E2)$ value obtained is very small, 0.01 $e^2\text{fm}^4$, much smaller than the experimental value shown in Table 1. We also calculated the 'neutron $B(E2)$' value ($B(E2)_n$ in Table 1) by exchanging the charges of neutrons and protons as $e_p = 0$ and $e_n = e$. The obtained value is relatively large; $B(E2; 2_1^+ \rightarrow 0_1^+)_n = 4.11$ $e^2\text{fm}^4$. From this we conclude that the neutron part of ^{16}C is deformed, and the proton part is spherical.

3.3. Clustering and role of the tensor force

According to the level order of the shell model, the ground state of ^{15}C should be $5/2^+$ ($0d_{5/2}$-orbit). However, the experimental ground state of ^{15}C is $1/2^+$. We conjectured that this level inversion may be accounted for in terms of the SMSO and the first-order tensor force contribution [14]. Therefore, we investigated the effect of the tensor correlation on the evolution of the clustering in the carbon isotopes, ^{15}C and ^{17}C. It was found that, for larger alpha-alpha distances, the repulsive effect of the first-order tensor correlation acts on the $5/2^+$-state strongly, and the $1/2^+$-state comes lower than the $5/2^+$-state in the ^{15}C, with the $R_1 \geq 1.5$ fm. However, such a large R_1 value does not yield the energy minimum. This indicates that other mechanisms, for example, the inclusion of the bare tensor interaction, are necessary.

4. Summary and discussion

We studied the cluster-shell competition by introducing a simple parameterization for the alpha-alpha distance in the core nucleus and for the dissolution of an alpha particle. In this way the characteristics of nuclei consisting of a few alpha particles and valence neutrons can be classified. The alpha particle is dissolved by the spin-orbit force, and the shell-model-like and cluster-model-like behaviour can be related to a dissolution parameter.

In this study, only one alpha particle was allowed to dissolve. We think that is not enough to describe the system realistically, and hence, in a future work, we proceed to breaking further alpha particles in the calculations.

References

[1] Itagaki N, Masui H, Ito M and Aoyama S 2005 *Phys. Rev.* C **71** 064307
[2] Itagaki N, Masui H, Ito M, Aoyama S and Ikeda K 2006 *Phys. Rev.* C **73** 034310
[3] Masui H and Itagaki N 2007 *Phys. Rev.* C **75** 054309
[4] Itagaki N, Cseh J and Płoszajczak M 2011 *Phys. Rev.* C **83** 014302
[5] Volkov A B 1965 *Nucl. Phys.* **74** 33
[6] Tamagaki R 1968 *Prog. Theor. Phys.* **39** 91
[7] Okabe S and Abe Y 1979 *Prog. Theor. Phys.* **61** 1049
[8] Kukulin V I and Krasnopol'sky V M 1977 *J. Phys.* G **3** 795
[9] Varga K and Suzuki Y 1995 *Phys. Rev.* C **52** 2885
[10] Itagaki N, Kobayakawa A and Aoyama S 2003 *Phys. Rev.* C **68** 054302
[11] Itagaki N, Otsuka T, Ikeda K and Okabe S 2004 *Phys. Rev. Lett.* **92** 142501
[12] Raman S *et al* 1987 *At. Data Nucl. Data Tabl.* **36** 1
[13] Ong H J *et al* 2008 *Phys. Rev.* C **76** 014308
[14] Otsuka T, Suzuki T, Fujimoto R, Grawe H and Akaishi Y 2005 *Phys. Rev. Lett.* **95** 232502

Journal of Physics: Conference Series **436** (2013) 012029 doi:10.1088/1742-6596/436/1/012029

Role of the tensor force in light nuclei with the tensor-optimized shell model

T Myo[1], A Umeya[2], H Toki[3] and K Ikeda[4]

[1] General Education, Faculty of Engineering, Osaka Institute of Technology, Osaka 535-8585, Japan
[2] Human Science and Common Education, Faculty of Engineering, Nippon Institute of Technology, Saitama 345-8501, Japan
[3] Research Center for Nuclear Physics (RCNP), Osaka University, Ibaraki, Osaka 567-0047, Japan
[4] RIKEN Nishina Center, Wako, Saitama 351-0198, Japan

E-mail: myo@ge.oit.ac.jp

Abstract. We investigate the structure of light nuclei using a bare nucleon-nucleon interaction. The tensor and short-range correlations are treated with the tensor-optimized shell model (TOSM) and the unitary correlation operator method (UCOM), respectively. In TOSM, the 2p2h states are fully optimized to describe the large tensor contribution and also the corresponding high-momentum components. We use the AV8$'$ interaction and show results for the excitation energy spectra and the radial properties of He and Li isotopes.

1. Introduction

The nucleon-nucleon (NN) interaction has strong tensor terms at long and intermediate distances caused by the pion exchange, which gives rise to large momentum transfer. The interaction also has strong central repulsions at short distances caused by the quark dynamics [1, 2]. It is important to investigate nuclear structure with these characteristics of the NN interaction treated properly. The tensor force is of intermediate range, and we are able to express the strong tensor correlation in a reasonable shell model space [3]. We call this method the Tensor-Optimized Shell Model (TOSM). In this model the wave function is constructed in terms of the shell model basis states with full optimization of the two-particle–two-hole (2p2h) states. The spatial shrinkage of the particle states is essential for the convergence of the tensor contribution involving high momenta, and we do not apply any truncation to the particle states in the TOSM. To treat the correlation coming from the short-range NN repulsion, we use the Unitary Correlation Operator Method (UCOM) [4]. We shall combine the two methods, TOSM and UCOM, to describe nuclei with a bare interaction.

In this paper, we present analyses for He and Li isotopes, and discuss their structures focusing on the effect of the tensor force on the energies and radii.

Journal of Physics: Conference Series **436** (2013) 012029 doi:10.1088/1742-6596/436/1/012029

2. Tensor-Optimized Shell Model (TOSM)

We begin with a many-body Hamiltonian having the bare nucleon-nucleon interaction AV8′ [1]. We write the TOSM wave function Ψ as

$$\Psi = \sum_{k_0} A_{k_0}|0p0h\rangle_{k_0} + \sum_{k_1} A_{k_1}|1p1h\rangle_{k_1} + \sum_{k_2} A_{k_2}|2p2h\rangle_{k_2}. \quad (1)$$

Here, the 0p0h, 1p1h and 2p2h states have various components, which are distinguished by the labels k_0, k_1 and k_2, respectively. The $\{A_k\}$ are variational coefficients. In the TOSM the hole states (h) are taken to be harmonic oscillator basis states. In the p-shell nuclei, the hole states are defined as the $0s$ and $0p$ shells; the higher shells are treated as particle states. Within the space of the 0p0h (hole) states, we allow the two-body interactions to excite at most two particles from the $0s$ orbit to the 0p orbits. From each configuration of the 0p0h states, up to two nucleons can be excited to the particle states as well to give rise to 1p1h and 2p2h states. For the radial wave functions involved in the particle states, we use the Gaussian expansion method [3]. In this way the radial factors of the particle states of each configuration are fully described. As many particle configurations are included as possible until convergence is reached. Thus, in this treatment, there is no truncation of the particle states. We used the short-range UCOM [4] to include the short-range correlation caused by the NN interaction in the TOSM basis states. We minimize the total energy by optimizing the length parameters of the Gaussian bases for all the hole states and particle states for each J^π state of each particular nucleus.

3. Results in TOSM+UCOM

We present and discuss the results for He and Li isotopes. The maximum orbital angular momentum of the particle states was taken to be 10 to get a sufficiently good convergence. The excitation energies of He isotopes and Li isotopes are shown in Fig. 1 and Fig. 2, respectively. For the excitation energies, we see good agreement with experiment, so that we can discuss the structure differences between energy levels. We also predict some energy levels which have not been observed experimentally yet or whose spin assignments are not yet settled, particularly on the neutron-rich side, for ^{7,8}He and ^{8,9}Li. In most of the indidvidual He and Li isotopes, the ground states have the largest contributions from the tensor force [5, 6]. The tensor contributions are mainly produced as 2p2h excitations to the particle states from the 0p0h hole states, i.e., from the $0s$ and $0p$ shells. The corresponding high-momentum components induced by the tensor force are demonstrated by enhancements of the kinetic energy with respect to the values produced by the naive shell model.

The resulting levels of the Li isotopes are spaced somewhat more compactly than the experimental spectra. For example, in ^{9}Li, the energy difference between the lowest $3/2^-$ and

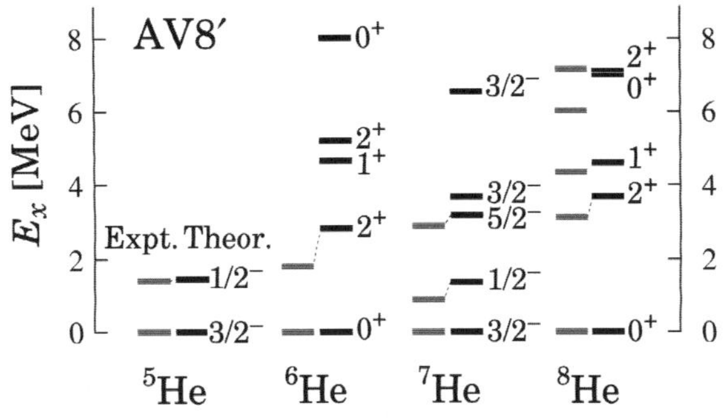

Figure 1. Excitation energies of He isotopes with TOSM+UCOM using AV8′.

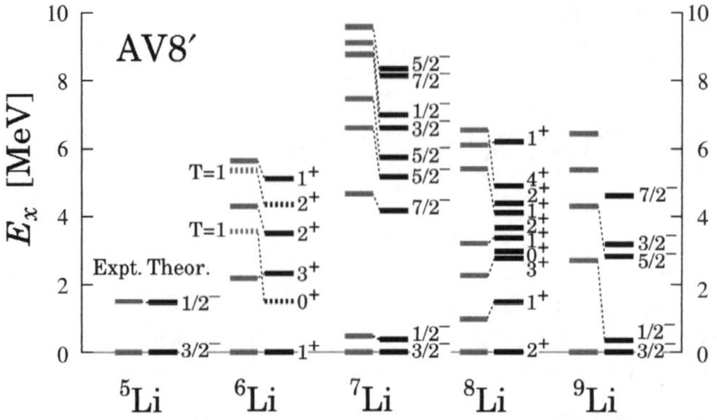

Figure 2. Excitation energies of Li isotopes with TOSM+UCOM using AV8′.

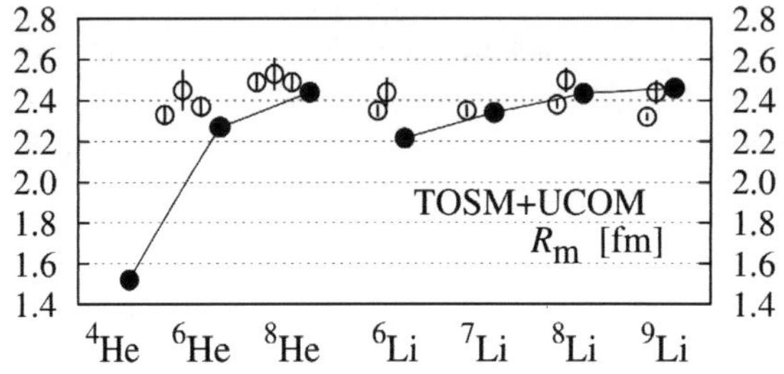

Figure 3. Matter radii of He and Li isotopes in fm (solid circles) comapred with experiment (open circles) [7, 8, 9].

$1/2^-$ states is too small. These characteristics are also common in the Green's Function Monte Carlo calculations performed without three-body terms in the interaction [1]. Thus a (partial) remedy may come from an additional genuine three-body interaction.

In TOSM, the length parameters of the basis states are determined variationally, hence we can discuss the radial properties of nuclei. The matter radii of He and Li isotopes are shown in Fig. 3 and the systematic trend agrees with the experiments. For ^{6}He and ^{8}He, the radii are larger than that of ^{4}He because of the loosely bound extra neutrons, they are however, slightly smaller than the experimental values. This difference could perhaps be recovered by including the continuum effects upon the extra neutrons. The radius of ^{6}Li is also smaller than experiment, which could be remedied by including the asymptotic α+d clustering component in addition to the shell model basis states in TOSM.

References

[1] Pieper S C and Wiringa R B 2001 *Annu. Rev. Nucl. Part. Sci* **51** 53
[2] Kamada H *et al* 2001 *Phys. Rev.* C **64** 044001
[3] Myo T, Toki H and Ikeda K 2009 *Prog. Theor. Phys.* **121** 511
[4] Feldmeier H, Neff T, Roth R and Schnack J 1998 *Nucl. Phys.* A **632** 61
[5] Myo T, Umeya A, Toki H and Ikeda K 2011 *Phys. Rev.* C **84** 034315
[6] Myo T, Umeya A, Toki H and Ikeda K 2012 *Phys. Rev.* C **86** 024318
[7] Tanihata I *et al* 1988 *Phys. Lett.* B **206** 592
[8] Dobrovolsky A *et al* 2006 *Nucl. Phys.* A **766** 1
[9] Kiselev O A *et al* 2005 *Eur. Phys. J.* A **25** Issue 1 Suppl. p. 215

10th International Conference on Clustering Aspects of Nuclear Structure and Dynamics IOP Publishing
Journal of Physics: Conference Series **436** (2013) 012030 doi:10.1088/1742-6596/436/1/012030

Microscopic description of $\alpha + \alpha$ bremsstrahlung from a realistic nucleon-nucleon interaction

Jérémy Dohet-Eraly[1], **Daniel Baye and Pierre Descouvemont**

Physique Nucléaire et Physique Quantique, CP229, Université Libre de Bruxelles (ULB), B-1050 Brussels Belgium.

E-mail: jdoheter@ulb.ac.be

Abstract. The $\alpha + \alpha$ bremsstrahlung is studied in a microscopic cluster approach with the Generator Coordinate Method associated with the Microscopic R-matrix Method. Two effective nucleon-nucleon interactions are considered: the Minnesota potential and a potential derived from the realistic Argonne AV18 potential by the Unitary Correlation Operator Method. The influence of the nucleon-nucleon interactions are discussed and the results are compared with experimental data.

1. Introduction

For a long time, microscopic approaches have been used to describe nuclear collisions [1, 2]. They were based on effective nucleon-nucleon (NN) interactions adapted to the cluster model and including one or two parameters fitted for each collision. In recent years, realistic NN interactions were developed and many efforts were done to base collision models on them. These realistic potentials have the feature to reproduce with a high precision the properties of two-nucleon systems. However, for three- or more-nucleon systems, a good agreement with experiment cannot be achieved without three-body forces and/or renormalization techniques.

In Ref. [3], we proposed to describe the α + N, $\alpha + {}^3$He, and $\alpha + \alpha$ elastic scatterings in a microscopic cluster model based on the UCOM-AV18 potential obtained from the two-body realistic potential Argonne AV18 [4] by the Unitary Correlation Operator Method (UCOM)[5, 6]. A rather good agreement with the experimental phase shifts was obtained. However, scattering data only depend on the asymptotic part of the scattering wave function. To probe the inner part of the scattering wave function, the study of bremsstrahlung is relevant, in particular for unbound systems.

Microscopic descriptions of the $\alpha + \alpha$ bremsstrahlung were previously performed [7, 8]. The scattering wave functions were obtained from a Schrödinger equation based on effective NN potentials: the Volkov [9] and Minnesota potentials [10]. In this contribution, we follow a very similar approach but based on the realistic effective UCOM-AV18 potential [3]. The obtained results are compared with the experimental data and with results from a microscopic cluster model based on the Minnesota effective NN interaction.

[1] F.R.S.-FNRS Research Fellow.

2. Microscopic bremsstrahlung model

An α particle moving in the z-direction with the energy E_α collides a second α particle initially at rest. Then, a photon is emitted in the direction $\Omega_\gamma = (\theta_\gamma, \varphi_\gamma)$ and the α particles are scattered in the directions $\Omega_1 = (\theta_1, \varphi_1)$ and $\Omega_2 = (\theta_2, \varphi_2)$. In the coplanar Harvard geometry which is considered here and in the experiments [11, 12], the α particles are detected only if $\varphi_1 = \varphi_2 = 0$.

Evaluating the differential bremsstrahlung cross sections requires calculating the matrix element of the photon-emission operator H_e between the incoming initial and outgoing final states. In this cluster approach, the wave function is assumed to be written as

$$\Psi = \mathcal{A}\phi_{\alpha_1}\phi_{\alpha_2}g(\boldsymbol{\rho}) \qquad (1)$$

where $\mathcal{A}$ is the antisymmetrization projector, ϕ_{α_1} and ϕ_{α_2} are the internal wave functions of the α particles, g is the relative wave function, and $\boldsymbol{\rho}$ is the vector between the c.m. of the nuclei. The α wave functions are obtained by removing a c.m. factor from the $(0s)^4$ state of the harmonic oscillator shell model. The initial and final relative wave functions $g(\rho)$ are calculated from the microscopic Schrödinger equation for eight nucleons

$$H\Psi(1, 2, \ldots, 8) = E_{\text{tot}}\Psi(1, 2, \ldots, 8), \qquad (2)$$

with

$$H = \sum_{i=1}^{8} t_i - T_{\text{c.m.}} + \sum_{i<j}^{8} \left[v_{ij}^{\text{N}} + v_{ij}^{\text{C}}\right], \qquad (3)$$

where E_{tot} is the energy of the nuclear system in the c.m. framework, t_i is the kinetic energy of nucleon i, $T_{\text{c.m.}}$ is the c.m. kinetic energy, v_{ij}^{N} is the nuclear interaction (the UCOM-AV18 potential or the Minnesota potential with the exchange parameter $u = 0.9474$) and v_{ij}^{C} is the Coulomb potential. Following the principle of the Microscopic R-matrix Method [13], the configuration space is divided in two regions at the channel radius a: an internal region ($\rho < a$) where the wave function is described by the Generator Coordinate Method [14, 2] and an external region ($\rho > a$) where the antisymmetrization between nuclei is neglected and the $\alpha\alpha$ interaction is considered purely Coulombic. The wave functions are evaluated by solving the Schrödinger equation (2) on the internal region requiring the continuity of the wave function at the channel radius a. They are insensitive to the value of a if a is chosen large enough.

The photon-emission operator is limited to the E2 transitions because they dominate greatly for the $\alpha + \alpha$ system. As in Ref. [7], we take only the convection current into account to evaluate the photon-emission operator. The long-wavelength approximation is not performed and the Siegert theorem is not used because it would lead to divergent integrals as the initial and final states are in the continuum, so not square-integrable [7]. Therefore, the spin-isospin-scalar part of the photon-emission operator reads [7]

$$H_e = \frac{ie\sqrt{\pi}}{2\sqrt{15}m_N c k_\gamma} \sum_{\mu=-2}^{2} \left[\mathfrak{D}^2_{\mu -1}(R_\gamma) + \mathfrak{D}^2_{\mu 1}(R_\gamma)\right] \sum_{i=1}^{8} \boldsymbol{\chi}_\mu(k_\gamma, \boldsymbol{r}_i - \boldsymbol{R}_{\text{c.m.}}) \cdot (\boldsymbol{p}_i - A^{-1}\boldsymbol{P}_{\text{c.m.}}), \qquad (4)$$

where m_N is the nucleon mass, k_γ is the photon wave number, $\mathfrak{D}$ is the rotation matrix element depending on the Euler angles $R_\gamma = (-\varphi_\gamma, -\theta_\gamma, 0)$, $\boldsymbol{r}_i$ and $\boldsymbol{p}_i$ are respectively the coordinate and momentum of nucleon i, $\boldsymbol{R}_{\text{c.m.}}$ and $\boldsymbol{P}_{\text{c.m.}}$ are the c.m. coordinate and momentum, and $\boldsymbol{\chi}_\mu$ is given by

$$\boldsymbol{\chi}_\mu(k, \boldsymbol{r}) = \left(k^2 \boldsymbol{r} + \nabla \frac{\partial}{\partial r} r\right) j_2(kr) Y_2^\mu(k\boldsymbol{r}). \qquad (5)$$

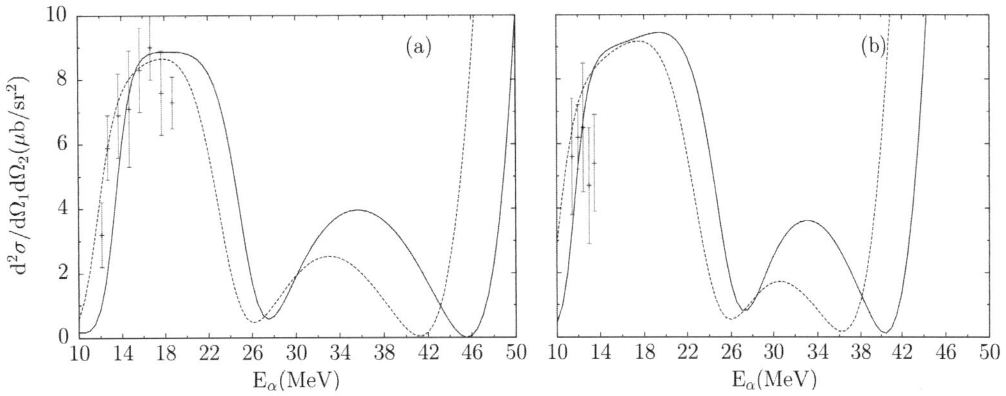

Figure 1. Differential bremsstrahlung cross sections obtained in the microscopic cluster models based on the UCOM-AV18 (solid lines) and Minnesota (dotted lines) potentials for $\theta_1 = \theta_2 = 35°$ (a) and $\theta_1 = \theta_2 = 37°$ (b) as a function of the incident α energy. Experimental data from [11, 12].

3. Results and discussions

The theoretical and experimental differential bremsstrahlung cross sections $d^2\sigma/d\Omega_1 d\Omega_2$ as a function of the incident α energy are displayed at Fig. 1 for the Harvard coplanar configurations $\theta_1 = \theta_2 = 35°$ and $\theta_1 = \theta_2 = 37°$. They are the only configurations for which experimental data are available.

In the range of energy where experimental data are available, differences between models are weak compared to the experimental error bars. At higher energies, differences become much larger. New bremsstrahlung experiments at higher energy are thus required to probe the models.

To conclude, let us note that differences between models come from the differences not only in the inner part of the wave functions but in their external part too since the phase shifts are not exactly equal for both models [3]. Indeed, for the lowest values of the relative orbital angular momentum $l = 0$ and $l = 2$, theoretical and experimental phase shifts are in a rather good agreement but from $l = 4$, the model based on the UCOM-AV18 potential does not reproduce anymore the experimental phase shifts as well as the model based on the Minnesota potential. This discrepancy between the models in the $\alpha + \alpha$ elastic phase shifts is partially responsible for their discrepancy in the description of $\alpha + \alpha$ bremsstrahlung. The contribution of the external part deserves to be investigated in further works.

References

[1] Wildermuth K and Tang Y C 1977 *A unified theory of the nucleus* (Braunschweig: Vieweg)
[2] Tang Y C 1981 *Topics in Nuclear Physics II* (Lecture Notes in Physics vol 145) (Berlin: Springer) p 571
[3] Dohet-Eraly J and Baye D 2011 *Phys. Rev.* C **84** 014604
[4] Wiringa R B, Stoks V G J and Schiavilla R 1995 *Phys. Rev.* C **51** 38
[5] Feldmeier H, Neff T, Roth R and Schnack J 1998 *Nucl. Phys.* A **632** 61
[6] Neff T and Feldmeier H 2003 *Nucl. Phys.* A **713** 311
[7] Baye D and Descouvemont P 1985 *Nucl. Phys. A* **443** 302
[8] Baye D, Descouvemont P and Kruglanski M 1992 *Nucl. Phys.* A **550** 250
[9] Volkov A B 1965 *Nucl. Phys.* A **74** 33
[10] Thompson D R, LeMere M and Tang Y C 1977 *Nucl. Phys.* A **286** 53
[11] Peyer U, Hall J, Müller R, Suter M and Wölfli W 1972 *Phys. Lett.* B **41** 151
[12] Frois B, Birchall J, Lamontagne C R, von Moellendorff U, Roy R and Slobodrian R J 1973 *Phys. Rev.* C **8** 2132
[13] Baye D, Heenen P-H and Libert-Heinemann M 1977 *Nucl. Phys.* A **291** 230
[14] Horiuchi H 1977 *Prog. Theor. Phys. Suppl.* **62** 90

Cluster structure in light nuclei studied with few-body models

W Horiuchi[1] and Y Suzuki[2,3]

[1] Department of Physics, Hokkaido University, Sapporo 060-0810, Japan
[2] Department of Physics, Niigata University, Niigata 950-2181, Japan
[3] RIKEN Nishina Center, RIKEN, Wako 351-0198, Japan

E-mail: whoriuchi@nucl.sci.hokudai.ac.jp

Abstract. We present our attempt at describing the low-lying states of $A = 16$ nuclei in a single scheme, a ^{12}C core-plus-few-nucleon model. Promising results are obtained for ^{16}C and ^{16}O, which suggest the coexistence of single-particle and cluster structure.

1. Introduction

Cluster structure is one of the unique and most interesting phenomena in nuclear many-body systems, and often appears in the excited spectrum of light nuclei. A famous and long-standing problem is the so-called mysterious 0^+ state in ^{16}O. In contrast to the shell-model-like ground state, the first excited 0^+ state is known to have a cluster structure of ^{12}C+α [1], which shows the coexistence of different structures in the spectrum but has not been reproduced microscopically even in modern large-scale calculations. A deep understanding requires a unified description. Since a fully microscopic calculation is not feasible at present, it is interesting to study ^{16}O in a ^{12}C core-plus-four-nucleon model, that is, without assuming the existence of an α cluster.

In this contribution, we present our calculations in a ^{12}C+4N five-body model. In the next section, we briefly explain our approach and in Section 3 we present preliminary results for ^{16}O as well as ^{16}C. A summary and the future prospects are outlined in Section 4.

2. A core-plus-few-nucleon model

The five-body system we consider here is characterized by the Hamiltonian which contains the nucleon-nucleon (N–N) potential for valence nucleons and the ^{12}C-nucleon (C$-N$) potential. For the N–N force the central Minnesota (MN) potential [2] is employed, while for the ^{12}C$-N$ potential, a Woods-Saxon-plus-derivative form is assumed with parameters determined so as to reproduce the low-lying spectrum of ^{13}C ($1/2^-$, $1/2^+$ and $5/2^+$ states). Our C$-N$ potential is deep enough to accommodate some redundant or Pauli-forbidden states. To eliminate such states we impose the orthogonality constraint for the motion of the valence nucleons, which is practically achieved by adding a pseudopotential to the Hamiltonian [3]. We assume the $0s_{1/2}$ and $0p_{3/2}$ harmonic oscillator wave functions for the orbits occupied in ^{12}C.

We express the wave function with total angular momentum J as a linear combination of many basis states of the following form

$$\Psi_{JM} = \mathcal{A}\left\{\left[[\Phi_L \chi_S]_{\mathcal{J}}\, \phi_I(^{12}\mathrm{C})\right]_{JM} \eta_{TM_T}\right\}, \tag{1}$$

where $\mathcal{A}$ is an antisymmetrizer which only acts on the valence nucleons. The $\phi_{IM_I}(^{12}\mathrm{C})$ is the core wave function but the core excitation is ignored here. The χ_{SM_S} (η_{TM_T}) is the spin (isospin) function of the valence nucleons. For the orbital part, Φ_L, we use a correlated Gaussian form reinforced with two global vectors, which enable us to obtain a precise solution of the many-body Schrödinger equation [4]. An advantage of this basis is that its functional form does not change under any coordinate transformation, and thus both cluster- and shell-model-like configurations can be expressed in a single scheme. We have applied this method not only to ground states but also to excited states [5]. In particular, the spectrum of ^{4}He is interesting as an analogue of the ^{16}O spectrum. Similarly to ^{16}O, the first excited state of ^{4}He has $J^\pi = 0^+$ and has been shown to be a cluster state of the $3N + N$ (^{3}H+p and ^{3}He+n) configuration [5, 6].

The variational parameters are optimized following the stochastic variational method [7]. Assuming that we already have some basis states, the parameters of the next one are determined by the following procedure: we generate several candidates with randomly chosen parameters, and select the state which gives the lowest energy among them when included in the basis, and repeat this procedure until the energy is converged.

3. Results

Figure 1 displays the calculated energy for ^{16}C and the ground and first excited state of ^{16}O as functions of the number of basis functions. The energy curve of ^{16}C converged at 7000 basis states and the energy is very close to experiment. The ^{15}C+n threshold is also well reproduced at -14.5 MeV (Exp. -14.3 MeV). The energy curves of ^{16}O exhibit slower convergence than that of ^{16}C due to the difficulty of eliminating the forbidden states. The case of ^{16}C is easier than that of ^{16}O because the valence neutrons do not form a strongly bound cluster. When the valence nucleons form a cluster, the wave function would tend to have a large overlap with the forbidden states, and a large basis is required to ensure orthogonality. However, the preliminary results for ^{16}O are promising. The ground state appears below the ^{12}C+α threshold and the first excited state appears just at a few MeV above the ground state, consistently with experiment. The calculation is still being in progress.

It is interesting to see how the cluster structure developes. Figure 2 displays the expectation values of the C$-N$ potential and of the N–N potential acting between the valence nucleons, as functions of the basis dimension. For ^{16}C, the contribution of the C$-N$ potential is very large compared to the N–N contribution. The state can be regarded as a shell-model-like state, in which the valence nucleons move almost independently around the core. Reflecting this simple

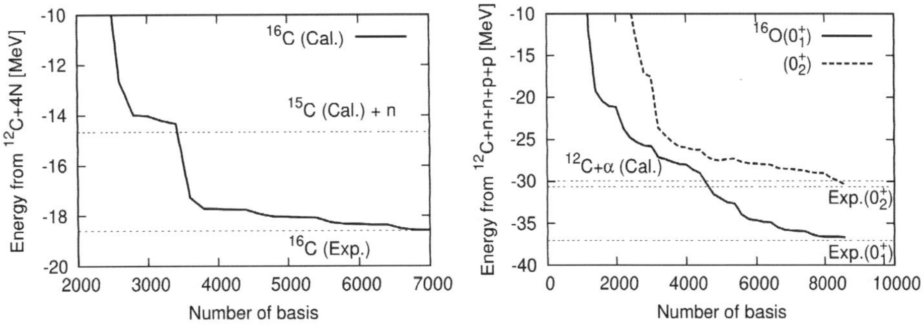

Figure 1. Energy curves for the ground state of ^{16}C (left) and for the ground and excited states of ^{16}O (right). The calculated ^{15}C($1/2^+$)+n (left) and ^{12}C(0_1^+)+α (right) thresholds and experimental energies are plotted with thin lines.

structure, the two energies of ^{16}C are almost independent of the basis. In contrast, the curves for the ground state of ^{16}O change drastically in the region of 4000 to 6000 basis states, where the energy crosses the ^{12}C+α threshold (See Fig. 1). We expect a phase transition from cluster- to shell-model-like structure at this basis size. Beyond 8000 basis states, where the energies are close to the convergence, the C$-N$ contribution obtained is similar to that of ^{16}C, whereas the N–N contribution is larger. The N–N energy is approximately half of the N–N contribution of the free alpha particle (-88.03 MeV) obtained with the MN potential [4]. In the ground state of ^{16}O, the α particle is distorted by the attraction of the core nucleus as well as by the forbidden states. In the first excited state of ^{16}O, the N–N contribution is dominant and is very close to that of the free alpha particle. This suggests that the first excited state has a well-developed ^{12}C+α structure as predicted by the cluster model [1]. It should be noted that the expectation value of the C$-N$ potential in the ground state is approximately equal to the N–N contribution in the first excited state of ^{16}O. Thus two different aspects coexist in the states of ^{16}O due to the balance of N–N and C$-N$ contributions.

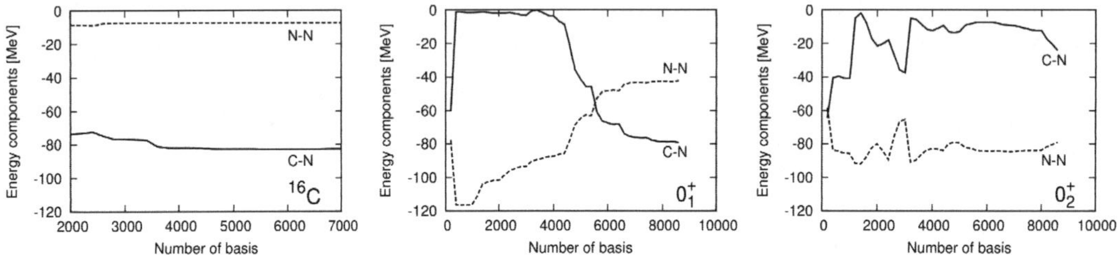

Figure 2. Expectation values of ^{12}C$-N$ (C$-N$) and nucleon-nucleon (N–N) potentials for the ground state of ^{16}C, the ground state of ^{16}O (0_1^+), and the first excited state of ^{16}O (0_2^+) as functions of the basis dimension.

4. Summary and future work

We described the low-lying spectrum of $A = 16$ systems with a ^{12}C+four-nucleon model. Solving the five-body equation is very hard mainly because we face with the difficulty of removing the forbidden states. Our preliminary results for ^{16}C and ^{16}O are yet promising: the energies of ^{16}C (0^+) and ^{16}O ($0_1^+, 0_2^+$) are reproduced in a single scheme. The ground state of ^{16}O seems to be a shell-model-like state, whereas the first excited state has a well-developed ^{12}C+α structure. To see the clustering more quantitatively, we will calculate the ^{12}C+α spectroscopic amplitude (factor). The calculation is still being in progress and will be reported elsewhere soon.

References

[1] Suzuki Y 1976 *Prog. Theor. Phys.* **55** 1751 (1976); *Prog. Theor. Phys.* **56** 111
[2] Thompson R, Lemere M and Tang Y C 1977 *Nucl. Phys.* A **286** 53
[3] Kukulin V I and Pomerantsev V 1978 *Ann. Phys. (N.Y.)* **111** 30
[4] Suzuki Y, Horiuchi W, Orabi M and Arai K 2008 *Few-Body Syst.* **42** 33
[5] Horiuchi W and Suzuki Y 2008 *Phys. Rev.* C **78** 034305
[6] Hiyama E, Gibson B F and Kamimura M 2004 *Phys. Rev.* C **70** 031001
[7] Varga K and Suzuki Y 1995 *Phys. Rev.* C **52** 2885

Journal of Physics: Conference Series **436** (2013) 012032 doi:10.1088/1742-6596/436/1/012032

Two-body Coulomb scattering and complex scaling

István Hornyák

University of Debrecen, Faculty of Informatics, PO Box 12, 4010 Debrecen, Hungary

E-mail: hornyak.istvan@atomki.mta.hu

Abstract. The two-body Coulomb scattering problem is studied with the complex scaling method. Our splitting method [1], on partial wave level, simplifies the scattering boundary condition. The scattered part of the wave function is square integrable. This property makes the numerical solution easier.

1. Introduction

The method of complex scaling (CS) has been successfully applied in many areas of quantum physics, but for scattering problems the CS procedure can only be applied to short-range potentials, which limits the applicability of CS in atomic and nuclear physics.

It has been shown in the Ref. [2] that scattering calculations with the exterior CS can be successfully performed for long-range interactions as well. Nevertheless, the exterior CS method has been under scrutiny since an artificial cutoff of some of the interaction may cause troubles. Recently, however, it has been shown [3] that the standard CS can be applied to scattering problems even in the presence of the pure Coulomb interaction. The method is based on the two-potential formalism. An alternative technique is presented below.

The full scattering solution is written in the form

$$\psi^{+}(\mathbf{k},\mathbf{r}) = \phi_0(\mathbf{k},\mathbf{r}) + \psi^{sc+}(\mathbf{k},\mathbf{r}), \tag{1}$$

where $\psi^{sc+}(\mathbf{k},\mathbf{r})$ is the scattered wave, and $\phi_0(\mathbf{k},\mathbf{r})$ is a known function. From the Schrödinger equation the so called driven Schrödinger equation can be derived for the scattered wave:

$$(E - \hat{H})\psi^{sc+}(\mathbf{k},\mathbf{r}) = S(\mathbf{k},\mathbf{r}), \tag{2}$$

where the source term is given by $S(\mathbf{k},\mathbf{r}) = (\hat{H} - E)\phi_0(\mathbf{k},\mathbf{r})$.

2. Theory

We have found that the choice of the so called Coulomb-modified plane wave (CMPW),

$$\phi_0(\mathbf{k},\mathbf{r}) = e^{i\mathbf{kr}}(kr - \mathbf{kr})^{i\gamma}, \tag{3}$$

does not simplify the boundary condition from the point of view of the complex scaling.

We rather split the wave function into incoming and scattered waves at the partial-wave level. The exact incoming wave is $\psi_{i,l}(k,r)$ [4] and we write

$$\psi_l^{+}(k,r) = \psi_{i,l}(k,r) + \psi_l^{sc+}(k,r), \tag{4}$$

where
$$\psi_{i,l}(k,r) = \omega_{i,l}(k,r) + \chi_l(k,r) \tag{5}$$
and
$$\omega_{i,l}(k,r) = e^{-ikr}e^{\gamma\pi/2}(-1)^{l+1}(2ikr)^l U(l+1-i\gamma, 2l+2, 2ikr), \tag{6}$$
$$\chi_l(k,r) = \frac{e^{ikr+\gamma\pi/2}}{2ikr}\frac{(-1)^l}{(2ikr)^l}\frac{\Gamma(2l+1)}{\Gamma(l+1-i\gamma)}\sum_{n=0}^{l}\frac{(-1)^n(i\gamma-l)_n}{(-2l)_n n!}(2ikr)^n, \tag{7}$$
with U being the confluent hypergeometric function. For the driven radial Schrödinger equation direct calculation gives the following source term
$$S_l(k,r) = \frac{e^{ikr+\gamma\pi/2}}{2r^2\Gamma(-i\gamma)}. \tag{8}$$

After CS by scaling angle θ, which satisfies the condition $0 < \theta < \pi$, and the transformation $h_{l,\theta}^{sc+}(k,r) = r^{l+1}\psi_{l,\theta}^{sc+}(k,r)$, which results in a regular function $h_{l,\theta}^{sc+}(k,r)$ at $r = 0$) we get the driven radial Schrödinger equation
$$\left[\frac{k^2}{2} + e^{-2i\theta}\frac{1}{2}\frac{d^2}{dr^2} - e^{-2i\theta}\frac{l}{r}\frac{d}{dr} - e^{-i\theta}\frac{\gamma k}{r} - V_s(re^{i\theta})\right] h_{l,\theta}^{sc+}(k,r) = r^{l+1}S_{l,\theta}^{tot}(k,r), \tag{9}$$
where the new source term reads
$$S_{l,\theta}^{tot}(k,r) = S_{l,\theta}(k,r) + e^{i3\theta/2}\psi_{i,l}(k,re^{i\theta})V_s(re^{i\theta}). \tag{10}$$
The boundary conditions are
$$h_{l,\theta}^{sc+}(k,0) = 0 \tag{11}$$
and
$$\lim_{r\to\infty} h_{l,\theta}^{sc+}(k,r) = 0. \tag{12}$$
One can show that the phase shift δ_l can be expressed as the $r \to \infty$ limit of a function $\sigma_l(r)$, which we call the local representation of the phase shift:
$$e^{2i\sigma_l(r)} = (2kre^{i\theta})^{i\gamma}\left[e^{-ikre^{i\theta}}2ikre^{-i\theta/2}\tilde{\psi}_{l,\theta}^{sc+}(k,r) + \frac{e^{\gamma\pi/2}(-1)^l(i\gamma-l)_l}{\Gamma(l+1-i\gamma)}\right] \to e^{2\delta_l} \quad (r\to\infty). \tag{13}$$

3. Numerical Results

In numerical calculations the boundary condition (12) is approximated by a similar expression for a large but finite value R of r, which in in the asymptotic region:
$$h_{l,\theta}^{sc+}(k,R) = 0. \tag{14}$$

The finite element method is used as a numerical technique for the solution of Eq. (9). In the calculations equally spaced finite elements of length 1 atomic unit are taken. The degree of the Lobatto shape functions is denoted by N, and the same N value is chosen for each element. A short-range potential $V_s = 7.5r^2\exp(-r)$ is added to the Coulomb potential.

In Figure 1 a case of $\theta = 0.1$ radian is shown, with the angular momentum $l = 0$ and $k = 3$. The Coulomb potential is given by $1/r$. The local representation of the phase shift is practically constant in a huge region, which can be identified with an approximate phase-shift value. In contrast the local approximation of the phase shift in [5] tends to the exact value by a persistent oscillation with decreasing order of amplitude.

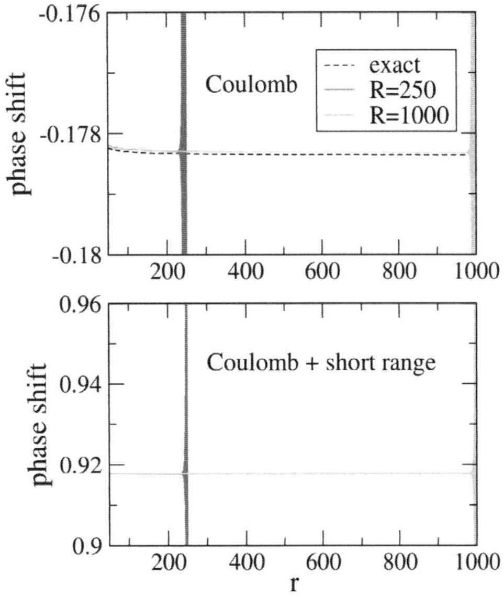

Figure 1. The local representation of the phase shift for the pure Coulomb potential (upper part) and for the Coulomb plus short-range potential (lower part). For the pure Coulomb case the dashed line displays the exact solution.

4. Conclusion

The two-body scattering problem has been solved for a pure Coulomb potential as well as for a Coulomb plus finite-range potential using the standard complex scaling method. No cutoff is applied to the tail of the long-range interaction. The CS makes it possible to observe the scattering boundary condition with great accuracy in a fairly simple manner.

Reference

[1] Hornyak I and Kruppa A T 2012 *Phys. Rev.* A **85** 022702
[2] Rescigno T N, Baertschy M, Byrum D and McCurdy C W 1997 *Phys. Rev.* A **55** 4253
[3] Kruppa A T, Suzuki R and Kato K 2007 *Phys. Rev.* C **75** 044602
[4] van Haeringen H 1976 *Il Nuovo Cimento* **34B** 53
[5] Volkov M V, Elander N, Yarevsky E and Yakovlev S L 2009 *Europhys. Lett.* **85** 30001

The geometric interpretation of the semimicroscopic algebraic cluster model and the role of the Pauli principle

H Yépez-Martínez[1], P O Hess[2,3,4] and G Lévai[5]

[1]Universidad Autónoma de la Ciudad de México, Prolongación San Isidro 151, Col. San Lorenzo Tezonco, Del. Iztapalapa, 09790 México D.F., Mexico
[2] Instituto de Ciencias Nucleares, UNAM, Circuito Exterior, C.U., A.P. 70-543, 04510 México, D.F., Mexico
[3] Frankfurt Institute for Advanced Studies (FIAS), J.W. von Goethe Universität, Ruth-Moufang-Str. 1, 60438 Frankfurt am Main, Germany
[4] GSI für Schwerionenforschung Helmholzzentrum GmbH,Max-Planck-Str. 1 64291 Darmstadt, Germany
[5]Institute of Nuclear Research of the Hungarian Academy of Sciences, Debrecen, Pf. 51, Hungary-4001

E-mail: hess@nucleares.unam.mx

Abstract. We study the properties of the Semimicroscopic Algebraic Cluster Model (SACM) under phase transitions. We show that the SACM can undergo first and second order phase transitions and a critical line appears. When the Pauli principle is not observed, the critical line appears, but other non-physical properties arise. At the end the meaning of the $SO(4)$ dynamical limit is discussed.

1. Introduction

In this contribution the structure of phase transition within the SACM is discussed. The same will be done when the Pauli exclusion principle is not taken into account. Some physical interpretation of the $SO(4)$ limit is also given. The discussion of phase transitions is enormously simplified when a geometrical mapping is applied. A semi-classical potential is obtained which depends on variables related to physical observables as the distance between two clusters. Phase transitions then manifest themselves by certain structural changes in the potential.

In section 2 the SACM is shortly reviewed, the Hamiltonian is defined and the method of geometrical mapping is explained. The semi-classical potential is obtained, which is used to investigate the phase transitions of the model. Changes are discussed when the Pauli exclusion principle is not observed. At the end some remarks are made concerning the physical meaning of the $SO(4)$ dynamical limit. In section 3 conclusions are drawn.

2. The SACM and the geometrical mapping

We shall shortly resume the main features of the SACM [1, 2]. The relative degrees of freedom are described by the oscillator boson operators $\pi^\dagger_m$ and π_m, with $m = 0, \pm 1$. To that the scalar boson operators $\sigma^\dagger$ and σ are also added: these have no physical meaning but are needed in order to introduce a cutoff in the theory. All combinations of any creation with any annihilation operator form the $U_R(4)$ group, where the R refers to the relative motion. The total number of bosons $N = n_\pi + n_\sigma$ is kept constant, i.e., the number of π-bosons ranges from 0 to N. The individual clusters are described via the $SU(3)$-model of Elliott [3]. In this way both the relative motion and the orbital structure of the clusters is characterized by an $SU(3)$ irreducible representation (irrep) (λ, μ).

The model space is constructed as follows. First, one determines the result of the product $(\lambda_1, \mu_1) \otimes (\lambda_2, \mu_2) \otimes (n_\pi, 0)$, which gives a sum of total $SU(3)$ irreps with a certain multiplicity. Second, all possible excitations of the nucleons within the shell model are constructed, for $n\hbar\omega$ excitation, with the center of mass removed. This list of shell model irreps, which by construction observe the Pauli exclusion principle is compared to the former list. Only those irreps are included in the SACM model space, which appear also in the shell model space. This method automatically reproduces the Wildermuth condition, which states that there is a minimal number of π-bosons (n_0) needed as a necessary condition to satisfy the Pauli exclusion principle. In this manner, the Pauli exclusion principle is observed. Since the model space is constructed microscopically, one can study in this framework the overlap of the cluster configurations with the states of the shell and collective models, e.g. the clusterization of the superdeformed (SD) and hyperdeformed (HD) states. It is remarkable that the predictions of the model seem to be justified by the experimental observation in some cases, like for the SD state of ^{28}Si [4] and for the HD state of the ^{36}Ar [5].

The Hamiltonian is phenomenological, indicated by "semi" in the acronym SACM. We consider the sum of Hamiltonians, each describing a particular dynamical symmetry limit. Due to lack of space we shall present here the Hamiltonian taking into account only the $SU(3)$ and $SO(4)$ dynamical symmetries. The first one is known as the *vibrational limit*, while the second one as the *deformed limit*, though, as will be shown below, within the SACM the potential is already deformed by construction.

The Hamiltonian has the form

$$\boldsymbol{H} = x\boldsymbol{H}_{SU(3)} + (1-x)\boldsymbol{H}_{SO(4)} \ , \tag{1}$$

where x is the parameter ranging from 0 to 1, determining the relative weight of each symmetry. In their most general form, the individual components are

$$\begin{aligned} \boldsymbol{H}_{SU(3)} &= \hbar\omega \boldsymbol{n}_\pi + a_C \boldsymbol{C}_2(\lambda_C, \mu_C) + (\bar{a} - \bar{b}\triangle \boldsymbol{n}_\pi)\boldsymbol{C}_2(n_\pi, 0) + (a - b\triangle \boldsymbol{n}_\pi)\boldsymbol{C}_2(\lambda, \mu) + \gamma \boldsymbol{L}^2 + t\boldsymbol{K}^2 \\ \boldsymbol{H}_{SO(4)} &= a_C \boldsymbol{C}_2(\lambda_C, \mu_C) + a_C {\boldsymbol{L}_C}^2 + a_R^{(1)} {\boldsymbol{L}_R}^2 + \gamma \boldsymbol{L}^2 + \frac{c}{4}\left[\left(\boldsymbol{\pi}^\dagger \cdot \boldsymbol{\pi}^\dagger\right) - \left(\boldsymbol{\sigma}^\dagger\right)^2\right]\left[(\boldsymbol{\pi}\cdot\boldsymbol{\pi}) - (\boldsymbol{\sigma})^2\right] . \end{aligned} \tag{2}$$

The $SU(3)$ Hamiltonian contains a third-order interaction. Because the interaction proportional, for example, to the $\boldsymbol{C}_2(n_\pi, 0)$ interaction is given by $\boldsymbol{n}_\pi(\boldsymbol{n}_\pi + 3)$, this term dominates for large n_π and destroys the mean field part $\hbar\omega\boldsymbol{n}_\pi$. In models with a fixed number of n_π this is avoided by limiting the model space to few $\hbar\omega$ excitations. However, we include mixing of n_π and this requires to test convergence for increasing N. This type of Hamiltonian was applied successfully in several studies, of which we cite here only a few [6, 7].

The semi-classical potential is obtained by choosing first a trial state $\mid\alpha\rangle$ and then calculating the expectation value of the Hamiltonian with respect to this normalized trial state, i.e., $V(\alpha) = \langle\alpha \mid \boldsymbol{H} \mid \alpha\rangle$. Here α is a short hand notation for the variables of the state.

As a trial state we use

$$\begin{aligned} |\alpha\rangle &= \mathcal{N}_{N,n_0}(\alpha\cdot\pi^\dagger)^{n_0}\left[\sigma^\dagger + \left(\alpha\cdot\pi^\dagger\right)\right]^N |0\rangle \\ &= \mathcal{N}_{N,n_0}\frac{N!}{(N+n_0)!}\frac{\mathrm{d}^{n_0}}{\mathrm{d}\gamma_1^{n_0}}\left[\sigma^\dagger + \gamma_1\left(\alpha\cdot\pi^\dagger\right)\right]^{N+n_0}|0\rangle \ , \end{aligned} \tag{3}$$

where, for convenience, we redefined the total number of relative oscillator quanta as $(N + n_0)$, while the γ_1 parameter has to be set equal to 1 after the differentiation, and the normalization factor $\mathcal{N}_{N,n_0}$ is given in [8]. Already here we obtain an essential difference with respect to the case when the Pauli principle is not observed. As shown in [8], the α variable is proportional to $\sqrt{n_0}$ and also proportional to the separation of the clusters. Thus the so-called vibrational limit ($\alpha = 0$) describes already a deformed system. Nevertheless, we shall keep the notation as it is. A special limit arises when n_0 is set to zero, i.e. no Pauli exclusion principle is observed. Models with this property we call PACM (*Phenomenologic Algebraic Cluster Model*).

The resulting potential has a surprisingly simple structure. The main part depends on only three parameters and is given by

$$\tilde{V}(\alpha) = \left(A(x,y)\alpha^2 \frac{F_{11}(\alpha^2)}{F_{00}(\alpha^2)} - B(x,y)\alpha^4 \frac{F_{22}(\alpha^2)}{F_{00}(\alpha^2)} + \alpha^6 \frac{F_{33}(\alpha^2)}{F_{00}(\alpha^2)} - C(x,y)\alpha^2 \frac{F_{20}(\alpha^2)}{F_{00}(\alpha^2)} \right) . \quad (4)$$

The expressions for $F_{pq}(\alpha^2)$ as well as for the coefficients $A(x,y)$, $B(x,y)$ and $C(x,y)$ can be found in Ref. [9]. They are functions of all parameters of the model. *It is important to note that although the Hamiltonians depends on many parameters, the semi-classical potential depends effectively only on three of them. Thus the discussion of phase transition properties is reduced to the variation of three parameters only!* The same potential could, in principle, be obtained in the PACM too, by paying the price of the introduction of non-polynomial interactions. This is similar to simulating Fermi repulsion in nucleus-nucleus collisions with a potential that is repulsive for small distances.

Phase transition is investigated, following the steps: i) The minima of the potential are determined numerically. There will appear only two possible minima, a deformed ($\alpha_2 > 0$) and a spherical one, the latter corresponding to $\alpha_1 = 0$. The point $\alpha_1 = 0$ is always an extremum. ii) The three parameters are varied and those points are determined, where the two minima coincide. These points, which form a surface in the space of the parameters, are at the phase transition boundary. iii) The corresponding α variables are substituted into the potential. For the spherical minimum $\alpha_1 = 0$ is always zero, which results in an expression for the potential which is *identical zero*, and therefore the derivatives of it with respect to the parameters is always zero. iv) The derivatives with respect to the parameters are determined and compared at the point of phase transition. Due to the structure of the potential a simple rule could be derived [9, 10]: If α_2 at the point of phase transition is greater than zero, then the first derivatives in the two potential minima are different and the transition is, therefore, of first-order. However, when α_2 tends to zero approaching the point of phase transition, then the phase transition is of second order. The new feature is that from a given point on, no phase transition appears and a critical line is obtained. As a conclusion, the SACM permits first- and second-order phase transitions. The situation is illustrated in fig. 1. In the right panel α_2 is plotted versus A and C (B is determined via the condition that only points at the surface of phase transition are considered). The area where α_2 is zero represents the surface of second-order phase transition, while the other area corresponds to the surface of a first-order phase transition. In the left panel this surface is plotted in the space of the parameters. Note the appearance of a dashed line (the solid line separates the first- from the second-order phase transitions). Beyond this line no phase transition appears and this represents a critical line.

The same considerations were applied, i.e. in the PACM, where the minimal number of π-oscillation quanta is zero. Though the Hamiltonian is the same, the semi-classical potential changes significantly. The important part has the form

$$\tilde{V}(\beta) = \left\{ A\beta^2 - B\beta^4 + \beta^6 \right\} , \quad (5)$$

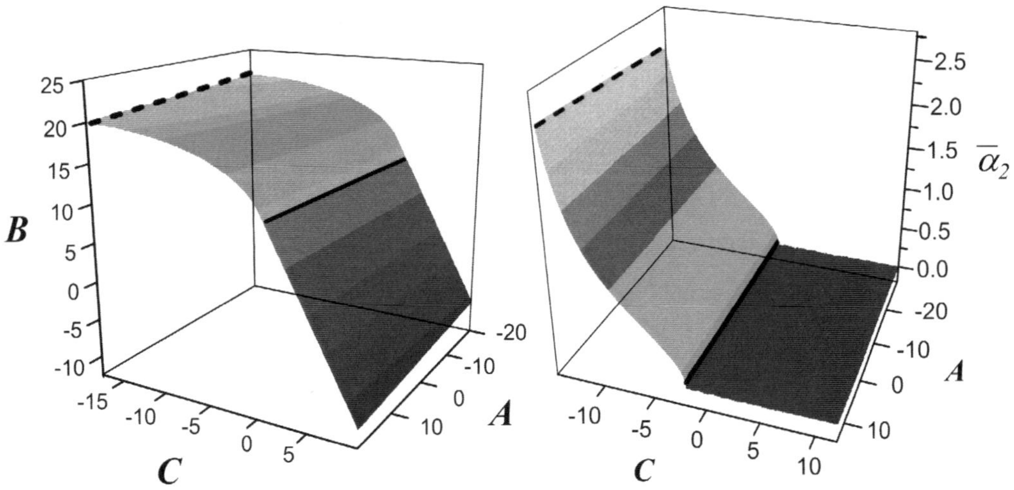

Figure 1. The phase space diagram of the SACM as a function of the independent parameters A, B and C (left panel) . The solid line marks the change from a second- to a first-order phase transition, the dashed line indicates a *critical line.* The variable $\bar{\alpha}_2$ of the deformed solution, as a function in A and C. B is fixed by the requirement that one is at a point of a phase transition (right panel).

with $\beta^2 = \frac{\alpha^2}{1+\alpha^2}$. The expression for the coefficients $A(x,y)$ and $B(x,y)$, can be found in Ref. [9]. The points of phase transitions now form a line. First- and second-order phase transition appear but not a critical point. The first-order phase transition appears due to the inclusion of a third-order interaction in the $SU(3)$ Hamiltonian, otherwise only second-order phase transitions are permitted.

In fig. 2 the space of phase transitions is plotted. There are only two relevant parameters A

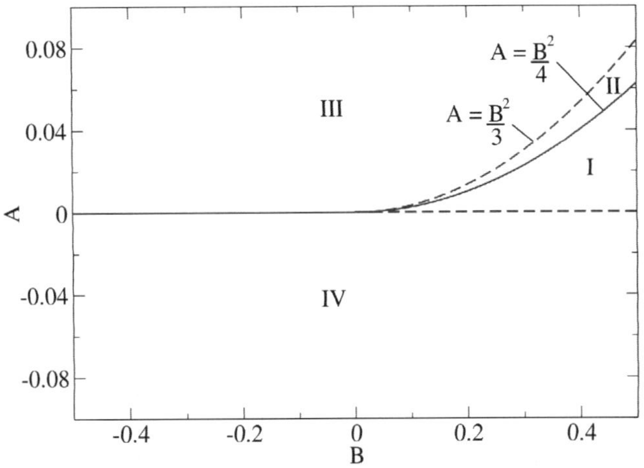

Figure 2. The parameter phase diagram for the PACM. The horizontal axis is B, while the vertical axis corresponds to A. In Region I and Region II two minima exist, one spherical and one deformed. In Region I the global minimum is the deformed one, while in Region II it is the spherical minimum. In Region III only a spherical minimum exists and in Region IV the only minimum is a deformed one.

Finally, we would like to point out severe problems for states at low energy, when the Pauli exclusion principle is not observed. As an example we consider the $^{16}O+\alpha \rightarrow {}^{20}Ne$ system. In the $SU(3)$ limit the ground state would be formed by a 0^+ state in the (0,0) $SU(3)$-irrep, a 2^+ state in (2,0), a 4^+ state in (4,0), etc. Even when one adds the interaction term $\boldsymbol{L}^2$, which gives a rotational $L(L+1)$ structure, simulating a rotational band, this band is not rotational because each state has a different intrinsic structure. Using the $SO(4)$ dynamical symmetry limit gives a quite good agreement to experiment (!), however, the resulting states jump between low n_π and hight n_π, near the cut-off N, making the result dubious. Another problem appears in the geometrical limit: the interaction $\left[\left(\hat{\pi}^\dagger \cdot \hat{\pi}^\dagger\right) - \left(\sigma^\dagger\right)^2\right] \left[(\hat{\pi} \cdot \hat{\pi}) - (\hat{\sigma})^2\right]$ tends to a potential of the type $[af(\alpha)\alpha^2 - bN]$, where $f(\alpha)$ is a simple expression of the order of one and a, b depend on N and n_0 (which is 0 for the PACM and > 0 for the SACM). Removing the cut-off ($N \to \infty$) the minimum of the deformed potential tends to infinity too, being more equivalent to a dissociation limit. This problem arises in both the SACM and PACM, shedding doubt on the physical meaning of the $SO(4)$ dynamical symmetry as a deformed limit.

3. Conclusions

In this contribution we showed that the SACM permits a very rich structure of phase transitions, including first and second order. A critical line also appears. When the Pauli exclusion principle is not observed, and a Hamiltonian of up to third order interactions is applied, then first- and second-order phase transitions still appear, but not a critical line. The results of the SACM could, in principle, be also reproduced within the PACM, but only paying the price of using a very complicated non-polynomial interaction that simulates the Pauli-exclusion principle. We also showed that neglecting the Pauli exclusion principle leads to dubious behavior for the states at low energy, even though the experimental spectrum is well reproduced. This puts some doubt on the argument that when the experiment is reproduced the model is fine. More has to be included, such as the structure of the states and the fulfillment of basic principles, as the Pauli exclusion principle is. These deficiencies may, of course, manifest themselves in failing to reproduce further experimental observables not discussed in the model.

Finally, we discussed the physical interpretation of the $SO(4)$ dynamical symmetry limit, which rather describes a dissociation of the cluster system than a deformed limit.

Acknowledgments

We gratefully acknowledge financial help from DGAPA-PAPIIT (no. IN103212), DGAPA, from the National Research Council of Mexico (CONACyT), OTKA (grant No. K72357), and from the MTA-CONACyT joint project. P.O.H. thanks the Frankfurt Institute for Advances Studies (FIAS) and the GSI for excellent working atmosphere and all the help received.

References

[1] Cseh J 1992 *Phys. Lett.* B **281** 173
[2] Cseh J and Lévai G 1994 *Ann. Phys. (N.Y.)* **230** 165
[3] Elliott J P 1958 *Proc. Roy. Soc.* A **245** 128
Elliott J P 1958 *Proc. Roy. Soc.* A **245** 562
[4] Darai J, Cseh J and Jenkins D G 2012 *Phys. Rev.* C **86** 064309
[5] Sciani W, Otani Y, Lépine-Szily A, Benjamim E A, Chamon L C, Lichtenthäler Filho R, Darai J and Cseh J 2009 *Phys. Rev.* C **80** 034319
Cseh J, Darai J, Sciani W, Otani Y, Lépine-Szily A, Benjamim E A, Chamon L C and Lichtenthäler Filho R 2009 *Phys. Rev.* C **80** 034320
[6] Cseh J and Darai J 2006 *Acta Univ. Debrecen, Ser. Phys. Chim.* **XL** 23
[7] Cseh J and Darai J 2008 *AIP Conf. Proc.* (Fusion08) **1098** 225
[8] Hess P O, Lévai G and Cseh J 1996 *Phys. Rev.* C **54** 2345
[9] Yépez-Martínez H, Fraser P R, Hess P O and Lévai G 2012 *Phys. Rev.* C **85** 014316
[10] Fraser P R, Yépez-Martínez H, Hess P O and Lévai G 2012 *Phys. Rev.* C **85** 014317

Journal of Physics: Conference Series **436** (2013) 012034 doi:10.1088/1742-6596/436/1/012034

Multicluster solutions of the Elliott SU(3) nucleonic Hamiltonian and alpha-cluster level density

I A Gnilozub, S D Kurgalin and Yu M Tchuvl'sky

Skobeltsyn Institute of Nuclear Physics, Lomonosov Moscow State University, 119991, Vorob'evy Gory, Moscow, Russia

E-mail: tchuvl@nucl-th.sinp.msu.ru

Abstract. Properties of the dense α-cluster spectra of even-even nuclei with masses falling in the range 30 – 40 are analyzed. The spectrum of ^{32}S nucleus is chosen as an example. The generalized Hamiltonian of the Elliott SU(3) model is invoked for the interpretation of the data.

The progress of the method of super-thick target [1] brings a bulk of experimental data on resonance α-cluster spectra of even-even nuclei. The spectra of ^{32}S [2], ^{34}S [3] and ^{40}Ca [4] are studied in detail. The properties of these spectra look enigmatic.

First, there is no chance to explain the density of the α-particle levels in the two-body model. To do this one needs an α-nucleus potential with a normal depth (~150 MeV) and a width of about 15 fm, or with a normal width (5.5 – 6.5 fm) and a depth of about 1.5 GeV. Second, the 'grouping' of the levels of one and the same spin J in certain energy domains is well-displayed. Third, the majority of the levels of a compound nucleus remains invisible. Indeed, a comparison of the α-particle level densities observed in the related domains ρ_α and the total densities of the levels of arbitrary nature calculated in the Gilbert-Cameron formalism ρ_{tot} demonstrates a significant discrepancy as it is presented in table 1. The α-decay characteristics of the proton resonances of the ^{32}S nucleus [5] provide a confirmation of this property.

Table 1. Level densities of the α-particle states and of the total ones in the ^{32}S nucleus

J^π	3^-	4^+	5^-	6^+	7^-	8^+	9^-
$\frac{\rho_\alpha}{2J+1}$ Mev^{-1}	8.0	2.9	5.4	6.7	8.6	5.5	7.6
$\frac{\rho_{tot}}{2J+1}$ Mev^{-1}	98	230	269	271	509	1029	1160

The second property looks more or less clear. The levels which may be observed are located in the domains related to the top of the potential barrier. The levels occuring lower are too small in width and the ones located higher are overlapped and thus shape a background. So this property can be explained by the 'window of observability' appearing at the top of the barrier.

But why the density of the levels populated in the α-nucleus collisions is small? To answer this question let us consider the approach developed in refs. [6, 7]. Briefly the formalism of the approach is the following.

Let us consider an A-nucleon system which is described by a Hamiltonian $\hat{H}$ characterized by sets of the eigenvalues (EV's) $\{E_\Delta\}$ and the eigenfunctions (EF's) $\{\Psi_\Delta\}$. The subset of A-nucleon eigenstates of the Hamiltonian $\hat{H}$ characterized by the EV's $\{E_\Delta\}$ and the EF's $\{\Psi_\Delta\}$ may be defined as a family of multicluster states of a partition $\{X_i\}$ ($\sum X_i = A$) if there exists a Hamiltonian $\hat{H}_{cl}$ acting in the reduced space of intercluster coordinates $\{\rho\}$ possessing the same EV's $\{E_\Delta\}$ and the EF's $\{\Psi_\Delta\}$ which fulfill the condition

$$\Psi_\Delta = \hat{A}\{\prod_{i=1}^{k}[\Psi_{X_i}]\hat{N}^{-1/2}\Phi_\Delta(\{\rho\})\} , \tag{1}$$

where $\hat{A}$ is the antisymmetrizer, Ψ_{X_i} are the wavefunctions (WFs) of clusters, $\hat{N}$ is the norm kernel of the multicluster resonating group model, while $\Phi_\Delta(\{\rho\})$ is the EF of the Hamiltonian H_{cl} related to the EV $\{E_\Delta\}$.

The problem of multicluster states possesses a precise solution if the A-nucleon system is described by the Hamiltonian of the Elliott SU(3) model [8] or a Hamiltonian of more general type, which can be expressed in terms of commutative Casimir operators of SU(3) group and its subgroups:

$$H = \hat{H}_{osc} + \beta(-1)^\pi + F_1(L^2) + F_2(\hat{g}_2) + F_3(\hat{g}_3) + F_4(\hat{\Omega}) , \tag{2}$$

where π is parity, $\hat{g}_i$ is the rank i Casimir operator of the SU(3) group, while $\hat{H}_{osc}$ and Ω are the oscillator Hamiltonian and the Bargman operator, respectively. They are also operators of the Casimir type. The Hamiltonian H_{cl} must also take the form (2) and such a multicluster system should be composed of oscillator magic clusters (SU(3)-scalars). The spectrum of such a Hamiltonian is determined by the quantum numbers (QN's) of the SU(3) group and of the related subgroups independently of the choice of the system, and in this case the QN's are determined by the symmetry of the relative motion WF $\Phi_\Delta(\{\rho\})$.

A two-body WF of the relative motion in the entrance channel of an α+A collision can be expanded onto the following series:

$$\phi_E^L(R) = \sum_{n=0}^{\infty} C_n^L(E)\varphi_{nL}(R) , \tag{3}$$

where $\varphi_{nL}(R)$ is an oscillator WF related to the $(n0)$ irreducible representation (IR) of the SU(3) group. If a WF of the target A Ψ_A is the WF of an $A/4$-α-particle state related to the IR of the SU(3) group $(\lambda_c\mu_c)$, then the component of WF (3) of the channel related to a certain value n turns out to be a superposition of some $(A/4{+}1)$-α-particle WF's characterized by the Elliott SU(3) symbols $(\lambda\mu) \subset (\lambda_t\mu_t) \times (n0)$. Obviously, the discussed channel is strongly coupled with multi-α-particle resonances with such QN's, whereas other resonances (ES's of H) are weakly coupled with it. So the natural qualitative explanation of the discussed properties is given.

To estimate the density of the α-particle states quantitatively, the spectrum of α-cluster states of the ^{32}S nucleus is calculated in the current work using the five-parameter linear Hamiltonian (2). The fitting procedure is based on the data for the lower levels of the ^{32}S nucleus and for two groups of 9^- levels observed in the experiment [2]. The oscillator frequency ω is not a fitted parameter. The results of the calculations are partially presented in table 2 together with the experimental data. The table is similar to the one contained in the paper [9], but the one at hand is based on the more recent experimental data [2]. For brevity we demonstrate only odd-parity levels here.

In this table M_J^{exp} denotes the number of the observed levels, ΔE_J the energy range in which these levels are observed, $(\lambda_c\mu_c)$ the SU(3) Elliott symbol of the set of degenerate levels with certain J^π with sub-index k labelling the values of the Bargman operator from minimal to

Table 2. Observed and calculated odd-parity α-particle levels of the ^{32}S nucleus.

J^π	M_J^{exp}	ΔE_J(MeV)	$(\lambda\mu)_k$	E(MeV)	η	M_J^Σ
3^-	25	11.3 – 14.4	$(47)_2$	13.7	15	30
			$(47)_1$	14.2	15	
5^-	17	14.1 – 17.3	$(58)_1$	14.7	5	34
			$(47)_3$	16.4	15	
			$(47)_2$	17.3	15	
7^-	21	17.4 – 19.8	$(96)_1$	17.6	1	26
			$(58)_2$	17.7	5	
			$(58)_1$	18.9	5	
			$(47)_3$	20.2	15	
9^-	18	21.0 – 23.4	$(69)_1$	20.6	1	13
			$(58)_3$	21.5	5	
			$(96)_3$	21.6	1	
			$(96)_2$	23.2	1	
			$(58)_2$	23.5	5	

maximal, finally E and η the energy of the centroid and multiplicity of the set. The multiplicity η appears due to the fact that there may be a number of shell-model states with the same global quantum numbers. The Hamiltonian (2) does not remove their degeneracy. Taking into account that the full-scale Hamiltonian may remove the degeneracy, the centroids falling into the range ΔE_J or deviating from it by less than 0.5 MeV are considered. M_J^Σ is the total number of calculated levels falling in such an interval.

The results of the numerical calculations are found to be in rather good agreement with the experimental data. There are few calculated levels with each J below the respective ΔE_J range and many levels above. The gaps between these levels and the levels presented in the table are not small, therefore they are invisible for the experiment. The discrepancies between the experimental M_J^{exp} and the theoretical multiplicities M_J^Σ probably appear due to the smallness of the α-spectroscopic factors (SF's) of some states (in the case $M_J^{exp} < M_J^\Sigma$), or due to the accidental proximity of levels resulting in the distribution of the α-particle component between few levels of the full-scale Hamiltonian (in the case $M_J^{exp} > M_J^\Sigma$). The even-parity spectrum possesses the same properties: $M_J^{exp} < M_J^\Sigma$ for $J = 4, 6$, and $M_J^{exp} > M_J^\Sigma$ for $J = 8$.

In summary, the selection rules inherent to the α-resonant scattering in the generalized Elliott SU(3) model adequately explain why the density of α-particle states is so small.

References

[1] Goldberg V Z and Pakhomov A E 1993 *Phys. At. Nucl.* **56** 406
[2] Lönnroth T *et al* 2010 *Eur. Phys. J.* A **46** 5
[3] Norrby M *et al* 2011 *Eur. Phys. J.* A **47** 73
[4] Norrby M *et al* 2011 *Eur. Phys. J.* A **47** 96
[5] Feng D 1987 *Proton Resonance Spectroscopy in* ^{32}S (PhD Thesis, Shanghai: Fuden Univ.)
[6] Gnilozub I A, Kurgalin S D and Tchuvil'sky Yu M 2006 *Phys. At. Nucl.* **69** 1014
[7] Gnilozub I A, Kurgalin S D and Tchuvil'sky Yu M 2007 *Nucl. Phys.* A **790** 687c
[8] Elliott J P 1958 *Proc. Roy. Soc.* A **245** 128
[9] Gnilozub I A, Kurgalin S D and Tchuvil'sky Yu M 2011 *AIP Conf. Proc.* **1377** 441

Journal of Physics: Conference Series **436** (2013) 012035 doi:10.1088/1742-6596/436/1/012035

Algebraic approach for α-cluster structure in ^{12}C and ^{16}O

T Yoshida[1], N Itagaki[2] and K Katō[3]

[1]Center for Nuclear Study, University of Tokyo, 7-3-1 Hongo, Bunkyo, Tokyo 113-0033, Japan,
[2]Yukawa Institute for Theoretical Physics, Kyoto University, Kyoto 606-8502, Japan,
[3]Nuclear Reaction Data Centre, Graduate School of Science, Hokkaido University, Sapporo 060-0810, Japan

E-mail: yoshida@cns.s.u-tokyo.ac.jp

Abstract. We investigate the properties of the states observed around the α-cluster breakup threshold region in ^{12}C and ^{16}O . We apply the orthogonality condition model (OCM), where the model space is given by the $SU3$ algebra. From bases of this $SU3$ configuration, two algebraic approaches of the symplectic (sp) method and μ-truncation method are introduced. In ^{12}C, we show that the sp-quanta are important to understand the mechanism of the appearance of the large $E0$ transition strength. With these methods, both strong- and weak-coupling configurations are taken into account at the same time. Therefore, the properties of gas-like and linear-chain states are investigated easily. In ^{16}O, a simpler truncation utilizing the μ quanta is applied. Here, the energy levels and the linear-chain (L. C.) configurations are investigated.

1. Introduction

Recently, the α-cluster structure of the states observed around the α-cluster breakup threshold energy region has been investigated from various viewpoints. For instance, the large iso-scalar monopole ($E0$) transition strength has been observed in the vicinity of the α-cluster threshold for many light nuclei such as ^{12}C, ^{16}O and ^{24}Mg [1, 2]. These states are the candidates of the cluster states. Some of them include dilute α-cluster configurations, in which each α-particle moves with a relatively weak interaction. One of the remarkable states is α-cluster gas state in the excited energy region of ^{12}C [3]. It is expected that such a kind of exotic state can be observed in wider mass and energy regions due to the development of experimental technique or theoretical framework.

From the theoretical viewpoint, the α-cluster structure like a gas-like state [3] and coherent α-cluster configuration associated with the large $E0$ transition strength have been studied [4, 5]. We present our recent analyses of the excited states of ^{12}C and ^{16}O by using many α-cluster model. The symplectic (sp) structure $Sp(2,R)_z$ [6, 7, 8, 5] has been found to be important in understanding the large $E0$ strength. Because the sp algebraic approach becomes complicated in the analysis when the number of α-particles increases, we introduce the μ-truncation method in the study of ^{16}O.

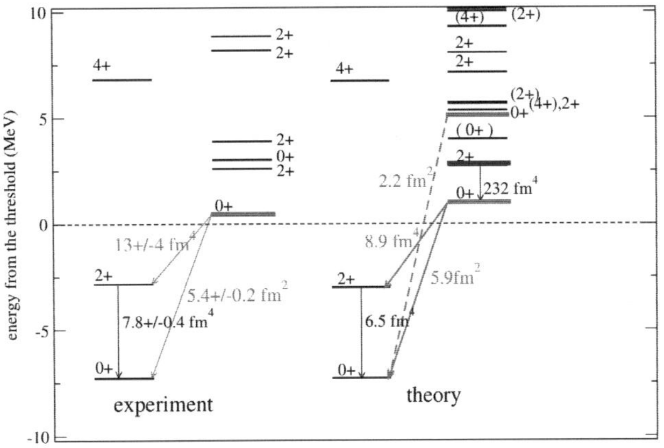

Figure 1. Comparison of the ^{12}C experimental data with the present calculation. The transition strength between 0^+ and 2^+ is also shown.

2. Results

In our three-α-cluster model, the $SU(3)$ model wave function is given by $V_{N_1,N_2}^{N(\lambda,\mu)J,\kappa}(\vec{r},\vec{R})$, where $\lambda = N_z - N_y$ and $\mu = N_y$ are evaluated from the total harmonic oscillator (H. O.) quanta for each 3D axis. The quantum number κ indicates the orthonormalized K quantum number for the intrinsic frame, and $(\vec{r},\vec{R})$ are Jacobi-coordinates to describe three-α-clusters. The use of the Pauli-allowed states can be performed in order to employ orthogonality condition model [9]. The Pauli-allowed state is obtained by the following transformation:

$$U_i(\vec{r},\vec{R}) \quad = \quad \sum_{N_1+N_2=N} A_{N_1,N_2}^{(\lambda,\mu)\rho} V_{N_1,N_2}^{N(\lambda,\mu)J,\kappa}(\vec{r},\vec{R}),$$

where i denotes a set of $(\lambda,\mu)\rho, J, \kappa$ quanta. The ρ shows the multiplicity of the Pauli-allowed state [10]. This ρ quanta is converted to the more physical basis state as follows.

$$w_\alpha(\vec{r},\vec{R}) = \sum_\rho C_\rho^\Lambda U_i(\vec{r},\vec{R}).$$

Here the coefficients C_ρ^Λ and the quantum number Λ are obtained as eigenvectors and eigenvalues, respectively of the Casimir operator of $Sp(2,R)_z$ [8]. In the transformed basis, each Λ is classified with respect to the band-head states that is connect to the specific Λ quantum numbers by the ladder operator.

In fig. 1 we show the energy levels obtained by experiments and the present calculation. Here we use the phenomenological inter-three-α interaction, which is optimized to reproduce the ground-band energy levels [11]. In the figure we show the energy levels, which are stable with respect the maximum H. O. quanta. The energy convergence of the Hoyle state is very slow as the number of H. O. quanta increases. The states in parentheses has slower energy convergence in comparison with that of the Hoyle state.

The $E2$ transition strength from the Hoyle state to the 2_1^+ state is in accordance with the large observed value. The calculated $E0$ and $E2$ transition are in a good agreement with observed values with error ranges. The other $E2$ strength from 2_2^+ to the Hoyle state becomes enormous (232 fm^4). From this analysis, we can see that the study to investigate the $E0$ transition strength can be expanded to other transition modes.

To investigate ^{16}O in a four-α model, we use the μ-truncation method. The Pauli-allowed basis state is obtained similarly to the three-α case by using three Jacobi-coordinates, $(\vec{r},\vec{R},\vec{\mathscr{R}})$ to describe four-α-clusters. In the μ-truncation procedure the eigenstate $u_k(\vec{r},\vec{R},\vec{\mathscr{R}})$ calculated

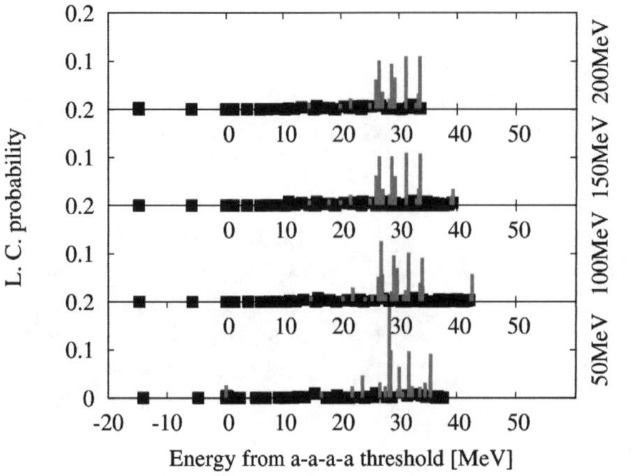

Figure 2. Probability of L. C. configuration for 0^+ (^{16}O) states for each eigenenergy. The right-hand vertical axis indicates the cutoff energy of the μ-truncation.

within subspace which is truncated by specific μ quanta and eigenenergies smaller than an appropriate cutoff energy. These eigenstates are expressed by using the original basis as follows,

$$u_k(\vec{r}, \vec{R}, \vec{\mathscr{R}}) = \sum_i D_i^k U_i(\vec{r}, \vec{R}, \vec{\mathscr{R}}) \ .$$

Here the model space is constructed with H. O. quanta, $N \leq 32$. We obtained good energy convergence with respect to the energy cutoff. In fig. 2, the L. C. state energy region is also estimated. The pure L. C. state is specified by the condition $\mu = 0$ and $N = \lambda$ in the case of four-α-cluster model. The horizontal axes show the 0^+ energy from the four-α threshold. The vertical axes show the probability of L. C. configuration. The cutoff parameter changes from 50 MeV (lower panel) to 200 MeV (upper panel). The colored bars present large L. C. strength. The strength is estimated by the overlap between the wavefunctions obtained by the μ-truncation with $\mu \leq 16$ quanta. The results show that the L. C. strength is about 10 % for each cutoff parameter. The strength is smaller than that for the L. C. states in ^{12}C, which have about 50 % pure L. C. configuration.

In both nuclei the algebraic model is found to work well. However, to confirm the existence of L. C. as the same accuracy as the gas-like states, larger N quanta are needed especially for ^{16}O.

References

[1] Kawabata T *et al* 2007 *Phys. Lett.* B **646** 6
[2] Bayman B F and Bohr A 1959 *Nucl. Phys.* **9** 596
[3] Tohsaki A, Horiuchi H, Schuck P and Röpke G 2001 *Phys. Rev. Lett.* **87** 192501
[4] Yamada T, Funaki Y, Horiuchi H, Ikeda K and Toshaki A 2008 *Prog. Theor. Phys.* **120** 1139
[5] Yoshida T, Itagaki N and Katō K 2011 *Phys. Rev.* C **83** 024301
[6] Arickx F 1976 *Nucl. Phys.* A **268** 347
[7] Suzuki Y, Arai K, Ogawa Y and Varga K 1996 *Phys. Rev.* C **54** 2073
[8] Katō K and Tanaka H 1989 *Prog. Theor. Phys.* **81** 841
[9] Saitō S 1960 *Prog. Theor. Phys.* **41** 705
[10] Horiuchi H 1989 *Prog. Theo. Phys.* **58** 204
[11] Schmidt E W and Wildermuth K 1961 *Nucl. Phys.* A **26** 463
Kurokawa C and Katō K 2007 *Nucl. Phys.* A **792** 87

Journal of Physics: Conference Series **436** (2013) 012036 doi:10.1088/1742-6596/436/1/012036

Comparing the group theoretical approaches to the description of dipole and quadrupole collectivity in nuclei

A I Georgieva

Institute of Nuclear Research and Nuclear Energy, Bulgarian Academy of Sciences, 72 bul. Tzarigradsko Chaussee, Sofia 1784, Bulgaria

E-mail: `anageorg@inrne.bas.bg`

Abstract. The comparison of the two types of collectivity in nuclear dynamics is based on the equal realizations of their basic symmetries, represented by the algebra-chain $u(3) \supset su(3) \supset so(3)$. We illustrate this on the example of two phenomenological algebraic models: the semimicroscopic algebraic cluster model (SACM) with one of the clusters considered as a scalar and the interacting two vector boson model (IVBM). In both of them the root group structure that leads to the above chain is based on the coupling of two different $U(3)$ representations, with different physical interpretations. These similarities allow for the use of the same mathematical tools in building the theory, but there are important differences that come from the different higher symmetries in which this equal substructures are embedded.

1. Introduction

The nucleus is a unique and sophisticated many-body system which at the present stage is impossible to be described with the direct application of the quantum mechanics. For this reason the accumulated experimental data is interpreted in the framework of different models, focusing its basic assumptions on specific properties of the nuclei. These models are based on different physical pictures. The shell model describes it as a small atom, considering microscopically the individual movements of the nucleons on their specific orbits. The collective model considers the nucleus as a liquid drop, introducing the quadrupole collectivity, which describes the deformation of the nuclear shape. The cluster model makes it similar to a molecule formed by two smaller systems and is based on the dipole degrees of freedom. Therefore, in order to really understand the nuclear structure it is of utmost importance to find the exact connection of these models.

In this contribution we discuss the interrelation of the two different types of collectivity, by comparing two models, exploring each of them: the Interacting Two Vector Boson model (IVBM) [1] and the Semimicroscopic Algebraic Cluster Model (SACM) with one of the clusters considered as a scalar [2]. This investigation is based on the comparison of the common algebraic structures used in both models, and on the comparison with the observed experimental spectra. As an example of the physical applications, the possible clusterizations of the shape isomers (superdeformed, hyperdeformed etc.) states of some nuclei are presented [3, 4].

2. Algebraic structures

2.1. *The basic SU(3) symmetry*

The basic connection between the quadrupole shape and clusterization was found in the fifties, via the bridge of the shell model. This relation is based on the reduction chain:

$$\begin{array}{ccccccc} U(3) & \supset & SU(3) & \supset & SO(3) & \supset & SO(2) \\ [n_1,n_2,n_3] & & (\lambda,\mu) & K & L & & M \end{array} \tag{1}$$

introduced by Elliott [5], and relates the invariant operators of the algebras in (1) to the observables – quadrupole-quadrupole interaction Q_m and angular momentum L_m, of the described system. The chain (1) also gives the harmonic oscillator basis, hence in this way the quadrupole deformation and collective rotation can be derived from the spherical shell model: the states belonging to a collective band are determined by their specific $SU(3)$ symmetry. Elliott's model has the advantage of allowing for a geometrical analysis of the eigenstates of a nuclear system via the relations between the microscopic parameters (λ, μ) (1) and the parameters (β, γ) of the collective model [6, 7]: $\beta^2 \sim \lambda^2 + \mu^2 + 3(\lambda + \mu) + \lambda\mu$ and $\gamma = \tan^{-1}\left(\frac{\sqrt{3}\mu}{(2\lambda+\mu+3)}\right)$.

Wildermuth and Kanellopoulos [8] established the relation between the shell and cluster models. They have shown that the Hamiltonians of the two models can be rewritten into each other exactly in the harmonic oscillator approximation. This relation results in a close connection between the corresponding eigenvectors, too: the wave function of one model is a linear combination of those of the other, which belong to the same energy. Later on this relation was interpreted by Bayman and Bohr [9] in terms of the $SU(3)$ symmetry. As a consequence, the cluster bands can be considered from the shell model viewpoint, like a set of states having special $SU(3)$ symmetry, just like the collective rotational bands.

2.2. *Direct product coupling*

The further elaboration of the algebraic collective nuclear models is based on the consideration of more realistic assumptions, considering two interacting systems of particles. The models presented in this contribution – the IVBM [1] and the SACM [2] are exactly of this type.

The IVBM is based on the introduction of two kinds of vector bosons (called p and n bosons), that 'build up' the collective excitations in the nuclear system. The creation operators of these bosons are assumed to be $SO(3)$ vectors and they transform according to two independent fundamental representations (1,0) of the group $SU(3)$. These bosons form an 'F-spin' doublet of the $U(2)$ group and differ in its projection. The introduction of this additional degree of freedom leads to the extension of the $SU(3)$ symmetry to $U(6) \supset U_p(3) \oplus U_n(3) \supset SU(3) \supset SO(3)$. The bilinear products of the creation and annihilation operators of the two vector bosons generate the noncompact symplectic group $Sp(12, R)$ [1]. The symplectic extension of the $u(6)$ algebra with the dynamical symmetry $Sp(12, R) \supset U(6) \supset U(2) \otimes U(3)$ [1] allows the change in the number of phonons needed to build the collective states, which results in larger model spaces, that can accommodate the more complex structural effects observed in the contemporary experiment.

The basic assumption of all the cluster models is that the relevant degrees of freedom of the nucleus are classified into two categories. Some of them account for the relative motion of the clusters, while others are related to the internal degrees of freedom of the clusters. We consider here binary cluster systems, with the interactions of the relative motion described by the vibron model of $U(4) \supset U(3)$ group structure [10]. The internal structure of the clusters is described by the Elliott model (1) [5], where $U(3)$ refers to the orbital part. From the viewpoint of the interaction the group structure simplifies to $U_{C_1}(3) \otimes U_{C_2}(3) \otimes U_R(3) \supset U(3) \supset O(3)$, where C_i stands for the i'th cluster, and R indicates relative motion. The Hamiltonian built up from the invariant operators of this chain has a $U(3)$ dynamical symmetry. The $U(3)$ limit of the algebraic cluster models corresponds to a soft vibrational-rotational motion in the collective model terms.

When one of the clusters is a scalar, like an α-particle for example, the $U_{C_2}(3)$ is neglected and practically the direct product coupling $U_{C_1}(3) \otimes U_R(3) \supset U(3) \supset O(3)$ is explored. From the microscopic viewpoint the $U(3)$ dynamical symmetry describes shell-model-like clusterization, when the cluster model state can be expressed in terms of a few (or a single) shell model basis states [11]. Furthermore, this symmetry is also the matching point with the quadrupole collective IVBM.

3. Application

The connection between the quadrupole and dipole (cluster-like) collectivity is revealed by the common symmetry characters, as discussed above. Both features are observed when the SACM is applied in the $SU(3)$ dynamical symmetry approximation. Then the cluster structure is evident from the model assumption, and the quadrupole shape of the state is uniquely determined by the $SU(3)$ quantum numbers. It was found that this dynamical symmetry gives a reasonable description of the detailed spectra in several cases, and it can reproduce the gross features especially well [11].

A very interesting phenomenon is the existence of the shape isomers, e.g. superdeformed (SD) and hyperdeformed (HD) states, and their possible clusterization [12, 13]. Symmetry-based methods can be very fruitful in their studies. Here we refer to an approach, which is closely related to the algebraic cluster model mentioned before. It is a two-step procedure [14]. First the shape isomers are obtained from a Nilsson-calculation by requiring the self-consistency of the quadrupole shape [15, 16, 17]. When doing so, the quasi-dynamical $U(3)$ symmetry is applied, which coincides with the real $U(3)$ symmetry in the simple cases when this latter one is valid too [18]. In the second step the $U(3)$ selection rule is used for the determination of the allowed clusterizations [19, 20, 21]. The model predictions can be compared directly to the experimental observation, and in some cases, like for the SD state of ^{28}Si [17] and for the HD state of the ^{36}Ar [22, 23, 24], they have been justified.

In conclusion, in this contribution it was established that the interrelation of the two approaches are due to the basic foundation of each of them by the microscopic shell model of nuclei, which also has an algebraic analogue: the $SU(3)$ model of Elliott. The comparison of the two models of nuclear dynamics is due to the common realization of their basic symmetries. The common intersection of the three basic structure models is the $U(3) \otimes U(3)$ dynamical symmetry. This finding is a generalization of the previously known result for the case of the single shell problem, when the intersection is the $U(3)$ dynamical symmetry.

Acknowledgments

This work is a result of a collaborative research with J. Cseh and J. Darai from the Insitute of Nuclear Research of the Hungarian Academy of Sciences and was supported by a JOINT RESEARCH PROJECT No. 7 'The atomic nucleus and neutron physics' between the Bulgarian Academy of Sciences and the Hungarian Academy of Sciences and the Bulgarian National Foundation for Scientific Research trough contract DID-02/16/17.12.2009.

References

[1] Georgieva A I, Ganev H G, Draayer J P and Garistov V P 2009 *Physics of Particles and Nuclei* **40** 461
[2] Cseh J 1992 *Phys. Lett.* B **281** 173
Cseh J, Lévai G 1994 *Ann. Phys. (NY)* **230** 165
[3] Darai J, Cseh J, Lépine-Szily A, Algora A, Hess P O, Antonenko N V, Jolos R V and W. Scheid 2010 *J. Phys.: Conf. Series* **205** 012022
[4] Darai J, Cseh J, Antonenko N V, Adamian G G and Georgieva A 2012 *J. Phys.: Conf. Series* **366** 012009
[5] Elliott J P 1958 *Proc. Roy. Soc. London* A **245** 128
Elliott J P 1958 *Proc. Roy. Soc. London* A **245** 562
[6] Bohr A, Mottelson B R and Pines D 1958 *Phys. Rev.* **110** 936

[7] Castaños O, Draayer J P and Leschber Y 1988 *Z. Phys.* A **329** 33
Leschber Y and Draayer J P 1987 *Phys. Lett.* **B190** 1
[8] Wildermuth K and Kanellopoulos Th 1958 *Nucl. Phys.* **7** 150
[9] Bayman B F and Bohr A 1958/59 *Nucl. Phys.* **9** 596
[10] Iachello F 1981 *Phys. Rev.* C **23** 2778
[11] Lévai G, Cseh J and Scheid W 1992 *Phys. Rev.* C **46** 548
Cseh J, Lévai G and Scheid W 1993 *Phys. Rev.* C **48** 1724
Cseh J 1994 *Phys. Rev.* C **50** 2240
Cseh J, Lévai G, Ventura A and Zuffi L 1998 *Phys. Rev.* C **58** 2144
Cseh J, Gupta R K and Scheid W 1992 *Phys. Lett.* B **299** 205
Lévai G and Cseh J 1996 *Phys. Lett.* B **381** 1
[12] Cseh J, Algora A, Darai J and Hess P O 2004 *Phys. Rev.* C **70** 034311
[13] Algora A, Cseh J, Darai J and Hess P O 2006 *Phys. Lett.* B **639** 451
[14] Darai J, Cseh J, Adamian G G and Antonenko N V 2012 *Eur. Phys. J. Web of Conf.* **38** 16001
[15] Cseh J, Darai J, Sciani W, Otani Y, Lépine-Szily A, Benjamim E A, Chamon L C and Lichtenthäler Filho R 2009 *Phys. Rev.* C **80** 034320
[16] Darai J, Cseh J, Antonenko N V, Royer G, Algora A, Hess P O, Jolos R V and Scheid W 2011 *Phys. Rev.* C **84** 024302
[17] Darai J, Cseh J and Jenkins D G 2012 *Phys. Rev.* C **86** 064309
[18] Hess P O, Algora A, Hunyadi M and Cseh J 2002 *Eur. Phys. J.* A **15** 449
[19] Cseh J and Scheid W 1992 *J. Phys.* G **18** 1419
[20] Cseh J, Darai J, Algora A, Yépez-Martinez H and Hess P O 2006 *Rev. Mex. Fis. S.* **52** 1
[21] Cseh J and Darai J 2009 *AIP Conf. Proc.* **1098** (Fusion08) 225
[22] Cseh J, Darai J, Algora A, Yépez-Martinez H and Hess P O 2008 *Rev. Mex. Fis. S.* **54** 30
[23] Sciani W, Otani Y, Lépine-Szily A, Benjamim E A, Chamon L C, Lichtenthäler Filho R, Darai J and Cseh J 2009 *Phys. Rev.* C **80** 034319
[24] Cseh J, Darai J, Algora A, Antonenko N V and Adamian G G 2011 *Eur. Phys. J. Web of Conf.* **17** 16001

Journal of Physics: Conference Series **436** (2013) 012037 doi:10.1088/1742-6596/436/1/012037

Cluster correlation in light nuclei

Y Kanada-En'yo[1], **F Kobayashi**[1] **and T Suhara**[2]

[1]Department of Physics, Kyoto University, Kyoto 606-8502, Japan

[2]Institute of Physics, University of Tsukuba, Tsukuba 305-8571, Japan

Abstract. Cluster aspects in light unstable nuclei as well as light stable nuclei are discussed. The cluster structures of ^{13}Be and ^{9}Li were investigated. The present calculation suggests that the breaking of the neutron magicity in ^{13}Be may occur due to the developed cluster structure as in the cases of ^{11}Be and ^{12}Be. Clustering in highly excited states of ^{9}Li was discussed in conjunction with those of ^{10}Be.

1. Introduction

The existence of a variety of cluster states has been suggested in neutron-rich Be, where the 2α core and surrounding excess neutrons play important roles (for example, Ref. [1] and references therein). In addition to such two-centre cluster structures conjectured for Be and Ne isotopes, three-centre cluster structures have been attracting great interest. For instance, a 3α cluster gas state identified with an excited ^{12}C state is one of the recent hot topics and it is often discussed in conjunction with alpha condensation [2]. In neutron-rich C, geometric 3α configurations, such as a linear-chain structure, were proposed to be found in some excited states [3].

Such rich cluster phenomena tell us that clustering is one of the essential features in unstable nuclei just as in stable nuclei. Various kinds of cluster structure emerge depending on the excitation energy, on the number and kinds of core clusters as well as on the number of excess neutrons. In low-lying states, clusters are in general tightly bound and clustering is characterized by the correlation between clusters and the actual way of cluster formation. On the other hand, in highly excited states, one may often see remarkable cluster structures in which clusters weakly couple to each other. Another cluster aspect peculiar to neutron-rich nuclei is the molecular orbital structure in which clusters are bonded by excess neutrons in molecular orbitals.

In the present work, we investigate the structure of low-lying states of ^{13}Be and discuss the breaking of neutron magicity. We also study the ^{6}He+t clustering of excited states of ^{9}Li and their analogy with and difference from ^{6}He+α cluster states in ^{10}Be. We also discuss how clustering properties change from low-lying states to highly excited states as a function of the inter-cluster distance.

2. Formulation

In the framework of antisymmetrized molecular dynamics (AMD) [4], an A-nucleon wave function is given by a Slater determinant of Gaussian wave packets:

$$\Phi_{\rm AMD}(\mathbf{Z}) = \frac{1}{\sqrt{A!}}\mathcal{A}\{\varphi_1, \varphi_2, ..., \varphi_A\}, \quad (1)$$

where the ith single-particle wave function is written as

$$\varphi_i = \phi_{\mathbf{X}_i}\chi_i\tau_i, \tag{2}$$

$$\phi_{\mathbf{X}_i}(\mathbf{r}_j) \propto \exp\left\{-\nu\left(\mathbf{r}_j - \frac{\mathbf{X}_i}{\sqrt{\nu}}\right)^2\right\}, \tag{3}$$

$$\chi_i = \left(\frac{1}{2}+\xi_i\right)\chi_\uparrow + \left(\frac{1}{2}-\xi_i\right)\chi_\downarrow. \tag{4}$$

Here $\phi_{\mathbf{X}_i}$ and χ_i are spatial and spin functions, and τ_i is the isospin function which is fixed to be up (proton) or down (neutron). We take an optimized value of the width parameter ν for each nucleus. Accordingly, an AMD wave function is expressed by a set of variational parameters, $\mathbf{Z} \equiv \{\mathbf{X}_1, \mathbf{X}_2, \cdots, \mathbf{X}_A, \xi_1, \xi_2, \cdots, \xi_A\}$, which indicate the Gaussian centres and the spin orientations for all nucleons. These parameters are determined by energy variation.

To study nuclear structure, energy variation and spin-parity projection are applied to the AMD wave functions. For the study of Be isotopes, the variation after spin-parity projection was performed in the AMD framework (VAP). By choosing specific configurations of $\mathbf{Z}$ and superposing several spin-parity projected AMD wave functions $P^{J\pm}_{M}K\Phi_{\rm AMD}(\mathbf{Z})$, one is able to perform GCM-type (GCM: generator coordiante method) cluster-model calculations in the AMD framework. To study the excited states of ^{9}Li and ^{10}Be, ^{6}He+t-type and ^{6}He+^{4}He-type GCM calculations are performed. We also do ^{12}Be+n-type calculations for ^{13}Be.

For more details of the present calculations, the reader is referred to Refs. [5, 6]. The general formulation of the AMD method is expounded in Refs. [7, 8].

The effective nuclear interactions used in the present calculations are the Volkov force and the MV1 force [9, 10] with a spin-orbit force supplemented by the Coulomb force.

3. Results

3.1. *Structure of ^{13}Be*

The structure of Be isotopes has been intensively studied by many groups. In Be, the formation of the 2α-cluster core and the surrounding neutrons play important roles. In ^{11}Be, the vanishing of the neutron magic number is known from the abnormal parity of the $1/2^+$ ground state. In recent years, the breaking of the neutron magicity was also found in ^{12}Be as well as ^{11}Be. The ^{12}Be ground state is a largely deformed state dominated by the intruder $2\hbar\omega$ configuration though ^{12}Be is an $N = 8$ nucleus. The lowering mechanism of these intruder states in ^{11}Be and ^{12}Be is successfully described by the molecular orbital picture as discussed by von Oertzen and many groups [1]. In the molecular orbital model, a 2α core is formed and the valence neutrons occupy the molecular orbitals around the 2α. In the intruder states, the molecular σ orbital is occupied by two valence neutrons. The σ orbital has two nodes along the α-α direction, and it gains kinetic energy as the 2α cluster develops. The σ orbital corresponds to the intruder sd-orbit which comes down into the p-shell in a developed cluster system. This means that the origins of the breaking of the neutron magicity in neutron-rich Be are the formation of the 2α-cluster core and the lowering of the σ orbital in the developed cluster state. Then a question to be answered is whether magicity is also broken in further neutron-rich Be isotopes. To answer this problem we applied the AMD+VAP method to calculate the structure of ^{13}Be.

Recently, the ^{13}Be spectrum has been measured. In the experimental work by Kondo *et al* [11], the low-lying $1/2^-$ state was reported at the energy much below the $5/2^+$ state. Since ^{13}Be is an $N = 9$ nucleus, the lowest state could be the $5/2^+$ if single-particle levels were ordered normally as would be naively expected from a simple shell model. The observation of the low-lying $1/2^-$ state in ^{13}Be is inconsistent with the normal single-particle level ordering and it suggests that $N = 8$ is not a magic number here since the sd-orbit intrudes in the p-shell.

Journal of Physics: Conference Series **436** (2013) 012037 doi:10.1088/1742-6596/436/1/012037

We applied the AMD+VAP method to ^{13}Be and calculated the states up to $J = 5/2$. The calculated energy levels of ^{13}Be are shown in Fig. 1 compared with the experimental energy levels reported in Ref. [11]. The interaction parameters are those adopted in Ref. [12] for ^{11}Be. In the present work, we used the set 1 interaction, which reproduces the abnormal parity ground state $1/2^+$, i.e., the parity inversion in ^{11}Be.

In the present result, the energy levels of ^{13}Be are found to be out of the normal order. The $1/2^-$ and $3/2^+$ states, almost degenerate, lie lowest, with a $5/2^+$ state above them. The $1/2^-$ and $3/2^+$ states are intruder states dominated by $1\hbar\omega$ and $2\hbar\omega$ excited configurations, respectively, while the $5/2^+$ state is described by a normal $0\hbar\omega$ configuration. The appearance of these intruder states at such low energies reveals that the $N = 8$ magicity is broken in ^{13}Be, just as in ^{11}Be.

As is discussed later, the structure of the low-lying ^{13}Be states obtained with AMD+VAP can be interpreted in a ^{12}Be+n picture, in which the ^{12}Be core is in the intrinsic state ^{12}Be(0_1^+), having an intruder $2\hbar\omega$ configuration, or in the state ^{12}Be(0_2^+), with a normal $0\hbar\omega$ configuration. In fact, ^{13}Be is an unbound nucleus and all its states are resonances above the ^{12}Be+n threshold. In such a case, the asymptotic behaviour of the valence neutron wave function in the outer region can be important, but the AMD method is not suitable to take that into account. To see the effects of the spatial extension of the last neutron on the excitation energies, we also calculated the energy levels in a ^{12}Be+n model [5]. In the ^{12}Be+n model the $1/2^-$ state is the lowest.

By analysing the intrinsic structures of the calculated ^{13}Be states, it is found that a 2α core is formed in most of the states just as in ^{10}Be, ^{11}Be and ^{12}Be. In particular, developed 2α cluster structures with large deformations are found in the negative-parity states ^{13}Be($1/2^-$), ^{13}Be($3/2^-$) and ^{13}Be($5/2^-$). In positive-parity states, ^{13}Be($1/2^+$) and ^{13}Be($5/2^+$) show relatively weaker cluster structures with smaller deformations while ^{13}Be($3/2^+$) shows a remarkable 2α structure similar to the negative-parity states. The intrinsic density distributions of the intruder states of ^{11}Be, ^{12}Be and ^{13}Be are shown in Fig. 2. We can say that the breaking of $N = 8$ magicity in ^{13}Be can be understood again as resulting from the lowering of the σ orbital in the developed cluster system just as in ^{11}Be and ^{12}Be.

It is also interesting how the ^{13}Be states can be interpreted as ^{12}Be+n states by adding a neutron on top of the ^{12}Be cores. The $N = 8$ magicity is broken in ^{12}Be. That is to say, the ground state of ^{12}Be is well deformed and it is considered to be dominated by the $2\hbar\omega$ excited configuration, while the second 0^+ state in the low-energy region is described dominantly by the normal $0\hbar\omega$ configuration. In the ^{12}Be+n picture, ^{13}Be($1/2^-$), ^{13}Be($3/2^-$) and ^{13}Be($5/2^-$) contain the ^{12}Be($2\hbar\omega$) core with one neutron in a p-orbit. ^{13}Be($3/2^+$) can be regarded as combined from the ^{12}Be($2\hbar\omega$) core and an sd-orbit neutron, while ^{13}Be($5/2^+$) is interpreted as the ^{12}Be($0\hbar\omega$) core and an sd-orbit neutron.

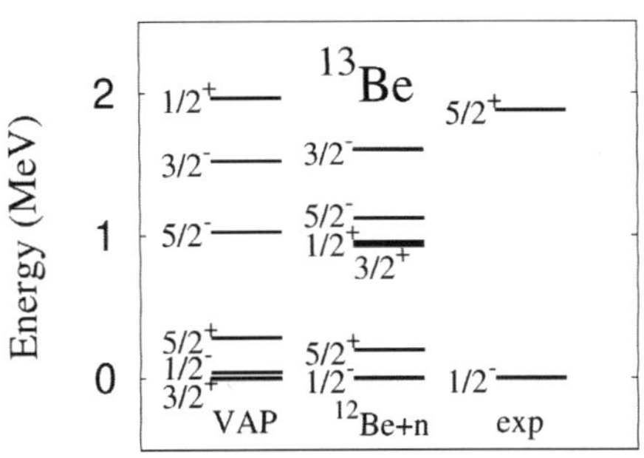

Figure 1. Energy levels of ^{13}Be calculated with the AMD+VAP method and the ^{12}Be+n model by using the set 1 interaction of Ref. [12]. The experimental data from Ref. [11] are also shown.

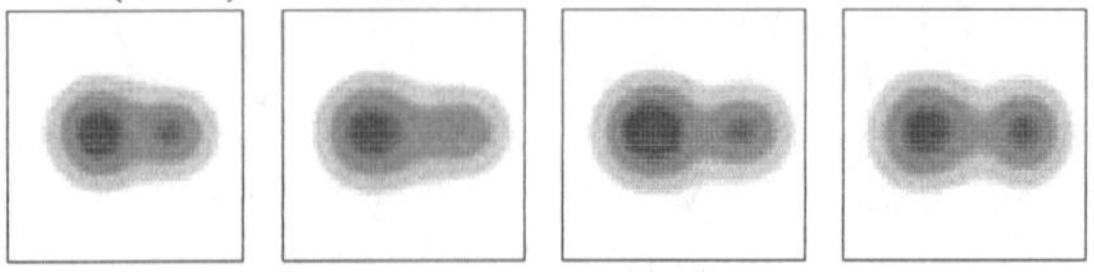

Figure 2. Density distribution in the intruder states of Be isotopes as obtained by AMD+VAP calculations. The box size is 10 fm×10 fm.

3.2. Cluster structure of $^9Li^*$

As described by the Ikeda diagram [13], cluster structures may appear in excited states near the threshold energy even if the ground state shows no prominent cluster structure. This threshold rule may be valid in unstable nuclei as well as in stable nuclei as is exhibited by various neutron-rich nuclei. Our aim here is to search for developed cluster states in excited states of ^{9}Li.

Let us first review the cluster features of ^{10}Be and then consider the analogy with ^{9}Li. The $K^\pi = 0_2^+$ band in ^{10}Be has been extensively studied in experimental and theoretical works. The 0_2^+ state at 6.18 MeV just below the ^{6}He+α threshold energy is considered to be the band-head state of the $K^\pi = 0_2^+$ band. The 2^+ state at 7.54 MeV and the 4^+ state at 10.2 MeV have been suggested by experiments to have developed ^{6}He+α-cluster structures and may belong to the $K^\pi = 0_2^+$ band [14, 15, 16, 17]. This 0_2^+ band of ^{10}Be is regarded as a rotational band constructed from the strong-coupling ^{6}He+^{4}He cluster state having an alpha cluster at a head-on position with respect to the deformed ^{6}He cluster. This state is also interpreted as a molecular orbital state with two neutrons in the σ orbital around the developed 2α core. In addition to the 0_2^+ state, the existence of a higher 0^+ ^{6}He+α resonance state has been recently proposed in theoretical and experimental works [18, 19].

An analogy with the ^{6}He+^{4}He cluster states in ^{10}Be suggests that the Ikeda threshold rule causes ^{6}He+t cluster states to appear in ^{9}Li near the ^{6}He+t threshold. In our previous study of ^{9}Li with the (β-γ)-constrained AMD [20], we have found a developed ^{6}He+t structure at a shallow local minimum in the energy surface on the β-γ plane. According to that calculation, the excitation energy of the ^{6}He+t cluster state in ^{9}Li is expected to be 2 MeV higher than that of the ^{10}Be(0_2^+), and hence it may be a resonance above the ^{6}He+t threshold.

To investigate the ^{6}He+t-cluster states near the ^{6}He+t threshold in ^{9}Li, we perform GCM calculations in a ^{6}He+t cluster model. We superpose the ^{6}He+t cluster wave functions with distances $d = 1, 2, \cdots, 8$ fm and obtain energy levels for ^{9}Li. The energy levels measured from the ^{6}He+t threshold energy are shown in Fig. 3. The developed ^{6}He+t cluster states are expected to be a few MeV above the ^{6}He+t threshold energy. The cluster states $J^\pi = 1/2_2^-, 3/2_3^-, 5/2_2^-$ and $7/2_2^-$ form a $K^\pi = 1/2^-$ rotational band, and show strong in-band $E2$ transitions. They show spatially well developed ^{6}He+t cluster structure. Compared with the $K^\pi = 0_2^+$ band of ^{10}Be, in which the ^{6}He cluster strongly couples to the neighbouring ^{4}He cluster, in the $K^\pi = 1/2^-$ band of ^{9}Li the ^{6}He cluster couples more weakly to the t cluster.

Unfortunately, there is no experimental information on the ^{6}He+t-type excited states of ^{9}Li. It is a challenging problem to search for such cluster states since they consist of two exotic clusters, ^{6}He and t.

4. Discussion

Recently, it was revealed that various kinds of cluster structures appear in light stable and unstable nuclei depending on proton and neutron numbers and on the excitation energy. Cluster states near the threshold energy are often characterized as weakly coupled clusters since the

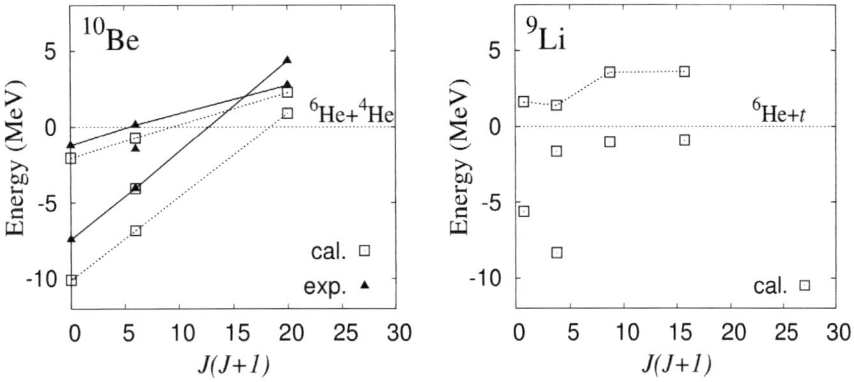

Figure 3. (Left) Energy levels of positive-parity states in ^{10}Be calculated by the ^{6}He+^{4}He-cluster GCM method with $d = 1, 2, \cdots, 8$ fm (open squares) and experimental ones (filled triangles). (Right) Energy levels of negative-parity states in ^{9}Li calculated by the ^{6}He+t-cluster GCM method. The energies measured from the ^{6}He+^{4}He and ^{6}He+t threshold are plotted as functions of spin $J(J+1)$. The points connected by lines are states which can be assigned to rotational bands: $K^\pi = 0_1^+$ and $K^\pi = 0_2^+$ for ^{10}Be, and $K^\pi = 1/2^-$ for ^{9}Li.

inter-cluster motion decouples from the cluster intrinsic degrees of freedom. On the other hand, in low-lying states we often see strongly coupled cluster structures in which clusters are tightly bound to form relatively compact states. Another type of cluster structure peculiar to neutron-rich nuclei is molecular orbital structure in which clusters are bonded by excess neutrons.

In ^{10}Be one may see such a change from a deeply bound cluster structure to a weakly coupled structure passing through the molecular orbital structure. Let us discuss the change of the two-cluster structure in ^{10}Be and ^{9}Li as a function of the inter-cluster distance (see the schematic picture in Fig. 4). How do the cluster structures of ^{10}Be and ^{9}Li compare?

The case of the small inter-cluster distance corresponds to the deeply bound cluster states. The ground state is obtained to be such a tightly bound compact cluster structure. With the increase of the inter-cluster distance, the structure of ^{10}Be changes into a molecular orbital structure, in which excess neutrons are moving around both clusters in the longitudinal σ orbital. With a further increase of the inter-cluster distance, the excess neutrons get localized around one of the α clusters to form a ^{6}He cluster. There, the ^{6}He cluster is still deformed and it may align toward the neighboring α cluster to gain potential energy. Finally, the ^{6}He cluster takes an energy eigenstate, and the system tends to a state with a weakly coupled pair of clusters, which is connected to the asymptotic ^{6}He+α cluster channel. In the realistic ^{10}Be system, the 0_2^+ state is considered to be intermediate between the molecular orbital structure and the aligned ^{6}He+α-cluster one, while a 0^+ state with a weakly coupled ^{6}He+α cluster pair was suggested to be a few MeV higher than the 0_2^+ state [18, 19].

In ^{9}Li one may also see a similar change of cluster structure from deeply bound clusters to the asymptotic ^{6}He+t channel. However, the cluster aspects of ^{9}Li may be somewhat different. One of the interesting properties of ^{9}Li different from ^{10}Be is that the t cluster in ^{9}Li easily decouples from the ^{6}He cluster because a t cluster gives a weaker attraction (or potential) than an α cluster. Therefore, the cluster states in ^{9}Li may show a trend to the weak coupling feature. Indeed, in the realistic ^{9}Li system, the ^{6}He+t cluster state in the $K^\pi = 1/2^-$ band, proposed to be near the threshold energy in the present work, is understood as an intermediate structure between the aligned ^{6}He+t and the weakly coupled ^{6}He+t cluster structures. We could not find molecular orbital structure in excited states of ^{9}Li. That is natural because the excess neutrons feel weaker attraction by the t cluster than by an α and they do not move around the whole $\alpha+t$ core but tend to localize around the α core. This is different from the case of ^{10}Be, in which a stable molecular structure is formed by σ-orbital neutrons around the developed 2α-cluster core and that is an essential property of the $K^\pi = 0_2^+$ band.

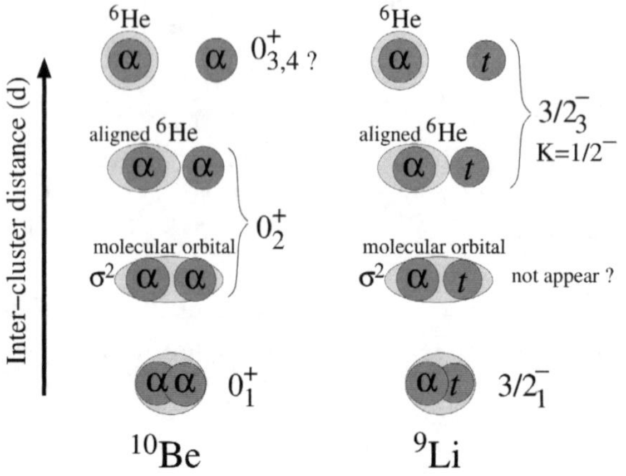

Figure 4. A schematic figure for the structure change of ^{10}Be and ^{9}Li toward the ^{6}He+^{4}He and ^{6}He+t channels with the increase of the inter-cluster distance.

5. Summary

Cluster structures in ^{13}Be were investigated with AMD+VAP calculations. Low-lying states having intruder $1\hbar\omega$ and $2\hbar\omega$ configurations were obtained, namely, a $1/2^-$ state with a dominant $1\hbar\omega$ excitation, and a $3/2^+$ state with $2\hbar\omega$. The result suggests that the neutron magicity in ^{13}Be has been broken because of the developed cluster structure, just as in the cases of ^{11}Be and ^{12}Be.

Highly excited cluster states of ^{9}Li were studied with cluster GCM calculations. The analogy and difference between the clustering properties of ^{9}Li and ^{10}Be were discussed.

It was found that various kinds of cluster structures appear depending on the neutron and proton numbers as well as on the excitation energy. In the neutron-rich nuclei ^{11}Be, ^{12}Be and ^{13}Be cluster structure develops even in the low-lying states. Their clusters are tightly bound or bonded by excess neutrons. On the other hand, there are also weakly coupled cluster states near the cluster thresholds in the same manner as in stable nuclei. The appearance of low-lying cluster states much below the cluster-decay threshold energy seems inconsistent with the Ikeda threshold rule. The weakly coupled cluster states do of course bear out the threshold rule.

The calculations reported in the present work suggest that there exist ^{6}He+t-type cluster resonance states in ^{9}Li. These states are peculiar because they consist of exotic clusters that are themselves unstable nuclei. It is a challenge to experimentalists to find them.

Acknowledgments

Discussions during YIPQS workshops held at YITP were helpful to advance this work. The computations were performed by using the supercomputers at YITP and at KEK. This work was supported by a Grant-in-Aid for Scientific Research from the Japan Society for the Promotion of Science (JSPS). It was also supported by the Grant-in-Aid for the Global COE Program "The Next Generation of Physics, Spun from Universality and Emergence" from the Ministry of Education, Culture, Sports, Science and Technology (MEXT) of Japan.

References

[1] Oertzen W, Freer M and Kanada-En'yo Y 2006 *Phys. Rep.* **432** 43
[2] Tohsaki A, Horiuchi H, Schuck P and Röpke G 2011 *Phys. Rev. Lett.* **87** 192501
[3] Suhara T and Kanada-En'yo Y 2010 *Phys. Rev.* C **82** 044301
[4] Kanada-En'yo Y, Horiuchi H and Ono A 1995 *Phys. Rev.* C **52** 628; Kanada-En'yo Y and Horiuchi H 1995 *Phys. Rev.* C **52** 647
[5] Kanada-En'yo Y 2012 *Phys. Rev.* C **85** 044320
[6] Kanada-En'yo Y and Suhara T 2012 *Phys. Rev.* C **85** 024303
[7] Kanada-En'yo Y and Horiuchi H 2011 *Prog. Theor. Phys. Suppl.* **142** 205
[8] Kanada-En'yo Y, Kimura M and Horiuchi H 2003 *Comptes Rendus Physique* Vol.**4** 497
[9] Volkov A B 1965 *Nucl. Phys.* **74** 33
[10] Ando T, Ikeda K and Tohsaki A 1980 *Prog. Theory. Phys.* **64** 1608
[11] Kondo Y *et al* 2010 *Phys. Lett.* B **690** 245
[12] Kanada-En'yo Y and Horiuchi H 2002 *Phys. Rev.* C **66** 024305
[13] Ikeda K, Tagikawa N and Horiuchi H 1968 *Prog. Theor. Phys. Suppl.* **Extra Number** 464
[14] Soić N *et al* 1996 *Europhys. Lett.* **34** 7
[15] Liendo J A, Curtis N, Caussyn D D, Fletcher N R and Kurtukian-Nieto T 2002 *Phys. Rev.* C **65** 034317
[16] Milin M *et al* 2005 *Nucl. Phys.* A **753** 263
[17] Freer M *et al* 2006 *Phys. Rev. Lett.* **96** 042501
[18] Kobayashi F and Kanada-En'yo Y 2012 arXiv:1208.0387 [nucl-th]
[19] Kuchera A N *et al* 2011 *Phys. Rev.* C **84** 054615
[20] Suhara T and Kanada-En'yo Y 2010 *Prog. Theor. Phys.* **123** 303

Journal of Physics: Conference Series **436** (2013) 012038 doi:10.1088/1742-6596/436/1/012038

Core excitations in exotic nuclei

Pierre Descouvemont[1], **Marianne Dufour**[2] **and Natalia Timofeyuk**[3]

[1] Physique Nucléaire Théorique et Physique Mathématique, C.P. 229,
Université Libre de Bruxelles (ULB), B 1050 Brussels, Belgium
[2] IPHC Bat27, IN2P3-CNRS/Université Louis Pasteur BP28,
F-67037 Strasbourg Cedex 2 - France
[3] Department of Physics, Faculty of Electronics and Physical Sciences,
University of Surrey, Guildford, GU2 7XH, UK

Abstract. The role of core excitations in exotic nuclei is discussed in the framework of a microscopic cluster model. This cluster approach is complemented by the R-matrix theory to take account of the long-range part of the wave functions. We briefly describe the model, and present two recent examples: the neutron-rich nucleus ^{16}B, described by a ^{15}B+n structure, and the proton-rich nucleus ^{17}Na, described by a ^{16}Ne+p structure. In both cases core excitations are shown to play an important role.

1. Introduction

Clustering is a well known effect in light nuclei (see Ref. [1] and references therein). This property is at the basis of cluster models [2], where the nucleus is described in terms of two (or more) cluster wave functions [3, 4, 5]. Several variants exist according to whether they exactly include antisymmetrization or not. A natural extension of cluster models to exotic nuclei is the description of halo states, where external nucleons are simply considered as clusters. Typical examples are the ^{6}He and ^{11}Li nuclei, which can be described by α+n+n [6] and ^{9}Li+n+n [7] configurations, respectively.

Simple variants of cluster models only include the ground state of each cluster and, in particular, of the core nucleus. However this approximation may be, in some cases, too simple, and the importance of core excitations needs to be addressed. Here we use a microscopic cluster model with an A-body hamiltonian. In the two-cluster variant, an approximate wave function of the system reads, in a schematic notation,

$$\Psi = \mathcal{A}\phi_1\phi_2 g(\boldsymbol{\rho}), \tag{1}$$

where $\mathcal{A}$ is the antisymetrization operator. In this definition, ϕ_1 and ϕ_2 are the internal wave functions of the two clusters and $g(\boldsymbol{\rho})$ the radial function depending on the relative coordinate $\boldsymbol{\rho}$. This model has been applied to many light nuclei and nucleus-nucleus reactions, including reactions of astrophysical interest (see references in Refs. [8, 5]). Definition (1) is the standard RGM (Resonating Group Method) wave function [9, 3], and is fully antisymmetric against exchanges of nucleons. Further simplification can be performed by neglecting the internal structure of the clusters, and by choosing an appropriate nucleus-nucleus potential [10]. Wave function (1) can be generalized to multicluster systems [5]. It is well adapted to exotic nuclei

[11], where the breakup threshold is low. In that case, the long-range part of the wave function must be accurately determined.

Here we focus on extensions of (1) to core excitations; in other words the total wave function can be generalized as

$$\Psi = \sum_c \mathcal{A}\phi_{1c}\phi_2 g_c(\boldsymbol{\rho}), \tag{2}$$

where c represents the excitation level of cluster 1 (cluster 2 is assumed to remain in its ground state). Previous applications to weakly bound nuclei and to unbound systems have shown that core excitations may play an important role [12, 13, 14]. This phenomenon was recently confirmed by an experimental study of ^{18}Ne+p elastic scattering, performed in parallel with the ^{18}Ne(p,p')^{18}Ne(2^+) inelastic scattering [15]. This experiment shows a clear evidence for ^{19}Na states with a dominant ^{18}Ne(2^+)+p structure. The importance of ^{6}He*+n configurations in the ^{7}He spectrum is also well established [16, 17].

The paper is organized as follows. In section 2, we briefly present the microscopic cluster model. Sections 3 and 4 are devoted to the recent examples of ^{16}B and ^{17}Na, respectively. These systems are typical neutron- and proton-rich nuclei. Concluding remarks are presented in section 5.

2. Microscopic cluster theories

Microscopic models are based on fundamental principles of quantum mechanics, such as the treatment of all nucleons, with exact antisymmetrization of the wave functions. Neglecting three-body forces, the Hamiltonian of an A-nucleon system is written as

$$H = \sum_{i=1}^{A} T_i + \sum_{i<j=1}^{A} V_{ij}, \tag{3}$$

where T_i is the kinetic energy of nucleon i, and V_{ij} an effective nucleon-nucleon interaction [3]. Effective forces, such as the Volkov [18] or the Minnesota [19] interactions are adapted to the model wave functions. Both contain one adjustable parameter which can be tuned to reproduce an important property of the nucleus, such as a threshold breakup energy.

In practice, the Schrödinger equation associated with this Hamiltonian cannot be solved exactly when $A > 4$. For very light systems ($A \sim 4-5$) efficient methods [20] exist, even for continuum states [21]. Recent developments of *ab initio* models (see for example Refs. [22, 23, 24]) are quite successful for spectroscopic properties of nuclei up to $A \approx 12$. These models make use of realistic interactions, which are fitted to many properties of the nucleon-nucleon system, and include three-body forces. However, a consistent description of bound and scattering states of an A-body problem remains a very difficult task [21], in particular for transfer reactions.

In cluster models, it is assumed that the nucleons are grouped in clusters [3, 13]. We present here the specific application to two-cluster systems. The internal wave functions of the clusters are denoted as $\phi_i^{I_i\pi_i\nu_i}(\boldsymbol{\xi_i})$, where (I_i, ν_i) and π_i are the spin/projection and parity of cluster i, and $\boldsymbol{\xi_i}$ represents a set of their internal coordinates. A channel function is defined as

$$\varphi_{\ell I}^{JM\pi}(\Omega_\rho, \boldsymbol{\xi_1}, \boldsymbol{\xi_2}) = \left[Y_\ell(\Omega_\rho) \otimes [\phi_1^{I_1\pi_1}(\boldsymbol{\xi_1}) \otimes \phi_2^{I_2\pi_2}(\boldsymbol{\xi_2})]^I\right]^{JM}, \tag{4}$$

where different quantum numbers show up: the channel spin I, the relative angular momentum ℓ, the total spin J and the total parity $\pi = \pi_1\pi_2(-)^\ell$.

The total wave function (1) of the A-nucleon system is therefore written in a more explicit way as

$$\begin{aligned}\Psi^{JM\pi} &= \sum_{\alpha\ell I} \Psi^{JM\pi}_{\alpha\ell I} \\ &= \sum_{\alpha\ell I} \mathcal{A}\, g^{J\pi}_{\alpha\ell I}(\rho)\, \varphi^{JM\pi}_{\alpha\ell I}(\Omega_\rho, \boldsymbol{\xi}_1, \boldsymbol{\xi}_2),\end{aligned} \tag{5}$$

where index α refers to different two-cluster arrangements. In most applications, the internal cluster wave functions $\phi_i^{I_i\pi_i\nu_i}$ are defined in the shell model. The relative wave functions $g^{J\pi}_{\alpha\ell I}(\rho)$ are to be determined from the Schrödinger equation which, in the RGM, is transformed into an integro-differential equation involving a non-local potential [9]. In most applications, this relative function is expanded over Gaussian functions [9, 3, 5], which corresponds to the Generator Coordinate Method (GCM). The wave function (5) is therefore rewritten as

$$\Psi^{JM\pi}_{\alpha\ell I} = \int f^{J\pi}_{\alpha\ell I}(R)\Phi^{JM\pi}_{\alpha\ell I}(R), \tag{6}$$

where $\Phi^{JM\pi}_{\alpha\ell I}(R)$ is a projected Slater determinant, and $f^{J\pi}_{\alpha\ell I}(R)$ the generator function, which must be determined. The GCM is equivalent to the RGM, but is better adapted to numerical calculations, as it makes uses of projected Slater determinants (see Refs. [9, 3, 5] for detail).

The main advantage of cluster models with respect to other microscopic theories is their ability to deal with reactions, as well as with nuclear spectroscopy. As mentioned before, the RGM radial wave functions are expanded over a Gaussian basis. The GCM is well adapted to numerical calculations, and to a systematic approach, but the Gaussian behaviour is not physical at large distances, and must be corrected. We use the Microscopic R-matrix Method (MRM) [25, 26] which is a direct extension of the standard R-matrix technique [27], based on the existence of two regions: the internal region (with channel radius a), where the nuclear force and the nucleus-nucleus antisymmetrization are important, and the external region where they can be neglected. In the external region, the Gaussian expansion of the RGM radial function is replaced by Coulomb functions. Matching the internal and external components provides either the collision matrix (for scattering states) or the binding energy (for bound states).

3. Application to the nucleus ^{16}B

The unbound nature of ^{16}B has been first proposed by Bowman *et al.* [28] and by Langevin *et al.* [29], and later confirmed by Kryger *et al.* [30] in an analysis of the ^{17}C breakup. The low-lying structure has been investigated by Kalpachieva *et al.* [31] in a heavy-ion multi-nucleon transfer reaction. The existence of a narrow peak above the ^{15}B+n threshold ($E = 40 \pm 40$ keV) with a width lower than 100 keV, together with a higher level at $E = 2.40$ MeV ($\Gamma = 0.15$ MeV) have been reported. More recently, the ^{16}B spectrum has been studied by single-proton removal from a 35 MeV/nucleon ^{17}C beam by Lecouey *et al.* [32]. In that experiment, a narrow resonant structure at $E = 85 \pm 15$ keV above the ^{15}B+n threshold, with a width of $\Gamma \ll 100$ keV, has been observed and confirms the experiment of Kalpachieva *et al.* This peak is interpreted as a narrow resonance which decays by d-wave neutron emission.

In this section, we investigate the ^{16}B nucleus in a ^{15}B+n model, involving several ^{15}B excited states (see detail in Ref. [13]). Shell-model wave functions are built from all configurations with 3 protons in the p shell, and 2 neutrons in the sd shell. This shell-model description involves 1320 Slater determinants and provides ^{15}B states from the diagonalization of several operators: the total angular momentum $\boldsymbol{J}^2$ (and its projection J_z), the orbital momentum $\boldsymbol{L}^2$ and the intrinsic spin $\boldsymbol{S}^2$ [5]. The lowest states can be considered as physical, and high-energy levels correspond to pseudostates, which simulate the ^{15}B distortion. To keep computer times and

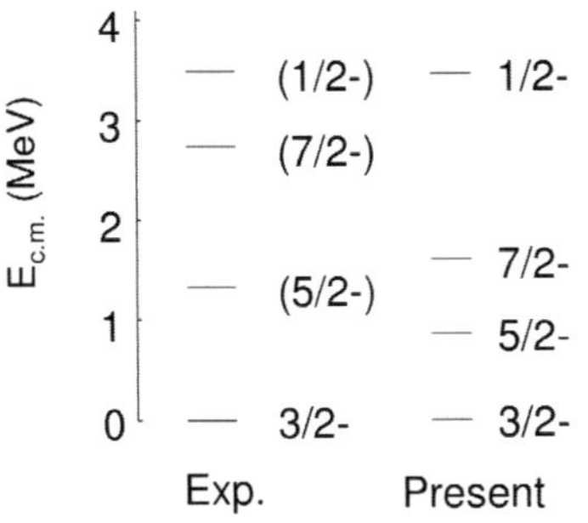

Figure 1. ^{15}B spectrum obtained with the present calculation, and compared to experiment [31, 33].

memory requirements within reasonable limits, we have limited the shell-model basis to the intrinsic spin $S = 1/2$; this provides 87 ^{15}B states.

In Figure 1, we present the low-lying ^{15}B spectrum obtained with the present conditions of calculation, i.e. with shell-model wave functions limited to the *sd* shell. This model predicts a $3/2^-, 5/2^-$ and $7/2^-$ level ordering consistent with the data. In particular, we suggest that the state at 3.48 MeV reported by Kalpachieva *et al.* [31] could be assigned $1/2^-$.

First we address the importance of core excitations, by analyzing the evolution of the ^{16}B spectrum as a function of the number of channels. Figure 2 shows low-lying states of ^{16}B for different numbers of channels. We present the single-channel ^{15}B$(3/2^-)$+n, two-channel ^{15}B$(3/2^-, 5/2^-)$+n, and four-channel ^{15}B$(3/2^-, 5/2^-, 7/2^-, 1/2^-)$+n calculations, as well as the full multichannel approach. Figure 2 is limited to the first states obtained for different J^π values from 0^- to 3^-.

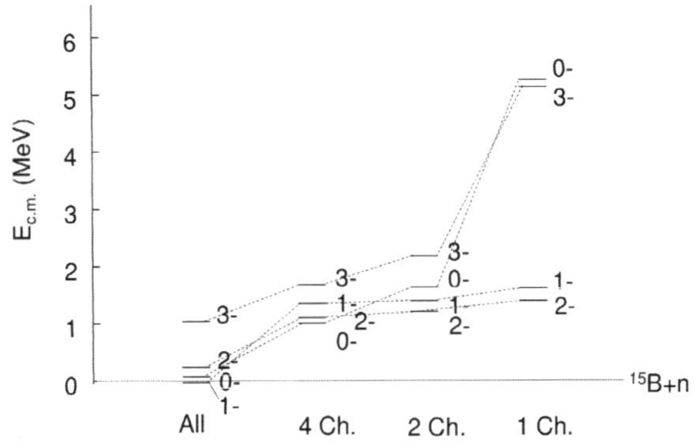

Figure 2. Evolution of ^{16}B low-lying states according to the number of channels. Labels '1 Ch.', '2 Ch.', '4 Ch.' and 'All' refer to the number of channels.

The resonance energies are clearly sensitive to the size of the variational basis. In particular, the 0^- level, assumed to be the ^{16}B ground state [31, 32], is strongly sensitive to excited configurations. In the multichannel calculation, the 1^- and 2^- states converge near the ^{15}B+n threshold. The 1_1^- state becomes the ^{16}B ground state, and is even slightly bound, when the full basis is considered. All these results support the importance of a multichannel framework to describe the ^{16}B nucleus.

Figure 3 provides the full GCM spectrum, compared with the available experimental data. The shell-model results of Ref. [32] are also shown for comparison. The existence of a narrow 0^- ($\ell = 2$) narrow resonance just above the threshold is confirmed by the GCM. The GCM predicts 1^- and 2^- states at low energies with a dominant component in the ^{15}B(g.s)+n channel. We find that the 1^- level is even slightly bound, although it presents a large reduced width. Of course a microscopic model cannot be expected to provide a precision of a few keV, but the existence of this low-lying state is likely. A similar *s*-wave resonance was predicted by the

GCM in ^{13}Be at low energies [34], and found experimentally later [35]. As a general conclusion the ^{16}B spectrum is predicted to have several additional states, which have not been observed experimentally yet.

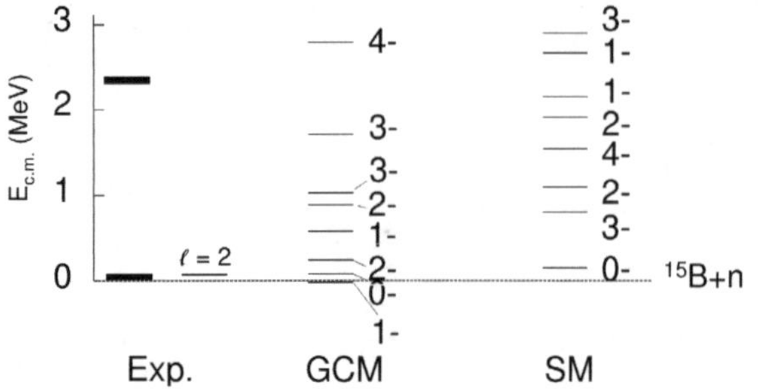

Figure 3. ^{16}B spectra. Experimental energies are taken from Refs. [31, 32]. Shell-model (SM) results are taken from Ref. [32].

4. Application to the nucleus ^{17}Na

Our aim here is to show that narrow states can also exist in the spectrum of the proton rich nucleus ^{17}Na, which is the mirror analogue of ^{17}C famous for its peculiar structure (see details in Ref. [12]). The neutron binding energy in the ground state ^{17}C($3/2^+$) is only 728 keV, typical of halo nuclei. However, knockout experiments have shown that the weakly bound ^{16}C(0^+)+n configuration is suppressed in ^{17}C($3/2^+$) and that this state is mainly based on the ^{16}C(2^+)+n configuration [36] where the neutron binding energy is 2.5 MeV. A similar structure should be expected in the mirror nucleus ^{17}Na($3/2^+$). Therefore, the decay branch ^{17}Na($3/2^+$) $\rightarrow$^{16}Ne(0^+)+p could be suppressed and, if energetically allowed, the main decay mode would be ^{17}Na($3/2^+$) $\rightarrow$^{16}Ne(2^+)+p. If its decay energy is below the Coulomb barrier then its width may be small. Since the decay product ^{16}Ne is unstable with respect to 2p emission, ^{17}Na should be a three-proton (3p) emitter.

The ^{17}C and ^{17}Na are studied in parallel with the Volkov and Minnesota interactions. The two ^{16}C valence neutrons occupy the $0d_{5/2}$, $1s_{1/2}$ and $0d_{3/2}$ orbitals, which gives many excitations in ^{16}C. We only consider the $2^+_{1,2}$, 3^+_1 and 4^+_1 states, motivated by neutron knockout experiments where they are strongly populated. The ^{17}C spectrum is fairly well known from experiment, and will be used to estimate the precision of the model applied to the mirror ^{17}Na nucleus.

Let us first discuss the ^{17}C spectrum shown in Fig. 4, compared to experiment. The parameters involved in the nucleon-nucleon interaction have been chosen to reproduce both the ^{16}C+n threshold in ^{17}C and the excitation energy of ^{17}C($1/2^+_1$). With these parameters, the energies of the three low-lying states are in very good agreement with experiment. In particular the level ordering is correctly reproduced. For comparison, the shell model spectrum is also shown in Fig. 4.

The decay scheme of the lowest part of the ^{17}Na spectrum is shown in Fig. 5. We find that the ^{17}Na($7/2^+_1$) level should be very narrow. It decays into the *d*-wave ^{16}Ne(2^+_1)+p and *s*-wave ^{16}Ne(4^+_1)+p channels with partial widths of $\Gamma(2^+_1) = 25$ keV and $\Gamma(4^+_1) = 98$ keV respectively. It should be noted that the theoretical value of the latter threshold is underestimated by 630 keV and, therefore, the energy in this channel is too large. Decreasing this energy by tunning the Majorana parameter m, we obtain a partial width $\Gamma(4^+_1) = 4$ keV. Thus, the ^{17}Na($7/2^+_1$) state should be as narrow as ^{17}Na($3/2^+_1$). A similar situation occurs for the ^{17}Na($9/2^+_1$) state. The partial width for the decay into ^{16}Ne(2^+_1)+p is predicted to be 211 keV. A similar width is expected for the decay into the ^{16}Ne(4^+_1)+p channel. Tuning the energy of this channel to reproduce the position of the ^{17}C($9/2^+_1$) state with respect to the ^{16}Ne(4^+_1)+p threshold, similar to what has been done in the case of $7/2^+_1$, we get a partial width $\Gamma(4^+_1) = 78$ keV.

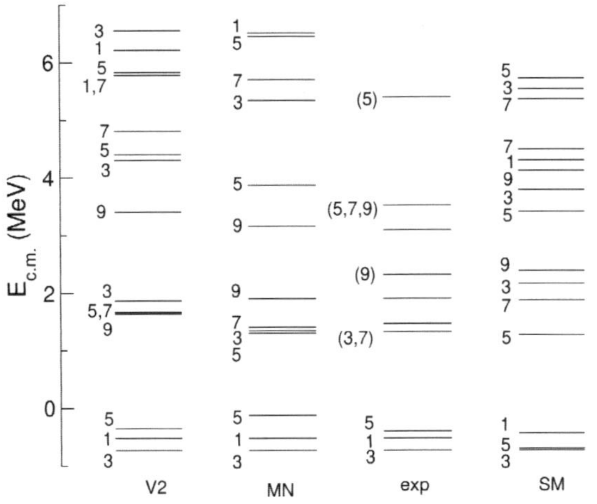

Figure 4. ^{17}C spectra calculated in the MCM with the V2 and MN interactions in comparison to the experimental spectrum and to shell model (SM) WBP predictions. Labels correspond to $2J$. The spin-parity assignment for observed unbound levels corresponds to that suggested by 3n transfer [37].

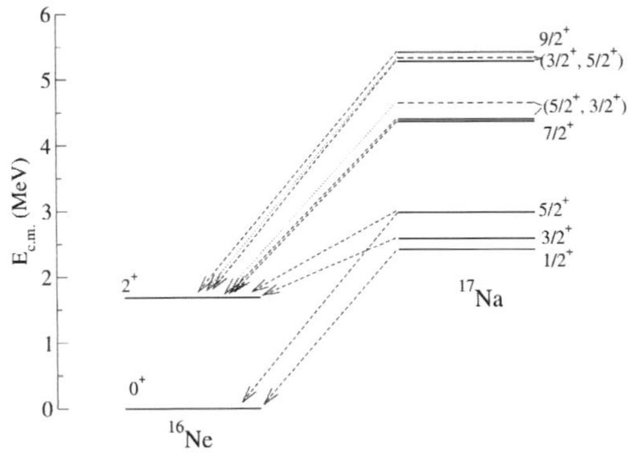

Figure 5. ^{17}Na decay scheme. The main decay branches are indicated by arrows.

There should be at least four narrow states, $3/2_1^+$, $5/2_1^+$, $7/2_1^+$ and $9/2_1^+$ in the ^{17}Na spectrum. The decay product of these states, ^{16}Ne, is unstable with respect to two-proton emission. Therefore, ^{17}Na is in fact a three-proton emitter with a decay path $^{17}\text{Na} \to {}^{16}\text{Ne}^*+\text{p} \to {}^{14}\text{O}+2\text{p}+\text{p}$. Consequently, ^{17}Na states can be identified by detecting ^{14}O+p+p+p events in coincidence. A multichannel algebraic study [38] confirms the importance of core excitations in ^{17}C and ^{17}Na, but finds some differences for the resonance properties of ^{17}Na.

5. Conclusion

Microscopic cluster models represent a powerful tool to investigate the structure of light nuclei. With the help of the R-matrix method, unbound systems can be studied with a rigorous treatment of the asymptotic wave function. This is particularly important for broad resonances showing up in many nuclei close to, or beyond, the driplines. Another advantage is that core excitations can be introduced without further parameters. The couplings between the channels are directly determined from the nucleon-nucleon interaction.

In most exotic nuclei, core excitations play an important role. This property is supported by several experiments, on ^{7}He or ^{19}Na, for example. The present work addresses core excitations in ^{16}B and ^{17}Na which are both unstable in their ground state. The microscopic calculation suggests that several states should exist in the ^{16}B spectrum. Most of them are very sensitive to core excitations, and the ground state should be close to particle stability. A similar conclusion holds for ^{17}Na. Due to the dominant ^{16}Ne*+p configurations, several states have narrow widths. These low-lying states have long lifetimes, and are suggested to be three-proton emitters.

References

[1] Horiuchi H, Ikeda K and Katō K 2012 *Prog. Theor. Phys. Suppl.* **192** 1
[2] Brink D M 2008 *J. Phys.: Conf. Series* **111** 012001
[3] Wildermuth K and Tang Y C 1977 *A Unified Theory of the Nucleus* (Braunschweig: Vieweg)
[4] Fujiwara Y *et al* 1980 *Prog. Theor. Phys. Suppl.* **68** 29
[5] Descouvemont P and Dufour M 2012 *Clusters in Nuclei, vol 2* (Springer) p 1
[6] Arai K, Suzuki Y and Lovas R G 1999 *Phys. Rev. C* **59** 1432
[7] Descouvemont P 1997 *Nucl. Phys. A* **626** 647
[8] Descouvemont P 2004 *Phys. Rev. C* **70** 065802
[9] Horiuchi H 1977 *Prog. Theor. Phys. Suppl.* **62** 90
[10] Buck B, Friedrich H and Wheatley C 1977 *Nucl. Phys. A* **275** 246
[11] Jonson B 2004 *Phys. Rep.* **389** 1
[12] Timofeyuk N K and Descouvemont P 2010 *Phys. Rev. C* **81** 051301
[13] Dufour M and Descouvemont P 2011 *Phys. Lett. B* **696** 237
[14] Timofeyuk N K *et al* 2012 *Phys. Rev. C* **86**(3) 034305
[15] Pellegriti M G *et al* 2008 *Phys. Lett. B* **659** 864
[16] Adahchour A and Descouvemont P 2006 *Phys. Lett. B* **639** 447
[17] Aksyutina Y *et al* 2009 *Phys. Lett. B* **679** 191
[18] Volkov A B 1965 *Nucl. Phys.* **74** 33
[19] Thompson D R, LeMere M and Tang Y C 1977 *Nucl. Phys. A* **286** 53
[20] Kievsky A, Rosati S, Viviani M, Marcucci L E and Girlanda L 2008 *J. Phys. G* **35** 063101
[21] Navrátil P, Roth R and Quaglioni S 2010 *Phys. Rev. C* **82** 034609
[22] Caurier E and Navrátil P 2006 *Phys. Rev. C* **73** 021302
[23] Neff T 2011 *Phys. Rev. Lett.* **106** 042502
[24] Navrátil P and Quaglioni S 2011 *Phys. Rev. C* **83** 044609
[25] Baye D, Heenen P H and Libert-Heinemann M 1977 *Nucl. Phys. A* **291** 230
[26] Descouvemont P and Baye D 2010 *Rep. Prog. Phys.* **73** 036301
[27] Lane A M and Thomas R G 1958 *Rev. Mod. Phys.* **30** 257
[28] Bowman J D, Poskanzer A M, Korteling R G and Butler G W 1974 *Phys. Rev. C* **9** 836
[29] Langevin M *et al* 1985 *Phys. Lett. B* **150** 71
[30] Kryger R A *et al* 1996 *Phys. Rev. C* **53** 1971
[31] Kalpakchieva R *et al* 2000 *Eur. Phys. J. A* **7**(4) 451
[32] Lecouey J L *et al* 2009 *Phys. Lett. B* **672** 6
[33] Stanoiu M *et al* 2004 *Eur. Phys. J. A* **22**(1) 5
[34] Descouvemont P 1994 *Phys. Lett. B* **331** 271
[35] Thoennessen M, Yokoyama S and Hansen P G 2000 *Phys. Rev. C* **63** 014308
[36] Maddalena V *et al* 2001 *Phys. Rev. C* **63** 024613(R)
[37] Bohlen H *et al* 2007 *Eur. Phys. J. A* **31** 279
[38] Amos K, Canton L, Fraser P, Karataglidis S, Svenne J and van der Knijff D 2012 *Nucl. Phys. A* **879** 132

Clustering aspects in highly excited states and in neutron-rich nuclei

Masaaki Kimura

Department of Physics, Hokkaido University, Sapporo 011-0021 Japan,

E-mail: masaaki@nucl.sci.hokudai.ac.jp

Abstract. A variety of exotic phenomena occurs at the limit of the nuclear binding. By use of Antisymmetrized Molecular Dynamics calculations shape coexistence and neutron-halo structure in the vicinity of the island of inversion is demonstrated. Furthermore, the presence of novel types of cluster states at very small excitation energy far from stability are predicted.

1. Introduction

The modern accelerator facilities and rapid development of the experimental technique have enabled to access the neutron drip-line nuclei with larger masses such as Ne and Mg isotopes. The S_n value of the "island of inversion" nuclei such as ^{32}Mg and ^{30}Ne are around 5 MeV, which is smaller than those of stable isotopes but not as small as to be called "weakly bound". Today, beyond them, the drip-line nuclei such as ^{31}Ne, ^{35}Mg and ^{37}Mg can also be accessed, where S_n is around or less than 1 MeV. At the limit of binding, naturally, we expect occurrence of exotic phenomena and have a chance to study the binding mechanism and new nuclear structure effects. As typical examples for weakly bound systems, we will discuss the structure of ^{35}Mg and ^{37}Mg using antisymmetrized molecular dynamics (AMD) calculations.

As another point of interest, clustering aspects of neutron rich-nuclei is also investigated. Owing to the theoretical and experimental developments, the domain of the cluster studies is rapidly expanding toward highly excited and neutron-rich nuclei. For example, the studies of unstable nuclei have revealed novel types of clustering represented by the Be isotopes [1, 2] and analogous structures have been discussed for heavier systems, too [3, 4]. Coexistence of the mean-field states and novel types of clustering composed of clusters and many valence neutrons are expected in F isotopes far from stability [5].

2. Theoretical Framework of the AMD

The framework of the AMD is given here briefly; readers are directed to the references [6, 7] for more detail. We start from the A-body Hamiltonian,

$$H = \sum_{i=1}^{A} t_i + \sum_{i<j}^{A} v_{NN}(ij) + \sum_{i<j \in p}^{Z} v_C(ij) - t_{c.m.} \tag{1}$$

Here, we have used Gogny D1S interaction [8] as an effective NN interaction, v_{NN}, and the Coulomb interaction is approximated by the sum of twelve Gaussians. $t_{c.m.}$ takes care of the center-of-mass kinetic energy. The AMD framework is free from the spurious center-of-mass motion. The variational wave function is represented by the parity-projected Slater determinant,

$$\Psi^{\pi} = \frac{1+\pi P_x}{2}\Psi_{\text{int}}, \tag{2}$$

$$\Psi_{\text{int}} = \mathcal{A}\{\varphi_1, \varphi_2, ..., \varphi_A\}, \tag{3}$$

where the ith single particle wave packet φ_i is a direct product of the spatial ϕ_i part with Gaussian form, spin χ_i and isospin η_i parts,

$$\varphi_i = \phi_i\chi_i\eta_i, \quad \phi_i = \exp\left\{-\sum_{\sigma=x,y,z}\nu_\sigma\left(r_\sigma - \frac{Z_{i\sigma}}{\sqrt{\nu_\sigma}}\right)^2\right\}, \tag{4}$$

$$\chi_i = \alpha_i\chi_\uparrow + \beta_i\chi_\downarrow, \quad \eta_i = \text{proton or neutron}. \tag{5}$$

The centroids of the Gaussian wave packets (Z_i) and spin directions (α_i, β_i) are the variational parameters. They are determined to minimize the total energy under the constraint on the matter quadrupole deformation parameter β. After the variational calculation, we obtain the optimized wave functions $\Psi^{\pi}(\beta)$ for each given value of β and project out the eigenstate of the total angular momentum,

$$\Psi^{J\pi}_{MK}(\beta) = \frac{2J+1}{8\pi^2}\int d\Omega D^{J*}_{MK}(\Omega)R(\Omega)\Psi^{\pi}(\beta), \tag{6}$$

where $D^J_{MK}(\Omega)$ and $R(\Omega)$ denote the Wigner D function and rotation operator, respectively. The integrals over three Euler angles Ω are performed numerically. Finally, the wave functions $\Psi^{J\pi}_{MK}(\beta)$ which have the same parity and angular momenta J and M, but have different K and β are superposed to take the configuration mixing and fluctuation of the quadrupole deformation into account,

$$\Psi^{J\pi}_{\alpha} = \sum_{iK} c_{iK,\alpha}\Psi^{J\pi}_{MK}(\beta_i). \tag{7}$$

The coefficients $c_{iK,\alpha}$ are determined by solving the Hill-Wheeler equation,

$$\sum_{jK'} H_{iKjK'}c_{jK',\alpha} = E_\alpha \sum_{jK'} N_{iKjK'}c_{jK',\alpha}, \tag{8}$$

$$\left\{\begin{array}{c} N_{iKjK'} \\ H_{iKjK'} \end{array}\right\} = \langle\Psi^{J\pi}_{MK}(\beta_i)|\left\{\begin{array}{c} 1 \\ H \end{array}\right\}|\Psi^{J\pi}_{MK'}(\beta_j)\rangle. \tag{9}$$

The practical procedure of the calculation is as follows. We first perform the variational calculation for positive- and negative-parity states with constraint on the matter quadrupole deformation to determine the variational parameters. It produces the energy curve for each parity as a function of the quadrupole deformation as shown in Fig. 1 (a) and (b) for the case of ^{31}Mg. We obtain several local energy minima on the energy curve which have different neutron configurations such as $0p1h$, $1p2h$ and $2p3h$. Here, the single particle configurations are identified by investigating the single particle energies and levels of each $\Psi^{\pi}(\beta)$ by using the AMD-HF method [10]. Figure 1 (c) and (d) show the neutron single particle energies. For example, in the spherical state of positive-parity the last neutron occupies the $0d_{3/2}$ orbit and no neutron is in the pf-shell, which confirms the neutron $0p1h$ configuration at the minimum around $\beta = 0.1$. When nuclear deformation increases, an intruder orbit, originating from the $0f_{7/2}$ configuration, comes down and will be occupied by two neutrons. Thus, we can determine that the local minimum

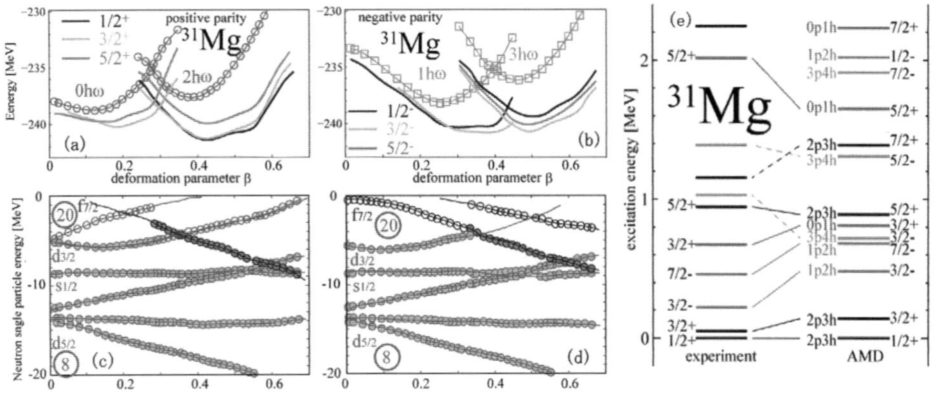

Figure 1. Results obtained for ^{31}Mg (see text). (a) and (b) Energy curves as a function of matter quadrupole deformation calculated before and after the angular momentum projection for the positive- and negative-parity states. (c) and (d) Neutron single particle energies. (e) Calculated and observed low-lying spectrum of ^{31}Mg. Experimental data are taken from the Ref. [9].

around $\beta = 0.4$ has a $2p3h$ configuration. After the variational calculation, we perform the angular momentum projection to obtain the energy curves for each angular momentum state. At this stage, the $1/2^+$ state with $2p3h$ configuration becomes the lowest in energy with a minimum around $\beta = 0.4$ (note that ^{31}Mg is located in the middle of the island of inversion). Finally, the wave functions on the energy curve are superposed to diagonalize the Hamiltonian, and we obtain the energy spectrum of ^{31}Mg (Fig. 1(e)) [11]. AMD predicted the coexistence of different particle-hole configurations with various deformations at very small excitation energy, which owes to the breakdown of the neutron magic number $N = 20$. Recently, the shape coexistence was confirmed by the various experiments [9]. In this way a nice agreement between the predicted and observed spectra has been presented suggesting that AMD has a predictive power in this mass region.

3. Parity inversion and halos in neutron-rich Mg isotopes

3.1. Possible parity inversion of ^{35}Mg

We now proceed toward drip-line Mg isotopes. Figure 2 shows the energy curve and neutron single particle energies for ^{35}Mg calculated by using the same procedure. Again, there are a couple of energy minima with different particle-hole configuration on the energy curves, which are identified as the neutron $(sd)^{-1}(fp)^4$ and $(sd)^{-3}(fp)^6$ configurations for the positive-parity minima and the neutron $(fp)^3$ and $(sd)^{-2}(fp)^5$ configurations for the negative-parity minima. Compared to ^{31}Mg, the level density around the Fermi surface is greatly increased, which simply owes to the increase of the neutron number and decrease of the neutron separation energy S_n. It can also be noted that an orbit intruding from the spherical $0p_{3/2}$ configuration appears at the largely deformed region of the positive- and negative-parity states, which generates the $(sd)^{-3}(fp)^6$ and $(sd)^{-2}(fp)^5$ configurations.

After the parity projection, the $3/2^+$ state with $(sd)^{-1}(fp)^4$ configuration becomes the lowest state, but other states with positive- and negative-parity also appear at very small excitation energies, which suggests shape coexistence in this nucleus. The GCM calculation does not change the lowest state, *i.e.* the $3/2^+$ state is the ground state and is followed by the $/5/2^+$ and so on to constitute the well deformed $K^\pi = 3/2^+$ band with $(sd)^{-1}(fp)^4$ configuration.

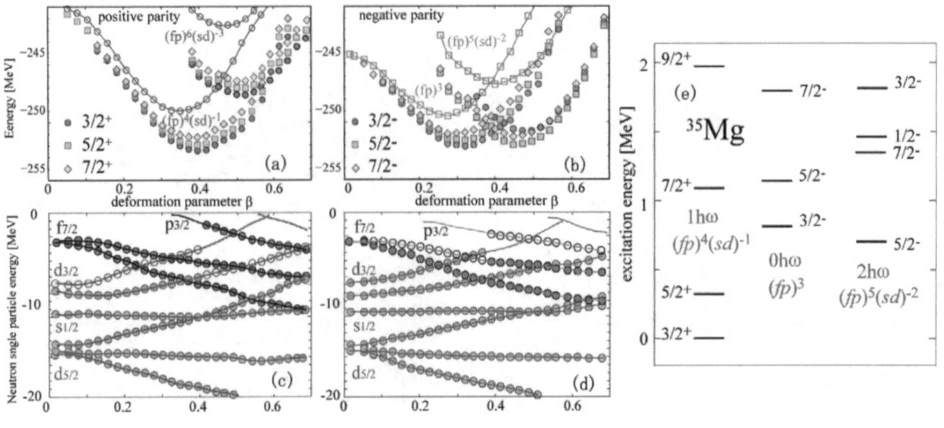

Figure 2. Same as Fig. 1, but for ^{35}Mg.

Negative parity states appear already below 1 MeV. The $3/2^-_1$ state which is the band head of the $K^\pi = 3/2^-$ band dominated by the $(fp)^3$ configuration (normal configuration) is located at around 800 keV. The $5/2^-_1$ state with a $(sd)^{-2}(fp)^5$ configuration appears around 700 keV. Thus, the ground state of ^{35}Mg is dominated by the intruder configuration and its parity is inverted to the positive-parity. It has to be noted that the ground state parity of the ^{35}Mg is sensitive to the effective interaction. For example, if we use the Gogny D1N parameter set [12], the $5/2^-$ state with a $(sd)^{-2}(fp)^5$ configuration becomes the ground state and the $3/2^+$ state appears as the first excited state. The shell model calculation [13] using SDPF-M interaction also predicts the $5/2^-$ ground state for ^{35}Mg, while the SDPF-U interaction predicts the $3/2^-$ ground state. Therefore the experimental information is essential for a more accurate knowledge on the nucleon-nucleon interaction.

3.2. Possible parity inversion and neutron halo of ^{37}Mg

^{37}Mg is the heaviest bound Mg isotope with odd mass number and is supposed to have very small S_n. Naturally, one-neutron halo structure may be expected. Indeed, quite recently, the total interaction cross section of ^{37}Mg has been measured at RIKEN and it was found that it has abnormally large cross section compared to the neighboring nuclei.

Figure 3 shows the AMD results for ^{37}Mg. Again, several local minima appear which are identified with neutron $(fp)^5$, $(fp)^6(sd)^{-1}$ and $(sd)^{-2}(fp)^6(g)$ configurations. After the angular momentum projection and GCM, the $5/2^-$ state with a $(fp)^5$ configuration becomes the ground state which constitute the head of the $K^\pi = 5/2^-$ rotational band. Note, that this ground state has normal configuration which means the end of the island of inversion, although the intruder configurations ($1/2^+_1$, $3/2^+_1$ and $5/2^+_1$ states) are below 1 MeV in excitation energies. Another notable point is the presence of an intruder orbit which originates from the spherical $g_{9/2}$-shell, appearing only at the largely deformed region of the positive-parity states (Fig. 3 (c)) with almost zero binding energy. This intruder orbit generates a strongly deformed $(sd)^{-2}(fp)^6(g)$ configuration around $\beta = 0.45$. Owing to the strong deformation of the system, this intruder orbit is not a pure g-wave but a linear combination of the $s, d, g, ...$ orbits. As discussed in Ref. [14], the s-wave component dominates over other partial waves due to the very weak binding energy. Indeed, after the angular momentum projection and the GCM, the $1/2^+_1$ state (s-wave state) has the lowest energy. Therefore, the $1/2^+_1$ state can be a candidate for a neutron-halo which is interpreted as an s-wave neutron orbiting around the deformed core (^{36}Mg).

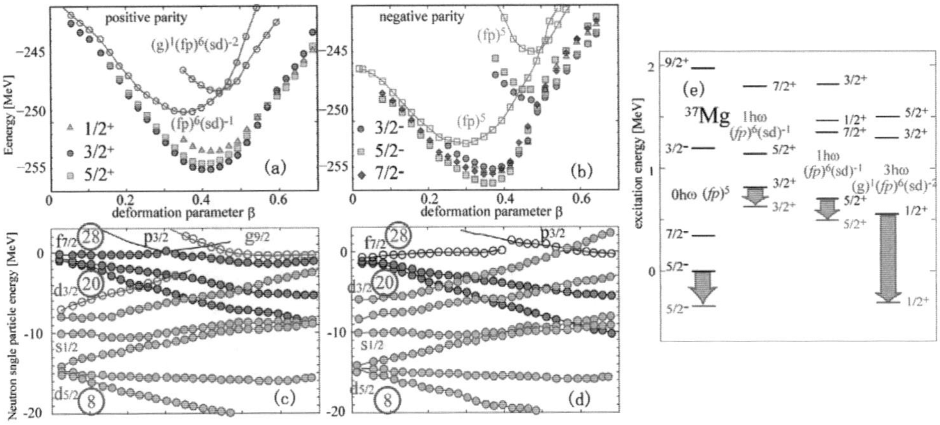

Figure 3. Same as Fig. 1, but for ^{37}Mg. Levels plotted by the red lines show the results obtained by the AMD-RGM calculation, while black lines show the AMD results.

Since the single particle wave function is represented by a Gaussian wave packet, AMD cannot describe properly the tail part of the halo wave function. This shortage can be overcome by combining AMD with the resonating group method (AMD+RGM). We used the RGM-type wave function for ^{37}Mg,

$$\Psi_{JM} = \sum_{\alpha l} c_{\alpha l} \mathcal{A}\left[R_{\alpha l}(r) Y_{lm}(\hat{r}) \phi_n \phi_{^{36}\mathrm{Mg}}(\alpha)\right]_{JM}, \tag{10}$$

where ϕ_n and $R_{\alpha l}(r)Y_{lm}(\hat{r})$ are the wave functions of the valence neutron and the relative motion between the valence neutron and ^{37}Mg, respectively. $\phi_{^{36}\mathrm{Mg}}(\alpha)$ is the internal wave function of ^{37}Mg solved by AMD, and α labels the ground and excited states. By using this wave function, the ^{37}Mg has been re-solved and the result is shown by red lines in Fig. 3 (e). Due to the improvement of the relative wave function, each band head state gains extra binding energy. Especially, the energy gain of the $1/2^+$ state is large; it amounts to more than 1 MeV, since the tail part of halo wave function is greatly improved by the AMD+RGM calculation. Now, $1/2^+_1$ is almost degenerate with the $5/2^-_1$ ground state and the observed abnormal reaction cross section can be explained.

4. Molecule-like states in the island of inversion

As a possible example of exotic phenomena driven by the shell erosion in the island of inversion, we discuss the molecule-like states of F isotopes, analogous to those of Be isotopes. As it is well known, the stable nucleus ^{19}F has a negative-parity state $(1/2^-_1)$ at 0.11 MeV, which is known to have an α+^{15}N cluster structure. Its extraordinarily small excitation energy is due to the strong α correlation or α clustering between four nucleons in the *sd* shell triggered by a proton excitation into this shell. Therefore, we may expect the presence of similar kinds of excited states in neutron-rich F isotopes which is also triggered by a proton excitation into the *sd* shell.

AMD calculation [5] predicts that simultaneous proton and neutron excitation across $Z = 8$ and $N = 20$ major shells will give rise to molecule-like states with α+N+"valence neutrons" structure throughout the F isotopes. Fig. 4 shows the band head states with a proton excitation across $Z = 8$ shell gap and various neutron configurations. Red lines show the states with a proton excitation but all neutrons are kept within the *sd* shell, which correspond to the low-lying

Journal of Physics: Conference Series **436** (2013) 012039 doi:10.1088/1742-6596/436/1/012039

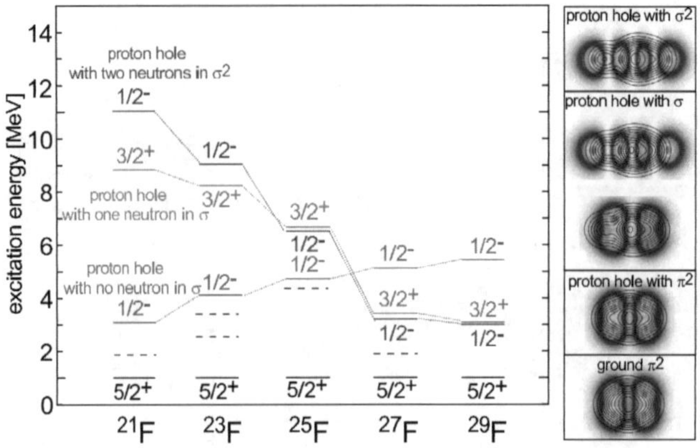

Figure 4. Left: Band-head energies of proton hole states of F isotopes. Red lines show the states with a proton excitation from p to sd shell and no neutrons in pf shell, while green (blue) lines show those with one (two) neutrons in pf shell that have molecule-like structure. Right: Density distribution of the core (^{19}F, contour lines) and two valence neutrons (color plot) of ^{21}F.

negative-parity states in lighter F isotopes. As the neutron number increases, their excitation energies gradually increase and the α clustering of these states diminishes since the increasing number of neutrons in sd shell reduces the deformation of the system. Therefore, a proton excitation is not enough to trigger the α clustering of neutron-rich F isotopes. On the contrary, a simultaneous excitation of a proton and neutrons across $Z = 8$ and $N = 20$ shell gap brings about the α clustering. These configurations are shown by green and blue lines, and they have a pronounced α cluster structure. For example, the density distribution of ^{21}F presented in Fig. 4 shows that the last neutron orbit is elongated along the symmetry axis and the last neutron pair is orbiting around the other nucleons which manifest the α+N cluster-like core structure. These configurations are analogous to the molecular structure of Be isotopes in which the 2α cluster core is surrounded by valence neutrons occupying the molecular orbitals around it.

It is important to note that a drastic reduction of the excitation energies toward neutron drip line is predicted for the molecular states. To understand this reduction, readers are reminded for the following points. (1) The last neutron's orbital originates in the pf shell, and its energy is lowered in the island of inversion due to the quenching of the $N = 20$ shell gap. (2) If the core has cluster structure, it induces the deformation of the system and it further lowers the energy of the last neutron's orbital as in the case of Be isotopes. Therefore, the neutron excitation into pf-orbital and α clustering of the core work in a cooperative way to reduce the excitation energies of molecular states in the island of inversion. Up to now, several candidates for the cluster states in the lighter F isotopes are experimentally known [15, 16] as shown in Fig. 4 with dashed lines. It is expected that more data on F isotopes near the drip line will be experimentally available in near future.

5. Summary

In this article we have reported our recent results. We have discussed the possible parity inversion and neutron-halo structure in ^{35}Mg, ^{37}Mg. The successful description of the low-lying states such as the shape coexistence and neutron-halo structure in the vicinity of the island of inversion was demonstrated. Furthermore, AMD prediction on the presence of the novel types of cluster states at very small excitation energy in F isotopes far from stability were also presented.

Journal of Physics: Conference Series **436** (2013) 012039 doi:10.1088/1742-6596/436/1/012039

References

[1] von Oertzen W 1996 *Z. Phys.* A **354** 37
[2] Kanada-En'yo Y, Horiuchi H and Doté A 1999 *Phys. Rev.* C **60** 064304
[3] Kimura M 2007 *Phys. Rev.* C **75**, 041302
[4] Furutachi N *et al* 2009 *Prog. Theor. Phys.* **122** 865
[5] Kimura M and Furutachi N 2011 *Phys. Rev.* C **83** 044304
[6] Kanada-En'yo Y, Kimura M and Horiuchi H 2003 *Comptes Rendus Physique* **4** 497
[7] Kanada-En'yo Y, Kimura M and Ono A 2012 *Prog. Theor. Exp. Phys.* **2012** 01A202
[8] Dechargé J and Gogny D 1980 *Phys. Rev.* C **21** 1568
[9] Neyens G 2011 *Phys. Rev.* C **84** 064301
[10] Doté A, Horiuchi H and Kanada-En'yo Y 1997 *Phys. Rev.* C **56** 1844
[11] Kimura M 2007 *Phys. Rev.* C **75** 034312
[12] Chappert F, Girod M and Hilaire S 2008 *Phys. Lett.* B **668** 420
[13] Gade A *et al* 2011 *Phys. Rev.* C **83** 044305
[14] Hamamoto I 2009 *Phys. Rev.* C **79** 014307
[15] Mairle G *et al* 1981 *Nucl. Phys.* A **363** 413
[16] Elekes Z, *et al* 2004 *Phys. Lett.* B **599** 17

Journal of Physics: Conference Series **436** (2013) 012040 doi:10.1088/1742-6596/436/1/012040

Mechanisms of direct reactions with halo nuclei

P Capel[1], F M Nunes[2], H Esbensen[3] and R C Johnson[4]

[1]Physique Nucléaire et Physique Quantique (CP 229), Université Libre de Bruxelles (ULB), 50 av. F. D. Roosevelt, B-1050 Brussels, Belgium

[2]National Superconducting Cyclotron Laboratory and Department of Physics and Astronomy, Michigan State University, East Lansing, MI 48824, USA

[3]Physics Division, Argonne National Laboratory, Argonne, IL 60436, USA

[4]Department of Physics, University of Surrey, Guildford GU2 7XH, UK

E-mail: pierre.capel@ulb.ac.be

Abstract. Halo nuclei are exotic nuclei which exhibit a strongly clusterised structure: they can be seen as one or two valence nucleons loosely bound to a core. Being observed at the shoulder of the valley of stability, halo nuclei are studied mostly through reactions. In this contribution the reaction models most commonly used to analyse experimental data are reviewed and compared to one another. A reaction observable built on the ratio of two angular distributions is then presented. This ratio enables removing most of the sensitivity to the reaction mechanism, which emphasises the effects of nuclear structure on the reaction.

1. Introduction

The advent of radioactive-ion beams has enabled the exploration of the nuclear landscape far from stability. This technological breakthrough has led to the discovery of exotic nuclear structures, such as halo nuclei [1]. Halo nuclei exhibit a large matter radius in comparison with their isobars. This peculiarity is qualitatively explained by the small binding energy of one or two valence nucleons, which then have a significant probability of presence at a large distance from the core of the nucleus. They thus exhibit a strongly clusterised structure in which one or two loosely-bound nucleons form a diffuse halo around a dense core. Examples of one-neutron halo nuclei are ^{11}Be and ^{15}C, whereas ^{6}He and ^{11}Li are the best known two-neutron halo nuclei. ^{8}B is a candidate one-proton halo nucleus.

Being far away from stability, halo nuclei cannot be studied through usual spectroscopic techniques, and we have to rely on indirect methods, such as reactions, to infer information about their structure. Breakup is probably the mostly used reaction to study halo nuclei [2, 3]. In that reaction, the halo dissociates from the core through the interaction with a target. In order to get valuable information from experimental data, various reaction models have been developed: the Continuum Discretised Coupled Channel model (CDCC) [4], the Time-Dependent approach (TD) [5], and the Dynamical Eikonal Approximation (DEA) [6]. These models are presented in Sec. 2. That section also includes a comparison between them for the breakup of ^{15}C on Pb at 68AMeV [7], which has been measured at RIKEN [3].

To reduce the sensitivity of the measurements to the reaction mechanism, one can also look for an observable that is independent of the reaction mechanism. Such an observable is presented in Sec. 3. It corresponds to the ratio of two angular distributions. According to the prediction of the Recoil Excitation and Breakup model (REB) [8], this ratio should depend only on the projectile structure. This prediction is confirmed by DEA calculations of reactions involving ^{11}Be, the archetypical one-neutron halo nucleus [9].

2. Reaction models

2.1. Theoretical framework

Let us consider the reaction involving a one-neutron halo nucleus, in which a valence neutron f is loosely bound to a core c. Such a two-cluster projectile is described by the Hamiltonian

$$H_0 = T_r + V_{cf}(\boldsymbol{r}), \tag{1}$$

where $\boldsymbol{r}$ is the neutron-core relative coordinate, T_r is the c-f kinetic-energy operator, and V_{cf} is a real potential that simulates the interaction between the halo neutron and the core. The parameters of this phenomenological potential are adjusted to reproduce the binding energy of the projectile and some of its low-lying levels. The structure of the target T is usually neglected and its interaction with the projectile components is simulated by optical potentials V_{cT} and V_{fT}. Within this framework, the theoretical study of reactions involving halo nuclei reduces to solving the three-body Schrödinger equation

$$\left[T_R + H_0 + V_{cT}(R_{cT}) + V_{fT}(R_{fT})\right]\Psi(\boldsymbol{r},\boldsymbol{R}) = E_T\Psi(\boldsymbol{r},\boldsymbol{R}), \tag{2}$$

where $\boldsymbol{R}$ is the projectile-target relative coordinate and T_R is the corresponding kinetic-energy operator. Eq. (2) is solved with the initial condition that the projectile is in its ground state ϕ_0

$$\Psi(\boldsymbol{r},\boldsymbol{R}) \underset{Z\to-\infty}{\longrightarrow} e^{iKZ+\cdots}\phi_0(\boldsymbol{r}), \tag{3}$$

where K is the wave number of the initial relative motion of the projectile to the target, which is assumed to be in direction $\widehat{\boldsymbol{Z}}$. The reaction models differ mostly in the way Eq. (2) is solved.

In the CDCC model, Eq. (2) is solved by expanding Ψ upon the eigenstates of H_0 [4]. Since breakup leads to the dissociation of the projectile, a tractable description of the projectile continuum must be included in this expansion. That is why the continuum is *discretised.* This is usually achieved by dividing the continuum into small energy intervals called *bins.* Each bin wave function is obtained by averaging the wave functions of the continuum states within the corresponding energy interval. This leads to a set of coupled equations, which we solve using FRESCO [10]. Apart from the continuum discretisation, the CDCC technique solves Eq. (2) exactly. It is a purely quantal model with no approximation on the projectile-target relative motion, and it does not imply any restriction on the beam energy. However, it is very expensive in a computational point of view, in particular at high beam energy.

A simpler way to solve Eq. (2) is the Time-Dependent (TD) approach, which relies on a semiclassical approximation. In that approximation, the projectile-target relative motion is simulated by a classical trajectory $\boldsymbol{R}(t)$. Along that trajectory, the projectile is affected by the time-varying Coulomb and nuclear fields of the target. The evolution of its internal structure is then obtained by solving a *time-dependent Schrödinger equation* [5]. That equation must be solved for each possible trajectory, with the condition that the projectile is initially in its bound state. Various computer codes have been developed to numerically solve the time-dependent equation. In this study, we will use the one detailed in Ref. [5]. Thanks to its semi-classical approximation, the time-dependent method is faster than CDCC. However, due to this approximation, quantal interferences between the trajectories cannot be taken into account.

Another way to avoid the heavy CDCC machinery is to use the eikonal approximation. At sufficiently high energy, one can assume that the projectile-target relative motion will be close to the initial plane wave (3). The main idea of the eikonal approximation is to factorise that plane wave out of the three-body wave function Ψ and to assume that the remaining factor is smoothly varying with $\boldsymbol{R}$. This leads to the Dynamical Eikonal Approximation (DEA), whose equation is mathematically equivalent to a time-dependent Schrödinger equation for straight-line trajectories [6]. As in the TD approach, the DEA equation must be solved for all the possible "trajectories". However, it remains a quantal approximation since it does not rely on any semiclassical approximation. This implies in particular that, contrarily to the TD approach, the DEA accounts for the quantal interferences between different "trajectories". However, since it relies on the eikonal approximation, the DEA is valid only at intermediate and high energies.

2.2. ^{15}C *Coulomb breakup*

To compare the three aforementioned models, we consider the breakup of the one-neutron halo nucleus ^{15}C on lead at 68AMeV [3]. The breakup cross section is plotted in Fig. 1 as a function of the relative energy E between the halo neutron and the ^{14}C core after dissociation. All three

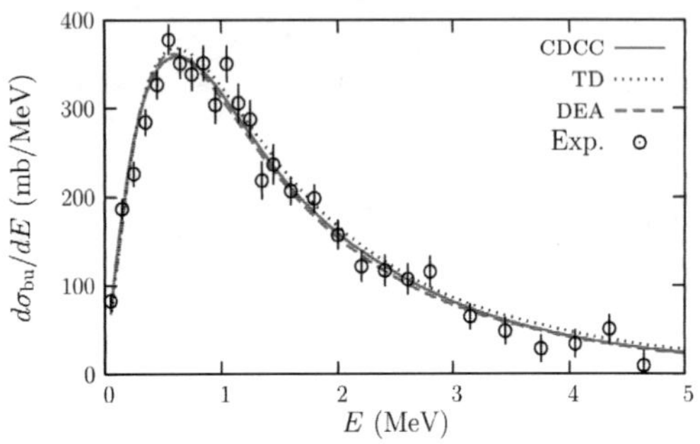

Figure 1. Energy distribution for the breakup of ^{15}C on Pb at 68AMeV. Comparison of CDCC, TD, and DEA [7]. Data from Ref. [3]

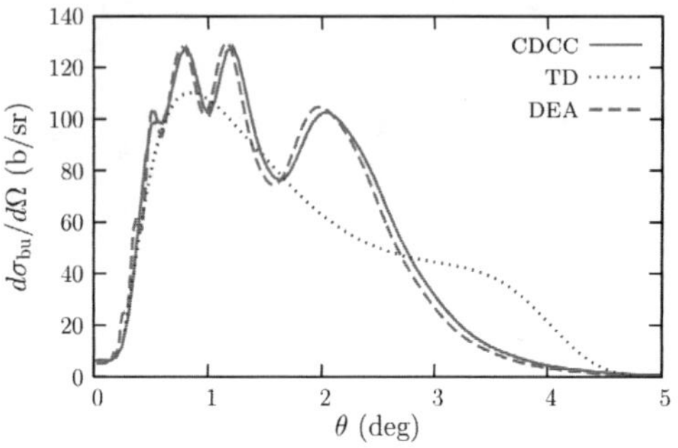

Figure 2. Angular distribution for the breakup of ^{15}C on Pb at 68AMeV. Comparison of CDCC, TD, and DEA [7].

models are in excellent agreement with each other [7] and with experiment [3]. This result shows that the reaction process is well understood, and that ^{15}C exhibits a two-cluster structure.

Fig. 2 shows the breakup angular distribution, i.e. the breakup cross section as a function of the scattering angle θ of the ^{14}C-n centre of mass [7]. CDCC and DEA lead to very similar cross sections. In particular they exhibit the same interference pattern, which confirms the validity of the DEA at these intermediate energies, even though it is much less time consuming than CDCC. As expected, the TD technique does not exhibit any quantal interference. Nevertheless, it reproduces the general trend of the quantal angular distributions, which explains why, once integrated over θ, it gives reliable results.

3. Ratio method

Albeit reliable, the aforementioned models remain sensitive to their input parameters. In particular, the choice the optical potentials V_{cT} and V_{fT} can affect significantly the results of the calculations [11]. An observable less sensitive to the reaction mechanism would naturally eliminate such sensitivity and improve the analysis of exotic nuclear structures from reactions.

It has been observed that angular distributions for elastic scattering and breakup exhibit very similar features: Coulomb rainbow, oscillatory patterns etc. [12]. This result can be qualitatively understood within the Recoil Excitation and Breakup model (REB) [8]. That model relies on two simplifying assumptions. First it includes an adiabatic treatment of the excitation of the projectile, i.e. the projectile excitation energy is neglected compared to the beam energy. Second, it neglects the interaction between the halo neutron and the target V_{fT}. Under these two assumptions an elegant factorisation for the elastic-scattering cross section is obtained [8]

$$\frac{d\sigma_{\rm el}}{d\Omega} = |F_{0,0}|^2 \left(\frac{d\sigma}{d\Omega}\right)_{\rm pt}, \tag{4}$$

which is the product of a cross section for the elastic scattering for a pointlike projectile $(d\sigma/d\Omega)_{\rm pt}$ times a form factor, which accounts for the halo extension

$$F_{0,0} = \int |\phi_0(\boldsymbol{r})|^2 e^{i\boldsymbol{Q}\cdot\boldsymbol{r}} d\boldsymbol{r}, \tag{5}$$

where $\boldsymbol{Q}$ is proportional to the transferred momentum. A similar factorisation can be performed for the angular distribution for the breakup of the projectile to energy E in the continuum

$$\frac{d\sigma_{\rm bu}}{dEd\Omega} = |F_{E,0}|^2 \left(\frac{d\sigma}{d\Omega}\right)_{\rm pt}, \tag{6}$$

where the form factor reads

$$|F_{E,0}|^2 = \sum_{ljm} \left| \int \phi_{ljm}(E,\boldsymbol{r})\phi_0(\boldsymbol{r}) e^{i\boldsymbol{Q}\cdot\boldsymbol{r}} d\boldsymbol{r}\right|^2. \tag{7}$$

It contains, besides the wave function of the initial bound state of the projectile ϕ_0, the wave functions describing the broken up projectile in partial wave ljm at energy E in the continuum.

The factorisations (4) and (6) explain the results of Ref. [12] since both expressions contain the same $(d\sigma/d\Omega)_{\rm pt}$, which is responsible for most of their angular dependence. In addition, the REB model also provides the idea for the *ratio method*: within that model, the ratio of cross sections (4) and (6) is the ratio of form factors (5) and (7), which depend only on the projectile structure. Accordingly, the ratio should be independent of the reaction process. In order to test this idea, we perform reaction calculations within the DEA [6], which does not include an adiabatic treatment of the projectile excitation, and which includes V_{fT}. As a testcase, we consider the collision of ^{11}Be with Pb at 69AMeV, which correspond to the conditions of a RIKEN experiment [2] with which the DEA calculations are in excellent agreement [13]. The result of this test is summarised in Fig. 3, which presents the DEA angular distributions, their ratio, and its REB prediction [9]. Note that instead of the mere elastic-scattering cross section (4), we use the summed cross section $d\sigma_{\rm sum}/d\Omega$, which corresponds to all quasi-elastic processes (elastic and inelastic scattering and elastic breakup). It is plotted here as a ratio to the Rutherford cross section (dotted line). This gives the best agreement between DEA and REB [9]. As already noted in Ref. [12], both angular distributions exhibit similar features. Interestingly their ratio (solid line) does not exhibit such features and it behaves smoothly with the scattering angle θ. In addition the DEA ratio is in excellent agreement with its REB prediction (thick grey line). This result suggests the ratio to be an excellent reaction observable to study loosely-bound systems, such as halo nuclei. Further tests have confirmed its independence to the reaction mechanism: the ratio computed for a carbon target is nearly superimposed to the one obtained on lead [9]. Moreover, the ratio is strongly sensitive to the binding energy, orbital, and radial wave function of the valence neutron [9], indicating that valuable information about the projectile structure can be inferred from it.

4. Summary and prospect

Halo nuclei are among the most peculiar nuclear structures discovered thanks to the development of radioactive-ion beams. They exhibit a strongly clusterised structure, which explains their large matter radius in comparison to their isobars. Being located on the shoulder of the valley

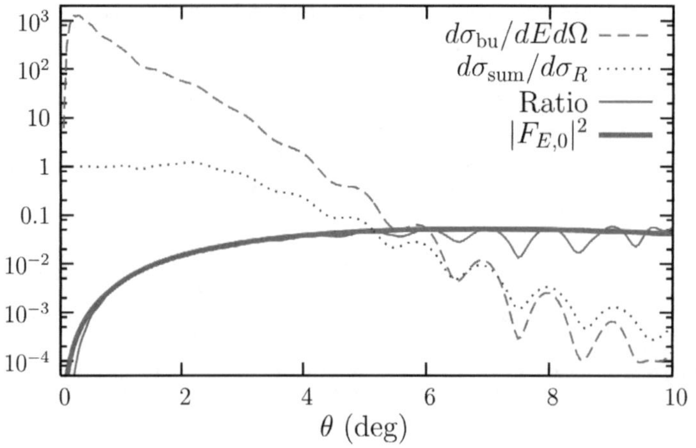

Figure 3. DEA calculations of ^{11}Be impinging on Pb at 69AMeV. Angular distribution for breakup at $E = 0.1$ MeV (dashed line) and the summed cross section (dotted line) are shown with their ratio (solid line) and the REB prediction (thick gray line) [9].

of stability, they are studied mostly through reactions, such as elastic scattering and breakup. Therefore, accurate reaction models must be available to extract valuable information from reaction data. In this contribution the mostly used reaction models have been presented and compared for the breakup of ^{15}C, a typical halo nucleus, at intermediate energy: the Continuum Discretised Coupled Channel model (CDCC) [4], the Time-Dependent approach (TD) [5], and the Dynamical Eikonal Approximation (DEA) [6]. Although based on different assumptions, they lead to nearly identical cross sections, but for the quantal interferences, which cannot be reproduced within TD, due to its relying on the semiclassical approximation.

To remove the influence of the reaction mechanism on observables, we suggest to measure the ratio of angular distributions, e.g. for scattering and breakup. As predicted by the REB model [8], this new observable is nearly independent of the reaction mechanism and provides detailed information on the halo structure. It seems that the idea could be easily applied to Borromean neutral systems like ^{11}Li. Extensions to charged valence nucleon, such as proton halœs may also be possible. We believe this ratio method will open a new era in the study of exotic systems.

References

[1] Tanihata I *et al* 1985 *Phys. Rev. Lett.* **55** 2676
[2] Fukuda N *et al* 2004 *Phys. Rev.* C **70** 054606
[3] Nakamura T *et al* 2009 *Phys. Rev.* C **79** 035805
[4] Tostevin J A, Nunes F M and Thompson I J 2001 *Phys. Rev.* C **63** 024617
[5] Esbensen H, Bertsch G F and Bertulani C A 1995 *Nucl. Phys.* A **581** 107
[6] Baye D, Capel P and Goldstein G 2005 *Phys. Rev. Lett.* **95** 082502
[7] Capel P, Esbensen H and Nunes F M 2012 *Phys. Rev.* C **85** 044604
[8] Johnson R C, Al-Khalili J S and Tostevin J A 1997 *Phys. Rev. Lett.* **79**, 2771
[9] Capel P, Johnson R C and Nunes F M 2011 *Phys. Lett.* B **705** 112
[10] Thompson I J 1988 *Comput. Phys. Rep.* **7** 167
[11] Capel P, Goldstein G and Baye D 2004 *Phys. Rev.* C **70** 064605
[12] Capel P, Hussein M and Baye D 2010 *Phys. Lett.* B **693** 448
[13] Goldstein G, Baye D and Capel P 2006 *Phys. Rev.* C **73** 024602

Journal of Physics: Conference Series **436** (2013) 012041 doi:10.1088/1742-6596/436/1/012041

Clustering in A=10 nuclei

G V Rogachev[1,2], L Baby[1], J Belarge[1], J C Blackmon[3], V Z Goldberg[4], E D Johnson[1], E Koshchiy[1], A N Kuchera[1], J Lai[3], L Linhardt[3], K Macon[3], M Matos[3], D Santiago-Gonzalez[1], I Wiedenhöver[1]

[1]Department of Physics, Florida State University, Tallahassee, FL, USA
[2]National Superconducting Cyclotron Laboratory, Michigan State University, MI, USA
[3]Department of Physics & Astronomy, Louisiana State University, Baton Rouge, LA, USA
[4]Cyclotron Institute, Texas A&M University, College Station, TX, USA

Abstract. We discuss the identification and properties of the states that belong to the highly clustered rotational band in A=10 nuclei, ^{10}Be, ^{10}B(T=1) and ^{10}C. The band is of interest because it may correspond to an exotic α:nn:α configuration.

1. Introduction

It has been suggested in several theoretical papers (see *e.g.* [1, 2, 3]) that highly clustered, double α-core structure, α:nn:α, may exist in A=10 T=1 nuclear systems. This structure would form a positive-parity rotational band (Fig. 1) based on the 0_2^+ state (0^+ at 6.17 MeV, 2^+ at 7.54 MeV and 4^+ at 10.15 MeV). Multiple experimental searches for the members of this band in ^{10}Be and ^{10}B have been performed [4, 5, 6, 7, 8]. We present new experimental data on the states in ^{10}Be populated in the ^{6}He+α interaction and discuss the spectrum of T=1 states in ^{10}B and ^{10}C in relation to the 0_2^+ rotational band.

2. Cluster States in ^{10}Be

Cluster states in ^{10}Be were searched for in the excitation function of ^{6}He+α elastic scattering which was measured in the excitation energy range between 9 and 18.5 MeV. Measurements were performed at the John D. Fox Superconducting Linear Accelerator Laboratory at Florida State University. The ^{6}He beam was produced by the in-flight rare isotope beam facility RESOLUT with intensity of about 10^5 pps. The new active target detector - Array for Nuclear Astrophysics and Structure with Exotic Nuclei (ANASEN) - was used to perform the measurements. ANASEN is a gas filled detector, which has an array of position sensitive silicon detectors that surrounds the beam axis and a cylindrical array of position sensitive proportional counters. The recoils that emerge from the place of interaction go through the proportional counter first. The location and energy loss in the active gas volume is measured. If the second hit is recorded by the silicon detector then the track of the recoil can be reconstructed using position information provided by the proportional counter and the silicon array. Particle ID is determined using energy loss in the active gas volume and the total energy measured by the silicon array. The ^{6}He+α elastic scattering events can be identified using the energy-angle kinematic dependence reconstructed from the tracks of the recoils.

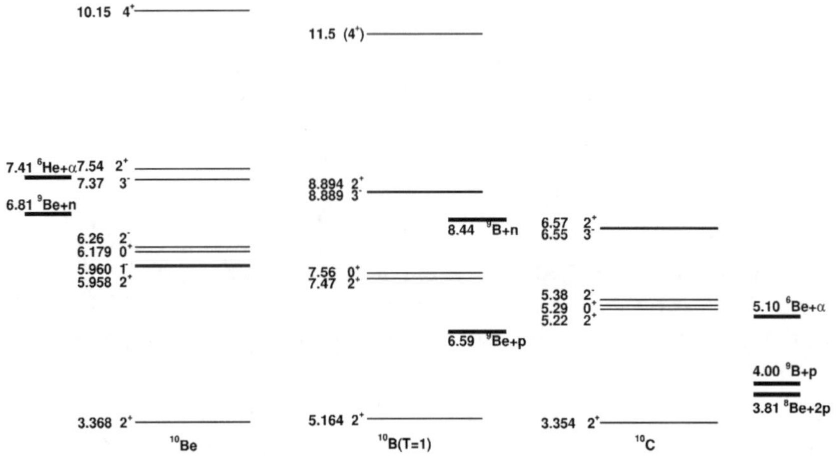

Figure 1. Level diagram for T=1 isobaric multiplet in A=10

We limit our discussion to the lower energy part of the ^{6}He+α elastic scattering excitation function. The energy range between 2 and 3.5 MeV in c.m.s. (9.5 and 11 MeV excitation energy) is shown in Fig. 2. The obvious feature of this excitation function is the peak at 2.75 MeV (10.16(2) MeV excitation energy). This is the state that was first observed in [4] and [5]. The 3^- spin-parity was assigned to it in [7]. However, this was contradicted by [8] where the ^{6}He+α angular distribution at 2.75 MeV energy in c.m. showed a maximum at 90 degrees, indicating positive parity. The 4^+ spin-parity assignment was made for the state in [8]. The excitation function was not measured in [8] and both the location and the width of the state were adopted from [7]. In this work we clearly observed the strong state at 10.16(2) MeV excitation energy in ^{10}Be with a width of 270(30) keV. The angular distribution for this state is shown in Fig. 2. Comparing this angular distribution to the squared Legendre polynomials give a unique spin-parity assignment for the state in question. We confirm the 4^+ spin-parity assignment for this state first made in [8]. The partial α width for the 10.16 MeV state is 150 keV. It was determined from the measured cross section and the total width of this state. This partial width exceeds the single-particle limit for the purely α-cluster states. Therefore, our data provide unambiguous evidence that the 10.16 MeV state is 4^+ and it has an extreme α-cluster character.

The 10.16 MeV state has been suggested as a possible member of the highly clustered 0_2^+ rotational band in [5]. Our data strongly support this suggestion. The degree of clustering for the other members of this rotational band in ^{10}Be, the 0^+ state at 6.18 MeV and the 2^+ state at 7.54 MeV is uncertain. The branching ratio Γ_α/Γ=3.5(12)$\times10^{-3}$ for the 2^+ state at 7.54 MeV reported in [9] would indicate a huge partial α width, 30-50 times above the single particle limit. Considering the experimental uncertainty of this measurement and unusually large value for the partial α width, independent confirmation is required.

3. The 0_2^+ band members in ^{10}B and ^{10}C

The spectrum of the T=1 states in ^{10}B has been studied recently in [10] using the ^{9}Be(^{1}H,α)^{6}Li(0^+; T=1) reaction. The partial α-width of the T=1 2^+ state at 8.89 MeV in ^{10}B, the isobaric analog of the 2^+ at 7.54 MeV in ^{10}Be was determined using R-matrix analysis of the ^{9}Be(^{1}H,α)^{6}Li(0^+; T=1) excitation function. Its value, 18±2.0±2.3(sys) keV corresponds

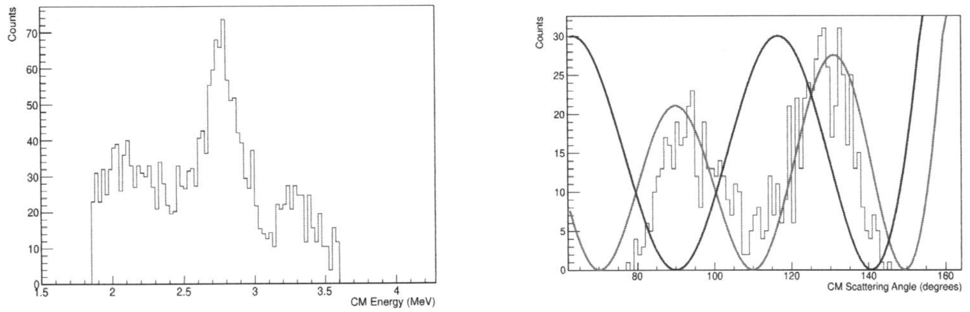

Figure 2. Left figure: Excitation function for the ^{6}He+α elastic scattering integrated over angular range from 90 to 140 degrees. Right figure: Angular distribution at 2.75±0.1 MeV in c.m.s. Red and blue curves corresponds to the square of L=4 and L=3 Legendre polynomials, respectively.

to the α single particle limit and clearly identifies this state as an extremely clustered structure. A similar conclusion can be found in [11]. The T=1 4^+ state in ^{10}B, the isobaric analog for the 10.16 MeV state in ^{10}Be is expected at excitation energy close to 11.5 MeV. No 4^+ state has been observed near that energy in [10]. However, in [12] a state at 11.5 MeV that decays by α emission to the ^{6}Li(0^+;T=1) and has small nucleon width has been observed. It is possible that because of the ^{9}Be+p entrance channel for the reaction used in [10] the 4^+ state was not populated. The reaction ^{11}B(^{3}He,α)^{10}B used in [12] does not have this limitation.

Spin-parity assignment of the states in ^{10}C is uncertain. However, recent analysis of the decay spectroscopy experiments [13] allows to make a suggestion that the state at 5.29 MeV and 6.57 MeV (Fig. 1) have large partial α widths (compared to α single particle limit) and small nucleon width. A particularly interesting feature of these states is their relatively large two proton decay branching ratio. It is argued in [14] that these states are highly clustered 0^+ and 2^+ states that belong to the 0_2^+ band. Confirming these spin-parity assignments and accurately measuring decay branching ratios would allow determination of the partial α width for the 0_2^+ state and, provided that it is large, firmly establish the highly clustered nature of the 0_2^+ band. The moment of inertia for this band is a factor of three larger than for the ground state band. This corresponds to the effective radius that is a factor of 1.8 larger than for the states in the ground state band. This was accurately predicted by the AMD calculations [1].

References

[1] Dote A *et al*1997 *Phys. Rev.* C **56** 1877
[2] Itagaki N and Okabe S 2000 *Phys. Rev.* C **61** 044306
[3] Ito M, Kato K and Ikeda K 2006 *Phys. Lett.* B **636** 293
[4] Hamada S *et al* 1994 *Phys. Rev.* C **49** 3192
[5] Soic N *et al*1996 *Europhys. Lett.* **34** 7
[6] Milin M *et al* 1999 *Europhys. Lett.* **48** 616
[7] Curtis N *et al* 2001 *Phys. Rev.* C **64** 044604
[8] Freer M *et al* 2006 *Phys. Rev. Lett.* **96** 042501
[9] Liendo J A *et al* 2001 *Phys. Rev.* C **63** 034301
[10] Kuchera A N *et al* 2011 *Phys. Rev.* C **84** 054615
[11] Fortune H T *et al* 2011 *Phys. Rev.* C **84** 024304
[12] Kirsebom O S *et al* 2012 *Phys. Rev.* C **85** 054308
[13] Charity R J *et al* 2009 *Phys. Rev.* C **80** 024306
[14] Goldberg V Z *et al* 2012 *Phys. Rev.* C **86** 044314

Two novel cluster states in ^{10}Be

Fumiharu Kobayashi and Yoshiko Kanada-En'yo

Department of Physics, Kyoto University, Kyoto 606-8502, Japan

E-mail: f-kobayashi@ruby.scphys.kyoto-u.ac.jp

Abstract. We have investigated the cluster structure of ^{10}Be by using ^{6}He+α cluster wave functions and 2α+$2n$ dineutron condensate wave functions. We found two novel cluster 0^+ states, one of which has an α+α+dineutron gas-like structure and the other contains ^{6}He and a well-separated α cluster. In this paper, we discuss their cluster structures.

1. Introduction

Cluster structure is a hot topic in nuclear physics, and it has been eagerly investigated both theoretically and experimentally. In Be isotopes, an α cluster, which is the most representative cluster, tends to be developed. As for the 0^+ states of ^{10}Be, the $0^+_{1,2}$ states have been confirmed experimentally, and their structures are explained well by a 2α core and two valence neutrons in the molecular orbits around the core. In addition to these 0^+ states, a new 0^+ state was proposed experimentally [1], which may consist of a ^{6}He and a well-developed α.

In conjunction with the cluster structure of finite nuclei, we are also interested in the dineutron correlation. Two neutrons coupled to spin-singlet have strong spatial correlation and become compact to be considered as a kind of cluster. Its universal properties are not well-known and we constructed a model, which we call the "dineutron condensate (DC) wave function", to investigate the dineutron correlation in various nuclei systematically [2].

In this work, we have investigated cluster states in ^{10}Be and put forward the possibility of a state which contains a compact dineutron and another state which contains ^{6}He and a well-separated α. We discuss their cluster structures in this paper.

2. Theoretical framework

In the present work, we superpose two kinds of cluster wave functions, namely, ^{6}He+α cluster wave functions used in Ref. [3] and 2α+$2n$ DC wave functions proposed in Ref. [2].

In the ^{6}He+α cluster wave function, the system is composed of a ^{6}He and an α cluster separated by a distance d_c. The single particle wave functions are described by the harmonic oscillator wave functions, and the configuration of an α is $(0s)^4$. The ^{6}He cluster is composed of an α core and two valence neutrons in the $(0p)^2$ configuration represented by Gaussians shifted with respect to the α core. By using such ^{6}He+α cluster wave functions, the $0^+_{1,2}$ states of ^{10}Be are well-described as shown in Ref. [3].

In addition to the ^{6}He+α wave functions, we superpose the 2α+$2n$ DC wave functions. In the 2α+$2n$ DC wave functions, we assume that the nuclear system is composed of a 2α core and two valence neutrons whose spins are coupled to singlet and their relative motion is an s-wave.

The relative and centre of mass (c.o.m.) wave functions of two neutrons are denoted by ψ_r and ψ_G, and the explicit form of the whole DC wave function is as follows.

$$\Phi_{\rm DC} = \frac{1}{\sqrt{10!}} \det \left[\psi_{\alpha 1}(\boldsymbol{r}_1) \cdots \psi_{\alpha 1}(\boldsymbol{r}_4) \; \psi_{\alpha 2}(\boldsymbol{r}_5) \cdots \psi_{\alpha 2}(\boldsymbol{r}_8) \; \psi_r(\boldsymbol{r}) \psi_G(\boldsymbol{r}_G) \right], \tag{1}$$

$$\psi_r(\boldsymbol{r}) \propto \exp\left[-\frac{r^2}{4b_n^2} \right], \qquad \psi_G(\boldsymbol{r}_G) \propto \exp\left[-\frac{r_G^2}{\beta^2} \right], \tag{2}$$

where $\psi_{\alpha i}$ $(i = 1, 2)$ are the single particle wave functions of the 2α core whose relative distance is d_n; $\boldsymbol{r}$ and $\boldsymbol{r}_G$ are the relative and c.o.m. coordinates of the two extra neutrons, respectively. The Gaussian widths in the relative and c.o.m. wave functions are b_n and β, and these parameters set the size and the spatial expansion of the dineutron around the core, respectively. In the DC wave functions, we fix the parameters β and b_n to various values to describe the system as a superposition of dineutron subsystems with various sizes and distributions.

In the present work, we include ^{6}He+α-type basis states with $d_c = 1, 2, \cdots, 8$ fm as well as 2α+$2n$ DC-type basis states with $d_n = 1, 2, \cdots, 6$ fm, $\beta = 2, 3, \cdots, 9$ fm and five b_n values for each β.

3. Results

By superposing the wave functions explained above, we obtain two kinds of excited states having distinct cluster structures above the 0_2^+ state. We explain the structures of two new 0^+ states below.

3.1. The α+α+dineutron gas-like state

We obtained a state, which we label 0_3^+ here, containing a rather compact dineutron separated from the core. To see its dineutron component, we calculated the overlaps with the DC wave functions whose parameters d_n, β and b_n are fixed to various values. A component with a small b_n corresponds to a state containing a compact dineutron, while a component with a large β corresponds to a state containing two neutrons far from the 2α core. We show the overlap of the 0_3^+ state obtained with DC wave functions whose d_n is fixed to 4.0 fm in the β-b_n plane in Figure 1 (a), and that whose b_n is fixed to 2.2 fm on the β-d_n plane in Figure 1 (b). It can be seen in Figure 1 (a) that the peak is located at $(\beta, b_n) \sim (4.0, 2.2)$ with a broad width in both directions of β and b_n, and in Figure 1 (b) that the peak is at $(\beta, d_n) \sim (4.0, 4.0)$ also

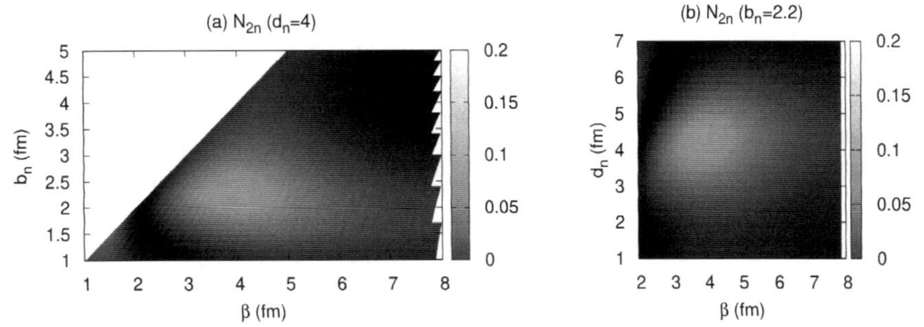

Figure 1. (a) The overlap of the 0_3^+ state with the DC wave functions with $d_n = 4$ fm. The horizontal axis is β and the vertical one is b_n. (b) The overlap of the 0_3^+ state with the DC wave functions with $b_n = 2.2$ fm. The horizontal axis is β and the vertical one is d_n. These figures are from Ref. [4]

with broad widths in both horizontal and vertical directions. This means that the dineutron is rather compact but its size is changeable to some extent, and such a soft dineutron and the two α clusters are weakly interacting with each other and spread broadly. We propose that this 0_3^+ state is a gas-like cluster state since it contains well-separated clusters loosely bound.

3.2. The state with a ^{6}He plus a well-separated α

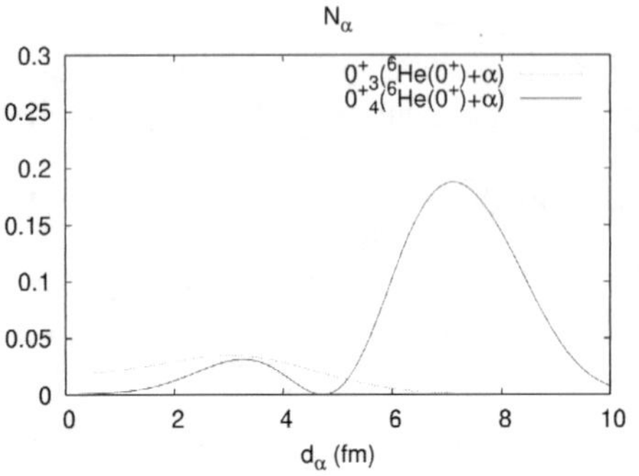

Figure 2. The overlaps of the $0_{3,4}^+$ states with the ^{6}He(0^+)+α wave functions. The horizontal axis is d_α. This figure is from Ref. [4]

We also obtained a ^{6}He+α state. We label it by 0_4^+. It lies energetically above the α+α+2n state 0_3^+. In order to see its components consisting of a ^{6}He and an α cluster, as a function of their distance, in Figure 2 we show the overlap of the 0_4^+state with ^{6}He(0^+)+α wave functions (the subsystem ^{6}He is projected to $J^\pi = 0^+$) whose relative distance is d_α. The 0_4^+ state has a large amplitude in the region of $d_\alpha \sim 7$ fm. This indicates that in the 0_4^+ state there is a well-developed α cluster displaced far from ^{6}He. On the other hand, the component of ^{6}He+α is minor in the 0_3^+ state because the two α clusters and the one dineutron in this state are weakly correlated, so that a ^{6}He cluster cannot be formed. Thus these two 0^+ states have cluster structures distinct from each other.

4. Summary

In the present work, we investigated excited cluster states in ^{10}Be and indicated that there may exist two novel 0^+ states with significantly different cluster structures. One of them has a gas-like structure of α+α+dineutron interacting weakly and extended spatially. Another consists of a ^{6}He cluster and a well-separated α cluster.

Acknowledgments

This work was supported by Grant-in-Aid for Scientific Research from Japan Society for the Promotion of Science (JSPS). It was also supported by the Grant-in-Aid for the Global COE Program "The Next Generation of Physics, Spun from Universality and Emergence" from the Ministry of Education, Culture, Sports, Science and Technology (MEXT) of Japan. A part of the computational calculations of this work was performed by using the supercomputers at YITP.

References

[1] Kuchera A N *et al* 2011 *Phys. Rev.* C **84** 054615
[2] Kobayashi F and Kanada-En'yo Y 2011 *Prog. Theor. Phys.* **126** 457
[3] Kanada-En'yo Y and Suhara T 2012 *Phys. Rev.* C **85** 024303
[4] Kobayashi F and Kanada-En'yo Y 2012 *Phys. Rev.* C **86** 064303

Journal of Physics: Conference Series **436** (2013) 012043 doi:10.1088/1742-6596/436/1/012043

Cluster structures of excited states in ^{11}B

Tadahiro Suhara[1] **and Yoshiko Kanada-En'yo**[2]

[1]Institute of Physics, University of Tsukuba, Tsukuba 305-8571, Japan[1]

[2]Department of Physics, Graduate School of Science, Kyoto University, Kyoto 606-8502, Japan

Abstract. The cluster structure of excited states in ^{11}B is investigated with a method of β-γ constrained antisymmetrized molecular dynamics in combination with the generator coordinate method. For negative-parity states, we identify a band with a $2\alpha+t$ cluster structure. This band starts from the $3/2_3^-$ state and may correspond to the experimental band observed recently. We find that the composision of the $3/2_3^-$ state s quite similar to that of the 0_2^+ state in ^{12}C.

1. Introduction

Recently, the cluster structure of ^{11}B has attracted much interest. Some theoretical and experimental studies have suggested that cluster structures develop well in the negative-parity states above or near $2\alpha+t$ the threshold energy [1, 2, 3]. In the recent experiment of α resonant scattering on ^{7}Li [4], Yamaguchi *et al* suggested a new negative-parity band consisting of 8.56 MeV ($3/2^-$), 10.34 MeV ($5/2^-$), 11.59 MeV ($7/2^-$) and 13.03 MeV ($9/2^-$) states, which have large α decay widths. It is interesting to know what kind of the cluster structure this band has.

Moreover, the analogy between the cluster features of ^{11}B and the three-α-cluster structure of ^{12}C is a fascinating problem to be clarified. In previous works [2, 5], the $3/2_3^-$ state was suggested to have a dilute cluster structure with a $2\alpha+t$ configuration, and to be an analogue of the 0_2^+ state in ^{12}C, which has a dilute 3α structure [6, 7, 8]. However, in a recent work [3], it was argued that ^{11}B($3/2_3^-$) cannot correspond to ^{12}C(0_2^+). The relation between the states ^{11}B($3/2_3^-$) and ^{12}C(0_2^+) is controversial and further studies are required.

In this article, we investigate the structures of some excited states in ^{11}B with a method of β-γ constrained antisymmetrized molecular dynamics (AMD) in combination with the generator coordinate method (GCM). To clarify the correspondence between the $3/2_3^-$ state of ^{11}B and the 0_2^+ state of ^{12}C, we compare the squared overlaps of their resulting wave functions with the basis functions on the β-γ plane.

2. Formulation

In this section, we briefly explain the β-γ constrained AMD + GCM [9]. In the method of AMD, a basis wave function $|\Phi\rangle$ is a Slater determinant of single-particle wave functions $|\varphi_i\rangle$. The ith single-particle wave function consists of the spatial part $|\phi_i\rangle$, spin part $|\chi_i\rangle$, and isospin part $|\tau_i\rangle$ of the following form

$$|\varphi_i\rangle = |\phi_i\rangle|\chi_i\rangle|\tau_i\rangle, \tag{1}$$

$$\langle \mathbf{r}|\phi_i\rangle = \left(\frac{2\nu}{\pi}\right)^{\frac{3}{4}} \exp\left[-\nu\left(\mathbf{r}-\frac{\mathbf{Z}_i}{\sqrt{\nu}}\right)^2+\frac{1}{2}\mathbf{Z}_i^2\right], \tag{2}$$

$$|\chi_i\rangle = \xi_{i\uparrow}|\uparrow\rangle+\xi_{i\downarrow}|\downarrow\rangle. \tag{3}$$

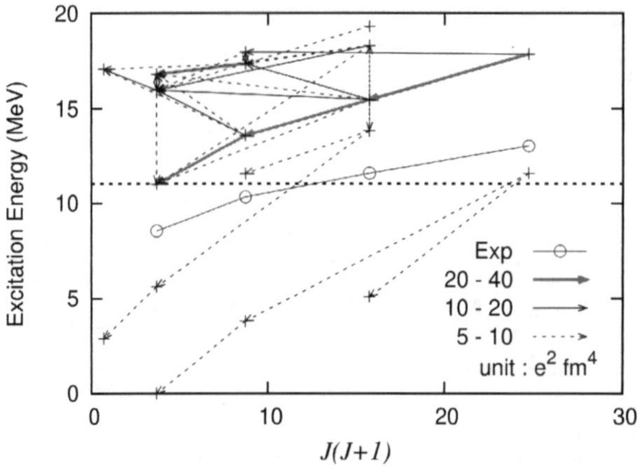

Figure 1. The black crosses are the calculated negative-parity states in ^{11}B. 5.0 e^2fm^4 $<$ $B(E2) \leq 10.0$ e^2fm^4, 10.0 e^2fm^4 $< B(E2) \leq 20.0$ e^2fm^4, and 20.0 e^2fm^4 $< B(E2) \leq 40.0$ e^2fm^4 transitions are denoted by broken, black solid and red bold solid arrows, respectively. The blue circles belong to the experimental band found in Ref. [4]. Dotted line: $2\alpha + t$ threshold.

The isospin part $|\tau_i\rangle$ is fixed to be up (proton) or down (neutron). In a basis wave function $|\Phi\rangle$, $\{\mathbf{Z}_1, \xi_1, \mathbf{Z}_2, \xi_2, \cdots, \mathbf{Z}_A, \xi_A\}$ are complex variational parameters and they are determined by the energy optimization with constraints on the quadrupole deformation parameters β and γ.

In the calculations of the energy levels, we superpose parity- and total-angular-momentum-projected AMD wave functions $\hat{P}^J_{MK}\hat{P}^{\pm}|\Phi(\beta,\gamma)\rangle$. Thus, the final wave function for the $J^{\pm}_n$ state is given by a linear combination of the basis wave functions, $|\Phi^{J\pm}_n\rangle = \sum_K \sum_i f_n(\beta_i, \gamma_i, K)\hat{P}^J_{MK}\hat{P}^{\pm}|\Phi(\beta_i,\gamma_i)\rangle$. The coefficients $f_n(\beta_i, \gamma_i, K)$ are determined by solving the Hill-Wheeler equation.

3. Results

In this section, we briefly show the results for negative-parity states. For more details, including the results for positive-parity states, the reader is referred to Ref. [10].

In Fig. 1, we show the calculated negative-parity energy levels as functions of the angular momentum $J(J+1)$ grouped according to the $E2$ transition strengths. We also show the experimental band [4]. At first, we explain the structures of ground and excited states. The calculated low-lying states have large overlaps with the basis wave functions that are in between the shell-model structure and the cluster structure. In the high-lying states near and above the $2\alpha + t$ threshold energy, we obtain various well-developed cluster states having significant overlaps with the basis wave functions which represent $2\alpha + t$ cluster structures. In particular, the $3/2^-_3$ state, which is found to have a dilute cluster structure with a $2\alpha + t$ configuration, is described by the linear combination of various $2\alpha + t$ spatial configurations.

For the $3/2^-_3$, $5/2^-_3$, $7/2^-_3$, and $9/2^-_3$ states, the $E2$ transition strengths are as large as 20–30 e^{2}fm^4, and therefore, we consider these states as members of a band. Although the calculated excitation energies are higher than the experimental ones of the band identified in Ref. [4] by 2.5–4.5 MeV, the systematics of the level structure corresponds well with the experimental ones. In both results, the higher angular momentum states satisfy the $J(J+1)$ rule. Only the $3/2^-_3$ state deviates from the straight line. Moreover, the present result suggests that clustering is well-developed in this band, and its members have large α decay widths. A confirmation of this

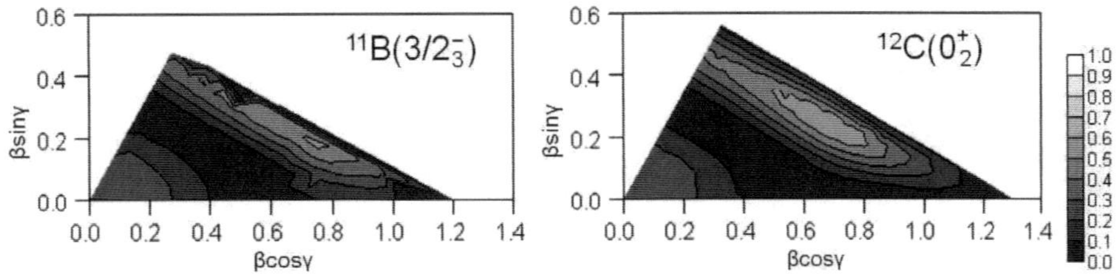

Figure 2. Squared overlaps between the GCM wave functions and basis wave functions for the $3/2_3^-$ state of ^{11}B (Left) and the 0_2^+ state of ^{12}C (Right).

conjecture would support the correspondence between the theoretical and experimental bands.

We discuss the relation between the $3/2_3^-$ state in ^{11}B and the 0_2^+ state in ^{12}C by comparing the squared overlaps between the GCM wave functions and the basis wave functions on the β-γ plane, $|\langle\Phi(\beta,\gamma)|\Phi_n^{J\pm}\rangle|^2$, in Fig. 2. The results for the 0_2^+ states in ^{12}C are calculations with the β-γ constrained AMD + GCM taken from Ref. [9]. The distributions of the squared overlaps between the GCM wave functions and the basis wave functions for these states are very similar to each other. Their comparison indicates that the $3/2_3^-$ state in ^{11}B is described by a linear combination of various $2\alpha + t$ configurations similar to the linear combination of the 3α configurations in the the 0_2^+ state of ^{12}C. That is, when an α cluster is replaced by a t cluster in the 0_2^+ state of ^{12}C, one obtains the $3/2_3^-$ state of ^{11}B.

4. Summary

We investigated the cluster structures of excited states in ^{11}B with the method of β-γ constrained AMD + GCM. By analysing the $E2$ transition strengths as well as the squared overlaps between the GCM wave functions and the basis wave functions, we discussed the negative-parity band starting from $3/2_3^-$, in which a $2\alpha+t$ structure is well developed. This band is a good theoretical candidate for a band which was identified in a recent experiment of alpha resonant scattering on ^{7}Li. To make the correspondence between the calculated and experimentally observed states more clear-cut, the partial decay widths of these excited states remain to be determined in the future. We found that the squared overlaps between the GCM wave functions and the basis wave functions of the $3/2_3^-$ state in ^{11}B are quite similar to those of the 0_2^+ state in ^{12}C.

Acknowledgments

The computations of this work were performed by supercomputers in YITP and KEK. This work was supported by the YIPQS program in YITP. It was also supported by the Grant-in-Aid for the GCOE Program "The Next Generation of Physics, Spun from Universality and Emergence" from MEXT of Japan, and a Grant-in-Aids for Scientific Research from JSPS.

References

[1] Nishioka H, Saito S and Yasuno M 1979 *Prog. Theor. Phys.* **62** 424
[2] Kanada-En'yo Y 2007 *Phys. Rev.* C **75** 024302
[3] Yamada T and Funaki Y 2010 *Phys. Rev.* C **82** 064315
[4] Yamaguchi H, Hashimoto T, Hayakawa S, Binh D N, Kahl D, Kubono S, Wakabayashi Y, Kawabata T and Teranishi T 2011 *Phys. Rev.* C **83** 034306
[5] Kawabata T *et al* 2007 *Phys. Lett.* B **646** 6
[6] Uegaki E, Okabe S, Abe Y and Tanaka H 1977 *Prog. Theor. Phys.* **57**, 1262
[7] Tohsaki A, Horiuchi H, Schuck P and Röpke G 2001 *Phys. Rev. Lett.* **87**, 192501
[8] Kanada-En'yo Y 2007 *Prog. Theor. Phys.* **117**, 655
[9] Suhara T and Kanada-En'yo Y 2010 *Prog. Theor. Phys.* **123**, 303
[10] Suhara T and Kanada-En'yo Y 2012 *Phys. Rev.* C **85**, 054320

Journal of Physics: Conference Series **436** (2013) 012044 doi:10.1088/1742-6596/436/1/012044

States in ^{10}Be populated in ^{6,7}Li(^{7}Li,^{3,4}He)^{10}Be* reactions at 30 and 52 MeV

M Uroić[1], M Milin[2], A Di Pietro[3], P Figuera[3], I Martel[4], M Fisichella[3], M Lattuada[3], Đ Miljanić[1], M G Pellegriti[3], L Prepolec[1], A M Sanchez Benitez[4], V Scuderi[3], N Soić[1], E Strano[3] and D Torresi[3]

[1]Ruđer Bošković Institute, Zagreb, Croatia
[2]Department of Physics, University of Zagreb, Zagreb, Croatia
[3]INFN-Laboratori Nazionali del Sud, Catania, Italy
[4]Departamento de Fisica Aplicada, Universidad de Huelva, Huelva, Spain

E-mail: milivoj.uroic@irb.hr

Abstract. In this contribution, an analysis of an experiment with the 30 and 52 MeV ^{7}Li beam and the ^{7}LiF and ^{6}LiF targets is given, emphasizing population of cluster states by triton transfer to ^{7}Li. Results concerning both reactions are given and discussed. Aside from the dominant $\alpha + d$ structure in ^{6}Li, a rather large overlap with ^{3}He+t cluster configuration populates cluster states in ^{10}Be by triton pick-up. A comparision with t pick-up from ^{7}Li, resulting with detected α instead of ^{3}He, with a considerably higher Q-value is given.

A special attention was paid to analysing and compensating non-uniformity of thicknes of standard 50 μm thin silicon detectors used in the measurement. Rather than measuring their thickness independantly, a method was derived to calculate and compensate the thickness variation directly from the measured data.

1. Introduction

Population of cluster states in light nuclei through low-energy nuclear reactions is highly dependent on the initial states of beam and target nuclei, as well as the transfer mechanism. Very suitable nuclei for $d-$ $t-$ and $\alpha-$transfer are both stable litium isotopes, where the ground state of ^{7}Li has well expressed $\alpha + t$ cluster structure, and ^{6}Li a dominant $\alpha + d$, but also ^{3}He+t structure. In this measurement, ^{7}Li beam was used to bombard isotopically enriched ^{6}LiF and ^{7}LiF targets, with one of the aims to study following reactions:

$$\begin{aligned}
&^{7}\text{Li}(^{7}\text{Li},\alpha)^{10}\text{Be}^*, && Q = 14.78\ \text{MeV}\\
&^{6}\text{Li}(^{7}\text{Li},\alpha)^{9}\text{Be}^*, && Q = 15.22\ \text{MeV}\\
&^{6}\text{Li}(^{7}\text{Li},^{3}\text{He})^{10}\text{Be}^*, && Q = 1.46\ \text{MeV}\\
&^{6}\text{Li}(^{7}\text{Li},t)^{10}\text{B}^*, && Q = 1.99\ \text{MeV}
\end{aligned}$$

Considering the ground state lithium structure, these reactions should be dominated by cluster transfer, namely $t-$transfer in the first and third, $d-$thansfer in the second, and ^{3}He$-$transfer in the fourth reaction. The well expressed ground state of the ^{7}Li, combined with the cluster transfer should populate excited cluster states in beryllium and boron, with ground-state recoil, thus enabling a very clean study of excited energy in single recoil spectra. Attention

Journal of Physics: Conference Series **436** (2013) 012044 doi:10.1088/1742-6596/436/1/012044

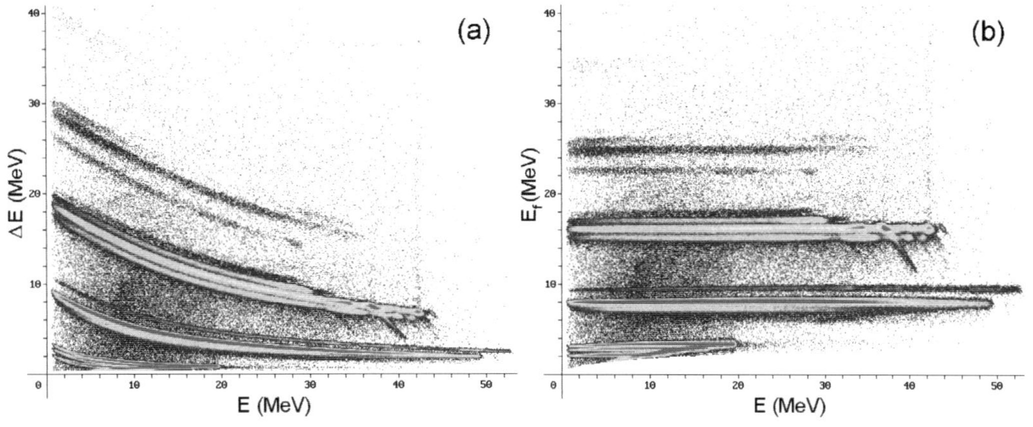

Figure 1. Standard ΔE (a) versus proposed E_f (b) plots for particle identification

will be given to the first and third reaction, producing ^{10}Be via triton transfer, studied through registering $\alpha-$ and ^{3}He$-$ recoil, respectively. In order to clearly distinguish α and ^{3}He particles, standard $\Delta E - E$ detector telescopes were used with somewhat unusual particle-identification criteria. It is discussed in the following section.

2. Technical remarks on particle identification

Instead of using standard graphical cut on a $\Delta E - E$ plot from raw detector readout, a slowly energy varying function E_f was used instead of ΔE:

$$E_f = \sqrt{\Delta E \cdot E + \alpha \Delta E^2} + \beta E,$$

with dimensionless constants $\alpha = 0.7$ and $\beta = 0.05$ chosen to give the smallest energy dependance [3]. Figure 1 ilustrates the comparision between standard $\Delta E - E$ plot (a) and $E_f - E$ plot (b), demonstrating nearly constant E_f values for specific nuclei.

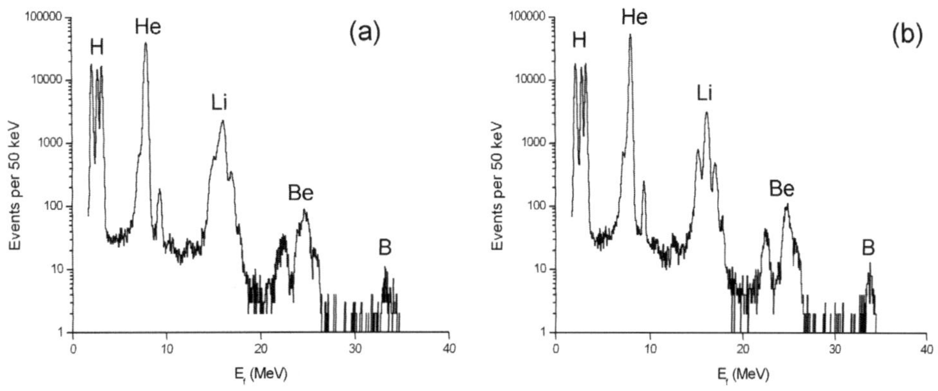

Figure 2. Projections of fig. 1b on E_f axis: raw data (a) and corrected (b)

Having reasonably large region (with high statistics in each DSSSD strip pixel) with practically constant E_f value (chosen for the most abundant α particles) enabled scaling the thin-detector energy *for each pixel* in thick detector to give exactly the same median E_f value, thus compensating the non-uniformity of the thin detector. As it can be seen from figure 2, resolution on *all* nuclei is improved, thus confirming the assumption that the spread has a thickness-based component.

It is evident that any cutting criteria based on corrected data (fig 2b) is more accurate than on uncorrected data (fig 2a). Specifically, the quality of ^{3}He-α separation is significantly improved.

3. Results and conclusion

The above listed procedure with standard callibration and event-by-event kinematical calculation gives the spectra plotted in figure 3. It is evident that different lithium isotopes in the target

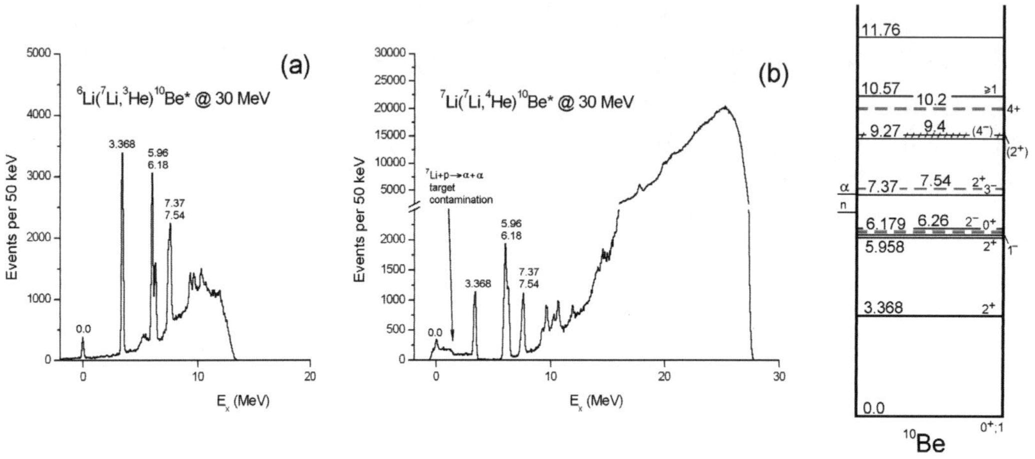

Figure 3. Excitations in ^{10}Be from triton transfer on two lithium isotopes. Three states marked red and dashed form a rotational band [1, 2]

populated the cluster states with similar intensities, thus confirming the assumption of the same $t-$ transfer mechanism and low energy dependance (Q-factors differ by more than 13 MeV). Data for the higher beam energy turned out to be unsuitable for these direct reactions, because the produced recoils had enough energy to pass through the whole detector telescope.

References

[1] Milin M, Zadro M *et al* 2005 *Nucl. Phys.* A **753** 263
[2] Freer M, Casarejos E *et al* 2006 *Phys. Rev. Lett.* **96** 042501
[3] Uroić M *et al* 2012 *AIP Conf. Proc.* **1491** 357

Journal of Physics: Conference Series **436** (2013) 012045 doi:10.1088/1742-6596/436/1/012045

Breakup of ^{11}Li in a three-cluster model

E. Pinilla[1], P. Descouvemont[1] and D. Baye[2]

[1] Physique Nucléaire Théorique et Physique Mathématique, C.P. 229,
Université Libre de Bruxelles (ULB), B 1050 Brussels, Belgium
[2] Physique Quantique, C.P. 165/82,
Université Libre de Bruxelles (ULB), B 1050 Brussels, Belgium

Abstract. The ^{11}Li breakup on a ^{208}Pb target is studied in the Coulomb corrected eikonal approximation by using a ^{9}Li+n+n three-body description of the projectile. The ^{11}Li wave functions are defined in the hyperspherical formalism for bound and scattering states, and are obtained from effective ^{9}Li+n and n+n interactions. The ^{9}Li+n+n three-body phase shifts suggest the existence of a narrow 1^- resonance near 0.5 MeV above threshold. This resonance shows up as a peak in the breakup cross section, and is supported by experimental data.

1. Introduction

Breakup reactions represent an efficient tool for the experimental investigation of exotic nuclei [1]. In particular, reactions involving a radioactive ^{11}Li beam on various targets have been carried out in recent years (see Ref. [2] and references therein). An accurate theoretical description of the breakup process requires a reaction theory complemented by a precise description of the projectile wave function.

The goal of the present work is to investigate recent breakup data of ^{11}Li on a ^{208}Pb target at 70 MeV/nucleon [2]. These cross section data present a maximum near 0.3 MeV above the ^{9}Li+n+n threshold, which could be associated with a dipole resonance in ^{11}Li [3]. We describe ^{11}Li in the three-body hyperspherical formalism [4, 5], which provides the ^{11}Li ground state, but also ^{9}Li+n+n three-body scattering states [6, 7]. This theoretical framework for the projectile can be implemented in the eikonal theory [8]. This approximation [9, 10] allows to derive elastic and breakup cross sections in a unified way. As the traditional eikonal method is known to diverge for the Coulomb potential, we use the Coulomb corrected version [11–13], which avoids the convergence problem in breakup cross sections. More detail can be found in Ref. [14].

2. Theoretical framework

The hamiltonian of the ^{11}Li+^{208}Pb system is defined as

$$H = H_0 - \frac{\hbar^2}{2\mu_{PT}}\Delta_R + V_{PT}(\boldsymbol{R}, \boldsymbol{x}, \boldsymbol{y}), \tag{1}$$

where hamiltonian H_0 is associated with the ^{11}Li projectile, μ_{PT} is the reduced mass, and $\boldsymbol{R}$ the relative coordinate. In Eq. (1), $\boldsymbol{x}$ and $\boldsymbol{y}$ are scaled Jacobi coordinates defining the ^{11}Li nucleus as a three-body ^{9}Li+n+n structure. The projectile-target interaction is taken as $V_{PT} = \sum_{i=1}^{3} V_{iT}(\boldsymbol{R}, \boldsymbol{x}, \boldsymbol{y})$, where $i = 1$ labels the ^{9}Li core, and $i = 2, 3$ the external neutrons. Those interactions are simulated by complex optical potentials.

The first step is to determine the ^{11}Li wave functions in the hyperspherical formalism. The ^{9}Li+n potential is taken from Ref. [15]. This interaction generates a core+n s wave scattering length of -5.7 fm. It also reproduces a $p_{1/2}$ resonance near 540 keV in agreement with Ref. [16]. A scaling factor 1.0051 is applied in order to reproduce the experimental binding energy of ^{11}Li. The n+n potential is the central part of the Minnesota interaction with the standard value $u = 1$ [17]. Three-body continuum wave functions and eigenphases are determined with the R-matrix theory [7].

In the second step, ^{11}Li+^{208}Pb scattering wave functions are determined at the eikonal approximation, i.e. the wave function is factorized as

$$\Phi(\boldsymbol{R}, \boldsymbol{x}, \boldsymbol{y}) = e^{iKZ}\hat{\Phi}(\boldsymbol{R}, \boldsymbol{x}, \boldsymbol{y}). \tag{2}$$

This factorization is introduced in the Schrödinger equation with Hamiltonian (1), with the adiabatic approximation that consists in replacing H_0 by the ^{11}Li ground-state energy E_0 [18]. The eikonal wave function, valid at high energies, is given by

$$\hat{\Phi}_{\text{eik.}}(\boldsymbol{R}, \boldsymbol{x}, \boldsymbol{y}) = \exp\Bigg(-\frac{i}{\hbar v}\int_{-\infty}^{Z} dZ'\, V_{PT}(\boldsymbol{b}, Z', \boldsymbol{x}, \boldsymbol{y})\Bigg)\Psi^{J_0M_0\pi_0}(\boldsymbol{x}, \boldsymbol{y}), \tag{3}$$

where v is the relative velocity between the target and the projectile, $\boldsymbol{b}$ the transverse component of $\boldsymbol{R}$, and $\Psi^{J_0M_0\pi_0}$ is the ground-state wave function. From wave function (3), elastic and breakup cross sections can be determined [8, 14]. The Coulomb potential leads to two divergence problems in the eikonal phase. They are solved by using the Coulomb-corrected eikonal approximation, which has been applied to describe breakup reactions involving halo nuclei in Refs. [8, 13].

For the n+^{208}Pb interaction, we consider the central part of the complex optical potential given in Ref. [19] at 70 MeV. For the ^{9}Li+^{208}Pb system, we use the α+^{208}Pb potential of Bonin et al. [20] at 699 MeV with scaled radii for the real and imaginary components ($R_r = 7.36$ fm and $R_i = 7.12$ fm).

3. Three-body phase shifts and breakup cross sections

Figure 1 displays the dominant 0^+, 1^- and 2^+ eigenphases [7] of ^{9}Li+n+n. The 0^+ and 2^+ curves exhibit a wide rise in energy with a resonant-like behavior. The 1^- eigenphase shows a sharp resonant behavior around 0.5 MeV. This energy is consistent with the maximum observed in the breakup experimental data, and suggests the existence of a low-lying resonance in ^{11}Li.

Figure 2 displays the total and 1^- breakup cross sections, convoluted with the detector response and compared with the experimental data of Ref. [2]. We observe a fair agreement for energies above 1.5 MeV, but the peak energy is slightly too high in the model. Including the 0^+ and 2^+ contributions increases the total cross section beyond 1 MeV, in better agreement with the experimental data. In the literature, calculations of breakup cross sections of halo nuclei often use the equivalent photon method [21]. This approximation assumes a dipole breakup process, and ignores other contributions. In contrast, the present eikonal description of the breakup reaction is more accurate, since it allows a quantitative evaluation of other partial wave contributions.

4. Conclusion

The ^{9}Li+n+n phase shifts suggest the existence of a narrow 1^- resonance near $E = 0.5$ MeV, corresponding to $E_x \approx 0.9$ MeV. Taking into account the $3/2^-$ spin of the ^{9}Li core nucleus, this resonance should correspond to $J = 1/2^+, 3/2^+$ or $5/2^+$ in ^{11}Li. Such a resonance is supported by a low-energy maximum in the experimental breakup cross section. The existence of a dipole resonance in the ^{11}Li nucleus seems now to be well established from experiment as well as from

Journal of Physics: Conference Series **436** (2013) 012045 doi:10.1088/1742-6596/436/1/012045

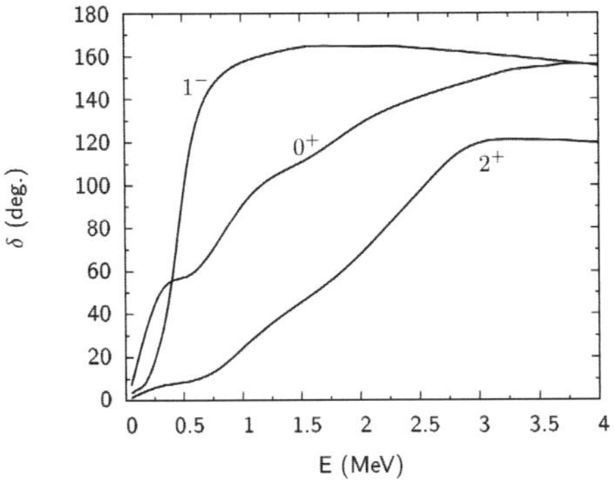

Figure 1. Dominant 0^+, 1^- and 2^+ eigenphases of the ^{9}Li+n+n system.

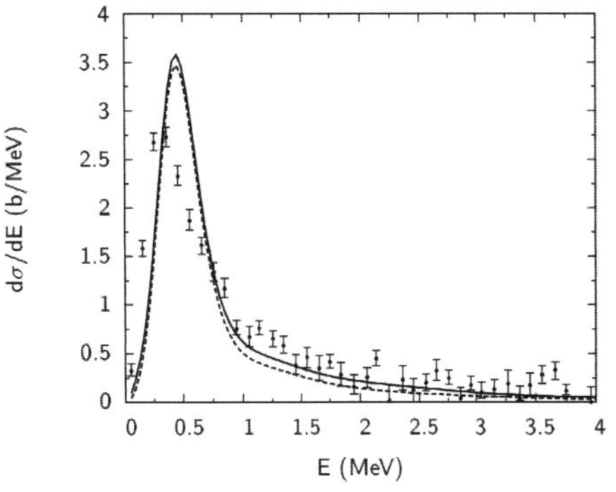

Figure 2. Total breakup cross section (full line) and 1^- contribution (dashed line) convoluted with the detector response. Experimental data are taken from Ref. [2].

theory. The situation is less clear in ^{6}He [8] where a 1^- resonance (broader than in ^{11}Li) is predicted by most theories, but not yet observed by experiment [22].

References

[1] Hansen P G, Jensen A S and Jonson B 1995 *Annu. Rev. Nucl. Sci.* **45** 591
[2] Nakamura T *et al* 2006 *Phys. Rev. Lett.* **96** 252502
[3] Hagino K and Sagawa H 2007 *Phys. Rev. C* **76** 047302
[4] Zhukov M V *et al* 1993 *Phys. Rep.* **231** 151
[5] Descouvemont P, Daniel C and Baye D 2003 *Phys. Rev. C* **67** 044309
[6] Thompson I J *et al* 2000 *Phys. Rev. C* **61** 024318
[7] Descouvemont P, Tursunov E M and Baye D 2006 *Nucl. Phys. A* **765** 370
[8] Baye D, Capel P, Descouvemont P and Suzuki Y 2009 *Phys. Rev. C* **79** 024607
[9] Glauber R J 1959 *High energy collision theory, in Lectures in Theoretical Physics, Vol. 1* (New York: Interscience)
[10] Yahiro M, Ogata K and Minomo K 2011 *Prog. Theor. Phys.* **126** 167
[11] Margueron J, Bonaccorso A and Brink D M 2003 *Nucl. Phys. A* **720** 337
[12] Abu-Ibrahim B and Suzuki Y 2004 *Prog. Theor. Phys.* **112** 1013
[13] Capel P, Baye D and Suzuki Y 2008 *Phys. Rev. C* **78** 054602
[14] Pinilla E C, Descouvemont P and Baye D 2012 *Phys. Rev. C* **85** 054610
[15] Esbensen H, Bertsch G F and Hencken K 1997 *Phys. Rev. C* **56** 3054
[16] Young B M *et al* 1994 *Phys. Rev. C* **49** 279
[17] Thompson D R, LeMere M and Tang Y C 1977 *Nucl. Phys. A* **286** 53
[18] Suzuki Y, Lovas R G, Yabana K and Varga K 2003 *Structure and Reactions of Light Exotic Nuclei* (London: Taylor & Francis)
[19] Koning A J and Delaroche J P 2003 *Nucl. Phys. A* **713** 231
[20] Bonin B, *et al* 1985 *Nucl. Phys. A* **445** 381
[21] Bertulani C and Baur G 1988 *Phys. Rep.* **163** 299
[22] Aumann T *et al* 1999 *Phys. Rev. C* **59** 1252

Mechanisms of Coulomb breakup reactions of ^{6}He and ^{11}Li

Y Kikuchi[1], **T Myo**[2,1], **K Katō**[3], **K Ikeda**[4]

[1] Research Center for Nuclear Physics (RCNP), Osaka University, Ibaraki 567-0047, Japan
[2] General Education, Faculty of Engineering, Osaka Institute of Technology, Osaka 535-8585, Japan
[3] Nuclear Reaction Data Centre, Faculty of Science, Hokkaido University, Sapporo 060-0810, Japan
[4] Nishina Center of Accelerator-Based Sciences, The Institute of Physical and Chemical Research (RIKEN), Wako 351-0198, Japan

E-mail: yuma@rcnp.osaka-u.ac.jp

Abstract. We investigate the three-body Coulomb breakup reactions of two-neutron halo nuclei and discuss the effect of binary subsystem correlations such as the of core-n and n-n. Furthermore, we calculate the invariant mass spectra. It is found that the final-state interactions of core-n and n-n binary subsystems dominantly determine the observed structures of the breakup cross sections, such as the low-lying enhancements.

1. Introduction

Since the discovery of neutron halo nuclei — such as ^{11}Be, ^{6}He, and ^{11}Li — extensive studies have been performed to understand their exotic structure caused by the weakly-bound nature of the valence neutrons. Theoretically, using the models based on the core + valence neutrons picture, the halo structure of the ground states and their excitations have been investigated [1, 2]. In two-neutron halo nuclei, it has been suggested that the correlation between two halo neutrons is important in reproducing the observed small two-neutron separation energies and large matter radii [1].

Experimentally, the Coulomb breakup reactions have been performed to expose the role of the n-n correlation in the two-neutron halo nuclei [3, 4]. The observed Coulomb breakup cross section provided us with interesting information on the weak-binding properties and the breakup mechanism of the halo nuclei. To investigate the properties and the breakup mechanism of halo nuclei, there are at least two problems to be solved: (i) clarifying the dominant breakup process such as the direct breakup to a noninteracting three-body continuum or sequential decay via the resonance of the binary subsystem, (ii) evaluating the influence of the final-state interactions (FSI) on the cross section.

In this study, we calculate the Coulomb breakup reaction of two-neutron halo nuclei, ^{6}He and ^{11}Li, and clarify the relation between the subsystem correlations of core-n and n-n and the Coulomb breakup cross section. To calculate the Coulomb breakup cross section, we use the complex-scaled solutions of the Lippmann-Schwinger equation (CSLS) [6, 7]. We calculate not only the breakup cross section with respect to the total energy of the system but also the invariant mass spectra of core-n and n-n by using CSLS, and discuss the effects of the binary subsystem correlations on the cross section.

2. Method

To evaluate the Coulomb breakup cross sections, it is necessary to obtain the scattering states, $\Psi^{(-)}(\mathbf{k},\mathbf{K})$, which is given as the function of the relative momenta, $\mathbf{k}$ and $\mathbf{K}$. Here, we obtain $\Psi^{(-)}(\mathbf{k},\mathbf{K})$ using CSLS [6, 7]. The incoming scattering state $\Psi^{(-)}$ in the bra-representation is described as

$$\langle\Psi^{(-)}(\mathbf{k},\mathbf{K})| = \langle\Phi_0(\mathbf{k},\mathbf{K})| + \langle\Phi_0(\mathbf{k},\mathbf{K})|\hat{V}\lim_{\varepsilon\to 0}\frac{1}{E-\hat{H}+i\varepsilon}, \tag{1}$$

where Φ_0 is a solution of an asymptotic Hamiltonian $\hat{H}_0$, and the interaction $\hat{V}$ is defined by subtracting $\hat{H}_0$ from the total Hamiltonian $\hat{H}$.

The Green's function in Eq. (1) is replaced with the complex-scaled Green's function in CSLS, and the relation between the Green's function in Eq. (1) and the complex-scaled one is given as

$$\lim_{\varepsilon\to 0}\frac{1}{E-\hat{H}+i\varepsilon} = U^{(-1)}(\theta)\frac{1}{E-\hat{H}^\theta}U(\theta) = \sum\!\!\!\!\!\!\int_n U^{(-1)}(\theta)|\chi_n^\theta\rangle\frac{1}{E-E_n^\theta}\langle\tilde{\chi}_n^\theta|U(\theta). \tag{2}$$

In the derivation of the right-hand-side of Eq. (2), we insert the complete set constructed with $\{\chi_n^\theta\}$, being the eigenstates of complex-scaled Hamiltonian $\hat{H}^\theta$. We obtain $\Psi^{(-)}(\mathbf{k},\mathbf{K})$ as

$$\langle\Psi^{(-)}(\mathbf{k},\mathbf{K})| = \langle\Phi_0(\mathbf{k},\mathbf{K})| + \sum\!\!\!\!\!\!\int_n\langle\Phi_0(\mathbf{k},\mathbf{K})|\hat{V}U^{-1}(\theta)|\chi_n^\theta\rangle\frac{1}{E-E_n^\theta}\langle\tilde{\chi}_n^\theta|U(\theta). \tag{3}$$

The scattering state in CSLS consists of two terms: the first term corresponds to the states of the noninteracting continuum, and the second is the effects of FSI. This decomposition of the scattering states is useful to estimate the effects of the correlations in the ground and final states on the Coulomb breakup cross section.

3. Results

In Fig. 1 we first show the Coulomb breakup cross section with respect to the total energy of ^{6}He and ^{11}Li in comparison with the observed ones [3, 4]. In both cases, the calculated cross sections show the low-lying enhancements, and well reproduce the observed data. We also estimate the effect of FSI on the cross sections by taking only the first term in Eq. (3) in the calculation, and it is found that FSI plays significant roles in reproducing the Coulomb breakup cross sections.

To discuss what kinds of FSI are important in the Coulomb breakup reactions, we calculate the invariant mass spectra for binary subsystems. In Fig. 2, we show the calculated invariant mass spectra for ^{11}Li. From the results, we confirm that both invariant mass spectra for ^{9}Li-n and n-n show the low-lying enhancements. It is also found that the virtual-state correlations of ^{9}Li-n and n-n subsystems have significant contributions in spectra. One the other hand, the p-wave resonances in ^{10}Li does not contribute to the invariant mass spectra for ^{9}Li-n subsystem since the resonance energies of the p-wave resonances, which are obtained at 275 keV and 506 keV in our calculation, are higher than the total energy of the peak position of the breakup cross section shown in the right panel of Fig. 1.

We also calculate the invariant mass spectra for ^{6}He. From the obtained results, the invariant mass spectra for the α-n subsystem shows the peak at around 0.7 MeV corresponding to the ^{5}He($3/2^-$) resonance. The α-n correlation is clearly confirmed in the invariant mass spectra of the Coulomb breakup cross section. For the n-n subsystem, the low-lying enhancement is seen near the zero-energy region, which indicates the importance of the n-n virtual state in the final states. (For details, see Refs. [6, 7])

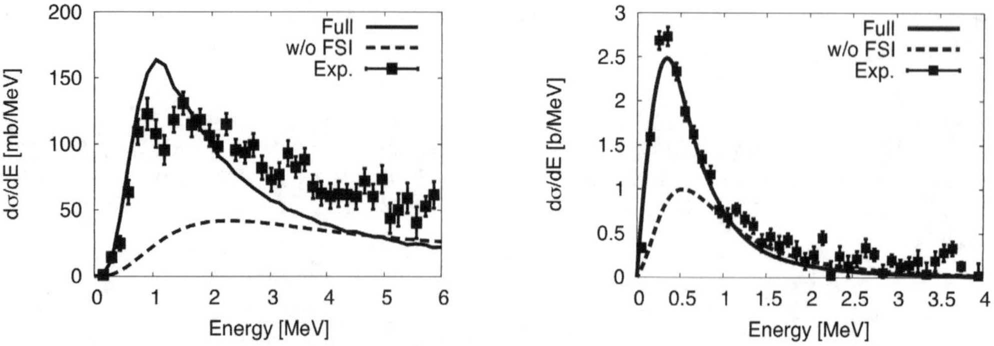

Figure 1. Obtained Coulomb breakup cross sections of ^{6}He and ^{11}Li, measured from the threshold energy of core + n + n. The left and right panels represent the cross sections for ^{6}He and ^{11}Li, respectively. The observed data, shown as solid squares, are taken from Ref. [3] for ^{6}He and from Ref. [4] for ^{11}Li.

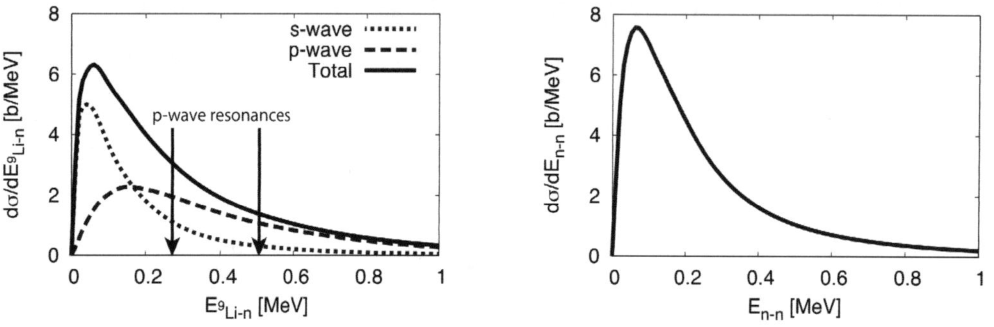

Figure 2. Invariant mass spectra of the Coulomb breakup reaction of ^{11}Li. The left and right panels present the invariant mass spectra for ^{9}Li-n and n-n, respectively. In the left panel, three types of spectra are presented: The total distribution (solid), s-wave component (dotted), and p-wave component (dashed). Two arrows in the left panel indicate the position of the p-wave resonances in our calculation.

From the results for ^{6}He and ^{11}Li, it is found that the invariant mass spectra for core-n subsystems reflects the characters of the core-n correlations in two-neutron halo nuclei, while the behaviors of the spectra for n-n subsystem are commonly dominated by the n-n virtual-state correlation in two nuclei.

References

[1] Aoyama S, Mukai S, Katō K and Ikeda K 1995 *Prog. Theor. Phys.* **93** 99
[2] Myo T, Kikuchi Y, Katō, K, Toki H and Ikeda K 2008 *Prog. Theor. Phys.* **119** 561
[3] Aumann T *et al* 1999 *Phys. Rev.* C **59** 1252
[4] Nakamura T *et al* 2006 *Phys. Rev. Lett.* **96** 252502
[5] Esbensen H and Bertsch G F 1992 *Nucl. Phys.* A **542** 310
[6] Kikuchi Y, Myo T, Takashina M, Katō K and Ikeda K 2009 *Prog. Theor. Phys.* **122** 499
[7] Kikuchi Y, Katō K, Myo T, Takashina M and Ikeda K 2010 *Phys. Rev.* C **81** 044308

Journal of Physics: Conference Series **436** (2013) 012047 doi:10.1088/1742-6596/436/1/012047

Breakup reactions of neutron drip-line nuclei near $N = 20, 28$

N Kobayashi[1], T Nakamura[1], Y Kondo[1], Y Satou[2] and the RIBF-55 Collaboration

[1] Department of Physics, Tokyo Institute of Technology, 2-12-1 O-Okayama, Meguro, Tokyo 152-8551, Japan

[2] Department of Physics and Astronomy, Seoul National University, Seoul 151-742, Korea

E-mail: kobayashi.n.aa@m.titech.ac.jp

Abstract. The cross sections of one-neutron removal via Coulomb and nuclear breakup from the very neutron-rich nuclei ^{29,31}Ne, ^{33,35,37}Mg and ^{39,41}Si and the momentum distributions of the fragments have been measured at around 240 MeV/nucleon. These reactions are combined to identify halo nuclei and obtain the spectroscopic factors, separation energies and spin-parities of ground states. The Coulomb breakup cross sections showed significant enhancement for ^{31}Ne and ^{37}Mg, which suggests halo formation. Further analysis is in progress.

1. Introduction

The present study aims at investigating quantitatively the single-particle structure of neutron-rich Ne, Mg, and Si isotopes, which are in the vicinity of the neutron drip line as well as of the conventional neutron magic numbers $N = 20$ and 28. Such investigations are also suitable for the search for new halo states. Two different reactions were utilized: Coulomb and nuclear breakup at energies around 240 MeV/nucleon. The Coulomb breakup is sensitive to the spatial distribution of a loosely bound valence neutron with low-ℓ orbital, and hence the reaction is especially sensitive to halo structure [3–7]. Nuclear breakup is sensitive to the single-particle configuration of the valence neutron [8, 9]. A combined analysis of the two processes is used for the first time. It has proved to be a powerful spectroscopic tool as, in particular, it may be used with very weak beams of nuclei very far from stability. In this context, a series of experiments were performed at the new-generation RI beam facility, the RI Beam Factory (RIBF) at RIKEN.

2. Experimental setup

The ^{29,31}Ne, ^{33,35,37}Mg, and ^{39,41}Si beams were produced through the projectile fragmentation of a ^{48}Ca beam at 345 MeV/u on a Be target. The intensity of a typical ^{48}Ca beam was about 100 p nA. The secondary beams were separated up to the F8 focal plane, which is at the end point of the projectile fragment separator BigRIPS [1, 2]. The secondary targets of lead (3.37 g/cm^2) and carbon (2.54 g/cm^2) were installed at F8. The details of the experimental setup and data analysis are described in Ref. [10, 11].

3. Results

Preliminary results for the inclusive one-neutron removal cross sections of the ^{29,31}Ne, ^{33,35,37}Mg, and ^{39,41}Si projectiles on the lead target, $\sigma^{\rm exp}_{-1n}$(Pb), and carbon target, $\sigma^{\rm exp}_{-1n}$(C), are shown in Fig. 1. The figure also shows the Coulomb breakup cross sections, $\sigma^{\rm exp}_{-1n}(E1)$, which are obtained from the cross sections $\sigma^{\rm exp}_{-1n}$(Pb) and $\sigma^{\rm exp}_{-1n}$(C). Although the Coulomb breakup cross section $\sigma^{\rm exp}_{-1n}(E1)$ of ^{29}Ne (S_{1n} = 0.95(15) MeV [13]) is much smaller than that of the one-neutron halo nucleus ^{31}Ne [14], it is significantly larger than that of the ordinary nuclei. This shows that ^{29}Ne has a moderate strength of soft $E1$ excitation. The cross section $\sigma^{\rm exp}_{-1n}(E1)$ of ^{37}Mg (S_{1n} = 0.25(1.03) MeV [12]) is almost as large as that of ^{31}Ne [14]. This result suggests that ^{37}Mg has halo structure. On the other hand, $\sigma^{\rm exp}_{-1n}(E1)$ of ^{33}Mg (S_{1n} = 2.22(3) MeV [12]) and ^{35}Mg (S_{1n} = 0.99(20) MeV [13]) are smaller than that of ^{37}Mg, although these $\sigma^{\rm exp}_{-1n}(E1)$ values are significantly larger than those of ordinary nuclei. We should note that, although the separation energy of ^{33}Mg is about twice that of ^{35}Mg, the cross sections are almost identical. Among Si isotopes, the cross section of ^{41}Si (S_{1n} = 1.38(44) MeV [13]) is larger than that of ^{39}Si (S_{1n} = 1.58(11) MeV [13]).

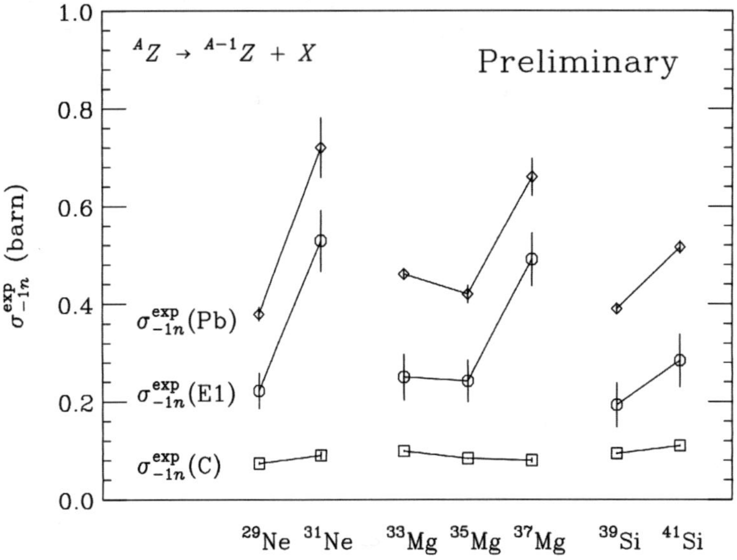

Figure 1. Experimental inclusive one-neutron cross sections for ^{29,31}Ne, ^{33,35,37}Mg and ^{39,41}Si. The diamonds, squares, and circles represent $\sigma^{\rm exp}_{-1n}$(Pb), $\sigma^{\rm exp}_{-1n}$(C) and $\sigma^{\rm exp}_{-1n}(E1)$, respectively.

Figure 2 shows the cross sections of the transitions to the 0^+ ground states of the fragments obtained from γ-ray analysis. By subtracting the cross sections corresponding to the transitions of $2^+_1 \rightarrow 0^+_{\rm g.s.}$ from the inclusive cross sections, we estimated the cross sections to $0^+_{\rm g.s.}$ state of the fragments. This estimation assumes that the direct decay from the states higher than the 2^+_1 state can be neglected. We found that the cross sections $\sigma^{\rm exp}_{-1n}(E1)$ to the 0^+ ground states of ^{29}Ne, ^{37}Mg, and ^{41}Si are particularly large, which suggests that in the relavant configurations of these nuclei the valence neutron with low-ℓ angular momentum is coupled to the ground state of the residual nucleus as in the case of ^{31}Ne. The Coulomb breakup cross sections of ^{33,35}Mg and ^{39}Si are consistent with 0 mb, which suggests that in these nuclei the valence neutron with low-ℓ angular momentum is coupled to the 2^+ or higher excited states, or the enhancement of $E1$ strength is due to mechanisms other than the soft $E1$ excitation. Further analysis concerning the above results and fragment momentum distributions is in progress.

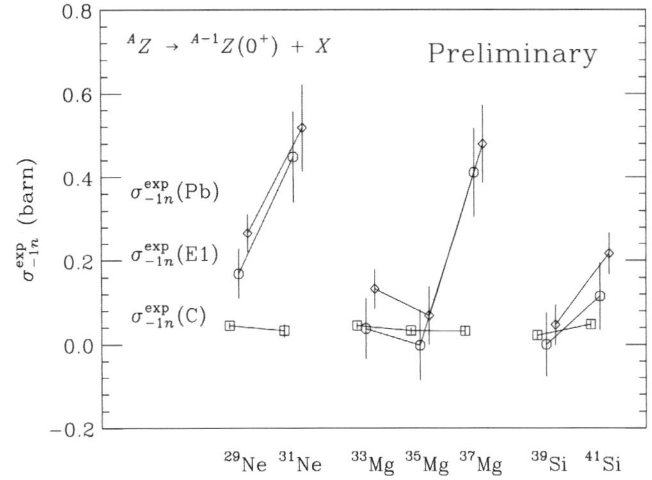

Figure 2. Experimental partial cross sections of transitions to the 0^{+} state of the heavy fragment for each reaction. The diamonds, squares and circles represent $\sigma^{exp}_{-1n}(\mathrm{Pb})$, $\sigma^{exp}_{-1n}(\mathrm{C})$ and $\sigma^{exp}_{-1n}(E1)$, respectively.

References

[1] Kubo T 2003 *Nucl. Instrum. Methods Phys. Res., Sect.* B **204** 97
[2] Ohnishi T *et al* 2008 *J. Phys. Soc.* **77** 083201
[3] Nakamura T *et al* 1994 *Phys. Lett.* B **331** 296; Fukuda N *et al* 2004 *Phys. Rev.* C **70** 054606
[4] Nakamura T *et al* 1999 *Phys. Rev. Lett.* **83** 1112
[5] Palit R *et al* 2003 *Phys. Rev.* C **68** 034318
[6] Aumann T *et al* 1999 *Phys. Rev.* C **59** 1252
[7] Nakamura T *et al* 2006 *Phys. Rev. Lett.* **96** 252502
[8] Kobayashi T *et al* 1988 *Phys. Rev. Lett.* **60** 2599
[9] Orr N A *et al* 1992 *Phys. Rev. Lett.* **69** 2050
[10] Kobayashi N *et al* 2011 *RIKEN Accel. Progr. Rep.* **44** 15
[11] Kobayashi N *et al* 2012 *Phys. Rev.* C **86** 054604
[12] Audi G *et al* 2003 *Nucl. Phys.* A **729** 337
[13] Jurado B *et al* 2007 *Phys. Lett.* B **649** 43
[14] Nakamura T, Kobayashi N *et al* 2009 *Phys. Phys. Lett.* **103** 011302

10th International Conference on Clustering Aspects of Nuclear Structure and Dynamics IOP Publishing
Journal of Physics: Conference Series **436** (2013) 012048 doi:10.1088/1742-6596/436/1/012048

^{8}He cluster structure studied by recoil proton tagged knockout reaction

Y Ye[1], Z Cao[1], J Xiao[1], D Jiang[1], T Zheng[1], H Hua[1], Y Ge[1], X Li[1], J Lou[1], Q Li[1], L Lv[1], R Qiao[1], H You[1], R Chen[1], H Sakurai[2], H Otsu[2], Z Li[2], M Nishimura[2], S Sakaguchi[2], H Baba[2], Y Togano[2], K Yoneda[2], C Li[2], S Wang[2,1], H Wang[2,1], K Li[2,1], Y Nakayama[3], Y Kondo[3], S Deguchi[3], Y Sato[4] and K Tshoo[4]

[1] School of Physics and State Key Laboratory of Nuclear Physics and Technology, Peking University, Beijing 100871, China

[2] RIKEN, 2-1 Hirosawa, Wako, Saitama, 351-0198, Japan

[3] Department of Physics, Tokyo Institute of Technology, Japan

[4] School of Physics, Seoul National University, Korea

E-mail: `yeyl@pku.edu.cn`

Abstract. Knockout reaction experiment for ^{8}He at 82.3 MeV/u on Hydrogen target was carried out at the RIPS beam line in RIKEN. Recoil protons were detected in coincidence with the forward moving core fragments and neutrons. The quasi-free knockout mechanism is identified through the polar angle correlation and checked by various kinematics conditions. The absolute differential cross sections for ^{6}He core cluster are obtained and compared with the simple Glauber model calculations. The extracted spectroscopic factor is close to unity and a shrinking of the cluster size is evidenced.

1. Introduction

Knockout reactions play an important role in studying the single-particle and cluster structure of stable nuclei [1, 2]. Since the advent of fast radioactive nuclear beams, knockout reactions with inverse kinematics have been used intensively for spectroscopic investigation of the exotic properties of unstable nuclei [3]. As indicated in many occasions, applicability of reaction tools to extract nuclear structure information depends sensitively on the handling of the reaction mechanisms [4, 5, 6]. It was suggested to use "a clean structure-less probe" like proton target in order to avoid the possible complex reaction processes [7].

^{8}He is an exotic nucleus with the largest neutron to proton ratio for any known particle-stable nucleus, and has attracted continuous attention experimentally as well as theoretically [8]. Based on the already established important properties ^{8}He provides an excellent test case to evaluate the reaction mechanisms. As demonstrated in a quasi-free scattering (QFS) experiment with ^{6,8}He beams impinging on a Hydrogen target [9], the core cluster knockout process can be isolated through the exclusive measurement of the recoil target protons in coincidence with the forward moving core fragments. This is of great importance since clustering structure seems growing in the vicinity of the neutron drip-line and spectroscopic investigation of this new degree of freedom is very demanding [10]. The reported experiment was carried out at very high energies and did not employ neutron detection [9]. It would be interesting to see if these reaction mechanisms could identified and applied at energies around 100 MeV/u where most knockout experiments for unstable nuclei have been performed.

2. The experiment

A detailed description of the experiment was given in Refs. [11, 12, 13]. The experiment was carried out at the RIKEN-RIPS beam line. In addition to the deflection magnet and neutron wall systems, two specially designed telescopes (D11 and D12) were installed, covering an angular range between 15° and 75° (for two setups) relative to the beam axis, to detect the recoil protons. Another telescope (D2) was installed beside the magnet acceptance, covering forward angles from 6° to 21°. Each of these telescopes is composed of one double-sided silicon strip detector (DSSD) of 1 mm in thickness, one large surface silicon detector of 1.5 mm in thickness, and one or two layers of thick CsI(Tl) crystals. The strip width of the DSSD is 2 mm at both X and Y directions.

3. Results and discussion

In Fig. 1 the polar angle correlation between the recoil protons and the forward moving ^{6}He fragments, for CH_2 target is plotted. At the upper right part of this figure, a component (in the frame F) appears clearly with relatively large proton and ^{6}He polar angles and follows very well the ^{6}He + p free scattering kinematics as described by the solid curve. It should be noted that this component does not exist for Carbon target [13]. According to earlier studies [9] this component arises from the core fragment knockout mechanism. For this component the upper limit of the proton angle is due to the angular coverage limit of the D2 telescope, whereas the lower limit at about 35° is due to the rapid decrease of the knockout cross section as demonstrated below.

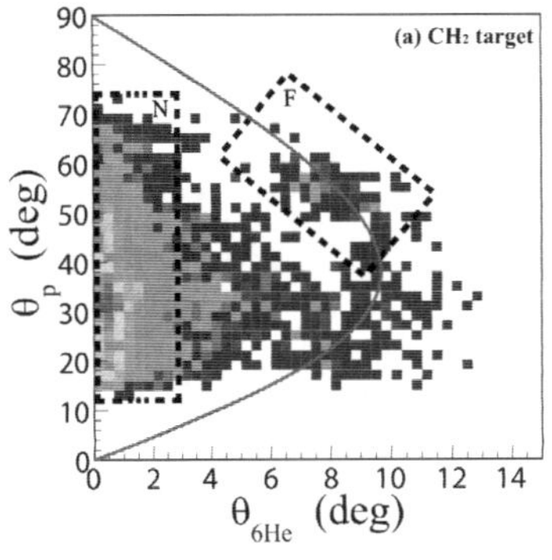

Figure 1. Polar angle correlations between the recoil protons and the forward moving ^{6}He fragments for CH_2 target. See the text for details.

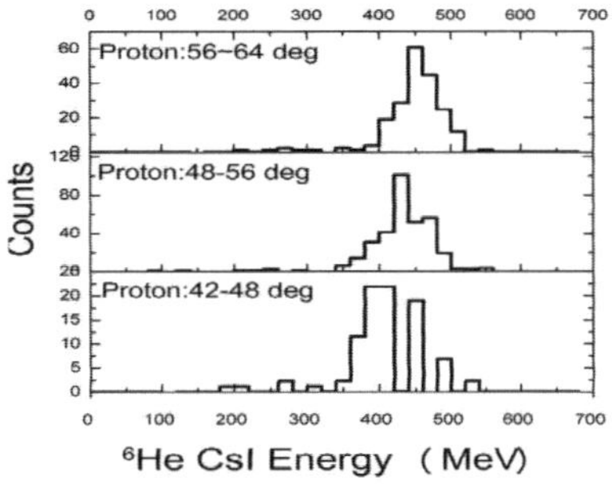

Figure 2. Energy distributions of the ^{6}He fragments in coincidence with the recoil protons in various angular bins.

The relation between the ^{6}He energy and its emission angle was further checked against the free scattering kinematics. It turns out that for each proton angular bin the energy distribution of the coincident ^{6}He fragments is well peaked and the peak moves as a function of the proton angle according to approximately the free scattering kinematics, as shown in Fig. 2. The azimuthal angle correlation between the core fragments and the recoil protons also satisfies well the free scattering condition. The quasi-free scattering mechanism between the ^{6}He core and the proton target is therefore well established for ^{8}He + p collision at about 80 MeV/u.

Based on the quasi-free scattering mechanism for ^{6}He core fragment, the absolute differential cross sections can be deduced accordingly. The solid angle for the coincident detection was determined by Monte Carlo simulation taking into account the two-body scattering kinematics, the realistic incident particle momentum distribution and the actual detector setup and efficiency. The results are shown in Fig. 3. Fig. 4 shows the differential cross sections for the knockout ^{6}He core fragments (filled diamonds) compared with the ^{6}He elastic scattering data (open circles) reported earlier [14]. A simple and approximate way to extract the SF for a cluster configuration is to compare the measured cluster knockout cross sections to the calculated elastic scattering ones [9]. As displayed in Fig. 4 by the dashed line, Glauber model calculations were performed for ^{6}He + p elastic scattering with the matter radii of ^{6}He to be 2.8 fm. A good description (solid curve) for the knockout data can also be obtained by just reducing the matter radii of ^{6}He to 2.2 fm. The shrinking of the the ^{6}He core cluster inside a ^{8}He nucleus was also find previously [9]. Since there is no need to substantially shift the absolute value of the calculated cross section to meet the experimental data, the SF of ^{6}He cluster configuration in ^{8}He should be close to 1.0. This is consist with previously reported result [9].

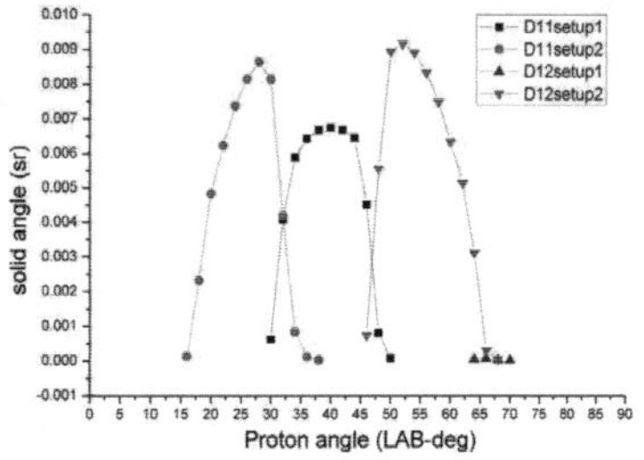

Figure 3. Absolute angular acceptance for ^{6}He + p free scattering, obtained from the simulation taken into account the real experimental setup.

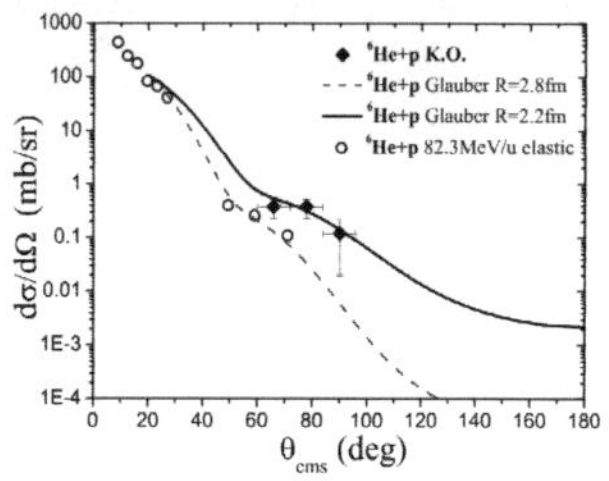

Figure 4. Differential cross sections of ^{6}He fragments from experiment at 82.3 MeV/u. See the text for details.

Acknowledgments

This work is supported by the 973 program of China (No. 2013CB834402) and NSFC projects (No.11035001 and 11275011)

References

[1] Sitenko A G, Translated by Kocherga O D 1990 *Theory of Nuclear Reaction* (Singapure: World Scientific) p 454
[2] Roos P G *et al* 1977 *Phys. Rev.* C **15** 69
[3] Hansen P G and Tostevin J A 2003 *Annu. Rev. Nucl. Part. Sci.* **53** 219
[4] Suzuki Y *et al* 2003 *Structure and Reaction of Light Exotic Nuclei* (London: Taylor & Francis)
[5] Louchart C *et al* 2011 *Phys. Rev.* C **83** 011601(R)
[6] Flavigny F *et al* 2012 *Phys. Rev. Lett.* **108** 252501
[7] Aksyutina Yu *et al* 2009 *Phys. Lett.* B **679** 191
[8] Lemasson A *et al* 2010 *Phys. Rev.* C **82** 044617
[9] Chulkov L V *et al* *Nucl. Phys.* A **759** 43
[10] Horiuchi H 2010 *J. Phys.* G **37** 064021
[11] Ye Y L, Cao Z and Jiang D X *et al* 2010 *Nucl. Phys.* A **834** 454c
[12] Lou J L, Ye Y L and Pang D Y *et al* 2011 *Phys. Rev.* C **83** 034612
[13] Ye Y L, Cao Z and Xiao J *et al* 2012 *Phys. Lett.* B **707** 46
[14] Faisal J Q, Lou J L and Ye Y L *et al* 2010 *Chin. Phys. Lett.* **27** 092501

Mystery of resonances in heavier systems

Eiji Uegaki

Graduate School of Engineering and Resource Science, Akita University, Akita 010-8502, Japan

E-mail: uegaki@phys.akita-u.ac.jp

Abstract. Studies of resonances in heavier systems are reviewed. A molecular model developed for heavy-ion resonances is described briefly. At high spins, stable dinuclear configurations are obtained as an equator-equator touching one for ^{28}Si + ^{28}Si (oblate-oblate system) and a pole-to-pole one for ^{24}Mg + ^{24}Mg (prolate-prolate system), respectively. Normal modes have been investigated around those equilibria. Those modes are expected to be the origin of a large number of the resonances observed. Furthermore, it is clarified that those different-type of equilibria give rise to their own characteristic rotational excitations. The importance of recent experimental data measured in Strasbourg, for both the ^{28}Si + ^{28}Si and the ^{24}Mg + ^{24}Mg systems are stressed. Analyses of physical quantities are made and compared with those data.

1. Introduction

The ^{28}Si + ^{28}Si and ^{24}Mg + ^{24}Mg systems exhibit a number of narrow and prominent resonances well above the Coulomb barrier, which are correlated among the elastic and inelastic channels [1, 2]. Maximum spins of both resonance systems are close to or over $40\hbar$, and the mass-asymmetric decays are much smaller than symmetric ones. In addition, high level densities over 'one per MeV' are observed. Those observations suggest rather long-lived compound systems in an extreme condition, but their structures as well as reaction mechanism are not known yet. Thus they remain a complete mystery. This is quite in contrast with lighter systems explained by the double resonance mechanism, such as band crossing model [3].

From the consideration of dinuclear structures with well-deformed constituent nuclei, the author and Y Abe proposed a new molecular model, in which internal degrees of freedom are introduced with the orientations of the deformation axes of the constituent nuclei. We investigated normal modes of motion at the equilibrium dinuclear configurations for high spins, which gave rise to various excitation modes such as butterfly and antibutterfly ones, and would be responsible for a large number of resonances observed [4, 5, 6, 7]. The analyses with a method of normal mode are briefly described in section 2.

The two systems exhibit rather different features. Namely, the ^{28}Si + ^{28}Si system shows a large number of sharp peaks on the bumps associated with the grazing angular momenta [8, 9], while the ^{24}Mg + ^{24}Mg system shows several prominent and isolated peaks [2]. Recently two important issues have appeared from Strasbourg. For ^{28}Si + ^{28}Si at $E_{\rm cm}$ = 55.8 MeV, the angular distributions exhibit disalignments between the orbital angular momentum and the fragment spins, and the particle-γ angular correlation measurements show characteristic features [10]. The other is the ^{24}Mg + ^{24}Mg experiment, which shows the important role of the $(4^+, 4^+)$ channel as well as the single and mutual 2^+ channels [11]. The analyses for those experimental data are given, for the $^{28}Si+^{28}Si$ system in section 2 and for the $^{24}Mg+^{24}Mg$ system in section 3, which clarify characteristic features of those systems.

2. Dinuclear molecular model for the $^{28}Si + ^{28}Si$ system

2.1. Stable molecular configurations at high spins

In fig. 1, stable molecular configurations at high spins are displayed, where the upper one is for $^{24}Mg - ^{24}Mg$ and the lower one for the $^{28}Si - ^{28}Si$ system. Those equilibria are obtaind by the molecular model, which will be explained later in this section. Because of centrifugal energy dominance, such elongated configurations appear with relatively lower energy, with contacts at the tips of the constituent nuclei. Normal mode analyses have been made around those equilibrium configurations with four internal degrees of freedom.

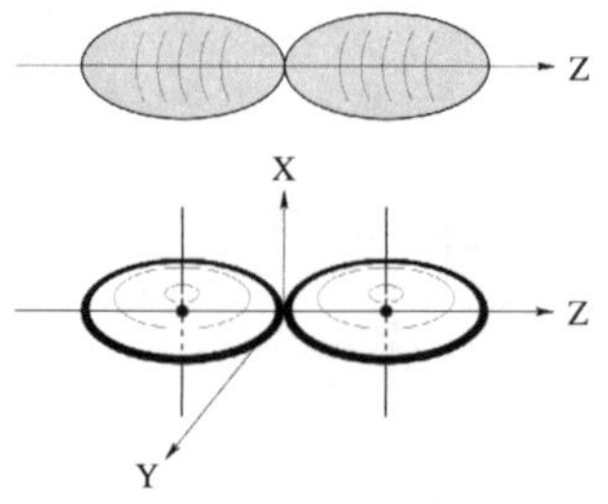

Figure 1. Equilibrium configurations of two dinuclear systems, $^{24}Mg - ^{24}Mg$ and $^{28}Si - ^{28}Si$.

2.2. Degrees of freedom and normal modes

For simplicity, assuming a constant deformation and an axial symmetry of the constituent nuclei, we have seven degrees of freedom, $(q_i) = (\theta_1, \theta_2, \theta_3, R, \alpha, \beta_1, \beta_2)$ illustrated in fig. 2, where (R, θ_2, θ_1) is the relative vector of the two ^{28}Si nuclei. As internal degrees of freedom, the orientations of the symmetry axes of the two constituent ^{28}Si nuclei are described with the Euler angles (α_i, β_i), which refer to the molecular axes. α_1 and α_2 are combined into $\theta_3 = (\alpha_1 + \alpha_2)/2$ and $\alpha = (\alpha_1 - \alpha_2)/2$.

The kinetic energy is given classically in terms of angular velocities, and then the quantization is done to describe the rotation of the whole system with the total angular momentum operator $\hat{J}$ and the internal motions referring to the molecular axes. The nucleus-nucleus interaction is given with folding potential using the nucleon-nucleon interaction DDM3Y. Expecting a stable configuration, the interaction potential is described with geometrical configurations specified by the internal degrees $(R, \alpha, \beta_1, \beta_2)$. Consistently with the coordinate system, at first we introduce a rotation-vibration type wave function as basis one, $\Psi_\lambda \sim D^J_{MK}(\theta_i)\chi_K(R, \alpha, \beta_1, \beta_2)$, where χ_K describes the internal motions. In fig. 3, a multidimensional energy surface including centrifugal energies is displayed. A local minimum is obtained around $\beta = 90°$ and $R = 7.6$ fm with the

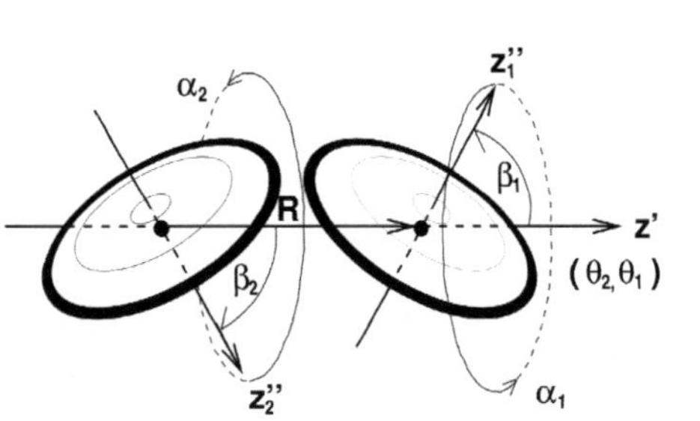

Figure 2. Dinuclear molecular coordinates; seven degrees of freedom of the $^{28}Si + ^{28}Si$ system are displayed,

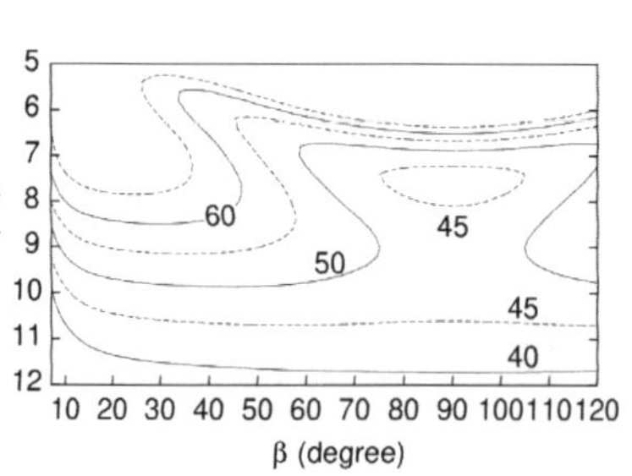

Figure 3. Contours of the effective potential energy V_{JK} with $J = 38$ and $K = 0$ for the $R - \beta(\beta_1 = \beta_2)$ degrees at $\alpha = \pi/2$.

Journal of Physics: Conference Series **436** (2013) 012049 doi:10.1088/1742-6596/436/1/012049

barrier at $\sim$ 9.5 fm, which is an *equator-equator* (E-E) configuration illustrated in the lower portion of fig. 1.

In order to solve the internal motions, we expand the effective potential at the energy minimum point and adopt the harmonic approximation to obtain the normal modes. Motions associated with β degrees are well confined to be vibrational, which are classified into new modes, *butterfly*: $\beta_+ = (\Delta\beta_1+\Delta\beta_2)/\sqrt{2}$ and *antibutterfly*: $\beta_- = (\Delta\beta_1-\Delta\beta_2)/\sqrt{2}$ around $\alpha = \pi/2$, respectively. In fig. 4(a) molecular normal modes of ^{28}Si + ^{28}Si with spin 38 are displayed, classified with quantum numbers K. Butterfly and antibutterfly modes are indicated with (b) under the levels. As for the α degree, $\alpha = (\alpha_1-\alpha_2)/2$, the confinements in the present folding potential appear to be unexpectedly weak, and hence the motion is close to a free rotation. We call it *twisting mode*, the associated levels of which are indicated by (t). All those are due to internal degrees of freedom, i.e., intrinsic excitations. These excitations exist over the range of spin $J = 34 - 40$ in which the E-E configuration is stable, and are expected to be the origin of the narrow resonances.

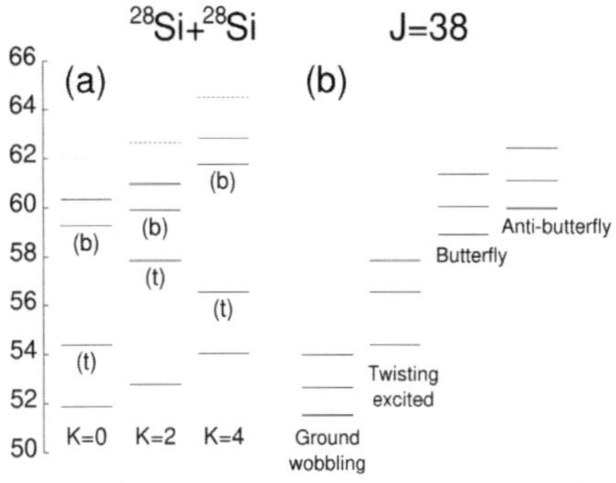

Figure 4. Energy spectra of the ^{28}Si+ ^{28}Si system for $J = 38$. (a) Molecular normal modes without K-mixing. (b) After K-mixing, with indication of the modes under the levels.

2.3. Wobbling motion with triaxial deformation

One of the characteristic features of the molecular spectrum is a series of low-energy K rotational excitation due to axial asymmetry around molecular z' axis, which is in contrast with the ^{24}Mg + ^{24}Mg case [4]. One can understand the reason immediately from fig. 1, where the upper configuration (^{24}Mg + ^{24}Mg) has axial symmetry as a total system, but the lower one for ^{28}Si + ^{28}Si has axial asymmetry. Thus K is not a good quantum number, namely, we expect that the eigenstates are K-mixed. In general, a triaxial system preferentially rotates around the axis with the largest moment of inertia. The oblate-oblate system, which can be seen as two pancake-like objects touching side-by-side in the lower panel of fig. 1, rotates around the X axis which is parallel to the normal to the reaction plane.

Those motions at high spins due to axial asymmetry give rise to mixing of quantum numbers K, which is known as *wobbling*. In the high spin limit ($K/J \sim 0$), the diagonalization in the K space is found to be equivalent to solving of the differential equation of the harmonic oscillator with parameters given by the moments of inertia. Thereby, the solution is a Gaussian, or a Gaussian multiplied by an Hermite polynomial, $F_n(K) = H_n(K/b)\exp[-(K/b)^2/2]$, with the width b given by the asymmetry of the moments of inertia. For the wobbling ground state, the total wave function is given with $F_0(K)$ as

$$\Psi_\lambda^{JM} \sim \sum_K \exp(-K^2/2b^2) D^J_{MK}(\theta_i)\chi_K(R,\alpha,\beta_1,\beta_2), \tag{1}$$

where in general, χ_K can be any molecular mode of triaxial deformations, such as the ground-state configuration or the butterfly modes. The resultant energy spectrum is displayed in fig.4(b) and compared with the spectrum without K-mixing in fig. 4(a).

In the experimental angle-averaged excitation functions, bumps are seen corresponding to the grazing angular momenta, the values of which are suggested by the Legendre fits [8]. Several sharp peaks are found on each bump [9], and thus those excited states with $J = 38$ are expected to the origin of those sharp peaks.

2.4. Disalignments: observed $L = J$ dominance

In the $^{28}Si + ^{28}Si$ system at 55.8 MeV, disalignments between the orbital angular momentum L and the fragment spins have appeared as a novel aspect of heavy-ion resonances [10], which has never been observed in any heavy-ion system so far and is considered mysterious.

Our molecular model naturally explains the mechanism of those disalignments, as is illustrated in fig. 5. Due to the characteristic rotational motion preferentially around the X axis, i.e., *wobbling*, the spins of the ^{28}Si fragments are approximately in the reaction plane, because they are in the YZ plane, as they are perpendicular to the symmetry axes of the constituent ^{28}Si nuclei. Thus the disalignments observed in the angular distributions, i.e., $L = J$ oscillation patterns both in the single and mutual 2^+ channels are well reproduced by the analyses with R-matrix theory [7].

2.5. Angular-correlation measurements: 'm=0' characteristics

Furthermore, fragment-fragment-γ angular correlations have been measured, which show a characteristic '$m = 0$' pattern of the γ distribution, and thus we have analyzed those data. Dots in fig. 6 display the angular correlation data [10], i.e., γ-ray intensity distributions of the $E2$ transition observed in the mutual 2^+ channel, and solid lines show theoretical results of the molecular ground state. Three different quantization axes are taken in panels (a)–(c), respectively: (a) beam direction, (b) z axis normal to the scattering plane, and (c) z axis perpendicular to those of (a) and (b). In panel (b) a typical '$m = 0$' pattern is obtained, and the results show good agreements in all the three panels. Again, figure 5 explains the origin of '$m = 0$', which means not only that the spins are in the reaction plane, but also that the spins in the Y and Z directions are almost the same in their strengths. Thus, the stable E-E configuration with the axial asymmetry rotates wobblingly around the axis normal to the reaction plane, which gives rise to the '$m = 0$' dominance [7].

In the angular correlations of the twisting excitation and the butterfly modes, clear differences from the molecular ground state are seen. For example, dashed lines in fig. 6 display the results of the twisting excitation, where a '$m = 1$' pattern appears in (b). The essential feature of the twisting excitation is '$m = 2$' dominance in panel (a), where the z axis is parallel to the beam direction. This means that the spin directions are not in B, but in A of fig. 5, due to the twisting motion around the molecular z' axis [7]. Such characteristics are also seen in the high excitation of the wobbling motion, where two ^{28}Si nuclei rotate around the molecular z' axis [12]. Thus the normal modes have their own characteristics in angular correlations, which give identifications of the normal-mode excitations, i.e. of the intrinsic nuclear structure of the resonances.

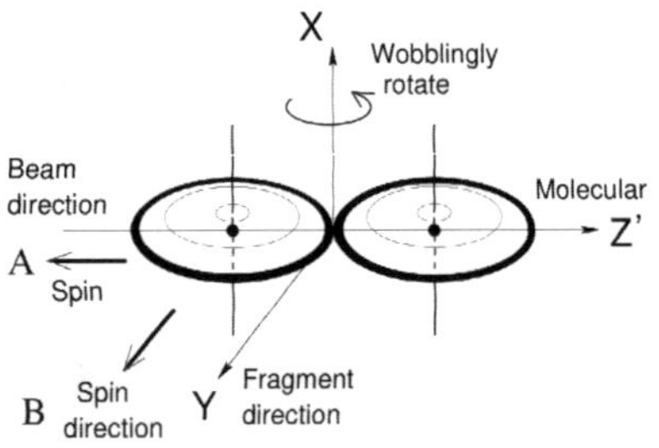

Figure 5. Spin directions of the molecular ground state with wobbling rotation, which give rise to disalignments.

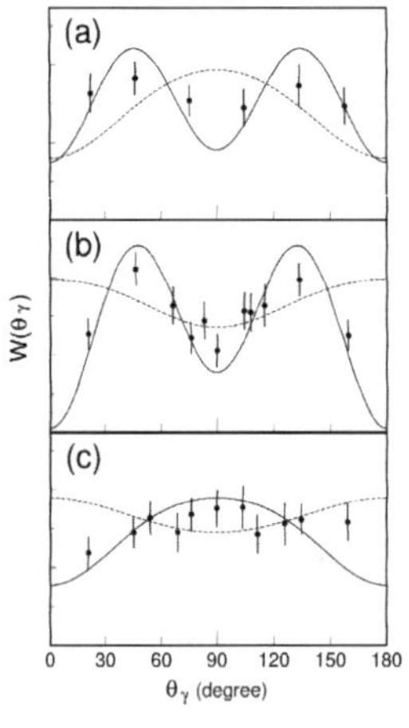

Figure 6. γ-ray intensities of particle-particle-γ angular correlations in the mutual 2^+ channel. Dots display the experimental data [10]. Solid lines show theoretical results of the molecular ground state, and dashed lines show those of the twisting rotational state. See text for the quantization axes of three panels.

3. Normal-mode spectrum for butterfly and antibutterfly; How they appear in the resonances of the ^{24}Mg + ^{24}Mg system?

3.1. Recent progress in the experimental study

Narrow and prominent resonance peaks are observed by Zurmühle *et al* in the ^{24}Mg + ^{24}Mg system around $E_{cm} = 46$ MeV, and four peaks starting from 45.7 MeV are assigned to be $J = 36$ [2]. We have analyzed the probability distributions in the decay channels and the partial widths for the normal mode excitations of the ^{24}Mg + ^{24}Mg system, and have showed that the excitation brings enhancements in the higher spin channels, such as 4^+ ones [13]. Recently, Salsac and Haas *et al* have made an experiment at $E_{cm} = 45.7$ MeV [11]. They showed that the 4^+ channels take much larger contributions to the resonance than those of the 2^+ channels, which is in good agreement with our analyses. In addition, they showed that the contribution from the 6^+ channel is quite small. Such a feature is seen in the angle averaged yields of Zurmühle *et al*, where strong $(4^+, 4^+)$ channel decays are observed [2]. Interesting is that enhancements of the $(4^+, 4^+)$ channel are seen to be rather different from those of the single and mutual 2^+ channels about excitation energies. Note that in the figure of Zurmühle's paper, the $(4^+, 4^+)$ channel is indicated as 6^+, since it is not possible to divide their contributions.

Now it is necessary to pay much attention to the mutual 4^+ channel. In table 1 data of the partial widths up to the $(4^+, 4^+)$ channel are collected [14] for the resonances at $E_{cm} = 45.66, 46.64$ and 47.25 MeV. In the resonance at 45.66 MeV, the contribution of the $(4^+, 4^+)$ channel is much larger than that of the $(2^+, 2^+)$ channel, while it is very small in the 46.64 MeV resonance, which suggests that the properties of those states are different.

Table 1. Partial widths (ratio $\Gamma_{\rm channel}/\Gamma$ in percent) of the resonances at $E_{cm} = 45.66, 46.64$ and 47.25 MeV observed in the elastic and inelastic scattering of ^{24}Mg + ^{24}Mg [14].

		Final state:	g.s.	2^+0^+	2^+2^+	4^+0^+	4^+2^+	4^+4^+
E_R(MeV)	Γ(keV)							
45.66	180 ± 25		1.75	3.80	4.40	2.10	2.15	11.00
46.64	230 ± 30		1.47	7.40	5.10	2.10	3.50	2.80
47.25	195 ± 30		1.39	4.90	2.70	0.65	1.10	5.40

3.2. Normal-mode analysis of the ^{24}Mg + ^{24}Mg system

In the ^{24}Mg + ^{24}Mg system the normal-mode energy spectrum is rather different from the ^{28}Si + ^{28}Si case. We can understand the basic aspect of the excitation in the pole motion from the equilibrium configuration of the ^{24}Mg + ^{24}Mg system, in which two pole tips are touching, as is seen in the upper panel of fig. 1. The vibrational motions of poles are available around the Z axis, both to the X and Y directions, which means that the dynamics of the single pole motion basically follows that of a two-dimensional harmonic oscillator. In addition, with two poles of prolate-deformed nuclei, their coupling should be taken into account. Thus we solved the single pole motion first, and then diagonalized the coupling between the two poles, with respect to the $(\alpha, \beta_1, \beta_2)$ degrees of freedom. In fig. 7, the normal-mode spectrum is displayed, where the levels displayed with short widths are those of the basic spectrum obtained at the first step. The levels C have two $\hbar\omega_\beta$ for the butterfly excitations with $K = 0$, while the level B exibits radial excitation. Due to the interaction energy difference between the butterfly and antibutterfly motions, which are illustrated in fig. 8, the degenerate levels split: for example, the levels C separate into two, about 2 MeV apart. The other levels D to L are also dissolved by such an interaction, which gives rise to the normal mode spectrum [4].

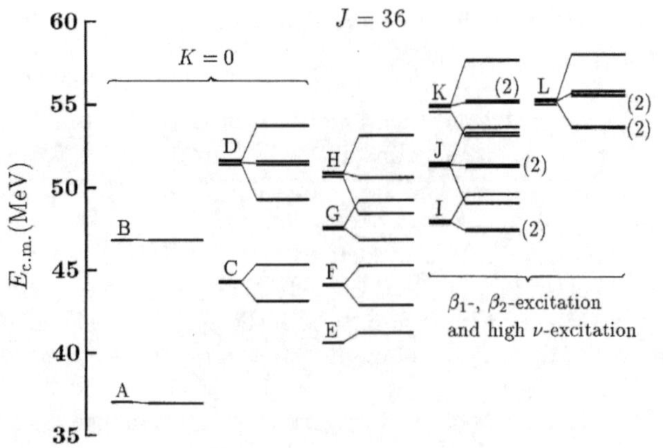

Figure 7. Energy spectra for the normal modes of the $^{24}Mg + ^{24}Mg$ system with $J = 36$ [4]. Levels displayed with short widths on the *l.h.s.* show the basic spectrum obtained without the orientation coupling between two poles. Level B shows the radial excitation. Levels A to D are of $K = 0$, and levels from E to H display $K = 1$ to $K = 4$ states, respectively. The degenerate levels of C split into two, i.e., butterfly and antibutterfly levels due to the interaction between pole tips.

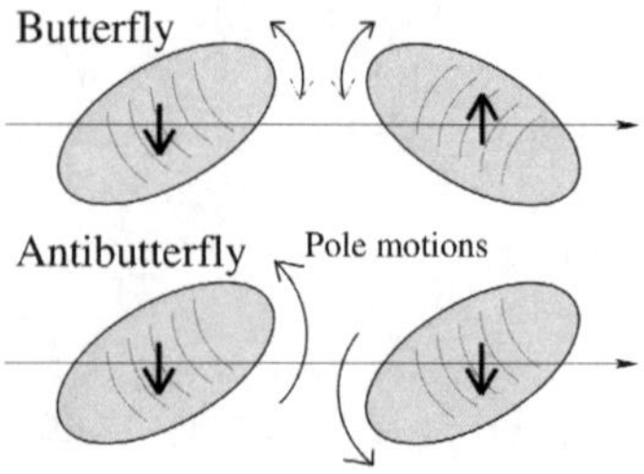

Figure 8. Illustrations of configurations for the butterfly mode (upper portion) and the antibutterfly mode (lower one), where the motions of poles of the ^{24}Mg nuclei are indicated by round allows. Bold arrows at the center of the ^{24}Mg nuclei show spin directions expected from those motions of the poles, respectively.

The physical properties of the butterfly modes have been investigated by the analyses of the partial widths. In fig. 9, probability distributions on the channel spins I are displayed for the $(4^+, 2^+)$ channel of the $^{24}Mg + ^{24}Mg$ system and the $(2^+, 2^+)$ channel of the $^{28}Si + ^{28}Si$ system, which are the dominant channels in the butterfly excitations. Characteristic features are seen for the dominant channel spins I in the two butterfly-type modes, i.e., $I \sim$ 'minimum' in butterfly, while $I \sim$ 'maximum' in antibutterfly [13]. We can easily understand such a mechanism of each mode from the spin directions caused by the butterfly motion, as is illustrated in fig. 8. Correspondingly to the channel spins, the dominant decay channels appear with quite different characteristics. In the antibutterfly state, the largest decay yield appears in the $(4^+, 2^+)$ channel, as expected. However, in the butterfly state the largest yield does not appear in the $(4^+, 2^+)$ channel, but appears in the $(2^+, 2^+)$ channel. This is due to dominant decaying angular momentum

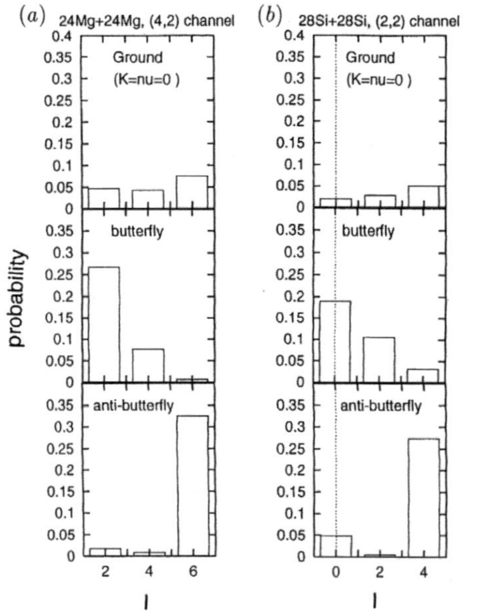

Figure 9. Probabilities in the channel spin I, for the systems (a) $^{24}Mg + ^{24}Mg$ with $J = 36$ and (b) $^{28}Si + ^{28}Si$ with $J = 38$, respectively.

values L, which are $L \sim 30$ with alignments of L and I in the former, while $L \sim 36$ with the small I value in the latter. The observation of the strong or weak enhancements of the $(4^+, 4^+)$ channel for each resonance at 45.66 MeV or 46.64 MeV discussed in section 3.1, are expected to correspond to such a mechanism.

4. Conclusions

Studies for resonances in heavier systems have been reviewed. By using a molecular model, the interaction between two deformed nuclei is described with the internal collective variables. Molecular normal modes have been obtained, such as butterfly and antibutterfly ones in addition to the radial excitation, which give rise to a variety of molecular states. Those modes are expected to be the origin of a large number of the resonance states.

For the ^{28}Si $-$ ^{28}Si resonance states the results show that the structure is an asymmetrically hyperdeformed molecule [5, 6]. By the R-matrix theory, analyses of the decay properties have been made [7]. Each normal mode exhibits its own different characteristics in the angular correlations. In other words, the angular correlation measurements provide information to distinguish among molecular intrinsic states of the resonances. Due to the axial asymmetry of the stable configuration, the system rotates wobblingly around the axis normal to the reaction plane, which gives rise to the '$m = 0$' dominance in the molecular ground state and good agreement with the experiment on the $E_{\rm cm} = 55.8$ MeV resonance.

For the ^{24}Mg + ^{24}Mg system, in the analysis on the partial widths [13] it is found that the butterfly mode has a small channel-spin value, while the antibutterfly mode has a large value. This mechanism is expected to give rise to disalignments and alignments, respectively.

With respect to the mutual 4^+ channel, its importance has been ignored, even though the contributions from this channel are very large. However, since Salsac and Haas have shown the importance of the 4^+ channels [11], the first data by Zurmühle *et al* [2] are reviewed. The statistical analyses exhibit strong or weak enhancements in the mutual 4^+ channel at different resonance energies [14]. Thus the above mechanism suggested theoretically is in agreement with the experimental data at present, and has to be confirmed by a further study.

As a whole, resonances in the heavier systems provide new, intriguing knowledge on nuclear structure and dynamics. The experimental challenges to uncover the characteristic features of the resonances are strongly desired.

Acknowledgments

The author heartily thanks to Dr Y Abe, for many discussions and encouragements in the collaboration for resonances in heavier systems.

References

[1] Betts R R 1985 *Clustering Aspects of Nuclear Structure* (Dordrecht: D Reidel) p 133
[2] Zurmühle R W, Kutt P, Betts R R, Saini S, Haas F and Hansen Ole 1983 *Phys. Lett.* **129B** 384
[3] Abe Y, Kondō Y and Matsuse T 1980 *Prog. Theor. Phys. Suppl. No.* **68** 303
[4] Uegaki E and Abe Y 1993 *Prog. Theor. Phys.* **90** 615
[5] Uegaki E and Abe Y 1994 *Phys. Lett.* B **340** 143
[6] Uegaki E and Abe Y 2012 *Prog. Theor. Phys.* **127** 831
[7] Uegaki E and Abe Y 2012 *Prog. Theor. Phys.* **127** 877
[8] Betts R R, DiCenzo S B and Petersen J F 1981 *Phys. Lett.* **100B** 117
[9] Betts R R, Back B B and Glagola B G 1981 *Phys. Rev. Lett.* **47** 23
[10] Nouicer R *et al* 1999 *Phys. Rev.* C **60** 041303
[11] Salsac M-D *et al* 2008 *Nucl. Phys.* A **801** 1
[12] Uegaki E and Abe Y 2013 *Proc. of this conference*
[13] Uegaki E 1998 *Prog. Theor. Phys. Suppl. No.* **132** 135
[14] Saini S, Betts R R, Zurmühle R W, Kutt P H and Dichter B K 1987 *Phys. Lett.* B **185** 316

Cluster and molecular structures observed in reactions of light nuclei

Matko Milin
Physics Department, Faculty of Science, University of Zagreb, Zagreb, Croatia
E-mail: matko.milin@phy.hr

Abstract. The experimental evidence for existence of nuclear molecular structures has been reviewed, starting with the simplest two-centre symmetric systems (^{9}Be and A=10 nuclei), then discussing the possibility of three-centre molecular states in carbon isotopes ^{13,14}C and finally giving experimental results for asymmetric cluster and molecular structures in oxygen isotopes $^{17-20}$O.

1. Introduction

The ^{8}Be nucleus in its three lowest lying states is a classical case of a clusterised system [1, 2] - its properties are completely consistent with configuration of two touching α-particles. Such a deformed structure has a simple experimental signature, *i.e.* a rotational band given in Fig. 1. ^{9}Be, with an additional neutron, also has a two-centre structure, with the neutron resembling the behaviour of the valence electrons in molecular physics. Two rotational bands (also given in Fig. 1) have been experimentally identified - one (K=3/2$^-$, based on the ground state) can be connected to a configuration in which the valence neutron is filling the molecular π-orbit, while the other (K=1/2$^+$, based on the state at E_x= 1.68 MeV) resembles the molecular configuration with the extra neutron in the molecular σ-orbit. As seen in Fig. 1, the K=1/2$^+$ band shows a "staggering" behaviour of the excitation energies, which is characteristic of the Coriolis decoupling.

The ^{9}Be nucleus thus seems to be a perfect example of a *nuclear molecule*, a two centre system which is given an extra binding through exchange of a valence particle (neutron). Several open questions remain to be answered to establish this picture - the most important one definitely being a clear identification of the K=1/2$^+$ band-head in the analog nucleus, ^{9}B. A number of experiments has recently been performed in that direction [3, 4, 5], so the final theoretical test of the consistency of molecular picture is to be expected soon. Another important unresolved problem is the question of the 9/2$^-$ member of the ground state rotational band - there is a long debate (see *e.g.* [6]) if the state seen in several experiments at $E_x\approx$ 11.3 MeV was indeed the missing member of the K=3/2$^-$ band.

2. Spectroscopy of A=10 nuclei

With the molecular structure of ^{9}Be almost established, the next step is an addition of further (valence) nucleon and systematic study of the A=10 nuclei. The best studied nucleus of the A=10 triplet is ^{10}Be - the rotational band associated with the α+α+n+n molecular structure

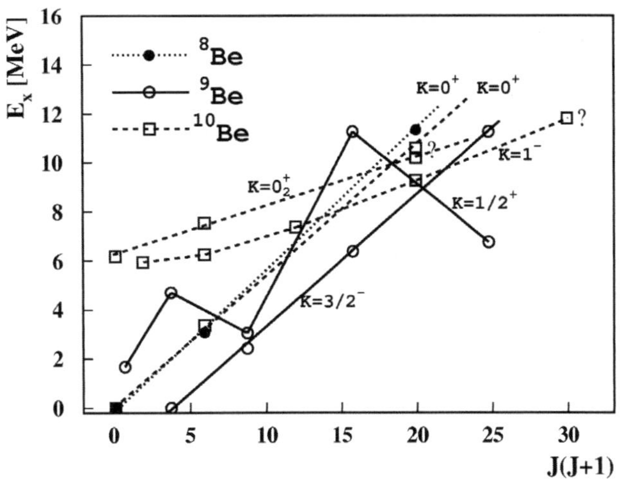

Figure 1. Rotational bands in ^{8}Be (full circles), ^{9}Be (open circles) and ^{10}Be (open squares).

has been clearly identified after a series of experiments with both stable and radioactive beams [7, 8, 9, 10, 11]. Three members of the band have been found: 0^+ at E_x= 6.18 MeV, 2^+ at E_x= 7.54 MeV and 4^+ at E_x= 10.15 MeV (given, together with the g.s. band in Fig. 1) - the deformation of the states seems to be one of the highest among all known states of all nuclei (and much larger than the deformation of the classic cluster example, ground state of ^{8}Be). This result is supported with a number of theoretical papers, *e.g.* [12, 13, 14], and it now seems to be well established.

A number of questions remain to be answered regarding the A=10 nuclei spectroscopy. First of all, although *all* experimentally seen states of ^{10}Be has been given a consistent theoretical explanation, theory predicts a number of states not yet seen in the experiment. Further experiments are needed to look for higher members of the molecular and g.s. rotational band [15], for states of the "atomic" ^{6}He+α character [16], for states with the α+t+t structure [17], and finally for states with the α+α+dineutron structure with an α-halo configuration [18].

Essential next step in establishing molecular picture as the important one for light nuclei would be the identification of states in ^{10}B and ^{10}C that are analog to the above mentioned members of the ^{10}Be molecular rotational band. Some experimental results in that direction have recently been published [19, 20, 21, 22, 23] and will be given in the following, but a clear picture is yet to be found.

For ^{10}B, results are obtained for different decay modes of the excited states, including proton, deuteron and α-particle emission. The α-decay is particularly interesting, when it leads to the excited states of ^{6}Li, especially the second one (0^+ T=1 at E_x= 3.56 MeV, denoted as ^{6}Li**), which is an analogue to the ^{6}He ground state. The ^{10}B states analogue to the two lowest members of the ^{10}Be molecular band are for long time known to be the ones at E_x= 7.56 and 8.89 MeV. The position of the 4^+ state is crucial for a complete understanding of the molecular configurations in A=10 nuclei, so several experiments have been performed to identify it. Results from the ^{11}B(^{3}He,^{6}Li** α)^{4}He measurements [19, 23] suggest that the 4^+ state is at $E_x\approx$ 11.5 MeV. It is clear that further work (*i.e.* clear spin and parity assignments) is definitely needed, especially because the analogue of the 9.56 MeV 2^+ state in ^{10}Be is also expected in this energy region (and the observed state at 11.5 MeV may actually be connected to it). The 9.56 MeV state in ^{10}Be (and its analogs in ^{10}B and ^{10}C) is interesting by itself, because that is the only

Journal of Physics: Conference Series **436** (2013) 012050 doi:10.1088/1742-6596/436/1/012050

^{10}Be state with significant and comparable decay strengths for both ^{9}Be+n and ^{6}He+α channels [7, 10], which makes it rather important for decay properties of high-lying cluster states in ^{11}Be [24, 25].

The experimental data for ^{10}C [20, 21] is still incomplete; nevertheless, good arguments [26] point to existence of the 0^+ state at E_x= 5.29 MeV, and 2^+ state at E_x= 6.57 MeV, both being relatively narrow and having the α+^{6}Be cluster structure. This suggestion is consistent with recent analysis within the shell-model picture [27, 28], but has to be checked in further experiments; our preliminary analysis of the ^{7}Be+^{6}Li measurement seems to confirm it, too. Crucial point is again the identification of the 4^+ member of the band, the analogue of the 10.15 MeV state in ^{10}Be.

The final recognition of the analogue states in ^{10}Be, ^{10}B and ^{10}C will allow direct insight in their structure by comparison of Coulomb effects. This will not at all be a trivial task, as witnessed by the fact that $B(E2)$ values measured for the $2_1^+ \to 0_1^+$ transition for the three isobars [29, 30] fail to have a simple interpretation and to agree with theory. Nevertheless, such result will definitely provide stringent tests for both *ab initio* and cluster calculations.

For systems with one nucleon more (*i.e.* the A=11 nuclei), measurements of the ^{4}He(^{7}Be,α)^{7}Be and ^{4}He(^{7}Be,p)^{10}B reactions [31] have been recently performed. A comparison of that experimental data with the one from the ^{7}Li+α resonant elastic scattering [32], reveals states of significant alpha partial widths in ^{11}B and ^{11}C, making them good candidates for being members of the deformed cluster rotational bands.

3. Clustering in ^{13}C and ^{14}C

Having in mind the three alpha-particle structure of the Hoyle state in ^{12}C, one may wonder what would happen when extra neutrons are added. Initial suggestion [33] of very deformed prolate structure in ^{13}C stabilised by valence neutron, seems to be too simple - subsequent measurements [34, 35, 36] have shown that some of the proposed members in the corresponding rotational band have properties that does not fit the suggested systematics. Further experimental work is needed to clarify the issue - in particular, one should try to identify ^{13}C states decaying to the Hoyle state by neutron emission. Equivalent study has been performed for the ^{13}N nucleus [37] and several states were found to have a significant branching ratio for decay by proton emission to the Hoyle state. Fully consistent picture of ^{13}C structure at E_x>10 MeV, and relation to theoretical predicted states in that energy region [38, 39], is yet to be found.

The situation seems to be a bit better for ^{14}C, where both oblate and prolate states were identified [40] starting with the 0^+ states at E_x= 6.59 and 9.75 MeV, respectively. Details of the proposed rotational bands yet has to be checked by measuring sequential decay reactions (*e.g.* [41]) or ^{10}Be+α resonant elastic scattering [42]. Theoretical interpretation in terms of three-centre α+α+α+n+n molecular structure and / or ^{10}Be+α cluster structure is also being built [43, 44].

A rather detailed experimental spectroscopy of states in ^{16}C [45] and ^{17}C [46] has been performed recently up to excitation energies of ≈16 MeV, with no clear signs of cluster and molecular states; further work is to be expected because calculations predict [47] that ^{16}C (at high excitation energies) should have the clearest signature of linear three-centre molecular structure stabilised by valence neutrons.

4. Cluster and molecular states of oxygen isotopes

Clustering in excited states of ^{16}O (starting already with the first excited state at E_x= 6.05 MeV) is another historical example - the rotational bands of both positive and negative parity states have been identified [48] long time ago and interpreted as parity splitting doublet due to the asymmetric ^{12}C+α cluster structure [49]. Again, one may wonder what happens when

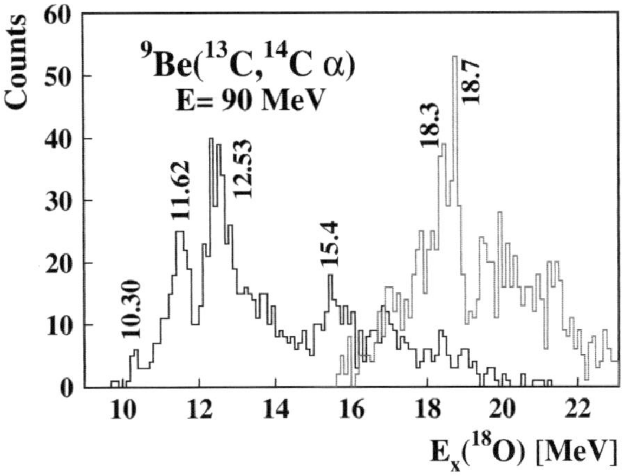

Figure 2. ^{18}O excitation energy spectrum reconstructed from the the $^{14}C_{g.s.}+\alpha$ coincidences. Two shown lines correspond to different combinations of detectors.

extra neutrons are added to such structure, and that question has been discussed in a number of recent papers [50, 51, 52, 53] for isotopes $^{17-20}$O.

Although with many unclear details, it seems that several cluster and molecular rotational bands can be found in these oxygen isotopes, compared in Table 1 (another classical example, the inversion doublet in ^{20}Ne, has been added for comparison).

nucleus	band-head	$\pi=+$ $\hbar^2/2\theta$ (keV)	$\pi=-$ $\hbar^2/2\theta$ (keV)	energy splitting (MeV)
^{16}O	0_2^+	230	210	2.9
^{18}O	0_2^+	200	200	4.0
^{18}O	0_4^+	110	110	2.4
^{19}O	$3/2_2^+$	200	200	3.8
^{20}O	0_2^+	160	150	5.2
^{20}O	0_4^+	90	100	1.7
^{20}Ne	0_1^+	170	190	4.8

Table 1. Cluster and molecular rotational bands in some oxygen isotopes. The slope parameters ($\hbar^2/2\theta$) are given for both positive and negative parity bands of a given structure. Last column gives the energy splitting ($2 \cdot \delta_E$) of the parity doublets.

From Table 1 it is evident that in both ^{18}O and ^{20}O the bands at higher excitation energies have also higher moments of inertia (*i.e.* lower slope parameters) and smaller energy splittings. For lighter well-studied systems, ^{10}Be [54] and ^{12}Be [55, 56], it was theoretically suggested that their higher-lying bands have more "ionic" character, while the lower-lying bands show molecular nature (*i.e.* with shared neutrons between two clusters), because the delocalisation of neutrons reduces significantly the excitation energy. Similar claims were made for ^{22}Ne [57] and finally for ^{20}O [58], where the lower-lying band was suggested to have mixed ^{14}C+^{6}He and ^{12}C+4n+α configurations, while the higher-lying one is suggested to correspond to the ^{16}C+α "ionic" structure.

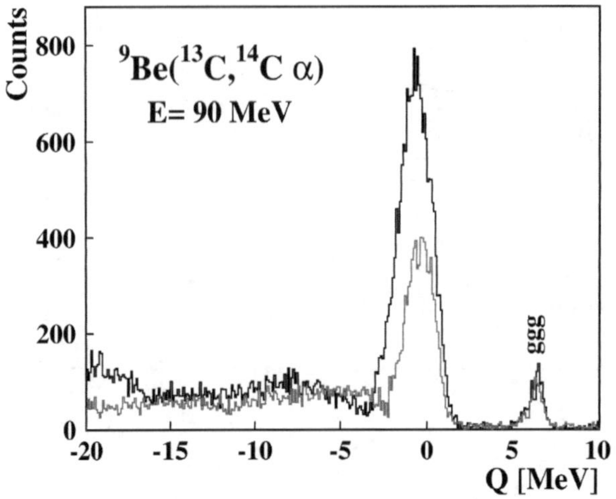

Figure 3. Q-value spectrum for the ^{14}C+α coincidences. Two shown lines correspond to different combinations of detectors. Gating on the peak at Q= 6.60 MeV gives the spectrum at Fig. 2.

Studies of decay modes of the observed states, (like the one performed for ^{17}O in Ref. [50] and for ^{18}O in Ref. [59]), should be done to give clear picture on the structure of high-lying states in oxygen isotopes. Figure 2 gives one such spectrum (previously unpublished). It is obtained measuring the ^{9}Be(^{13}C,^{14}C α) reaction with the E= 90 MeV beam. The details of the experimental set-up can be found in Ref. [50], which also discusses the results obtained in the same measurement for the ^{17}O nucleus.

Though statistics is rather low (especially when compared to the similar ^{17}O spectrum [50]), several peaks are clearly seen in the ^{18}O excitation energy spectrum reconstructed from the the ^{14}C$_{g.s.}$+α coincidences. Some of these peaks correspond to the states previously strongly populated in the ^{12}C(^{18}O,^{14}Cα) reaction at E= 82 MeV [60, 61], and clearly populated in the ^{14}C(^{6}Li,d) reaction at 34 MeV [62, 63] They are also strong in the ^{14}C(^{18}O,^{14}Cα)^{14}C reaction [59] and already there suggested to be members of the deformed cluster rotational band. The state at $E_x\approx$ 10.30 MeV is also the strongest state seen in the ^{12}C(^{7}Li,p) reactions, and as such suggested to have 6p-4h configuration [64, 65].

Similar discussion can be made for every state seen in Fig. 2; obviously the studied sequential decay reaction selectively populates the states with large alpha-strengths. Those states were, along with others, grouped into rotational bands in Ref. [51]. The measurement of the ^{14}C+α coincidences actually showed two strong peaks in the Q-value spectrum (Fig. 3). The peak marked with "ggg" and corresponding to the ^{14}C(g.s.)+2α exit channel is indeed clearly seen (Q= 6.60 MeV). There is another strong and broad structure at $Q\approx -0.5$ MeV; it corresponds to the ^{14}C nucleus left in its excited (unresolved) states at E_x= 6-8 MeV. Gating on the latter, the corresponding ^{18}O excitation energy spectrum shows only one strong (and broad) peak, at $E_x\approx$ 18.5 MeV.

Nevertheless, a measurement with better resolution is needed to resolve the ^{14}C states in the broad bump at $Q\approx -0.5$ MeV in Fig. 3 and thus study a decay of the ^{18}O states into different ^{14}C states (some of which have well developed cluster structure, like the 0^+ state at 6.59 MeV). A better statistics is also needed, as well as the detector set-up which will allow the clear identification of the particles, thus enabling a study of other decay channels, like ^{6}He emission (to be expected for states with molecular character). We have performed such a measurement recently at the Orsay accelerator facility, so new experimental results for ^{18}O and other oxygen isotopes are to be expected soon.

5. Summary

Recent experimental results regarding cluster and molecular states in the A=10 nuclei, ^{13,14}C and $^{17-20}$O, have been reviewed and corresponding open problems listed. The A=10 isobars are suggested to provide the best test of consistency of molecular picture, in which two clusters are additionally bound through exchange of valence nucleons. The possible existence of three-centre structures in ^{13,14}C and asymmetric configurations in $^{17-20}$O is also discussed, and current knowledge on them reviewed. The Zagreb group has recently performed a number of experiments studying all the three discussed topics, so new results are to be expected soon.

Acknowledgments

This work is based on experiments done in collaborations between groups from Zagreb, Berlin, Catania, Birmingham, Edinburgh and Huelva - thanks are due to all the collaborators.

References

[1] von Oertzen W, Freer M and Kanada-En'yo Y 2006 *Phys. Rep.* **432** 43
[2] Freer M 2007 *Rep. Prog. Phys.* **70** 2149
[3] Akimune H *et al* 2001 *Phys. Rev.* C **64** 041305(R)
[4] Scholl C *et al* 2011 *Phys. Rev.* C **84** 014308
[5] Baldwin T D, Catford W N *et al* 2012 *Phys. Rev.* C **86** 034330
[6] Soić N *et al* 1998 *Europhys. Lett.* **41** 489
[7] Soić N *et al* 1996 *Europhys. Lett.* **34** 7
[8] Milin M *et al* 1999 *Europhys. Lett.* **48** 616
[9] Milin M *et al* 2004 *Phys. Rev.* C **70** 044603
[10] Milin M *et al* 2005 *Nucl. Phys.* A **753** 263
[11] Freer M *et al* 2006 *Phys. Rev. Lett.* **96** 042501
[12] Kanada-En'yo Y, Horiuchi H and Dote A 1999 *Phys. Rev.* C **60** 064304
[13] Itagaki N and Okabe S 2000 *Phys. Rev.* C **61** 044306
[14] Ogawa Y, Arai K, Suzuki Y and Varga K 2000 *Nucl. Phys.* A **673** 122
[15] Rogachev GV 2012, contribution to these proceedings
[16] Ito M 2006 *J. Phys.: Conf. Ser.*, **49** 206
[17] Itagaki N *et al* 2008 *Phys. Rev.* C **77** 067301
[18] Kobayashi F and Kanada-En'yo Y 2012 *Phys. Rev.* C **86** 064303
[19] Uroić M, Miljanić Đ *et al* 2008 *Int. J. Mod. Phys.* E **17** 2345
[20] Curtis N *et al* 2008 *Phys. Rev.* C **77** 021301(R)
[21] Charity R J *et al* 2009 *Phys. Rev.* C **80** 024306
[22] Kuchera A N, Rogachev G V *et al* 2011 *Phys. Rev.* C **84** 054615
[23] Kirsebom O S *et al* 2012 *Phys. Rev.* C **85** 054308
[24] Majer M, Raabe R, Milin M *et al* 2010 *Eur. Phys. J.* A **43** 153
[25] Madurga M, Borge M J G *et al* 2008 *Nucl. Phys.* **810** 1
[26] Goldberg V Z and Rogachev G V 2012 *Phys. Rev.* C **86** 044314
[27] Fortune H T and Sherr R 2006 *Phys. Rev.* C **73** 064302
[28] Barker F C 2009 *Phys. Rev.* C **79** 017302
[29] McCutchan E A, Lister C J *et al* 2012 *Phys. Rev.* C **86** 014312
[30] McCutchan E A, Lister C J *et al* 2012 *Phys. Rev.* C **86** 057306
[31] Freer M *et al* 2012 *Phys. Rev.* C **85** 014304
[32] Yamaguchi H *et al* 2011 *Phys. Rev.* C **83** 034306
[33] Milin M and von Oertzen W 2002 *Eur. Phys. J.* A **14** 295
[34] Sasamoto Y *et al* 2006 *Mod. Phys. Lett.* A **21** 2393
[35] Freer M *et al* 2011 *Phys. Rev.* C **84** 034317
[36] Wheldon C *et al* 2012 *Phys. Rev.* C **86** 044328
[37] Fujimura H *et al* 2004 *Phys. Rev.* C **69** 064327

[38] Itagaki N, von Oertzen W and Okabe S 2006 *Phys. Rev.* C **74** 067304
[39] Furutachi N and Kimura M 2011 *Phys. Rev.* C **83** 021303(R)
[40] von Oertzen W *et al* 2004 *Eur. Phys. J.* A **21** 193
[41] Milin M *et al* 2004 *Nucl. Phys.* **A 730** 285
[42] Freer M 2012 contribution to these proceedings
[43] Itagaki N *et al* 2004 *Phys. Rev. Lett.* **92** 142501
[44] Suhara T and Kanada-En'yo Y 2011 *Phys. Rev.* C **84** 024328
[45] Bohlen H G *et al* 2003 *Phys. Rev.* C **68** 054606
[46] Bohlen H G *et al* 2007 *Eur. Phys. J.* A **31** 279
[47] Itagaki N, Okabe S, Ikeda K and Tanihata I 2001 *Phys. Rev.* C **64**, 014301
[48] Bauhoff W *et al* 1984 *Phys. Rev.* **C 29** 054604
[49] Horiuchi H and Ikeda K 1968 *Prog. Theor. Phys. (Japan)* **40** 277
[50] Milin M *et al* 2009 *Eur. Phys. J.* A **41** 335
[51] von Oertzen W *et al* 2010 *Eur. Phys. J.* A **43** 17
[52] von Oertzen W, Milin M *et al* 2010 *Eur. Phys. J.* A **46** 345
[53] Bohlen H G, von Oertzen W, Milin M *et al* 2011 *Eur. Phys. J.* A **47** 44
[54] Ito M 2006 *Phys. Lett.* B **636** 293
[55] Ito M, Itagaki N, Sakurai H and Ikeda K 2008 *Phys. Rev. Lett.* **100** 182502
[56] Ito M and Itagaki N 2008 *Phys. Rev.* C **78** 011602(R)
[57] Kimura M 2007 *Phys. Rev.* C **75** 034312
[58] Furutachi N *et al* 2008 *Prog. Theor. Phys.* **119** 403
[59] Curtis N *et al* 2002 *Phys. Rev.* C **66** 024315
[60] Bhowmik R K, Rae W D M and Fulton B R 1984 *Phys. Lett.* B **136** 149
[61] Rae W D M, Cole A J, Harvey B G, and Stokstad R G 1984 *Phys. Rev.* C **30** 158
[62] Cunsolo A *et al* 1981 *Phys. Rev.* C **24** 476
[63] Cunsolo A *et al* 1982 *Phys. Lett.* B **112** 121
[64] Fortune H T 1978 *Phys. Rev.* C **18** 1053
[65] Fortune H T and Sherr R 2003 *Phys. Rev.* C **68** 034307

Electromagnetic decay of molecular states

S Courtin, A Goasduff and F Haas

Institut Pluridisciplinaire Hubert Curien, UMR 7178, Université de Strasbourg / CNRS-IN2P3. 23, rue du Loess, 67037 Strasbourg, France

E-mail: sandrine.courtin@iphc.cnrs.fr

Abstract. Electromagnetic transitions from deformed structures based on α configurations or on heavier clusters are discussed, drawing the link between multiparticle-multihole excited bands and cluster structures. Enhanced E2 and E1 transitions are reviewed in the light nuclei, ^{8}Be, ^{10}Be, ^{12}C, ^{16}O, ^{18}O and heavier ones like ^{212}Po. Connections between cluster structures and superdeformed configurations in ^{36}Ar and ^{40}Ca are discussed. What the cluster states based on heavier substructures like ^{12}C and ^{16}O are concerned, recent results on the resonant radiative capture reaction ^{12}C(^{16}O,γ)^{28}Si are presented, in particular the strong decay mode involving the feeding of low-lying ^{28}Si 1^+ and 2^+ T=1 states by enhanced M1 isovector transitions.

1. Introduction

Throughout the nuclear chart, there exist well identified regions of nuclei with large deformations. For such nuclei, the characteristic signature of deformation is the grouping of excited states into rotational bands with large moment of inertia. The deformed feature of an excited band member generally manifests itself through collective, *i.e.* enhanced, E2 intraband transitions. In light nuclei, as we will show later, deformation is very often connected with clusterization. To assess such connection, it is thus highly desirable to search for the electromagnetic decay mode of cluster states, sometimes also called molecular states. The 'basic' cluster states in light nuclei are composed of α particles and we will now briefly recall some 'highlight' cases for which enhanced E2 and sometimes also E1 γ transitions have been observed. We will then discuss the occurrence of heavier clusters in the *sd* shell, and in particular the ^{12}C-^{16}O resonances in ^{28}Si that we have explored via resonant radiative capture experiments.

2. Alpha cluster bands probed by E1 and E2 transitions

Among the light even-even N=Z nuclei, ^{8}Be is certainly the 'archetype' example of an α-α nucleus showing a band structure composed of the 0^+ ground state and excited 2^+ and 4^+ states at 3.04 and 11.4 MeV, respectively. These three states mainly decay into 2α particles, but a strong $4^+ \rightarrow 2^+$ γ transition has been observed with a reduced B(E2) strength of 25.8 $\pm$ 8.4 W.u. [1]. This very important result in the context of γ decay of molecular states, has been obtained through the radiative capture reaction $\alpha(\alpha,\gamma)$ and an α-α-γ coincidence selection in the exit channel [1]. The E2 $2^+ \rightarrow 0^+$ γ transition remains to be measured and is clearly a challenging goal.

In ^{9}Be, there is a well developed α-α-n cluster band built on the $J^{\pi} = \frac{3}{2}^-$ ground state. The excited band members with $J^{\pi} = \frac{5}{2}^-$ and $\frac{7}{2}^-$ are located at E_x = 2.43 and 6.38 MeV, respec-

Journal of Physics: Conference Series **436** (2013) 012051 doi:10.1088/1742-6596/436/1/012051

tively. The intra cluster band transitions $\frac{5}{2}^- \rightarrow \frac{3}{2}^-$ and $\frac{7}{2}^- \rightarrow \frac{5}{2}^-$ are highly collective with B(E2) = 24.4 ± 1.8 W.u. and 8.5 ± 3.6 W.u. [2], confirming thus the cluster, *i.e.* deformed, structure of this nucleus.

^{10}Be is one of the best example for clustering among light nuclei. However in this case, the α-α-n-n clustering is mainly concentrated in two excited bands, one with $K^\pi = 0^+$ and $J^\pi = 0^+$, 2^+ and 4^+ band members at 6.18, 7.54 and 10.15 MeV, respectively and another one with $K^\pi = 1^-$ and $J^\pi = 1^-$, 2^-, 3^- and 4^- at E_x = 5.96, 6.26, 7.37 and 9.27 MeV, respectively [3, 4, 5]. The first excited 0^+ state at 6.18 MeV which is the band-head of the $K^\pi = 0^+$ α-α-n-n cluster band cannot be reproduced by conventional shell model calculations which always place the first 0^+ excited state of ^{10}Be at excitation energies above 10 MeV. This cluster 0^+ state decays by a very moderate E2 transition of B(E2) = 2.6 ± 0.9 W.u. to the first ^{10}Be 2^+ state. But it also decays to the cluster 1^- state of the $K^\pi = 1^-$ band by a γ transition of E_γ = 219 keV and a very large E1 strength of $4.3 \pm 1.6 \times 10^{-2}$ W.u. [2], indeed one of the strongest E1 transitions for nuclei in the p shell. We will come back to these strong E1 transitions between cluster states a little later. These two positive and negative parity α-α-n-n cluster bands have a much larger moment of inertia than the cluster bands in ^{8}Be(α-α) and ^{9}Be(α-α-n). In this case, the intraband E2 transitions should be searched for in the $K^\pi = 0^+$ band between the 7.54 MeV, $J^\pi = 2^+$ state and the 6.18 MeV, $J^\pi = 0^+$ state and also in the $K^\pi = 1^-$ band between the 7.37 MeV, $J^\pi = 3^-$ state and the 5.96 MeV, $J^\pi = 1^-$ state.

What α clustering is concerned, ^{12}C is certainly a textbook example. The first excited 0^+ state is located just above the ^{8}Be+α threshold, it has an excitation energy of 7.65 MeV and an almost pure α-α-α cluster configuration. This so-called 'Hoyle' state decays mainly through the 3α channel but has also small decay branches to the 0^+_{gs} (E0 decay) and to the first 2^+ state at 4.43 MeV (E2 decay). The E2 strength of 8.1 ± 1.1 W.u. [2] corresponds to a moderately collective transition between the very deformed 0^+ α-α-α 'Hoyle' state and the almost pure 2^+ shell model 'p' state. Between the slightly oblate deformed ^{12}C first 2^+ and ground state, we have B(E2) = 4.7 ± 0.2 W.u. [2]. Possible excited states built on the 'Hoyle' state are actually actively searched for and candidates for the 2^+ and 4^+ members of the 'Hoyle' band are reported at excitation energies of ∼ 9.8 and ∼ 13.3 MeV, respectively [6, 7]. AMD model predictions of E2 transitions inside the 'Hoyle' band have been performed by Kanada-En'Yo [8] and values of 61.3 and 376.6 W.u. are obtained for the $2^+ \rightarrow 0^+$ and $4^+ \rightarrow 2^+$ E2 transitions, respectively. It turns out that it is very difficult to locate experimentally and characterize the excited 2^+ and 4^+ 'Hoyle' states which are at high excitation energies and are surrounded by several other states which generally have a large natural width. It will thus probably be experimentally difficult, if not impossible, to measure and eventually confirm these predicted very enhanced E2 transitions between these strongly deformed α-α-α cluster 'Hoyle' states.

^{16}O is a doubly-magic nucleus with N=Z=8 and with a first excited 0^+ state at 6.05 MeV. This state is the band head of a cluster α-^{12}C band with the 2^+ and 4^+ band members at 6.92 and 10.36 MeV, respectively. These α cluster states are strongly excited in Li induced α transfer reactions on ^{12}C [2], the unbound 4^+ state at 10.35 MeV has also been observed as a strong resonance in the ^{12}C(α,γ) capture reaction. This α-^{12}C cluster band shows remarkably strong E2 strengths of 27 ± 3 W.u. for the $6.92(2^+) \rightarrow 0(0^+)$ transition and of 65 ± 6 W.u. for the $10.36(4^+) \rightarrow 6.92(2^+)$ transition [2]. These E2 strengths are among the strongest in the whole *sd* shell. Generally, cluster states cannot be described by shell model (SM) calculations. In the case of ^{16}O, the SM calculations have been rather successful and the cluster α band has been clearly identified with a 4p-4h structure using the ZBM interaction with a ^{12}C core [9]. ^{16}O was the first case where the correspondence between α clustering and the SM deformed 4p-4h quadrupole correlations was clearly established. More recently, such features were also seen in the case of ^{40}Ca, another doubly magic nucleus (see below).

As seen previously, α cluster states in light even-even nuclei with N=Z are generally connected by strong E2 transitions of several tenths of W.u. But it has also been shown that in nuclei with N different from Z, the E1 transitions between cluster states could be particularly strong. We have mentioned previously the case of ^{10}Be where the 0^+ and 1^- cluster states at 6.18 and 5.96 MeV are connected by a strong E1 transition of $4.3 \pm 1.6 \times 10^{-2}$ W.u. The case of ^{18}O is even more remarkable, α-^{14}C cluster states with positive and negative parity have been clearly identified through the α transfer reaction ^{14}C(^{7}Li,t) and the resonant radiative capture reaction ^{14}C(α,γ) [10]. It has been demonstrated that the α cluster states with large spectroscopic factors in the transfer reactions have also enhanced E1 decay transitions. While for the 'normal' ^{18}O states with SM ^{16}O+2n structure, the B(E1) transition strength is in the 10^{-3} to 10^{-4} W.u. range, this B(E1) strength for transitions between the ^{18}O 'pear-shaped' α-^{14}C cluster states is larger than 10^{-2} W.u. [10]. It is worth mentioning that in the heavy nucleus ^{212}Po, states populated selectively by the α transfer reaction ^{208}Pb(^{18}O,^{14}C) also show strong E1 decay strengths between 10^{-3} and 2×10^{-2} W.u. [11]. The ^{212}Po states populated are probably of α-^{208}Pb cluster structure.

In the last decade, one of the most remarkable discovery in nuclear structure of light nuclei has been the observation of superdeformed (SD) rotational bands in the N=Z nuclei ^{36}Ar and ^{40}Ca close to the end of the *sd* shell. These bands were fed by fusion-evaporation reactions: ^{24}Mg(^{20}Ne,2α) in the case of ^{36}Ar [12] and ^{28}Si(^{20}Ne,2α) in the case of ^{40}Ca [13]. The 0^+ band heads of these bands are located at E_x = 4.33 MeV in ^{36}Ar and at E_x = 5.21 MeV in ^{40}Ca. For these very deformed prolate bands, the deformation parameter β_2 is 0.46 ± 0.03 in ^{36}Ar and 0.59 ± 0.09 in ^{40}Ca. In the two SD bands, the B(E2) strength of the $4^+ \rightarrow 2^+$ transition is 53 ± 9 W.u. for ^{36}Ar and 170 ± 40 W.u. for ^{40}Ca. In the case of ^{40}Ca, the SD band is nicely described by SM calculations [14] using a kind of ZBM interaction [9] with a ^{28}Si core and allowing up to 8 jumps of nucleons from the *sd* to *pf* shell. In ^{40}Ca, there are three 0^+ states below an excitation energy of 5.5 MeV, their structure in terms of SM np-nh states is given in Table 1 [14].

Table 1. Table extracted from Ref.[14] which shows the percentage of np-nh shell model wave function components and the energy (in MeV) of the first three 0^+ states (GS, ND and SD) of ^{40}Ca.

	0p-0h	2p-2h	4p-4h	6p-6h	8p-8h	E(th)	E(exp)
0^+_{GS}	65	29	5	-	-	0	0
0^+_{ND}	1	1	64	25	9	3.49	3.35
0^+_{SD}	-	-	9	4	87	4.80	5.21

The ^{40}Ca ground state 0^+_{GS} is spherical with 65% 0p-0h component in its wave function, the 0^+ excited state at E_x = 3.35 MeV has a main 4p-4h component (64%) and is the band head of a normal deformed (ND) band. The second 0^+ excited state at 5.21 MeV, is the band head of the SD band with a 87% 8p-8h component in its wave function. In a sense, the structure of the 0^+_{ND} at 3.35 MeV and of the 0^+_{SD} at 5.21 MeV can be interpreted as α-^{36}Ar and ^{8}Be-^{32}S (or α-α-^{32}S) cluster states. These two states are indeed strongly excited in the α transfer reactions induced by Li beams on ^{36}Ar targets [2]. It is quite interesting to note here that very deformed cluster bands are observed in doubly-magic, in principle closed-shell, ^{16}O and ^{40}Ca nuclei. These cluster bands are of np-nh structure and built on excited and highly deformed 0^+ states. The 'physics' involved here is very similar to what is called the island of inversion where N$\sim$20 Ne, Na, and Mg isotopes show strong deformations in their ground states. In this case, the sometimes called shell erosion is nothing else than the appearance in the ground states of strong deformations due to quadrupole np-nh correlations. It is quite possible that nuclei in the island of inversion have ground states with a cluster structure which calls for future experiments using radioactive neutron-rich beams.

3. ^{12}C–^{16}O clusters in ^{28}Si

Alpha clustering plays an important role in the description of the ground state and excited states of light nuclei in the *p* shell. For heavier nuclei, in the *sd*-shell, cluster configurations may be based on heavier substructures like ^{12}C and ^{16}O. These have been discussed to appear in ^{24}Mg(^{12}C–^{12}C) and ^{28}Si(^{12}C–^{16}O) both theoretically and experimentally [15, 16, 17]. The case of the mid-*sd* shell nucleus ^{28}Si is of particular interest as it shows the coexistence of deformed and cluster structures at rather low excitation energies. Its ground state is oblate, with a partial α-^{24}Mg structure, two prolate normal deformed bands are found, one built on the 0_2^+ state at 4.98 MeV and on the 0_3^+ one at 6.69 MeV. Recently, a candidate superdeformed band with a pronounced α-^{24}Mg structure has been suggested by Jenkins *et al* [18]. In this band, the 2^+ (9.8 MeV), 4^+ and 6^+ members have been identified. The 0^+ band head remains to be discovered. We shall discuss here a resonant cluster band which is predicted to start close to the Coulomb barrier of the ^{12}C+^{16}O collision, *i.e.* around 25 MeV excitation energy in ^{28}Si [15, 16]. This band could be related to the breakup band of ^{28}Si in ^{12}C+^{16}O observed some time ago by the CHARISSA collaboration with a dedicated setup [19]. This was observed in the ^{12}C(^{20}Ne,^{12}C^{16}O)α reaction up to 50 MeV excitation energy. Spin systematics has been deduced from the corresponding angular correlation measurement and the breakup states are grouped in a band which moment of inertia is close to a very deformed, molecular like ^{28}Si. Breakup states with low spins (≤ 6) corresponding to low excitation energy have not been measured in this study due to experimental limitations. Our experimental work aimed at measuring resonant states in this low spin region, possibly the low-lying members of this breakup band. We have studied the ^{12}C(^{16}O,γ)^{28}Si radiative capture reaction at 5 resonant energies around the Coulomb barrier (CB). The ^{12}C+^{16}O CB energy is $\sim$ 7.9 MeV [20]. This reaction was already investigated in the 1980ies close to the CB in an experimental campaign at Brookhaven National laboratory by Sandorfi *et al* [17]. Narrow resonances were identified and significant radiative capture cross-sections to some low-lying members of the ^{28}Si ground state band were measured making use of a large volume NaI detector coupled, in one of their experiments, to a Wien velocity filter for better recoil selection [21]. Unlike the Sandorfi measurement [17], our data do not show the piling-up of low energy transitions due to fusion-evaporation events, and thus our spectra extends to lower energies. This may have led Sandorfi to miss some flux mediated via low-energy transitions in the decay of the resonances and thus to miss some part of the capture cross-section.

We have decided to measure the full electromagnetic decay of some resonances in the energy region explored by Sandorfi *et al*, making use of the state-of-the-art recoil 0° spectrometer Dragon installed at Triumf, Vancouver, to identify the ^{28}Si recoils with an incident beam rejection factor better than 10^{12} and the associated 30 BGO array for the coincident high efficiency γ-ray detection. Details about the setup can be found in [22]. For our two ^{12}C+^{16}O radiative capture studies, five ^{16}O bombarding energies have been used between 21.1 and 15.4 MeV, three above the CB, on- and off-resonances observed by Sandorfi, and two below CB: the first on a previously observed resonant structure and the lower one below all previous radiative capture measurements was chosen on the maximum of a structure observed in the fusion excitation function [23]. A very efficient selection of the radiative capture channel out of the five orders of magnitude dominant fusion-evaporation events was performed by setting a gate on the time-of-flight between the DSSSD detector at the Dragon focal plane and signals of the 21 m upstream BGO array.

Monte Carlo GEANT calculations have been performed to simulate the response function of the experimental setup to different entrance channel conditions, *i.e.* energy and spin of the resonance.

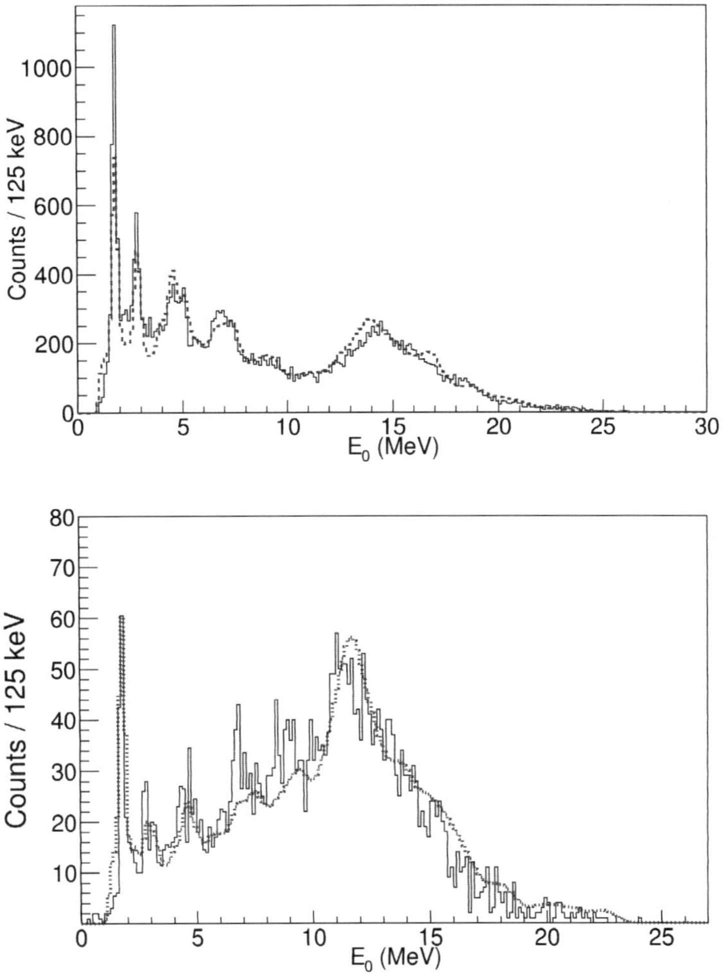

Figure 1. (Color online) Gamma spectra of the highest energy γ-ray in the cascade (E_0) at $E_{c.m.}$ = 9 MeV (top) and at $E_{c.m.}$ = 6.6 MeV (bottom) and Monte-Carlo simulations (blue dashed line) of the decay spectrum for a 6^+ entrance resonance (top) and for a 2^+ entrance resonance (bottom).

Figure 1 shows the γ spectra of the highest energy γ-ray (E_0) in the cascade measured in the BGO array at two bombarding energies of our study, the highest energy E_{Lab} = 21.1 MeV ($E_{c.m.}$

= 9 MeV) at the top and the lowest bombarding energy E_{Lab} = 15.4 MeV ($E_{c.m.}$ = 6.6 MeV) at the bottom. Below 10 MeV, the spectra show γ transitions between ^{28}Si low lying states and above an unexpected bump around E_0 = 14 and 12 MeV. This bump results from the feeding of intermediate states in the decay in an energy region above the α emission threshold (at 9.98 MeV). Due to the BGO array resolution, individual peaks are not resolved in this bump. Data is compared to numerical simulations (blue dashed curves on Fig. 1). These simulations take into account the entire experimental setup, *i.e.* the BGO array, the Dragon 0° spectrometer optical elements and focal plane detection. The ^{28}Si recoils are fully tracked down through the Dragon spectrometer. Characteristics of the ^{28}Si compound nucleus are used in the simulations, such as entrance excitation energy, bound and quasi-bound states from the literature and their γ branching ratios [2]. The entrance angular momentum is the main parameter of the simulations and the γ spectra obtained take into account the electromagnetic transition properties in the ^{28}Si mass region [24]. The best fit of the data is obtained for :

- A 6^+ entrance resonance at the highest bombarding energy (E_{Lab} = 21.1 MeV / $E_{c.m.}$ = 9.0 MeV). The data clearly show the direct feeding of the prolate deformed 4_3^+ state at 9.16 MeV and of the octupole deformed 3^- at 6.88 MeV. This state is the band head of an octupole band which mainly decays to the ^{28}Si oblate ground state with a strong E3 transition of reduced width B(E3) = 20 W.u. Our results are very similar to what has been measured for the ^{12}C+^{12}C radiative capture reaction above the CB in a previous Triumf Dragon experiment [25] where the enhanced feeding of the ^{24}Mg prolate band has been measured from a 4^+-2^+ resonance at $E_{c.m.}$ = 8.0 MeV near the CB.
- A 2^+ entrance resonance at the lowest bombarding energy (E_{Lab} = 15.4 MeV / $E_{c.m.}$ = 6.6 MeV). Our data clearly show the enhanced feeding, directly from the resonance, of $J^\pi=2^+$ and 1^+ T=1 states around 11 MeV. The main flux is going to the 1^+ state at 11.45 MeV which decays mainly to the ^{28}Si ground state by an isovector M1 transition. This scenario is confirmed by experimental angular distributions which show that dipole transitions are emitted from the 2^+ entrance resonance, which feed the intermediate T=1 states around 11 MeV, which then also decay via dipole transitions to the ^{28}Si ground state (see simplified view of this scenario on Fig. 2). At this energy, the decay is again quite similar to what occurs in the ^{12}C+^{12}C system at the lowest explored energy of Ref. [25]. Here as well, $J^\pi=2^+$ has been assigned to the entrance resonance and an enhanced decay has been measured via intermediate 1^+ T=1 states around 11 MeV in ^{24}Mg. A definitive scenario for the decay of the resonances at these low bombarding energies in both systems would come from the measurement of the γ decay spectra with a γ spectrometer with better resolution than BGO but still rather good efficiency. This type of array, based on $LaBr_3$ crystals is under construction by the PARIS collaboration [26] and should be available in the near future for such radiative capture studies.

Moreover, it would be interesting to determine if the decay scenario involving M1 ΔT=1 transitions is specific to self conjugate nuclei. This could be investigated by performing for instance the ^{12}C+^{14}C radiative capture reaction.

Interestingly enough, our spin assignments around the CB agree very well with the previously mentioned breakup band spin systematics measured by the CHARISSA collaboration and our resonant states could be the low lying members of the very deformed molecular-like ^{12}C-^{16}O band they have identified [19].

What higher bombarding energies are concerned, it would be interesting to search for the evolution of the observed feeding of the ^{28}Si intermediate states in the decay of the resonances. A pioneering campaign of experiments has been performed by Harakeh *et al* measuring the isospin mixing in the ^{28}Si excitation energy region around 35 MeV via the ^{12}C(^{16}O,γ)^{28}Si radiative capture reaction [27]. The authors discuss their γ spectra in terms of the feeding of the giant

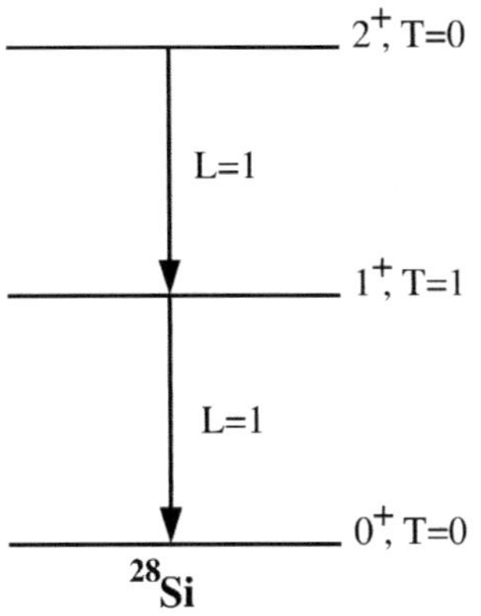

Figure 2. Schematic view of the decay scenario of the lowest energy 2^+ resonance ($E_{c.m.} = 6.6$ MeV / $E^*(^{28}Si) = 23.4$ MeV) via isovector M1 transitions feeding an intermediate T=1 state.

dipole resonance built on ^{28}Si excited states. It could be possible that in our studies of the electromagnetic decay of $^{12}C+^{16}O$ resonances near CB, the giant quadrupole resonance (GQR) is excited, built on the ^{28}Si prolate band starting at ~7 MeV. Such a mechanism would thus populate states located ~7 MeV above the maximum of the GQR, measured at 18.8 MeV [28], i.e. a region around 26 MeV, close to our resonance excitation energies.

References

[1] Datar V M *et al* 2005, *Phys. Rev. Lett.* **94** 122502
[2] http://www.nndc.bnl.gov/nudat2, and references therein
[3] Milin M *et al* 2005, *Nucl. Phys.* A **753** 263
[4] Freer M *et al* 2006, *Phys. Rev. Lett.* **96** 042501
[5] Milin M 2011, *Int. J. Mod. Phys.* E **20** 750
[6] Freer M *et al* 2011, *Phys. Rev.* C **83** 034314
[7] Itoh M *et al* 2011, *Phys. Rev.* C **84** 054308
[8] Kanada-En'yo Y. 2007, *Prog. Theor. Phys. (Kyoto)* **117** 655
[9] Zuker A P, Buck B and Mc Grory J B 1968, *Phys. Rev. Lett.* **21** 39
[10] Gai M *et al* 1987, *Phys. Rev.* C **36** 1256; 1989, *Phys. Rev. Lett.* **62** 874; 1991, *Phys. Rev.* **C 43** 2127; 1992, *Phys. Rev.* C **45** R2548
[11] Astier A *et al* 2010, *Phys. Rev. Lett.* **104** 042701; 2010 *Eur. Phys. J.* **A 46** 165
[12] Svensson C *et al* 2000, *Phys. Rev. Lett.* **85** 2693
[13] Ideguchi E *et al* 2001, *Phys. Rev. Lett.* **87** 222501
[14] Caurier E, Menendez J, Nowacki F and Poves A 2007, *Phys. Rev.* C **75** 054317
[15] Ichikawa T, Kanada-En'Yo Y and Horiuchi H 2011, *Phys. Rev.* C **83** 054319, and references therein
[16] Kanada-En'yo *et al* 2004, *Nucl. Phys.* A **738** 3
[17] Sandorfi A M 1984, *Treatise on Heavy-Ion Science*, Vol.2, p.52, Ed. Allan Bromley, Plenum Press, New-York, and references therein
[18] Jenkins D G *et al* 2012, *Phys. Rev.* C **86** 064309
[19] Metelko C J *et al* 2003, *J. Phys.* G **29** 697
[20] Lebhertz D *et al* 2012, *Phys. Rev.* C **85** 034333
[21] Collins M T *et al* 1982, *Phys. Rev Lett* **49** 1553
[22] Hutcheon D A *et al* 2003, *Nucl. Instrum. Methods* A **498** 190
[23] Christensen P, Switkowski Z and Dayras R 1977, *Nucl. Phys.* A **280** 189
[24] Endt P M 1993, *At. Data Nucl. Data Table* **55** 171
[25] Jenkins D G *et al* 2007, *Phys. Rev.* C **76** 044310
[26] Maj A *et al* 2009, *Acta Phys. Polonica* B **40** 565
[27] Harakeh M N *et al* 1986, *Phys. Lett.* B **176** 297
[28] Youngblood D H, Lui Y W and Clark H L 2007, *Phys. Rev.* C **76** 027304, and references therein

Journal of Physics: Conference Series **436** (2013) 012052 doi:10.1088/1742-6596/436/1/012052

Power series expansion method for determining collective quantities

Yoritaka Iwata
GSI Helmoltzzentrum für Schwerionenforschung, Darmstadt, Germany
E-mail: y.iwata@gsi.de

Abstract. A method of extracting the collective profiles from the microscopic many-body quantum dynamics is presented. In this paper a power series expansion method is introduced with the aim of utilizing the spline interpolation method.

1. Introduction
The master equation of the collective dynamics is

$$\mu(t)\ddot{R}(t) + \tfrac{1}{2}\tfrac{d\mu}{dR}\dot{R}(t)^2 + \gamma(t)\dot{R}(t) + \tfrac{dV(t)}{dR} = 0. \tag{1}$$

Only trajectories $R(t)$, $\dot{R}(t)$ and $\ddot{R}(t)$ are the input, where $R(t)$ is the center-of-mass motion obtained in microscopic many-body time-dependent Schrödinger-type equations, and $\dot{R}(t)$ and $\ddot{R}(t)$ are calculated by $R(t)$. Solving this equation, the collective quantities such as mass parameter of relative motion $\mu(t)$, friction coefficient $\gamma(t)$ and collective potential $V(t)$ are identified. In this sense the problem is a kind of inverse problem profiling the collectivity; clarifying the relation between the single-particle and collective degrees of freedom. In this paper, following [1], a power series expansion method for solving this inverse problem is presented in which each function is assumed to be analytic. In general this analysis requires a number of similar trajectories $R_i(t)$, where it can be assumed that $\mu(t)$, $\gamma(t)$ and $V(t)$ are the same in their dependence on the spatial coordinate.

2. Power series expansion
Let $R(t)$ be expanded as

$$R(t) = \textstyle\sum_{k=0}^{\infty} r_k t^k,$$

where r_k are known real numbers. It follows that

$$\dot{R}(t) = k\textstyle\sum_{k=1}^{\infty} r_k t^{k-1}, \quad \ddot{R}(t) = k(k-1)\textstyle\sum_{k=2}^{\infty} r_k t^{k-2}.$$

Let $\mu(t)$, $d\mu(t)/dR(t)$, $\gamma(t)$ and $dV(t)/dR(t)$, which vary depending on time, be expanded as

$$\begin{aligned} \mu(t) &= \textstyle\sum_{k=0}^{\infty} m_k t^k, \quad \tfrac{d\mu}{dR}(t) = \textstyle\sum_{k=0}^{\infty} n_k t^k, \\ \gamma(t) &= \textstyle\sum_{k=0}^{\infty} g_k t^k, \quad \tfrac{dV}{dR}(t) = \textstyle\sum_{k=0}^{\infty} v_k t^k, \end{aligned} \tag{2}$$

respectively, where m_k, n_k, g_k and v_k are unknown real numbers. As is discussed in the following section, it is not necessary to introduce n_k if μ and therefore m_k are introduced. Substituting them into Eq. (1), we have

$$k(k-1)\{\textstyle\sum_{k'=0}^{\infty} m_{k'}t^{k'}\}\{\sum_{k=2}^{\infty} r_k t^{k-2}\} + \frac{kk'}{2}\{\sum_{k=0}^{\infty} n_{k''}t^{k''}\}\{\sum_{k'=1}^{\infty} r_{k'}t^{k'-1}\}\{\sum_{k=1}^{\infty} r_k t^{k-1}\}$$
$$+k\{\textstyle\sum_{k'=0}^{\infty} g_{k'}t^{k'}\}\{\sum_{k=1}^{\infty} r_k t^{k-1}\} + \sum_{k=0}^{\infty} v_k t^k = 0$$

resulting in

$$\begin{aligned}&\{2m_0r_2 + (6m_0r_3 + 2m_1r_2)t + (12m_0r_4 + 6m_1r_3 + 2m_2r_2)t^2\cdots\}\\&+\tfrac{1}{2}\{n_0r_1^2 + 2(2n_0r_1r_2 + n_1r_1^2)t + (3n_0r_1r_3 + 4n_0r_2^2 + 2n_1r_1r_2 + n_2r_1^2)t^2\cdots\}\\&+\{g_0r_1 + (2g_0r_2 + g_1r_1)t + (3g_0r_3 + 2g_1r_2 + g_2r_1)t^2\cdots\} + \{v_0 + v_1t + v_2t^2 + \cdots\} = 0.\end{aligned} \quad (3)$$

This expansion method for solving the ordinary differential equations is well-known and legitimate.

3. Power series analysis

3.1. Power series expansion

According to the relation between m_k and n_k, we have

$$\frac{d\mu}{dR}\frac{dR}{dt} = \frac{d\mu}{dt} \quad \Leftrightarrow \quad (\textstyle\sum_{k'=0}^{\infty} n_{k'}t^{k'})(k''\sum_{k''=1}^{\infty} r_{k''}t^{k''-1}) = k\sum_{k=1}^{\infty} m_k t^{k-1}$$

therefore

$$\begin{aligned}&n_0 = m_1/r_1, \quad n_1 = 2(m_2 - m_1r_2/r_1)/r_1,\\&n_2 = \{3(m_3 - m_1r_3/r_1) - 4(m_2r_2/r_1 - m_1r_2^2/r_1^2)\}/r_1, \quad \cdots\end{aligned} \quad (4)$$

Taking up to the second order of t, we have

$$\begin{aligned}&2m_0r_{2,i} + (\tfrac{1}{2}m_1 + g_0)r_{1,i} + v_0 = 0\\&6m_0r_{3,i} + 2(m_1 + g_0)r_{2,i} + (2m_2 + g_1)r_{1,i} + v_1 = 0\\&12m_0r_{4,i} + (6m_1 + 3g_0)r_{3,i} + 2m_1r_{2,i}^2/r_{1,i} \quad + 2(m_2 + g_1)r_{2,i} + (\tfrac{3}{2}m_3 + g_2)r_{1,i} + v_2 = 0\end{aligned} \quad (5)$$

where $r_{j,i}$ is the j-th coefficient of the i-th trajectory ($i = 1, 2, 3, \cdots$). It implies that there should be several trajectories to solve the problem. Let $m_0 = \mu(0)$ and $g_0 = \gamma(0)$ be given constants. Using the first equation of Eq. (5), we have

$$\tfrac{1}{2}m_1 = 2m_0\frac{r_{2,2}-r_{2,1}}{r_{1,1}-r_{1,2}} - g_0, \quad v_0 = 2m_0\frac{r_{1,2}r_{2,1}-r_{1,1}r_{2,2}}{r_{1,1}-r_{1,2}}.$$

Using the second equation of Eq. (5), we have

$$2m_2 + g_1 = \frac{6m_0(r_{3,2}-r_{3,1})+2(m_1+g_0)(r_{2,2}-r_{2,1})}{r_{1,1}-r_{1,2}}, \quad v_1 = \frac{6m_0(r_{3,1}r_{1,2}-r_{1,1}r_{3,2})+2(m_1+g_0)(r_{2,1}r_{1,2}-r_{1,1}r_{2,2})}{r_{1,1}-r_{1,2}}.$$

Using the third equation of Eq. (5), we have

$$\begin{aligned}2(m_2 + g_1) &= \frac{(a_1-a_3)(r_{1,1}-r_{1,2})-(a_1-a_2)(r_{1,1}-r_{1,3})}{(r_{2,1}-r_{2,2})(r_{1,1}-r_{1,3})-(r_{2,1}-r_{2,3})(r_{1,1}-r_{1,2})},\\ \tfrac{3}{2}m_3 + g_2 &= \frac{(a_1-a_3)(r_{2,1}-r_{2,2})-(a_1-a_2)(r_{2,1}-r_{2,3})}{(r_{2,1}-r_{2,2})(r_{1,1}-r_{1,3})-(r_{2,1}-r_{2,3})(r_{1,1}-r_{1,2})},\\ v_2 &= \frac{a_1(r_{2,3}r_{1,2}-r_{2,2}r_{1,3})+a_2(r_{2,1}r_{1,3}-r_{2,3}r_{1,1})+a_3(r_{2,2}r_{1,1}-r_{2,1}r_{1,2})}{(r_{2,1}-r_{2,2})(r_{1,1}-r_{1,3})-(r_{2,1}-r_{2,3})(r_{1,1}-r_{1,2})},\end{aligned}$$

where $a_i = 12m_0r_{4,i} + 3(2m_1 + g_0)r_{3,i} + 2m_1r_{2,i}^2/r_{1,i}$. Although it is necessary to consider the third order equation to identify m_3 and g_2, v_i; i = 0, 1, 2 is fully identified within this framework.

$$\frac{dV}{dt} = \frac{dV}{dR}\frac{dR}{dt} = (\textstyle\sum_{k'=0}^{\infty} v_{k'}t^{k'})(k''\sum_{k''=1}^{\infty} r_{k''}t^{k''-1})$$

leading to the potential V

$$V(t) = V_0 + \tfrac{1}{2}v_0r_1t + \tfrac{1}{3}(v_0r_2 + v_1r_1)t^2 + \tfrac{1}{4}(v_0r_3 + v_1r_2 + v_2r_1)t^3 + \cdots$$

where $V(0) = V_0$ is a constant.

3.2. Potential as a function of spatial coordinate

Among μ, γ and V, we pay attention to the potential V. If t can be represented by the spatial coordinate ($t = \tau(R)$) using the relation $R = r_0 + r_{1,1}t + r_{2,1}t^2 + \cdots$, it is possible to have the potential representation

$$V(R) = V_0 + \tfrac{1}{2}v_0 r_1 \tau(R) + \tfrac{1}{3}(v_0 r_2 + v_1 r_1)\tau(R)^2 + \tfrac{1}{4}(v_0 r_3 + v_1 r_2 + v_2 r_1)\tau(R)^3 + \cdots$$

as a function of spatial coordinate R. If it is not easy to determine the functional form of $t = \tau(R)$, numerical methods are utilized to find the relation between t, $R(t)$ and $V(t)$ instead.

4. Simple cases

The most simple case is the case with a constant reduced mass $\mu(t) = m_0$ and no friction force $\gamma(t) = 0$. Using the first, second and third equations of Eq. (5), we have

$$v_0 = -2m_0 r_2, \quad v_1 = -6m_0 r_3, \quad v_2 = -12m_0 r_4, \tag{6}$$

respectively, where only one trajectory is sufficient to determine v_0, v_1 and v_2 in such a limited case.

Another simplified case is the case with a constant reduced mass $\mu(t) = m_0$ with fully operational friction force. Using the first, second and third equations of Eq. (5), we have

$$\begin{aligned}
v_0 &= \frac{2m_0(r_{1,2}r_{2,1} - r_{1,1}r_{2,2})}{r_{1,1} - r_{1,2}}, \\
v_1 &= \frac{6m_0(r_{3,1}r_{1,2} - r_{1,1}r_{3,2}) + 2g_0(r_{2,1}r_{1,2} - r_{1,1}r_{2,2})}{r_{1,1} - r_{1,2}}, \\
v_2 &= \frac{12m_0(r_{4,1}r_{1,2} - r_{1,1}r_{4,2}) + 3g_0(r_{3,1}r_{1,2} - r_{1,1}r_{3,2}) + 2g_1(r_{2,1}r_{1,2} - r_{1,1}r_{2,2})}{r_{1,1} - r_{1,2}},
\end{aligned}$$

with

$$\begin{aligned}
g_0 &= \frac{2m_0(r_{2,2} - r_{2,1})}{r_{1,1} - r_{1,2}}, \\
g_1 &= \frac{6m_0(r_{3,2} - r_{3,1}) + 2g_0(r_{2,2} - r_{2,1})}{r_{1,1} - r_{1,2}}, \\
g_2 &= \frac{12m_0(r_{4,2} - r_{4,1}) + 3g_0(r_{3,2} - r_{3,1}) + 2g_1(r_{2,2} - r_{2,1})}{r_{1,1} - r_{1,2}},
\end{aligned}$$

where two trajectories are sufficient to determine v_0, v_1 and v_2 in this case.

5. Concluding remark

This method is designed for the trajectories expressed as segmented spline interpolation functions. This representation provides a high precision analysis when using an adapted variable segmentation interval.

Acknowledgments

This work was supported by the Helmholtz alliance HA216/EMMI. The author thanks Prof. J. A. Maruhn for reading this manuscript carefully.

References

[1] Iwata Y Feldmeier H and Maruhn J A 2011 *AIP Conf. Proc.* **1377** 365

Wobbling excitation of triaxial nuclear molecule ^{28}Si − ^{28}Si

E Uegaki[1] and Y Abe[2]

[1]Graduate School of Engineering and Resource Sci., Akita University, Akita 010-8502, Japan
[2]Research Center for Nuclear Physics, Osaka University, Ibaraki 567-0047, Japan

E-mail: uegaki@phys.akita-u.ac.jp

Abstract. High-spin resonances observed in the ^{28}Si + ^{28}Si collisions are investigated with a molecular model. At high spins, a stable dinuclear configuration is found to be an equator-equator touching one. Since the E-E configuration is slightly triaxial, rotations of the total system induce mixing of K quantum numbers, called wobbling motion, which clearly explains the particle-γ angular correlations observed as well as the disalignments in a simple and natural way. Furthermore, predictions are given for the angular correlations of the wobbling excited states. The first excited state of wobbling shows strong alignments, which is quite different from the molecular ground state.

1. Introduction

High-spin resonances well above the Coulomb barrier in the ^{28}Si + ^{28}Si system exhibit a number of sharp peaks correlated among the elastic and inelastic channels. In their angle-averaged excitation functions, bumps are seen corresponding to the grazing angular momenta, and several sharp peaks are found on each bump [1]. The high level density of the resonance states indicates activation of some internal degrees of freedom other than the relative motion. Hence the rotational or the orientation degrees of freedom of the constituent deformed nuclei have to be taken into account. As a first step for that, we solved normal modes of motion in multi-dimensional space around the equilibrium configurations of interacting two oblate nuclei [2].

Recently new experimental data were obtained for ^{28}Si + ^{28}Si at $E_{\rm cm}$ = 55.8 MeV [3]; the angular distributions exhibit disalignments between the orbital angular momentum and the fragment spins, and the particle-γ angular correlation measurements show characteristic features. We analyzed the correlations by using the normal modes of the equilibrium configuration, and found that the molecular ground state reproduces those characteristic features of the angular correlations very well [4]. Furthermore, we expect that some of those sharp peaks on the bump correspond to wobbling excited states, and thus we have analyzed their angular correlations. The results for a couple of the excited states are compared between them, which show quite different characteristics, respectively.

2. Dinuclear molecular structure of ^{28}Si − ^{28}Si

2.1. *Normal mode analyses*

Assuming a constant deformation and axial symmetry of the constituent nuclei for simplicity, we have seven degrees of freedom $(q_i) = (\theta_1, \theta_2, \theta_3, R, \alpha, \beta_1, \beta_2)$, as illustrated in fig. 1, where

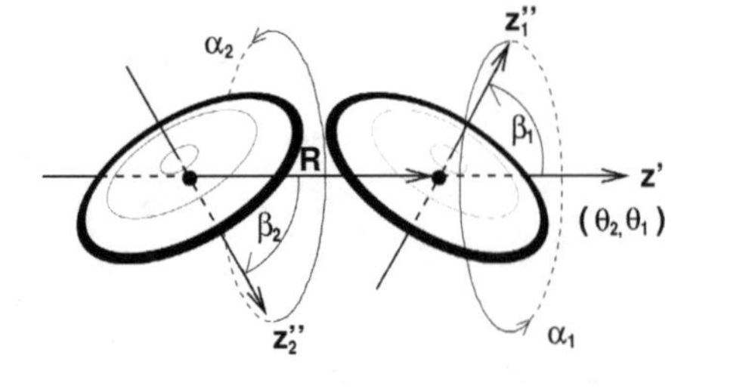

Figure 1. Dinuclear molecular coordinates; seven degrees of freedom.

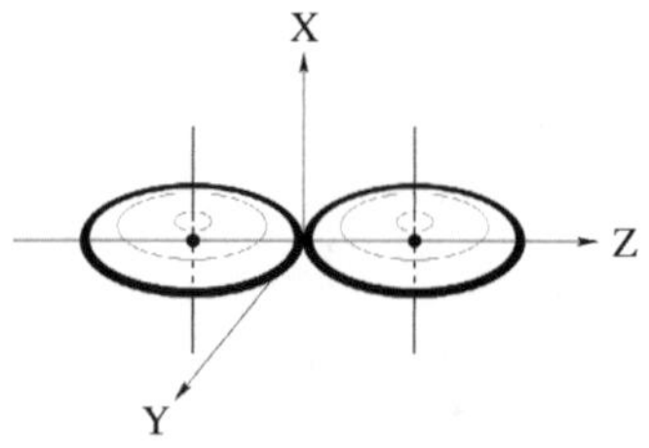

Figure 2. Equilibrium configuration of two interacting oblate nuclei.

(R, θ_2, θ_1) is the relative vector of two ^{28}Si nuclei. The orientations of the symmetry axes of the ^{28}Si nuclei are described with Euler angles (α_i, β_i) referring to the molecular axes, where α_1 and α_2 are combined into $\theta_3 = (\alpha_1 + \alpha_2)/2$ and $\alpha = (\alpha_1 - \alpha_2)/2$. Consistently with the coordinate system, at first, we introduce a rotation-vibration type wave function $D^J_{MK}(\theta_i)\chi_K(R, \alpha, \beta_1, \beta_2)$, as basis one, where χ_K describes internal motions. Details are given in ref. [5].

By the molecular model the authors investigated an equilibrium configuration of two interacting oblate nuclei, and showed it to be the *equator-equator* (E-E) one, illustrated in fig. 2. Couplings among various molecular configurations are taken into account by the method of normal mode around the equilibrium, which gives rise to the molecular modes of excitation, such as *butterfly* with $\beta_+ = (\Delta\beta_1 + \Delta\beta_2)/\sqrt{2}$ and *anti-butterfly* with $\beta_- = (\Delta\beta_1 - \Delta\beta_2)/\sqrt{2}$ around $\alpha = \pi/2$, respectively [2, 5]. In fig. 3(a) molecular normal modes of $^{28}\text{Si} + ^{28}\text{Si}$ with spin 38 are displayed, classified with quantum numbers K, for $K = even$.

2.2. Wobbling motion with triaxial deformation

Since the E-E configuration is triaxial, rotations of the total system induce mixing of quantum numbers K at high spins, which is known as *wobbling*. Such a motion is consistent with the measured angular correlations as well as the disalignments. At a given angular momentum J, this configuration rotates in a triaxial way preferentially about the axis corresponding to the largest moment of inertia in the state with the lowest energy. Therefore the whole system rotates about the normal to the plane defined by the two pancake-like nuclei. With such a characteristic rotation, the spins of the ^{28}Si fragments are preferentially in the reaction plane, because they are perpendicular to the symmetry axes of ^{28}Si. After K-mixings due to the axial asymmetry, we obtain a new energy spectrum, displayed in fig. 3(b).

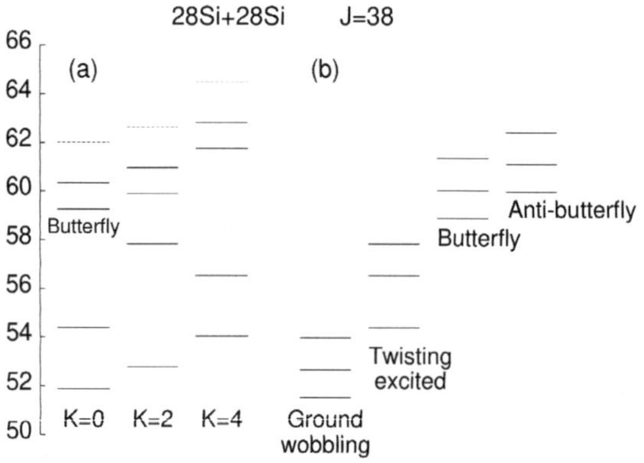

Figure 3. Energy spectra of the $^{28}\text{Si} + ^{28}\text{Si}$ system for $J = 38$. (a) Molecular normal modes without K-mixing. (b) After K-mixing, with indications of the modes under the levels.

In the high-spin limit ($|K|/J \sim 0$) we obtain analytical wobbling (K-mixing) solutions as $F_n(K) = H_n(K/b)\exp(-K^2/2b^2)$, with the Hermite polynomial H_n for the n-th excited state, which provide a good perspective [5, 4]. The width b is given by the asymmetry of the moments of inertia.

3. Angular correlations of the wobbling excitations

Figure 4 shows γ-ray intensities emitted from the first excited 2^+ state of ^{28}Si fragments, obtained for the mutual 2^+ channel decays in the particle-γ angular correlation measurements [3]. Data on 4π solid angle are classified with three different quantization z axes: (a) beam direction, (b) normal to the plane and (c) fragment direction taken perpendicular to the axes (a) and (b), respectively. Solid lines show theoretical results of the molecular ground state with $J = 38$, which have dominant $m = 0$ (disalignment) in (b). It is found that the molecular ground state reproduces the measured correlations very well [3, 4]. Furthermore, we have predicted the first and second excited states of the wobbling motion (for the states denoted 'Ground wobbling' in fig. 3(b)). Apparently the excited states exhibit different characteristics from that of the molecular ground state. For the first excited state (dashed lines), with a dominant $m = 2$ pattern in (b), strong alignments are indicated. The second excited state (dotted lines) gives dominant $m = 2$ pattern in (a). Due to high K-rotation, the fragment spins are almost parallel to the molecular z' axis. Those states are possible origin of the resonance peaks observed on the same bump of the 55.8 MeV resonance.

Decay properties have also been investigated; due to the strong K-mixing, the wobbling excited states show strong suppression in the elastic channel, which is consistent with the data around $E_{\mathrm{cm}} = 57$ MeV [1].

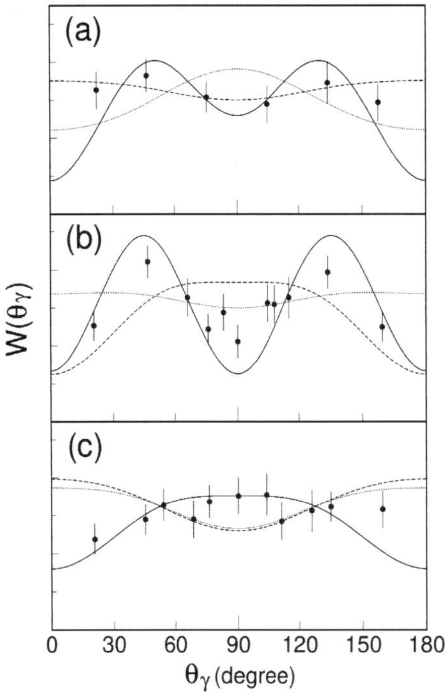

Figure 4. Angular correlations for three quantization z axes (see text for the axes). Solid lines are those of the molecular ground state, dashed lines for the first excited state and dotted lines for the second excited state.

4. Conclusions

The molecular ground state with $J = 38$ is a good candidate for the resonance at $E_{\mathrm{cm}} = 55.8$ MeV [4, 6]. In addition, we have predicted angular correlations of the wobbling excited states. Contrary to the molecular ground state with the property of disalignments, the first excited state of wobbling motion shows strong alignments. In brief, it has turned out that particle-γ angular correlations provide decisive information on orientations or motions of the constituent nuclei in resonance states. Experiments on the nearby resonances are strongly called for, which unveil the long-standing mystery of the high-spin resonances in heavy-ion collisions.

References

[1] Betts R R, Back B B and Glagola B G 1981 *Phys. Rev. Lett.* **47** 23
[2] Uegaki E and Abe Y 1994 *Phys. Lett.* B **340** 143
[3] Nouicer R *et al* 1999 *Phys. Rev.* C **60** 041303
[4] Uegaki E and Abe Y 2012 *Prog. Theor. Phys.* **127** 877
[5] Uegaki E and Abe Y 2012 *Prog. Theor. Phys.* **127** 831
[6] Uegaki E 2013 *Proc. of this conference*

Journal of Physics: Conference Series **436** (2013) 012054 doi:10.1088/1742-6596/436/1/012054

Clustering effects in ^{48}Cr composite nucleus produced via the reaction ^{24}Mg + ^{24}Mg at the excitation energy of 60 MeV

E Vardaci[1,2], A Di Nitto[1,2,5], P Nadtochy[2,6], A Brondi[1,2], G La Rana[1,2], R Moro[1,2], A Vanzanella[2], M Cinausero[3], G Prete[3] and N Gelli[4]

[1] Dipartimento di Fisica, Universitá degli Studi di Napoli "Federico II", 80126 Napoli, Italy
[2] Istituto Nazionale di Fisica Nucleare, 80126 Napoli, Italy
[3] Laboratori Nazionali di Legnaro, INFN, Legnaro (Padova), Italy
[4] Istituto Nazionale di Fisica Nucleare, Firenze, Italy

E-mail: Emanuele.Vardaci@unina.it

Abstract. The reaction ^{24}Mg + ^{24}Mg was used to produce the composite nucleus ^{48}Cr at 60 MeV of excitation energy where a narrow resonance (170 KeV) has been found by measuring the elastic and inelastic channels. To determine the occurrence of deformation of this compound nucleus and its possible connection with the resonances and the hypothetical cluster structure, evaporative Light Charged Particles (LCP) were measured and compared to Statistical Model (SM) predictions. The experiment was performed at LNL using the 8πLP apparatus. The comparison of the evaporation residue-LCP coincidence angular distributions and LCP energy spectra with SM calculations supports the presence of a very large deformation of the composite nucleus ^{48}Cr that scales with the angular momentum.

1. Introduction

In many reactions involving light α-like reacting partners there is a consistent occurrence of narrow resonances in the elastic and inelastic channels [1]. Examples of such systems are ^{24}Mg + ^{24}Mg, ^{12}C + ^{12}C, ^{28}Si + ^{28}Si or ^{12}C + ^{28}Si. Furthermore, these resonances are linked to the formation of an intermediate highly deformed system with a lifetime comparable to the time scale of evaporation from compound nuclei at moderate excitation energy. The origin and characteristic features of these resonances and their connection with the cluster structure of the two reacting partners is a matter of discussion.

One recurring hypothesis is that the intermediate resonant states, with very large deformations, may act as doorway states to fusion. In this respect, entrance channel dynamics may contribute to the selective population of highly deformed configurations with a lifetime comparable to the evaporative time scale. The memory of the entrance channel asymmetry may therefore persist in the Light Charged Particles (LCP) evaporative angular distributions and energy spectra. Consequently, evaporative LCP can be a good probe to independently estimate the deformation of these states.

[5] Present address: *Johannes Gutenberg – Universität Mainz Institut für Kernchemie*, Mainz, Germany
[6] Present address: *Omsk State University*, Mira prospekt 55-A, Omsk 644077, Russia

In this framework, we have undertaken an exploratory study on the system ^{24}Mg + ^{24}Mg by using LCP evporative emission. A narrow resonance with J=36$\hbar$ at E_x=60 MeV was observed [2] for this system. Surprisingly, the resonance spin is higher than the grazing angular momentum in the entrance channel (L_{graz} = 32$\hbar$) by 4$\hbar$, raising many questions on this matter. Furthermore, the width of 170 KeV implies a lifetime of 4x10^{-21}s, ten times longer than a typical nuclear lifetime. A study [3] has been recently carried out on the J=36$\hbar$ resonance in order to investigate its decay into the inelastic and fusion-evaporation channels. The properties of the studied resonance are in agreement with the molecular model in [4] and the intermediate system's shape predicted is very close to a highly prolate deformed ^{48}Cr after a Jacobi shape transition as calculated by the Lublin-Strasbourg-Drop (LSD) model [5] between J=28$\hbar$ and 32$\hbar$ at E_x = 60 MeV. To estimate the extent of deformation of the excited intermediate system, evaporative LCP are measured in coincidence with Evaporation Residues (ER) and compared with the SM predictions which include a treatment of a spin-dependent compound nucleus shape.

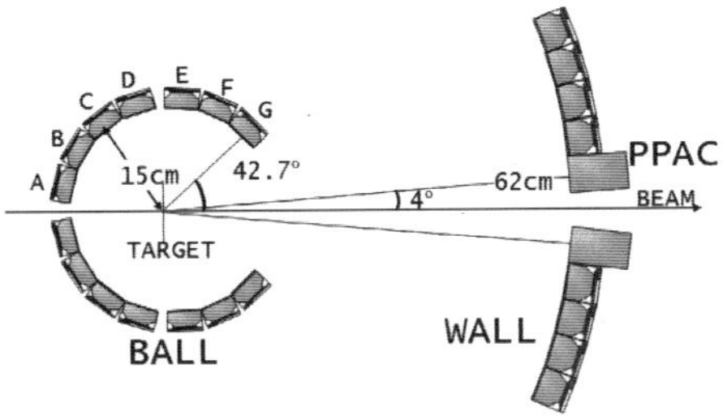

Figure 1. Schematic layout of the experimental apparatus. The letters from A to G label the 7 rings of the BALL section.

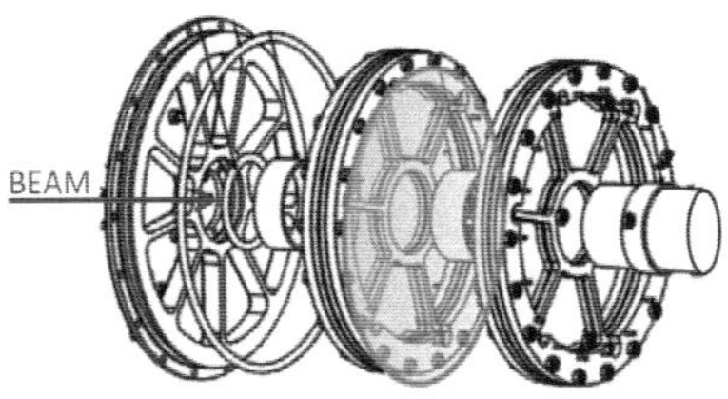

Figure 2. Scheme of the PPAC system.

2. Experimental procedure and observables

The experiment was performed at the Tandem accelerator of Laboratori Nazionali di Legnaro. A pulsed beam of ^{24}Mg of intensity of about 5-10 enA was used to bombard a ^{24}Mg target $60\mu g/cm^2$ thick on $15\mu g/cm^2$ C backing. The chosen beam energy of 91.4 MeV corresponds to the resonance $J^{\pi}=36\hbar^{+}$ at E_x=60 MeV in the ^{48}Cr compound nucleus (CN). The light charged particles emitted from the CN de-excitation were measured with the 8πLP apparatus [6].

The detector system, shown schematically in Fig. 1, is constituted by about 250 telescopes made by a ΔE stage of Si detectors followed by a second stage of CsI detectors, geometrically arranged in two sections: the BALL and the WALL. Particle identification is carried out by using the ΔE-E, Pulse Shape Discrimination and time of flight techniques. The BALL consists of 7 rings of detectors, labeled from A to G each containing 18 telescopes. ER were detected by a circular PPAC system (Fig. 2). The LCP and the ER events were collected by requiring the OR mode of singles and coincidence mode, including up to four-fold events between LCP. This triggering scheme was very effective for the evaluation of the multiplicities because it provided, in the same run, for the measurement of the singles and coincidence yields.

3. Experimental data and Statistical model simulations

SM calculations were performed with the code LILITA_N97 [7] which models the multistep evaporative decay of a compound nucleus by using the Hauser-Feshbach formulation of the SM. The program produces laboratory frame energy spectra and angular distributions for both ER and LCP, once the constraints imposed by the experimental apparatus are included. In this study we simulated the LCP decay from a normal and highly deformed ^{48}Cr composite system using two sets of deformability parameters δ_1 and δ_2 to modulate the deformation. These parameters are usually adopted in the statistical model to calculate the moment of inertia $\Im$ of the emitting nucleus as a function of the angular momentum J through the expression:

$$\Im = \Im_0 \cdot (1 + \delta_1 J^2 + \delta_2 J^4) \quad (1)$$

In the Rotating Liquid Drop Model (RLDM) calculation a set of deformability parameters corresponding to a deformation with an axis ratio b/a=1.51 for J=36 ħ were used. In the calculation indicated as Highly Deformed for the angular momenta in the window J=28ħ − 36ħ (around the angular momentum of the resonance) very large deformations were used (b/a= 3.0) whereas for lower angular momenta the RLDM deformation was assumed.

Our goal was to find an indication of the occurrence of a large deformation in this α-like system, and possibly its link with the observed resonances by studying the LCP evaporation at the excitation energy corresponding to a resonance. From Fig. 3 and 4 the experimental data are consistent with a deformation of the compound nucleus larger than the one predicted by the RLDM. The large deformation found for the nucleus ^{48}Cr at the resonance energy is, however, a common feature of non α-like system as well, where no resonances appear, though the deformation appears to be less pronounced. The cases of the reactions ^{28}Si + ^{28}Si [8] and ^{30}Si + ^{30}Si [9] are examples. This means that the large deformation of ^{48}Cr compound nucleus is not necessarily connected with the presence of resonances. Consequently, the resonances might be more strictly related to the α cluster composition of the projectile and target and to the mass symmetry of the entrance channel. To clearify on this point a measurement on the system ^{24}Mg + ^{24}Mg at an off-resonance energy would be very important.

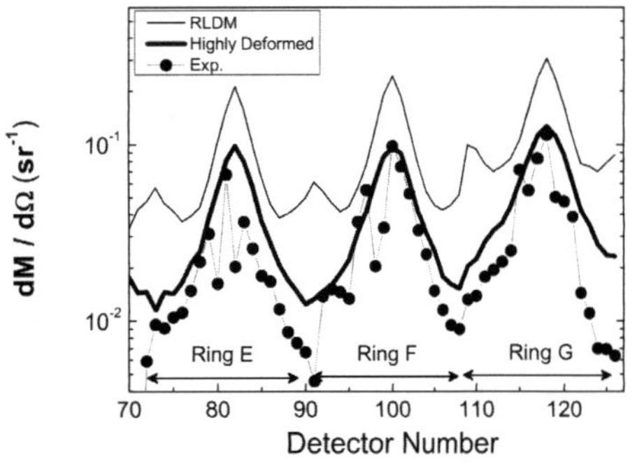

Figure 3. Experimental angular multiplicity distributions (full dots) of α particles in coincidence with ER versus the identification number of the BALL detectors in rings E, F and G compared to the simulations.

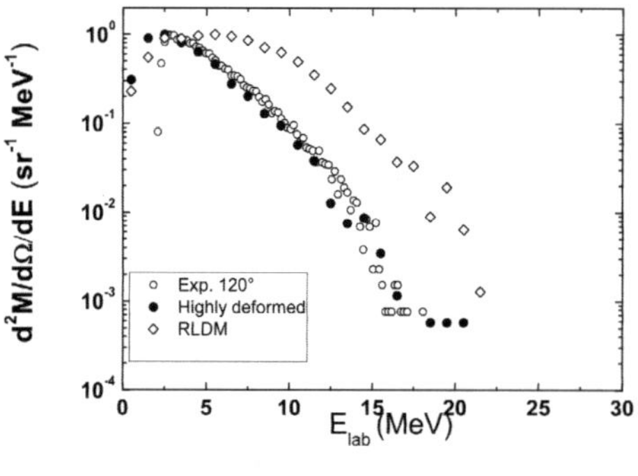

Figure 4. Laboratory energy spectrum of α particles at 120° in coincidence with ER compared to simulations

References

[1] Sanders S J, Szanto De Toledo A and Beck C 1999 *Phys. Rep.* **311** 487
[2] Wuosmaa A H *et al* 1990 *Phys. Rev.* C **41** 2666
[3] Salsac M D *et al* 2008 *Nucl. Phys.* A **801** 1
[4] Uegaki E and Abe Y 1989 *Phys. Lett.* B **231** 28
[5] Pomorski K and Dudek J 2003 *Phys. Rev.* C **67** 044316
[6] Fioretto E *et al* 1997 *IEEE Trans. Nucl. Sci.* **44** 1017
[7] Extensively modified version of the original code LILITA written by Gomez del Campo J and Stockstad R G 1981 Oak Ridge National Laboratory Report No. TM7295 (unpublished)
[8] Bhattacharya C *et al* 2001 *Phys. Rev.* C **65** 014611
[9] La Rana G *et al* 1988 *Phys. Rev.* C **37** 1920

Journal of Physics: Conference Series **436** (2013) 012055 doi:10.1088/1742-6596/436/1/012055

Cluster correlations in largely deformed states

Yasutaka Taniguchi
Center for Computational Sciences, University of Tsukuba
1-1-1, Tennodai Tsukuba-city, Ibaraki 305-8577, Japan
E-mail: yasutaka@ccs.tsukuba.ac.jp

Abstract. The coexistence of various deformed states in ^{42}Ca and α–^{38}Ar correlations have been investigated using deformed-basis antisymmetrized molecular dynamics. Wavefunctions of the low-lying states are obtained via parity and angular momentum projections and the generator coordinate method (GCM). Basis wavefunctions of the GCM calculation are obtained via energy variation with constraints on the quadrupole deformation parameter β and the distance between α and ^{38}Ar clusters. The rotational band built on the $J^\pi = 0^+_2$ (1.84 MeV) state as well as the $J^\pi = 0^+_3$ (3.30 MeV) state are both reproduced. The coexistence of two additional $K^\pi = 0^+$ rotational bands is predicted; one band is shown to be built on the $J^\pi = 0^+_3$ state. Members of the ground-state band and the rotational band built on the $J^\pi = 0^+_3$ state contain α–^{38}Ar cluster structure components. The α–^{38}Ar higher-nodal $J^\pi = 0^+$ state is also obtained, and the $J^\pi = 0^+$ state has large E0 transition strength from the ground state.

1. Introduction

Drastic structural changes initiated by low excitation energies are a significant characteristic of nuclear systems, and the coexistence of deformed states and cluster structures is a typical phenomenon. In the mass number region $A \sim 40$, low-lying normal-deformed (ND) and superdeformed (SD) bands with many-particle-many-hole (mp-mh) configurations have been confirmed experimentally. Coupling of the cluster structure components in deformed states such as the α-cluster structure in the ND band of ^{40}Ca [1, 2, 3, 4] and the ground-state band in ^{44}Ti [1, 2, 5] has also be investigated. Recently, enhancement of E0 and IS0 strengths by cluster structure components was discussed [6, 7].

In ^{42}Ca, deformed states with mp-mh configurations and clustering behavior have been observed experimentally, and the rotational band built on the $J^\pi = 0^+_2$ (1.84 MeV) state ($K^\pi = 0^+_2$ band) has been observed [8, 9]. With regard to α-cluster structures, strong population to the $J^\pi = 0^+_1$ and 0^+_3 (3.30 MeV) states has been observed in α-transfer reactions to ^{38}Ar, and the ratios of the cross sections of α and $2n$ transfer reactions suggest that the $J^\pi = 0^+_2$ and 0^+_3 states have configurations of $6p4h$ and $4p2h$, respectively [10]. Theoretically, the α+^{38}Ar orthogonality condition model (OCM) describes $4p2h$ states with α–^{38}Ar cluster structures, but a rotational band with a $6p4h$ configuration is not obtained in low-lying states [11]. To understand the structures in ^{42}Ca, various deformations with mp-mh configurations and clustering should be taken into account, but such a study has never been performed. The structures of low-lying states in ^{42}Ca have not been clarified yet.

This paper aims at clarifying the structures of low-lying positive-parity states in ^{42}Ca by focusing on the coexistence of rotational bands with mp-mh configurations. The α–^{38}Ar cluster correlations in low-lying deformed states are also discussed.

2. Framework

The wavefunctions in low-lying states are obtained using the parity and angular momentum projection (AMP) and the generator coordinate method (GCM) with deformed-basis antisymmetrized molecular dynamics (AMD) wavefunctions. A deformed-basis AMD wavefunction $|\Phi\rangle$ is the Slater determinant of Gaussian wave packets that can deform triaxially such that

$$|\Phi\rangle = \hat{\mathcal{A}}|\varphi_1, \varphi_2, \cdots, \varphi_A\rangle, \quad (1)$$

$$|\varphi_i\rangle = |\phi_i\rangle \otimes |\chi_i\rangle \otimes |\tau_i\rangle, \quad (2)$$

$$\langle \mathbf{r}|\phi_i\rangle = \pi^{-3/4}(\det \mathsf{K})^{1/2} \exp\left[-\frac{1}{2}(\mathsf{K}\mathbf{r} - \mathbf{Z}_i)^2\right], \quad (3)$$

$$|\chi_i\rangle = \chi_i^{\uparrow}|\uparrow\rangle + \chi_i^{\downarrow}|\downarrow\rangle, \; |\tau_i\rangle = |\pi\rangle \text{ or } |\nu\rangle, \quad (4)$$

where $\hat{\mathcal{A}}$ denotes the antisymmetrization operator, and $|\varphi_i\rangle$ denotes a single-particle wavefunction. The $|\phi_i\rangle$, $|\chi_i\rangle$, and $|\tau_i\rangle$ states denote the spatial, spin, and isospin components, respectively, of each single-particle wavefunction $|\varphi_i\rangle$. The real 3×3 matrix K denotes the wave number of the Gaussian single-particle wavefunction that can deform triaxially, which is common to all nucleons. The $\mathbf{Z}_i = (Z_{ix}, Z_{iy}, Z_{iz})$ vectors are complex parameters to denote the centroid of each single-particle wavefunction in phase space. The complex parameters $\chi_i^{\uparrow}$ and $\chi_i^{\downarrow}$ denote the spin directions. Neither time reversal nor axial symmetry are assumed. The basis wavefunctions of the GCM are obtained via the energy variation with two types of constraints: the quadrupole deformation parameter β of the total system and the distance between α and ^{38}Ar clusters, which is defined by the positions of the wave packet centroids [12]. The energy variational calculations are performed after the projection onto positive-parity states. For the energy variation with β, harmonic oscillator (HO) quanta for the protons and neutrons, N_π and N_ν, respectively, relative to the lowest allowed state are also constrained. The variational parameters are K, $\mathbf{Z}_i$, and $\chi_i^{\uparrow,\downarrow}$ $(i = 1, ..., A)$. The isospin component of each single-particle wave function is fixed as a proton (π) or a neutron (ν). The Gogny D1S force is used as the effective interaction. Details of the framework are provided in refs. [13, 14, 12].

3. Results and discussions

Figure 1 shows the energy curves obtained from the energy variation with the constraints on β and the HO quanta N_π and N_ν as functions of β. The HO quanta (N_π, N_ν) are constrained to (0, 0), (2, 0), (2, 2), and (4, 2), which appear on the β–energy curve obtained with the β-constrained energy variation without the HO constraint. The (0, 0), (2, 0), (2, 2), and (4, 2) configurations correspond to the $[(pf)^2]_\nu$, $[(sd)^{-2}(pf)^2]_\pi[(pf)^2]_\nu$, $[(sd)^{-2}(pf)^2]_\pi[(sd)^{-2}(pf)^4]_\nu$, and $[(sd)^{-4}(pf)^4]_\pi[(sd)^{-2}(pf)^4]_\nu$ structures, respectively, which, in total are the $2p$, $4p2h$, $6p4h$, and $8p6h$ configurations. The $4p2h$, $6p4h$, and $8p6h$ states have local minima at $\beta \sim 0.3$, 0.4, and 0.5, respectively. Deformations of the protons and neutrons are similar across the whole β region. The $4p2h$, $6p4h$, and $8p6h$ states form triaxially deformed structures. Through the AMP, largely deformed states gain higher binding energies, and the energy of the local-minimum state for the $6p4h$ configuration projected onto the $J^\pi = 0^+$ state becomes lower than that for $4p2h$ states. The order of the local-minimum energies are the reverse of that before the AMP.

Two kinds of configurations obtained by using energy variation with the intercluster distance constraint are adopted as the GCM basis, and are labeled as A and B. The difference between them is in the orientation of the ^{38}Ar cluster relative to the direction of the α cluster. ^{38}Ar clusters have two proton holes in the sd-shell, and the direction of the proton holes is parallel and vertical to the direction of α clusters for configurations A and B, respectively. Superposing those configurations, the rotation of the ^{38}Ar clusters is taken into account partly. Figures 2 and 3

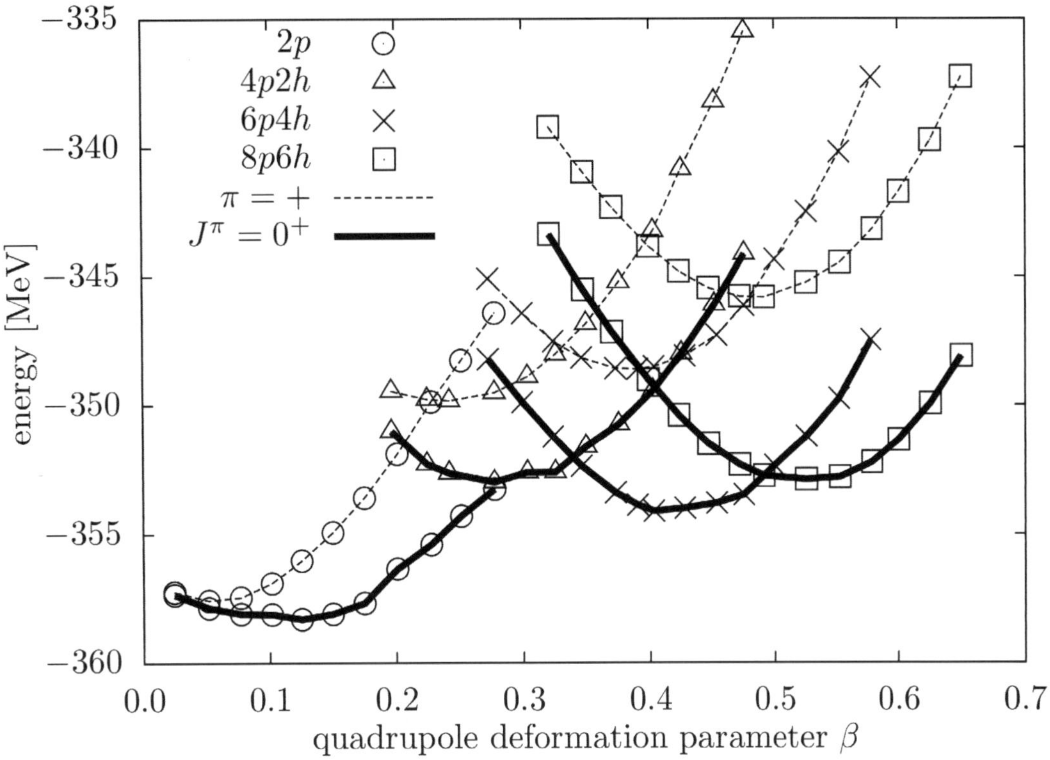

Figure 1. The energy curves as functions of the quadrupole deformation parameter β for positive-parity (solid lines) and $J^\pi = 0^+$ (dashed lines) states. Circles, triangles, crosses, and squares indicate the $2p$, $4p2h$, $6p4h$, and $8p6h$ configurations, respectively (see text).

show the energy and HO quanta, respectively, for the α–^{38}Ar wavefunctions with configuration A and B. In small intercluster distance region the energy of configuration A wave functions go to the ground-state energy, but configuration B is still excited (fig. 2). The excitation of configuration B is caused by $2\hbar\omega$ excitation of the protons. In configuration B the sd-shell of protons is fully occupied in the direction of the α cluster, and the protons in the α cluster should be excited to the pf shell due to the Pauli blocking effect (fig. 3).

Figure 4 shows the level scheme of the positive-parity states in ^{42}Ca up to the $J^\pi = 8^+$ states obtained via the AMP and the GCM. The GCM bases are deformed-structure wavefunctions with configurations of $2p$, $4p2h$, $6p4h$ and $8p6h$ obtained via energy variation with the β and HO quanta constraints, and α–^{38}Ar cluster structure wavefunctions obtained via energy variation with the α–^{38}Ar intercluster distance constrained to a maximum of 8.5 fm. Convergence of the GCM calculation was confirmed by a comparison with a restricted set of basis wavefunctions (the energies of states listed in fig. 4 change by less than 0.25 MeV when the number of basis wavefunctions is halved). Three $K^\pi = 0^+$ rotational bands coexist in the excited states, labeled as ND1, ND2, and SD. The dominant components of these states have $6p4h$, $4p2h$, and $8p6h$ configurations, respectively, and the quadrupole deformation parameters of their dominant components are $\beta = 0.40$, 0.28 and 0.53, respectively. The ground-state (GS) band has a $2p$ configuration. The theoretical level spacings of the GS band are underestimated, although the GS band is considered to have a simple $[(f_{7/2})^2]_\nu$ structure. The present framework does not

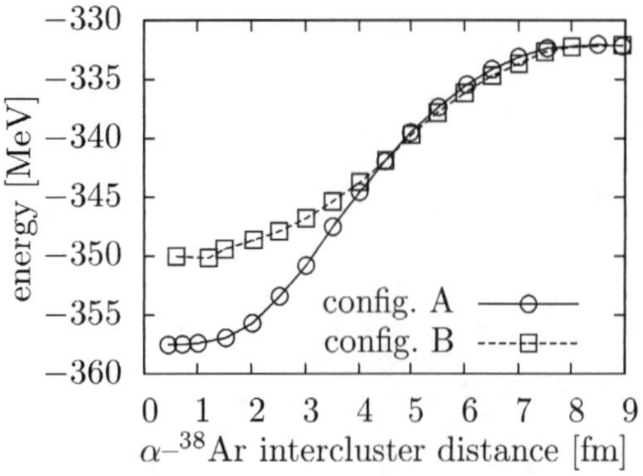

Figure 2. Energy of the α–^{38}Ar wavefunction as the function of the intercluster distance between the α and ^{38}Ar clusters. Solid and dotted lines show configurations A and B, respectively (see text).

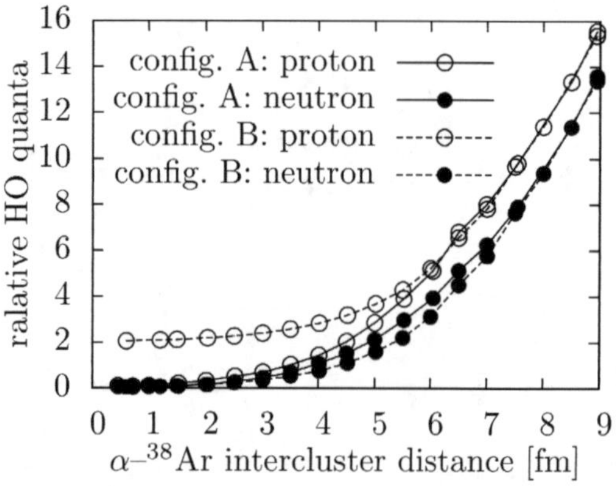

Figure 3. Particle-hole configurations of protons and neutrons relative to the lowest allowed state as the function of the intercluster distance between the α and ^{38}Ar clusters. Solid and dotted lines show configurations A and B, respectively (see text), while open and closed symbols indicate protons and neutrons.

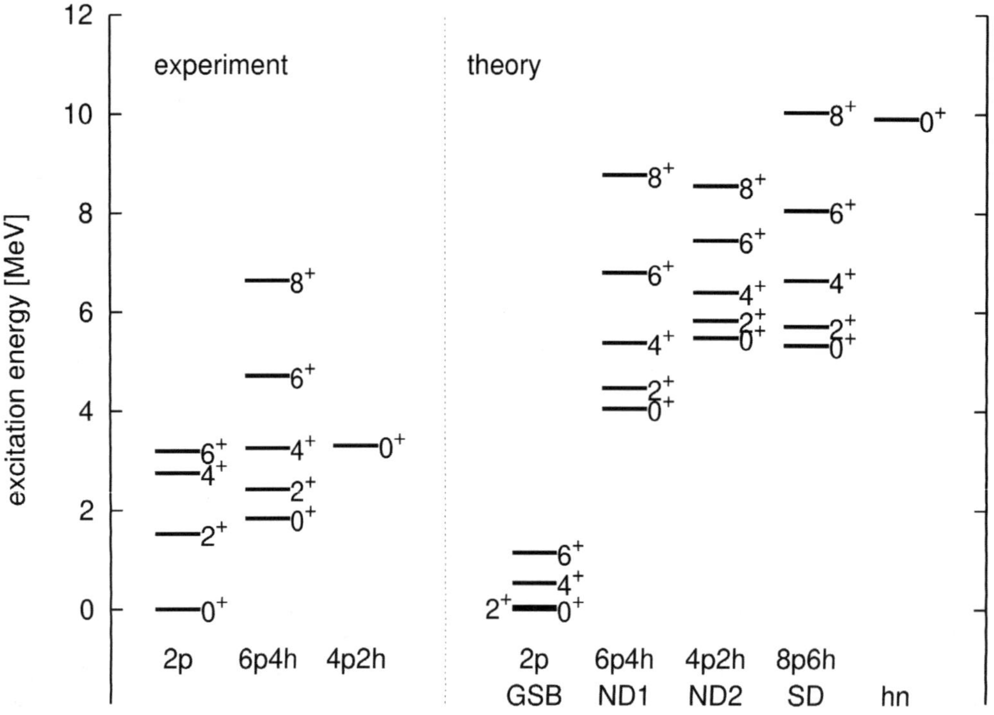

Figure 4. The experimental and theoretical level schemes in ^{42}Ca.

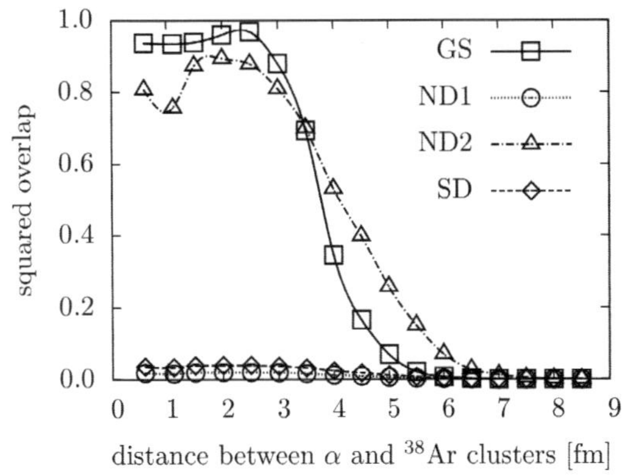

Figure 5. Squared overlaps of the $J^\pi = 0^+_{\rm GS}$ (solid), $0^+_{\rm ND1}$ (dotted), $0^+_{\rm ND2}$ (dot-dashed), and $0^+_{\rm SD}$ (dashed) states as a function of the distance between the α and ^{38}Ar clusters.

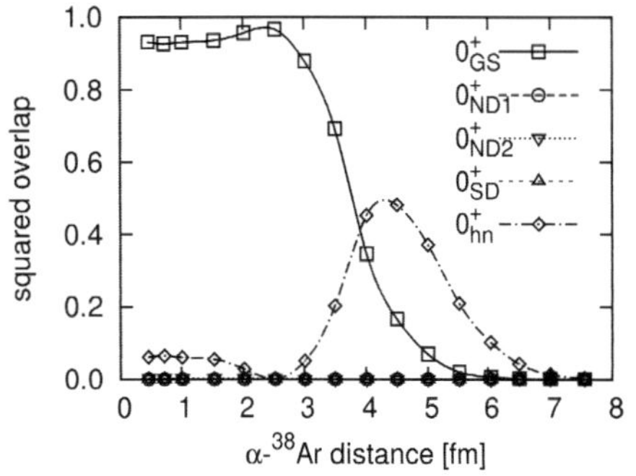

Figure 6. Squared overlaps of the configuration A α–^{38}Ar wavefunctions for the $J^\pi = 0^+_{\rm GS}$ (solid), $0^+_{\rm ND1}$ (dotted), $0^+_{\rm ND2}$ (dot-dashed), and $0^+_{\rm SD}$ (dashed) states as a function of the distance between the α and ^{38}Ar clusters.

include a time-reversal symmetry assumption, and the 2^+ and 4^+ states contain low-energy magnetized components, which leads to the underestimation. In the highly excited region a 0^+ state that contains large amount of α–^{38}Ar cluster structure components is obtained, which has α–^{38}Ar higher-nodal (hn) structure as discussed below.

To analyze the α–^{38}Ar cluster structure correlations in the low-lying rotational bands, squared overlaps of the band-head states and α–^{38}Ar cluster structure components were calculated, as shown in fig. 5. The squared overlaps of the $J^\pi = 0^+_{\rm GS}$, $0^+_{\rm ND1}$, $0^+_{\rm ND2}$, and $0^+_{\rm SD}$ states, which are band heads of the GS, ND1, ND2, and SD bands, respectively, are shown with the space spanned by α–^{38}Ar cluster structure wave functions obtained via energy variation with the α–^{38}Ar cluster distance constraint. In the calculations, two types of α–^{38}Ar cluster structure wavefunctions are obtained that differ in the orientation of the ^{38}Ar clusters. The $J^\pi = 0^+_{\rm GS}$ and $0^+_{\rm ND2}$ states have large amount of α–^{38}Ar cluster structure components at large intercluster distances as well as at small distances. The $J^\pi = 0^+_{\rm ND1}$ and $0^+_{\rm SD}$ states have small amount of α–^{38}Ar cluster structure components for any intercluster distance.

B(E2) values of transitions within the theoretical ND1, ND2 and SD bands, as well as within the experimental $K^\pi = 0^+_2$ band are listed in table 1 in Weisskopf units. The B(E2) values of ND1 and ND2 are similar, while those of the SD band are much larger. The theoretical values for the ND1 and ND2 bands are also similar to the experimental values for the $K^\pi = 0^+_2$ band.

The amount of α–^{38}Ar cluster components (fig. 5) and the in-band transition B(E2) values (table 1) indicate that the ND1 band and the band head of the ND2 band correspond to the experimental $K^\pi = 0^+_2$ band and $J^\pi = 0^+_3$ state, respectively. The large amount of α–^{38}Ar cluster components in the $J^\pi = 0^+_{\rm ND2}$ state reveals that this state corresponds to the experimental $J^\pi = 0^+_3$ state because of strong population to the $J^\pi = 0^+_3$ state by α-transfer reactions to ^{38}Ar [10], which is sensitive to the α–^{38}Ar cluster structure components. The ND1 states have small amount of α–^{38}Ar cluster structure components and similar in-band B(E2) values to those of the experimental $K^\pi = 0^+_2$ band, which indicates that the ND1 band corresponds to the experimental $K^\pi = 0^+_2$ band. The particle-hole configurations of the ND1 ($6p4h$) and ND2 ($4p2h$) bands are consistent with those of the $J^\pi = 0^+_2$ and 0^+_3 states, respectively, as suggested by the results of an α-transfer experiment [10]. The members of the ND2 band apart from the band head and

Table 1. Theoretical and experimental B(E2) values in Weisskopf units (I_i and I_f indicate initial and final states, respectively). Experimental values are taken from refs. [8] and [15].

	I_i	I_f	B(E2)
theory	$2^+_{\rm ND1}$	$0^+_{\rm ND1}$	28.55
	$4^+_{\rm ND1}$	$2^+_{\rm ND1}$	33.12
	$6^+_{\rm ND1}$	$4^+_{\rm ND1}$	38.45
	$2^+_{\rm ND2}$	$0^+_{\rm ND2}$	29.02
	$4^+_{\rm ND2}$	$2^+_{\rm ND2}$	24.70
	$6^+_{\rm ND2}$	$4^+_{\rm ND2}$	24.61
	$2^+_{\rm SD}$	$0^+_{\rm SD}$	82.12
	$4^+_{\rm SD}$	$2^+_{\rm SD}$	107.99
	$6^+_{\rm SD}$	$4^+_{\rm SD}$	130.18
experiment	4^+_2	2^+_2	57 ± 42
($K^\pi = 0^+_2$ band)	6^+_2	4^+_2	50^{+35}_{-16}

Table 2. Theoretical and experimental values of electric monopole strengths $M(\mathrm{E0})_{\rm th}$ and $M(\mathrm{E0})_{\rm exp}$, respectively, from the ground state. The unit is in e fm^2. Theoretical values are taken from Ref. [15].

		in e fm^2
transition	$M(E0)_{\rm th}$	$M(E0)_{\rm exp}$
$0^+_{\rm ND1}$	0.09	5.24 ± 0.39
$0^+_{\rm ND2}$	0.52	
$0^+_{\rm SD}$	0.11	
$0^+_{\rm hn}$	4.48	

those of the SD band have not been observed. The members of the ND2 band could possibly be observed by a combination of α-transfer reactions and γ-spectroscopy experiments because the ND2 band contains large amount of α–^{38}Ar cluster structure components and has large in-band B(E2) values.

Table 2 shows E0 transition strengths [M(E0)] from the ground state. The $0^+_{\rm hn}$ has large E0 value compared with other states. It is because the $0^+_{\rm hn}$ state and the ground state contain large amount of the configuration A α–^{38}Ar cluster structure components. Figure 6 shows squared overlaps of the configuration A α–^{38}Ar cluster wavefunctions as a function of the intercluster distance. The ground state and the $0^+_{\rm hn}$ state contains large amount of the configuration A α–^{38}Ar cluster structure components with small and large intercluster distance, respectively. It shows that the $0^+_{\rm hn}$ state has a α–^{38}Ar higher-nodal structure. The α–^{38}Ar cluster structures in the ground state and the $0^+_{\rm hn}$ state enhance E0 transition strengths because E0 transitions excite intercluster motion [6, 7].

The M(E0) of the $0^+_{\rm ND1}$ state underestimates the experimental data. A possible reason is that cluster structure components other than the α–^{38}Ar cluster structure are insufficient in the present calculation. The structure of α + ^{36}Ar with two neutrons in molecular orbitals is a candidate cluster structure coupled in the $0^+_{\rm ND1}$ state. The structure of α + ^{36}Ar with two neutrons in molecular orbitals probably contains α–^{38}Ar cluster structure components, and the

$0^+_{\rm ND1}$ state is also populated by the α-transfer reactions to ^{38}Ar, although the population is weaker than that of the ND2 states.

4. Conclusions

The structures of the deformed states in ^{42}Ca have been investigated using deformed-basis AMD and the GCM by focusing on the coexistence of various rotational bands with mp-mh configurations and α–^{38}Ar clustering. In the excited states, three $K^\pi = 0^+$ bands, ND1, ND2 and SD were obtained, which have dominant $6p4h$, $4p2h$, and $8p6h$ configurations, respectively. The ND1 band corresponds to the experimental $K^\pi = 0^+_2$ band, while the ND2 and SD bands have not yet been observed. The band head of the ND2 band corresponds to the experimental $J^\pi = 0^+_3$ state. The B(E2) values of the in-band transitions of the ND1 band are consistent with experimental data. The members of the GS and ND2 bands contain α–^{38}Ar cluster structure components, which is consistent with results that show the $J^\pi = 0^+_1$ and 0^+_3 states are strongly populated by ^{38}Ar(^{6}Li, d) reactions. Particle-hole configurations of the dominant components of the GS and ND2 bands are consistent with the suggestions of the α-transfer to ^{38}Ar experiment. The α–^{38}Ar higher-nodal state is also obtained, which has large E0 strength from the ground state. It is necessary to consider both clustering and various deformations with mp-mh configurations for understanding the structures in ^{42}Ca.

Acknowledgments

This work was supported by the University of Tsukuba Research Infrastructure Support Program. Numerical calculations were conducted on the RIKEN Integrated Cluster of Clusters (RICC) and the "Interdisciplinary Computational Science Program" in Center for Computational Sciences, University of Tsukuba.

References

[1] Michel F, Ohkubo S and Reidemeister G 1998 *Prog. Theor. Phys. Suppl.* **132** 7 and references therein
[2] Yamaya T, Katori K, Fujiwara M, Kato S and Ohkubo S 1998 *Prog. Theor. Phys. Suppl.* **132** 73 and references therein
[3] Sakuda T and Ohkubo S 1998 *Prog. Theor. Phys. Suppl.* **132** 103 and references therein
[4] Taniguchi Y, Kimura M, Kanada-En'yo Y and Horiuchi H 2007 *Phys. Rev.* C **76** 044317
[5] Kimura M and Horiuchi H 2006 *Nucl. Phys.* A **767** 58
[6] Kawabata T *et al* 2007 *Phys. Lett.* B **646** 6
[7] Yamada T, Funaki Y, Horiuchi H, Ikeda K and Tohsaki A 2008 *Prog. Theor. Phys.* **120** 1139
[8] Betz P, Bitterwolf E, Busshardt B and Röpke H 1976 *Z. Phys.* A **276** 295
[9] Lach M *et al* 2003 *Eur. Phys. J.* A **16** 309
[10] Fortune H, Betts R, Bishop J, AL-Jadir M and Middleton R 1978 *Nucl. Phys.* A **294** 208
[11] Sakuda T and Ohkubo S 1995 *Phys. Rev.* C **51** 586
[12] Taniguchi Y, Kimura M and Horiuchi H 2004 *Prog. Theor. Phys.* **112** 475
[13] Kanada-En'yo Y and Horiuchi H 1995 *Prog. Theor. Phys.* **93** 115
[14] Kimura M 2004 *Phys. Rev.* C **69** 044319
[15] Cameron J A and Singh B 2004 *Nuclear Data Sheets* **102** 293

Journal of Physics: Conference Series **436** (2013) 012056 doi:10.1088/1742-6596/436/1/012056

Heavy-particle radioactivity

D N Poenaru[1,2], R A Gherghescu[1,2] and W Greiner[2]

[1]Horia Hulubei National Institute for Physics and Nuclear Engineering
PO Box MG-6, 077125 Bucharest-Magurele, Romania
[2]Frankfurt Institute for Advanced Studies, Ruth-Moufang-Strasse 1, 60438 Frankfurt am Main, Germany

E-mail: D.Poenaru@fias.uni-frankfurt.de

Abstract. The competition of heavy particle radioactivity (HPR) and α decay is investigated in the region of superheavy (SH) nuclei with atomic numbers $Z = 104-124$. Calculations of half-lives within analytical superasymmetrical fission (ASAF) model are performed by using different theoretical mass tables to determine Q, the energy released. For α decay the calculations are made using semi-empirical Fission (semFIS) model and AME11 mass table. A trend toward shorter half-lives and larger branching ratios relative to alpha decay for heavier SHs is observed.

1. Introduction

Superheavy (SH) elements with atomic numbers $Z = 104 - 118$ (see the review papers [1, 2] and the presentations at this Conference by Ch. E. Düllmann [3] and N. Schunck [4]) have been synthesized with cold fusion reactions [5, 6] or with hot fusion induced by ^{48}Ca projectiles [7, 8]. The names and symbols for $Z = 104-112, 114, 116$ are: Rf, Db, Sg, Bh, Hs, Mt, Ds, Rg, Cn, Fl, and Lv, respectively. The elements 113, 115, 117, and 118 are still waiting to be named. Many of them are identified through α decay chains. Previously we also discussed the competition of α decay and heavy particle radioactivity (HPR) [9], which may be important [10, 11] in the region of the heaviest SHs. In this process, from one parent nucleus ^AZ, one obtains an emitted particle $^{A_e}Z_e$, and a daughter $^{A_d}Z_d$:

$$^AZ \rightarrow {}^{A_e}Z_e + {}^{A_d}Z_d \tag{1}$$

Alternative theories are available [12, 13, 14, 15, 16]. A universal decay law based on R-matrix theory was recently introduced for α emission and HPR [17].

Starting with 1984 [18] many HPR have been experimentally confirmed [19, 20] in heavy parent nuclei with $Z = 87$ to 96: ^{14}C, ^{20}O, ^{23}F, $^{22,24-26}$Ne, ^{28,30}Mg, and ^{32,34}Si. The measured half-lives are in good agreement with predicted values within the ASAF model (see the review [21] and references therein). The largest branching ratio relative to α decay, $b_\alpha = T_\alpha/T_c$, of $10^{-8.9}$ was observed for ^{14}C radioactivity of ^{223}Ra. Usually in this region of the nuclear chart HPR is a rare process in a huge background of α particles.

The measurable quantities are: the kinetic energy of the emitted cluster $E_k = Q_cA_d/A$ or the released energy Q_c and the half-life T_c or the branching ratio $b_\alpha = T_\alpha/T_c$. With the existing techniques it is possible to measure half-lives shorter than the limit $T_c < 10^{32}$ s and branching ratios large enough $b_\alpha > 10^{-17}$. In the present work we give new details concerning the use of different mass tables to estimate the competition of α decay and HPR.

2. Calculations

A strong shell effect of the doubly magic daughter $^{208}_{82}Pb_{126}$ was observed. In order to study his importance in the region of SHs with $Z > 110$ we changed the concept of HPR, previously [22] associated with a maximum $Z_e^{max}|_{old} = 28$. Now we allow

$$Z_e^{max} = Z - 82 \tag{2}$$

The accuracy of half-life calculation is essentially dependent on the precision with which we know the Q-value

$$Q = [M - (M_e + M_d)]c^2 \tag{3}$$

obtained as a difference between the parent, M, and the two decay product masses, M_e and M_d, in units of energy; c is the light velocity. Q-value should be positive in order to allow a spontaneous decay.

The decay constant

$$\lambda = \ln 2/T_c = \nu S P_s \tag{4}$$

is expressed by a product of three model dependent quantities ν, S and P_s where ν is the frequency of assaults on the barrier per second, S is the preformation probability and P_s is penetrability of external barrier. According to our method [23] the preformation in a fission theory is given by the penetrability of the internal part (overlapping stage of the two fragments) of the barrier.

We developed our ASAF model starting with the Myers-Swiatecki liquid drop model [25] adjusted with a phenomenological correction. The half-life is given by

$$T = [(h \ln 2)/(2E_v)]exp(K_{ov} + K_s) \tag{5}$$

It is calculated by using the WKB quasiclassical approximation

$$K = \frac{2}{\hbar}\int_{R_a}^{R_b} \sqrt{2B(R)E(R)}dR \tag{6}$$

with $B = \mu$ (the reduced mass), $K = K_{ov} + K_s$, and $E(R)$ replaced by $[E(R) - E_{corr}] - Q$ where E_{corr} is a correction energy similar to the Strutinsky shell correction. The turning points of

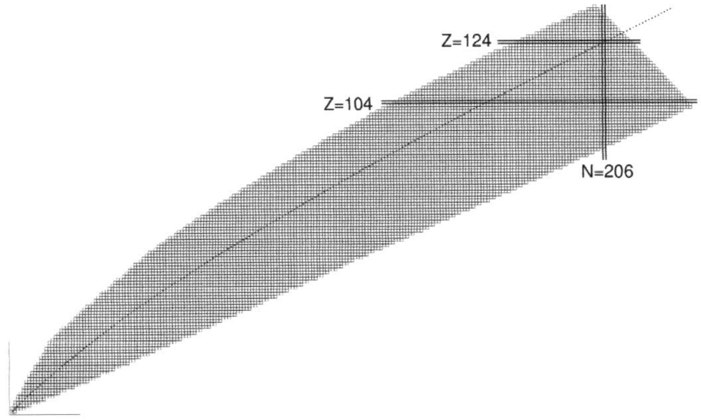

Figure 1. Chart of nuclides for which calculated masses are available according to the MySw94 mass tables. 8824 masses for $Z = 1 - 136, N \leq 236$. The Green approximation of the beta stability is marked by full squares.

the WKB integral are: $R_a = R_i + (R_t - R_i)[(E_v + E^*)/E_b^0]^{1/2}$ and $R_b = R_t E_c\{1/2 + [1/4 + (Q + E_v + E^*)E_l/E_c^2]^{1/2}\}/(Q + E_v + E^*)$ where E^* is the excitation energy concentrated in the

Journal of Physics: Conference Series **436** (2013) 012056 doi:10.1088/1742-6596/436/1/012056

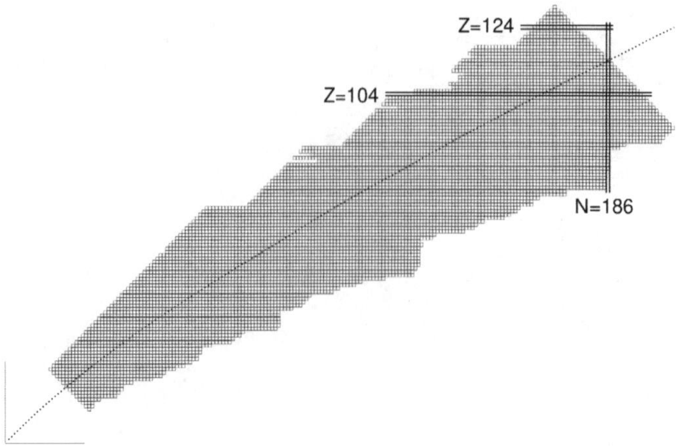

Figure 2. Chart of nuclides for which calculated masses are available according to the APDT95 mass tables. 7508 masses for $Z = 10 - 130, N \leq 206$. The Green approximation of the beta stability is marked by full squares.

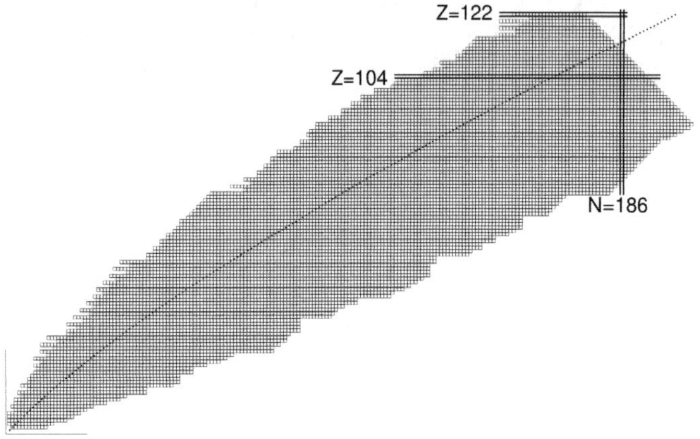

Figure 3. Chart of nuclides for which calculated masses are available according to the DuZu96 mass tables. 9321 masses for $Z = 2 - 122, N \leq 207$. The Green approximation of the beta stability is marked by full squares.

separation degree of freedom, $R_i = R_0 - R_e$ is the initial separation distance, $R_t = R_e + R_d$ is the touching point separation distance, $R_j = r_0 A_j^{1/3}$ ($j = 0, e, d$; $r_0 = 1.2249$ fm) are the radii of parent, emitted, and daughter nuclei, respectively, and $E_b^0 = E_i - Q$ is the barrier height before correction. The interaction energy at the top of the barrier, in the presence of a non-negligible angular momentum, $l\hbar$, is given by $E_i = E_c + E_l = e^2 Z_e Z_d / R_t + \hbar^2 l(l+1)/(2\mu R_t^2)$. The two terms of the action integral K, corresponding to the overlapping K_{ov} and separated K_s fragments, are calculated analytically [11, 22]. The potential barrier shape similar to that which we considered within the ASAF model was calculated by using the macroscopic-microscopic method [24].

3. Mass tables

Half-life calculations are very sensitive to the Q-values. The closest to reality are the updated table of evaluated experimental masses **AME11** [26], but many masses are still not available for new SHs in this table. We have also used calculated masses:

- Macroscopic-Microscopic: **FRDM95** (Möller, Nix, Myers, Swiatecki) [27],
- Thomas-Fermi: **MySw94** [28],
- Phenomenological/Hybrid: **KTUY05** (Koura, Tachibana, Uno, Yamada) [29],

- Hartree-Fock: **APDT95** (Aboussir, Pearson, Dutta, Tondeur) [30],
- Semiempirical Shell Model: **LiMaZe01** (Liran, Marinov, Zeldes) [31, 32],
- Shell Model: **DuZu96** (Duflo, Zuker) [33].

We show in figures 1, 2, and 3 the nuclides for which calculated masses are available according to the tables MySw94, APDT95, and DuZu96, respectively. For the region of interest $Z = 104 - 124$ the beta stability line goes through $N_g = 106$ for $Z = 104$ and $N_g = 206$ for $Z = 124$. It is clearly seen that the most complete mass table of these three is MySw94. The other two will allow us to make calculations only for some of the neutron deficient heaviest SHs.

When using calculated masses for parent and daughter nuclei we take into account the nuclides stable against one proton, two protons, one neutron and two neutrons spontaneous emissions which leads to a smaller number of parent nuclei than those shown in figures 1-3.

4. A trend toward shorter half-lives for heavy superheavies

Besides the emitted clusters with $Z_e \leq 28$ (Be, Ar, Ca, Ti, V, Cr, Mn, Fe, and Ni), many other types of new HPR with $Z_e > 28$ appear when we use the MySw94 calculated mass table to determine the Q-values: Cu, Zn, Ga, Ge, As, Se, Br, Kr, Rb, Sr, Y, Zr, Nb, and Mo (see figure 4). Many of the SH nuclides are ^{8}Be emitters, but they have a very low branching ratio b_α. Most frequently occurs the doubly magic ^{78}Ni radioactivity.

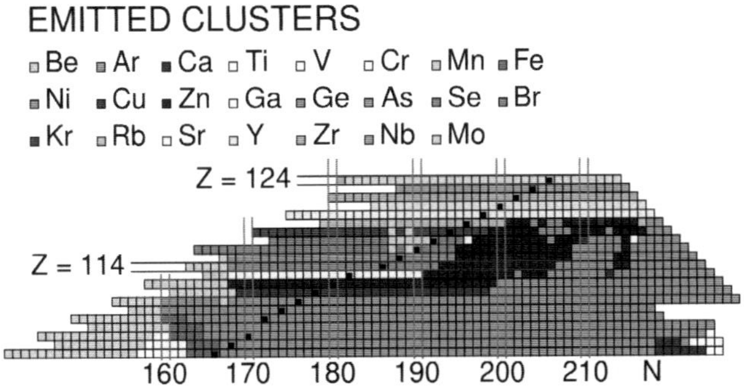

Figure 4. (color online) Chart of superheavy cluster emitters with atomic numbers $Z = 104 - 124$. The Q values are calculated using the MySw94 mass table. Black squares mark the Green approximation of the line of beta stability.

An even-odd staggering of HPR half-lives was observed [11] leading to shorter T_c for even N nuclides compared to the neighboring odd N ones. In order to avoid such a complication we consider in figure 5 for HPR only the odd N isotopes. Two important trends are observed: (1) both T_c and T_α are shorter for SHs with larger atomic number, and (2) for some of the isotopes of $Z = 122, 123$ and $Z = 124$ elements cluster decay half-life may be shorter than that of α decay: $T_c < T_\alpha$ (or the branching ratio $b_\alpha = T_\alpha/T_c$ becomes larger than unity when the atomic number of the parent nucleus increases over 121). Very large values of $\log_{10} T_\alpha$ (over 13) occurring at $N = 195$ for $Z = 121 - 124$ and at $N = 193, 195$ for $Z = 124$, due to very low value of Q_α have been removed from the plot.

We observe in table 1 that large differences in Q-values for HPR calculated with various mass tables occur very frequently and make an important contribution to the broad range of branching ratios b_α for any particular nucleus. In this table we kept for α decay the Q_α-value determined using AME11 experimental masses and the half-life T_α calculated within semFIS model. The semFIS calculations are in good agreement with experimental values for $\log_{10} T_\alpha$ of ^{292}116, ^{293}117, ^{294}118 which are −1.74 s, −1.84 s, and −3.05 s, respectively. By taking as the optimum AME11, the closest values of Q_c (and consequently of T_c, and b_α) are obtained when we use the KTUY05 mass table. The most optimistic results are based on the LiMaZe01 table and the most pessimistic on the FRDM95 masses. The sensitivity of half-life values to a

Journal of Physics: Conference Series **436** (2013) 012056 doi:10.1088/1742-6596/436/1/012056

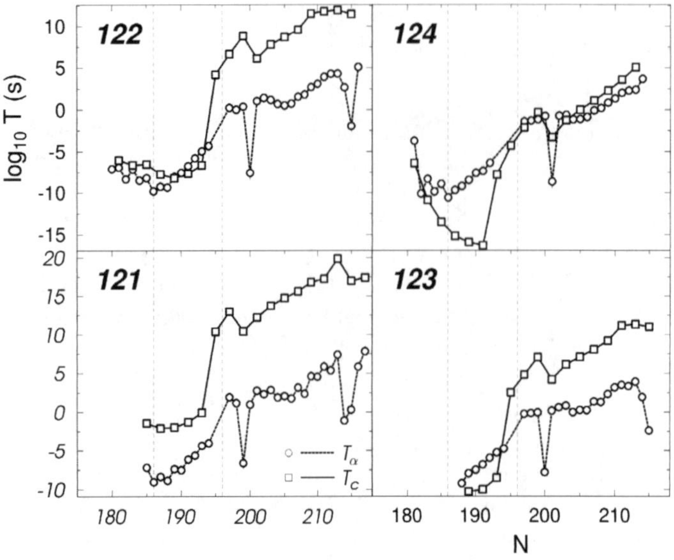

Figure 5. Decimal logarithm of the half-lives of superheavy nuclei with atomic numbers 121 − 124 against α decay (open circles) and HPR for odd-neutron isotopes (open squares) versus the neutron number of the parent nucleus. Q values are calculated using the MySw94 mass tables. Vertical lines correspond to $N = 186, 196$.

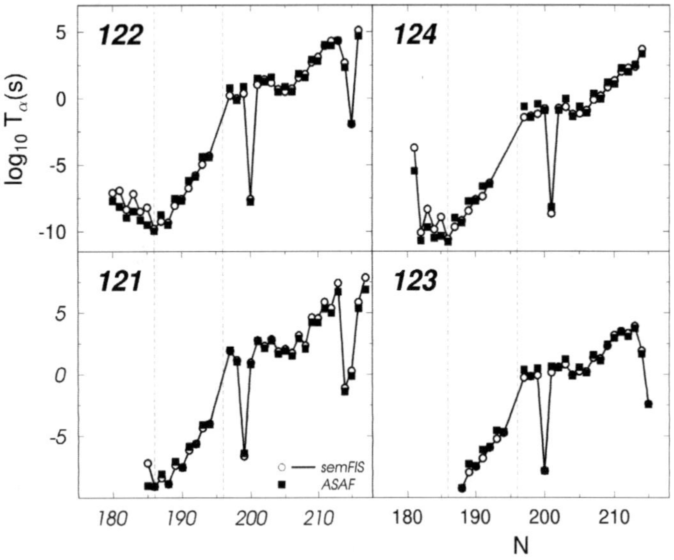

Figure 6. Comparison of alpha-decay half-lives calculated with ASAF (full squares) and semFIS models vs. the neutron number of the parent nucleus. Both even and odd N values are considered. Vertical lines correspond to $N = 186, 196$. Q-values are calculated using the MySw94 mass tables.

small variation of the Q_c value is also evident. Let us take for example the case of $^{296}Lv_{116}$: the decrease of Q_c with 1.17 MeV from 284.90 (AME11) to 283.73 (APDT95) leading to an increase of 1.32 orders of magnitude in T_c, whereas a decrease with 4.19 MeV (FRDM95) gives 4.72 orders of magnitude increase in T_c.

More elaborate models should be used (see, e.g., [34, 4]) in order to estimate the competition of spontaneous fission.

An estimation of the accuracy gives the standard rms deviation of $\log_{10} T$ values:

$$\sigma = \left\{ \sum_{i=1}^{n} [\log_{10}(T_i/T_{exp})]^2/(n-1) \right\}^{1/2} \quad (7)$$

Table 1. Qvalues in MeV, half-lives and branching ratios for the most probable HPR of SH nuclei with $Z = 116, 117, 118$, $N = 176$ obtained by using seven mass tables: AME11, DuZu96, LiMaZe01, APDT95, KTUY05, MySw94, and FRDM95.

Parent	Mass table	Q_α	$\log_{10} T_\alpha(s)$	Emitted cluster	Q_c	$\log_{10} T_c(s)$	$\log_{10} b_\alpha$
$^{292}116$	AME11	10.809	-1.55	^{84}Se	284.90	0.53	-2.08
	DuZu96				281.09	4.82	-6.37
	LiMaZe01				288.20	-3.21	1.66
	APDT95				283.73	1.85	-3.40
	KTUY05				284.39	1.10	-2.66
	MySw94				282.02	3.77	-5.32
	FRDM95				280.71	5.25	-6.80
$^{293}117$	AME11	11.183	-2.03	^{85}Br	293.73	1.16	-3.19
	DuZu96				290.40	4.90	-6.93
	LiMaZe01				296.73	-2.24	0.21
	APDT95				292.57	2.47	-4.50
	KTUY05				293.52	1.39	-3.42
	MySw94				291.58	3.57	-5.60
	FRDM95				290.04	5.31	-7.34
$^{294}118$	AME11	11.811	-3.31	^{86}Kr	303.69	-2.01	-1.30
	DuZu96				300.22	1.86	-5.17
	LiMaZe01				306.27	-4.91	1.60
	APDT95				302.23	-0.39	-2.93
	KTUY05				303.11	-1.37	-1.94
	MySw94				302.06	-0.20	-3.12
	FRDM95				300.29	1.78	-5.09

Compared to calculations within ASAF model, lower values of σ for α decay half-lives may be obtained [35] within our UNIV (universal curve) [36] and semFIS (semiempirical) models. For 44 even-even nuclei we obtained $\sigma = 0.164$ within semFIS, $\sigma = 0.267$ within UNIV and $\sigma = 0.402$ within ASAF model. For 25 odd-odd nuclei $\sigma = 0.451, 0.456$ and 0.795, respectively.

A comparison of $\log_{10} T_{ASAF}$ with $\log_{10} T_{semFIS}$ is shown in figure 6. Very large values of $\log_{10} T_\alpha$ (over 13) occurring at $N = 195, 196$ for $Z = 121 - 124$ and at $N = 193 - 196$ for $Z = 124$, due to very low value of Q_α, have been removed from the plot. We assume that semFIS calculations are closer to reality, particularly in the vicinity of the neutron and proton magic numbers. From figure 6 one can see that odd-even staggering are more pronounced for semFIS than ASAF model for odd-N values in the neighborhood of $N = 184$ when $N \leq 186$. The agreement between ASAF and semFIS values is rather good.

In conclusion we found that calculated half-lives T_c against HPR and the branching ratios relative to α decay are showing a trend toward shorter T_c and larger b_α for the heaviest SHs. In the vicinity of neutron magic number $N = 184$ the half-lives for α decay calculated within ASAF model are shorter than those determined with semFIS which takes into account the influence of closed shells. There is a need of extending atomic mass models in the region of heaviest neutron-rich superheavy nuclides. The accuracy of calculated masses in the region of heaviest SHs should be improved in order to make reliable predictions of half-lives.

Acknowledgments
This work was partially supported within the IDEI Programme under Contracts No. 43/05.10.2011 and No. 42/05.10.2011 with UEFISCDI, Bucharest.

References
[1] Sobiczewski A 2011 *Radiochimica Acta* **99** 395
[2] Greiner W and Poenaru D N 2010 *Cluster Structure of Atomic Nuclei* ed Brenner M (Research Signpost, Trivandrum, India) chap 5, pp 119
[3] Düllmann C E 2012 Search for superheavy elements at GSI, Invited talk at this Conference
[4] Schunck N 2012 Density functional theory approaches to fission, Talk at this Conference
[5] Hofmann S 2011 *Radiochim. Acta* **99** 405
[6] Morita K *et al* 2007 *J. Phys. Soc. Jpn.* **76** 045001
[7] Oganessian Y T 2011 *Radiochimica Acta* **99** 429
[8] Düllmann C E *et al* 2010 *Phys. Rev. Lett.* **104** 252701
[9] 2011 Encyclopaedia Britannica Online. Web http://www.britannica.com/EBchecked/topic/465998/
[10] Poenaru D N, Gherghescu R A and Greiner W 2011 *Phys. Rev. Lett.* **107** 062503
[11] Poenaru D N, Gherghescu R A and Greiner W 2012 *Phys. Rev.* C **85** 034615
[12] Blendowske R, Fliessbach T and Walliser H 1987 *Nucl. Phys.* A **464** 75
[13] Lovas R G, Liotta R J, Insolia A, Varga K and Delion D S 1998 *Phys. Rep.* **294** 265
[14] Adamian G G, Antonenko N V and Zubov A S 2005 *Phys. Rev.* C **71** 034603
[15] Kuklin S, Adamian G and Antonenko N 2005 *Phys. Rev.* C **71** 014301
[16] Denisov V Y and Khudenko A A 2009 *Atomic Data Nucl. Data Tables* **95** 815
[17] Qi C, Xu F R, Liotta R J and Wyss R 2009 *Phys. Rev. Lett.* **103** 072501
[18] Rose H J and Jones G A 1984 *Nature* **307** 245
[19] Bonetti R and Guglielmetti A 2007 *Rom. Rep. Phys.* **59** 301
[20] Poenaru D N, Nagame Y, Gherghescu R A and Greiner W 2002 *Phys. Rev.* C **65** 054308; Erratum: C **66** 049902
[21] Poenaru D N and Greiner W 2010 *Clusters in Nuclei Vol. 1. Lecture Notes in Physics* **818** ed Beck C (Springer:Berlin) chap 1, pp 1
[22] Poenaru D N, Schnabel D, Greiner W, Mazilu D and Gherghescu R 1991 *Atomic Data Nucl. Data Tables* **48** 231
[23] Poenaru D N and Greiner W 1991 *Physica Scripta* **44** 427
[24] Poenaru D N, Gherghescu R A and Greiner W 2006 *Phys. Rev.* C **73** 014608
[25] Myers W D and Swiatecki W J 1966 *Nucl. Phys.* A **81** 1
[26] Audi G and Meng Wang 2011 Private communication
[27] Möller P, Nix J R, Myers W D and Swiatecki W J 1995 *Atomic Data Nucl. Data Tables* **59** 185
[28] Myers W D and Swiatecki W J 1994 Table of nuclear masses according to the 1994 Thomas-Fermi model, Report LBL-36803, Lawrence Berkeley Laboratory
[29] Koura H, Tachibana T, Uno M and Yamada M 2005 *Prog. Theor. Phys.* **113** 305
[30] Aboussir Y, Pearson J M, Dutta A K and Tondeur F 1995 *Atomic Data Nucl. Data Tables* **61** 127
[31] Liran S, Marinov A and Zeldes N 2000 *Phys. Rev.* C **62** 047301 (*Preprint* `nucl-th/0006045`, `nucl-th/0102055v1`)
[32] Liran S, Marinov A and Zeldes N 2002 *Phys. Rev.* C **66** 024303 (*Preprint* `nucl-th/0110010`)
[33] Duflo J and Zuker A P 1995 *Phys. Rev.* C **52** R23 and Private Communication to AMDC, February 1996
[34] Smolanczuk R, Skalski J and Sobiczewski A 1995 *Phys. Rev.* C **52** 1871–1880
[35] Poenaru D N, Plonski I H and Greiner W 2006 *Phys. Rev.* C **74** 014312
[36] Poenaru D N, Gherghescu R A and Greiner W 2012 *J. Phys.* G **39** 015105

Journal of Physics: Conference Series **436** (2013) 012057 doi:10.1088/1742-6596/436/1/012057

Two-proton radioactivity: 10 years of experimental progresses

J Giovinazzo[1], P Ascher[1*], L Audirac[1†], B Blank[1], C Borcea[2], G Canchel[1], C E Demonchy[1], C Dossat[1], S Grvy[1,4], S Leblanc[1], I Matea[3], F de Oliveira-Santos[4], J L Pedroza[1], L Perrot[3], J Pibernat[1], L Serani[1], C Stodel[4] and J C Thomas[3]

[1] Centre d'Etudes Nucléaires de Bordeaux-Gradignan, CNRS/IN2P3 Univ. Bordeaux1, Le Haut Vigneau BP 120, 33175 Gradignan, France

[2] National Institute for Physics and Nuclear Science, P.O. Box MG6, Bucharest-Margurele, Romania

[3] Institut de Physique Nucléaire, CNRS/IN2P3 - Univ. Paris Sud, 15, rue G. Clmenceau, F-91406 Orsay, France

[4] Grand Accélérateur National d'Ions Lourds, CEA/DSM-CNRS/IN2P3, BP 55027, F-14076 Caen cedex 5, France

E-mail: giovinaz@cenbg.in2p3.fr

Abstract. The two-proton radioactivity has been observed experimentally in 2002, at projectile fragmentation facilities, more than 40 years after the first theoretical prediction of this process. First observations were indirect measurements, using standard silicon detector devices. Since then, a new generation of experiments allowed for a direct observation, and opened the field of more detailed studies, using tracking devices for the detection of the emitted protons.

1. Introduction

In the early 60s, Goldanskii proposed a first description of two decay processes at the proton drip-line: the 1-proton and the 2-proton radioactive decays [1], for odd- and even-Z nuclei respectively. These decay modes may occur for very proton-rich isotopes, where the nuclear strong interaction should not bind the last protons anymore. The existence of such isotopes is due to the Coulomb barrier, and in the case of the 2-proton decay, to the extra binding energy provided by the pairing of the last protons.

A first description of the emission process was proposed, based on the tunnelling of the 2-proton sub-system, that breaks-up into 2 individual protons in the final state since the diproton system is unbound. In addition to this unique quantum mechanics process, the 2-proton radioactivity studies allow for the characterisation of the masses and the structure of isotopes beyond the proton drip-line. The best candidates for this new decay mode were predicted in A=50 mass region, which was confirmed by the mass models in the 90's [2].

Experimentally, the 2-proton radioactivity has been evidenced in the decay of ^{45}Fe in two independent experiments in 2002, and in the decay of ^{54}Zn in 2005, using standard implantation-decay technique in silicon detectors. After these experiments, an effort has been made in order to

* Max Planck Institut für Kernphysik, Saupfercheckweg 1, D-60117 Heidelberg, Germany † CEA Saclay, DSM/Irfu/SPhN, Orme des Merisiers, F-91191 Gif-sur-Yvette, France

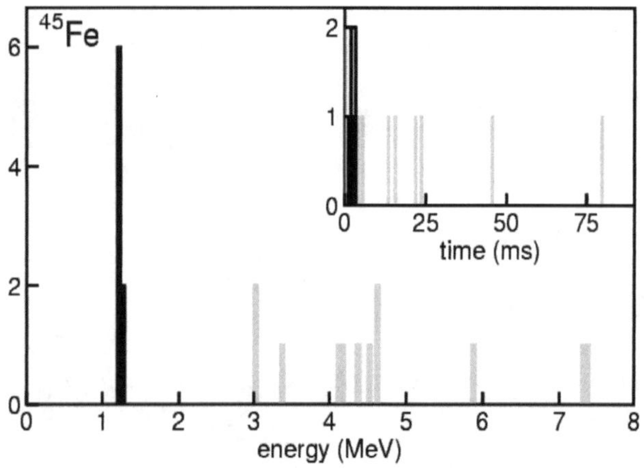

Figure 1. Energy and time distribution of decay events in the implantation silicon detector after implantation and identification of a ^{45}Fe (left) and ^{54}Zn (right) implantation. The events represented in black correspond to the 2-proton decay peak, while the grey ones are beta-decay events (most of them being the second decay after implantation, i.e. the daughter decay).

develop detection tools that allows to observe and characterise individually the emitted protons, in order to study in more details the decay process.

2. The standard silicon experiments

The first evidence of the 2-proton radioactivity could be obtained from indirect measurements, in the decay of ^{45}Fe, in projectile fragmentation experiments at GANIL [3] and GSI [4]. The ^{45}Fe nuclei, identified by standard energy loss and time-of-flight technique, were implanted in a silicon detector. The decay occur from the implantation position. From estimated masses, there are 2 possibilities for the decay: either the 2-proton decay, either beta delayed proton(s) emission. Protons up to few MeV are stopped in the detector, leaving their full energy, while beta particles (positrons) may escape it and only deposit a variable fraction of their energy.

In the GANIL experiment, the 2-proton radioactivity could be concluded from several observations. A narrow peak around 1.1 MeV was observed in the silicon detector energy distribution of the decay events after ^{45}Fe implantation. The implantation detector was surrounded in a close geometry by other silicon detectors, where no beta was observed in coincidence with the peak. The second decay events (after the events from the peak), are only compatible with ^{43}Cr decay, which is the ^{45}Fe daughter for 2-proton emission. This decay process was thus the only decay mode consistent with the observations. The GSI experiment leaded to the same conclusion, from the absence of positron annihilation gamma-rays that should have been observed in case of beta decay.

Using the same type of measurement, another experiment has been performed at GANIL, were the 2-proton decay of ^{54}Zn could be observed [5] and the previous results for ^{45}Fe could be confirmed [6]. The time and energy distribution of the corresponding decay events are shown in Figure 1. In addition, a first indication that ^{48}Ni could be a 2-proton emitter was proposed.

Those first experiments allowed to measure the global characteristics of the decay process: the transition energy Q_{2p}, the half-life $T_{1/2}$ and the branching ratio of the 2-proton decay (in competition with beta decay). The results could be compared with available theoretical descriptions from R-matrix formalism [7], from shell model embedded in the continuum (SMEC)

Journal of Physics: Conference Series **436** (2013) 012057 doi:10.1088/1742-6596/436/1/012057

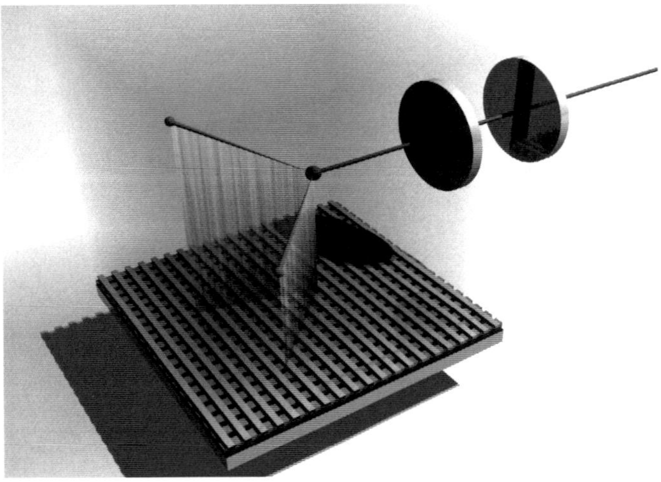

Figure 2. Schematic view of the TPC: the ions produced by projectile fragmentation and identified with silicon detectors, are implanted in a gas volume, where the decay occurs. When the protons are emitted, they ionise the gas, and due to a vertical electric field in the chamber, the ionisation electrons drift towards a double stripped X-Y detection plane. The energy signal on each strip is collected to provide the X and Y projections of the energy distributions. The time that charges arrive on strips gives the Z position where ionisation occurred.

[8], or from a 3-body model [9].

3. The tracking experiments

In order to get a more detailed description of the decay mechanism, in particular with comparison with the 3-body model that predicts angular and energy distributions for the emitted protons, the second generation of experiments was designed to measure the energy and the emission direction of both protons individually. Since the 2-proton ground-state emitters have half-lives in the order of few milliseconds, they can only be produced and studied in fragmentation experiments. To detect individually each emitted proton, an experimental option is to use an active implantation material.

For this purpose, a new detection device was developed, on the principle of a time-projection chamber (TPC) [10], as described in Figure 2. The purpose of this device is to reconstruct the tracks of the 2 protons, that are emitted from nuclei implanted in a gas chamber. The 2-dimension X-Y is composed of 2 sets of orthogonal strips, which collects the ionization signal from the protons along their track in the gas volume. The amplitude of the collected signal is used for the reconstruction of the energy loss distributions of the protons along the X and Y directions. The collection time on each strip is related to the drift time of the ionization electrons, and thus allows to reconstruct the vertical information associated to the previous distributions.

This TPC has been successfully used for 2 experiments performed at GANIL, the first one for the 2-proton radioactivity of ^{45}Fe [11], and the second one for ^{54}Zn [12]. An example of the tracks reconstruction of the emitted protons is presented in Figure 3, in the case of ^{45}Fe decay. The result from this experiment is the first direct observation of the 2-proton radioactivity, and confirms the previous conclusions from indirect measurements. Nevertheless, only few events have been observed and fully reconstructed in these experiments: 7 events in both cases, which considerably limits the comparison with theory.

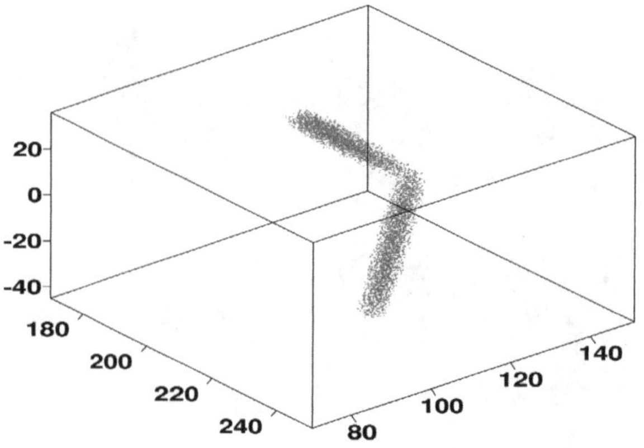

Figure 3. Example of the tracks reconstruction in 3 dimensions, for the protons emitted in the decay of ^{45}Fe. The X and Y coordinates are the strip numbers (the strips pitch is 0.4 mm), and the Z axis is expressed in the same units. The full reconstruction is used for proton-proton angle estimates.

Another TPC device, based on the same principle but with a different 2D detection technique has been developed by a Warsaw group [13]. It has been used in an experiment at MSU, where a first angular distribution of the emitted protons could be obtained for the decay of ^{45}Fe [14]. This result opens new structure studies at the drip-line, since the angular distribution is sensitive to the wave function of the parent nucleus, according to the 3-body model. In another experiment, the 2-proton radioactivity of ^{48}Ni could be established [15].

4. Experimental and instrumental perspectives

Up to now, three isotopes have been identified as ground-state 2-proton emitters: ^{45}Fe (the most studied one), ^{54}Zn and ^{48}Ni. To test our understanding of this new decay process, with comparison with theoretical descriptions, the known decays must still be characterised more precisely. In addition, we also need to extend the observation to other emitters, in order to compare experiment and theory on a wider range of nuclear configurations.

Any isotopes at the drip-line, with an even proton number, is a potential candidate for the 2-proton radioactivity. According to mass models [2], new candidates could be found for higher proton numbers, with Q_{2p} values between 1 and 2 MeV, and for which the Coulomb barrier could be responsible for a half-life in an observable range. The next isotopes that should be studied are ^{59}Ge, ^{63}Se and ^{67}Kr. An experiment has been proposed [16] to determine whether these isotopes are bound, and in such case, to search for a 2-proton radioactivity branch in their decay. This study will first be performed with standard experimental techniques (with silicon detectors).

Another perspective for the 2-proton radioactivity studies is the development of the detection tools for energy and angular correlation measurements. The TPC used for the tracking experiments suffer from various instrumental limitations, that limit the tracks reconstructions, and for some cases, even make it impossible. For the TPC used in GANIL experiments, some limitations comes from the 2D detection plane: the 2 sets of strips provide X and Y one-dimension distributions for energy and time. The 3D reconstruction has to be performed from the combination of the 1D distributions. If protons are emitted in the same X or Y directions,

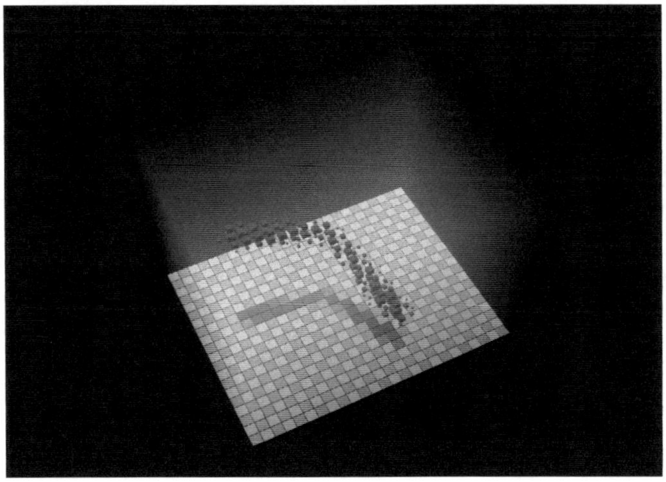

Figure 4. Schematic view of the ACTAR-TPC device: as the previous detectors, it works on the principle of a TPC. In this case, the X-Y detector for charges collection is a pad plane, which measures a digitised 2D projection of charged particles tracks. Since the signal on each pad is sampled in time, this sampling corresponds to a Z digitisation of the charges created (by ionisation) along the vertical axis above the pads. As a result, the detector should provide a 3D digitisation of the ionisation along the particles tracks. In this picture, the size of the boxes represents the amplitude of the signal in the 3D cells.

the signals from the 2 protons are mixed, and the unfolding of the contribution of each is difficult.

We started to design a new generation of TPC, which purpose is to be able to obtain a real 3D image of the ionisation of charged particles along their tracks. The 2D plane for ionisation charges collection is a pad detector (for a real 2D projection), with time sampling of the signal on each pad, to get a digitisation of the signal along the vertical axis above the pad (since the drift time is related to the vertical position where charges are created). The principle of this new detector is shown in Figure 4.

The final detector will have 16384 pads, each pad been 2x2mm2, with a micromegas [17] for amplification of the ionisation signal. A small size demonstrator is currently under construction in the context of the ACTAR-TPC collaboration [18], to test the feasibility of the detector.

The readout electronics is developed by the GET collaboration [19]. The input signal from collected charges is shaped and stored into analog memories, and when the system is trigged, the samples are digitised and the full event is build. The trigger has several levels, based on the multiplicity of pads hits. The amount of data to process is huge for an event, which will result in important dead-time, limiting the running capabilities around 1000 events per second. For the case of 2-proton radioactivity, where the decay event occurs few milliseconds after the implantation event, a special mode is requested: the implantation signal is stored in half the memory, and the system waits for the decay signals (stored in the 2nd half of the memory) before the full data is processed.

Acknowledgments
Authors would like to thank the GANIL / LISE staff and the electronics and data acquisition group for their constant support during the experiments. We are also grateful to the CERN PCB workshop for their collaboration in the technical development of the new TPC detector, to the ACTAR-TPC collaboration, and to the GET collaboration for readout electronics development.

References

[1] Goldanskii V I 1960 *Nucl. Phys.* **19** 482
[2] Brown B A 1991 *Phys. Rev.* C **43** R1513; Cole B J 1996 *Phys. Rev.* C **54** 1240; Ormand W E 1997 *Phys. Rev.* C **55** 2407
[3] Giovinazzo J *et al* 2002 *Phys. Rev. Lett.* **89** 102501
[4] Pftzner M *et al* 2002 *Eur. Phys. J.* A **14** 279
[5] Blank B *et al* 2005 *Phys. Rev. Lett.* **94** 232501
[6] Dossat C *et al* 2005 *Phys. Rev.* C **72** 054315
[7] Brown B A and Barker F C 2003 *Phys. Rev.* C **67** 041304(R)
[8] Rotureau J, Okolowicz J and Ploszajczak M 2005 *Phys. Rev. Lett.* **95** 042503
[9] Grigorenko L V and Zhukov M V 2003 *Phys. Rev.* C **68** 054005
[10] Blank B *et al* 2010 *Nucl. Instr. Meth.* A **613** 65
[11] Giovinazzo J *et al* 2007 *Phys. Rev. Lett.* **99** 102501
[12] Ascher P *et al* 2011 *Phys. Rev. Lett.* **107** 102502
[13] Miernik K *et al* 2007 *Nucl. Instr. Meth.* A **581** 194
[14] Miernik K *et al* 2007 *Phys. Rev. Lett.* **99** 192501
[15] Pomorski M *et al* 2011 *Phys. Rev.* C **83** 061303(R)
[16] Blank B *et al* 2012 RIKEN PAC proposal NP0702-RIBF04R1
[17] Giomataris Y *et al* 1996 *Nucl. Instr. Meth.* A **376** 29
[18] ACTAR-TPC, http://pro.ganil-spiral2.eu/spiral2/instrumentation/actar-tpc
[19] Pollaco E *et al* 2012 *Phys. Procedia* **37** 1799

10th International Conference on Clustering Aspects of Nuclear Structure and Dynamics IOP Publishing
Journal of Physics: Conference Series **436** (2013) 012058 doi:10.1088/1742-6596/436/1/012058

Microscopic description of induced fission

Nicolas Schunck
Physics Division, Lawrence Livermore National Laboratory, CA 94551, USA
E-mail: schunck1@llnl.gov

Abstract. Selected aspects of the description of neutron-induced fission in ^{240}Pu in the framework of the nuclear energy density functional theory at finite temperature are presented. In particular, we discuss aspects pertaining to the choice of thermodynamic state variables, the evolution of fission barriers as function of the incident neutron energy, and the temperatures of the fission fragments.

1. Introduction

Nuclear fission remains one of the most complex physics phenomena in nature. Powerful phenomenological models, finely tuned to a wealth of precise experimental data, have given us a good qualitative, and at times quantitative, understanding of the phenomenon. However, these models of fission lack a firm connection with the theory of the nuclear force. This jeopardizes their ability to supply the high-precision predictions currently needed to understand the later stages of the formation of elements in supernova, or to simulate new prototypes of nuclear reactors.

Some of the basic concepts of a microscopic theory of fission were already laid out in the eighties by the theory group at CEA Bruyères-le-Châtel in France [1, 2]. Starting from an effective interaction between nucleons that embeds (in a phenomenological way) in-medium many-body correlations, the nuclear self-consistent mean-field theory provides a generic framework to construct ever more accurate representations of the nuclear wavefunction [3]. To describe fission, a number of collective variables must be introduced. The potential energy surface in the collective space thus defined serves as a basis to perform a time evolution of the nuclear wave packet in the framework of the time-dependent generator coordinate method. As evidenced from the few applications reported, the method is very promising [4].

However, at the time of its inception, the necessary computing power was lacking so that this approach could not yet compete with more empirical models. Today, the fast development of leadership class computers and interdisciplinary collaborations such as the NUCLEI project open new perspective [5]. In addition, recent progress in deriving quality energy functionals for fission applications [6] and of understanding the quantum mechanics of scission [7] suggests the microscopic approach to fission has enough potential to become predictive.

In this work, we discuss some aspects of the microscopic theory of fission in the poster-child example of the neutron-induced fission of ^{240}Pu at low energies. After a brief reminder of the theoretical framework, we show the evolution of the fission barriers as function of the nuclear temperature, and discuss some of the practical aspects related to such simulations.

2. Theoretical framework

Our formalism is based on the nuclear energy density functional theory (DFT) with local energy functionals of the Skyrme type [8] for the particle-hole channel, and contact density-dependent interactions with mixed volume and surface character for the particle-particle channel [9]. The nucleus is described at the Hartree-Fock-Bogoliubov approximation: its ground state is a quasi-particle vacuum that depends on a number of collective variables $\boldsymbol{q} = \{q_i\}_{i=1,\dots,N}$, such as, e.g. the expectation value of multipole moments or angular momentum. The set of the N collective variables defines a point $\boldsymbol{q}$ in the collective space. For each point in the collective space, the requirement that the energy be minimal with respect to variations of the generalized density leads to the HFB equations; solving the latter defines the one-body density matrix and the pairing tensor, and thereby all observables of interest.

The finite-temperature extension of the HFB formalism (FT-HFB) was introduced in nuclear physics more than 30 years ago [10]. We only recall that in the finite-temperature HFB theory, the ground-state is a statistical superposition of pure quantum states, characterized by a density operator $\hat{D}$. Adopting the HFB approximation for the density operator leads to the finite-temperature HFB equations: they are formally equivalent to their form at zero-temperature, only the expression of the density matrix and pairing tensor is different. Most importantly, the Wick theorem still applies, meaning that all observables are computed from the trace (in the given representation) of the corresponding operator and the one-body density matrix.

In addition to the density operator, macroscopic thermodynamical concepts are also used to describe the system at finite temperature. In particular, several thermodynamical potentials are available to describe the nucleus, each coming with its set of state variables. The two most useful are (i) the internal energy $E(V, S, X)$, where V is the (constant) nuclear volume, S the entropy and X any additional extensive state variable needed to characterize the nucleus (expectation value of multipole moment, for example) and (ii) the Helmholtz free energy $F(V, T, X)$ with $F = E - TS$ and T the temperature. The FT-HFB equations are obtained by minimization of the grand potential at temperature T, and naturally yield the free energy in the isothermal representation (constant T). Passage to an isentropic description (constant S) is sometimes needed, as discussed below.

3. Induced fission and finite-temperature formalism

In the description of induced fission in the finite-temperature DFT framework, the first step is to compute the potential energy surface of the nucleus as a function of the chosen collective coordinates and the nuclear temperature. In the following, we work in a four-dimensional collective space characterized by the expectation values of the axial, $\hat{Q}_{20}$, and triaxial, $\hat{Q}_{22}$, quadrupole moment, mass octupole moment $\hat{Q}_{30}$ and hexadecapole moment $\hat{Q}_{40}$. All calculations have been performed with the SkM* parametrization of the Skyrme functional [11]. In the particle-particle channel, the proton and neutron pairing strengths were fitted to the experimental 3-point formula for the odd-even mass differences in ^{240}Pu. A cutoff of $E_{\text{cut}} = 60$ MeV limits the number of quasi-particles taken into account in the calculation of the densities.

Calculations were performed with the DFT solvers HFBTHO [12] and HFODD [13]. In both codes the solutions to the HFB equations are expanded in the one-center harmonic oscillator (HO) basis. In HFBTHO axial and time-reversal symmetry are assumed so that solutions are labeled by the projection of the angular momentum on the z-axis. By contrast, HFODD is fully symmetry-unrestricted, and can, in particular, describe triaxial and parity-breaking shapes. The two programs have been benchmarked against one another and agree within a few eV for an axial configuration [12]. In practice, the HFB wavefunctions were expanded onto $N_{\text{states}} = 1100$ HO basis states with contributions from up to $N_{\text{max}} = 30$ oscillator shells.

3.1. Evolution of fission barriers

In figure 1, we show the evolution of the free energy $F = E - TS$ as function of the expectation value Q_{20} of the axial quadrupole moment. All three other degrees of freedom are locally minimized: in practice, the first barrier has a non-zero value of $\hat{Q}_{22}$ while the octupole moment is non-zero past the second barrier. Note the gradual disappearance of the barriers as the temperatures increases.

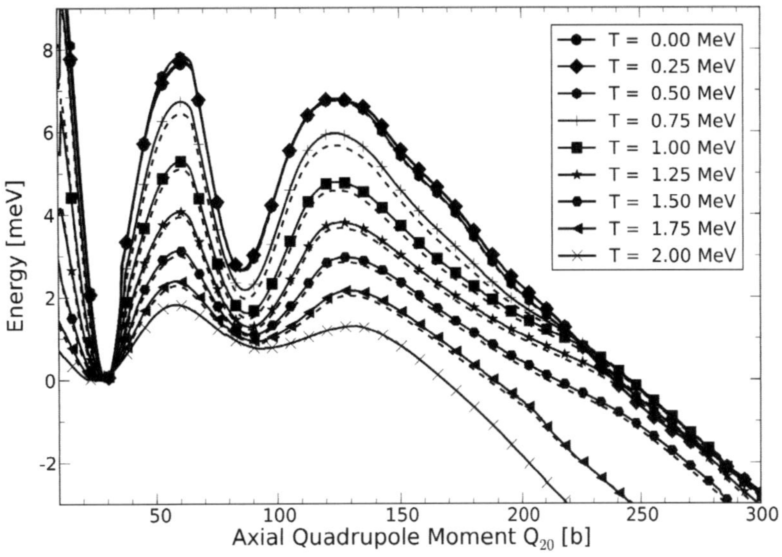

Figure 1. Evolution of fission barriers in ^{240}Pu as a function of temperature along the most probable fission path. Plain lines curves with a symbol show the free energy at constant temperature (isothermal process), the dashed-line curves next to them show the corresponding internal energy at constant entropy (isentropic process).

In applications of induced fission, the primary motivation has to do with the properties of the fission fragments rather than those of the fissioning compound nucleus. Of particular importance are the charge and mass distributions of the fragments, their total kinetic energy (TKE), and their excitation energy. The latter is a major input in the reaction codes that model the neutron and gamma spectrum generated during fission. Estimates of TKE and excitation energy require, in principle, to work in the isentropic representation. The Maxwell relations of thermodynamics state that, for any (extensive) collective coordinate q_i,

$$\left.\frac{\partial F}{\partial q_i}\right|_{T,q_{j\neq i}} = \left.\frac{\partial E}{\partial q_i}\right|_{S,q_{j\neq i}}. \tag{1}$$

Therefore, variations of internal energy with respect to deformation at constant entropy are equal to the variations of free energy at constant temperature [14].

From the set of curves $\{F(\boldsymbol{q}, T_k)\}_k$ obtained directly from the FT-HFB calculations, one can, by numerical interpolation, reconstruct the $E(\boldsymbol{q}, S)$ for any value S of the entropy. The procedure is based on a standard spline interpolation. Finite-temperature HFB calculations produce at each point $\boldsymbol{q}$ in the collective space the quantities $E(T)|_{\boldsymbol{q}}$ and $S(T)|_{\boldsymbol{q}}$, where from we easily obtain $E(S)|_{\boldsymbol{q}}$ for any desired value of S. When normalized to the same deformation point, the curves $E(\boldsymbol{q})|_S$ and $F(\boldsymbol{q})|_T$ should be strictly equivalent within the interpolation errors. This is verified in figure 1, where all dashed lines are obtained for the values of entropy at the top of the first barrier.

3.2. Temperature of the fragments

Until now, macroscopic-microscopic approaches assumed either the same temperature in the two fragments [15], or a ratio of temperatures proportional to the ratio of the fragment masses $T_1/T_2 \approx A_1/A_2$ [16]. The validity of the Maxwell thermodynamical relations in a practical case suggests a method to predict microscopically the nuclear temperature of each individual fragment after scission.

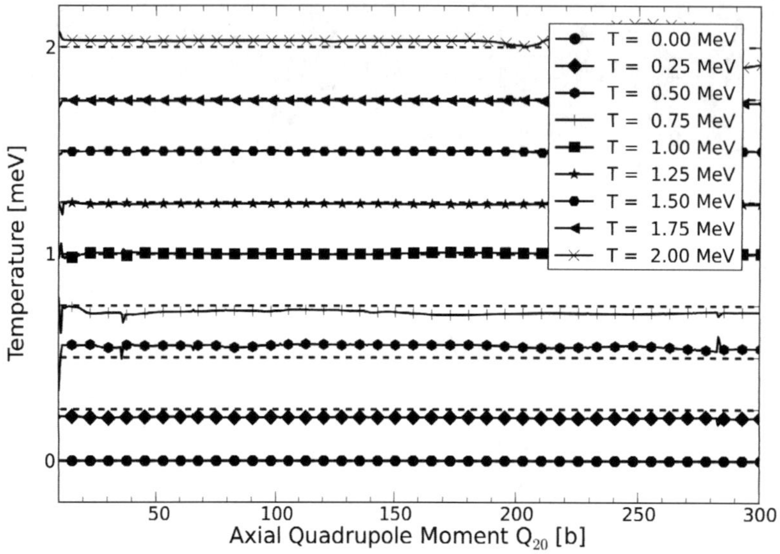

Figure 2. Temperature along the most probable fission path of ^{240}Pu as extracted from the entropy as function of the internal energy.

As seen in the previous section, the isothermal and isentropic representations are equivalent. The advantage of the latter is that it involves only extensive variables, and is, therefore, more amenable to the description of the splitting of the compound nucleus into two subsystems. The scission of the compound nucleus, characterized with an internal energy E and an entropy S, yields two fragments, with respective internal energy and entropy (E_1, S_1) and (E_2, S_2). Since the entropy is an extensive variable, we have $S = S_1 + S_2$ (here, we have to assume implicitly that the interaction energy between the two fragments is 'small enough'). The next step is to take advantage of the thermodynamical relation

$$\left.\frac{\partial S}{\partial E}\right|_{q_i} = \frac{1}{T}. \tag{2}$$

For each fragment $n = 1, 2$, we can perform a separate FT-HFB calculation constrained on the deformations $q_i^{(n)}$ of the fragment at scission. This gives us the functions $S^{(n)}(E^{(n)})$. Taking the derivative with respect to $E^{(n)}$ gives the temperature $T^{(n)}$ of the fragment n. While the value of both the internal energy and entropy in this separate FT-HFB calculation should be the same as the ones extracted from the compound nucleus (again: if the interaction energy can be neglected), the temperature may be different, as it is related to the derivative $\partial S/\partial E$.

The method just outlined will be applied to estimating the temperatures of the fragments in the induced fission of ^{240}Pu in a forthcoming publication. As a preliminary step, we estimate in figure 2 its numerical accuracy by comparing the exact value of the temperature, as set in the FT-HFB calculation, with the numerical estimates obtained from the $S(E)$ function. Overall, the accuracy is of the order of 50 keV at worse, owing to interpolation errors.

4. Conclusions

Recent advances in the nuclear energy density functional theory, combined with the constant increase of computing power, have enabled significant progress toward the development of a microscopic theory of nuclear fission. In this short article, we have presented early results related to the evolution of fission barriers with nuclear temperature and the equivalence between isentropic and isothermal representations.

Acknowledgments

Enlightening discussions with W. Younes and D. Gogny are very warmly acknowledged. This work was supported by the U.S. Department of Energy under Contract Nos. DE-AC52-07NA27344 (Lawrence Livermore National Laboratory) and DE-SC0008499 (NUCLEI SciDAC Collaboration). An award of computer time was provided by the Innovative and Novel Computational Impact on Theory and Experiment (INCITE) program. This research used resources of the Oak Ridge Leadership Computing Facility located in the Oak Ridge National Laboratory, which is supported by the Office of Science of the Department of Energy under Contract DE-AC05-00OR22725. It was also supported by the National Energy Research Scientific Computing Center supported by the Office of Science of the US Department of Energy under Contract No. DE-AC02-05CH11231.

References

[1] Berger J-F, Girod M and Gogny D 1984 *Nucl. Phys.* A **428** 23c
[2] Berger J-F, Girod M and Gogny D 1989 *Nucl. Phys.* A **502** 85c
[3] Bender M, Heenen P-H and Reinhard P-G 2003 *Rev. Mod. Phys.* **75** 121
[4] Goutte H, Berger J F, Casoli P and Gogny D 2005 *Phys. Rev.* C **71** 024316
[5] http://computingnuclei.org.
[6] Kortelainen M, McDonnell J, Nazarewicz W, Reinhard P-G, Sarich J, Schunck N, Stoitsov M V and Wild S M 2012 *Phys. Rev.* C **85** 024304
[7] Younes W and Gogny D 2011 *Phys. Rev. Lett.* **107** 132501
[8] Perlińska E, Rohoziński S G, Dobaczewski J and Nazarewicz W 2004 *Phys. Rev.* C **69** 014316
[9] Dobaczewski J, Nazarewicz W and Stoitsov M V 2002 *Eur. Phys. J.* A **15** 21
[10] Goodman A L 1981 *Nucl. Phys.* A **352** 30
[11] Bartel J, Quentin P, Brack M, Guet C and Håkansson H-B 1982 *Nucl. Phys.* A **386** 79
[12] Stoitsov M, Schunck N, Kortelainen M, Michel N, Nam H A, Sarich J and Wild S 2012 *Preprint* nucl-th/1210.1825
[13] Schunck N, Dobaczewski J, McDonnell J, Satuła W, Sheikh J A, Staszczak A, Stoitsov M and Toivanen P 2012 *Comp. Phys. Comm.* **183** 166
[14] Pei J C, Nazarewicz W, Sheikh J A and Kerman A K 2009 *Phys. Rev. Lett.* **102** 192501
[15] Wilkins B D, Steinberg E P and Chasman R R 1976 *Phys. Rev.* C **14** 1832
[16] Schmidt K H and Jurado B 2010 *Phys. Rev. Lett.* **104** 212501

Cluster aspects of binary fission

A V Andreev, G G Adamian and N V Antonenko
Joint Institute for Nuclear Research, 141980 Dubna, Russia
E-mail: antonenk@theor.jinr.ru

Abstract. With the improved scission-point model the mass distributions are calculated for induced fission of different Hg isotopes with even mass numbers A =180, 184, 188, 192, 196, 198. The calculated mass distribution and mean total kinetic energy of fission fragments are in a good agreement with the existing experimental data. The change in the shape of the mass distribution from asymmetric to more symmetric is revealed with increasing A of the fissioning ^{A}Hg nucleus, and the reactions are proposed to verify this prediction experimentally.

1. Introduction

The asymmetric shape of the mass distribution is well known in the spontaneous, neutron induced and β-delayed fission of most actinide isotopes. Such asymmetric shape was theoretically explained by taking into account the shell structure of the fragments [1, 2, 3, 4, 5]. The studies of the Coulomb-excited fission of radioactive nuclei revealed the predominance of symmetric fission in the light thorium to astatine region [6]. In the fission of stable targets with mass numbers $A = 185 - 210$ induced by protons and ^{3,4}He, the mass distribution was also found to be symmetric in most cases [7]. However, for several nuclei with $A \approx 200$ the mass distribution looks symmetric, but with a small dip on the top at small excitation energies [7]. Based on the most of experimental data, one could conclude that the asymmetric shape of mass distribution in low-energy fission changes to symmetric one with decreasing mass number of the fissioning nucleus. It was unexpected that in the recent experiment [8] on β-delayed fission of ^{180}Tl the shape of the mass distribution was found to be clearly asymmetric. The explanation of this interesting result is a challenge for nuclear theory and a good test for the existing models of nuclear fission.

2. Cluster approach to the last stage of fission

The statistical scission-point model [1] relies on the assumption that statistical equilibrium is established at scission and the observable characteristics of the fission process are formed near the prescission configurations of the fissioning nucleus. In spite of the criticism raised in more sophisticated models, this model remains to be unique for describing all observable characteristics of fission. With the modified scission-point model [9] we can describe the experimental data on fission of actinides: mass, charge, kinetic energy distributions and neutron multiplicity distributions. With this model a new explanation of a bimodality effect in fission of heavy actinides and fine structure of mass-energy distribution in fission of ^{236}U have been proposed. The model has also been extended to the description of ternary fission [10].

Here we give a short description of the model [9, 10]. The fissioning nucleus at the scission point is modeled by the two nearly touching coaxial spheroids – fragments of a dinuclear system with the masses (charges) A_L (Z_L) and A_H (Z_H) of the light (L) and heavy (H) fragments, respectively. $A = A_L + A_H$ ($Z = Z_L + Z_H$) is the mass (charge) number of fissioning nucleus. Taking into account the volume conservation, the shape of the system is defined by the mass and charge numbers of the fragments, deformation parameters of the fragments β_i ($i = L, H$) and the interfragment distance R. The deformation parameter of each fragment is the ratio of the major and minor semi-axes of the spheroid $\beta_i = c_i/a_i$. Here and further i denotes the light or heavy fragments of the dinuclear system. The case $\beta_i = 1$ corresponds to the spherical shape of the fragment. For small values of $\beta_i - 1$, the following equality is valid: $\beta_i \approx \beta_{2i} + 1$, where β_{2i} is the parameter of quadrupole deformation of the i'th fragment in the multipole expansion of the fragment shape.

The potential energy of the system

$$U(A_i, Z_i, \beta_i, R, l) = U^{macro}(A_i, Z_i, \beta_i, R, l) + \delta U^{shell}(A_i, Z_i, \beta_i), \quad (1)$$

$$\begin{aligned} U^{macro}(A_i, Z_i, \beta_i, R, l) &= U_L^{LD}(A_L, Z_L, \beta_L) + U_H^{LD}(A_H, Z_H, \beta_H) \\ &+ V^C(A_i, Z_i, \beta_i, R) + V^N(A_i, Z_i, \beta_i, R) \\ &+ V^{rot}(A_i, Z_i, \beta_i, R, l) + U^{zpv}(A_i, Z_i), \end{aligned}$$

$$\delta U^{shell}(A_i, Z_i, \beta_i) = \delta U_L^{shell}(A_L, Z_L, \beta_L) + \delta U_H^{shell}(A_H, Z_H, \beta_H)$$

is the sum of the liquid drop energies U_i^{LD} of each fragment, the energy of interaction of the fragments $V^C + V^N$, the rotational energy V^{rot}, the energy U^{zpv} of zero-point vibrations, and the shell correction terms δU_i^{shell}. The shell corrections are calculated with the Strutinsky method and two-center shell model [11]; the damping of the shell corrections with excitation energy and angular momentum l is introduced in our model. The interaction energy consists of the Coulomb interaction V^C of two uniformly charged spheroids and nuclear interaction V^N in the form of a double folding of nuclear densities and density-dependent Skyrme-type nucleon-nucleon forces [12]. For β-delayed and induced fission we use zero and non-zero angular momenta l, respectively (see below). We related the energy $U^{zpv} = E_i^{2^+} \coth[E_i^{2^+}/T(l)]$ of zero-point vibrations with the energies $E_i^{2^+}$ of the first 2^+ excited states of the fragments from [13]. The definition of the temperature $T(l)$ is given below. Here, we use a simplification that for all deformations the zero-point vibration energies at zero temperature are equal to the E_i^2+ energy at the ground state deformation. However, our calculations show that the nucleus being stiffer in the ground state in comparison to the neighboring nuclei remains stiffer at large deformations. The shape dependence of zero-point vibrations of the nascent fragments is taken effectively into account through the shape dependent temperature or excitation energy. At high excitation energy $U^{zpv} = T(l)$ for all fragments.

All terms in (1) excepting U^{zpv} depend on deformations of the fragments. For given deformations of the fragments the interaction potential has a potential minimum (pocket) as a function of the interfragment distance R. For calculation of the potential energy we take the value of interfragment distance $R = R_m$ corresponding to this minimum $[U(A_i, Z_i, \beta_i, l) \equiv U(A_i, Z_i, \beta_i, R_m, l)]$. Depending on the masses of the fragments and their deformations the calculated distance between the tips of the spheroids is 0.5–1 fm.

Because the thermodynamical equilibrium is postulated at the scission point, the excitation energy of the nuclear system at scission is calculated as a difference between the potential energy $U_{CN}(A, Z, \beta, l)$ of the compound nucleus (fissioning nucleus) and the potential energy $U(A_i, Z_i, \beta_i, l)$ of the dinuclear system at the scission point plus the initial excitation energy $E^*_{CN}(l)$ of the compound nucleus: $E^*(l) = U_{CN}(A, Z, \beta, l) - U(A_i, Z_i, \beta_i, l) + E^*_{CN}(l)$. The temperature is calculated as $T(l) = \sqrt{E^*(l)/a}$, where $a = A/12$ is the level density parameter in the Fermi-gas model. The yield of a particular scission configuration with given mass and charge numbers and deformation parameters of the fragments is proportional to the exponential Boltzmann-factor:

$$Y(A_i, Z_i, \beta_i, l) \sim \exp\left\{-\frac{U(A_i, Z_i, \beta_i, l)}{T(l)}\right\}. \tag{2}$$

To obtain the relative mass distribution as a function of the mass number of one of the fragments in fission of a compound nucleus with mass and charge numbers A and Z, one should integrate the expression (2) over Z_L, β_L, and β_H, sum over l and take into account that $A_H = A - A_L$ and $Z_H = Z - Z_L$:

$$Y(A_L) = \frac{\sum\limits_{l=0}^{l_{max}} (2l+1) \int \exp\left\{-\frac{U(A_i,Z_i,\beta_i,l)}{T(l)}\right\} \mathrm{d}Z_L \mathrm{d}\beta_L \mathrm{d}\beta_H}{\sum\limits_{l=0}^{l_{max}} (2l+1) \int \exp\left\{-\frac{U(A_i,Z_i,\beta_i,l)}{T(l)}\right\} \mathrm{d}A_L \mathrm{d}Z_L \mathrm{d}\beta_L \mathrm{d}\beta_H}. \tag{3}$$

The value of angular momentum l is limited by either the kinetical angular momentum $l_{kin} = R_b\sqrt{2\mu(E_{c.m.} - V_b)}$ ($R = R_b$ is the position of the Coulomb barrier with the height $V_b = V^C(A_i, Z_i, \beta_i = 0, R = R_b) + V^N(A_i, Z_i, \beta_i = 0, R = R_b)$ in the entrance channel, $\mu = m_0 A_L A_H/(A_L + A_H)$ is the reduced mass parameter, and m_0 is the nucleon mass) or by the calculated critical angular momentum l_{cr} in the entrance channel, depending on which one is smaller: $l_{max} = \min\{l_{kin}, l_{cr}\}$.

We calculate the total kinetic energy (TKE) supposing that all interaction energy at the scission point transforms after fission into the kinetic energy of the fission fragments. Therefore, the value of the TKE strongly depends on the deformation of the fragments at the scission point. The smaller the deformations of the fragments, the larger the Coulomb repulsion, the larger the TKE. The mean value of the total kinetic energy for particular binary splitting is calculated by averaging over deformations of the fragments:

$$\langle TKE\rangle(A_i, Z_i) = \frac{\sum\limits_{l=0}^{l_{max}} (2l+1) \int TKE(A_i, Z_i, \beta_i, l) \exp\left\{-\frac{U(A_i,Z_i,\beta_i,l)}{T(l)}\right\} \mathrm{d}\beta_L \mathrm{d}\beta_H}{\sum\limits_{l=0}^{l_{max}} (2l+1) \int \exp\left\{-\frac{U(A_i,Z_i,\beta_i,l)}{T(l)}\right\} \mathrm{d}\beta_L \mathrm{d}\beta_H}, \tag{4}$$

where

$$TKE(A_i, Z_i, \beta_i, l) = V^C(A_i, Z_i, \beta_i, R_b) + V^N(A_i, Z_i, \beta_i, R_b) + V^{rot}_{rel}(A_i, Z_i, \beta_i, l), \tag{5}$$

$V^{rot}_{rel}(A_i, Z_i, \beta_i, l) = \hbar^2 fl(fl+1)/(2\mu R_b^2)$, $f = \mu R_m^2/(\Im_L + \Im_H + \mu R_m^2)$. Here, $\Im_L$ and $\Im_H$ denotes the moment of inertia of the fragments.

3. Fission of Hg isotopes

For β-delayed fission of ^{180}Tl [8] the excitation energy $E^*_{CN}(l = 0)$ of the fissioning nucleus ^{180}Hg does not exceed 10.44 MeV. The mass distribution of fission fragments is presented in fig. 1. We obtained clearly asymmetric mass distribution with the average masses of the light and heavy fragments about 80 and 100, respectively, that is in agreement with the experimental data [8]. The calculated $\overline{TKE} = 136$ MeV is also in good agreement with the experiment [8].

It is convenient to analyze the obtained results for fission of ^{180}Hg by comparison of the potential energy surfaces of different mass/charge splits (fig. 2). If one excludes the shell corrections, the potential energy surfaces will have a minimum at the deformations of the

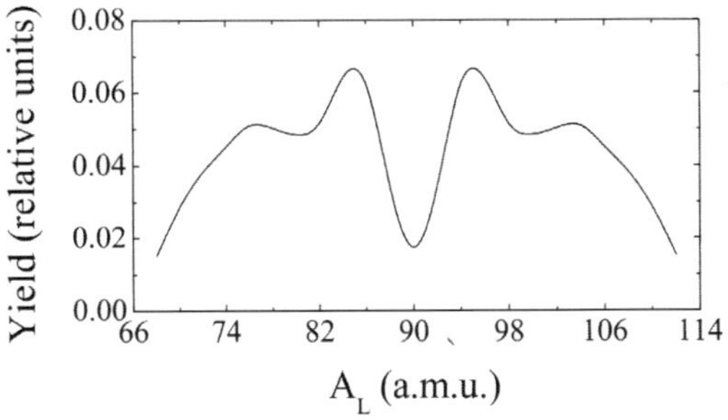

Figure 1. Calculated mass distribution of fission fragments for β-delayed fission of ^{180}Tl (fissioning nucleus is ^{180}Hg).

fragments of about β_i=1.6. These β_i are larger than the ground-state deformations of the corresponding nuclei because of the polarization effect. In fission of ^{180}Hg the symmetric scission configuration is ^{90}Zr+^{90}Zr. The shell correction for ^{90}Zr has a negative value $\delta U_i^{shell} \approx -2$ MeV near β_i=0, at larger deformations it becomes positive, at β_i=1.6 it is equal to $\delta U_i^{shell} \approx 1$ MeV, then grows further and at β_i=1.85 reaches $\delta U_i^{shell} \approx 4$ MeV. On the contrary, the shell corrections for non-magic nuclei in the scission configurations Kr+Ru and Se+Pd are usually positive at small deformations ($\delta U_L^{shell} \approx 2.5$ MeV, $\delta U_H^{shell} \approx 1.5$ MeV) and have zero or slightly negative values in the region around β_i=1.6. Because of these shell effects, the minimum is narrow for ^{90}Zr+^{90}Zr, while for ^{76}Se+^{104}Pd, where the shell effects are weaker, it is wide.

Figure 3 shows a change in the shape of the mass distribution from asymmetric to more symmetric with increasing mass number A of the fissioning nucleus ^{A}Hg. While the mass distribution is well asymmetric for ^{180}Hg and ^{184}Hg, for ^{188}Hg the asymmetry is less pronounced, and for ^{192}Hg and ^{196}Hg the mass distribution looks more symmetric but with a dip on the top similar to that observed experimentally in the fission of ^{198}Hg [7]. In our model, with increasing mass of the fissioning nucleus, the fragments of symmetric scission configurations deviate from the magic ^{90}Zr, and the role of strong shell effects at symmetric splits decreases. Thereby, in the heavy isotopes of Hg the shape of the mass distribution is generally defined by the liquid-drop part of the energy, and we obtain more symmetric mass distributions.

The excitation energy reduces the shell effects and smooths out the shape of the mass distribution (fig. 3). However, for some isotopes the influence of the excitation energy is rather weak. For example, in the case of ^{180}Hg the mass distribution has a pronounced asymmetric shape even at the excitation energy E^*_{CN}=64.2 MeV [8].

To study the fission properties of all considered isotopes ^{180}Hg, ^{184}Hg, ^{188}Hg, ^{192}Hg and ^{196}Hg (fig. 3), we propose the induced fission reactions at bombarding energies 10 MeV and 30 MeV above the corresponding Coulomb barriers V_b for the spherical nuclei: ^{36}Ar+^{144}Sm→^{180}Hg (V_b=126.2 MeV, l_{max}=44 and 61); ^{40}Ar+^{144}Sm→^{184}Hg (V_b=124.55 MeV, l_{max}=46 and 71); ^{40}Ar+^{148}Sm→^{188}Hg (V_b=123.9 MeV, l_{max}=46 and 66); ^{32}S+^{160}Gd→^{192}Hg (V_b=114.4 MeV, l_{max}=42 and 63); ^{36}S+^{160}Gd→^{196}Hg (V_b=112.8 MeV, l_{max}=45 and 70).

4. Summary

Based on the cluster approach to the last stage of fission, we got a good description of the recent experiment [8] where the asymmetric mass distribution in fission of ^{180}Hg was observed. The results of our calculations confirm the importance of the shell structure and, correspondingly, the deformation effects in the fission process.

Acknowledgments

We thank Dr. A. N. Andreyev for fruitful discussions. This work was supported by DFG (Bonn) and RFBR (Moscow).

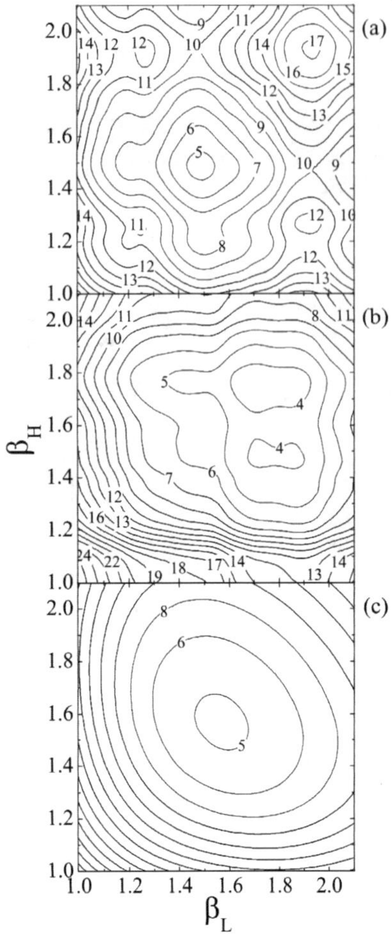

Figure 2. Calculated potential energy at the scission point as a function of deformations of the fragments in the binary systems $^{90}Zr+^{90}Zr$ (a) and $^{76}Se+^{104}Pd$ (b). The energy is given in MeV relative to the energy of the fissioning nucleus ^{180}Hg. For the binary system $^{90}Zr+^{90}Zr$, the potential energy calculated without shell corrections is shown (c).

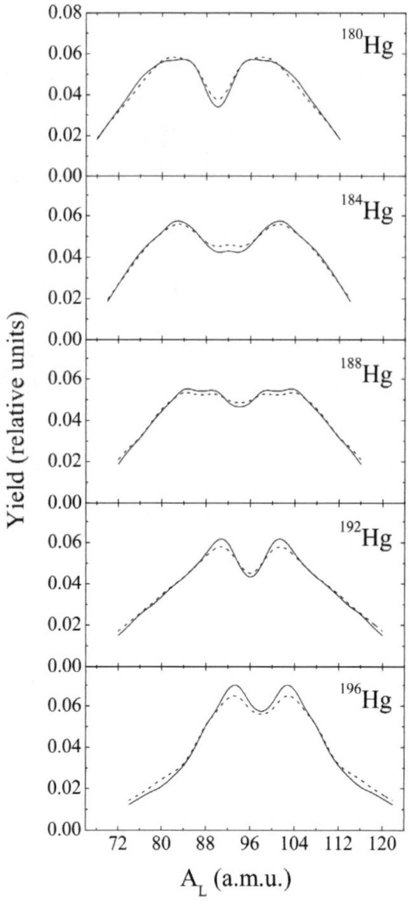

Figure 3. Calculated mass distributions of fission fragments for induced fission of $^{180,184,188,192,196}Hg$ with the impact energies of 10 MeV (solid lines, $E^*_{CN}(l=0)$=44.2, 43.9, 49.7, 62.4, 56.0 MeV for $^{180,184,188,192,196}Hg$, respectively) and 30 MeV (dashed lines, $E^*_{CN}(l=0)$=64.2, 63.9, 69.7, 82.4, 76.0 MeV for $^{180,184,188,192,196}Hg$, respectively) above the corresponding Coulomb barriers for the spherical nuclei.

References

[1] Wilkins B D, Steinberg E P and Chasman R R 1976 *Phys. Rev.* C **14** 1832
[2] Möller P, Madland D G, Sierk A J and Iwamoto A 2001 *Nature* **409** 785
[3] Gönnenwein F 1991 in *Nuclear Fission Process*, ed C. Wagemans (Boca Raton, FL: CRC Press)
[4] Hall H L and Hoffman D C 1990 *J. Radiol. Nucl. Chem.* **142** 53
[5] Oganessian Yu Ts 2007 *J. Phys.* G **34** R165
[6] Schmidt K-H, Benlliure J and Junghans A R 2001 *Nucl. Phys.* A **693** 169
[7] Itkis M G, Kondrat'ev N A, Mul'gin S I, Okolovich V N, Rusanov A Yu and Smirenkin G N 1990 *Sov. J. Nucl. Phys.* **52** 601

Itkis M G, Kondrat'ev N A, Mul'gin S I, Okolovich V N, Rusanov A Yu and Smirenkin G N 1991 *Sov. J. Nucl. Phys.* **53** 757
[8] Andreyev A N *et al* 2010 *Phys. Rev. Lett.* **105** 252502
[9] Andreev A V, Adamian G G, Antonenko N V, Ivanova S P and Scheid W 2004 *Eur. Phys. J.* A **22** 51
Andreev A V, Adamian G G, Antonenko N V and Ivanova S P 2005 *Eur. Phys. J.* A **26** 327
[10] Andreev A V, Adamian G G, Antonenko N V, Kuklin S N, Ivanova S P and Scheid W 2006 *Eur. Phys. J.* A **30** 579
[11] Maruhn J and Greiner W 1972 *Z. Physik* **251** 431
[12] Adamian G G, Antonenko N V, Jolos R V, Ivanova S P and Melnikova O I 1996 *Int. J. Mod. Phys.* E **5** 191
[13] Raman S, Nestor JR C W, Tikkanen P 2001 *At. Data and Nucl. Data Tables* **78** 1

Formation of strongly deformed states in entrance channels

A S Zubov[1], **V V Sargsyan**[1,2], **G G Adamian**[1], **N V Antonenko**[1] **and W Scheid**[3]

[1]Joint Institute for Nuclear Research, 141980 Dubna, Russia
[2]International Center for Advanced Studies, Yerevan State University, 0025 Yerevan, Armenia
[3]Institut für Theoretische Physik der Justus–Liebig–Universität, D–35392 Giessen, Germany

E-mail: adamian@theor.jinr.ru

Abstract. The high-spin hyperdeformed (HD) nuclear states treated as dinuclear or quasi-molecular configurations are suggested to be directly populated in heavy ion-collisions at near Coulomb barrier energies.

1. Introduction

The evidence of low-spin HD states in actinides has been experimentally established in induced fission reactions (n,f), (t,pf) and (d,pf) [1]. The question of experimental indications of high-spin HD states is still open. No convincing evidence for the discrete-line HD rotational bands has been found so far. The question arises how to obtain high-spin HD nuclei. In the high-spin HD studies one can use the experimental and theoretical knowledge about high-spin superdeformed (SD) states and low-spin SD and HD isomers. As shown in shell model calculations [2], the third minimum in the potential energy surface of actinide nuclei belongs to a quasi-molecular configuration of two touching nuclei (clusters). Based on the results of refs. [3, 4, 5] one can be convinced that certain quasi-molecular configurations with dumb-bell shapes have the same quadrupole momenta and momenta of inertia as those measured for the high- and low-spin SD states and low-spin HD isomer states. In light α-particle nuclei the similarity between the HD and cluster-type states was mentioned in refs. [6, 7, 8]. The HD configurations of ^{36}Ar, ^{40}Ca and ^{56}Ni seem to be the dinuclear systems (DNS) formed in the reactions ^{24}Mg+ ^{12}C, ^{28}Si+ ^{12}C and ^{28}Si+^{28}Si [7, 9, 10], respectively.

2. Cluster model

According to the cluster interpretation the HD state can be treated as a cold rotating DNS (such that the internal excitation energy of the DNS is zero). The relative distance R between the center of two touching clusters corresponds to the minimum of the potential energy surface and, correspondingly, to the minimum of the pocket of nucleus-nucleus interaction potential. The large overlap of the DNS nuclei is hindered by a repulsive nucleus-nucleus interaction potential at smaller relative distances. The pocket of the nucleus-nucleus potential at a given angular momentum contains the quasi-bound states with energies below the potential barrier and with quite long half-lives. The lowest quasi-bound state, which is identical to the HD state can

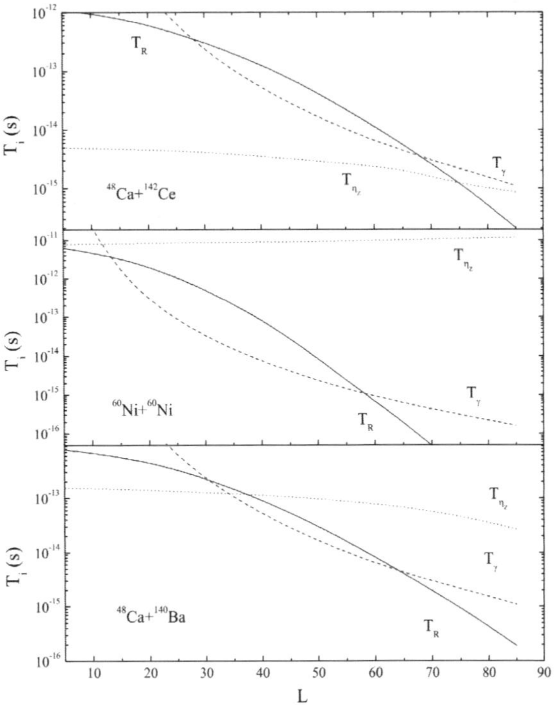

Figure 1. Time T_γ of $E2$ transition and tunneling times T_{R,η_Z} through the barriers in the R and the η_Z coordinates for the HD states formed in the entrance channel of indicated reactions as the function of angular momentum L.

be populated in capture process either by de-excitation (neutron emission) of the initial DNS formed at collision energy above the entrance Coulomb barrier [11], or directly by tunneling through this barrier [5, 12]. Note that the entrance barrier contains the centrifugal term.

2.1. HD state formation cross section

The partial cross section $\sigma_{\rm HD}(E_{\rm c.m.}, L)$ for the formation of the HD state at the bombarding energy $E_{\rm c.m.}$ in the center-of-mass system depends on the capture probability $P_{\rm cap}$, which is related to the formation of the initial excited DNS and to the probability $P_{\rm HD}$ of transformation of this DNS into a HD state by neutron emission [11]:

$$\sigma_{\rm HD}(E_{\rm c.m.}, L) = \frac{\pi\hbar^2}{2\mu E_{\rm c.m.}}(2L+1)P_{\rm cap}(E_{\rm c.m.}, L)P_{\rm HD}(E_{\rm c.m.}, L)\ . \qquad (1)$$

For each reaction there is a certain interval of angular momentum L, where the value $\sigma_{\rm HD}$ is maximal and there is a chance to observe the γ quanta from the formed HD states. Outside this interval the identification of HD states seems to be difficult. The probability $P_{\rm HD} = \sum_{k=1}^{2} P_{n_k} w_{n_k}$ of forming the HD state depends on the probability P_{n_k} of neutron emission from the k'th nucleus of the DNS and on the probability w_{n_k} of emitting a neutron with large enough kinetic energy to cool the excited DNS to zero internal excitation energy. We assume that the neutron can be evaporated from each fragment of the DNS and the neutron spectrum has Maxwellian form. The probability P_{n_k} depends on the competition between neutron emission and the DNS decay either in the relative distance coordinate R or via evolution in the charge

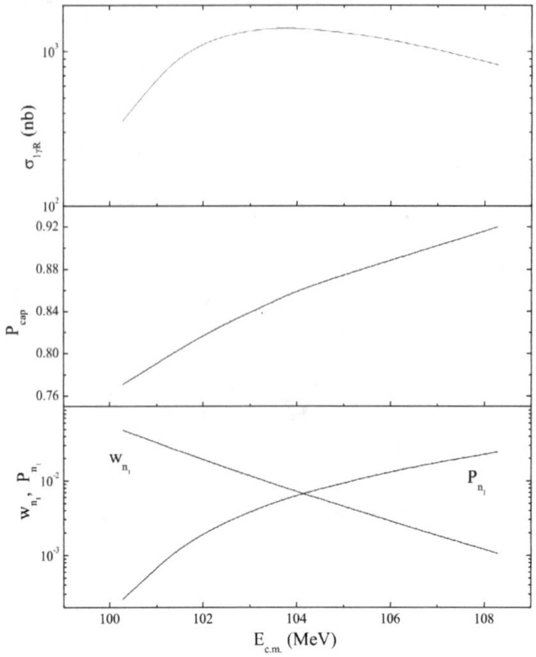

Figure 2. Probability P_{n_1} of neutron emission, probability w_{n_1} of cold system formation (lower panel), probability of capture P_{cap} (middle panel) and identification cross section $\sigma_{1\gamma R}$ of the HD state formed in the ^{60}Ni+^{60}Ni reaction at L=30–40 $\hbar$ (upper panel) as the functions of the bombarding energy $E_{\rm c.m.}$ in the center-of-mass system.

(mass) asymmetry $\eta_Z(\eta)$. The widths of these different processes are calculated within the statistical approach [11].

2.2. HD state identification cross section

One can propose the experimental method of identifying the HD states by measuring rotational γ quanta in the HD band in coincidence with decay fragments of the DNS constituting the HD state [5, 11, 12]. This means that the properties of the formed cold system must fulfill the following conditions:

$$T_\gamma < T_R < T_{\eta Z}, \tag{2}$$

where T_R and $T_{\eta Z}$ are the tunneling time of the HD DNS decay and that of the transformation in charge (mass) asymmetry, respectively. Condition (2) sufficiently restricts the interval of angular momenta in which it is possible to identify HD states, because the characteristic times of different processes depend on L. The value of T_γ mainly depends on L through E_γ. The angular momentum dependence of T_R and $T_{\eta Z}$ is defined by the angular momentum dependence of the corresponding barriers. While the barrier in η_Z is weakly affected by the change of L, the value of the decay barrier in R decreases much strongly with increasing contribution of the repulsive centrifugal part of the nucleus-nucleus potential. For example, the values T_γ , T_R and $T_{\eta Z}$ are presented in fig. 1 as the functions of L for the HD states formed in the entrance channel of reactions ^{48}Ca+^{142}Ce, ^{60}Ni+^{60}Ni, and ^{48}Ca+^{140}Ba. The condition $T_\gamma \leq T_R$ is satisfied only in some interval of L. At very small angular momentum the time of $E2$ transition becomes enormous due to the small values of $E_\gamma(L \to L-2)$, while at very large L the quasi-fission (the DNS decay) barrier vanishes. In addition, the minimum in η_Z in this interval of L should

Journal of Physics: Conference Series **436** (2013) 012060 doi:10.1088/1742-6596/436/1/012060

Table 1. The momenta of inertia $\Im$, electric quadruple momenta $Q_2^{(c)}$, range of angular momenta L, optimal bombarding energy $E_{\rm c.m.}$ in the center-of-mass system and cross sections σ_{HD} and $\sigma_{x\gamma R}$ ($x = 1-3$) calculated for the HD states formed in the entrance channel of the indicated reactions.

Reactions	$\Im$ $\hbar^2/MeV$	$Q_2^{(c)}$ $10^2(e\ {\rm fm}^2)$	$L_{min} < L < L_{max}$	$E_{\rm c.m.}$ MeV	σ_{HD} μb	$\sigma_{1\gamma R}$ nb	$\sigma_{2\gamma R}$ nb	$\sigma_{3\gamma R}$ nb
^{40}Ca+^{40}Ca	38.1	13.9	$20 < L < 30$	69.7	10	123	122	120
			$30 < L < 40$	76.7	7.3	110	100	96
			$40 < L < 50$	86.0	2.8	250	130	75
^{48}Ca+^{48}Ca	51.5	15.6	$40 < L < 50$	66.5	59	23	4.6	1.1
			$50 < L < 60$	75.1	32	2400	970	390
			$60 < L < 70$	85.2	11	1300	270	83
^{58}Ni+^{58}Ni	83.0	31.3	$20 < L < 30$	104.4	2.2	1700	1200	760
			$30 < L < 40$	107.6	1.1	450	350	270
			$40 < L < 50$	111.8	0.6	340	210	140
			$50 < L < 60$	116.9	0.3	59	17	6.2
^{60}Ni+^{60}Ni	89.3	32.9	$20 < L < 30$	99.9	2.2	1400	1200	980
			$30 < L < 40$	103.3	2.0	1400	1400	1200
			$40 < L < 50$	106.7	1.0	770	650	570
			$50 < L < 60$	111.5	0.5	280	180	120
^{48}Ca+^{140}Ba	138	36.8	$40 < L < 50$	134	0.08	35	19	10
			$50 < L < 60$	137.1	0.07	37	23	14
			$60 < L < 70$	140.8	0.02	11	5.3	2.9

be quite deep to provide the condition $T_R \le T_{\eta_Z}$. Thus, from the analysis of fig. 1 one can see that the HD states can be identified by measuring the rotational γ quanta in coincidence with the decay fragments at $20 < L < 60$ and $40 < L < 70$ in the reactions ^{60}Ni+^{60}Ni and ^{48}Ca+^{140}Ba, respectively, and cannot be identified by γ quanta at any angular momenta in the reaction ^{48}Ca+^{142}Ce.

Using condition (2), one can estimate the probability of emitting x γ quanta from the HD state just before its decay in R as

$$P_{x\gamma R} = \frac{\Lambda_R(L-2x)}{\Lambda_{tot}(L-2x)} \prod_{k=0}^{x-1} \frac{\Lambda_\gamma(L-2k)}{\Lambda_{tot}(L-2k)} \, . \tag{3}$$

Here $\Lambda_{\gamma,R,\eta} = \hbar/T_{\gamma,R,\eta_Z}$ are the rates of different competing processes (rotational γ quantum emission and tunneling in R and η_Z), while $\Lambda_{tot} = \Lambda_\gamma + \Lambda_R + \Lambda_{\eta_Z}$. Then one can estimate the cross section of emitting at least x γ quantum from the HD state just before its decay in R

$$\sigma_{x\gamma R} = \sum_{L=L_{min}}^{L_{max}} \sigma_{\rm HD}(E_{\rm c.m.}, L) \sum_{x'=x}^{L/2} P_{x'\gamma R} \, . \tag{4}$$

The values of L_{min} and L_{max} are defined by eq. (2). Since the detection of γ quanta in coincidence with the decay products of HD state provides the identification of HD state, the cross section

$\sigma_{x\gamma R}$ is also the identification cross section. This should be taken into account in planning the experiments. Measuring more than one γ quantum will provide the experimentalists an opportunity to determine the momenta of inertia of the analyzed HD states. One can compare the values of $\sigma_{x\gamma R}$ with the cross section of the formation of the HD state calculated for the same interval of angular momenta $\sigma_{HD} = \sum_{L=L_{min}}^{L_{max}} \sigma_{\rm HD}(E_{\rm c.m.}, L)$.

3. Results of calculations

Table 1 displays the calculated values of various characteristics of HD states formed in the entrance channel of several reactions. These characteristics include the moment of inertia, electric quadrupole moment, optimal bombarding energy, range of angular momenta and the cross section of formation with emitting up to 3 γ quanta before decaying in R. With increasing angular momentum the value of $\sigma_{\rm HD}$ decreases due to the increase of the DNS decay in R. The choice of the reaction is influenced by several reasons. In the entrance channel of the reaction the DNS should have a local potential minimum which is populated, and after neutron emission leads to th formation of the cold quasi-bound state, treated here as a HD state. There should exist a range of angular momentum satisfying conditions (2). As one can see from table 1, symmetric reactions lead to larger $\sigma_{x\gamma R}$ if more neutron-rich nuclei are used. One can compare, for example, the reactions ^{40}Ca+^{40}Ca (^{58}Ni+^{58}Ni) and ^{48}Ca+^{48}Ca (^{60}Ni+^{60}Ni). Comparing the calculated values of $\sigma_{1\gamma R}$, $\sigma_{2\gamma R}$ and $\sigma_{3\gamma R}$ one can see that for some reactions these cross sections decrease rather slowly with the number of emitted γ quanta. This means that an articulate rotational band of the formed HD state can be observed experimentally. For example, in the ^{60}Ni+^{60}Ni reaction at initial $L = 40 - 50$ the calculated value of $\sigma_{10\gamma R}$=260 nb. Note that the $1n$-emission channel is the optimal evaporation channel for the formation of the HD state.

An example for the dependence of $\sigma_{1\gamma R}$ and that of the factors P_{cap}, P_{n_1} and w_{n_1} on $E_{\rm c.m.}$ is shown in fig. 2. The capture probability slowly increases with $E_{\rm c.m.}$ and does not affect much the location of the maximum of the function $\sigma_{1\gamma R}(E_{\rm c.m.})$. This location is mainly determined by P_{n_1} and w_{n_1}. While P_{n_1} increases, w_{n_1} decreases with increasing $E_{\rm c.m.}$. The final curve representing the function $\sigma_{1\gamma R}(E_{\rm c.m.})$ has a maximum, which is usually about 8–10 MeV higher than the value of the entrance Coulomb barrier. The location of this maximum gives us the optimal bombarding energy for each considered reaction.

4. Summary

We propose to consider the reactions ^{48}Ca+^{48}Ca and ^{58,60}Ni+^{58,60}Ni as good candidates for the production and experimental identification of the HD states. The estimated identification cross sections $\sigma_{\gamma R}$ for the HD states formed in these reactions are of the order of 1 μb. We suggested a new experimental method for the identification of the high-spin HD state by measuring the consecutive rotational $E2$ γ-transitions in the HD band in coincidence with decay fragments constituting the HD state. Observing these signatures would be a unique proof of the idea that HD states are cluster-type states, and further that quasi-molecular configurations also exist in heavier nuclei. Then the concept of nuclear molecular states would not be restricted to light and medium nuclei, but would spread over whole mass region.

References

[1] Buck B B, Britt H C, Garett J D and Hansen O 1972 *Phys. Rev. Lett.* **28** 1707
Baumann F F and Brinkmann K Th 1989 *Nucl. Phys.* A **502** 271
Krasznahorkay A *et al* 1998 *Phys. Rev. Lett.* **80** 2073
Krasznahorkay A *et al* 1999 *Phys. Lett.* B **461** 15
[2] Cwiok S, Nazarewicz W, Saladin J X, Plociennik W and Johnson A 1994 *Phys. Lett.* B **322** 304
[3] Royer G and Haddad F 1995 *J. Phys.* G **21** 339
[4] Shneidman T M, Adamian G G, Antonenko N V, Ivanova S P and Scheid W 2000 *Nucl. Phys.* A **671** 119
[5] Adamian G G, Antonenko N V, Nenoff N and Scheid W 2001 *Phys. Rev.*C **64** 014306

Adamian G G, Andreev A V, Antonenko N V, Nenoff N, Scheid W and Shneidman T M 2003 *Heavy Ion Phys.* **19** 87
Adamian G G, Andreev A V, Antonenko N V, Nenoff N, Scheid W and Shneidman T M 2003 *Acta Phys. Pol.* B **34** 2147
Adamian G G, Antonenko N V and Scheid W 2003 *Fizika* B **12** 21
Kuklin S N, Adamian G G, Antonenko N V and Scheid W 2008 Int. J. Mod. Phys. E **17** 2020
[6] Cindro N (1978) *J. Phys.* G **4** L23 (1978)
Greiner W, Park J Y and Scheid W *Nuclear Molecules* (World Scientific, Singapore, 1995).
[7] Sanders S J, Szanto de Toledo A and Beck C 1999 *Phys. Rep.* **311** 487
[8] Cseh J, Lévai G, Ventura A and Zuffi L, 1998 *Phys. Rev.* C **58** 2144;
Darai J, Cseh J and Jenkins D G 2012 *Phys. Rev.* C **86** 064309
[9] Beck C *et al* 2000 *Phys. Rev.* C **63** 014607
Rousseau M *et al* 2002 *Phys. Rev.*C **66** 034612
[10] Sciani W, Otani Y, Lépine-Szily A, Benjamim E A, Chamon L C, Lichtenthäler Filho R, Darai J and Cseh J (2009) *Phys. Rev.* **80** 034319
Cseh J, Darai J, Sciani W, Otani Y, Lépine-Szily A, Benjamim E A, Chamon L C and Lichtenthäler Filho R 2009 *Phys. Rev.* C **80** 034320
[11] Zubov A S, Sargsyan V V, Adamian G G, Antonenko N V and Scheid W 2010 *Phys. Rev.* C **81** 024607
[12] Zubov A S, Sargsyan V V, Adamian G G, Antonenko N V and Scheid W 2010 *Phys. Rev.* C **82** 034610

Journal of Physics: Conference Series **436** (2013) 012061 doi:10.1088/1742-6596/436/1/012061

Complex-energy shell model description of alpha decay

R Id Betan

Department of Physics and Chemistry FCEIA(UNR) - Physics Institute of Rosario (CONICET), Av. Pellegrini 250, 2000 Rosario, Argentina
Department of Physics and Astronomy, University of Tennessee, Knoxville, Tennessee 37996, USA
Physics Division, Oak Ridge National Laboratory, P.O. Box 2008, Oak Ridge, Tennessee 37831, USA

E-mail: idbetan@ifir-conicet.gov.ar

Abstract. The alpha decay width has been successfully described in the framework of a combined shell and cluster model. It has been found that the shell model does not provide enough alpha clustering. The cluster model cures this deficiency by including the alpha cluster in the model from the very beginning. In the complex-energy shell model attempt is made to produce the clustering needed by a large model space which involves particles in the continuum. An advantage of such a formulation is that the tail of the formation amplitude may be consistent with the decaying boundary condition. The formulation can be applied to systems, like ^{104}Te, decaying to the continuum.

1. Introduction

The theoretical description of the alpha decay in ^{212}Po was a long standing unsolved problem. From the first successful description of the relative decay given in a semi-microscopic description by Gamow in 1928 [1, 2] up to the satisfactory description of the absolute width given by Varga *et al* [3] 64 years passed. The theoretical enterprise started with an outstanding progress made by Gamow by dividing the decay into two steps and by using the very new idea that Quantum Mechanics can describe the tunnelling of the alpha particle through the Coulomb barrier, providing its probability. Gamow assumed that the alpha particle (formed by four protons and two electrons) is pre-formed inside the nucleus. Thomas [4] made another major step by contributing with a new element: a way to calculate microscopically the formation of the alpha particle by using the then newly invented formalism of nuclear reactions: R-matrix theory. Using time dependent perturbation theory Mang [5] elaborated the connection of the formation amplitude with the Shell Model. Mang's approach happens to be equivalent to Thomas' approach [6]; the absolute width is given by $\Gamma = 2P(R)\gamma^2(R)$ [4, 5]. The formation stage is incorporated in the reduced width $\gamma^2(R)$ while the decay stage is represented by the penetrability $P(R)$. Alternatively, the absolute width can be calculated from the spectroscopic factor S and the single-particle width Γ_{sp} as $\Gamma = S\Gamma_{sp}$ [4, 7].

The first Shell Model calculation of the absolute width involved a single Harmonic Oscillator (HO) configuration [5]. The absolute width obtained was several orders of magnitude far from the experimental one. This failure was caused by the unrealistic rapid falloff of the HO wave function tail. Harada [8] showed that a mixing of configurations improves the absolute width calculation due to the constructive interference of the HO states in the tail region. But the enhancement was not enough to reproduce the experimental result. Even the more extended HO shell model calculation carried out by Tonozuka and Arima [9] did not produce enough clustering. In order to improve the tail of the formation amplitude, Woods-Saxon wave functions expanded in an HO basis were used [10, 11]. A schematic calculation in the Berggren basis showing the importance of the continuum states was done in ref. [12]. But all the improvements mentioned were not enough to reproduce the experimental ^{212}Po decay.

Fliessbach [13] noticed for the first time that the normalization of the relative wave function in the decay channel implies a normalization to be interpreted in terms of a resonating-group-like theory. As a consequence, the formation amplitude has to be modified by a renormalization. The first calculation [14] in a single-configuration model space gave a promising result by showing that the absolute width increased by a factor of 100. Unfortunately the systematic calculation done by Tonozuka and Arima in the extended shell model proved that that hope was unjustified.

The final word to explain theoretically the absolute alpha decay width in the nucleus ^{212}Po came when the Shell Model, Cluster Model and the renormalization were all combined. In ref. [3] Varga *et al* reproduced the experimental absolute width within 10% and predicted 30% of clusterization for the alpha particle in the mother nucleus. It was also suggested that the low-lying yrast spectrum of ^{212}Po could also be explained in terms of an α+^{208}Pb cluster model [15].

The goal of this contribution is to show that the experimental absolute width of the elusive alpha decay of ^{212}Po can also be described within the Shell Model Framework. The way to get at this goal involves two improvements: (1) use a realistic basis of Woods-Saxon eigenfunctions without an HO expansion, and (2) use $\Gamma = S\Gamma_{sp}$ to calculate the width, where Γ_{sp} is the width obtained in a daughter + alpha single-particle model of the parent nucleus. The effect of (1) is to improve the tail of the shell-model formation amplitude while (2) reduces the sensitivity of the final result to the model space and makes the result independent of a channel radius R. Other formulations which avoid the use of a channel radius can be found in references [16, 17, 18, 19].

2. Formalism

The first ingredient in any many-body calculation is the basis. An extensive shell-model space has to be used in order to gain sufficient amount of clustering. This implies that one has to represent the continuum in the basis somehow. For the Woods-Saxon mean-field used here, this implies that we have to deal with positive-energy states. We will use a complex-energy representation taking a truncated Berggren basis [20]. Since we are not using HO states, we have to deal with nucleon single-particle coordinates and daughter–alpha relative coordinates at the same time. For this reason the six-dimensional integral involved in the formation amplitude must be calculated numerically.

We are going to show results for the alpha decay absolute width calculated by (1) the R-dependent formula $\Gamma = 2P(R)\gamma^2(R)$ (R the channel radius) coming from the R-matrix theory and (2) the R-independent formula $\Gamma = S\Gamma_{sp}$. Both expressions split the decay process into a formation stage, represented by γ and S, and a decay stage, represented by $P(R)$ and Γ_{sp}, respectively. The reduced width amplitude $\gamma(R) = (\hbar^2 R/2\mu)^{1/2} G(R)$ is proportional to the modified formation amplitude $G(R)$, which gives the probability amplitude that an alpha particle is formed inside the mother nucleus,

$$G(R) = \int \mathcal{N}^{-1/2}(R, R')\, g(R')\, R'^2\, dR' \qquad (1)$$

where $\mathcal{N}$ is the norm kernel [21] of the alpha-particle + daughter nucleus problem and g the

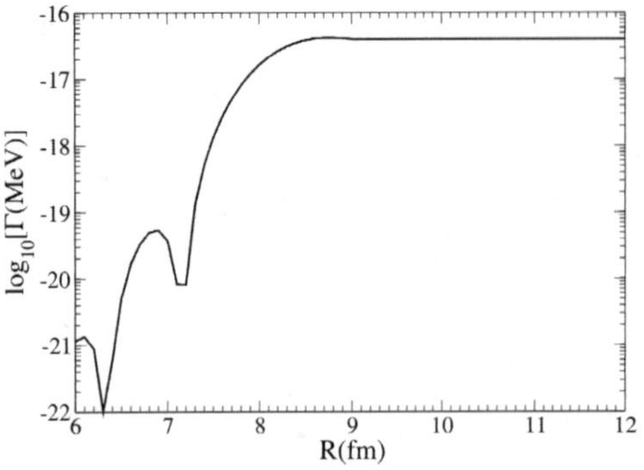

Figure 1. Alpha decay absolute width for ^{212}Po using the R-matrix expression $\Gamma = 2\,P\,\gamma^2$. Beyond 9 fm the tail of Γ was replaced by $H_0^+(kR)$ with $\frac{\hbar^2 k^2}{2\mu} = 8.986$ MeV.

(conventional) formation amplitude

$$\mathcal{N}(R, R') = \left\langle \mathcal{A}\frac{\delta(R_\alpha - R)}{R^2}\phi_\alpha Y_{00}\Psi_D \middle| \mathcal{A}\frac{\delta(R_\alpha - R')}{R'^2}\phi_\alpha Y_{00}\Psi_D \right\rangle,$$

$$g(R) = \int d\Omega_R \int d\xi_\alpha \int d\xi_d\, \Phi_P^*\, \phi_\alpha(\xi_\alpha)\, \Psi_D(\xi_d)\, Y_{00}(\hat{R}). \tag{2}$$

Here Φ_P and Ψ_D are the parent and daughter wave functions, respectively, both in the ground state; ϕ_α is the alpha intrinsic wave function.

The spectroscopic factor $S = \int dR\, R^2\, G^2(R)$ gives the probability that an alpha particle and the daughter ground state is formed. The penetration factor $P(R) = kR/|H^+(\chi, kR)|^2$ is the probability that a pre-formed alpha particle can penetrate through the section of the Coulomb barrier between R and ∞ (H^+ being the outgoing spherical Coulomb function), while the single-particle absolute width Γ_{sp} gives the width of the alpha decay of a system formed by a structureless daughter and a structureless alpha particle. To determine Γ_{sp}, one needs the daughter-alpha relative wave function $u(R)$, which is obtained as a solution of the daughter-alpha Schrödinger equation with outgoing boundary condition, and the single-particle absolute width is obtained from [22]

$$\Gamma_{sp} = \frac{\hbar^2 Re(k)}{\mu}\frac{|u(R)|^2}{|H_0^+(kR)|^2}. \tag{3}$$

3. Results

The neutron and proton single-particle model spaces needed for the calculation of the formation amplitude, were defined as the two major valence shells with widths smaller than 1 MeV for both ^{212}Po and ^{104}Te. The ^{212}Po wave function was taken as the seniority zero part of the Tonozuka and Arima [9] wave function from their $N7$ model space. (But the Tonozuka–Arima basis was spanned by HO states, while here we use Woods–Saxon states with outgoing boundary condition.) The wave function for the nucleus ^{104}Te was constructed as the product of the two-proton and two-neutron wave functions calculated using a separable force. The relative wave function for the calculation of the penetrability and the single-particle absolute width (3) was calculated with the program [23] using $Q_\alpha(^{212}\text{Po}) = 8.986$ MeV and $Q_\alpha(^{104}\text{Te}) = 5.151$ MeV. Figure 1 shows the absolute width $\Gamma = 2\,P\,\gamma^2$ for the ^{212}Po alpha decay with the tail substituted by $H_0^+(kR)$. The asymptotic value of Γ is 36 times smaller than the experimental one.

Table 1 shows the absolute width $\Gamma = S\,\Gamma_{sp}$ for ^{212}Po and ^{104}Te alpha decays. The respective single-particle absolute width used are $\Gamma_{sp}(^{212}\text{Po}) = 0.1247 \times 10^{-12}$ MeV and

$\Gamma_{sp}(^{104}\text{Te}) = 0.162 \times 10^{-12}$ MeV. The norm kernel was represented by its eigenvalues and eigenfunctions. The eigenvalue problem was solved in an equidistant shifted Gaussian basis, with Gaussian centres at $R_k = k\,\Delta R$ and $\Delta R = 0.56$ fm (for details see Ref. [24].) The terms belonging to eigenvalues larger than 0.001 were taken into account. The agreement with experiment for ^{212}Po is much better with this method.

Table 1. Alpha decay absolute width calculated using the expression $\Gamma = S\Gamma_{sp}$.

	Γ (MeV$\times10^{14}$)	Γ_{exp} (MeV$\times10^{14}$)
^{212}Po	0.14	0.153
^{104}Te	0.083	—

4. Discussion and conclusions

The absolute width obtained from $\Gamma = 2\,P(R)\,\gamma^2(R)$ should not depend on R around and beyond the nuclear surface where the interaction is dominated by the Coulomb interaction. But the approximations involved in the R-matrix formalism make this ideal condition unrealizable [25]. Another alarming fact is that the penetrability $P(R)$ and the reduced width amplitude γ are calculated using different procedures, which makes the matching inconsistent. On the other hand the absolute width from $\Gamma = S\,\Gamma_{sp}$ has no R dependence because none of S and Γ_{sp} depends on R. Another important advantage of calculating the absolute width using $\Gamma = S\,\Gamma_{sp}$ is that this formula proves to be less sensitivive to the model space [24]. The R-matrix expression uses the formation amplitude, which is very sensitive to the collective behaviour of the wave function at the nuclear surface, which depends on the model space, while the spectroscopic factor involves an integral which strongly diminishes the sensitivity to the model space. This explains the difference between the results obtained for the ^{212}Po decay in the two approaches.

A microscopic formulation of the formation amplitude for alpha particle decay in an extended Woods-Saxon basis including the continuum spectrum (Berggren bases) was presented. To preserve the correct tail of the wave function, no harmonic oscillator expansion was used. The absolute width calculated using the spectroscopic factor for ^{212}Po is consistent with the experimental result. This shows the feasibility of the Shell Model for calculating the alpha width.

As a byproduct of the single-particle complex energy shell model presented here, we calculated the alpha decay width of the nucleus ^{104}Te, whose all valence orbits lie completely in the continuum. An upper bound for the life-time of this nucleus was calculated and found to be 0.55×10^{-6} s. The Γ value obtained in this work shows that the decay model involving the mother wave function has to be treated as in the case of ^{212}Po.

Finally, let us mention the intriguing 1% found in this Shell Model calculation for the amount of clusterization of the α particle compared with 30% obtained using the Cluster plus Shell Model formalism [3].

Acknowledgments

Useful discussions with Kiyoshi Kato, Rezső Lovas and Witold Nazarewicz are gratefully acknowledged. This work was supported by the Office of Nuclear Physics, U.S. Department of Energy under Contract No. DE-FG02-96ER40963 and by the National Council of Research PIP-77 (CONICET, Argentina).

Journal of Physics: Conference Series **436** (2013) 012061 doi:10.1088/1742-6596/436/1/012061

References

[1] Gamow G 1928 *Z. Physik* **51** 204
[2] Condon E U and Gurney R W 1928 *Nature* **122** 439
[3] Varga K, Lovas R G and Liotta R J 1992 *Nucl. Phys.* A **550** 421
[4] Thomas R G 1954 *Prog. Theor. Phys.* **12** 253
[5] Mang H J 1957 *Z. Phys.* **148** 582
[6] Zeh H D 1962 *dissertation* (Univ. Heidelberg)
[7] Arima A and Yoshida S 1972 *Phys. Lett.* B **40** 15
[8] Harada K 1962 *Prog. Theor. Phys.* **27** 430
[9] Tonozuka I and Arima A 1979 *Nucl. Phys.* A **323** 45
[10] Insolia A, Liotta R J and Maglione E 1988 *Europhys. Lett.* **7** 209
[11] Delion D S and Suhonen J 2000 *Phys. Rev.* C **61** 024304
[12] Lenzi S M, Dragún O, Maqueda E E, Liotta R J and Vertse T 1993 *Phys. Rev.* C **48** 1463
[13] Fliessbach T and Mang H J 1976 *Nucl. Phys.* A **263** 75
[14] Fliessbach T, Mang H J and Rasmussen J O 1976 *Phys. Rev.* C **13** 1318
[15] Astier A, Petkov P, Porquet M-G, Delion D S and Schuck P 2010 *Phys. Rev. Lett.* **104** 042701
[16] Harada K and Rauscher E A 1968 *Phys. Rev.* **169** 169
[17] Kadmensky S G and Kalechits V E 1970 *Yad. Fiz.* **12** 70
[18] Jackson D F and Rhoades-Brown M 1977 *Nature* **267** 593
[19] Jackson D F and Rhoades-Brown M 1977 *Ann. Phys.* (N. Y.) **105** 151
[20] Berggren T 1968 *Nucl. Phys.* A **109** 265
[21] Fliessbach T 1976 *Z. Physik* A **277** 151
[22] Kruppa A T and Nazarewicz W 2004 *Phys. Rev.* C **69** 054311
[23] Ixaru L Gr, Rizea M and Vertse T 1995 *Comp. Phys. Commun.* **85** 217
[24] Id Betan R and Nazarewicz W 2012 *Phys. Rev.* C **86** 034338
[25] Descouvemont P and Baye D 2010 *Rep. Prog. Phys.* **73** 036301

Population and properties of superdeformed bands in $A \sim 150$ region

A S Zubov[1], V V Sargsyan[1,2], G G Adamian[1,3] and N V Antonenko[1]
[1]Joint Institute for Nuclear Research, 141980 Dubna, Russia
[2] Yerevan State University, 0025 Yerevan, Armenia
[3]Institute of Nuclear Physics, Tashkent 702132, Uzbekistan
E-mail: azubov@theor.jinr.ru

Abstract. The production of superdeformed nuclei, treated as cluster configurations in the $A \sim 150$ region is described in the framework of statistical approach and the dinuclear system concept. The competition of particle emission from the dinuclear system and quasifission is taken into account. The calculated nuclear characteristics of the superdeformed band in ^{152}Dy are close to the experimental values.

Among nuclei in which the effect of superdeformation (SD) occurs, the region $A \sim 150$ is of theoretical and experimental interest for certain reasons. The known SD states of several isotopes of Dy, Gd and Tb [1] are produced in different evaporation channels of fusion evaporation reactions. In this region of the nuclear chart it is possible to observe the rotational γ-transitions from very high ($\sim 60\hbar$) to lower ($\sim 20\hbar$) spins and, in some cases, to determine the energy of the lowest state in the SD band. Such experimental basis allows one to describe theoretically not only the structure of SD states, but also the mechanism of their population.

According to the cluster interpretation, the SD state can be treated as a cold rotating dinuclear system (DNS). The relative distance between the centers of two touching clusters corresponds to the minimum of the pocket of the nucleus-nucleus interaction potential. The large overlap of the DNS nuclei is hindered by a repulsive nucleus-nucleus interaction potential at smaller relative distances. The validity of the cluster interpretation of strongly deformed states was investigated in ref. [2]

The scheme of formation of the SD state in fusion-evaporation reaction is shown in figure 1. The potential energy U of the DNS is calculated as a sum of the mass excesses of the fragments at their ground states and the nucleus-nucleus potential [3]. Due to the shell and even-odd effects, it has its minima at different DNS configurations. After the colliding nuclei pass over the Coulomb barrier and come to the touching configuration, the initial excited DNS is formed. This DNS can evolve by diffusion in the relative distance R between the centers of nuclei and in the charge $\eta_Z = (Z_1 - Z_2)/(Z_1 + Z_2)$ and mass η asymmetry coordinates (Z_1 and Z_2 are the charge numbers of the DNS nuclei). It can also be de-excitated by particle emission during this evolution. The residue daughter DNS configuration can be cold enough to be trapped in some potential minimum in both the R and the η_Z coordinates and live a rather long time to be interpreted as a SD state. Such a system may emit $\gamma-$quanta between collective rotational states.

One can identify a DNS configuration, which corresponds to the experimentally observed

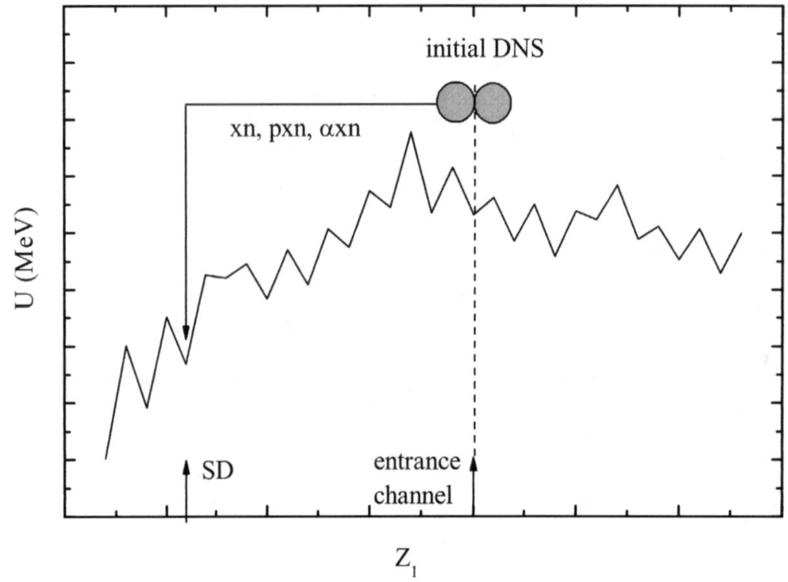

Figure 1. The scheme of formation of the SD state in fusion-evaporation reaction. Z_1 is the charge number of the lighter DNS nucleus.

SD state by comparing the calculated nuclear characteristics for different DNS configurations with the measured ones for the SD state. For example, the SD states of ^{152}Dy with measured moment of inertia $\Im_{SD}^{exp}$=85±3 $\hbar^2$/MeV, quadrupole moment Q_2^{exp}=18±3×10^2(e fm^2) and energy of the lowest state in SD band $E_{SD}^{exp}(24^+)$=10.64 MeV can be treated as the DNS configuration ^{12}C+^{140}Nd, for which our calculations give the values $\Im_{SD}^{th}$=76 $\hbar^2$/MeV, $Q_2^{th} = 11 \times 10^2$(e fm^2) and $E_{SD}^{th}(24^+)$=9.94 MeV.

The capture process is described using the quantum diffusion approach, based on the formalism of reduced density matrix [4]. The probability of particle emission from the evolving DNS is calculated using the statistical approach. At each evaporation step the competition between particle emission and the diffusion of DNS to large R (quasifission) is considered for all possible DNS configurations, with their statistical weights given by Bolzmann distribution.

The emission of rotational γ-quanta from the cold dinuclear configuration corresponding to the SD state competes with tunneling through barriers in R (quasifission) and η_Z (fusion). The rates of these processes $\Lambda_{\gamma,R,\eta_Z}$ can be expressed through their times T_{γ,R,η_Z} as $\Lambda_{\gamma,R,\eta_Z} = \hbar/T_{\gamma,R,\eta_Z}$, and can be calculated, as in Refs. [5, 6].

Measured [1, 7] and calculated intensities of the SD band in ^{152}Dy are shown in figure 2. The results of our calculation represent the dependence of this value on angular momentum. The strong fall of intensity at $L \geq 50$ can be explained by the significant decrease of quasifission barriers in this region due to the increase of centrifugal forces. At $L \leq 25$ $E2$-transitions are suppressed, as the time of these processes is too large, and the transition to the normally deformed (ND) rotational band (complete fusion, in cluster interpretation) becomes more favorable. Measured [8, 9] and calculated intensities of the SD band normalized to those of the ND band are presented in figure 3 as functions of the bombarding energy. This dependence is similar to the excitation functions for the compound nucleus reactions, but has much broader shape.

The agreement of our results with the experimental data proves the validity of the cluster

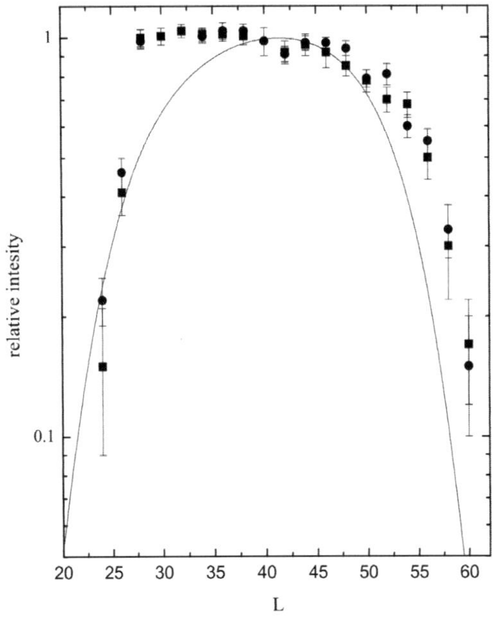

Figure 2. Measured [1] (circles), [7] (squares) and calculated (line) intensities of the SD band in ^{152}Dy produced in the reaction ^{108}Pd(^{48}Ca,4n).

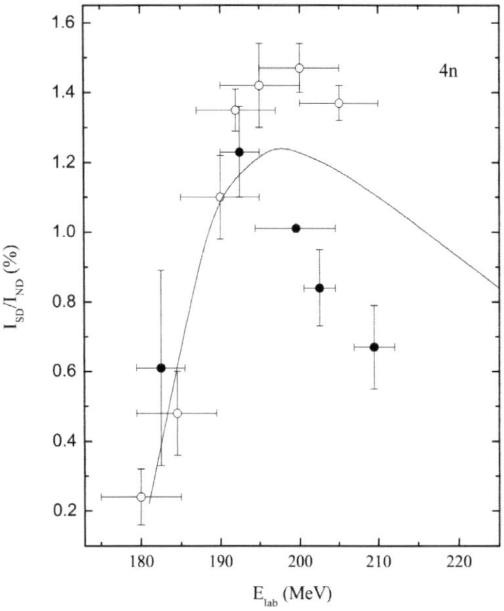

Figure 3. The ratio of SD and ND band intensities as a function of bombarding energy for the reaction ^{108}Pd(^{48}Ca,4n). The results of calculations and the experimental values taken from refs. [8] and [9] are presented by line, full and open circles, respectively.

interpretation of strongly deformed nuclear states and supports the predictions concerning the possible formation of hyperdeformed states in the entrance channel of heavy-ion reactions [5, 6], based on the same theoretical approach.

Acknowledgments
This work was supported by RFBR.

References

[1] Twin P J 1990 *Nucl. Phys.* A **520** 17
[2] Shneidman T M, Adamian G G, Antonenko N V, Ivanova S P and Scheid W 2000 *Nucl. Phys.* A **671** 119
[3] Adamian G G, Antonenko N V and Scheid W 2000 *Nucl. Phys.* A **678** 24
[4] Sargsyan V V, Adamian G G, Antonenko N V and Scheid W 2010 *Eur. Phys. J.* A **45** 125
[5] Zubov A S, Sargsyan V V, Adamian G G, Antonenko N V and Scheid W 2010 *Phys. Rev.* C **81** 024607
[6] Zubov A S, Sargsyan V V, Adamian G G, Antonenko N V and Scheid W 2010 *Phys. Rev.* C **82** 034610
[7] Lauritsen T *et al* 2007 *Phys. Rev.* C **75** 064309
[8] Twin P J 1988 *Proc. Int. Conf. on Contemporary Topics in Nuclear Structure Physics (Cocoyoc, Mexico)* (Singapore: World Scientific) p. 445
[9] Smith G *et al* 1992 *Phys. Rev. Lett.* **68** 158

Journal of Physics: Conference Series **436** (2013) 012063 doi:10.1088/1742-6596/436/1/012063

Charge distribution in disintegration of uranium isotopes

N Eren, N Buyukcizmeci and R Ogul

Department of Physics, University of Selcuk, TR-42075 Konya, Turkey

E-mail: eren@selcuk.edu.tr

Abstract. Charge distributions of the fragments emitted from the disintegration of uranium isotopes ^{235}U and ^{238}U at various excitation energies up to 10 MeV per nucleon were investigated on the basis of a statistical approach. The calculations reproduce the fundamental properties of the fission at low excitation energies for $E^* \leq 0.5$ MeV/A. The variation of the average value of the maximal charge, $< Z_{max} >$, of the emitted fragments with excitation energy is determined.

1. Introduction

Nuclear fragmentation reactions are of great interest in the context of nuclear physics and astrophysical processes such as supernova simulations and neutron star models [1]. In theoretical approaches to investigate the nuclear disintegration process, statistical models were found to be very successful to reproduce the experimental data [2, 3, 4, 5]. The mass and charge distributions at various excitation energies exhibit very different and interesting behavior. In our previous paper [10], we demonstrated that the evolution of the mass yield distribution is determined by the excitation energy. In the excitation energy range for $E^* \leq 1$ MeV per nucleon, one may observe fission-like distributions, while in the region of 1–3 MeV per nucleon an inverted U-shaped distribution is observed. The region of 3–6 MeV per nucleon is called the liquid-gas phase transition region, where the hot, compressed matter tends to expand in a thermodynamical equilibrium.

2. Theoretical calculations and results

There are several models to study nuclear fragmentation at various excitation energies. At low excitation energies up to 1 MeV per nucleon, disintegration of heavy nuclei are very well described by the compound nucleus model. At higher energies, beside the dynamic and kinetic models, statistical models have been found to be very suitable to reproduce the experimental data, especially for the description of the fragment production. The present calculations have been carried out on the basis of a statistical approximation, the so called statistical multifragmentation model (SMM) [6]. For a comparison of the statistical approach with experimental data we refer the reader to Ref. [10], where it has been shown that the results we obtained on the basis of the present statistical model are in good agreement with existing theoretical and experimental data.

In this short paper, we report on the analysis of the fission process of ^{235}U and ^{238}U nuclei in the excitation energy range of 0.05 – 0.5 MeV/nucleon. Fission of the compound nucleus

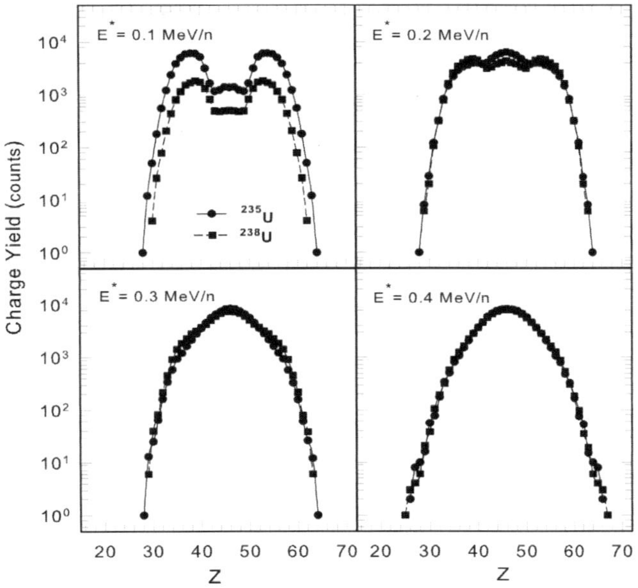

Figure 1. Charge yield distributions for the fission of ^{235}U and ^{238}U isotopes at low excitation energies, over 50000 events.

is investigated in the framework of the liquid-drop model taking into account shell effects at various deformation modes using the Bohr-Wheeler approach [7]. We used the fission barriers determined by Myers and Swiatecki [8], and the results of Ref. [9] for the level densities at the saddle point. The fragmentation processes of these isotopes were also examined for higher energies up to 10 MeV/nucleon. The shell effects on the level densities of the nucleus can be neglected at these high energies [11].

In Fig.1, we plotted the predicted charge yield distributions for the fission of ^{235}U and ^{238}U

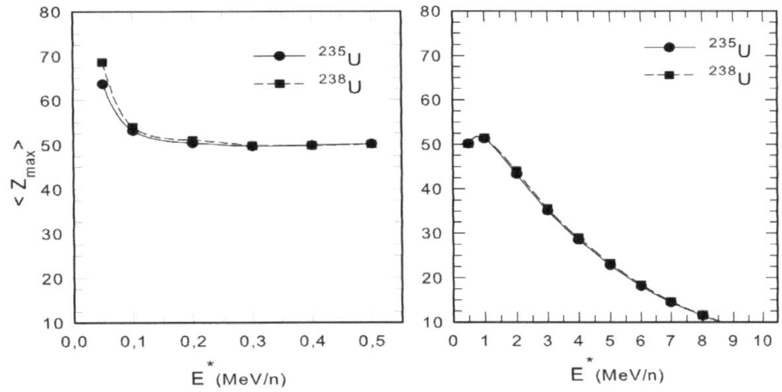

Figure 2. Average values of maximum charge of the fragments $< Z_{max} >$ as a function of the excitation energy in fission channel (left panel), and for higher energies (right panel).

nuclei in the low excitation energy region. From the yield differences shown in this figure, the rigidity of the nuclei can also be deduced, namely, the even-odd ^{235}U nucleus is more unstable against fission than ^{238}U. At higher energies this difference is washed out.

Fig.2 shows the variation of the average value of maximum charge yield as a function of excitation energy. To get the average value $< Z_{max} >$, we determined the maximal Z value in each simulated event and averaged it over the 50.000 simulated events produced at each excitation energy point. We observed a strong reduction in the average maximum charge, $< Z_{max} >$ with increasing excitation energy, E* (left panel). It is interesting to mention that in the fission excitation energy range, especially at energies ≤ 0.1MeV/nucleon, there is a difference in the $< Z_{max} >$ value between the two isotopes, in agreement with Fig. 1, where it has been shown that the fission products of the ^{235}U isotope have a wider Z distribution.

3. Conclusion

It is shown that $< Z_{max} >$ may be an indicator for the way of disintegration of heavy nuclei from fission to multifragmentation. In the low energy part of the fission zone (approximately E*= 0.02 – 0.2 MeV/A) one may distinguish between the average maximum charge values, $< Z_{max} >$, obtained for different isotopes, which may be correlated with the stability of these heavy nuclei against fission.

References

[1] Buyukcizmeci N *et al* 2010 *Journal of Physics: Conf. Series* **202** 012003.
[2] Ogul R and Botvina A S 2002 *Phys. Rev.* C **66** 051601.
[3] Buyukcizmeci N *et al* 2011 *Acta Physica Polonica* B **42** 697.
[4] Ogul R, *et al* 2009 *J. Phys.* G **36** 115106.
[5] Buyukcizmeci N *et al* 2012 *J. Phys.* G **39** 115102.
[6] Bondorf J P, Botvina A S, Iljinov A S, Mishustin I N and Sneppen K 1995 *Phys. Rep.* **257** 133
[7] Bohr N and Wheeler J 1939 *Phys. Rev.* **56** 426
[8] Myers W D and Swiatecki W J 1966 *Nucl. Phys.* **81** 1
[9] Iljinov A S *et al.* 1992 *Nucl. Phys.* A **543** 517
[10] Eren N, Buyukcizmeci Nand Ogul R 2007 *Physica Scripta* **76** 657.
[11] Cherepanov E A, Iljinov A S and Mebel M V 1983 *J. Phys.* G **9** 1397

Alpha-decay quantum-tunnelling calculations based on a folded Woods-Saxon potential

F R Xu, S M Wang, Z J Lin and J C Pei

State Key Laboratory of Nuclear Physics and Technology, School of Physics, Peking University, Beijing 100871, China

E-mail: frxu@pku.edu.cn

Abstract. Assuming that the α particle is a structureless point particle with two protons and two neutrons, we construct a mean-field-type cluster potential based on the Woods-Saxon potential with a folding factor which is to satisfy the quantization condition of a quasibound cluster state. The folded Woods-Saxon cluster potential has been successfully applied to the calculations of α-particle decay in light and superheavy nuclei. The standard values of the Woods-Saxon parameters were used without any adjustment. The calculated α-decay widths or lifetimes agree generally with experiment. Such a cluster potential leads to a consistent description of single-particle and cluster motions.

1. Introduction

Alpha-particle emission is the most important decay mode of nuclei heavier than Pb. Also, α decay has been observed widely from the excited states of light nuclei. In light nuclei, α-cluster structures are favoured when nuclei are excited to the vicinity of the α-decay threshold [1]. Molecular structures with two or more α particles and covalent neutrons in light nuclei have become a hot topic currently in both experiment [2, 3] and theory [4, 5]. In heavy nuclei, α decay can happen from the ground states as well. For superheavy nuclei, detecting α decay is a unique method to identify new superheavy elements.

Theoretically, α-particle emission can be considered a process of quantum tunnelling of an α particle through a potential barrier, which is called the Gamow decay model [6]. A reasonable barrier is crucial for the calculation of decay width or lifetime. Several phenomenological α-cluster potentials have been proposed for the calculations of α-decay half-lives and spectroscopic properties [7, 8, 9]. Double-folding microscopic cluster potentials have also been successfully applied to α-decay and α-scattering calculations [10, 11, 12, 13, 14].

2. The model

In our previous works [15, 16, 17, 18, 19, 20], we constructed folded mean-field-type cluster potentials based on microscopic Skyrme-Hartree-Fock [15, 16, 17] or phenomenological Woods-Saxon potentials [18, 19, 20]. These cluster potentials have been successfully applied to the calculations of various cluster decays including α and heavier cluster decays [15, 16, 19, 20] and molecular structures as well [16, 18]. In this paper, we review the Woods-Saxon-potential-based calculations with focusing on α decay from excited states of the light nucleus ^{24}Mg and from the ground states of superheavy nuclei.

In the spherical case, the cluster potential can be decomposed as (e.g., [8])

$$V(r) = V_N(r) + V_C(r) + \frac{\hbar^2}{2\mu r^2} l(l+1), \tag{1}$$

which contains the nuclear potential $V_N(r)$, the Coulomb potential $V_C(r)$ and the centrifugal potential with l and μ for the angular momentum carried by the α particle and the reduced mass of the α-core system, respectively [15]. The general folding procedure to derive a cluster potential can be written as [10]

$$V_N(r) = \lambda \int\int \rho_1(\mathbf{r}_1)\rho_2(\mathbf{r}_2) v_{\text{eff}}(|\mathbf{r}+\mathbf{r}_1-\mathbf{r}_2|)\mathrm{d}\mathbf{r}_1\mathrm{d}\mathbf{r}_2, \tag{2}$$

where λ is the folding factor, $v_{\text{eff}}(|\mathbf{r}+\mathbf{r}_1-\mathbf{r}_2|)$ is an effective nucleon-nucleon interaction, and $\rho_1(\mathbf{r}_1)$ and $\rho_2(\mathbf{r}_2)$ are the densities of the daughter and the cluster, respectively. To simplify the equation above, we consider the cluster as a structureless point particle, then obtain

$$V_N(r) = \lambda \int [Z_c \rho_1^p(\mathbf{r}_1) v_{\text{eff}}^p(|\mathbf{r}+\mathbf{r}_1|) + N_c \rho_1^n(\mathbf{r}_1) v_{\text{eff}}^n(|\mathbf{r}+\mathbf{r}_1|)]\mathrm{d}\mathbf{r}_1, \tag{3}$$

where the superscripts 'p' and 'n' indicate the proton and neutron, respectively, and N_c and Z_c are the neutron and proton numbers of the cluster, respectively. For the α particle, $N_c = 2$ and $Z_c = 2$. In Eq. (3) the integral gives the mean-field single-particle potential. Therefore, the nuclear potential between the cluster and the remaining core can be simplified further,

$$V_N(r) = \lambda[N_c v_n(r) + Z_c v_p(r)], \tag{4}$$

where $v_n(r)$ and $v_p(r)$ are single-neutron and single-proton potentials (excluding the Coulomb potential) respectively, generated by the core. Single-particle potentials can be obtained by mean-field models, such as the Skyrme-Hartee-Fock (SHF) [15, 16, 17] or the simple Woods-Saxon potential [18, 19, 20]. The Coulomb potential $V_C(r)$ need not be calculated by a folding procedure; we take its usual form [21] with assuming a homogeneous charge distribution in the daughter.

In this paper, we adopt the Woods-Saxon potential for the single-particle mean-field, that is

$$v(r) = \frac{V_0}{1 + e^{\frac{r-R}{a}}}, \tag{5}$$

with

$$V_0(r) = -V_{00}\left(1 \pm \kappa \frac{N_d - Z_d}{N_d + Z_d}\right), \tag{6}$$

where the sign is + (−) for the proton (neutron). The index 'd' indicates the daughter.

The folding factor λ is determined by the Bohr-Sommerfeld quantization condition, which, for the ground state, looks like

$$\int_{r_1=0}^{r_2} \sqrt{\frac{2\mu}{\hbar^2}|Q_0 - V(r)|}\mathrm{d}r = (2n+1)\frac{\pi}{2} = (G+1)\frac{\pi}{2}, \tag{7}$$

where r_1, r_2 (and r_3 later) are classical turning points obtained from $V(r_i) = Q_l^*$ (the decay energy). The global quantum number $G = 2n$ (n is the node number in the radial wave function of the cluster tunnelling motion) is determined by the Wildermuth rule [22], giving $G = \sum_{i=1}^{A_c} g_i$ where A_c is the nucleon number of the cluster and g_i is the oscillator quantum number of a

Journal of Physics: Conference Series **436** (2013) 012064 doi:10.1088/1742-6596/436/1/012064

cluster nucleon orbiting the core. The g_i numbers are those of the single-particle states occupied in the parent nucleus by the nucleons constituting the α particle to be emitted.

The partial α-decay width is calculated by [8, 16, 19]

$$\Gamma = P\frac{\hbar^2}{4\mu}\frac{\exp\left[-2\int_{r_2}^{r_3} k(r)\mathrm{d}r\right]}{\int_{r_1}^{r_2} \mathrm{d}r/2k(r)}, \tag{8}$$

where $k(r) = \sqrt{(2\mu/\hbar^2)|Q_l^* - V(r)|}$ is the wave number, and P is the preformation factor of the α-particle being formed in the mother. For even-even nuclei, it has been well established that the $P = 1$ assumption under the use of the Bohr-Sommerfeld condition can well reproduce the experimental half-lives of various cluster decays [8, 15]. The decay half-life is calculated by $T_{1/2} = \hbar \ln 2/\Gamma$. In the equations above, Q_l^* is the α-decay energy from an excited state,

$$Q_l^* = Q_0 + E_{J_i}^* - E_{J_f}^*, \tag{9}$$

where Q_0 is the α-decay energy of the ground state, and $E_{J_i}^*$ and $E_{J_f}^*$ are the excitation energies of the mother with spin J_i and the daughter with spin J_f. The orbital angular momentum l carried by the α particle can be determined from the vector coupling of $\mathbf{l}$ and $\mathbf{J}_f$ to $\mathbf{J}_i$ and from parity conservation. Since the decay calculation is very sensitive to the Q_l^* value, experimental values have been used for Q_0 and the excitation energies.

3. Alpha-decay calculations for the excited states of ^{24}Mg and the ground states of superheavy nuclei

In α-decay calculations for ^{24}Mg, we take the Chepurnov parameters [23] of the Woods-Saxon potential, which work well for light nuclei. For the ^{24}Mg ground state, nucleons belonging to the α particle should occupy orbits immediately above the Fermi surface of the daughter ^{20}Ne, i.e., the $d_{5/2}$ shell, which gives a value of $G = 8$ [19]. Using the Bohr-Sommerfeld condition with taking the experimental $Q_0 = -9.316$ MeV for the α+^{20}Ne channel, we obtain a folding factor of $\lambda = 0.608$ for the α-cluster potential in ^{24}Mg. It has been known experimentally that the ^{24}Mg excited states in the energy range of $\approx 10 - 15$ MeV decay dominantly by α emission [24]. The occurrence of α decay requires decay energies satisfying $Q_l^* = Q_0 + E_{J_i}^* - E_{J_f}^* > 0$. Therefore, the excited states in this energy range decay mainly into the 0_1^+ (g.s.) and 2_1^+ (1.63 MeV) states of ^{20}Ne. Assuming a preformation factor of $P \approx 1$ for every α-decay transition, we calculated the α-decay widths of some excited states of ^{24}Mg. The widths obtained agree, in most cases, with the experimental data within two orders of magnitude, see [19] for detailed results. The preformation probabilities of an α particle in the excited states of ^{24}Mg range from 10^{-2} to 1 [25], which might explain the discrepancies between the calculated and experimental widths.

The folded Woods-Saxon cluster potential has also been applied in the α-decay calculations of the ground states of superheavy nuclei [20]. For superheavy nuclei, however, we took another set of the Woods-Saxon parameters, which has been widely used in cranking shell-model calculations of high-spin states in heavy and superheavy nuclei (see, e.g., [26] and references therein). They are

$$\begin{aligned} V_{00} &= 53.754 \text{ MeV}, \\ \kappa &= 0.791, \\ a &= 0.637 \text{ fm}, \\ r_0 &= 1.19 \text{ fm}. \end{aligned} \tag{10}$$

The main difference between this set of parameters and the Chepurnov parameters for light nuclei is in the radius parameter r_0. The Chepurnov parameterization takes $r_0 = 1.24$ fm [23].

Table 1 lists the calculated half-lives of α decays for even-even superheavy nuclei. They agree with the experimental values within one order of magnitude. For superheavy nuclei, we took the global quantum number $G = 22$ [17], which is consistent with the Wildermuth rule. The discrepancies between calculations and data might come from two main factors: 1) deformation effects (most superheavy nuclei are deformed, while our calculations were limited to spherical shapes); 2) possible large uncertainties in experimental half-lives due to poor statistics of decay events.

Table 1. Calculated half-lives of observed α decays for even-even superheavy nuclei. Data are from [27, 28, 29].

Nuclei	λ	$T^{\rm calc}_{1/2,\alpha}$ (second)	$T^{\rm expt}_{1/2,\alpha}$ (second)
^{246}Fm	0.764	7.33×10^{-1}	$1.1\times10^{+0}$
^{248}Fm	0.763	$1.28\times10^{+1}$	$3.6\times10^{+1}$
^{250}Fm	0.762	$5.16\times10^{+2}$	$1.8\times10^{+3}$
^{252}Fm	0.761	$2.02\times10^{+4}$	$9.1\times10^{+4}$
^{254}Fm	0.757	$4.32\times10^{+3}$	$1.2\times10^{+4}$
^{256}Fm	0.755	$5.99\times10^{+4}$	$9.5\times10^{+3}$
^{252}No	0.758	$1.01\times10^{+0}$	$2.4\times10^{+0}$
^{254}No	0.757	$1.13\times10^{+1}$	$5.1\times10^{+1}$
^{256}No	0.751	6.95×10^{-1}	$2.9\times10^{+0}$
^{256}Rf	0.755	3.30×10^{-1}	3.6×10^{-1}
^{260}Sg	0.747	2.27×10^{-3}	3.6×10^{-3}
^{266}Sg	0.744	$1.94\times10^{+0}$	$2.6\times10^{+1}$
^{264}Hs	0.742	1.88×10^{-4}	1.0×10^{-4}
^{266}Hs	0.740	7.59×10^{-4}	2.3×10^{-3}
^{270}Hs	0.747	5.60×10^{-1}	
^{270}Ds	0.734	2.55×10^{-5}	1.0×10^{-5}
^{284}Cn	0.730	$9.39\times10^{+0}$	$9.8\times10^{+0}$
^{288}114	0.725	4.37×10^{-1}	$1.9\times10^{+0}$
^{292}116	0.720	1.82×10^{-2}	3.3×10^{-2}

In summary, we have proposed a folded Woods-Saxon α-cluster potential. The advantage of this cluster potential is that no new free parameter is to be introduced, which makes the predictions reliable. Moreover, it leads to a consistent description of single-particle and cluster motions. Within the framework of the quantum-tunnelling picture, the potential has been successfully applied to the α-decay of some excited states of ^{24}Mg with excitation energies of $E_x \approx$ 10–15 MeV and to the ground states of superheavy nuclei. The calculated decay widths or half-lives agree reasonably with experimental values.

Acknowledgments
This work has been supported by the National Key Basic Research Program of China under Grant 2013CB834400, and the National Natural Science Foundation of China under Grant Nos. 11235001 and 10975006.

References

[1] Ikeda K, Takigawa N and Horiuchi H 1968 *Prog. Theor. Phys. Suppl. (Japan)* **Extra Number** 464
[2] Freer M *et al* 1999 *Phys. Rev. Lett.* **82** 1383
[3] Itoh M *et al* 2011 *Phys. Rev.* C **84** 054308
[4] Kanada-En'yo Y and Horiuchi H 2003 *Phys. Rev.* C **68** 014319
[5] Ito M, Itagaki N, Sakurai H and Ikeda K 2008 *Phys. Rev. Lett.* **100** 182502
[6] Gamow G 1928 *Z. Phys.* **51** 204
[7] Buck B, Merchant A C and Perez S M 1992 *Phys. Rev.* C **45** 2247
[8] Buck B, Merchant A C and Perez S M 1993 *At. Data Nucl. Data Tables* **54** 53
[9] Xu C and Ren Z 2004 *Phys. Rev.* C **69** 024614
[10] Mohr P 2000 *Phys. Rev.* C **61** 045802
[11] Atzrott U, Mohr P, Abele H, Hillenmayer C and Staudt G 1996 *Phys. Rev.* C **53** 1336
[12] Xu C and Ren Z Z 2005 *Nucl. Phys.* A **760** 303
[13] Ren Z Z, Xu C and Wang Z 2004 *Phys. Rev.* C **70** 034304
[14] Chowdhury P R, Samanta C and Basu D N 2006 *Phys. Rev.* C **73** 014612
[15] Xu F R and Pei J C 2006 *Phys. Lett.* B **642** 322
[16] Pei J C and Xu F R 2007 *Phys. Lett.* B **650** 224
[17] Pei J C, Xu F R, Lin Z J and Zhao E G 2007 *Phys. Rev.* C **76** 044326.
[18] Xu C, Qi C, Liotta R J, Wyss R, Wang S M, Xu F R, and Jiang D X 2010 *Phys. Rev.* C **81** 054319
[19] Wang S M, Xu C, Liotta R J, Qi C, Xu F R and Jiang D X 2011 *Sci. China Ser. G-Phys. Mech. Astron.* **54** s130
[20] Lin Z J, Pei J C and Xu F R 2012 *Phys. Scr.* **T150** 014022
[21] Buck B, Hopkins P D B and Merchant A C 1990 *Nucl. Phys.* A **513** 75
[22] Wildermuth K and Tang Y C 1977 *A Unified Theory of the Nucleus* (New York: Academic Press)
[23] Chepurnov V A 1967 *Yad. Fiz.* **6** 955
[24] Firestone R B 2007 *Nucl. Data Sheets* **108** 2319
[25] Hodgson P E and Betak E 2003 *Phys. Rep.* **374** 1
[26] Xu F R, Satula W and Wyss R 2000 *Nucl. Phys.* A **669** 119
[27] Oganessian Y T *et al* 2004 *Phys. Rev.* C **70** 064609
[28] Oganessian Y T *et al* 2004 *Phys. Rev.* C **74** 044602
[29] Audi G *et al* 2003 *Nucl. Phys.* A **729** 3

Alpha-particle condensation in nuclear systems: present status and perspectives

P Schuck[1,2,3]

[1] Institut de Physique Nucléaire, CNRS, UMR8608, Orsay, F-91406, France
[2] Université Paris-Sud, Orsay, F-91505, France
[3] Laboratoire de Physique et de Modélisation des Milieux Condensés, CNRS et, Université Joseph Fourier, 25 Av. des Martyrs, BP 166, F-38042 Grenoble Cedex 9, France

Abstract. Alpha clustering and α condensation in lighter nuclei are presently strongly and increasingly discussed in the literature both from the experimental and from the theoretical side. A discussion of the present status of the theory as well as outlooks for future developements will be presented.

1. Introduction

For about ten years, since the idea of the possible existence of α condensate type of states in $n\alpha$ nuclei was formulated for the first time [1], many exciting new theoretical and experimental results have been produced. In this contribution, we would like to assess where we stand and what possible future extensions may be.

Let us start with the reminder that nuclear clustering and, in particular, α clustering would not exist, if we did not have in nuclear physics *four* different types of fermions (proton/neutron spin up/down), all attracting one another. We should be aware of the fact that this is a rather singular situation in fermionic many-body systems. However, the possibility of future trapping of four different kinds of cold fermionic atoms may open a new field of cluster physics with similar features. In a mean field description of an isolated α particle (which, with, e.g., Skyrme forces, gives reasonable results, if the c.o.m. motion is treated correctly) the four fermions can occupy the lowest (0S) level. Were there only neutrons, only two of them could be in the 0S-level, and the other two neutrons would have to be in the energetically very penalising 0P-state. That is why α particles exist, tetra-neutrons not. The ensuing fact is that α particles are very strongly bound ($E/A \sim 7$ MeV), almost as strongly as the most strongly bound nucleus, which is ^{56}Fe ($E/A \sim 8$ MeV). In addition the first excited state of the α particle (~ 20 MeV) is by factors higher than that of any other nucleus. The α particle can, therefore, be considered as an almost inert ideal bosonic particle. As we will see in the discussion below, in spite of its strong binding, α particle condensation can only exist in the so-called BEC (Bose-Einstein Condensation) phase, which implies low density. There is no analogue to the BCS phase of pairing where the Cooper pairs can have very large extensions, strongly overlapping with one another, still being fully antisymmetrised. This is the reason why α condensation can only be present at low densities where the α particles do not overlap strongly. (This holds, if the system consists of protons and neutrons and α's. If other clusters as t, ^{3}He, d are around, the situation may change, see below.) These considerations apply to nuclear matter as well as to finite nuclei. The Hoyle state in ^{12}C which can, to a good approximation, be described as a product of three α particles occupying all the 0S state of their bosonic mean field has a density which is by a factor 3-4 lower than the one of the ground state of ^{12}C. In the ground state there exist α-type of correlations but there is no condensation phenomenon. Let us start our considerations with infinite matter.

2. Alpha particle condensation in infinite matter

The in-medium four-body equation can be written in the following form

$$(E_{\alpha,\mathbf{K}} - \varepsilon_1 - \varepsilon_2 - \varepsilon_3 - \varepsilon_4)\Psi^{\alpha,\mathbf{K}}_{1234} = (1 - f_1 - f_2)v_{121'2'}\Psi^{\alpha,\mathbf{K}}_{1'2'34} + (1 - f_1 - f_3)v_{131'3'}\Psi^{\alpha,\mathbf{K}}_{1'23'4} + ... \quad (1)$$

In total, there are six terms coming from permutations. The ε_i are kinetic energies plus mean field corrections; v_{1234} are the matrix elements of the two-body interaction, and f_i is a Fermi-Dirac distribution of the uncorrelated nucleons accounting for phase space blocking. Repeated indices are summed over and index numbers comprise momenta and spins. The above equation describes *one* quartet in a gas of uncorrelated nucleons at temperature T. The analogous two-body equation can be used to determine the critical temperature T_c for the onset of superfluidity or superconductivity, where T_c has to be determined so that the eigenvalue comes at two times the chemical potential μ. This is the famous Thouless criterion of BCS theory. In analogy with pairing, one has to find the critical temperature T^α_c so that the eigenvalue of the four-body equation (1) comes at 4μ. The in-medium four-body equation is very difficult to solve. None the less, the solution has been found employing the Faddeev-Yakubovsky equations and using the Malfliet-Tjon bare nucleon-nucleon interaction which yields realistic nucleon-nucleon phase shifts and properties of an isolated α particle [2]. To simplify the problem, we made in addition a very easy-to-handle variational ansatz of the four-body wave function in (1). It consists of a mean field ansatz for the α particle projected on good total momentum. In momentum space this is

$$\Psi_{1234} \propto \delta(\mathbf{K} - \mathbf{k}_1 - \mathbf{k}_2 - \mathbf{k}_3 - \mathbf{k}_4)\varphi(\mathbf{k}_1)\varphi(\mathbf{k}_2)\varphi(\mathbf{k}_3)\varphi(\mathbf{k}_4) \quad (2)$$

Inserting this ansatz into (1), one obtains a nonlinear HF-type equation for the S-wave function $\varphi(\mathbf{k})$. Of course, for quartet condensation, we choose $\mathbf{K} = 0$. With the mean field ansatz (2), one cannot use a bare force. We adjusted an effective separable force with two parameters which are chosen to reproduce the binding energy and radius of the free α particle. The full Faddeev-Yakubovsky solution of (1) is shown for symmetric matter in Fig. 1 (crosses). We see

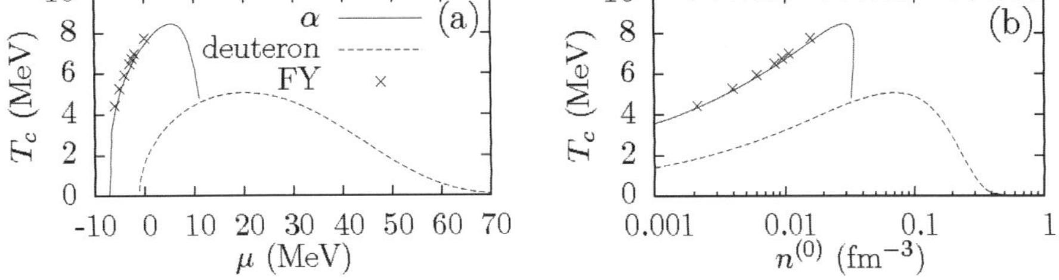

Figure 1. Critical temperatures for α particle and deuteron condensation in symmetric nuclear matter as a function of μ (a) and density $n^{(0)}$ (b).

that the ansatz (2) which very much eases the otherwise difficult solution of (1) works very well (continuous line). Also shown is the critical temperature for deuteron condensation. The striking feature is that α particle condensation abruptly breaks down already at very low density which approximately coincides with the point where the α's start to overlap appreciably (this fact was already found in [3] using a somewhat different variational ansatz for Ψ_{1234}). On the other hand deuteron condensation goes on up to very high densities and the limit is only triggered by the range of the effective force (which has been readjusted to reproduce the deuteron properties). This is so for symmetric nuclear matter. For strong asymmetry, deuteron condensation breaks down earlier than α condensation because the α particle according to its much stronger binding is less sensitive to asymmetry [4]. This is the afore-mentioned phenomenon that α condensation only exists in the BEC phase, i.e., at low density, whereas the deuteron continuously goes from negative to positive chemical potentials where for the latter the deuterons turn into large-size Cooper pairs. More on this can be found in [2, 3]. We should mention that our calculation of T_c^α is only reliable rather close to the breakdown point. For lower densities, T_c^α should join the one for condensation of ideal bosons (α's). To describe this feature, one should extend our theory to the so-called Nozières–Schmitt-Rink (NSR) theory [5] for pairing, see also [6], to α particle condensation. This, however, has not been worked out so far and remains a task for the future.

At zero temperature, there are many α's which go into the condensate phase. For this, we have to set up an approach analogous to the nonlinear BCS theory. Equation (1) corresponds to the linearised version and only describes *one* α particle in an otherwise uncorrelated gas (at finite T) of fermions. In finite nuclei, there may exist such a situation even at zero temperature. This is the case of ^{212}Po which can, to a certain extent, be viewed as an α particle sitting on top of the doubly magic core of ^{208}Pb which can be well described by a HF mean-field approach, i.e. a Fermi gas in a container. We will come back to this later when we discuss finite nuclei.

After having considered the linearised version of the equation for the quartet order parameter at the critical temperature, let us now try to write down, in analogy to the BCS case, the fully non-linear system of equations for the quartet order parameter. To clearly see the analogy to the BCS case, let us repeat the latter equations in a slightly unusual way. The pairing order parameter, allowing for non-zero c.o.m. momentum of the pairs, reads

$$(\varepsilon_{k_1} + \varepsilon_{k_2})\kappa_{\mathbf{k}_1\mathbf{k}_2} + (1 - n_{k_1} - n_{k_2})\Delta_{\mathbf{k}_1\mathbf{k}_2} = 2\mu\kappa_{\mathbf{k}_1\mathbf{k}_2} \tag{3}$$

where $\Delta_{\mathbf{k}_1\mathbf{k}_2} = \sum v_{k_1k_2k_{1'}k_{2'}}\kappa_{\mathbf{k}_{1'}\mathbf{k}_{2'}}$ and $\kappa_{\mathbf{k}_1\mathbf{k}_2} = \langle \mathrm{BCS}|c_{\mathbf{k}_1}c_{\mathbf{k}_2}|\mathrm{BCS}\rangle = u_{\mathbf{k}_1}v_{\mathbf{k}_2}$ is the usual pairing tensor (with spin indices suppressed) and $n_k = v_k^2 = 1 - u_k^2$ are the BCS occupation numbers. The ε_k's are, as before, the kinetic energies, eventually including a mean field correction. The occupation numbers can be obtained from the Dyson equation

$$G_k^\omega = G_k^0 + G_k^0 M_k^\omega G_k^\omega, \tag{4}$$

with $M_k = \Delta_k\Delta_k^*/(\omega + \varepsilon_k)$ the BCS mass operator where Δ_k is the diagonal part of the gap for cases where the pairs are at rest. From the single-particle Green's function, obviously we can calculate the occupation numbers thus closing the typical BCS self-consistency cycle. Inspired by the BCS case, we then write for the quartet order parameter [see (1)]

$$(4\mu - \varepsilon_1 - \varepsilon_2 - \varepsilon_3 - \varepsilon_4)\kappa_{1234} = (1 - n_1 - n_2)v_{121'2'}\kappa_{1'2'34} + (1 - n_1 - n_3)v_{131'3'}\kappa_{1'23'4} + \ldots \tag{5}$$

with $\kappa_{1234} = \langle c_1c_2c_3c_4\rangle$. Again, we have to close the equation with the Dyson equation for the occupation numbers. However, the mass operator now contains the quartet order parameter

$$M_1^\alpha = \sum_{234} \frac{\Delta_{1234}[\bar{n}_2^0\bar{n}_3^0\bar{n}_4^0 + n_2^0n_3^0n_4^0]\Delta_{1234}^*}{\omega + \varepsilon_2 + \varepsilon_3 + \varepsilon_4}\delta(\mathbf{k}_1 + \mathbf{k}_2 + \mathbf{k}_3 + \mathbf{k}_4) \tag{6}$$

with $\bar{n}^0 = 1 - n^0$ and n_i^0 being the uncorrelated Fermi function, i.e., the Fermi step and $\Delta_{1234} = \sum V_{123'4'}\kappa_{3'4'34}$ where V_{1234} is an effective coupling vertex linking the single-particle

Journal of Physics: Conference Series **436** (2013) 012065 doi:10.1088/1742-6596/436/1/012065

motion to the order parameter (for more details and derivation, see [7]). Before trying to solve this equation, let us discuss the differences between the pairing and the quartet case. The first thing which strikes is that the three 'holes' only have to have total momentum $\mathbf{k}_2+\mathbf{k}_3+\mathbf{k}_4 = -\mathbf{k}_1$ and, therefore, we have a remaining sum over momenta. In the pairing case with only one 'hole', there is no sum. Furthermore in the pairing case the hole propagator has no phase space factor because the 'forward' and 'backward' going parts add up to one: $\bar{n}^0 + n^0 = 1$. In the quartet case there are three hole propagators and the corresponding sum of phase space factors does *not* add up to one, i.e. $\bar{n}_1^0\bar{n}_2^0\bar{n}_3^0 + n_1^0 n_2^0 n_3^0 \neq 1$! This makes a dramatic difference from the pairing case. In order to understand this a little better, let us compare the level density of a single hole with the one of three holes:

$$g^{1h}(\omega) = \sum_k [\bar{n}_k^0 + n_k^0]\delta(\omega + \varepsilon_k) = \sum_k \delta(\omega + \varepsilon_k) \tag{7}$$

$$g^{3h}(\omega) = \sum_{\mathbf{k}_1\mathbf{k}_2\mathbf{k}_3} [\bar{n}_{k_1}^0\bar{n}_{k_2}^0\bar{n}_{k_3}^0 + n_{k_1}^0 n_{k_2}^0 n_{k_3}^0]\delta(\omega + \varepsilon_{k_1} + \varepsilon_{k_2} + \varepsilon_{k_3}) \tag{8}$$

The 3h level density is shown in Fig.2. We see that for positive μ there is a striking difference

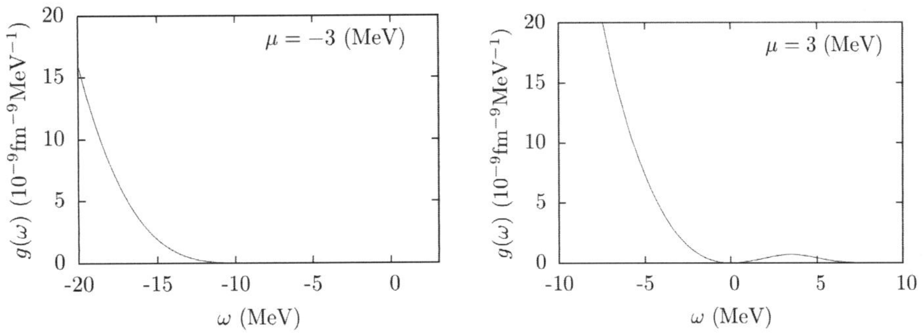

Figure 2. 3h level densities for negative and positive chemical potentials, respectively. Note that on the horizontal axis, the origin corresponds to $\mu = 0$.

from the 1h level density. At positive μ, $g^{1h}(\omega = \mu)$ is obviously finite (not shown), whereas $g^{3h}(\omega = 3\mu)$ goes through zero. This is because phase space constraints and energy conservation cannot be fullfilled simultaneously at the Fermi surface in the latter case as is easily verified. That is, in the quartet case, exactly at the point where the correlations should be built up, namely at the Fermi level, there is no level density! As a consequence, no quartet condensation is possible for positive μ. On the contrary, for negative μ, $n^0 = 0$ and thus the phase space factor in the case of g^{3h} is also equal to one and then there is no qualitative difference from the 1h case. This explains in a natural way why quartet condensation is not possible at positive chemical potentials. It is, by the way, well known that any mp–nh level density, besides the single-particle case, goes through zero at the Fermi level. For example, the origin of fermions at the Fermi energy having an infinite free mean path stems from the fact that the 2p-1h (2h-1p) level density is zero at the Fermi level and that is the equivalent to the imaginary part of the optical potential being zero at that energy. Also the 2p-2h level density, which plays an important role for the damping of zero sound modes, is zero at zero frequency. As first pointed out by Landau, it approaches zero as ω^2. In conclusion, for positive μ only pairing survives, whereas quartetting breaks down; it only exists in the BEC phase with negative μ. The situation may be different when other light clusters are present, i.e. in a mixed gas of, e.g., nucleons, tritons (^{3}He) and deuterons. Then a nucleon with momentum $\mathbf{k}$ may, eventually, directly pair with, e.g., a triton of momentum $-\mathbf{k}$

(or the other way round), rather similarly to the standard pairing situation apart from the fact that now two fermions with different masses have to pair up. Similar considerations hold for the pairing of two deuterons (pairing of 'bosons'). In compact star physics such situations may exist when the star is cooling down. The extension of our theory to this scenario is a task for the future. The full solution of the nonlinear set of equations (6) and (7) is again very much eased in taking for the order parameter the factorisation ansatz (2). The most interesting result is that the occupation number, e.g., for μ around zero is far from being close to saturation. At $k = 0$ it is only approximately $n_{k=0} \sim 0.30$. This scenario is analogous to pairing in the BEC regime. More results can be found in [7].

3. Finite Nuclei

As we know from pairing, a direct observation of condensation phenomena is only possible in finite nuclei. Of course, in such small systems, there cannot exist a condensation in the macroscopic sense. Nevertheless, as we know very well, only a handful of Cooper pairs suffices to show clear signatures of pairing in nuclei. For α particle condensation it is the same story. We can only expect that there exists about a handful of α particles, essentially in lighter $n\alpha$ nuclei, in a gaseous phase at low density. It is, indeed, surprising that such states at low density with about $\rho = \rho_0/3 - \rho_0/4$ with ρ_0 the density at saturation do exist as quite long lived excited states in those nuclei. The most famous example is the Hoyle state in ^{12}C at 7.65 MeV, just about 300 keV above the 3α threshold. We will not dwell much on the successful theoretical description of this state (and others, e.g., in ^{16}O) with the THSR wave function [1], since this has been presented in the recent literature a great number of times [8]. Let us only make a couple of remarks. The THSR wave function is schematically written for a finite number of quartets as

$$\Psi_{n\alpha}^{\mathrm{THSR}} = \mathcal{A}[\Phi_\alpha \Phi_\alpha ... \Phi_\alpha] \tag{9}$$

where the single-α wave function Φ_α depends on four spatial coordinates, the spin-isospin part being suppressed. This wave function is fully antisymmetrical due to the antisymmetriser $\mathcal{A}$ and is analogous to the number-projected BCS wave function

$$\Psi^{\mathrm{BCS}} = \mathcal{A}[\phi_{\mathrm{pair}} \phi_{\mathrm{pair}} ... \phi_{\mathrm{pair}}] \tag{10}$$

where ϕ_{pair} is the Cooper pair wave function depending on two spatial coordinates. The calculus with the α condensate wave function is very much facilitated by a variational ansatz where Φ_α is split into a product of a c.o.m. gaussian with a large width parameter B times another, intrinsic, gaussian depending only on the relative coordinates of the α particle and having a width parameter b which corresponds to the size of an isolated α particle. The first remark to be made is that this THSR wave function contains two important limits: if $B = b$ then it corresponds to a pure harmonic oscillator Slater determinant. If $B \gg b$, then the α's are so distant from one another that the Pauli principle among the different α's can be neglected and, thus, the antisymmetriser be dropped. The THSR wave function is then a pure product state of α particles, i.e., a condensate state of ideal bosons. Reality is, of course, in between those limits and one important task is to find out whether reality is closer to a Slater determinant or to a Bose condensate. That this question must be carefully investigated, can also be deduced from the fact that a number-projected BCS wave function always leads to a non-trivial pairing solution. For example for ^{208}Pb, one obtains a non-trivial BCS solution in spite of the fact that ^{208}Pb is certainly not superfluid. Only when the original particle-number-breaking BCS theory with $|\mathrm{BCS}\rangle = e^{\sum z_{kk'} c_k^+ c_{-k}^+} |\mathrm{vac}\rangle$ has a non-trivial solution, can we speak of a superfluid nucleus. For ^{208}Pb there is no such solution. One way to analyse whether the THSR approach leads primarily to an α condensate or to a Slater determinant, is to investigate the bosonic occupation numbers. An ideal Fermi gas has occupation numbers which are either one or zero.

In an ideal Bose condensate the bosons will occupy the lowest single-particle state with 100%. Of course, in real nuclei, neither the fermionic nor the bosonic occupation numbers agree with those of an ideal gas. For nucleons, we know that occupation numbers are around 70-80% down from 100%. For the α occupations in the condensate, we also have found a number around 70%, all the other occupancies being down by at least a factor of ten, whereas for the ground state the occupation numbers are almost equally distributed according to the SU(3) shell model scheme [8]. This is a typical condensate situation, though not totally the one of an ideal condensate. Residual antisymmetry effects between the slightly overlapping α particles makes that, with a certain probability, the α's are scattered out of the condensate. In addition there are also other correlations at work. One of the most important ones is certainly the formation of ^{8}Be clusters out of two α's. The answer is unclear even to the question whether α gas does not in reality consist of a gas of ^{8}Be's. The latter are, of course, also bosons and the old question arises whether in an attractive Bose gas the bosons condensate as singles or as molecules [9]. This is certainly a very interesting question which deserves further studies in the future.

Many other extensions of α condensation are presently discussed theoretically and experimentally. An intersting aspect is whether on top of the α condensate states excited α gas states exist. A long debate has recently been closed about the nature and existence of the second 2^+ state close to 10 MeV excitation energy in ^{12}C. Very nice experimental results by M. Freer, M. Itoh, and M. Gai have recently shown that this state is there and that it is a member of a family of α gas states [10, 11, 12]. Many more results, for instance, in ^{16}O are to be expected. One of the most exciting aspects is that one may be able to dismantle $n\alpha$ nuclei, with n a rather large number like $n = 10$ or more, into n α-particles. The first results in this direction have been reported at this conference on ^{56}Ni, a nucleus with 14 α-particles, by H. Akimune. A dream would be that all α particles be just excited to the Ikeda threshold and then they disintegrate in a very slow motion as a kind of coherent state driven by the Coulomb force, i.e., it would be some kind of soft Coulomb explosion. This would then be rather close to what happens with a trapped Bose condensate of cold atoms upon switching off the trapping potential. Other exciting perspectives are that α particles could exist in a gaseous phase on top of an inert core. For example four α's on top of ^{16}O in ^{32}S, or other variants, even in quite heavy nuclei. Because of space restrictions, we, unfortunately, cannot enlarge further upon these exciting aspects of α-gas type of states in nuclei.

Before closing, we would like to discuss, however, the question of a possible preformation of α particles in heavy nuclei. As is well known, this question is of great importance for the description of α decay rates. Let us take the example already alluded to in the infinite matter section, namely ^{212}Po. Since the lead core is doubly magic, one can treat it in mean field, i.e. as a Slater determinant with a suitable Skyrme- or Gogny-type force. That is we may view ^{212}Po as an α particle on top of a finite Fermi gas. If such a configuration exists, the α decay rate of ^{212}Po suggests, and it is clear from our experience with the Hoyle state as well, that such a cluster state cannot be described within the shell model alone, see [13, 14]. In the infinite matter section, we have already considered a situation where a single α particle is embedded in an uncorrelated Fermi gas. This was the case with respect to the critical temperature. For ^{212}Po we can consider our treatment of the single-α case even at zero temperature. Of course we cannot use the ansatz (2) as it is, since it corresponds to a c.o.m. wave function which is a plane wave $e^{i\mathbf{KR}}$. In position space (2) reads $\Psi_{1234} \propto e^{i\mathbf{KR}}\psi_{int.}(|\mathbf{r}_i - \mathbf{r}_j|)$ where $\psi_{int.}$ is the intrinsic wave function depending only on the relative coordinates. In a finite nucleus, instead of the plane wave, one would have to use a wave packet considering the following wave function

$$\Psi_{1234} \propto \Phi(\mathbf{R})\psi_{int.}(|\mathbf{r}_i - \mathbf{r}_j|) \tag{11}$$

where $\psi_{int.}$ and Φ are to be determined variationally upon inserting (11) into Eq. (1) where all indices and ingredients correspond now to the mean field of ^{208}Pb. For instance, ε_i correspond to the HF energies and the f_i are to be replaced by the HF occupation numbers at $T = 0$. In addition, for $\psi_{int.}$, one could use, as for the study of the Hoyle state, a gaussian wave function with the width parameter b. Then the single unknown would be the spherical wave function $\Phi(R)$ to be determined variationally where R is the c.o.m. coordinate of α with respect to the centre of the Pb core. Our knowledge that α's can only exist at very low density incites us to believe that this Φ wave function should be peaked rather far out in the surface of the Pb nucleus. If true, this would be a nice explanation of a preformed α particle in the nuclear surface. Adding more than one α to the Pb core may suggest that there could exist some sort of α condensate in the surface in a fluctuating state. However, as we know, adding more α's to the Pb core leads to deformed nuclei and the treatment of α's in the surface becomes a much more delicate subject.

4. Conclusions

As we have seen, the existence of α gas and α condensate states in nuclear systems where the α's play practically the role of elementary bosons, is fascinating. Nuclear physics is at the forefront of this kind of physics. In the future, experiments with cold atoms trapping four (or more) different kinds of fermions may also open wide perspectives in the field of cluster physics. For more reading on α cluster states, we invite the reader to consult our review article [8].

Acknowledgments

This contribution is in collaboration with, Y. Funaki, H. Horiuchi, G. Röpke, T. Sogo, A. Tohsaki, and T. Yamada.

References

[1] Tohsaki A, Horiuchi H, Schuck P and Röpke G 2001 *Phys. Rev. Lett.* **87** 192501
[2] Sogo T, Lazauskas R, Röpke G and Schuck P 2009 *Phys. Rev.* C **79** 051301
[3] Röpke G, Schnell A, Schuck P and Nozières P 1998 *Phys. Rev. Lett.* **80** 3177
[4] Sogo T, Röpke G and Schuck P 2010 *Phys. Rev.* C **82** 034322
[5] Nozières P and Schmitt-Rink S 1985 *J. Low Temp. Phys.* **59** 195
[6] Jin M, Urban M and Schuck P 2010 *Phys. Rev. C* **82** 024911
[7] Sogo T, Röpke G and Schuck P 2010 *Phys. Rev.* C **81** 064310
[8] Yamada T, Funaki Y, Horiuchi H, Röpke G, Schuck P and Tohsaki A 2011 *Nuclear Alpha-Particle Condensates* arXiv:1103.3940; *Clusters in Nuclei vol 2* (Lecture Notes in Physics vol 848) ed C Beck (Berlin: Springer) pp 229–298
[9] Nozières P and Saint-James D 1982 *J. Physique* **43** 1133
[10] Gai M for the UConn-Yale-Duke-Weizmann-PTB-UCL Collaboration 2011 *J. Phys.* C **267** 012046; 2011 *Acta Phys. Pol.* B **42** 775
[11] Itoh M *et al* 2011 *Phys. Rev.* C **84** 054308
[12] Freer M *et al* 2009 Phys. Rev. C **80** 041303
[13] Varga K, Lovas R G and Liotta R J 1992 *Phys. Rev. Lett.* **69** 37; 1992 *Nucl. Phys.* A **550** 421
[14] Astier A, Petkov P, Porquet M-G, Delion D S and Schuck P 2010 *Eur. Phys. J.* A **46** 165

Nuclear matter and surface clustering

Q N Usmani,[1] N Abdullah[1], K Anwar[1] and Z Sauli[2]

[1]Institute of Engineering Mathematics, University Malaysia Perlis, Malaysia
[2]School of Microelectronics Engineering, University Malaysia Perlis, Malaysia

Email:qamar@unimap.edu.my

Abstract. We demonstrate that the binding energy per nucleon of symmetric nuclear matter (SNM) (with Coulomb interaction switched off and $N = Z$) in the limit of zero density approaches to its value, u_v, at the saturation density, where u_v is the volume term of the Weizsäcker mass formula. This phenomenon is a direct result of the clustering of nuclei in the low density region of nuclear matter. We study the implications of this result on the properties of nuclei. We also study the properties of asymmetric nuclear matter. Because of clustering a provocative interpretation of the equation of state of asymmetric nuclear matter emerges which is at considerable variance at low densities with hitherto all the previous calculations. For nuclei, as a framework, an extended version of Thomas-Fermi theory is invoked. Calculations are performed for 2149 nuclei with N, $Z \geq 8$. The present scheme leads to a forceful interpretation of the low density asymmetry energy data of Natowitz *et al.* [1].

1. Introduction

In this paper we discuss the implication of clustering in nuclear matter at low densities. Properties of the nuclear matter play a very important role in shaping the structure of a nucleus, its energy and other properties. Here, we consider a hypothetical symmetric nuclear matter ($N = Z$) where the Coulomb interaction is switched off. Large number of binding energy calculations of such a system have been carried out in the past with two and three-nucleon realistic or simplified potentials using either variational or different versions of Bruckner g-matrix techniques. A common feature of most of these calculations is that the binding energy per nucleon (*BE*/*A*) of the SNM approaches zero as the density approaches zero. The following thought experiment demonstrates that this feature is at a gross variance with clustering aspect of nuclear matter at low densities. We demonstrate that, based upon a quite general consideration, that in the limit of zero density the *BE*/*A* approaches u_v ($\approx$ 16 MeV), where u_v is the volume term of the Weizsäcker mass formula which corresponds to the value at the saturation density ρ_0 ($\approx$ 0.16 fm^{-3}). In [2], it was demonstrated that an idealized picture of α-matter of SNM leads to *BE*/*A* $\approx$ 7.3 MeV, which is roughly the binding energy of α-particle per nucleon without the Coulomb interaction. But why α-particles? Why not heavier nuclei since they have lower *BE*/*A* since the Coulomb interaction is switched off? One may thus consider SNM, in the limit of zero density, as an aggregate of heavier nuclei or large chunks of SNM itself. In the latter case the *BE*/*A* will then tend to its value at the saturation density which is u_v. We thus obtain the exact relation

$$E(\rho \to 0) \to E(\rho = \rho_0) = -u_v, \tag{1}$$

where, E is the energy per nucleon of SNM. The above considerations are based upon the concept of ideal cluster matter which do not interact in the limit of zero density. However, at somewhat higher but low densities the SNM will be a highly correlated system with complicated structure. We have shown elsewhere [3], that the equation of state (EOS) of SNM will have one maximum between 0 and

Journal of Physics: Conference Series **436** (2013) 012066 doi:10.1088/1742-6596/436/1/012066

ρ_0. This claim along with the identity (1) leads to the possibility of the presence of clusters at low densities, which in turn then offers a force interpretation of the asymmetry energy data of Natowitz *et al* [1]. In addition, within the framework of an extended Thomas-Fermi model [4], we demonstrate that clustering is fully consistent with the static properties of nuclei, namely, the binding energies and rms radii. Because of the clustering the EOS of asymmetric nuclear matter (NM) is also found to be qualitatively different than all the previous studies.

In section 2, we give the formulations and show how the above presented picture of nuclear matter is incorporated in the theory to calculate the properties of the nuclei. Section 3 gives the results and discussion. Section 4 contains the conclusions.

2. Asymmetry Energy, EOS of Nuclear Matter and an Extended Tomas Fermi Model

The asymmetry energy $E_{\text{asym}}(\rho)$ is defined as the difference between the $E(\rho)$ of pure neutron matter (PNM) and the SNM

$$E_{\text{asym}}(\rho) = E(\rho, \delta = 1) - E(\rho, \delta = 0) \quad (2)$$

where the isospin asymmetry parameter is defined as $\delta = (\rho_n - \rho_p)/\rho$ with ρ_n and ρ_p as the densities of neutrons and proton respectively. Neutron matter is a gas of neutrons at all densities, we obtain the exact result $E_{\text{asym}}(\rho \to 0) \to u_v$, since $E(\rho \to 0,\ \delta = 0) \to -u_v$. Also, the EOS of NM with arbitrary δ can be expanded by Taylor series where we ignore the small charge symmetry breaking component:

$$E(\rho, \delta) = E(\rho, \delta = 0) + S(\rho)\delta^2 + Q(\rho)\delta^4 \quad (3)$$

In the above expansion, we have retained terms up to the fourth power of δ. This was found necessary to obtain an adequate description of the static properties of nuclei when we use the EOS of PNM calculated using Diffusion Monte Carlo technique with realistic interactions [5]. We have no knowledge about the density dependence of $Q(\rho)$. We make the simplifying assumption that $Q(\rho)$ is proportion to $S(\rho)$. We thus replace $S(\rho)$ by $(1\text{-}q)\ E_{\text{asym}}$ and $Q(\rho)$ by qE_{asym}, where q is an input parameter which is determined from the fits to the static properties of nuclei.

For the EOS of SNM we use the following phenomenological form [3,6] which satisfy the identity (1) with a one maximum between 0 and ρ_0.

$$E(\rho_{\geq}) = -u_v + \frac{K}{18}\left(\frac{\rho - \rho_0}{\rho_0}\right)^2 + M\left(\frac{\rho - \rho_0}{\rho_0}\right)^3, \qquad \rho \geq \rho_x \quad (4a)$$

$$E(\rho_{\leq}) = -u_v + A\rho + B\rho^2 + C\rho^3 + D\rho^4 + \frac{3\hbar^2(3\pi^2/2)^{2/3}}{5m_N}\rho^{2/3}, \qquad \rho \leq \rho_x \quad (4b)$$

where ρ_x is a parameter between 0 and ρ_0. It may be noted from (4a) and (4b) that as the density approaches ρ_0 and 0, E approaches $-u_v$ in accordance with the identity (1). The constant terms A, B, C, and D are determined by equalizing $E(\rho_{\geq})$ and $E(\rho_{\leq})$ and their first three derivatives at $\rho = \rho_x$. (4a) essentially contains two parameters, M and ρ_x. The other parameters are consistent with generally accepted values, namely, $u_v \approx 16$MeV, the compression modulus $K \approx 230$ MeV, and $\rho_0 \approx 0.16$ fm^{-3}.

To calculate the properties of nuclei, we use an extended version of the Thomas-Fermi theory [3, 4] which is a precursor of density functional theories recently being developed under the SciDAC program [7]. Density Functional theories consist of terms pertaining to homogenous part of the nuclear matter and non-homogenous parts which controls the surface properties of the nucleus [4, 7]. We write the energy of a nucleus as a functional of neutron and proton densities:

$$\mathcal{E}[\rho_n, \rho_p] = \int \left[E(\rho,\delta)\rho + \frac{\hbar^2}{2m}\left(\tau_2(\rho_n,\rho_p) + \tau_4(\rho_n,\rho_p)\right) + a_\rho(\nabla\rho)^2 - a_{np}\left(\nabla\rho_n - \nabla\rho_p\right)^2 \right] d\vec{r}$$

$$+ \frac{1}{2}e^2 \int \frac{\rho_p(\vec{r})\rho_p(\vec{r}\,')}{|\vec{r} - \vec{r}\,'|} d\vec{r}\,' d\vec{r} - \frac{3}{4}\left(\frac{3}{\pi}\right)^{1/3} e^2 \int \rho_p^{4/3}(\vec{r})\, d\vec{r} + Shell + a_{pair}A^{-1/3}\Delta_{np} + E_W \quad (5)$$

The first line in (5) corresponds to volume and surface terms. The first and the second integrals in the second line are, respectively the direct and exchange Coulomb terms. The expressions for the kinetic energy densities can be found in Ref. [3]. The *shell* contribution is taken from Ref. [8], the pairing Δ_{np} is taken from Ref. [9]. The Wigner contribution is taken from Refs. [10]:

$$E_W = V_W \exp\left\{-\lambda\left(\frac{N-Z}{A}\right)^2\right\} + W_W\,|N-Z|\exp\left\{-\left(\frac{A}{A_0}\right)^2\right\} \quad (6)$$

For neutron and proton densities we employ a three parameter modified Fermi distribution [2], we thus have a total of six variational parameters which were obtained by minimizing the density functional (5) for each nucleus. Fits to energies are obtained by minimizing the rms deviation

$$\sigma(E) = \sqrt{\sum_i \left(E^i_{theo} - E^i_{\exp}\right)/N} \quad (7)$$

We similarly define the rms deviations $\sigma(R)$, $\sigma(S1)$, $\sigma(S2)$, and $\sigma(Q_\beta)$ for the point proton radii, one neutron, two neutron and β-decay energies, respectively. We minimize $\sigma(E)$ by varying M, u_v, ρ_o a_ρ, q, a_{pair} and the Wigner parameters V_W, λ, W_W and A_0 for fixed values of ρ_x. The value of the parameter ρ_x is sensitive to the asymmetry energy data of Natowitz *et al.*, [1]. It is found to be around 0.05–0.06 fm^{-3}. The parameter searches were made through an automated procedure.

3. Results and Discussion

Neutron Matter	Clustering	$\sigma(E)$ *MeV*	$\sigma(R)$ *fm*	$\sigma(S1)$ *MeV*	$\sigma(S2)$ *MeV*	$\sigma(Q_\beta)$ *MeV*	*q*
AV8'+UIX	Yes	0.664	0.044	0.391	0.519	0.510	0.171
AV8'+UIX	No	0.671	0.042	0.419	0.564	0.557	0.151
AV8'	Yes	0.637	0.043	0.381	0.508	0.506	0.097

Table I: Results with 2149 nuclei. $\rho_x = 0.06$ fm^{-3}.

We have considered a total of 2149 nuclei [11]. Results are displayed in Table I. $\sigma(R)$, $\sigma(S1)$, $\sigma(S2)$, $\sigma(Q_\beta)$, are for the measured 799 [12], 1988, 1937, and 1868 nuclei, respectively. The column "Clustering" indicates that if the clustering has been taken into account or not. If we use the EOS (4) as it is we have the clustering indicated by "Yes". If we make $u_v = 0$ in (4b), it will correspond to no clustering in the surface region indicated by "No". The non-zero values of q signify that the quartic isospin term is of some importance in the expansion (3). These results are discussed in details in Ref. [3].

rms deviations	Present	Möller *et al.*, Ref.[13]	Brussels-Montreal	
			HFB-14 Ref.[10]	HFB-17 Ref.[10]
$\sigma(E)$ *MeV*	0.637	0.669	0.729	0.581
$\sigma(R)$ *fm*	0.043	–	0.031	0.030
$\sigma(S1)MeV$	0.380	0.411	–	0.506
$\sigma(S2)MeV$	0.505	–	–	–
$\sigma(Q_\beta)MeV$	0.504	–	–	0.583
$\sigma(S_{nr})MeV$	0.561	0.910	0.833	0.729
No. of Nuclei	2149	1654	2149	2149

Table II: Root mean square deviations in various approaches.

In Table II, we compare our results with some of the earlier authors where the clustering aspect demonstrates improvement, particularly in the one neutron separation energies $\sigma(S1)$ and $\sigma(S_{nr})$, where $\sigma(S_{nr})$ is the one neutron separation energies which are less than 5 MeV (a subset of 186 neutron rich nuclei). These indeed are sensitive to surface properties. In Fig. 1, we plot the calculated and experimental differences for energies of 2149 nuclei for $N, Z \geq 8$ in the left panel. The right panel gives the differences of proton point rms radii as a function of A. The points are for AV8' from table I.

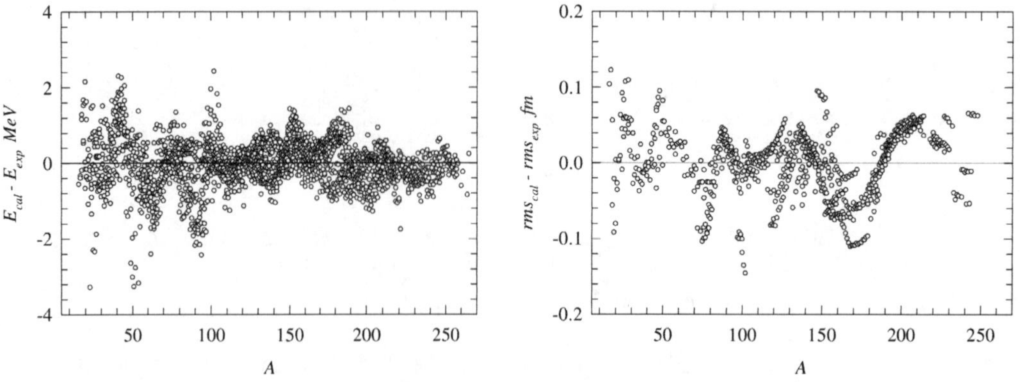

Fig1 : The calculated and experimental differences in energies and point proton rms radii as a function of *A*.

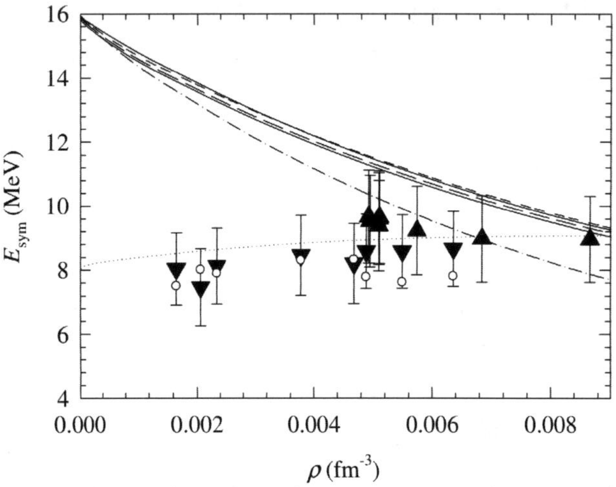

Fig2: Comparison of the calculated and experimental asymmetry energy data. For details see text.

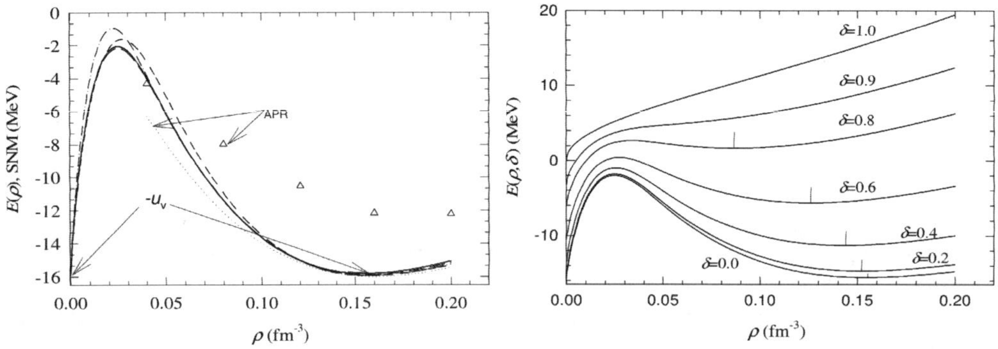

Fig3: EOS of symmetric (left panel) and asymmetric (right panel) nuclear matter. For details see text.

In Fig. 2, we compare our asymmetry energy results with the experimental results of Natowitz *et al.*, [1]. The up triangles represent the results after the medium modification is taken into account on the data represented by down triangles. The open circles and the dotted line are the results of Quantum Statistical calculations. The curves (with the same legend as in Fig3, see next paragraph) are the results of our calculations.

In the left panel of Fig 3, we plot the EOS for SNM. The long dashed line represents the results for AV8'+UIX, the solid line shows for AV8'. These curves are for ρ_x = 0.06 fm^{-3}. The dashed-dot and short-short curves are for AV8'+UIX with ρ_x = 0.05 fm^{-3} and ρ_x = 0.07 fm^{-3}, respectively. It is seen that for a change of ρ_x by 0.02 fm^{-3}, the change in the location of maximum, E_{max}, in the EOS of SNM is only 0.005 fm^{-3}. It is fixed around $\rho \approx 0.025\ fm^{-3}$. Also shown is the EOS of Akmal *et al.*, (APR) [14]. The triangles are the results with $AV18 + UIX$ with relativistic boost corrections. The dotted line is obtained by invoking heuristic corrections to account for the known empirical values of ρ_0, K and u_v. Below the spinodal density (0.05 fm^{-3}) most of the many body calculations of SNM become unreliable.

In the right panel, we plot the EOS of asymmetric NM for different values of the asymmetry parameter δ. The vertical short lines represent the location of equilibrium densities. Note that all the curves meet the y-axis [$E(\rho, \delta)$-axis] at nonzero negative values for $\delta < 1$. It is interesting to note that these negative values of energies are always less than the corresponding values of the minimum energies at the equilibrium densities. In fact they represent the ground state of the asymmetric NM; not the energies at the equilibrium densities. In earlier NM calculations all the energies approach zero as the density approaches zero for all values of δ.

4. Conclusions

In a phenomenological way cluster of all sizes, shapes along with medium modifications are included. This modifies the equation of state (EOS) of nuclear matter at low densities significantly. It is demonstrated that clustering is consistent with the static properties of nuclei. Also, a forceful interpretation of the experimental asymmetry data emerges and the exact result that $E_{asym}(\rho \to 0) \to u_v$.

References

[1] Natowitz J B, Röpke G, Typel S, Blaschke D, Bonasera A, Hagel K, Klähn T, Kowalski S, Qin L, Shlomo S, Wada R and Wolter H H 2010 *Phys. Rev. Lett.* **104** 202501
[2] Johnson M T and Clark J W, 1980 *Kinam* **2** 3
[3] Usmani Q N, Nooraihan A, Anwar K and Zaliman S 2011 *Phys. Rev.* C **84** 064313
[4] Brack M, Guet C and Håkansson H-B 1985 *Phys. Rep.* **123** 275
[5] Gandolfi S, Illarionov A Yu, Schmidt K E, Pederiva F and Fantoni S, 2009 *Phys. Rev.* C **79** 054005;
Gandolfi S, Carlson J and Reddy S, arXiv:1101.1921v1
[6] Usmani Q N, Bodmer A R and Sauli Z, 2008 *Phys. Rev.* C **77** 034312;
Usmani Q N, Sauli Z and Abdullah N 2011 *J. Phys. Conf. Ser.* **312** 092060;
Sauli Z, Abdullah N, Anwar K and Usmani Q N 2012 *J. Phys. Conf. Ser.* **381** 012102
[7] http://www.unedf.org/content/papers_db.php
[8] Myers W D and Swiatecki W J 1994 *LBL Report* 36803;
Myers W D and Swiatecki W J 1996 *Nucl. Phys.* A **601** 141
[9] Mendoza-Temis J, Hirsch J G and Zuker A P 2010 *Nucl. Phys.* A **843** 14
[10] Goriely S, Samyn M, Bender M and Pearson J M 2003 *Phys. Rev.* C **68** 054325
Goriely S, Chamel N and Pearson J M 2009 *Phys. Rev. Lett.* **102** 152503
[11] Audi G, Wapstra A H and Thibault C 2003 *Nucl. Phys.* A **729** 337
[12] Angeli I 2004 *At. Data Nucl. Data Tables* **87** 18
[13] Möller P, Nix J R, Myers W D and Swiatecki W J 1985 *At. Data Nucl. Data Tables* **59** 185
[14] Akmal A R, Pandharipande V R and Ravenhall D G 1998 *Phys. Rev.* C **58** 1804

Dynamical description of exotic structures at subnuclear densities

V de la Mota[1], F Sébille[1], M Novak[1] and Ph Eudes[1]
[1] SUBATECH, Ecole des Mines, Université de Nantes, CNRS/IN2P3, Nantes, France
E-mail: delamota@subatech.in2p3.fr

Abstract. The dynamics of infinite nuclear matter in the conditions of density and temperature expected in the outermost layers of neutron stars is studied in the framework of a microscopic time-dependent mean-field approach around zero temperature. Dynamical processes in inhomogeneous nuclear matter are studied using a large number of nucleons in numerical simulations without any assumptions on the morphology of nuclear matter. The occurrence of exotic structures when varying internal conditions as densities, nuclear species and elementary cell symmetries is investigated. The corresponding structures are studied in terms of a phase diagram in density space evidencing some sensitivity to the isospin-dependent part of the equation of state.

1. Introduction

Neutron stars (NS) have interested physicists and astronomers for long time insofar as they can shed light on our understanding of fundamental properties of matter under extreme conditions. The study of neutron stars has gained a renewed interest on the last decade by means of the new generation of launched observatories which are enabling new observations and providing us with a large amount of new data.

Modern nuclear physics input as well as many-body theories are valuable tools to improve our comprehension of the physics of compact objects. Astrophysics and nuclear physics are inextricably intermingled disciplines. Any advance in these two areas leads to a progress in the other and this interplay enables us to deepen our knowledge of the universe. In the nuclear domain, new experimental programs with radioactive beams are nowadays emerging (SPIRAL2, FRIB, FAIR). Their corresponding goal is to produce those rare isotopes playing a relevant role in astrophysical phenomena, aiming at providing answers to open questions about the nature of NS and about nucleosynthesis processes.

In most theoretical models of NS the outermost layers have the structure of a crystal lattice of neutron rich nuclei immersed in a degenerate electron gas and, as density increases, it is essentially a fluid composed of dripped neutrons. In this region, where the density varies from 0.1 to 0.5 ρ_∞, the saturation value of nuclear matter, the occurrence of exotic structures as rods, slabs and other complex shapes has been predicted by theoretical approaches in the framework of the liquid drop model [1] a few decades ago. More recently other static [2] and dynamical [3] models have confirmed this early prediction.

The existence of these structures is assumed to be a consequence of the competition between Coulomb and nuclear forces, this phenomenon being called "frustration", and to be sensitive to the equation of state (EOS) of nuclear matter. Their presence may have important astrophysical consequences on macroscopic characteristics, as masses and radii [4], as well as on the cooling processes of the star [5]. In particular, they should affect the neutrino opacity [6] and the transport [7] and elasticity [8] properties of the crust.

Aiming at investigating the microscopic dynamics of matter in the crust, we have developed a dynamical model [10] beyond the Wigner-Seitz approximation [11]. Since these structures involve a large number of low energy configurations, a pure mean field description of the nuclear dynamics has been performed. In this approach, starting from initial crystalline lattices of nuclei, with different symmetries, non spherical structures occur as the result of microscopic self-organization processes. The survival of those meta-stable equilibrium structures has been checked over thousands fm/c. In this work, the influence of the EOS, of the isotopic composition and of the lattice symmetries on the formation of exotic structures are studied.

This paper is organized as follows, in Section 2 the essential ingredients of the model are presented. In Section 3 a selection of relevant results are shown and in Section 4 the conclusions and perspectives are given.

2. The model: Dynamical Wavelets in Nuclei for Neutron Stars (DYWAN-NS)

In the range of characteristic temperatures and densities of NS, namely $T \ll 1$ MeV and $10^6\text{g/cm}^3 \leq \rho \leq 10^{13}\text{g/cm}^3$, the stellar matter is correctly described in terms of a system of interacting nucleons in a uniform background of electrons insuring the neutrality of matter. In this context we have developed a dynamical approach which is based on the DYWAN model [12] describing heavy ion collisions.

In the present approach nuclei are initially located on the sites of a crystalline lattice with periodic boundary conditions. A static self-consistent procedure is implemented in order to solve the Hartree-Fock (HF) equation for nuclear composites, which are prepared either in their ground states or in excited states according to mechanical or thermal constraints. The HF equation is numerically solved by spanning the one-body density matrix in a convenient basis. The elements of this basis are wavelets [13], which are functions of a set of four parameters: $\{\langle x\rangle, \langle p_x\rangle, \chi, \gamma\}$ in x coordinate, with similar expressions in y and z coordinates. Here $\chi = \langle x^2\rangle - \langle x\rangle^2$ and γ is defined as:

$$\gamma = \frac{\sigma^2}{2\chi} \qquad \text{with} \qquad \sigma = \langle[(x - \langle x\rangle), (p_x - \langle p_x\rangle)]_+\rangle$$

For this calculation a density-dependent zero-range effective interaction has been chosen with the following self-consistent field:

$$\begin{aligned} V_q^{HF}(\rho,\xi) \; = \; & \frac{t_0}{\rho_\infty}\,\rho + \frac{t_3}{\rho_\infty^{7/6}}\,\rho^{7/6} + \frac{c}{{\rho_\infty}^2}\xi^2 + \frac{4qc}{{\rho_\infty}^2}\rho\xi + \\ & \frac{\Omega}{3{\rho_\infty}^2}\xi^2 + \frac{4q\Omega}{3{\rho_\infty}^2}(\rho-\rho_\infty)\xi + V_q^C \end{aligned} \qquad (1)$$

In Eq. (1) ρ_n and ρ_p stand for neutron and proton densities, $\rho = \rho_n + \rho_p$, $\xi = \rho_n - \rho_p$, q=1/2 for neutrons and -1/2 for protons, ρ_∞=0.145 fm^{-3} is the saturation density of infinite nuclear matter and V_q^C is the Coulomb interaction. With the current values of the parameters, namely c=20 MeV, Ω= -100 MeV, $t_0/\rho_\infty = -356$ MeV fm^3 and $t_3/\rho_\infty^{7/6} = 303$ MeV fm$^{7/2}$, the principal static characteristics of nuclei and of infinite neutron matter are reproduced.

The Coulomb term for protons is calculated using Ewald summation techniques [14], which are adapted to the calculation of long range potentials in periodic systems.

The parameters c and Ω are related to the coefficients of the mass formula J and L corresponding to, respectively, the volume-symmetry and to the density-dependent symmetry energy as follows:

$$c = J - \frac{h^2}{6M}k_F^2 \qquad\qquad L = 6c + \Omega + L_{kin}$$

where L_{kin} is the kinetic energy contribution to L (see Eq. (5) below). These parameters are still uncertain because they are not completely constrained from nuclear data. It seems then worthwhile to perform a study of their influence on the overall dynamical behavior of the crust. To this end c has been fixed to the value above defined but Ω was varied in the range -120 $\leq \Omega \leq$ -75 MeV in order to analyze the sensitivity of the dynamics to this interaction. The density dependence of the symmetry energy has been determined here in a phenomenological way according to current estimates at the saturation value in pure neutron matter [9]. The values of the parameters reproduce the principal static characteristics of nuclei, as binding energies, radii and equilibrium densities. The associated incompressibility modulus in symmetric matter K_∞= 200 MeV corresponds to a "soft" EOS.

The energy density per baryon ω_q is defined as:

$$\omega_q = \frac{\varepsilon}{\rho} = \frac{\int V_q^{HF}\, d\rho}{\rho} + \omega_{kin}, \tag{2}$$

where ω_{kin} corresponds to the kinetic contribution. According with Ref. [15] the isovector part of the energy density per baryon is:

$$\omega_\delta = \frac{1}{2} \lim_{\delta \to 0} \frac{\partial^2 \omega_q}{\partial \delta^2} \tag{3}$$

with $\delta = \xi/\rho$, and the corresponding contribution to the incompressibility modulus is:

$$K_{sym} = 9\rho^2 \frac{\partial^2 \omega_\delta}{\partial \rho^2}\Big|_{\rho=\rho_\infty}. \tag{4}$$

The kinetic contribution to the density-dependent symmetry energy is defined as:

$$L_{kin} = 3\rho \frac{\partial \omega_{kin}}{\partial \rho}\Big|_{\rho=\rho_\infty} \tag{5}$$

In Fig. 1 we have plotted the values of L and of K_{sym} as a function of J. The squares A, B and C are the results obtained with the implemented force for Ω= -75, -100 and -120 MeV, respectively. The remaining symbols are extracted from Skyrme Hartree-Fock (SHF) calculations from Ref. [15] and references therein. The numbers in the figure denote the different parameter sets for SHF:1 for SI, 2 for SIII, 3 for SIV, 4 for SVI, 5 for Skya, 6 for SkM, 7 for SkM*, 8 for SLy4, 9 for MSkA, 10 for SkI3, 11 for SkI4, 12 for SkX, 13 for SGII. In this picture the results corresponding to EOS A and B are compatible to the reference theoretical calculations, while the soft EOS C value is too low for both correlations. Nevertheless, we recall that the effective force implemented here is a simplified version of a Skyrme-like interaction. It depends on a reduced set of parameters thus providing efficient computational formula. It should be underlined here that in spite of the simplicity of the force, the global trends of the calculated physical quantities displayed in this section are in good accordance with other theoretical approaches.

The dynamical evolution of the system is governed by the time-dependent Hartree-Fock (TDHF) equation for the one body density matrix. As shown elsewhere [10] the TDHF equation is solved by a variational principle which gives the evolution of the 12 parameters of wavelets (4 for each dimension). The evolution is ruled then by the mean-field given by Eq. (1), the influence of which is felt by all nuclei belonging to the periodic box.

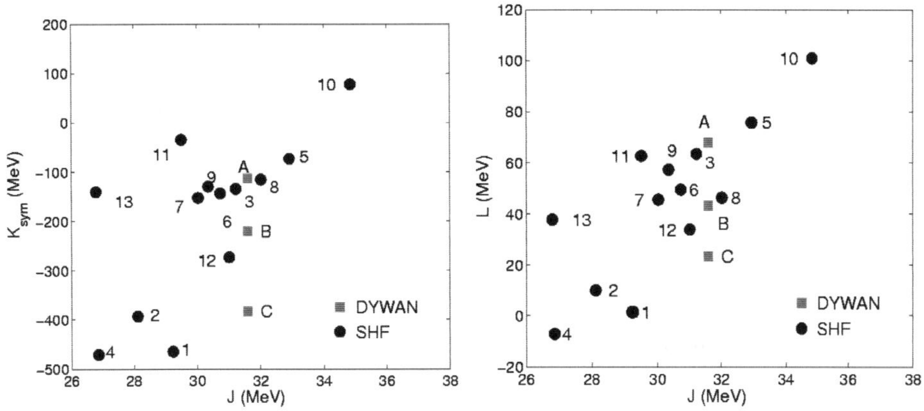

Figure 1. Correlations between macroscopic properties of asymmetric nuclear matter K_{sym} vs J (right) and L vs J (left). Full dots are Skyrme Hartree-Fock (SHF) calculations form Ref. [15] and squares are DYWAN-NS results with Ω= -75 MeV (A), -100 MeV (B) and -120 MeV (C).

3. Results

Let us consider the simplest crystalline arrangement given by a simple cubic cell (SCC) box of oxygen isotopes with proton fraction x=$\langle\rho_p\rangle/\langle\rho\rangle$=0.2, $\langle\rho_p\rangle$ and $\langle\rho\rangle$ being proton and total average densities, respectively. The lattice sites are located at the corners of a cube. Each nuclei at a given site is then shared equally between 8 adjacent cubes. These nuclei have been coherently excited with the same small quadrupole deformation at the initial time. In light of the lattice level density distribution it is possible to select spatial regions where the density is higher than a given threshold, called threshold density ρ_t. The morphological characteristics of the selected matter distribution can be analyzed by means of Morphological Image Analysis (MIA) technics [16] in order to characterize the underlying structures. In Fig. 2 is represented the structure diagram observed for the neutron density as a function of the mean density, normalized to the saturation value, in terms of ρ_t which is represented in the vertical axis.

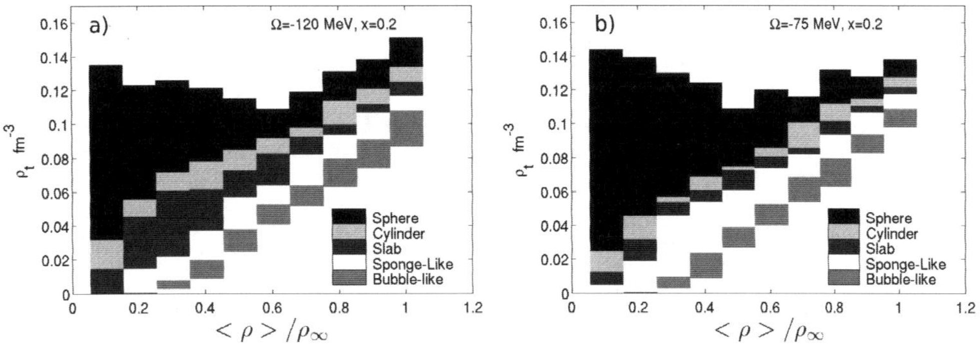

Figure 2. Neutron phase diagram in the threshold density versus the mean density plane for two values of Ω: -120 MeV (a) and -75 MeV (b), for proton fraction x=0.2. The different structures are represented in grey scale.

Two choices of the asymmetry parameter have been provided for: Ω= -120 MeV (a) and -75 MeV (b). In both cases, for a given threshold density, the five standard types of pasta phases emerge naturally and in the same order they should appear with increasing density as predicted by static models [1] and confirmed by recent calculations. Some sensitivity to the asymmetry-dependent contribution to the effective force also emerges from Fig. (2). The stiffer force, for Ω=-75 MeV, favors the occurrence of spherical and sponge-like structures, while restraining that of cylinder and slabs. This is consistent with the fact that the stiff potential is more repulsive, namely at high mean densities. The consequence is that more and more neutrons are dripped or localized at the surface of clusters. In the last case, quasi-free neutrons contribute either to build heavy spherical nuclei or to link neighboring clusters in all directions, giving rise to the so-called sponge-like structures.

As underlined by other authors [17] the inner crust of a neutron star is an extremely complex system in which the appearance of disordered phases can be favored. There is then a special interest to investigate the response of nuclear matter to the introduction of incoherent perturbations at the initial time. The behavior of the system under incoherent perturbations strongly depends on the symmetry of the crystal. It is well known from solid state physics that SCC are unstable against Coulomb interactions. In the NS context, where the nuclear field plays a preponderant role, the system is shown to loose the fingerprints of the initial lattice symmetry [18] all along the dynamical evolution.

The feasibility of those disordered phases in a different, more complex, kind of crystalline arrangement has been studied. To this end face centered cubic (FCC) lattices of symmetric oxygen nuclei have been perturbed incoherently, by shifting at random their position from the lattice sites. In addition to the 8 corner lattice sites are included here the centers of the 6 faces, giving a total of 4 lattice points per unit cell. Some snapshots of the corresponding neutron density evolution are shown in Fig. 3. In this case, the average density is $0.4\rho_\infty$ and the threshold

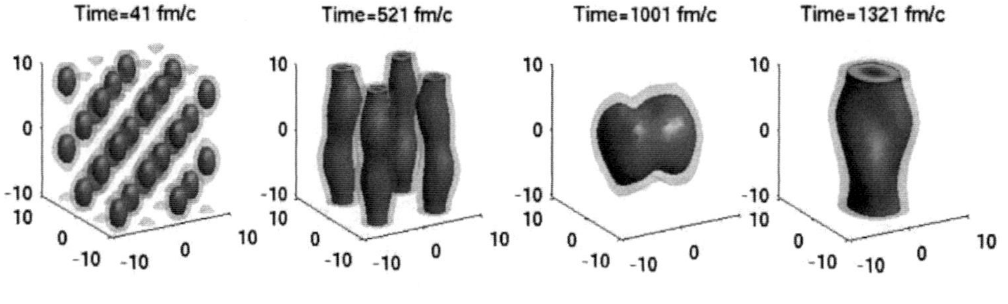

Figure 3. Neutron density profiles of a perturbed supercell of 0.5 proton fraction oxygen isotopes in face centered cubic lattices at different times. The mean and threshold densities are $\langle\rho\rangle = 0.4\rho_\infty$ and 0.04 fm^{-3}, respectively.

density for the 3-dimensional plot is ρ_t=0.04 fm^{-3}. The stability conditions of the FCC initial lattice is such that despite of the random perturbations imposed to the location of nuclei, the system attempts to preserve symmetries exhibiting stringy or cylindrical structures. At sufficiently long times the dominant arrangement corresponds to a single infinite rod-like structure which is periodically replicated.

In order to investigate the influence of the charge and the mass of nuclear species on the onset and on the dynamical evolution of structures, iron nuclei with different isotopic compositions have been considered. In Fig. 4 the neutron phase diagram corresponding to the particular case of SCC iron lattices with proton fraction x=0.5 is shown. The overall mean density is in this

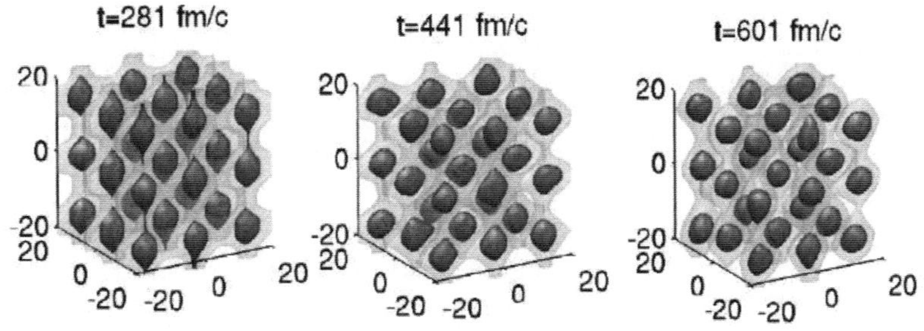

Figure 4. Neutron density profiles of a perturbed simple cubic cell iron lattice with 0.5 proton fraction as a function of time. The mean and threshold densities are $\langle\rho\rangle/\rho_\infty$=0.1 and 0.05 fm^{-3}, respectively.

case $\langle\rho\rangle$=0.1ρ_∞ and the threshold density is $\rho_t = 0.05$ fm^{-3}. In the same way as for the system represented in Fig. 3, the positions of nuclei have been randomly shifted from the lattice sites at the initial time. In this case, the observed structures are the individual iron nuclei which undergo shape oscillations around their initial deformed states. In this case, the stability and symmetry of non-equilibrium structures seem not being substantially modified by incoherent perturbations. This behavior can be observed similarly for other kind of cells as face centered cells, not shown here, and can be interpreted as the consequence of a balance in Coulomb and nuclear forces relative weight between neighboring nuclei. For iron lattices in the symmetric $x = 0.5$ case clusters are smaller than those found in perturbed oxygen systems and are stable since perturbations do not break the initial symmetries. From this result we can expect that at low densities matter could exist as a stable crystalline arrangement of iron nuclei without breaking up of symmetries as a consequence of the microscopic dynamics.

4. Conclusions

In this work we presented an overview of the microscopic DYWAN-NS model, which has been recently developed in order investigate the dynamics of matter as it should exist in the outermost layers of neutron stars. In this framework, the occurrence of self-organized structures in nuclear matter emerges as a natural consequence of the microscopic many-body dynamics. The structure of the stellar crust is self-consistently build up starting from a crystal of slightly deformed nuclei under periodic boundary conditions. The lattice is immersed in a degenerate electron gas insuring the neutrality of the system. The principal static properties of nuclear matter have been compared with other theoretical calculations. Despite the fact that a simplified effective force has been implemented, the considered macroscopic properties of nuclear matter are shown to be in close agreement with reference calculations. In the current version of the model the evolution of the system is ruled by the nuclear mean-field owing to the fact that in this work we focus the investigation on the crust at around zero temperature .

A variety of structural phases, the so-called "pasta phases", are shown to be built self-consistently from the microscopic nuclear motion. The complex morphology of these structures has been analyzed using integral geometry techniques. Beyond spherical nuclei, rod-like structures, slabs, cylindrical bubbles, slabs with holes, connected slabs and sponges have been found. Transitions between these dynamical phases take place at constant energy.

The morphology of the dominant phases has been represented in structure diagrams on the plane defined by the threshold density ρ_t and the overall mean density $\langle\rho\rangle$. A sensitivity to the

EOS has been evidenced in SCC cells with 0.2 proton fraction.The neutron diagram plots exhibit an increase of the regions corresponding to spherical and sponge-like structures at expense of slabs and cylinders with the stiffening of the effective interaction. In this case more neutrons are propelled toward the surface regions of initial clusters.

The response of the system to the effect of random perturbations on the initial lattice has been analyzed in order to investigate the possibility of the occurrence of disordered phases. The breaking of initial symmetries observed previously in oxygen SCC systems is not seen in FCC lattices, nevertheless a non-equilibrium steady cylindrical structure is formed in the particular case of oxygen isotopes with proton fraction 0.5 at $0.4\rho_\infty$.

Finally the sensitivity to the nuclear masses composition has been considered by studying the evolution of a randomly perturbed iron lattice in a SCC configuration. The corresponding proton fraction and mean density are, respectively, 0.5 and $0.1\rho_\infty$. In this case the system performs shape oscillations around the initial quasi-spherical state, which is consistent with the hypothesis of crystalline structures composed by iron nuclei at the low-density external layers.

In the present version of the model individual nucleons, either bound or unbound, evolve in a self-consistent mean-field holding in the overall box under periodic boundary conditions. The number of nuclei considered in a given box depends on the nature of the elementary cell. The occurrence of heavy aggregates and uniform phases at low proton fraction is deeply related with the ability of the model to describe the spreading of wave functions widths, corresponding to unbound nucleons. This fact is a consequence of the time dependence of wavelet parameters. Nevertheless, the formation and transitions between these non-equilibrium structures can be strongly modified using more sophisticated non-local effective forces and beyond a pure mean field description. These investigations are currently in progress.

References

[1] Ravenhall D G, Pethick C J and Wilson J R 1983 *Phys. Rev. Lett.* **50** 2066; Hashimoto M, Seki H and Yamada M 1984 *Prog. Theor. Phys.* **71** 320
[2] Magierski P, Bulgac A and Heenen P-H 2003 *Nucl. Phys.* A **719** 719; Montani F, May C and Muther H 2004 *Phys. Rev.* C **69** 0645801; Onsi M, Dutta A K, Chatri H, Goriely S, Chamel N and Pearson J M 2008 *Phys. Rev.* C **77** 065805; Newton W G and Li B-A 2009 *Phys, Rev.* C **80** 065809
[3] Horowitz C J, Perez-Garcia M A and Piekarewicz J 2004 *Phys. Rev.* C **69** 045804; Sonoda H, Watanabe G, Sato K, Takiwaki T, Yasuoka K and Ebisuzaki T 2007 *Phys. Rev.* C **75** 042801(R); Gögelein P, van Dalen E N E, Fuchs C and Müther H 2008 *Phys. Rev.* C **77** 025802
[4] Haensel G, Potekhin A Y and Yakovlev D G 2007 *Neutron Stars* vol 1 (Springer)
[5] Page D, Geppert U and Weber F 2006 *Nucl. Phys.* A **777** 497
[6] Horowitz C J, Perez-Garcia M A and Piekarewicz J 2004 *Phys. Rev.* C **69** 045804
[7] Zhang H F, Lombardo U and Zuo W 2010 *Phys. Rev.* C **82** 015805
[8] Gearheart M, Newton W G, Hooker J and Li B-A 2011 arXiv:1106.4875v1 [astro-ph.SR]
[9] Friedman B and Pandharipande V R 1981 *Nucl. Phys.* A **361** 501
[10] Sébille F, Figerou S and de la Mota V 2009 *Nucl. Phys.* A **822** 51
[11] Chamel N, Naimi S, Khan E and Margueron J 2007 *Phys. Rev.* C **75** 055806
[12] Jouault B, Sébille F and de la Mota V 1998 *Nucl. Phys.* A **628** 119
[13] Unser M and Aldroubi A 1992 *Polynomial splines and wavelets- A signal processing perspective in Wavelets: A tutorial in Theory* ed C K Chui (New York: Academic Press) p 91
[14] Ewald P 1921 *Ann. Phys.* **64** 253; Toukmaji A Y and Board J A Jr. 1996 *Comp. Phys. Comm.* **95** 73
[15] Yoshida S and Sagawa H 2006 *Phys. Rev.* C **73** 044320
[16] Michielsen K and De Raedt H 2001 *Phys. Rep.* **347** 461
[17] Magierski P and Heenen P-H 2002 *Phys. Rev.* C **65** 045804
[18] Sébille F, de la Mota V and Figerou S 2011 *Phys. Rev.* C **84** 055801

10th International Conference on Clustering Aspects of Nuclear Structure and Dynamics IOP Publishing
Journal of Physics: Conference Series **436** (2013) 012068 doi:10.1088/1742-6596/436/1/012068

Nuclear matter properties in fragmentation reactions

Akira Ono

Department of Physics, Tohoku University, Sendai 980-8578, Japan

Abstract. Nuclear multifragmentation is closely related to the properties of nuclear matter. In particular, the possibility of strong cluster correlations is explored based on the antisymmetrized molecular dynamics calculations. We introduce the two-nucleon collision process to the final states in which the scattered nucleons may form clusters (such as deuterons, tritons and α particles). This cluster correlation has strong impacts on the whole collision dynamics. The multifragmentation data at 50 MeV/nucleon are well reproduced with this strong cluster correlation if pairs of clusters with small relative velocities are assumed to form light nuclei. Effects of the density dependence of symmetry energy are discussed by analyzing the compression-expansion dynamics and the emission of nucleons, clusters and fragment nuclei.

1. Introduction

In medium-energy heavy-ion collisions, the system is compressed at an early stage and then expands so that the whole system disintegrates into many intermediate mass fragments and light particles. The reaction dynamics can be regarded as a manifestation of nuclear matter properties, such as the equation of state and the liquid-gas phase transition. In addition to the mean-field aspect, many-body correlations are also an important aspect of matter properties.

The experimental data shows that the clusters are actually an important part of multi-fragmenting systems. In central Xe + Sn collisions at 50 MeV/nucleon [1], only about 10 % of the total protons in the system are emitted as free protons, while the other protons are bound in light particles (d, t, ^{3}He, α) and heavier fragments in the final state. The number of protons in α particles (20 %) is more than the number of free protons (10 %). In theoretical studies, there are large discrepancies in the number of emitted nucleons predicted by different transport models (see Ref. [2] for example), and most transport models tend to overestimate the experimental data of the proton multiplicity. Therefore the cluster correlations may be much stronger than usually assumed by transport models.

We are recently extending the antisymmetrized molecular dynamics (AMD) approach in order to properly incorporate cluster correlations in the final states of two-nucleon collisions induced by the residual interaction. This model can explain the multiplicities of protons and clusters (such as α particles) as well as the fragment size distribution in central collisions at 50 MeV/nucleon. Currently there is a lot of interest in the equation of state of asymmetric nuclear matter at various densities [3, 4]. The model is also applied to the study on how the density dependence of symmetry energy is reflected in the collision dynamics and final observables such as the triton/^{3}He spectra and the isotopic yields of fragments.

2. Model

AMD [5] solves the time evolution of many-nucleon system starting with two boosted nuclei with a given impact parameter and a suitable distance between them. To describe the state at each time, AMD employs a single Slater determinant of Gaussian wave packets for A nucleons

$$\langle \mathbf{r}_1 \dots \mathbf{r}_A | \Phi(Z) \rangle \propto \det_{ij} \Big[\exp\Big\{ -\nu(\mathbf{r}_i - \mathbf{Z}_j/\sqrt{\nu})^2 \Big\} \chi_{\alpha_j}(i) \Big], \tag{1}$$

where χ_{α_i} are the spin-isospin states with $\alpha_i = p\uparrow, p\downarrow, n\uparrow$, or $n\downarrow$. Thus the many-body state $|\Phi(Z)\rangle$ is parametrized by a set of complex variables $Z = \{\mathbf{Z}_i\}_{i=1,\dots,A}$. The width parameter $\nu = (2.5 \text{ fm})^{-2}$ is treated as a constant parameter common to all the wave packets. The time evolution of the wave packet parameters Z is determined, up to the extensions described later, by applying the time-dependent variational principle from which the equation of motion for Z is obtained,

$$i\hbar \sum_{j\tau} C_{i\sigma,j\tau} \frac{dZ_{j\tau}}{dt} = \frac{\partial \mathcal{H}}{\partial Z_{i\sigma}^*}, \tag{2}$$

where $\sigma, \tau = x, y, z$. The Hamiltonian $\mathcal{H}$ is the expectation value of the Hamiltonian operator with an additional correction for the spurious zero-point energies of the center-of-mass motions of fragments. A positive definite Hermitian matrix $C_{i\sigma,j\tau}$ appears in the equation of motion, suggesting that the variables Z are not canonical coordinates. The effective interactions for the mean-field calculations, such as the Gogny force and the Skyrme force, have been usually employed. The calculated results presented in this paper were obtained with the Skyrme SLy4 force [6] unless otherwise mentioned. The equation of motion can be interpreted intuitively as representing the motion of individual wave packets in the mean-field potential.

In addition to the mean-field effect, the residual interaction plays important roles in collisions at energies of more than 10 MeV/nucleon. This effect of two-nucleon collision is treated as a stochastic transition form an AMD state to one of the possible other AMD states. The probability is decided based on the differential cross section $(d\sigma/d\Omega)_{\text{NN}}$ of the two-nucleon scattering, which depends on the scattering energy and may be modified in nuclear medium. In the calculations presented in this paper, we employ the in-medium density-dependent total cross sections by Li and Machleidt [7] for the scattering energies $E > 35$ MeV in the nucleon-nucleon center-of-mass system, while the cross sections at $E = 35$ MeV are employed at lower energies. Angular distributions are assumed to be similar to the experimental data in the free space.

As the AMD variables Z do not have an intuitive meaning of nucleon coordinates due to the effect of antisymmetrization, we have introduced physical coordinates $W = \{\mathbf{W}_1, \dots, \mathbf{W}_A\}$ which are nonlinear functions of Z [5]. Let us consider a two-nucleon collision between the 1st and the 2nd nucleons. The assumption, which is usually taken for granted but is essential for the discussion in this paper, is that only the two coordinates are changed by the collision with a randomly generated scattering angle, so that the physical coordinates after the collision are written as $W' = \{\mathbf{W}'_1, \mathbf{W}'_2, \mathbf{W}_3, \dots, \mathbf{W}_A\}$. The Pauli-blocking for the final state of the collision is considered with the physical coordinates.

The two-nucleon collision process in AMD is a transition from an AMD wave function $|\Psi_i\rangle$ to one of possible final states $|\Psi_f\rangle$ which is assumed to occur instantaneously and to conserve the energy expectation value $\langle\Psi_i|H|\Psi_i\rangle = \langle\Psi_f|H|\Psi_f\rangle$. As the transition is induced by the residual interaction V between the two colliding nucleons, the transition rate may be expressed as

$$W_{i\to f} = \frac{2\pi}{\hbar} |\langle\Psi_f|V|\Psi_i\rangle|^2 \delta(E_f - E_i) \tag{3}$$

which is similar to Fermi's golden rule. However, some care should be taken since the transition is considered here between states which are not eigenstates of the unperturbed Hamiltonian.

An important question here is how to choose the complete basis $\{|\Psi_f\rangle\}$ of final states. If the correlations exist in the final states between the scattered nucleons and other nucleons, it is reasonable to construct the final states $\{|\Psi_f\rangle\}$ taking account of the correlations.

Cluster correlations can exist in nuclear medium under some conditions [8, 9]. We assume here that clusters with $A = 2, 3$ and 4 can propagate in the medium if it is allowed by the Pauli principle. In AMD, if several wave packets (with different spins and isospins) are placed at the same phase space point, these wave packets will tend to move together as a cluster by the equation of motion. On the other hand, if the wave packets are placed randomly in the phase space, the chance for these nucleons to form a cluster after propagation is small. Therefore, in order to respect the possibility of forming a cluster, the set of final states $\{|\Psi_f\rangle\}$ should be suitably constructed.

To obtain the transition probabilities to clusterized states, we employ an approximation similar to Ref. [8]. As an example of possible final states $|\Psi_f\rangle$ for a collision of two nucleons N_1 and N_2 with the initial relative velocity v_{NN}, let us consider a case that N_1 (N_2) forms a cluster C_1 (C_2) with another nucleon B_1 (B_2) in the final state. The partial differential cross section to this final channel is given by

$$v_{\mathrm{NN}} d\sigma(\mathrm{N}_1\mathrm{B}_1\mathrm{N}_2\mathrm{B}_2 \to \mathrm{C}_1\mathrm{C}_2) = \frac{2\pi}{\hbar}|\langle\varphi_1'|\varphi_1^{+\mathbf{q}}\rangle|^2 \, |\langle\varphi_2'|\varphi_2^{-\mathbf{q}}\rangle|^2 \, |M|^2 \, \delta(\mathcal{H} - E)\frac{p_{\mathrm{rel}}^2 dp_{\mathrm{rel}} d\Omega}{(2\pi\hbar)^3}, \tag{4}$$

where M is the matrix element for the two-nucleon scattering to the final state with the relative momentum p_{rel} and the scattering angle Ω in the two-nucleon center-of-mass system. The overlap matrix $\langle\varphi_1'|\varphi_1^{+\mathbf{q}}\rangle$ is taken between $|\varphi_1^{+\mathbf{q}}\rangle = e^{i\mathbf{q}\cdot\mathbf{r}_1}|\varphi_1\rangle$ and $|\varphi_1'\rangle$, where $|\varphi_1\rangle$ and $|\varphi_1'\rangle$ are the initial and final states of the $\mathrm{N}_1 + \mathrm{B}_1$ system, respectively, and the operator $e^{i\mathbf{q}\cdot\mathbf{r}_1}$ gives the momentum transfer to the nucleon N_1. The clusterized states $|\varphi_1'\rangle$ and $|\varphi_2'\rangle$ are approximated by the simple harmonic oscillator $(0s)^n$ configuration with the oscillator constant associated with the wave packet width ν in AMD, so that any final state of the collision is represented by an AMD wave function. By identifying the two-nucleon matrix element M with that of the usual two-nucleon scattering, the cluster-forming cross section can be expressed by using the two-nucleon collision cross section that is an input to the calculation.

The actual situation of a two-nucleon collision requires more considerations because there are many possible ways of forming a cluster with different nucleons (B's) in the system for each of the scattered nucleons N_1 and N_2. It is important to take care of the non-orthogonality of the final states corresponding different ways of cluster formation.

The procedure is repeated for the cluster formation with nucleons (B's) with other spin and isospin directions until formation of an α cluster is considered. The particle N should be regarded as a cluster, instead of a nucleon, if a (sub)cluster has been already formed in previous steps of the repetition.

There is a different direction of extending AMD by introducing wave packet splitting [10, 11]. The extension for cluster correlations can be introduced together with wave packet splitting as reported in Ref. [12]. In this paper, however, the results without wave packet splitting are presented below. Surprisingly, conclusion does not depend on wave packet splitting.

3. Effects of cluster correlations in multifragmentation

It is known that the early version of AMD with usual two-nucleon collisions does not reproduce multifragmentation data except for collisions of light nuclei. In the left part of Fig. 1, the fragment charge distribution calculated with the usual AMD without special care of cluster correlations is compared with the INDRA data [1]. The number of fragments with $Z > 20$ is seriously overestimated while the number of intermediate mass fragments is underestimated. The proton multiplicity $M_p = 40.2$ is too large while the α-particle multiplicity $M_\alpha = 2.5$ is too small compared to the experimental data ($M_p = 8.4$ and $M_\alpha = 10.1$) [1].

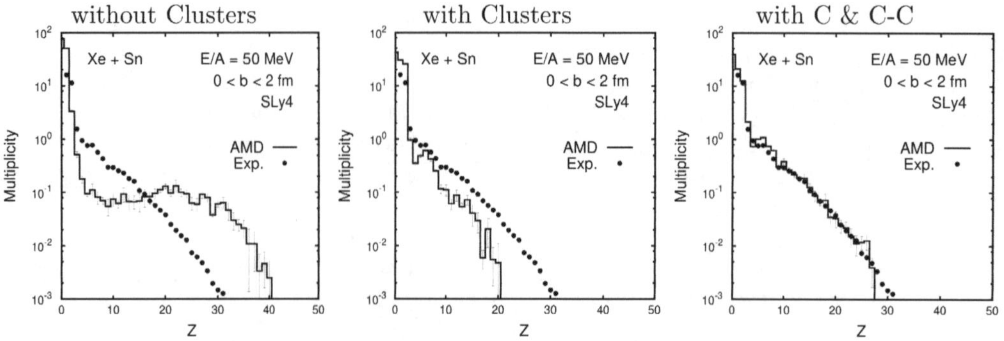

Figure 1. The charge distribution of fragments produced in central Xe + Sn collisions at 50 MeV/nucleon. The red histogram was calculated by the AMD with usual two-nucleon collisions without explicit treatment of cluster correlations (left), by the AMD with cluster correlations (middle), and by the AMD with cluster and inter-cluster correlations (right). The blue points show the experimental data by INDRA [1].

When cluster formation is turned on in the final states of two-nucleon collisions, many clusters are created in heavy-ion collisions as soon as two nuclei overlap and hard two-nucleon collisions occur. The fragment charge distribution has been drastically changed by cluster correlations as shown in the middle part of Fig. 1. The proton multiplicity has decreased from $M_p = 40.2$ to 10.9 and the α-particle multiplicity has increased from $M_\alpha = 2.5$ to 23.2 by the incorporation of cluster correlations. Although the proton multiplicity is now consistent with the experimental data [1], the α particles are produced too much and heavier fragments are not sufficient. In short, the tendency of turning into cluster gas is too strong compared to experimental data.

The reason of the unphysically strong tendency of cluster gas may be due to the fact that the motion of clusters is almost classical in AMD calculations when the equation of motion for the wave packet centroids is solved. In reality, a few clusters can form a bound state corresponding to the ground state of a light nucleus such as the Li and Be isotopes. The binding of clusters in many of these nuclei is weak so that there is only one bound intrinsic state. Even though the AMD wave function can describe the ground state, the phase space of classically bound configurations does not necessarily have a correct volume corresponding to a single quantum state. When the binding is weak, the classical phase space is much smaller than the quantum phase space, which suppresses the binding probability of clusters in AMD.

To test this idea, some inter-cluster correlation is introduced as a stochastic process of bonding clusters. The relative momentum between clusters is replaced by zero if moderately separated clusters ($R_{\text{rel}} < 5$ fm) are moving away from each other with a small relative momentum ($\mathbf{R}_{\text{rel}} \cdot \mathbf{P}_{\text{rel}} < 0$ and $P_{\text{rel}}^2/2\mu < 8$ MeV). Thus this cluster bonding process works to select bound states of two or several clusters during the time evolution of reactions.

This inter-cluster correlation certainly reduces the tendency of cluster gas and increases the yields of relatively heavy fragments as shown in the right part of Fig. 1. The α-particle multiplicity is reduced from $M_\alpha = 23.2$ to 10.7 by the inter-cluster correlation, while the the proton multiplicity changes only slightly from $M_p = 10.9$ to 10.8. This calculated result is now consistent with the data in many aspects.

4. Symmetry energy effects in multifragmentation

The results in the previous section demonstrate that the global dynamics and fragmentation are reasonably described by AMD with cluster and inter-cluster correlations, at least for central collisions of $A \approx 120$ nuclei at 50 MeV/nucleon. Based on it, we investigate here how the collision

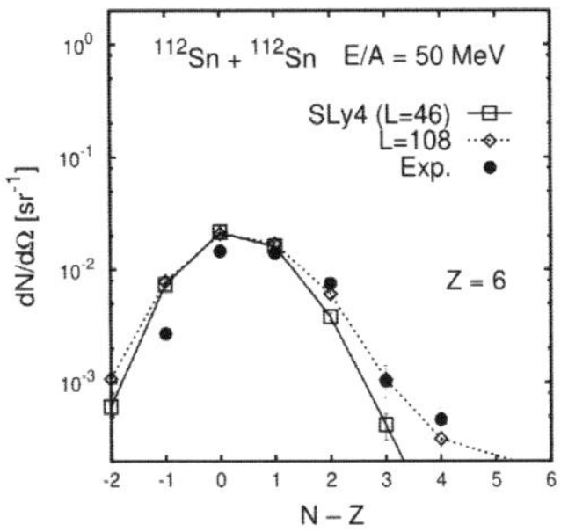

Figure 2. Yield distribution of carbon isotopes in the ^{124}Sn + ^{124}Sn central collision at 50 MeV/nucleon.

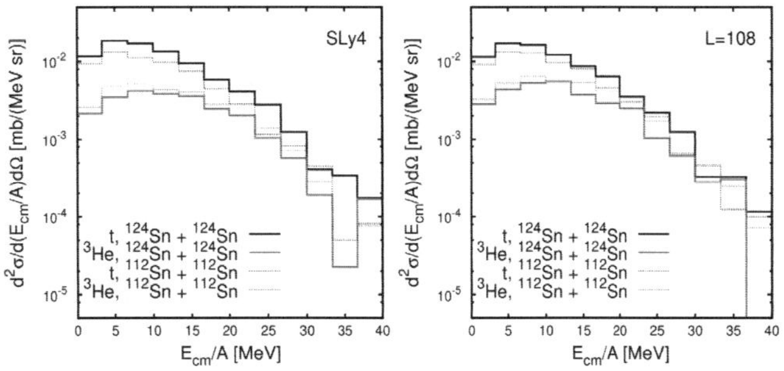

Figure 3. Spectra of tritons and ^{3}He in central collisions of Sn isotopes at 50 MeV/nucleon, calculated with the symmetry energy parameter $L = 46$ (left) and 108 (right) MeV.

dynamics and final observables are reflecting the equation of state of asymmetric nuclear matter. The density dependence of the symmetry energy $E_{\rm sym}(\rho)$ can be characterized by a parameter $L = 3\rho_0(dE_{\rm sym}/d\rho)_{\rho=\rho_0}$. We compare the calculated results with two different Skyrme-type interactions corresponding to $L = 46$ and 108 MeV.

At the early stage of the central collisions of Sn isotopes at 50 MeV/nucleon, the difference of the calculated neutron and proton densities in the high density region show a very clear influence of the density dependence of the symmetry energy. In the late stage of the reaction, the calculation clearly shows that the symmetry energy influences the degree of fractionation, i.e., how much the gas part is neutron rich compared with the liquid part of the system in case of collisions of neutron-rich nuclei. The effects in some final observables are shown in Figs. 2 and 3. The width of the isotope distribution is sensitive to the symmetry energy parameter L, as well as the mean value of the distribution (for carbon isotopes produced in ^{124}Sn + ^{124}Sn collisions, $\langle N - Z\rangle = 0.75$ for $L = 46$ MeV, and $\langle N - Z\rangle = 0.99$ for $L = 108$ MeV). Comparison with data [13, 14] for these observables suggests that the interior (and therefore low energy) part of the expanding system should be more neutron rich even than in the case of $L = 46$ MeV.

Acknowledgments
This work was supported by JSPS KAKENHI Grant Number 21540253. The numerical calculations were partly carried out on SR16000 at YITP in Kyoto University.

References
[1] Hudan S *et al* 2003 *Phys. Rev.* C **67** 064613
[2] Colonna M, Ono A and Rizzo J 2010 *Phys. Rev.* C **82** 054613
[3] Li B A, Chen L W and Ko C M 2008 *Phys. Rep.* **464** 113
[4] Tsang M B *et al* 2012 *Preprint* arXiv:1204.0466v1 [nucl-ex]
[5] Ono A, Horiuchi H, Maruyama T and Ohnishi A 1992 *Prog. Theor. Phys.* **87** 1185
[6] Chabanat E, Bonche P, Haensel P, Meyer J and Schaeffer R 1998 *Nucl. Phys.* A **635** 231; Chabanat E, Bonche P, Haensel P, Meyer J and Schaeffer R 1998 **643** 441(E)
[7] Li G Q and Machleidt R 1993 *Phys. Rev.* C **48** 1702;Li G Q and Machleidt R 1994 *Phys. Rev.* C **49** 566
[8] Danielewicz P and Bertsch G F 1991 *Nucl. Phys.* A **533** 712
[9] Röpke G 2011 *Nucl. Phys.* A **867** 66
[10] Ono A and Horiuchi H 1996 *Phys. Rev.* C **53** 2958
[11] Ono A, Hudan S, Chbihi A and Frankland J D 2002 *Phys. Rev.* C **66** 014603
[12] Ono A 2012 *EPJ Web of Conferences* **31** 00023
[13] Liu T X *et al* 2004 *Phys. Rev.* C **69** 014603
[14] Liu T X, Lynch W G *et al* 2012 *Preprint* arXiv:1208.3108v1 [nucl-ex]

A new step of microscopic cluster model

The present status of the THSR wave function

Akihiro Tohsaki[1]

[1] Research Center for Nuclear Physics, Osaka University, Osaka 567-0047, Japan

E-mail: tohsaki@rcnp.osaka-u.ac.jp

Abstract. In this talk, we present new developements concerning the cluster-gas wave function of a fully microscopic description.

1. Introduction

A whole cycle of the Oriental Zodiac has passed since we proposed a specific form of the wave function describing α-cluster-gas-like states. Thanks to the fully microscopic description, this form of the wave function can naturally describe some ground states as well as other ordinary cluster-model states. The microscopic model of nuclei takes account of the Pauli principle correctly and is based on an effective nuclear force. We call this model wave function the THSR wave function, which comes from the names of the authors (Tohsaki, Horiuchi, Schuck, Röpke) of our first article [1]. Until now the idea has been applied to the aggregates of 2 to 5 α-clusters. First of all, the Hoyle state of ^{12}C has been conjectured to be a 3α gas state which can be regarded as an α-condensate.

The idea is characterised by the following:

(i) Some kind of container for an α-aggregate is introduced.

(ii) The behaviour of the container is not *a priori* given but self-consistently determined by the Hill-Wheeler equation.

(iii) At the beginning, the container was assumed to be spherical, but then, it became possible to have it deformed. The wave function has been applied to positive parity states [2].

(iv) The competition of the attractive nuclear force and the repulsive Coulomb force around the rarefied density of the α-gas plays an important role in garanteeing its stability.

In the new microscopic cluster model defined by the THSR wave function one can vary the variational parameters so as to change the distances between clusters as well as the density of clusters. This enables one to reduce the number of variational parameters drastically. The α-gas state is an excited state orthogonal to the ground one. The Hill-Wheeler equation results in a density which is about $1/3$ of that of the ground state. The THSR wave function has the remarkable advantage of being constructed from α-clusters free from the c.m. energy, which we could not achieve in a conventional cluster model written as a geometrical configuration. Unfortunately, at that moment, the container is still restricted to a kind of spheroid, which excludes the treatment of negative-parity states.

2. Present status

The idea of the container has been recently applied to systems that contain other clusters in addition to the α-aggregate. For instance, we have already investigated the ^{16}O-α system in a container [3]. Introducing the distance parameter between ^{16}O and α, we have succeeded in estimating the negative-parity state as well as the positive-parity ones. Surprisingly enough, the stationary energy of the negative-parity state is obtained to be at the limit of zero distance. This shows that the idea of the container is more important in showing the cluster structure than that of the distance between the clusters.

The present status can be summarised as follows:

(i) The description of negative-parity states is also possible by using the THSR wave function.

(ii) The door is open to the study of a wide range of nuclear systems which can be regarded as including various kinds of clusters in a container.

(iii) The interaction between α-aggregates can be described within the scope of the THSR wave function. The suprious c.m. coordinate is naturally removed by going back to the idea of the resonating group method which is free from he c.m. variable.

(iv) A clarification of the mathematical structure of the formalism and the development of computer facilities allow us to tackle larger nuclear systems with THSR wave functions.

The idea of the cluster density is not appropriate for the nuclear systems consisting of different kinds of clusters. At this stage, we should see the nucleon density and the mean deviation caused by cluster conformation.

3. Conclusion and remaining problem

We introduced the idea of the cluster-gas in the nuclear system, in which the complete inclusion of the Pauli principle is absolutely important in describing the cluster structure. This is because the Pauli principle is indispensable in producing the ground state and the excited states simultaneously. The distances between clusters are replaced by the nuclear container which confines nuclear clusters. It is of utmost importance that it is self-consistently determined by the Hill-Wheeler equation.

It is, however, unfortunate that we do not have a suitable effective nuclear force that reproduces the saturation properties of a wide range of nuclei.

Acknowledgments

The speaker thanks Y Funaki, B Zhou, H Horiuchi, T Yamada, G Röpke, P Schuck, Z Ren and C Xu for their good collaboration with him.

References

[1] Tohsaki A, Horiuchi H, Schuck P and Röpke G 2001 *Phy. Rev. Lett.* **87** 192501
[2] Funaki Y *et al* 2003 *Phys. Rev.* C **67** 051306
[3] Zhou B *et al* 2012 *Phys. Rev.* C **86** 014301

Journal of Physics: Conference Series **436** (2013) 012070 doi:10.1088/1742-6596/436/1/012070

Quantum statistical calculation of cluster abundances in hot dense matter

G Röpke[1]

[1]Universität Rostock, Institut für Physik, 18051 Rostock, Germany

E-mail: gerd.roepke@uni-rostock.de

Abstract. The cluster abundances are calculated from a quantum statistical approach taking into account in-medium corrections. For arbitrary cluster size the self-energy and Pauli blocking shifts are considered. Exploratory calculations are performed for symmetric matter at temperature $T = 5$ MeV and baryon density $\varrho = 0.0156$ fm^{-3} to be compared with the solar element distribution. It is shown that the abundances of weakly bound nuclei with mass number $4 < A < 12$ are strongly suppressed due to Pauli blocking effects.

The investigation of thermodynamic properties of hot nuclear matter for parameter values temperature $k_{\rm B}T \approx 1 - 100$ MeV, baryon density $\varrho \leq \varrho_0 \approx 0.166$ fm^{-3} is recently of interest with respect to astrophysical and cosmological questions as well as the investigation of heavy ion collisions. In particular, the problem of the calculation of abundances of clusters $\{AnP\}$ (nuclei with mass number A, excitation state n and center of mass momentum P) within a systematic quantum statistical approach is of relevance for the equation of state (EoS) and the thermodynamic properties. We will not give here an exhaustive review of the quantum statistical treatment of hot nuclear matter, see [1, 2]. We give only some details and results in calculating the cluster abundances, in particular the treatment of heavier clusters at low densities.

We start from a cluster decomposition of the EoS obtained from a self-consistent ladder Hartree-Fock approximation for the self-energy. Neglecting continuum correlations,

$$\varrho(\beta,\mu) = \sum_{A=1}^{\infty} A\varrho_A(\beta,\mu) \approx \sum_{A=1}^{\infty} A \sum_{n}^{\rm bound} g_{A,n} \int_{P>P^{\rm Mott}_{A,n}} \frac{{\rm d}^3P}{(2\pi)^3} f_A(E_{AnP}),$$
$$f_A(E) = \left[\exp\beta(E - A\mu) - (-1)^A\right]^{-1}, \tag{1}$$

where $\varrho_A(\beta,\mu)$ denotes the contribution of clusters A to the total baryon density, $g_{A,n}$ is the degeneration factor of the state n. The energy eigenvalue E_{AnP} of the cluster $\{AnP\}$ is obtained from a Bethe-Goldstone equation that takes into account the influence of the surrounding matter on the quantum state of the considered cluster. In ladder Hartree-Fock approximation it can be decomposed as follows:

$$E_{AnP} = E^0_{AnP}(\varrho,T) + \frac{P^2}{2AM} + \Delta E^{\rm SE}_{AnP} + \Delta E^{\rm Pauli}_{AnP}, \tag{2}$$

where $E^0_{AnP}(\varrho,T)$ describes the dependence of the energy of the cluster because of the Coulomb interaction with the surrounding charged particles, $\Delta E^{\rm SE}_{AnP}$ is the self-energy shift, and $\Delta E^{\rm Pauli}_{AnP}$

denotes the shift due to phase space occupation (Pauli blocking). The summation in Eq. (1) concerns only the bound states. This means that the condition

$$E_{AnP} < A\Delta(P/A) \tag{3}$$

holds, since the lower edge of the continuum of scattering states for each of the A cluster constituents is shifted by the single particle self-energy shift $\Delta(p)$ at momentum $p = P/A$. Above the so-called Mott density introduced in [2], the bound state merges in the continuum of scattering states and disappears. The condition (3) excludes the interval $P < P^{\rm Mott}_{A,n}(\varrho, T)$ for the integral over P in Eq. (1) if the density exceeds the Mott density.

To evaluate the EoS (1) we have to determine the energies E_{AnP}. We give some details considering as example the parameter values $T = 5$ MeV and $\varrho = 0.0156$ fm^{-3}. In addition to the ladder-Hartree Fock approximation and the approximation for the evaluation of the Pauli shift given below, that the surroundings of the cluster under consideration is described as a nondegenerated, uncorrelated gas of nucleons, we assume symmetric nuclear matter where protons and neutrons are considered as particles interacting in the same manner. Within the exploratory model calculations given here for symmetric matter, we make the approximation that we attribute to a system of A nucleons the charge $eA/2$, this means, each nucleon (mass M) has the charge $e/2$. For symmetric matter, we have only one chemical potential μ (for asymmetric matter see [1] where the two-component case is treated). Further assumptions refer to empirical quantities (radii of nuclei, density distribution within a cluster, single-particle Hartree-Fock shift, density of states of excited nuclei, and so on) that can be improved in a systematic way. Our main interest is the calculation of the energy shift of heavier elements, in particular $4 < A < 12$.

The relevance of Coulomb interaction is increasing with increasing cluster size A and leads to two effects: The nucleon density near a cluster is reduced due to the Debye-Thomas-Fermi theory of screening of a charge distribution by the surrounding plasma, whereas the density of the strongly degenerated electrons (as compensating background) remains nearly constant. At the surface of the cluster A, given by the radius $R_A = r_0 A^{1/3}$ with $r_0 = 1.2$ fm, we have the density $\varrho^*_A(\varrho, T) < \varrho$. Furthermore, the energy of a cluster is shifted to the value $E^0_{AnP}(\varrho, T)$ due to the screening of the Coulomb potential.

The screening radius $1/\kappa$ is given by

$$\kappa^2 = \pi e^2 \frac{\mathrm{d}}{\mathrm{d}\mu}\varrho(\mu), \tag{4}$$

where the value μ in Eq. (4) is obtained from the relation for the ideal Fermi gas ($\Lambda^2 = 2\pi\hbar^2/Mk_BT$)

$$\varrho(\mu) = \frac{2}{\pi^2\Lambda^3}\int \mathrm{d}x \frac{x^2}{\exp(x^2/4\pi - \mu/k_BT) + 1}. \tag{5}$$

For the parameter set given above, $\mu/k_BT = 0.892$; $k_BT/\varrho \times \mathrm{d}\varrho/\mathrm{d}\mu = 0.668$ results. According to the Debye theory follows

$$\varrho^*_A(\varrho, T) = \varrho \exp\left\{-\frac{\kappa^2 A}{4\pi\varrho R_A}\frac{1+\gamma}{1+\kappa R_A}\right\}, \tag{6}$$

$$E^0_{AnP}(\varrho, T) - E^0_{AnP}(0) = \frac{e^2}{4r_0}A^{5/3}\left\{\frac{3}{5}(1+\gamma)^2 - \frac{3}{5} - \frac{1}{2}(1+\gamma)^2\frac{\kappa R_A}{(1+\kappa R_A)^2}\left(\frac{3}{2} + \kappa R_A\right)\right\}, \tag{7}$$

$$\gamma = \frac{\varrho}{\varrho_0}\left[\exp\left\{-\frac{e^2}{4}\frac{A}{R_A k_B T}(1+\gamma)\left(\frac{1}{1+\kappa R_A}+\frac{1}{2}\right)\right\}-1\right]. \tag{8}$$

Results for different values A are given in Tabs. 1, 2 of Ref. [3] where for $E^0_{AnP}(0)$ the value of the empirical binding energy for the most abundant isobar is taken; the dependence of the energy $E^0_{AnP}(\varrho,T)$ from the center of mass momentum P has been neglected.

For the parameter values considered here, the values $\varrho^*_A(\varrho,T)$ for $A \leq 11$ exceeds the Mott density so that for these cluster the value $P^{\rm Mott}_A \neq 0$ has to be determined where the cluster breaks up into free nucleons or smaller clusters (with energies shifted according to the density $\varrho^*_A(\varrho,T)$, Eq. (6), as shown below). This way, the interval of values P for that the cluster can exist, is reduced.

The single-particle self-energy shift $\Delta(\varrho)$ can be determined from a Skyrme interaction with zero range so that it is not dependent of temperature and momentum. According to [4] we have (units MeV, fm)

$$\Delta(\varrho) = -792.975\varrho + 2711.9\varrho^2 + 125.226\varrho^{5/3}. \tag{9}$$

For light clusters $A < 12$, the self-energy shift is taken as $\Delta E^{\rm SE}_{AnP} = A\Delta(\varrho^*_A)$. For $A \geq 12$ we use within the Fermi gas model

$$\Delta E^{\rm SE}_{AnP} = \int d^3r \Delta(\varrho_A(r))\rho^*_A, \tag{10}$$

where $\rho_A(r)$ is the nucleon density as function of the radius r for the cluster A. According to [5] we can use the density distribution

$$\rho_A(r) = \frac{3A}{4\pi R^3}\frac{1}{1+(\pi b/R)^2}\left[\frac{1}{1+{\rm e}^{\frac{r-R}{b}}}+\frac{1}{1+{\rm e}^{\frac{-r-R}{b}}}-1\right] \tag{11}$$

with parameter values $b = 0.57$ fm and $R = 1.05$ fm $A^{1/3}$.

The Pauli shift has been determined for light clusters ($A \leq 12$) approximating the cluster wave function by an antisymmetrized product of single-particle wave functions $\varphi_\nu(i)$:

$$\Delta E^{\rm Pauli}_{AnP} = -\sum_\nu \sum_1 |\varphi_\nu(1)|^2 \left(E_\nu - \frac{\hbar^2 p_1^2}{2M}\right)\frac{\rho^*_A \Lambda^3}{4}{\rm e}^{-\frac{\hbar^2(p_1-P/A)^2}{2Mk_BT}}. \tag{12}$$

This expression is evaluated for $4 < A < 12$ using, according to Ref. [6], the eigenstates of the harmonic oscillator with potential $V(r) = V_0 + M\omega^2 r^2/2$, $\hbar\omega = 40A^{-1/3}$ MeV. The single-particle Hartree-Fock energies E_ν and the binding energy $E^0_{A0P}(\varrho,T)$ lead to the relation

$$V_0 = \frac{2}{A}\left(E^0_{A0P}(\varrho,T) + 3\hbar\omega\right) - \frac{15}{4}\hbar\omega. \tag{13}$$

For $A \leq 4$ we also can use a product of Gauss functions $\varphi_\nu(p) \propto \exp(-p^2/2k_A^2)$ with the parameter

$$k_A^2 = 4M|E^0_{A0P}|/A\hbar^2. \tag{14}$$

As before, we find the relation

$$E_\nu = 2E^0_{A0P}(\varrho,T)/A - 3\hbar^2 k_A^2/4M = 5E^0_{A0P}(\varrho,T)/A. \tag{15}$$

The Pauli blocking shifts for $A \leq 4$ follow from Eq. (12) as

$$\Delta E^{\rm Pauli}_{A0P} = A\frac{\varrho^*_A\Lambda^3}{4}\frac{|E^0_{A0P}(\varrho,T)|}{\delta^{3/2}}\left\{5+\frac{3}{\delta}+\frac{x^2(\delta-1)}{2\pi\delta^2}\right\}{\rm e}^{-\frac{x^2}{4\pi\delta}}$$

$$\delta = 1 + 2|E^0_{A0P}(\varrho,T)|/AT, \qquad x = P\Lambda/A. \tag{16}$$

For $4 < A < 12$ we have according Eq. (12)

$$\Delta E_{A0P}^{\rm Pauli} = \frac{\varrho_A^* \Lambda^3}{4} {\rm e}^{-\frac{x^2}{4\pi\delta}} \frac{\hbar\omega}{\delta^{3/2}} \left\{ \frac{3}{\delta} - 4\left(\frac{V_0}{\hbar\omega}+\frac{3}{2}\right) + \frac{x^2(\delta-1)\hbar\omega}{2\pi\delta^2} \right.$$
$$-2(A-4)\frac{x^2(\delta-1)}{12\pi\delta^2}\left[\left(\frac{V_0}{\hbar\omega}+\frac{5}{2}\right) - \frac{x^2(\delta-1)}{8\pi\delta^2}\right]$$
$$\left. -2(A-4)\frac{1}{3\delta}\left[-\frac{5x^2(\delta-1)}{8\pi\delta^2} + \frac{3}{2}\left(\frac{V_0}{\hbar\omega}+\frac{5}{2}\right)\right] + 2(A-4)\frac{5}{8\delta^2} \right\},$$
$$\delta = 1 + \frac{\hbar\omega}{2k_BT}. \tag{17}$$

For large cluster, $A \geq 12$, the Pauli blocking shift is calculated within a Fermi gas model so that the compensation between self-energy shift and Pauli blocking shift at low temperatures is correctly reproduced. The total shift results as

$$\Delta E_{A0P} = \varrho_A^* \int {\rm d}^3 r \Delta(\varrho_A(r)) \int_{\Lambda p_F(\varrho_A)}^{\infty} \frac{y {\rm d}y A}{2\pi P\Lambda} \left\{ {\rm e}^{-(y-P\Lambda/A)^2/4\pi} - {\rm e}^{-(y+P\Lambda/A)^2/4\pi} \right\},$$
$$p_F(\varrho) = \left(\frac{3\pi^2}{2}\varrho\right)^{1/3}. \tag{18}$$

The integral over y depends on $\varrho_A(r)$ an can be approximated by $\exp(-\delta\varrho^{2/3})$. To fit the value at ϱ_0, the value $\delta = 20.926$ results for $P\Lambda/A = 0.3$.

The sum over the excited states n was performed using the density of states [6]

$$D_A(E) = \frac{1}{12}\left(\frac{\pi^2}{a}\right)^{1/4} E^{-5/4} \exp\left(2\sqrt{aE}\right) \tag{19}$$

with $a = A/15$ MeV^{-1}. This way, the abundance

$$X_A = \frac{\varrho_A}{\varrho} = \frac{A^3}{2\pi^2 \varrho \Lambda^3} \int {\rm d}x x^2 \int {\rm d}E D_A(E)$$
$$\times \exp\left\{-(E_{A0P}^0(\varrho,T) + E + \Delta E_{A0P}(x,\varrho_A^*) + ATx^2/4\pi - A\mu)/k_BT\right\} \tag{20}$$

results for cluster with $A \geq 12$. It has been assumed that the shifts are independent of n. The evaluation of X_A is simplified if the shift ΔE_{AnP} is taken for a representative value $x = P\Lambda/A = 0.3$ considering the dependence on P.

For the light cluster $A < 12$, the density considered here exceeds the Mott density. We take into account only the ground state with degeneration factors $g_{2,0} = 3, g_{3,0} = 4, g_{A,0} = 1$ else, calculating the abundances X_A. The result for the cluster abundances for the parameter values $T = 5$ MeV and $\varrho = 0.0156$ fm^{-3} exemplarily considered here is shown in Fig. 1. These parameter values have been chosen to reproduce the global behavior of the solar element distribution. The results are of interest considering the evaporation of matter from the surface of a dense hot nuclear system, see [3].

The approach given here considers in particular for the first time the abundance of the light clusters $4 < A < 12$ that are weakly bound so that they are strongly suppressed in hot dense matter. Some approximations can be improved. The inclusion of scattering states in Eq. (1) is considered in the cluster-virial expansion of the EoS [7]. A more detailed calculation of the Pauli blocking shift for the light clusters $A \leq 4$ is given in [8], and the EoS that contains only

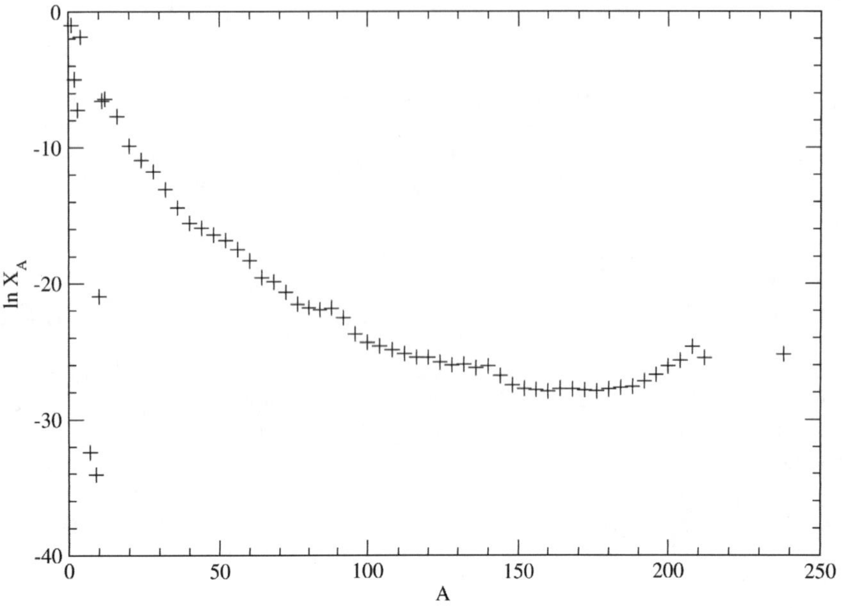

Figure 1. Cluster abundances $\ln X_A$ for hot and dense symmetric matter as function of mass number A, parameter values $T = 5$ MeV and $\varrho = 0.0156$ fm^{-3}. Most abundant isobars are considered for $A < 12$. For $A \geq 12$ abundances of clusters with $A = 4n$ are shown.

these light clusters was discussed in Ref. [9]. The inclusion of arbitrary clusters calculating the nuclear matter EoS was considered in [10] where the medium effects are taken into account using the phenomenological concept of the excluded volume. The strong influence of medium effects on the abundance of elements with $4 < A < 12$ is caused by the week binding of these cluster, similar to the deuteron that is also strongly influenced by medium effects.

References

[1] Röpke G 1983 *Phys. Lett.* B **121** 223; Röpke G 1984 *Wiss. Z. Univ. Rostock* **33** 33
[2] Röpke G, Münchow L, Schmidt M and Schulz H 1982 *Nucl. Phys.* A **379** 536; 1983 *Nucl. Phys.* A **399** 587; 1984 *Nucl. Phys.* A **424** 594; 1982 *Phys. Lett.* B **110** 21
[3] Röpke G 1983 *Wiss. Z. Univ. Rostock* **32** 30
[4] Vautherin D and Brink D M 1972 *Phys. Rev.* C **5** 626
[5] Burov V V, Eldyshev Yu N, Lukyanov V K and Pol Yu S 1974 *JINR preprint* E4-8029 Dubna
[6] Bohr A and Mottelson B 1963 *Nuclear Structure* (New York: Benjamin)
[7] Röpke G, Bastian N-U, Blaschke D, Klähn T, Typel S and Wolter H H 2013 *Nucl. Phys.* A **897** 70
[8] Röpke G 2011 *Nucl. Phys.* A **867** 66
[9] Typel S, Röpke G, Klähn T, Blaschke D and Wolter H H, 2010 *Phys. Rev.* C **81** 015803
[10] Hempel M and Schaffner-Bielich J 2010 *Nucl. Phys.* A **837** 210

Journal of Physics: Conference Series **436** (2013) 012071 doi:10.1088/1742-6596/436/1/012071

Role of clusters in nuclear astrophysics with Cluster Nucleosynthesis Diagram (CND)

S Kubono[1,5,13], Dam N Binh[2], S Hayakawa[3], H Hashimoto[4], D Kahl[5], H Yamaguchi[5], Y Wakabayashi[1], T Teranishi[6], N Iwasa[7], T Komatsubara[8], S Kato[9], A Chen[10], S Cherubini[3], S H Choi[11], I S Hahn[12], J J He[13], Le Hong Khiem[14], C S Lee[15], Y K Kwon[16], S Wanajo[17] and H-T Janka[18]

[1] RIKEN Nishina Center, 2-1 Hirosawa, Wako, Saitama 351-0198, Japan
[2] 30MeV Cyclotron Center, Tran Hung Dao Hospital, Hanoi, Vietnam
[3] Universita di Catania and INFN-LNS, Catania, Italy
[4] Research Center of Nuclear Physics, Osaka University, Osaka, Japan
[5] Center for Nuclear Study, University of Tokyo, 2-1 Hirosawa, Wako, Saitama 351-0198, Japan
[6] Department of Physics, Kyushu University, Japan
[7] Department of Physics, Tohoku University, Aoba, Sendai, Miyagi, Japan
[8] Department of Physics, Tsukuba University, Japan
[9] Faculty of Science, Yamagata University, Japan
[10] Department of Physics, University of McMaster, Canada
[11] Department of Physics, Seoul National University, Seoul, Korea
[12] Ewha Womans' University, Seoul, Korea
[13] Institute of Modern Physics, Chinese Academy of Sciences, Lanzhou, China
[14] Institute of Physics, VAST, Hanoi, Vietnam
[15] Department of Physics, Chung Ang University, Seoul, Korea
[16] RISP, Basic Science Institute, Daejoun, Korea
[17] National Astronomical Observatory Japan, Mitaka, Tokyo, Japan
[18] Max-Planck-Institut für Astrophysik, D-85748 Garching, Germany

E-mail: kubono@riken.jp

Abstract. The role of nuclear clustering in stellar reactions is discussed, with Cluster Nucleosynthesis Diagram (CND) proposed before, for nucleosynthesis in stellar evolution and explosive stellar phenomena. Special emphasis is placed on α-induced stellar reactions. We report here the first experimental evidence that α cluster resonances dominate the (α,p) stellar reaction cross sections that is crucial for the νp-process in core-collapse supernovae.

1. Introduction

Nuclear processes play a crucial role in evolution of the universe as well as in various stellar phenomena. Especially, nuclear reactions lead to synthesis of almost all elements in our universe, and large energy generation which drives stars to evolve. Hydrogen burning is the main energy source and the burning takes place hydrostatically or explosively. After the hydrogen burning, the ashes of the hydrogen burning, helium, become the fuel for the next nuclear burning. Subsequently, carbon and oxygen burn to produce heavier elements, mostly Si. Along the stellar evolution, these lead to Si burning to produce the iron core in the middle of massive

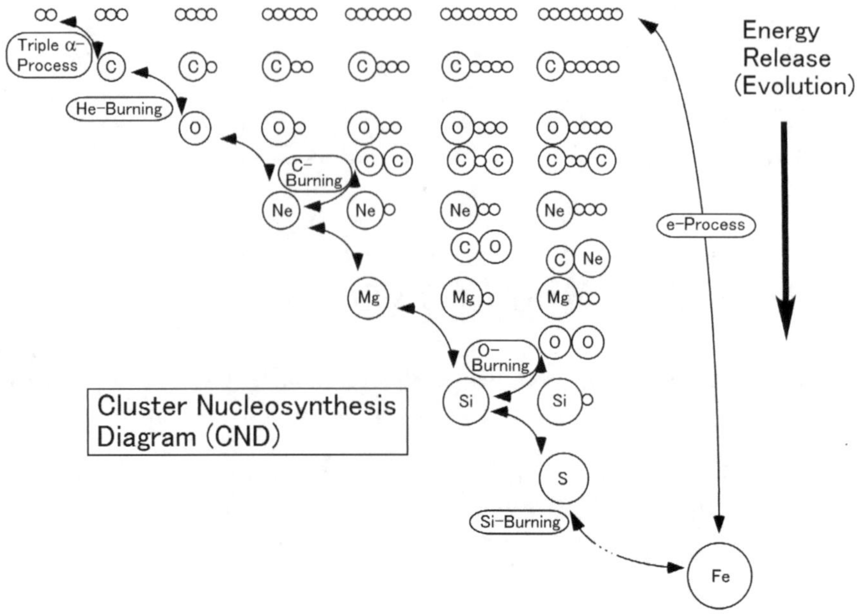

Figure 1. Cluster Nucleosynthesis Diagram (CND) [2] for nucleosynthesis along the stellar evolution. Small circles indicate α-clusters. The processes to go to lower levels imply energy releases.

stars. Thus, nucleosynthesis after hydrogen burning involves α, ^{12}C, etc., which can be via the well-known, very interesting clusters in nuclei.

2. Cluster Nucleosynthesis Diagram (CND)

The successive nucleosynthetic processes after hydrogen burning can be summarized in Fig. 1 [1], where the smallest circles indicate alpha particles. As mentioned above, after hydrostatic hydrogen burning, the helium ashes become the fuel in the next stage, called helium burning. Subsequent burning processes involve C, O and Si as fuels.

Figure 1. depicts the idea of these nuclear burning processes in stars till the epoch of the iron-core formation. This idea also applies to other burning phenomena as well. This diagram was proposed in 1992 [2], to give a natural way of understanding the stellar nucleosynthesis. This is very much like the Ikeda diagram for cluster physics, but it includes the way of stellar evolution in terms of production of energies and elements. The most important point here is that this diagram is derived by physics point of view. Since the relevant scattering energy is small in stellar environments, resonances near the cluster threshold are crucial in nucleosynthesis. This is actually where one should expect states that have a large parentage of the cluster configuration from the cluster threshold rule.

The first step of the helium burning is the synthesis of ^{12}C by capture of an α particle on ^{8}Be mainly through the Hoyle state in ^{12}C. The second step is the ^{12}C(α,γ)^{16}O reaction. Hydrostatic helium burning process is considered to terminate at this point. After the helium burning, the ashes of helium burning, ^{12}C and ^{16}O, become the fuel and lead to C and O burning, which go through the fusion reactions of ^{12}C + ^{12}C and ^{16}O + ^{16}O, respectively, at very low energies mainly emitting α particles, together with subsequent α-induced reactions on even-even sd-shell nuclei. Eventually, the silicon burning begins from ^{28}Si with successive (α,γ) and (α,p) reactions

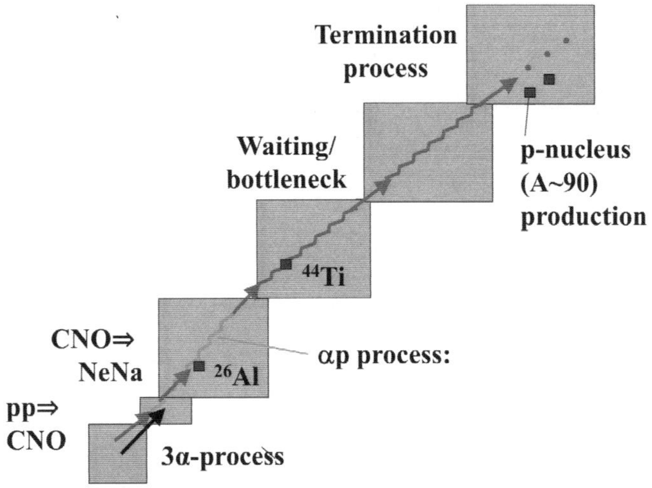

Figure 2. Sketch of the νp-process and the problems.

together with photo-disintegration of ^{28}Si, etc. Here, an interesting observation is that in nuclear physics, one may add excitation energy to the nuclear system to see the evolution of clusters. For instance, in ^{24}Mg there appears an α-cluster state, then a 2α-cluster state or ^{12}C+^{12}C molecular state, etc., whereas nucleosynthesis in nature goes in other way around. Nucleosynthesis here is a series of successive processes of crushing clusters to form a one-body system, gaining the difference in binding energies as thermal energy. Thus, the vertical axis in the figure should be regarded as the energy release during the progression of stellar evolution.

Here, the most important fact is that the CND diagram arises naturally from "the cluster threshold rule", as mentioned above. Namely, the rule says that there is a good chance that cluster states play an important role for nucleosynthesis. There have been many efforts for studying nuclear reactions of hydrogen burning, but not much for burning of helium, C and heavier clusters. Especially, the reactions that involve unstable nuclei for these reactions are very scarce. Since ^{4}He is the second most abundant isotope, there are many environments where α induced reactions play a role in the universe. They include (α,γ), (α,p) and (α,n) reactions. Because of the charge of α particle, α-induced stellar reactions play usually a role at higher energies than proton induced reactions, but in the light mass regions.

In this talk, I will concentrate my discussion on α-induced stellar reactions in the νp-process [3, 4, 5], which supposed to take place in the very early epoch in type II supernovae.

3. The νp-Process in Type II Supernovae

We discuss here the role of α clusterization in the νp-process [3, 4, 5] which is considered to be one of the challenging subjects for investigating the mechanism of type II supernovae, because it involves a variety of wide-range problems for nucleosynthesis.

The νp-process [3, 4, 5] was proposed in 2006 which would take place at the very early epoch of type II supernovae in the ejecta near the inner core due to the neutrino processes there. Here, it can be proton-rich, because of the high-intensity neutrino flux through the neutrino processes. This process also has been discussed as a source of p-nuclei near A=90, which have anomalously large isotopic abundances among the p-nuclei [6]. If the νp-process runs up to the mass 100 region, it will contribute to the p-nuclei productions. See Fig. 2.

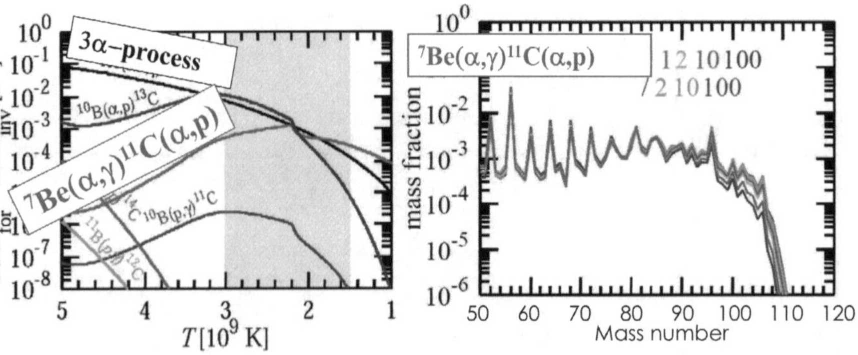

Figure 3. Nuclear reaction rate sensitivities for heavy element synthesis in the νp-process [7].

Recently, we made analysis on the uncertainties among the νp-process extensively in ref. [7]. There are many unknown nuclear physics problems in the νp-process, which involves many proton capture reactions at extremely high temperatures, but very few of them are known yet. In addition, because the site has certain fractions of neutrons and alpha particles, the νp-process involves also neutron-induced reactions as well as alpha induced reactions, which will involve high-lying states of proton-rich unstable nuclei. The latter reactions would be relevant to α cluster resonances and the former reactions giant resonances. An interesting consequence of neutron induced reactions will be discussed later in the last section.

Along the stellar evolution, nuclear burning around the N=Z line or proton rich side is important. At higher temperature and higher density condition of hydrogen and helium rich material, (α,p) reactions dominate the burning process in the light mass regions, which is called the αp-process. This process would play a crucial role in various stage of the evolution and phenomena. For instance, the ^{14}O(α,p)^{17}F [8] and ^{18}Ne(α,p)^{21}Na [9] reactions are important for ignition of X-ray bursts, which still await experimental investigation. Similarly, (α,n) reactions play a crucial role in the early stage of the r-process in neutron-rich nuclear regions.

A detailed discussion on the nuclear reactions in the νp-process will be made in the following sections including recent experimental investigations on the alpha-induced stellar reaction relevant.

4. (α,p) reactions in the νp process

The α induced reactions, especially the (α,p) reactions play a crucial role in the νp-process like in the rp-process. At high-temperature and high-density sites, this type reaction will dominate and possibly bypass the waiting point of the nucleosynthesis flow and the breakout from one region to the next in the proton-rich mass region. However, the role of α cluster states just above the α-threshold are not studied yet, although we expect an important role of α clusters there from the CND.

The (α,p) reactions are of course favored in the proton-rich nuclear region because of the positive Q values. A resonant contribution to the reaction rate in the (α,p) reaction can be written as follows;

$$\langle \sigma v \rangle \propto \frac{\Gamma_\alpha \Gamma_p}{\Gamma_{tot}} ,$$

where Γ_α, Γ_p are the α and p decay widths of the resonance, respectively. Thus, α resonant elastic scattering study is important together with the (α,p) cross section measurement for the

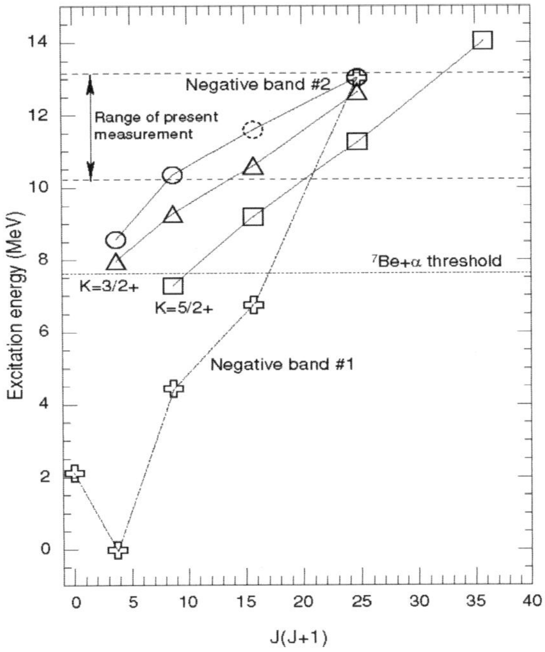

Figure 4. A new rotational band of α clusters observed by a thick target method of ^{4}He + ^{7}Li [15].

present study.

A series of experiments on (α,p) reactions as well as resonant elastic scattering of (α,α) have been made at the low-energy RI beam facility, CRIB [10, 11], of Center for Nuclear Study, the University of Tokyo. This is a part of RIBF in RIKEN.

I will first discuss here the pathway from the pp-chain region to the CNO region. The main process for synthesis of CNO elements is considered to be the triple alpha process. However, in addition, α-induced reactions on proton-rich nuclei would have led to synthesis of CNO elements. These reactions are not known well and thus need to be investigated experimentally. There are possible three major side flows that reach CNO elements, ^{7}Be(p,γ)^{8}B(p,γ)^{9}C(α,p)^{12}N, ^{7}Be(α,γ)^{11}C(α,p)^{14}N, and ^{7}Be(α,p)^{10}B(α,p)^{13}C.

The ^{7}Be+α and ^{7}Li+α resonant elastic scattering was studied recently by a thick target method [12, 13] in order to learn the entrance channel of ^{7}Be + α for these reaction chains. Here, an intense ^{7}Be beam [14] was obtained at CRIB. Some new resonances were identified in the ^{7}Li+α scattering [15], suggesting a rotational band which could have a cluster structure of 2α + t. See Fig. 4. Similar results are expected to be seen in the study of ^{7}Be+α, which will be published soon.

One of the crucial branching points for the breakout from the pp-chain region should be the nucleus ^{11}C, where the three processes are competing; ^{11}C(α,p)^{14}N, ^{11}C(p,γ)^{12}N and the beta decay to ^{11}B. The least known process among them was the ^{11}C(α,p)^{14}N reaction, which was investigated previously only by the time reverse reaction with an activation method [16]. Recently, a successful experiment was performed for the first time using a low-energy, high intensity ^{11}C beam at CRIB [17]. The cross sections were measured at E_{cm}= 1.0 $\sim$ 4.5 MeV, which covers an effective temperature range of T = 1 – 5$\times$ 10^9 K. The ^{11}C(α,p)^{14}N

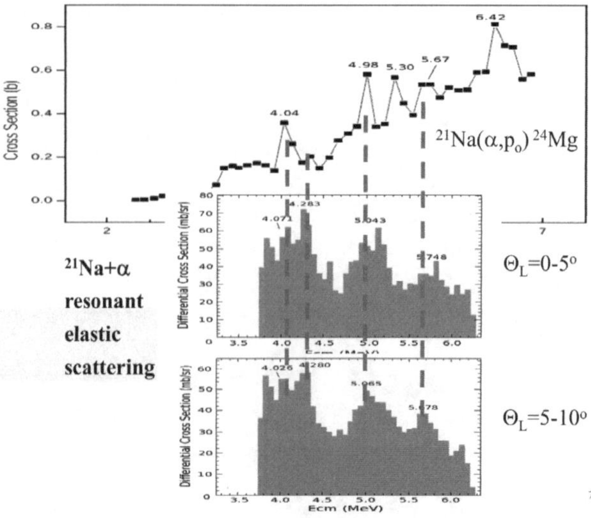

Figure 5. α resonances observed in the thick target experiment of ^{4}He + ^{21}Na [19], compared with the ^{21}Na(α,p$_0$) reaction cross sections measured by the time reverse process [20].

reaction for the low-lying excited states in ^{14}N were also successfully observed [17], but with a large uncertainty at the lowest energy. This experiment now provides reliable reaction rate at the temperature range for the νp-process. The experiment also has confirmed the previous data of ^{11}C(α,p$_0$)^{14}N(g.s.) by the activation method. The observed excitation functions of the ^{11}C(α,p)^{14}N cross sections were characterized by individual resonances, especially those with large α widths, and considerably deviate from the statistical model calculations. The experimental cross sections of the ^{11}C(α,p$_1$) reaction are also smaller than the statistical model prediction roughly by a factor of two. The present results demonstrate that one should study directly the (α,p) reaction cross sections identifying each level in these low mass regions.

After production of CNO elements in the αp-process, the next step of the nucleosynthesis is a flow out from the CNO region to the heavier element region. The first half of the sd shell nuclear region is of great interest because of the observation of nuclear gamma rays from long-lived nuclei ^{26}Al and ^{22}Na [18].

A successful experimental result was obtained recently in a study of the ^{21}Na(α,p) stellar reaction [19], which was suggested to play an important role in the αp-process [7]. A ^{21}Na beam was obtained from the CRIB separator. As can be seen in Fig. 5, the four prominent peaks appear to correspond to the peaks in the excitation function of the ^{21}Na(α,p)^{24}Mg reaction, which was measured by the time reverse reaction using an activation method [20], although the correspondences are not perfect in detail. The R-matrix analysis revealed that all these resonances have quite large α widths which exhausted large fractions of the Wigner limits. This result implies that α-resonances have a major role for the ^{21}Na(α,p) reaction rate, as expected by the CND [2]. We also have succeeded to measure directly the cross sections of ^{21}Na(α,p)^{24}Mg, which show much larger cross sections than those obtained by the activation method [20].

5. Scope

The Cluster Nucleosynthesis Diagram (CND), introduced before [1], was re-discussed here in order to emphasize importance of nuclear clusterization for nucleosynthesis in the universe. It gives a natural way of understanding of the evolution of nucleosynthesis and also suggests interesting stellar reactions which need to be investigated in nuclear astrophysics.

Specifically, we discussed here the role of alpha clusters for the αp-process, which use low-energy RI beams from the CRIB facility. Especially, our main efforts have been placed for the moment on (α,p) stellar reactions which set in under high-temperature and high-density hydrogen burning conditions. A crucial role of α resonances to the (α,p) stellar reactions has been demonstrated experimentally for the first time, as suggested by the CND.

In this talk, the (α,p) stellar reactions on proton-rich unstable nuclei were mainly discussed. However, there are other interesting nuclear physics problems in order to understand fully the mechanism of type II supernovae. Since there are significant amount of neutrons in the environment of the νp-process, neutron induced reactions on proton-rich unstable nuclei are of great interest and challenging. The (n,p) and (n,γ) reactions would accelerate the nucleosynthesis flow at the waiting point nuclei of ordinary hydrogen burning, which will eventually affect greatly the production rate of heavy nuclei at around mass 100 [5]. A sensitivity test of the ^{56}Ni(n,p) rate indicates quite large effect to the yields of the nuclei there. Here, the n-induced reactions involve high-lying states of unstable neutron-deficient nuclei, which are totally unknown yet. They are treated by the Hauser-Feshbach model together with E1 giant resonances. Recent theoretical works on giant resonances for wide-range unstable nuclei would improve the estimate for the rate. This subject is very important but it is one of the toughest challenges for nuclear astrophysics experiments. One possibility would be the Trojan Horse Method with proton-rich beams.

References

[1] Proc. Origin of Matter and Evolution of Galaxies 2011 ed Kubono S *et al* 2012 *AIP Conf. Proc.* **1484**
Proc. Origin of Matter and Evolution of Galaxies 2010 ed Tanihata I, Shima T, Ong H J, Tamii A, Kishimoto T, Toki H, Kajino T and Kubono S 2010 *AIP Conf. Proc.* **1269**
[2] Kubono S 1994 *Z. Phys.* A **349** 237
Kubono S *et al* 2010 *Nucl. Phys.* A **834** 647c
[3] Fröhlich C *et al* 2006 *Phys. Rev. Lett.* **96** 142502
[4] Pruet J, Hoffman R D, Woosley S E, Buras R and Janka H-T 2006 *Astrophys. J.* **644** 1028
[5] Wanajo S 2006 *Astrophys. J.* **647** 1323
[6] Arnould M and Goriely S 2003 *Phys. Rep.* **384** 1
[7] Wanajo S, Janka H-T and Kubono S 2011 *Astrophys. J.* **729** 46
[8] Notani M *et al* 2004 *Nucl. Phys.* A **746** 113
[9] He J J *et al* 2008 *Eur. Phys. J.* A **36** 1
[10] Kubono S, Yanagisawa Y, Teranishi T, Kato S, Kishida Y, Michimasa S, Ohshiro Y, Shimoura S, Ue K, Watanabe S and Yamazaki N 2002 *Eur. Phys. J.* A **13** 217
[11] Yanagisawa Y, Kubono S, Teranishi T, Ue K, Michimasa S, Notani M, He J J, Ohshiro Y, Shimoura S, Watanabe S, Yamazaki N, Iwasaki H, Kato S, Kishida T, Morikawa T and Mizoi Y 2005 *Nucl. Instr. Meth.* A **539** 74
[12] Artemov K P *et al* 1990 *Sov. J. Nucl. Phys.* **52** 408
[13] Kubono S 2001 *Nucl. Phys.* A **693** 221
[14] Yamaguchi H *et al* 2008 *Nucl. Instr. Meth.* A **589** 150
[15] Yamaguchi H *et al* 2011 *Phys. Rev.* C **83** 034306
[16] Ingalls P D, Schweitzer J S, Anderson B D and Rios M 1976 *Phys. Rev.* C **13** 524
[17] Hayakawa S private communication.
[18] Diehl R *et al* Proc. Origin of Matter and Evolution of Galaxies 2010 ed I. Tanihata *et al* 2010 *AIP Conf. Proc.* **1269** 144
[19] Dam G Binh private communication.
[20] Gruhle W and Kober B 1977 *Nucl. Phys.* A **286** 523

^{12}C and the triple-α reaction rate

Oliver Sølund Kirsebom
TRIUMF, 4004 Wesbrook Mall, Vancouver BC, V6T 2A3, Canada
E-mail: oliverk@triumf.ca

Abstract. We briefly review the role of the triple-α reaction in astrophysics and discuss the uncertainties associated with the determination of the reaction rate. We summarize the results of three recent experimental studies of the breakup of the Hoyle state in ^{12}C into three α particles, and we show how these results eliminate an often overlooked source of uncertainty in the determination of the triple-α reaction rate. Finally, we contemplate whether improved studies of the breakup of the Hoyle state can teach us something about the structure of the Hoyle state.

1. Introduction

The cross section of a nuclear reaction $A + B \rightarrow X + Y + \ldots$ can be enhanced by many orders of magnitude if the compound nucleus, $C = AB$, has a state with the appropriate quantum numbers at an energy corresponding to the combined energy of the reactants, A and B; we then speak of *resonance* in the channel $A+B$. Resonances play a very important role in astrophysics by speeding up nuclear reactions that otherwise would proceed very slowly. Furthermore, because thermal energies in stars tend to be small compared to nuclear excitation energies, the astrophysically important resonances are usually the ones nuclear physicists consider to be "low energy", *i.e.* for a nuclear state in the compound nucleus, C, to be astrophysically important, it must be close to the combined rest mass of A and B, *i.e.* close to the decay threshold.

At the same time, it is close to threshold that we find the best examples of cluster structures in nuclei. Thus, there exists a remarkable connection between stellar phenomena and the clustering aspects of nuclear structure. The famous Hoyle state in ^{12}C provides arguably the best example of this connection. Were it not for the tendency of nucleons to cluster into α particles, the Hoyle state would not exist, and without the Hoyle state little carbon would be produced in stars [1]. (And without carbon, we humans would not exist.)

2. Astrophysical importance

The triple-α reaction, *i.e.* the fusion of three α particles into a ^{12}C nucleus, plays a role in many astrophysical processes, but its great importance in astrophysics is to be attributed to two circumstances in particular: Firstly, it is the main mechanism behind stellar synthesis of carbon, essential to life as we know it on earth, and secondly, it is the mechanism by which the first generation of stars, created from the hydrogen and helium ashes of the Big Bang, were able to bridge the $A = 5$ and $A = 8$ mass gaps, thus allowing nucleosynthesis to proceed.

Carbon is mainly created during two phases of stellar evolution: the horizontal branch (HB) phase where helium burns in the core of the star, and the asymptotic giant branch (AGB) phase where helium burns in a shell surrounding the core. During these phases the star is in hydrostatic equilibrium and helium burns at temperatures between 0.1 GK and 2 GK (1 GK = 10^9 K). The triple-α reaction and the ^{12}C(α, γ) reaction operate side by side, and their relative rates determine the relative abundances of carbon and oxygen created. The triple-α reaction proceeds as a two-step process, strongly enhanced by two low-lying s-wave ($\ell = 0$) resonances,

$$\alpha + \alpha \to {}^{8}\text{Be(gs)} , \quad \alpha + {}^{8}\text{Be(gs)} \to {}^{12}\text{C(Hoyle)} , \tag{1}$$

the first being due to the ground state in ^{8}Be and the second being due to the Hoyle state in ^{12}C.

Furthermore, the triple-α reaction plays an important catalytic role in various explosive stellar processes in which the peak temperature exceeds 2 GK. Examples are given in Table 1. At such elevated temperatures, higher-lying resonances begin to contribute, and the triple-α reaction is no longer completely dominated by the low-lying s-wave resonances.

At temperatures somewhat below 0.1 GK, the thermal energy is too small for the triple-α reaction to proceed sequentially via the s-wave resonances; instead the reaction proceeds directly by some three-body mechanism. The rate of the triple-α reaction below 0.1 GK is important for the evolution of the first generation of stars. Since they were created from the hydrogen and helium ashes of the Big Bang, they lacked the carbon necessary for the CNO-I cycle to operate, which therefore had to be produced by the triple-α reaction, but at the relatively low temperature sustained by the pp chain [2].

Table 1 gives an overview of the three temperature domains of the triple-α reaction, including the most important astrophysical sites.

Table 1. Temperature domains of the triple-α reaction.

T (GK)	Burning	Reaction mechanism	Astrophysical sites
0.1–2	hydrostatic	s-wave resonances	– helium core burning in HB stars – helium shell burning in AGB stars
< 0.1	hydrostatic	direct three-body	– first-generation stars
> 2	explosive	higher-lying resonances	– shock fronts of core-collapse supernovae – X-ray bursts – helium flashes during AGB phase

3. Determination of the reaction rate

In the intermediate-temperature domain (0.1–2 GK) the rate of the triple-α reaction is fully determined by the properties of the Hoyle state, which can be measured in laboratory experiments. Precisely for this reason the rate is known relatively well, even though it cannot be directly measured. Currently, the rate is known with 10% precision, but efforts are underway to reduce this to 5% [3]. The determination of the reaction rate is far more complicated in the low- and high-temperature domains. In the high-temperature domain (> 2 GK) the rate depends on the properties of not one, but several states in ^{12}C, which are not well established. This is especially true for the possible 2^+ state around 9.6 MeV [4, 5, 6, 7, 8]; see also the contributions of M. Itoh and M. Gai. In the low-temperature domain (< 0.1 GK) sophisticated three-body models are necessary to calculate the rate [9].

When the triple-α reaction proceeds through the two low-lying s-wave resonances, the reaction rate is given by,

$$R \propto \frac{\Gamma_{\alpha 0}\Gamma_{\rm rad}}{\Gamma} T^{-3/2} \exp\left(-\frac{Q}{kT}\right) , \tag{2}$$

where Q is the energy of the Hoyle state relative to the $\alpha+\alpha+\alpha$ threshold, Γ_{α_0} is the partial α-decay width to the ground state of ^{8}Be, $\Gamma_{\rm rad}$ is the radiative width and $\Gamma = \Gamma_\alpha + \Gamma_{\rm rad}$ is the total width, Γ_α being the total α-decay width. The current evalution of the reaction rate assumes that $\Gamma_\alpha = \Gamma_{\alpha 0}$, i.e. the breakup of the Hoyle state into three α particles proceeds exclusively as a sequential two-step process via the ground state of ^{8}Be. Futher using that $\Gamma_{\rm rad}/\Gamma_\alpha \ll 1$, we have

$$\frac{\Gamma_{\alpha 0}\Gamma_{\rm rad}}{\Gamma} = \frac{\Gamma_\alpha \Gamma_{\rm rad}}{\Gamma_\alpha + \Gamma_{\rm rad}} \approx \Gamma_{\rm rad} , \tag{3}$$

showing that the rate depends on two quantities only, Q and $\Gamma_{\rm rad}$. The latter is determined from the measurement of three quantities,

$$\Gamma_{\rm rad} = \Gamma_\gamma + \Gamma_\pi = \frac{\Gamma_\gamma + \Gamma_\pi}{\Gamma} \frac{\Gamma}{\Gamma_\pi} \Gamma_\pi . \tag{4}$$

The current error budget is,

$$\begin{array}{rcll} Q & = & 379.38 \pm 0.20 \text{ keV} & (1.2\%) \;\; [10] \\ (\Gamma_\gamma + \Gamma_\pi)/\Gamma & = & (4.12 \pm 0.11) \times 10^{-4} & (2.7\%) \;\; [11] \\ \Gamma_\pi/\Gamma & = & (6.8 \pm 0.7) \times 10^{-6} & (10\%) \;\; [12] \\ \Gamma_\pi & = & 62.3 \pm 2.0 \;\mu\text{eV} & (3.2\%) \;\; [3] \end{array} \tag{5}$$

with experiments underway to reduce the uncertainty on Γ_π/Γ to 5%.

4. An often overlooked assumption

Let $\Gamma_{3\alpha} = \Gamma_\alpha - \Gamma_{\alpha 0}$ denote the combined width of all the direct α-decay branches of the Hoyle state bypassing the ground state of ^{8}Be. As we have seen, the assumption that $\Gamma_{3\alpha} = 0$ is of critical importance to the determination of the triple-α reaction rate. Nevertheless, the assumption had received relatively little scrutiny until recently. In 1994 an upper limit of $\Gamma_{3\alpha}/\Gamma_\alpha < 0.04$ was obtained by Freer *et al.* [13]. However, in 2011 Raduta *et al.* identified two direct α-decay branches with a combined branching ratio of $\Gamma_{3\alpha}/\Gamma_\alpha = 0.17(5)$ [14]. If correct, this would imply $\Gamma_{\alpha 0}/\Gamma = 0.83(5)$ and, as seen from Eq. (2), a corresponding reduction in the reaction rate in the temperature range 0.1–2 GK. The reduction is larger than the current estimate of the uncertainty on the rate and would have consequences for a number of astrophysical processes in which the triple-α reaction plays a role [15, 16, 17, 18].

The consequences would be far more dramatic in the low-temperature domain, where, due to the different energy dependences of $\Gamma_{\alpha 0}$ and $\Gamma_{3\alpha}$ at very low energy [19], the findings of Raduta *et al.* [14] would imply an *increase* of several orders of magnitude in the reaction rate.

5. New constraints on $\Gamma_{3\alpha}$

In Ref. [20] we reported on a new measurement of the breakup of the Hoyle state into three α particles. We used the ^{11}B(^{3}He, d) reaction at 8.5 MeV to populate the Hoyle state, and we used a compact detection system consisting of four segmented ΔE-E telescopes to measure the momenta of the deuteron and the three α particles, see Ref. [21] for details. The complete kinematics information allowed us to obtain very clean data and allowed us to improve the resolution using a mathematical procedure known as kinematic fitting. We used the Dalitz-plot technique to analyse the momentum distribution of the three α particles. The distribution of the experimental data was fully consistent with the distribution expected for sequential decay via the ground state of ^{8}Be (SD). We were able to put a stringent limit on a hypothetical phase-space distributed component (DDΦ) and even more stringent limits on the two components identified by Raduta *et al.*, namely, three α particles of equal energy (DDE) and one α particle at rest and the other two with equal energy (DDL). The results are summarized in Table 2 along with the results of another recent experiment by Manfredi *et al.* [22].

Table 2. Experimentally determined magnitudes of various Dalitz-plot components. SD denotes sequential decay via the ground state of ^{8}Be. DDE, DDL and DDΦ are associated with direct decay branches, see text for explanation. Upper limits are given at 95% confidence level. The uncertainties quoted by Raduta *et al.* are presumably 1σ uncertainties, *i.e.* 68% confidence level.

Dalitz-plot component	Freer *et al.* [13]	Raduta *et al.* [14]	Manfredi *et al.* [22]	Kirsebom *et al.* [20]
SD	1	0.830(50)	1	1
DDE	...	0.075(40)	$< 0.3 \times 10^{-2}$	$< 0.9 \times 10^{-3}$
DDL	...	0.095(40)	...	$< 0.9 \times 10^{-3}$
DDΦ	< 0.03 [a]	...	$< 3 \times 10^{-2}$	$< 5 \times 10^{-3}$

[a] Given as 0.04 at 99.5% confidence level in Ref. [13].

6. Link to structure

The structure of the Hoyle state is a subject of long-standing interest, which has received much theoretical attention recently (lattice effective field theory [23, 24], no-core shell model [25], fermionic molecular dynamics [26] and α-cluster models [27, 28, 29]).

Raduta *et al.* make the premise that observation of a certain distribution of α-particle energies provides evidence for corresponding structural features of the Hoyle state. Specifically, decay with three α particles of equal energy (DDE) is linked to α-condensate structure [30, 31] and decay with one α particle at rest and the other two with equal energy (DDL) is linked to the long-discussed linear-chain structure [32]. Raduta *et al.* thus conclude that the observation of a DDE component constitutes direct evidence for α-condensate structure of the Hoyle state. While we dispute the suggested direct link between energy distribution and structure, because the former must be strongly influenced by the tunneling through the Coulomb barrier, it seems natural to expect *some* connection between energy distribution and structure, and therefore a precise measurement of the energy distribution, *viz.* the Dalitz-plot distribution, should provide a sensitive test of structure models if the latter is combined with a sophisticated decay model such as that described in Refs. [27, 28].

7. A comment on semantics

In the scientific literature it is commonplace to refer to the breakup mechanism as if it were an observable ammendable to measurement. This is an unprecise use of terms, which can lead to confusion in discussions between theorists and experimentalists. The observable is not the breakup mechanism, but the momentum distribution of the α particles, *viz.* the Dalitz-plot distribution. The task of the experimentalist is to measure this distribution; that of the theorist is to calculate it. Any *interpretation* in terms of a breakup mechanism is only meaningful in the context of the model adopted by the theorist. That said, the interpretation of the experimental data presently available is straightforward and unambiguous: the breakup proceeds sequentially via the ground state of ^{8}Be in at least 99.5 of every 100 occurences.

8. Summary

The triple-α reaction plays a very important role in astrophysics. Under the conditions prevailing during helium burning in HB and AGB stars, the rate of the triple-α reaction is fully determined by the properties of the Hoyle state in ^{12}C. A central assumption in the determination of the rate is that the breakup of the Hoyle state into three α particles always proceeds sequentially via the ground state of ^{8}Be. Recently, Raduta *et al.* have reported experimental results that challenge this view: They claim observation of two direct α-decay branches bypassing the ground state of ^{8}Be with a combined branching ratio of 17(5)% [14]. This would imply a corresponding reduction in the reaction rate with significant astrophysical consequences. The results of Raduta *et al.* are, however, not verified in two more recent studies by Manfredi *et al.* [22] and Kirsebom *et al.* [20]. On the contrary, neither finds any evidence for the proposed direct branches. The most stringent upper limit is obtained by Kirsebom *et al.*: 5×10^{-3} at 95% confidence level.

While the currently available experimental data is consistent with a purely sequential decay, future experiments with higher sensitivity may potentially uncover other, very rare, decay branches. Besides having significant implications for the triple-α reaction rate at low temperatures, such a discovery would also have implications for our understanding of the structure of the Hoyle state.

[1] Hoyle F, Dunbar D N F, Wenzel W A and Whaling W 1953 *Phys. Rev.* **92** 1095
[2] Siess L, Livio M and Lattanzio J 2002 *Astrophys. J.* **570** 329
[3] Chernykh M, Feldmeier H, Neff T, von Neumann-Cosel P and Richter A 2010 *Phys. Rev. Lett.* **105** 022501
[4] Itoh M *et al.* 2004 *Nucl. Phys. A* **738** 268–272
[5] Freer M *et al.* 2009 *Physical Review C* **80** 041303
[6] Gai M 2011 *A. Phys. Pol. B* **42** 775
[7] Zimmerman W R, Destefano N E, Freer M, Gai M and Smit F D 2011 *Phys. Rev. C* **84** 027304
[8] Itoh M *et al.* 2011 *Phys. Rev. C* **84** 054308
[9] Nguyen N B, Nunes F M, Thompson I J and Brown E F 2012 *Phys. Rev. Lett.* **109** 141101
[10] Nolen J A and Austin S M 1976 *Phys. Rev. C* **13**(5) 1773–1779
[11] Markham R, Austin S M and Shahabuddin M 1976 *Nucl. Phys. A* **270** 489–500
[12] Alburger D E 1977 *Phys. Rev. C* **16**(6) 2394–2400
[13] Freer M *et al.* 1994 *Phys. Rev. C* **49**(4) R1751–R1754
[14] Raduta A *et al.* 2011 *Phys. Lett. B* **705** 65–70
[15] Fynbo H O U *et al.* 2005 *Nature* **433** 136–139
[16] Herwig F, Austin S M and Lattanzio J C 2006 *Phys. Rev. C* **73** 025802
[17] Tur C, Heger A and Austin S M 2007 *Astrophys. J.* **671** 821
[18] Tur C, Heger A and Austin S M 2010 *The Astrophys. J.* **718** 357
[19] Garrido E, de Diego R, Fedorov D and Jensen A 2011 *Euro. Phys. J. A* **47** 1–14
[20] Kirsebom O S *et al.* 2012 *Phys. Rev. Lett.* **108** 202501
[21] Kirsebom O S 2010 *^{8}B Neutrinos and ^{12}C Resonances* Ph.D. thesis Aarhus University
[22] Manfredi J *et al.* 2012 *Phys. Rev. C* **85** 037603
[23] Epelbaum E, Krebs H, Lähde T, Lee D and Meißner U G 2012 *arXiv:1208.1328v2 [nucl-th]*
[24] Epelbaum E, Krebs H, Lee D and Meißner U G 2011 *Phys. Rev. Lett.* **106** 192501
[25] Roth R, Langhammer J, Calci A, Binder S and Navrátil P 2011 *Phys. Rev. Lett.* **107**(7) 072501
[26] Chernykh M, Feldmeier H, Neff T, von Neumann-Cosel P and Richter A 2007 *Phys. Rev. Lett.* **98** 032501
[27] Álvarez-Rodríguez R, Jensen A S, Fedorov D V, Fynbo H O U and Garrido E 2007 *Phys. Rev. Lett.* **99** 072503
[28] Álvarez-Rodríguez R, Jensen A S, Garrido E, Fedorov D V and Fynbo H O U 2008 *Phys. Rev. C* **77** 064305
[29] S I Fedotov O I Kartavtsev A V M 2010 *JETP Lett.* **92** 647
[30] Tohsaki A, Horiuchi H, Schuck P and Röpke G 2001 *Phys. Rev. Lett.* **87** 192501
[31] Zinner N T and Jensen A S 2008 *Phys. Rev. C* **78** 041306
[32] Morinaga H 1956 *Phys. Rev.* **101** 254–258

Low-energy d+d fusion via the Trojan Horse Method

A Tumino[1,2], C Spitaleri[1,3], A M Mukhamedzhanov[4], S Typel[5], M Aliotta[6], V Burjan[7], M Gimenez del Santo[8], G G Kiss[1,9], V Kroha[7], Z Hons[7], M La Cognata[1], L Lamia[3], J Mrazek[7], R G Pizzone[1], S Piskor[7], G G Rapisarda[1,3], S Romano[31], M L Sergi[1], R Spartà[1,3]

[1]Laboratori Nazionali del Sud - INFN, Catania, Italy
[2]Università degli Studi di Enna "Kore", Enna, Italy
[3]Dipartimento di Fisica e Astronomia, Università di Catania, Catania, Italy
[4]Cyclotron Institute Texas A&M University - College Station, Texas, USA
[5]GSI Helmholtzzentrum für Schwerionenforschung GmbH - Theorie Darmstadt Germany
[6]School of Physics and Astronomy - University of Edinburgh; SUPA, UK
[7]Nuclear Physics Institute of ASCR - Rez near Prague, Czech Republic
[8]Departamento de Fisica Nuclear - Universitade de Sao Paulo, Sao Paulo, Brasil
[9]ATOMKI - Debrecen, Hungary

E-mail: tumino@lns.infn.it

Abstract. The ^{2}H(d,p)^{3}H and ^{2}H(d,n)^{3}He reactions have been recently investigated from E_{dd}=1.5 MeV down to 2 keV, by means of the Trojan Horse Method (THM) applied to the Quasi Free ^{3}He+d interaction at 18 MeV [1]. The knowledge of their fusion cross section at low energies is of interest for pure and applied physics. Both reactions belong to the network of processes to fuel the first inertial confinement fusion reactors in the range of kT= 1 to 30 keV. These energies overlap with the burning temperatures of deuterium in the Pre-main sequence of stellar evolution. They are key processes in the Standard Big Bang Nucleosynthesis (SBBN), in an energy region from 50 to 300 keV and experimental data at least up to 1 MeV are required for an accurate calculation of the reaction rate. Providing experimental data for both channels from a single experiment and over the entire energy range of interest is crucial for an accurate calculation of the reaction rates. This is what has been obtained from the present Trojan Horse (TH) investigation with new reaction rates which deviate by more than 20% from available direct data. This represents also the first pioneering experiment in quasi free regime where the charged spectator is detected.

1. Introduction

Low energy ^{2}H(d,p)^{3}H and ^{2}H(d,n)^{3}He reactions belong to the Standard Big Bang Nucleosynthesis (SBBN) network, synthesizing D, ^{3}He, ^{4}He and ^{7}Li in the early universe.They are the most important and still critical points at the SBBN relevant energies ($d-d$ relative energies E_{dd} from 50 to 350 keV). All published measurements for both channels have been mainly focused either on the E_{dd} region below 200 keV [2, 3, 4, 5] or at E_{dd} above 1 MeV [6], while an accurate calculation of the reaction rate needs the cross section to be known from E_{dd}=30 keV to at least 1 MeV. All the reaction rates quoted so far rely either on polynomial [7, 8] or in the best case on R-Matrix fits [9], whose different slopes bring inside the calculations deviations of more than 15%. In a recent work [10] the total cross section for both ^{2}H(d,p)^{3}H and ^{2}H(d,n)^{3}He reactions was measured at E_{dd} from 55 to 325 keV in

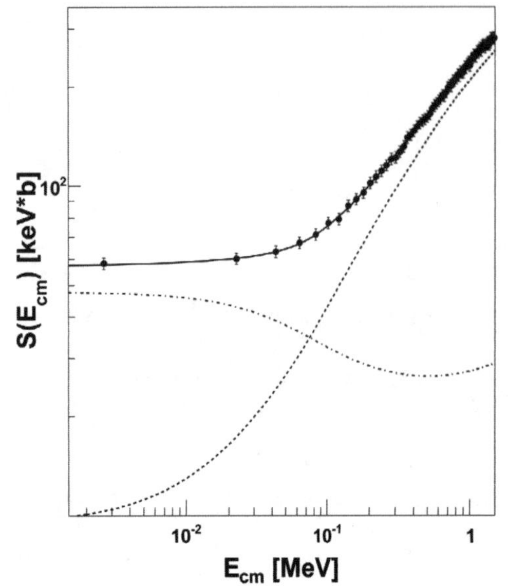

Figure 1. $S(E)$ factor for the ^{2}H(d,p)^{3}H reaction: black dots=THM results; The black solid lines represents the new TH parameterization as sum of the dashed-dotted (l=0) and dashed (l=1) lines. See text for details.

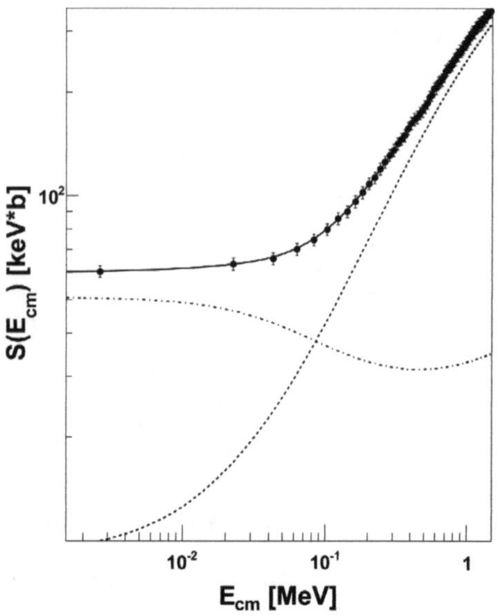

Figure 2. $S(E)$ factors of the ^{2}H(d,n)^{3}He reaction: black dots=THM data; Lines have the same meaning as in Fig.1.

steps of about 50 keV. Although uncertainties of better than 4% are quoted, there are some concerns about the procedure followed to calculate the reaction rates. For this reason, a new experimental campaign was recommended by the astrophysical community, to provide new data in the SBBN energy range of interest. Moreover, the importance of the ^{2}H(d,p)^{3}H and ^{2}H(d,n)^{3}He reactions in the ultra-low energy region below 30 keV is twofold: it overlaps with the burning temperatures (0-10 keV) of deuterium in Pre Main Sequence (PMS) stars and corresponds to the thermal energy range of future fusion power plants. However, the measured cross section below 10 keV, available only for the ^{2}H(d,p)^{3}H channel with a gas target [4], has shown a clear enhancement attributed to electron screening. The deduced electron screening potential of U_e= 25±5 eV is significantly larger than the adiabatic limit (14 eV) provided by atomic physics. Unfortunately, nuclear reactions in the laboratory are affected by a different mechanism of electron screening than in a plasma [11], either astrophysical or laboratory fusion with inertial confinement. Hence, the laboratory screening effects of the bound electrons should be removed from the data to assess the reaction rate correctly. This makes the knowledge of the bare nucleus cross section essential for several purposes, from basic to applied research. Usually, a simple extrapolation of available unscreened higher energy data [12] is taken as an approximation of the bare nucleus cross section, and compared to measured low-energy data to extract the electron screening potential. However, this approach can lead to large uncertainties. One can consider that available theoretical extrapolations of the bare nucleus ^{2}H(d,p)^{3}H and ^{2}H(d,n)^{3}He cross sections provide quite different S(0) values which vary by 10-15%. A valid alternative approach is represented by the Trojan Horse Method (THM) ([13, 14, 15] and references therein), that provides at present the only way to measure the energy dependence of the bare nucleus cross section down to the relevant ultra-low energies, overcoming the main

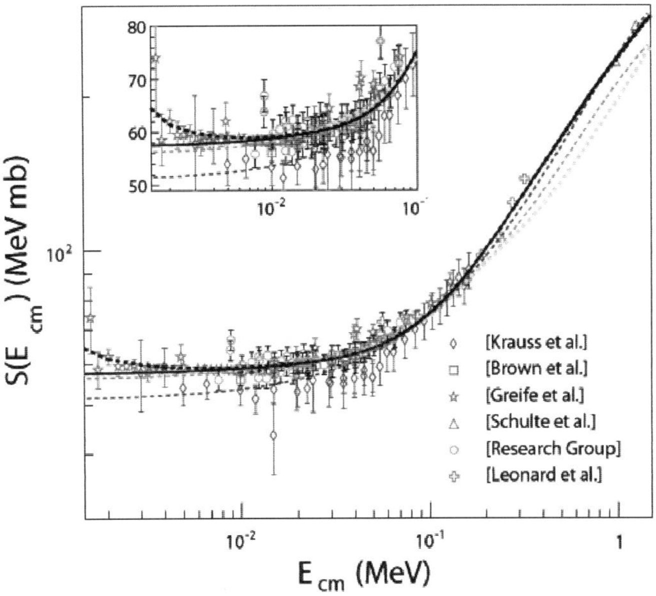

Figure 3. (Color online) $S(E)$ factor for the ^{2}H(d,p)^{3}H reaction: black solid line=THM result; colored symbols=direct data. See text for the description of the other lines.

problems of direct measurements [16, 17, 18]. The THM has already been applied several times to reactions connected with fundamental astrophysical problems [19-35] and references therein). In the present paper we report on a novel investigation of the deuterium depletion reactions ^{2}H(d,p)^{3}H and ^{2}H(d,n)^{3}He [1, 36] throughout the energy range relevant for pure and applied physics by means of the THM. The method was applied by choosing the ^{2}H(^{3}He,p^{3}H)^{1}H and ^{2}H(^{3}He,n^{3}He)^{1}H reactions in QF kinematics with ^{3}He as the Trojan horse. For the first time in the application of the THM, we have detected the particle that acts as a spectator to both ^{2}H(d,n)^{3}He and ^{2}H(d,p)^{3}H binary processes, namely the proton. This technique was mandatory for the ^{3}He+n fusion channel to avoid the limitation of standard neutron detectors. However, it was well suited also for the ^{3}H+p channel, preventing unwanted quasi free (QF) events from target break-up to be detected. A detailed account of the measurement and related data analysis is given in [1, 36]. Here, we briefly report on the main steps of analysis and on the final results.

2. Discussion and results

In previous R-matrix calculations [9], it was necessary to include the l = 0 and l = 1 components for both ^{2}H(d,p)^{3}H and ^{2}H(d,n)^{3}He reactions for the determination of the $S(E)$ factor throughout the investigated E_{cm} region. Provided this information, data analysis follows the standard procedure as reported in [21, 22, 31] and references therein. With the deduced scaling ratio of the s- and p-wave contributions, the $S(E)$ factors for both reactions were then extracted as a function of E_{cm} in steps of 20 keV, with a statistical error of 4%. The normalization of the S(E) factors to direct data has to be carried out in a region where electron screening effects are negligible (E_{cm} ≥15 keV). To account for the different accuracies and large scatters of direct data [2, 3, 4, 5, 6, 10], a weighted normalization was carried out to all of them from E_{cm}= 1.5 MeV down to 15 keV, each with their quoted total errors (both statistical and

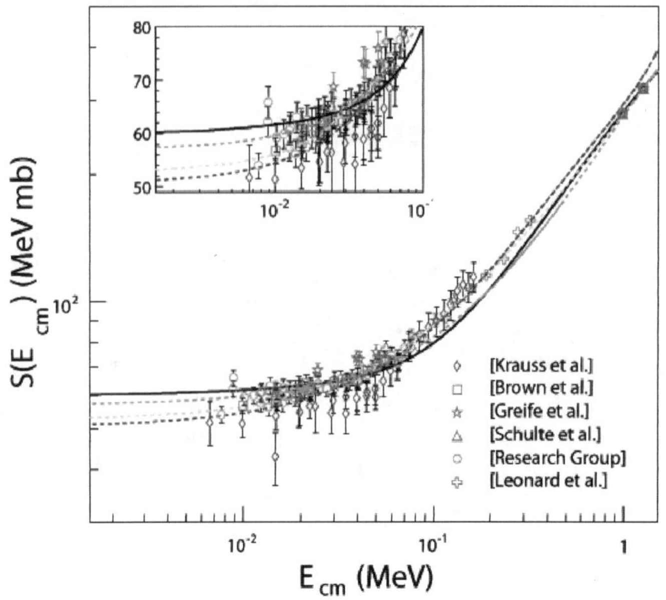

Figure 4. (Color online) $S(E)$ factors of the ^{2}H(d,n)^{3}He reaction: black solid line=THM result; colored symbols=direct data. See text for the description of the other lines.

systematic combined in quadrature). The weighted normalization procedure clearly favors the most accurate direct data, some of which have total uncertainties better than 1% [4, 6]. A total of 43 THM data points were used in the energy regions where direct data are available. The procedure leads to an overall normalization error of about 1%, obtained from the standard error combination in quadrature. The TH $S(E)$ factors of the ^{2}H(d,p)^{3}H and ^{2}H(d,n)^{3}He reactions are shown as black full dots in Figs. 1 and 2, respectively, with uncertainties accounting for statistical and normalization errors. We provide a new parameterization of the THM S(E) factor (black solid line in both figures) as a sum of the theoretical $l = 0$ and $l = 1$ contributions (red and blue solid lines, respectively), with relative weight fixed from the fit to the measured three-body coincidence yield. Direct data from [2, 3, 4, 5, 6] are shown as colored symbols in Figs. 3 and 4 (the legend in both figures associates a symbol and a color to each ref.), with the black TH solid line superimposed on them. The very regular trend of TH data contrasts with that of direct data taken as a whole, both in energy dependence and absolute values, with deviations of more than 15%. The polynomial/R-matrix fits to direct data, usually taken as reference, are shown as green [7], blue [8] and yellow [9] dashed lines. They barely overlap with most of the direct low-energy data, in particular in the case of the ^{3}He+n channel, probably due to their larger scatter. Conversely, the green line seems to agree with low-energy TH data, while for E_{cm} >200 keV they seem to be better reproduced by the blue line. However, none of them correctly reproduces the slope of the THM S(E)-factor in the entire energy region investigated, thus calling for a revision of the previous theoretical descriptions based on these new high quality data.

The THM parameterizations of the S(E) factors lead to new values of S(0) = 57.4±1.8 keVb for ^{3}H+p and 60.1±1.9 keVb for ^{3}He+n, with uncertainties including the 1% normalization error and 3% coming from the theory, combined in quadrature. The insets in the upper panel of both Figs. 3 and 4 help to compare these values with previous ones usually taken as reference [7, 8, 9].

The deviations are of 15%-20%. A comparison between the THM S(0) factors to the ^{2}H(d,p)^{3}H direct data below 15 keV provides further insight into the electron screening effect. Low-energy direct data at 14.95 keV from [4] were first normalized to the THM S_{bare}(E) (black solid line) and then fitted with the screening function [1] leaving U_e as free parameter. This provides a value of $U_e = 13.2 \pm 1.8$ eV, not exceeding the adiabatic limit (14 eV) for a molecular deuteron target (gas target), but covering it with its uncertainty. Additional low energy direct data for both channels would help to fix the electron screening potential.
Preliminary estimates of reaction rates based on the new THM data show that they experience relevant changes. For the ^{3}H+p (^{3}He+n) channel, the rate increases (decreases) by up to 15% at the SBBN temperatures with respect to previous calculations [7, 8, 9], while a stronger increase of more than 25% for both channels is found at the temperatures relevant for PMS and future fusion power plants.

Acknowledgments
This work was supported in part by the Italian MIUR under Grant (RBFR082838).

References

[1] Tumino A *et al* 2011 *Phys. Lett.* B **700** 111
[2] Krauss A *et al* 1987 *Nucl. Phys.* A **465** 150
[3] Brown R E and Jarmie N 1990 *Phys. Rev.* C **41** 1391
[4] Greife U *et al* 1995 *Z. Phys.* A **351** 107
[5] The First Res. Group 1985 *Chin. J. Nucl. Phys.* **9** 723
[6] Schulte R L *et al* 1972 *Nucl. Phys.* A **192** 609
[7] Angulo C *et al* 1999 *Nucl. Phys.* A **656** 3
[8] Cyburt R H 2004 *Phys. Rev.* D **70** 023505
[9] Descouvemont P *et al* 2004 *At. Data Nucl. Data Tab.* **88** 203
[10] Leonard D S *et al* 2006 *Phys. Rev.* C **73** 045801
[11] Shaviv G and Shaviv N 2000 *Astroph. J.* **529** 1054
[12] Bosch H S and Hale G M 1992 *Nucl. Fusion* **32** 611
[13] Spitaleri C *et al* 2011 *Phys. of Atomic Nuclei* **74** 12 1725
[14] Tumino A *et al* 2012 *Few-Body Syst.* 1 DOI 10.1007/s00601-012-0407-1
[15] Tumino A *et al* 2012 *AIP* **1488** 87
[16] Tumino A *et al* 2007 *Phys. Rev. Lett.* **98** 252502
[17] Mukhamedzhanov A M *et al* 2008 *J. Phys.* G **35** 014016
[18] Tumino A *et al* 2008 *Phys. Rev.* C **68** 064001
[19] Spitaleri C *et al* 2000 *Eur. Phys. Journ.* A **7** 181
[20] Aliotta M *et al* 2000 *Eur. Phys. Journ.* A **9** 435
[21] Spitaleri C *et al* 2001 *Phys. Rev.* C **63** 005801
[22] Tumino A *et al* 2003 *Phys. Rev.* C **67** 065803
[23] Pizzone R G *et al* 2003 *A&A* 2003 **398** 423
[24] Tumino A *et al* 2005 *Eur. Phys. J.* A **25** 649
[25] Pizzone R G *et al* 2005 *A&A* **438** 779
[26] Pizzone R G *et al* 2005 *Phys. Rev.* C **71** 058801
[27] Tumino A *et al* 2006 *Eur. Phys. J.* A **27** 243
[28] Wen Q *et al* 2008 *Phys. Rev.* C **78** 035805
[29] Lamia L *et al* 2009 *Il Nuovo Cimento* **31** 423
[30] Gulino M *et al* 2010 *J. Phys.* G **37** 125105
[31] La Cognata M *et al* 2010 *Astrophysical J.* **708** 796
[32] Sergi M L *et al* 2010 *Phys. Rev.* C **R82** 032801
[33] La Cognata M *et al* 2011 *Astrophysical J. Lett.* **739** L54
[34] Wen Q *et al* 2011 *J. Phys.* G **38** 085103
[35] Lamia L *et al* 2012 *J. Phys.* G **39** 015106
[36] Tumino A *et al* 2011 *Phys. Lett.* B **705** 546

Journal of Physics: Conference Series **436** (2013) 012074 doi:10.1088/1742-6596/436/1/012074

Study of ^{24}Mg resonances relevant to carbon burning nucleosynthesis

V Tokić[1], N Soić[1], S Blagus[1], S Fazinić[1], D Jelavić-Malenica[1], Đ Miljanić[1], L Prepolec[1], N Skukan[1], S Szilner[1], M Uroić[1], M Milin[2], A Di Pietro[3], P Figuera[3], M Fisichella[3,5], M Lattuada[3,4], V Scuderi[3], E Strano[3], D Torresi[3,8], N Ashwood[6], N Curtis[6], M Freer[6], V Ziman[6], I Martel[7], A M Sánchez-Benítez[7] and L Acosta[7]

[1]Ruđer Bošković Institute, Bijenička cesta 54, 10000 Zagreb, Croatia
[2]Faculty of Science, University of Zagreb, Bijenička cesta 32, 10000 Zagreb, Croatia
[3]INFN-Laboratori Nazionali del Sud, via S.Sofia 62, 95125 Catania, Italy
[4]Department of Physics, University of Catania, Piazza Universita, 2 I, 95124 Catania, Italy
[5]Department of Physics, University of Messina, Piazza Pugliatti 1, 98122 Messina, Italy
[6]School of Physics and Astronomy, University of Birmingham, Birmingham B15 2TT, UK
[7]University of Huelva, Department of Applied Physics, Avda. Fuerzas Armadas, s/n, 21071 Huelva, Spain

E-mail: vtokic@irb.hr

Abstract. We have studied the decays of the resonances of ^{24}Mg at excitation energies 1–6 MeV above the ^{12}C+^{12}C decay threshold, using the ^{12}C(^{16}O,α)^{24}Mg* reaction at E(^{16}O) = 94 MeV. Some preliminary results are presented and further analysis is in progress.

1. Introduction

Carbon-carbon burning has an important role in many stellar systems, which include super AGB stars, SN-type Ia systems and superbursts. To see how likely this fusion process is in these environments, we need to know the cross section in the energy range that occurs in these systems. The relevant energy range of cross section measurements for super AGB stars and SN-type Ia is $E_{c.m.}$ = 1.5–3.3 MeV for the ^{12}C+^{12}C reaction, while for the superbursts this energy range extends up to 5.7 MeV [1]. The available experimental data for the cross section cover the energy range down to $E_{c.m.}$ = 2.14 MeV [2, 3, 4]. Since there are resonances in the cross section at energies higher than the Gamow peak (E_G = 1.5 MeV for temperature $5{\cdot}10^8$ K), an extrapolation to the energy range required is uncertain, therefore, new data are needed. As this region is far below Coulomb barrier, it is extremely difficult to obtain these results.

2. Experiment

Instead of the direct cross section measurements used in [2, 3, 4], we decided to take a different approach. We have studied the ^{12}C(^{16}O,α)^{24}Mg* reaction at E(^{16}O) = 94 MeV. The decay channels of ^{24}Mg* and some threshold energies are presented in Table 1. The main idea was to

[8] Present address: INFN-Sez. di Padova, Via Marzolo 8, 35131 Padova, Italy

Table 1. Decay channels of ^{24}Mg* and threshold energies E_{th}.

decay channel	$^{12}C+^{12}C$	$^{8}Be+^{16}O$	$\alpha+^{20}Ne$	$p+^{23}Na$
E_{th} (MeV)	13.93	14.14	9.31	11.69

study the resonance structures of ^{24}Mg at excitation energies from 1 up to 6 MeV above the $^{12}C+^{12}C$ decay threshold and to determine their parameters by detecting decay fragments and recoil nuclei. The experiment was performed at INFN-LNS, using a Tandem accelerator and a large Si detector array. Our experimental setup consisted of 6 Si detector telescopes, each of them included a thin ΔE detector (their thickness was 19 or 20 μm), and one thick Position-Sensitive Silicon Strip Detector (PSSSD) or a Double-Sided Silicon Strip Detector (DSSSD).

3. Results

In calibrating the detectors we have noticed that the thin detectors are nonuniform in thickness (Fig. 1). The energy loss of the detected ions was calculated using the SRIM-2010 code. Similar results were obtained in Ref. [5].

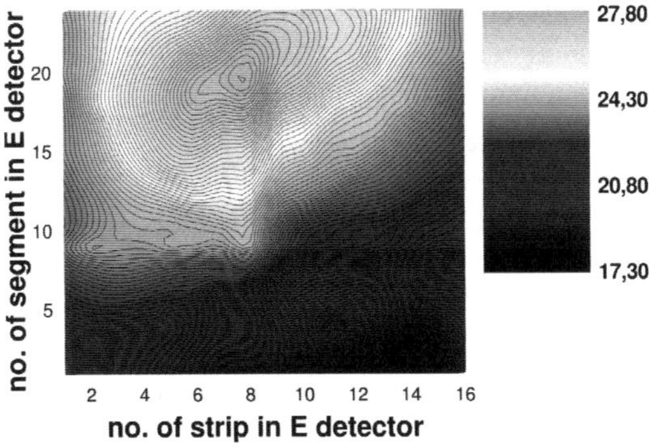

Figure 1. Thickness variation in μm of a 20 μm detector, made by Micron Semiconductor Ltd.

The Romano plot [6] for the coincidence of ^{12}C and α in one of the reaction channels is presented in Fig. 2. From these data it is inferred that the third, undetected particle has $A = 12$, while the Q-value of this reaction is –7.6 MeV (Fig. 2, inset 1), and $A = 12.03$ with a Q-value of –12.1 MeV (Fig. 2, inset 2). This difference in Q-value is about 4.4 MeV which corresponds to the energy of the first excited state of ^{12}C. This indicates that the third, undetected particle is also ^{12}C indeed, and in one case both nuclei ^{12}C are in their ground state, while in the other case one of them is in its first excited state. The Q-value as determined at various angles θ of the detected ^{12}C is presented in Fig. 3.

Further analysis is in progress.

4. Acknowledgments

This work was supported by the Croatian Science Foundation Project 'Experimental nuclear physics inputs for thermonuclear runaway' in the ESF EuroGENESIS program. The speaker (the first author) would like to thank the staff of INFN-LNS for their support and assistance.

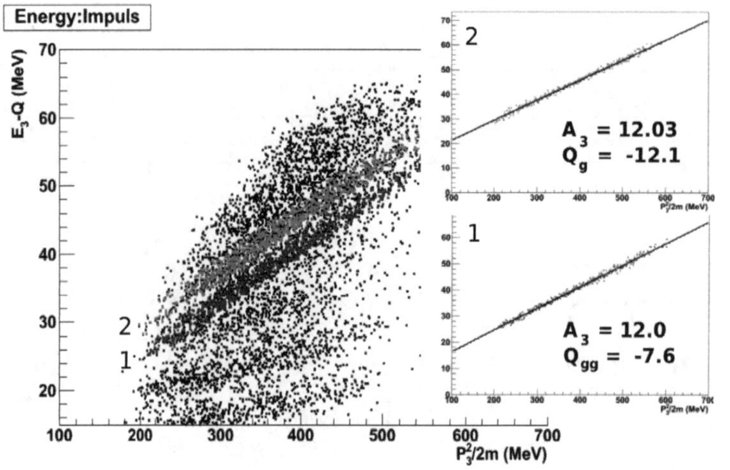

Figure 2. The difference $E_3 - Q$ of the energy of the third, undetected particle and the Q-value as a function of the $P_3^2/2\mathrm{m}$ (P_3 being the momentum of the third particle and m the nucleon mass) for the measurement of the coincidence of a ^{12}C and an α. From the crossing of the perpendicular axis and the slope of each 'stripe' we get $-Q$ and $1/\mathrm{A}_3$.

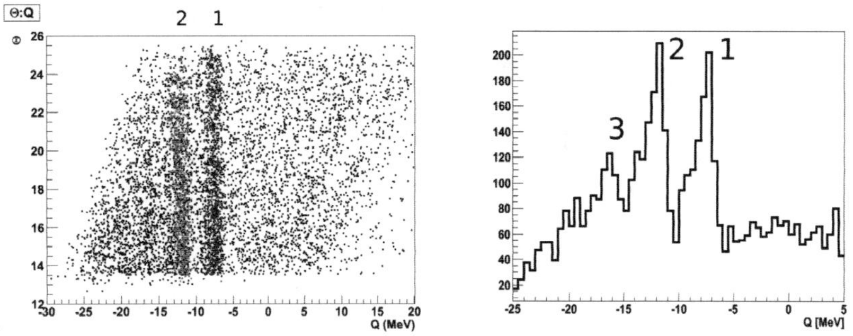

Figure 3. Detection angle θ of ^{12}C versus the Q-value (left) and the number of detected particles as a function of the Q-value (right). The Q-value peaks are at –7.6 MeV (1) and –12.1 MeV (2), corresponding to two ^{12}C in ground state and one in the ground state and the other in the first excited state, respectively. There is also an indication for a third peak (3).

References

[1] Gasques L R *et al* 2007 *Phys. Rev.* C **76** 035802
[2] Spillane T *et al* 2007 *Phys. Rev. Lett.* **98** 122501
[3] Aguilera E F *et al* 2006 *Phys. Rev.* C **73** 064601
[4] Notani M *et al* 2012 *Phys. Rev.* C **85** 014607
[5] Davin B *et al* 2001 *Nucl. Instr. and Meth.* A **473** 302
[6] Costanzo E *et al* 1990 *Nucl. Instr. and Meth.* A **295** 373

10th International Conference on Clustering Aspects of Nuclear Structure and Dynamics IOP Publishing
Journal of Physics: Conference Series **436** (2013) 012075 doi:10.1088/1742-6596/436/1/012075

Electron screening effects in (p,α) reactions induced on boron isotopes studied via the Trojan Horse Method

L Lamia[1], C Spitaleri[1,2], N Carlin[3], S Cherubini[1,2], M Gameiro Munhoz[3], M Gimenez Del Santo[3], M Gulino[1,2], G G Kiss[4], V Kroha[5], S Kubono[6], M La Cognata[2], C Li[7], A Mukhamedzhanov[8], R G Pizzone[2], S M R Puglia[1,2], Qungang Wen[7], G G Rapisarda[1,2], S Romano[1,2], M L Sergi[1,2], E Somorjai[4], F A Souza[3], A Szanto de Toledo[3], A Tumino[2,9], Y Wakabayashi[6], H Yamaguchi[6]

[1] Dipartimento di Fisica e Astronomia, Università degli Studi di Catania, Italy
[2] Laboratori Nazionali del Sud, INFN, Catania, Italy
[3] Departamento de Fisica Nuclear, Universitade de São Paulo, São Paulo, Brasil
[4] Atomki, Debrecen, Hungary
[5] Institute for Nuclear Physics, Prague-Rez,Czech Republic
[6] CNS-University of Tokyo-Japan
[7] China Institute of Energy, Beijing, China
[8] Cyclotron Institute, Texas A&M University, College Station, TX 77843,Usa
[9] Università degli Studi di Enna "Kore", Enna, Italy

E-mail: llamia@lns.infn.it

Abstract. The Trojan Horse Method is a powerful indirect technique allowing one to measure the bare nucleus S(E)-factor and the electron screening potential for astrophysically relevant reactions without the needs of extrapolations. The case of the (p,α) reactions induced on the two boron isotopes ^{10,11}B is here discussed in view of the recent Trojan Horse (TH) applications to the quasi-free ^{10,11}B+^{2}H reactions. The comparison between the TH and the low-energy direct data allowed us to determine the electron screening potential for the ^{11}B(p,α) reaction, while preliminary results on the ^{10}B(p,α) reaction have been extracted.

1. Method and results

The Trojan Horse Method (THM) [1-4] is an indirect technique allowing one to measure the low-energy S(E)-factor of a charged particle induced reaction $a+x \to c+C$ in correspondence of the Gamow energy, at which the direct cross section measurements exhibit a drastic exponential decrease because of the Coulomb barrier. The method selects the quasi-free (QF) contribution on a suitable 2$\to$3 reaction $a + A \to c + C + s$, performed at energies well above the Coulomb barrier, thus ensuring the $a + x$ interaction without any influence of the Coulomb penetration effects [5-6]. In that framework, the Trojan Horse nucleus A is chosen because of its large amplitude for the $A = x \oplus s$ cluster configuration and, once the Impulse Approximations (IA) hypothesis are fulfilled, x represents the *participant* while s is the *spectator* [7]. The THM has been applied for shedding light on different astrophysical problems, such as the study of

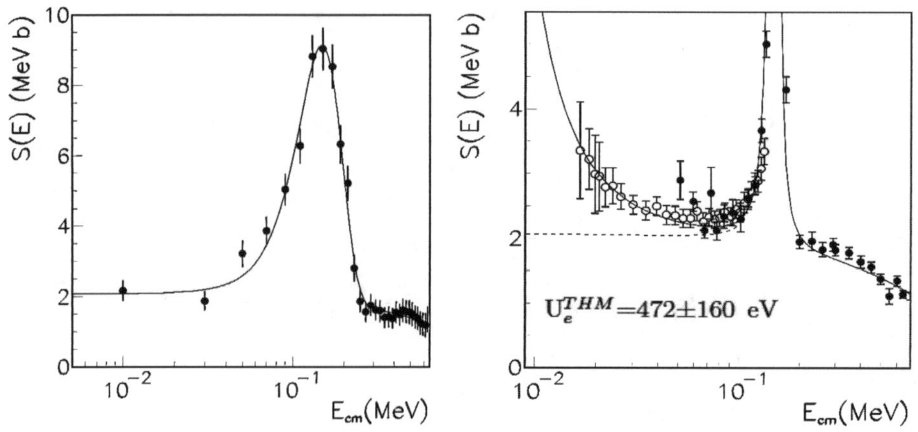

Figure 1. Left panel: the TH ^{11}B(p,α)^{8}Be data (black points) with the corresponding fit. Right panel: the low-energy direct data [21] have been fitted by means of the enhancing factor $\exp(\pi\eta U_e/E)$ [26], with a value of the screening potential of U_e=472 eV.

light element burning reactions ([8-15]), CNO reactions (see [16-19]), and removing/producing neutron reactions ([20-22]).
Recently in [23] we have reported on the study of the astrophysically relevant ^{11}B(p,α)^{8}Be reaction, via the THM application to the ^{2}H(^{11}B,α^{8}Be)n reaction induced at the boron beam energy of 27 MeV in the laboratory, where the ^{2}H has been used as TH-nucleus thanks its obvious p$\oplus$n structure. Once the contribution of the QF reaction mechanism has been carefully selected, the experimental momentum distribution for the p-n relative motion has been extracted and it was found to be in agreement with the expected Hulthén wave function in momentum space. In addition, in a further work, we investigated also the possible influence of the d-state component in the p-n relative motion obtaining a contribution of less than 1% with respect the s-state one [24]. The ^{11}B(p,α)^{8}Be have been then studied from ~600 keV down to ~10 keV, the last corresponding to the Gamow energy window for quiescent boron burning. In particular the zero-energy S(E)-factor measurements leads, after a normalization to the direct data (see [25] and references therein), to the value of S(0)=2.07±0.41 MeV b (left panel of Fig. 1) and, from the comparison with the available direct measurements, a corresponding electron screening potential of 472±160 eV has been extracted (right panel of Fig. 1).

In order to complement the information about electron screening gained in the previous case, the TH has been also applied for studying the ^{10}B(p,α)^{7}Be reaction. In that case, the ^{2}H(^{10}B,α^{7}Be)n experiment have been performed, by using a 24 MeV boron beam on a CD_2 target at LNS, Catania. Left panel of Fig.2 shows the experimental Q-value for the three-body reaction, centered at ~-1.1 MeV in excellent agreement with the theoretical one. Right panel of Fig. 2 shows the preliminary TH S(E)-factor in arbitrary units. The astrophysical S(E) factor shows up the resonant l=0 contribution centered at about E_{cm}~10 keV, i.e. in correspondence to the Gamow peak. The behavior of the S-factor is strongly dominated by the ~10 keV resonant state, falling just in the Gamow peak energy region where, up to now, the NACRE compilation [25] reports only an R-matrix calculation. Even if the experimental uncertainties affecting the TH data do not allow for definitive conclusions, these results encourage further investigations. The U_e^{THM} obtained for the ^{11}B(p,α)^{8}Be is in agreement with the value of 430±80 eV (see [25] and references therein), though it is higher than the upper limit of 340 eV predicted by the adiabatic limit. The same conclusions apply to the lithium-burning reactions for which a value

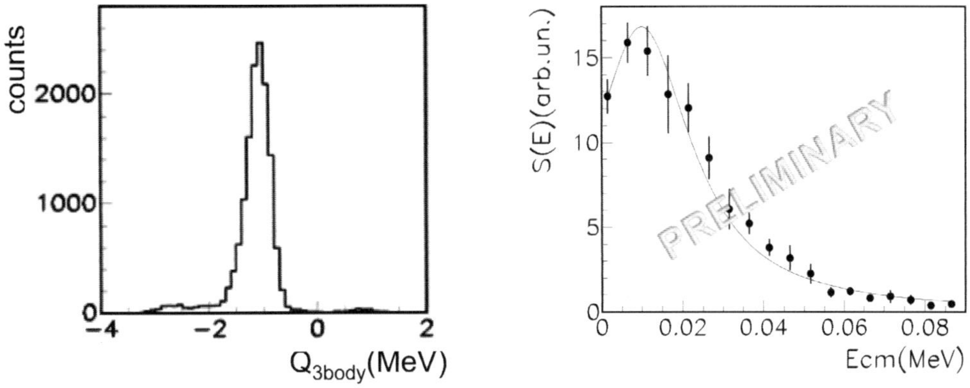

Figure 2. Left panel: the experimental Q-value for the 3body ^{2}H(^{10}B,α^7Be)n reaction. Right panel: preliminary TH ^{10}B(p,α)^{7}Be S(E)-factor in arbitrary units.

of 425±60 eV has been recently deduced in [15], in contradiction with the theoretical one of 175 eV. Thus, the deviation from the adiabatic limit confirms once again the systematic discrepancy between experimental and theoretical values for the electron screening potential, thus leaving the electron screening understanding as an important open question for nuclear physics and astrophysics, as well as atomic physics or condensed matter physics [26].

References

[1] Baur G *et al* 1986 *Phys. Lett.* B **178** 135
[2] Spitaleri C *et al* 1999 *Phys. Rev.* C **60** 055802
[3] Spitaleri C *et al* 2004 *Phys. Rev.* C **69** 055806
[4] Spitaleri C *et al* 2011 *Phys. of Atomic Nuclei* **74** 12 1763-1777
[5] Tumino A *et al* 2007 *Phys. Rev. Lett.* **98** 252502
[6] Tumino A *et al* 2008 *Phys. Rev.* C **78** 064001
[7] Pizzone R G *et al* 2005 *Phys. Rev.* C **71** 058801
[8] Tumino A *et al* 2006 *Eur. Phys. J.* A **27** s01 243-248
[9] Romano S *et al* 2006 *Eur. Phys. J.* A **27** s01 221-225
[10] Lamia L *et al* 2007 *Nucl. Phys.* A **787** 309c
[11] Lamia L *et al* 2008 *Nuovo Cimento della Società Italiana di Fisica-C* **31** 423
[12] Tumino A *et al* 2011 *Phys. Lett.* B **700** 111-115
[13] Tumino A *et al* 2011 *Phys. Lett.* B **705** 546
[14] Pizzone R G *et al* 2011 *Phys. Rev. C* **83** 04580
[15] Lamia L *et al* 2012 *Astron. & Astrophys.* **541** A158
[16] La Cognata M *et al* 2006 *Eur. Phys. J.* A **27** s01 249-254
[17] Sergi M L *et al* 2010 *Phys. Rev.* C **82** 032801(R)
[18] La Cognata M *et al* 2010 *Astr. Phys. Journ.* **708** 796
[19] La Cognata M *et al* 2011 *Astr. Phys. Lett.* **739** L54
[20] Tumino A *et al* 2005 *Eur. Phys. J.* A **25** s01 649-650
[21] Gulino M *et al* 2010 *J. Phys.* G **37** 125105
[22] La Cognata M *et al* 2012 *Phys. Rev. Lett.* **109** 232701
[23] Lamia L *et al* 2012 *J. Phys.* G **39** 015106
[24] Lamia L *et al* 2012 *Phys. Rev.* C **85** 025805
[25] Angulo C *et al* 1999 *Nucl. Phys.* A **656** 3
[26] Rolfs C & Rodney W *Cauldrons in the Cosmos* 1988 (Chicago: The University of Chicago press)

Nuclear (multi-)fragmentation

Adriana R Raduta[1]
[1]National Institute for Nuclear Physics and Engineering (IFIN-HH), RO-077125 Bucharest, Romania

E-mail: araduta@nipne.ro

Abstract. Nuclear (multi-)fragmentation, defined as the nuclear decay mechanism in which at least three intermediate mass fragments ($Z \geq 3$) are produced, is the disassembly phenomenon specific to hot nuclear matter produced in nuclear collisions at beam energies of 20-100 MeV/nucleon. Considered as the manifestation of the liquid-gas phase transition predicted by nuclear matter mean-field models or an instrument to address the nuclear equation of state, it enjoys for thirty years a vivid scientific interest and motivates the construction of more and more high-performance 4π-detectors allowing for almost complete reaction characterisation on an event-by-event basis. Over the years, many experimental signatures of first- and second-order phase transitions have been proposed and identified in both central and peripheral collisions. Despite their seemingly puzzling messages, coherent understanding of multifragmentation is possible accounting for statistical ensemble inequivalence and finite size effects. Two particular situations will be discussed in detail. The first one corresponds to a multifragmenting medium size nucleus where the expected first-order phase transition is blurred by surface effects. The second one corresponds to an infinitely large system, the clusterized nuclear matter thought to constitute the main sector of baryonic matter in (proto-)neutron stars, where the first-order phase transition is quenched by Coulomb effects. In both cases statistical models with cluster degrees of freedom have been employed.

1. Introduction

Due to the short-range repulsive and finite-range attractive features of the nucleon-nucleon interaction, the idea that nuclear matter exhibits a phase diagram similar to the liquid-gas one was advanced already thirty years ago [1, 2]. Detailed studies performed within mean-field approaches have confirmed the presence of first and second-order phase transitions and determined the evolution of the phase coexistence domain with respect to temperature and isospin asymmetry [3, 4, 5, 6]. Moreover, the existence of a density-driven instability domain where density fluctuations are spontaneously and exponentially amplified in order to achieve phase separation lead to the idea that over the corresponding sub-saturation density range the homogeneous matter scenario cease to stand valid and matter organizes itself in clusters.

Clusterized structures of dilute nuclear matter potentially stemming from the spinodal instabilities were experimentally evidenced in nuclear multifragmentation, that is the decay mechanism taking place at excitation energies of the order of 2-10 MeV/nucleon [7], and an impressive experimental and theoretical effort was done with the aim to answer the challenging question on whether this decay channel really corresponds to a phase transition or not. The difficulties of this task were both conceptual and technical. From the conceptual point of view, the finite size of a nucleus makes it lie outside the domain of validity of the phase transition theory which is the thermodynamical limit (*i.e.* infinite size). This means that in the most optimistic situation in which a nucleus would behave like the infinite, electrically uncharged, homogeneous system that nuclear matter is, none of the known first or second order phase transition signals could survive undistorted in multifragmentation. The situation is even more dramatic if one takes into account that, for a charged system, the thermodynamical limit is, by construction, not achievable because of the diverging Coulomb energy. Moreover, Coulomb and surface interaction energies which scale with Z^2 and, respectively, A^σ, violate extensivity property and might be responsible for statistical ensemble in-equivalence, that is the dependence of the observed physics on the constraints externally imposed to the system. From the experimental perspective, the situation is equally difficult given that the experimentally accessible information concerning fragment partitions and kinetic properties does not correspond to the thermodynamically relevant break-up stage, but to the so called asymptotic one obtained after the break-up: hot fragments suffered a sequence of particle evaporation processes and propagation in the mutual Coulomb field. Not less important, most intensive thermodynamical observables (temperature, pressure, density) which are essential in plotting the phase diagram or characterize the path followed by the system in the phase space are subject to large uncertainties or impossible to constrain.

All these issues have been thoroughly addressed over the last two decades and led, on one hand, to the re-formulation of phase transition theory in the case of finite systems [8] and, on the other hand, to the construction of performant 4π-detectors allowing for almost complete reaction characterization on an event-by-event basis [7, 9].

What is interesting is that the efforts paid in order to understand the thermodynamics associated to multifragmentation pay their dividends in helping us to interpret the thermal and phase properties of dilute unhomogeneous baryonic matter produced in the explosion of core-collapsing supernova and neutron star crust. At variance with a fragmenting nucleus, star matter is - because of the electrons - neutral and, practically, at the thermodynamical limit. Though, phase separation is quenched as the huge electron gas incompressibility prevents the electrons to follow on a macroscopic scale the baryonic matter density fluctuations such as to compensate the proton charge. As such, dishomogeneities are of finite size and the whole can be viewed as an admixture of nuclear clusters and nucleons [10]. Mastery of both the thermodynamical response of baryonic matter, commonly termed as equation of state (EOS), and its chemical composition is of major astrophysical interest as it influences the dynamics of core-collapsing supernovae and cooling phase of proto-neutron stars.

In the following we shall spot the thermodynamics of multifragmenting nuclei and dilute star matter within a statistical model with cluster degrees of freedom, commonly known under the generic name of nuclear statistical equilibrium (NSE) models. This kind of approaches rely on the hypotheses that nuclear correlations are entirely exhausted by clusterization and offer the remarkable advantage of treating all nuclear bound and continuum states via empirical parameterizations. Their major inconvenience is the neglection of in-medium effects. In order to approach the crust-core transition and account for nucleon-nucleon interactions in the vicinity of normal nuclear density and beyond that, in the case of star matter the loosely interacting gas of clusters is considered to coexist with an uniform gas of interacting nucleons treated within the mean-field approach.

We shall first show that an infinite neutral clusterized system manifests first and second order phase transitions. Then, we shall demonstrate that, because of surface, for a multifragmenting nucleus in which Coulomb interaction is switched off the first order phase transition is quenched but it can be traced back from the back-bending behavior of any $Y_i(X_i)|_{X_{j,j\neq i}}$ curve or, equivalently, bi-modal pattern of probability distributions of the order parameters of the phase transition. Y_i and X_i stand for an intensive and, respectively, the conjugated extensive variable. Further on we shall show that provided that Coulomb effect is not too strong, the fossil signal

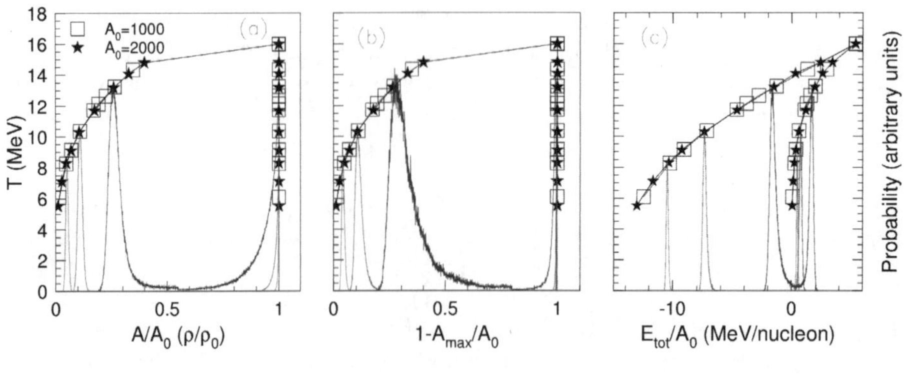

Figure 1. Probability distributions of the total number of particles (left), largest fragment in each event (middle) and total energy (right) for a one-component cluster model with standard Fisher fragment definition in the grand-canonical ensemble characterized by V=14476.4 fm^3 (A_0=2000) and $(T, (\mu/T))$ = (8.3, -2.40), (10.35, -2.20) and (13.19, -2.05), where T is expressed in MeV. The solid star symbols mark the phase diagram. Open squares mark the phase diagram of a system half the size, (V=7238.2 fm^3 and A_0=1000). Figure taken from Ref. [12].

of the phase transition can be found also in real (charged) nuclei. Finally, we shall show that the non-saturating Coulomb interaction is responsible for statistical ensemble in-equivalence in non-homogeneous dilute star matter and that correct thermodynamical characterization can be achieved only in the ensemble in which the order parameter of the phase transition, *i.e.* total baryonic density, is controlled, *i.e.* the canonical ensemble.

2. Infinite neutral clusterized matter

Let us first consider a one-component clusterized gas of non-interacting clusters whose energy density functional $B(A) = a_v A - a_s A^{2/3}$ consists, similarly to the one of nuclei, of a bulk and a surface term and deduce, under the equilibrium hypotheses, the associated thermodynamics.

Probability distributions of the total number of particles and energy, the expected order parameters of a liquid-gas (LG) type phase transition, of a system composed of an arbitrary number of A_0=2000 particles occupying an arbitrary volume V=14476.4 fm^3 at T = $8.3, 10.35, 13.19$ MeV as obtained out of a grand-canonical Monte-Carlo Metropolis simulation are plotted in Fig. 1 (left and right panels). As one may note, both variables present a bi-modal pattern meaning that the system manifests a first-order phase transition. It comes out that the dense phase is characterized, whatever the temperature $T < T_C$, by a density equal to the normal nuclear matter density. At contrast, the dilute phase frontier evolves monotonically with the temperature such that the phase coexistence region shrinks (open symbols). At a certain value of the temperature, $T_C \approx 16$ MeV, it gets reduced to a unique point, the second-order phase transition. As such one may conclude that, as expected given the van der Waals-type of the interaction, clusterized nuclear matter behaves like macroscopic fluids. A difference is nevertheless worthwhile to be noticed: the limiting point of the phase diagram is characterized by $\rho = \rho_0$ which suggests that it does not correspond to the critical point of the Fisher model [11]. This fact is confirmed by the study of critical exponents, as demonstrated in Ref. [12]. What is more remarkable and encouraging for experimentalists aiming to pin down the phase diagram out of multifragmentation data is that the largest fragment bi-modal distribution correctly signals phase coexistence (middle panel in Fig. 1): the different temperatures for which the two peaks

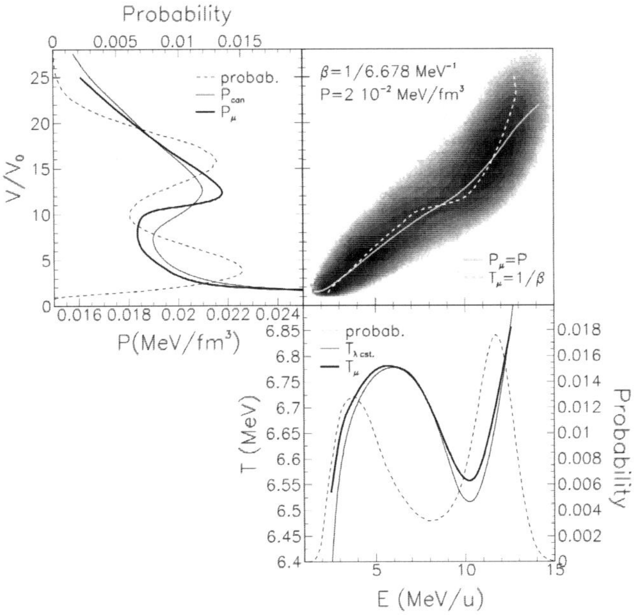

Figure 2. Right-upper panel: probability distribution (in log scale) of a canonical ensemble at constant pressure corresponding to the un-charged nucleus (200, 82). The thick lines correspond to microcanonical isobaric and isothermal paths corresponding to $P_\mu = P$ and $T_\mu = 1/\beta$. The lateral panels correspond to projections (dashed lines) of the probability distributions on V/V_0 (left) and excitation energy E (down) axes. Canonical pressure (left panel) and constant-λ temperature are figured with thin lines. The corresponding microcanonical curves are plotted with thick lines. Figure taken from Ref. [14].

corresponding to the two phases have the same height differ by less than 0.1 MeV from the ones corresponding to equal height peaks in $P(E)$ and $P(\rho)$. This is obviously due to the correlation between A_{max} and the system's total energy.

3. Un-charged multifragmenting nucleus

To what extend the large system phase transition phenomenology stands valid while decreasing the system size is a non-trivial question taking into account that surface effects and conservation laws play a non-negligible role. Surface effects, which do not scale with the system size, are expected to violate extensivity. Finite size and conservation laws are expected to induce statistical ensemble in-equivalence meaning that, in order to pin down the phase coexistence, one has to choose a signal compatible with the way in which the system is constrained. To account as realistically as possible for the particularities of an isolated system exploding into vacuum, it is clear that the most appropriate statistical ensemble is the microcanonical one. The details on how microcanonical multifragmentation is modeled and numerically solved may be found in Ref. [13]. But, constraining the energy, volume and particle number, that is the order parameters, does not allow the system to explore the two phases provided that the external constraints are such that it lies in the phase coexistence region. To overcome this inconvenient, one has to work in a constant-pressure canonical ensemble. One possibility to do that is to start from microcanonical configurations and alter accordingly their statistical weights,

$$\mathcal{P}(E,V) = \frac{W(E,V)\exp(-\beta E - \beta PV)}{Z_{can}(\beta,P)}, \tag{1}$$

where $W(E,V)$ stands for the microcanonical weight of a micro-state defined by the excitation energy E and freeze-out volume V, P is the externally imposed pressure, $\beta = 1/T$ is the inverse temperature and $Z_{can}(\beta,P)$ is the canonical partition function.

The right-upper panel of Fig. 2 depicts (in log scale) the probability distribution $\mathcal{P}(E,V)$ corresponding to the un-charged nucleus (200, 82) at $T = 6.678$ MeV and $P = 2 \cdot 10^{-2}$ MeV/fm^3.

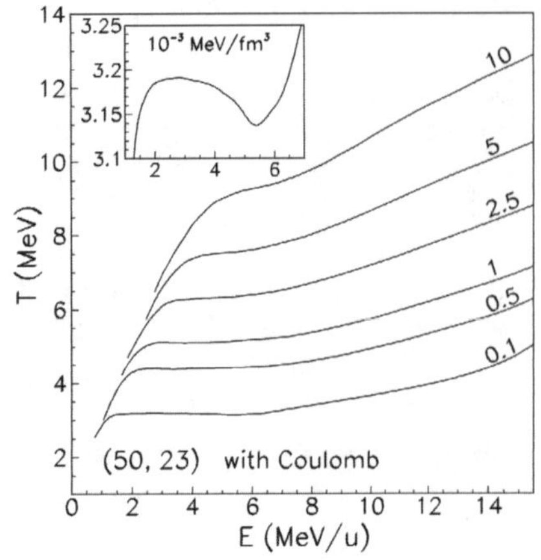

Figure 3. Constant-pressure caloric curves for the (50, 23)-nucleus. The numbers accompanying the curves mark the pressure expressed in 10^{-2} MeV/fm^3. Figure taken from Ref. [15].

The obvious bi-modal pattern as a function of both excitation energy and volume stems from the convexity anomaly of the associated thermodynamical potential $\bar{S}[T, P]$, where $\bar{S}[Y_i]$ represents the Legendre transformation of the entropy with respect to the fixed intensive Y_i, and signals phase coexistence. In addition to how this two-dimensional probability distribution looks like while projected on the two axes E and V (dotted lines), the lateral panels show a constant-pressure caloric curve $T(E)|_P$ (lower panel) and, respectively, a constant-temperature pressure versus volume curve $P(V)|_T$ (left panel). For the sake of completeness, also constant $\lambda = \beta P$ caloric curves and pressure versus volume curves are represented. All these curves back-bend in the region situated between the peaks of the probability distribution which corresponds, in the mean-field approach, to the spinodal instability domain.

In conclusion, 1) a neutral finite system keeps the memory of the LG phase transition taking place in infinite matter, 2) first-order phase transition is quenched by surface effects and 3) back-bending behavior of $T(E)|_P$ and $P(V)|_T$ is a signature of phase coexistence.

4. Multifragmenting nucleus

Despite the achieved progress in understanding in what regards a finite system differs from its infinite counterpart, our main question is still not answered because, as already stressed, real nuclei are also electrically charged. At first glance this should imply, in the most optimistic scenario, a decrease of the limiting temperature and, in a more pessimistic one, the complete suppression of phase coexistence.

Using the previously discussed back-bending behavior of isobar caloric curves, Fig. 3 shows that at least for relatively small nuclei, where Coulomb field is not too strong, the phase transition survives.

5. Dilute star matter

Clusterized nuclear matter is not only created in small amounts in laboratories employing heavy ion collisions at bombarding energies of the order of several tens of MeV/nucleon but is abundantly populated during the post-bounce phase of a core-collapsing supernovae, as well as over the much longer Kelvin-Helmholtz epoch during which the remnant changes from a hot lepton-rich proto-neutron star to a cold deleptonized neutron star. The duration of the first process is of about several hundreds of ms while the one of the latter is of several tens of seconds. In both cases temperatures $10^9 < T < 2 \cdot 10^{11}$ K, nuclear sub-saturation densities

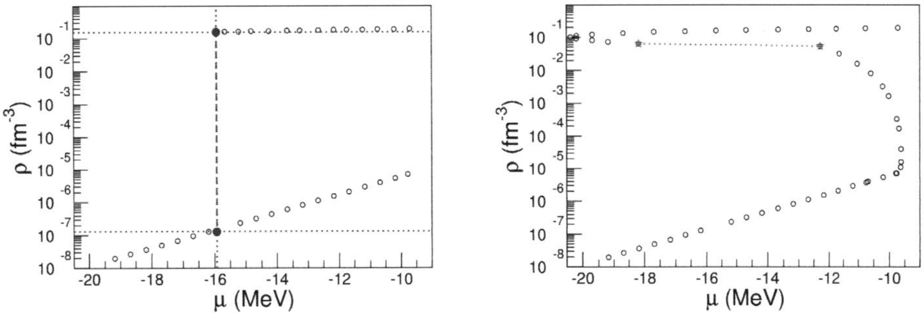

Figure 4. Symbols: Grand-canonical (left) and canonical (right) $\mu_{tot}(\rho_{tot})$ at $T = 1.6$ MeV and $\mu_I = 1.68$ MeV. Vertical dashed line illustrates the Gibbs construction; horizontal dot-dashed lines mark the frontiers of the phase coexistence domain. Figure taken from Ref. [17].

$10^{10} < \rho < 10^{14}$ g/cm^3 and electron fractions $0 < Y_e < 1$ are spanned.

Adding to the previously discussed loosely interacting gas of clusters an extra component made out of an uniform distribution of interacting nucleons such as to correctly describe the crust-core solid-liquid transition phenomenology, we investigate whether phase coexistence really occurs in star matter [16, 17]. The fragment component is described by a liquid drop energy density functional whose parameters are fitted to experimental data but no restriction is imposed with regard to the cluster size or isospin-asymmetry and the Coulomb term is altered such as to account for the screening effect of the electrons. The latter is estimated within the Wigner-Seitz approximation. For the uniform gas of nucleons we use a mean-field approach and Skyrme-like interaction potentials. The assumed lack of interaction between the two components otherwise than via the excluded volume allows the total partition function be factorized in a free nucleon and, respectively, a cluster gas component such that the total entropy writes

$$\sigma^{can}_{\beta,\mu_I}(\rho) = \ln \mathcal{Z}^{a=1}_{\beta,\mu_I}(\rho_f) + \lim_{V\to\infty} \frac{1}{V} \ln \mathcal{Z}^{a>1}_{\beta,\mu_I}(V\rho_{cl}). \quad (2)$$

Taking into account that 1) in a two-component system a standard procedure of reducing the mathematically difficult Gibbs construction to a much easier Maxwell construction is to control one of the two chemical potentials and the other particle density and 2) at constant temperature the phase separation direction in NM is largely dominated by the total density, we choose to fix the iso-vector chemical potential $\mu_I = (\mu_n - \mu_p)/2$.

The evolution of total baryonic density with respect to the conjugated total chemical potential $\mu = (\mu_n + \mu_p)/2$ as obtained in a grandcanonical ensemble is represented with open symbols in the left panel of Fig. 4 for T=1.6 MeV and μ_I =1.68 MeV. As one may note, there is an important density range which is jumped over and, for a certain domain of μ, two kinds of mixtures exist. One of them entirely consists of clusters. The second one consists of uniform NM at normal or supra-saturation nuclear density. The impossibility to access the phase coexistence density range while controlling the intensive conjugated to the order parameter is a notorious issue in the case of first-order transitions. Which of the multiple mixture solutions obtained for $-16 < \mu < -9.5$ MeV or a linear combination of them corresponds to equilibrium can be learnt by seeking in which case the associated energetic thermodynamical potential gets minimized. The result is plotted with a vertical dashed line on the figure and corresponds to the Maxwell construction performed for $\mu \approx 16$ MeV. This means that the dilute clusterized phase obtained for larger values is meta-stable.

Taking into account that 1) charged matter can occur on macroscopic scales if and only if net charge neutrality is strictly obeyed, 2) because the symmetry energy, a density fluctuation leads to a proton, *i.e.* charge, fluctuation, 3) electron gas incompressibility might not allow them to follow the baryonic matter fluctuations such as to compensate the proton charge, one may wonder whether phase separation may really exist.

The answer can be given by following how the system behaves while constraining the density, that is in a canonical ensemble. If the thermodynamics was that of a first-order phase transition, a plateau would be obtained. If, by contrary, the system preferred to organize itself in an admixture of free nucleons and nuclear clusters, the transition between the extremities of the Maxwell construction would be continuous. The right panel of Fig. 4 confirms the last scenario meaning that the dis-homogeneities happen at microscopic scale. From the statistical physics point of view, the situation is a paradigm of ensemble in-equivalence and represents, to our knowledge, one of the few cases in which it could have significant phenomenological consequences. The narrow density interval presenting the most sudden evolution of $\mu(\rho)$ and whose extremities are marked by stars corresponds to the so-called pasta phase region disregarded by the spherical symmetry assumed in our simple model.

6. Conclusions

Clusterized structures in fragmented excited nuclei and dilute star matter are discussed from the perspective of statistical models with cluster degrees of freedom. It was shown that multifragmentation does really correspond to a phase transition but phase separation is quenched by finite size effects. In the case of dilute stellar matter, phase separation is also quenched and the first-order phase transition experienced by nuclear matter is replaced by a continuous one.

Acknowledgments

Participation to this conference was in part supported by IUPAP via a Travel Grant. Enlightening discussions held over many years with Bernard Borderie, Francesca Gulminelli, Alexandru Raduta and Marie-France Rivet are gratefully acknowledged.

References

[1] Finn J E *et al* 1982 *Phys. Rev. Lett.* **49** 1321
[2] Bertsch G and Siemens P J 1983 *Phys. Lett.* **126** 9
[3] Glendenning N K, Csernai L P and Kapusta J I 1986 *Phys. Rev.* C **33** 1299
[4] Muller H and Serot B D 1995 *Phys. Rev.* C **52** 2072
[5] Lee S J and Mekjian A Z 2001 *Phys. Rev.* C **63** 044605
[6] Ducoin C, Chomaz Ph and Gulminelli F 2006 *Nucl. Phys.* A **771** 68
[7] Borderie B and Rivet M-F 2008 *Progr. in Part. and Nucl. Phys.* **61** 551
[8] Gulminelli F 2004 *Ann. Phys. Fr.* **296** 1
[9] *Dynamics and Thermodynamics with Nuclear Degrees of Freedom* Gulminelli F, Chomaz P *et al.* (Eds) 2006 *Eur. Phys. J.* A **30**
[10] Lattimer J M, Pethick C J, Ravenhall D G and Lamb D Q 1985 *Nucl. Phys.* A **432** 646
[11] Fisher M E 1967 *Physics (NY)* **3** 255
[12] Raduta Ad R and Gulminelli F 2009 *Phys. Rev.* C **80** 024606
[13] Raduta Al H and Raduta Ad R 1997 *Phys. Rev.* C **55** 1344
[14] Raduta Al H and Raduta Ad R 2002 *Nucl. Phys.* A **703** 876
[15] Raduta Al H and Raduta Ad R 2001 *Phys. Rev. Lett.* **87** 202701
[16] Raduta Ad R and Gulminelli F 2010 *Phys. Rev.* C **82** 065801
[17] Gulminelli F and Raduta Ad R 2012 *Phys. Rev.* C **85** 025803

Journal of Physics: Conference Series **436** (2013) 012077 doi:10.1088/1742-6596/436/1/012077

Collective and thermal signatures in π^4He interaction in the Δ excitation energy region

I Gnesi [1,2,3]

[1]Dipartimento di Fisica Generale "A. Avogadro", Università di Torino, via Pietro Giuria 1, 10125 Torino, Italy (present address)
[2]INFN, Sezione di Torino
[3]Centro Studi e Ricerche "Enrico Fermi", Roma, Italy

E-mail: gnesi@to.infn.it

Abstract. Measurement of the in-medium modifications of the Δ resonance has been performed at a PAINUC experiment. Signatures of the excitation of nuclear collective states and the first experimental evidence for thermal emission of photons have been obtained from the analysis of the $\pi^{\pm}$ ^{4}He data taken at T_π =106 MeV.

1. Introduction

The study of low and intermediate energy pion-induced reactions is of great interest. The PAINUC experiment has collected new data on elastic and inelastic $\pi^{\pm}$ ^{4}He interaction at T_π =106 MeV where the maximum of the Δ resonance production on ^{4}He has been observed. The events have been collected using a triggerable self shunted streamer chamber filled with helium at atmospheric pressure, installed in a magnetic field and exposed to the pion beam of the Joint Institute for Nuclear Research (JINR) in Dubna (Russia). The apparatus makes possible the observation of all the charged secondaries, down to 1 MeV protons and other helium fragments. The particle identification efficiency has been improved by the use of the track brightness method [1]. At low energies, where other detectors start to show their limitations, by use of this technique the PAINUC collaboration was able to collect information in different fields starting from the physics of nuclei and pion-nucleus interactions, with interesting observations of nuclear collective states, to the physics of phase transitions, with evidences for thermal emission of photons from the so called *nuclear matter*.

2. Experimental findings in $\pi^{\pm 4}$He interactions at T_π =106 MeV

2.1. The thermal emission of photons

The analysis of the $\pi^{\pm 4}$He scattering events at T_π ~106 MeV revealed the existence of a channel with emission of a high energy photon in the final state (see [2, 3]). Several hypotheses have been tested in order to identify the γ emission mechanism, namely the initial or final state radiation (external bremsstrahlung), the internal radiation (internal bremsstrahlung), the Δ^{++} magnetic dipole radiation, the Δ radiative decay and, finally, the thermal emission. The energy spectra of the external radiation has shown a bad agreement with the experimental data, both in the π^+ and π^- cases. The internal radiation spectrum shows a good agreement in the high energy γ region, but is totally in disagreement with the experimental spectra at low γ-ray energies. However, the main evidence against the bremsstrahlung radiation, as the mechanism responsible for the observed γ emission, is the high branching ratio of the channel, $\sim$0.1 relative to the elastic scattering, which is a factor of 10 to 20 higher than expected.

The hypothesis of the emission of a γ via an intermediate Δ^{++} magnetic dipole radiation has also been considered. This mechanism is expected to occur only in π^+-proton and π^+-nucleus interactions, producing a clear bump at 80-100 MeV in the energy distribution of the emitted γ-rays. With the current statistics the presence of the peak can neither be excluded nor confirmed. It has also to be stressed that the presence of the Δ magnetic dipole radiation has not been confirmed even on a proton target and at higher pion energies. ([4]). The free Δ radiative decay ($\Delta \rightarrow$Nγ) has a too low branching ratio and it cannot be observed with the present statistics. In addition, one has also to bear in mind that modifications are expected both in Δ^{++} magnetic dipole radiation and in radiative Δ decay taking place in a nucleus, compared to the free nucleon case, which would make their identification more difficult.

Finally, the γ energy distributions have been found in good agreement with a Planck black body radiation distribution (see fig. 1), assuming about 16 MeV temperature for both π^+ ^{4}He and π^- ^{4}He interactions. The hypothesis of a thermal radiation is also in agreement with other physical features of the observed radiative events: a) the high energy tail of the gamma energy distributions, which exceeds 100 MeV, is compatible with the low energy available per nucleon (20 MeV) and with the extracted nuclear temperature ($\sim$16 MeV); b) the thermal emission is consistent with the isotropic differential cross section in the pion scattering angle and c) with the high branching ratio observed, unlike the bremsstrahlung radiation mechanism.

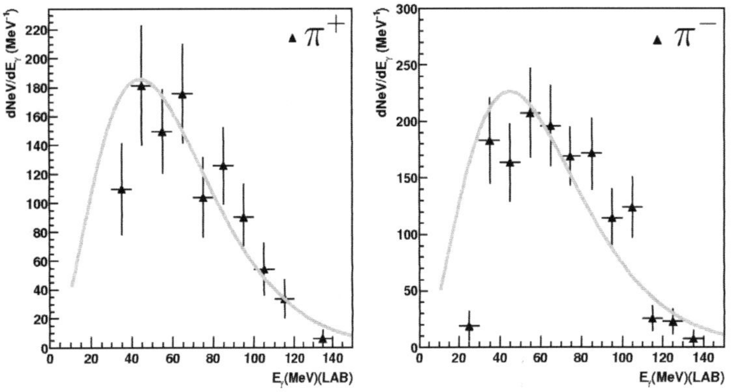

Figure 1. Photon energy distributions from $\pi^\pm$-^{4}He$\rightarrow$ $\pi^\pm$-^{4}Heγ reactions. The green curves are fits with Planck black-body radiation distributions. The extracted temperature of the corresponding black-bodies is 16 MeV.

2.2. Signatures of Collective Resonances

The hypothesis of the existence of a giant (I,S)=(3/2,3/2) nuclear resonance was presented by Dillig and Huber in 1974 [5] while the first experimental observations of modifications in the various resonant elastic π-nucleus cross sections were done by F. Balestra et al. during the 80's [6].

From our analysis of the π^-^{4}He$\rightarrow$ π^-^{3}He n reaction at T$_\pi$ =106 MeV, the first experimental evidence on the excitation of the Δ^- resonance has been obtained in an inelastic channel and below the pion production threshold. The mass of the resonance turns out to be M$_\Delta$ =(1157$\pm$14) MeV/c^2 and the measured width is Γ =(38$\pm$2) MeV/c^2, thus smaller and narrower with respect to the corresponding values of the Δ excitation on a free-nucleon (see fig. 2).

The observed parameter modifications are similar to those observed in ref. [6] on the Δ excitation in elastic scattering of pions on several nuclei. The peaks of the Δ excitation functions show a shift towards lower pion energies, while their widths undergo a narrowing: the phenomenon is more pronounced in the case of pion scattered at high angle (>120°). The modifications of the observed Δ^- mass and width could be caused by the activation of an isobaric collective resonance, involving more than one nucleon in the excitation of the resonant state, as it was proposed for the explanation of the effects observed in Ref. [6]. Both in elastic ([6]) and inelastic (this work) π-nucleus interactions, the modified Δ resonance seems to be produced at high momentum transfer and low Q^2, in support of the hypothesis that several nucleons are involved in the energy transfer (low Q^2 means large probing wavelength). The observed excitation energy shift and the width shrink seem to be mainly related to the increase of the binding energy coming from the nucleons participating in the resonance (see [3]). In order

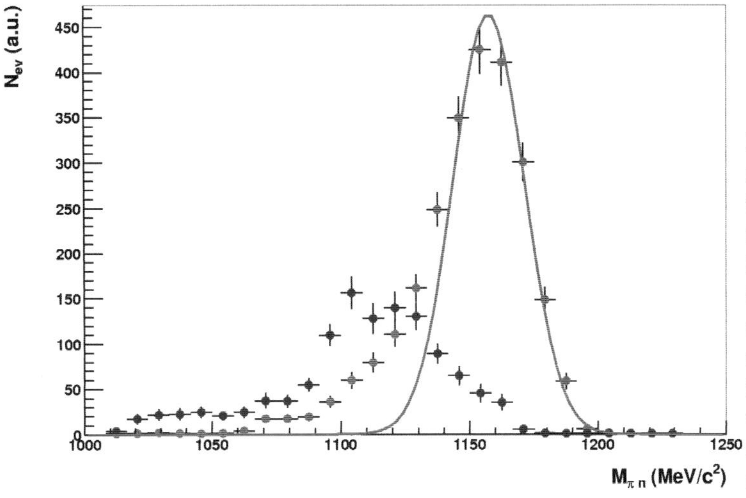

Figure 2. π^-n invariant mass distribution in neutron knock-out reaction for high (red points) and low (blue points) transferred momentum. The resonant (red) distribution reveals a mass lowering and width narrowing with respect to the values for a Δ excited on a free nucleon.

to explain both the observations of the present work and those of [6], a semi-empirical model has been developed, which assumes that a Δ resonance excited in a nucleus interacts with the remaining nucleons, giving rise to a collective state; the parton density function (p.d.f) of the collective state has been formulated as a composition of a spherical symmetric and uniform central part and of an exponentially decreasing peripheral part. The semi-empirical model has been used to fit data on several nuclei in Ref. [6]; the fit is based on the simple idea that each additional nucleon involved in the collective resonance gives a contribution in terms of binding energy, thus, narrowing the width and increasing the life time of the resonance.

The results of the fit describe the collective state as a cluster of nucleons and a Δ, where the total number of nucleons goes from 1 for H to 1.7 and 2.7 for deuteron and ^{4}He, respectively. The number saturates at 3.5 for nuclei with A>12-16 (carbon and oxygen). The size of the uniform central part turns out to be 1.15 nucleon radius and the exponentially decreasing peripheral part of the p.d.f. turns out to fall in 0.3-0.4 fm. The contribution to the binding energy per additional nucleon is E_B ~50 MeV (see fig. 3). Therefore, the described collective resonance seems to be a strongly bound state, with binding energies per nucleon 7 times the standard binding energy of a nucleon in a nucleus. The fast drop of the p.d.f. suggests that the state is well confined within the uniform central part, thus loosely interacting with the remaining nucleons in the nucleus (see [3, 8]). With the present model the Fermi momentum contribution to the lowering of the excitation energy cannot be evaluated, however its presence cannot be excluded. At the same time multiple scattering effects, producing slight broadening of the resonance, can also be present. As it has been demonstrated by several works ([7, 9]) neither Fermi momentum of nucleons nor multiple scattering effects are able to explain the width narrowing, which can only

Journal of Physics: Conference Series **436** (2013) 012077 doi:10.1088/1742-6596/436/1/012077

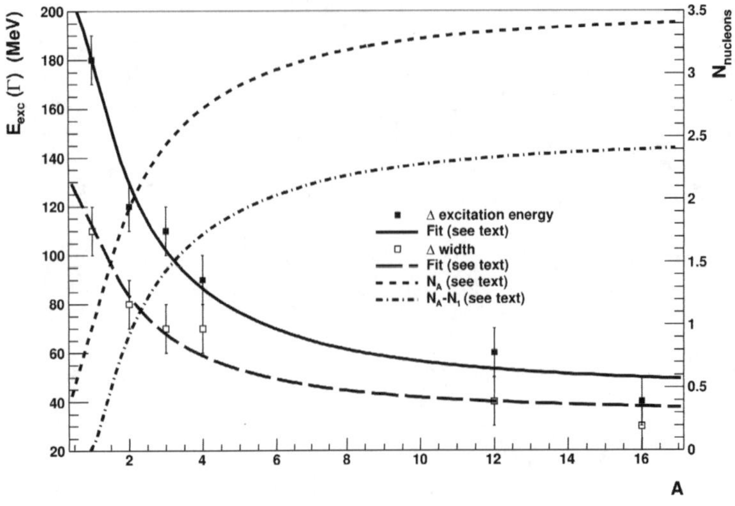

Figure 3. Full and empty squares: peaks and widths of Δ excitation functions for several nuclei, respectively; solid and dashed curves are fits with the semi-empirical model (see text). Fine-dashed and dotted-dashed curves: number of all and of additional nucleons (out of the one destroyed by Δ excitation) participating in the formation of the collective state.

be explained by the excitation of a collective resonant state. Finally, the contribution coming from the Δ excitation on peripheral single nucleons to the total Δ width and amplitude has also to be taken into account. However, in the present data at T_π =106 MeV, this contribution has to be small, since the excitation on a peripheral, thus quasi-free, nucleon occurs at higher pion energies (180 MeV).

It clearly turns out that the excitation of resonances in nuclear matter at intermediate energies is far from being well understood and the present experimental observations clearly show that there are strong motivations to deepen the study of the cold *nuclear matter*, both on experimental and theoretical sides.

2.3. 3-body correlation in pion absorption reactions

The pion absorption channel in the region of excitation of the Δ resonance in nuclei is of interest because it can give information on the multinucleon-pion absorption mechanisms and on the role of the in-medium excitation of the resonance. The analysis of the π^{+4}He→3pn absorption reactions at T_π ~106 MeV allowed the obsrvation of 2-3 nucleon absorption signatures, as well as signals of the formation of a collective state. It has to be mentioned that the complete phase space has been measured down to 1 MeV protons.

The study of the two-nucleon correlations has been performed by using a set of two-nucleon kinematical variables, namely: the opening angle, the relative momentum and the invariant mass of the two-nucleon systems. The bidimensional distributions of these variables have been compared with those obtained from the Monte Carlo simulation of the phase space. The analysis of the two-nucleon kinematical correlations, for all the six possible nucleon pairs in the final state, reveals that none of the observed absorption events, in the region where strong correlations have been observed, can unambiguously be identified as a pure two-nucleon absorption (2NA); feeble signatures of Hard and Soft Final State Interactions ((H)SFSI) or Initial State Interactions (ISI), as well as signatures of three-nucleon absorption are present.

The study of three-nucleon correlation has been performed by using a set of variables in order to identify the three-nucleon system under study, namely: the bidimensional distributions of the two opening angles θ_{N1N2}-θ_{N2N3}, the distribution of three-nucleon plane angle ξ (see [3]) and the kinetic energy of the three-nucleon system. The analysis reveals that a fraction of ~14% of the absorption events occurs on the three final state protons, in agreement with an absorption on a pd cluster in the initial state: the absence of strong (H)SFSI suggests the absorption occurs

Journal of Physics: Conference Series **436** (2013) 012077 doi:10.1088/1742-6596/436/1/012077

on a pd cluster rather than on the deuteron with a proton as a mere spectator. The result is in agreement with that obtained at PSI [10], even if, in that case, protons with energy below 20 MeV were not measured and the observations showed agreement with a true three-nucleon absorption mechanism. According to our analysis, another ∼42% cannot be unambiguously identified as a 2NA + (H)SFSI/ISI or 3NA process; the result is again in agreement with the findings of Matheos [11], which gives ∼37% for the same uncertainty. In addition, from the analysis of the three-nucleon kinetic energy distributions, the excitation of a dibarionic d' in the pion absorption, via the channel $(\pi^+nn)p \rightarrow d'N \rightarrow 3N+190$ MeV, cannot be excluded.

The behavior of the differential cross section in the three-nucleon plane angle ξ, for all four possible three-nucleon systems in the final state, has been fitted with Legendre polynomials, revealing the presence of a P wave contribution to all the final state three-nucleon systems. This is again in agreement with the observations obtained at PSI [10]. Further statistics is needed in order to perform a global fit with a weighted sum of models (3NA, 2NA, 2NA+(H)SFSI, 3NA(d')), that would allow to simultaneously extract the different contributions of different absorption processes to the whole population of pion absorption events. However it is clear from several pion absorption studies ([3, 10, 11]) that new theoretical models, taking into account a real 3-4NA process, have to be developed in order to reliably explain the absorption interactions in the Δ resonance energy region.

In order to correctly understand the meaning of these experimental findings, some comments are needed. First, the analyzed pion absorption events are measured at T_π =106 MeV, where the maximum cross section of pion-induced excitation of the Δ resonance on ^{4}He is expected. Signatures of the excitation of a collective resonance, involving up to 3-4 nucleons, on ^{4}He have been obtained. Second, the well known change of the slope of the pion absorption cross section as a function of the mass number of the target nucleus, from a fast $\sigma_{abs} \propto A^3$ if $A< 4$ to a slow $\sigma_{abs} \propto A^{2/3}$ if $A> 4$ change (see [11] and references therein), is again in support of a collective excitation; the change occurs precisely on the nucleus (the ^{4}He) with the number of nucleons equal to the maximum involved in the collective resonance activation, according to the semi-empirical model. In this view, the observation of signatures of 3NA processes strongly suggests that the collective resonance takes a fundamental role also in the pion absorption channel in the Δ energy region.

The physical parameters of the collective resonance extracted (number of nucleons involved, binding energy per additional nucleon, ΔN potential size and fall, contribution to the mean life time), can be useful physical quantities to improve our knowledge on the collective resonant states and to develop static and dynamic models to be compared with future measurements.

References

[1] Andreev E M *et al* 2002 *Nucl. Instr. and Meth.* A **489** 99
[2] Angelov N *et al* 2007 *Eur. Phys. Jour.* A **34** 255
[3] Gnesi I 2010 *Pion Induced Reactions in the Delta Resonance Energy Region*, PhD Thesis, University of Turin, copy available from gnesi@to.infn.it.
[4] Sober D I *et al* 1975 *Phys. Rev.* D **11** 1017
[5] Dillig M and Huber M G 1974 *Phys. Lett.* B **48** 417
[6] Balestra F *et al* 1980 *N. Cim.* A **55** 273
[7] Landau R H, Phatak S C and Tabakin F 1973 *Ann. Phys.* **78** 299
[8] Gnesi I *et al* 2011 *Eur. Phys. Jour.* A **47** 3
[9] Sedlak J E and Friedman W A 1977 *Phys. Rev.* C **16** 2306
[10] Lehmann A *et al* 1997 *Phys. Rev.* C **55** 2931
[11] Matheos A O *Multinucleon Pion Absorption on ^{4}He in the Δ resonance region* 1991 PhD Thesis, Princeton University, available at MIT http://hdl.handle.net/1721.1/37014.

Journal of Physics: Conference Series **436** (2013) 012078 doi:10.1088/1742-6596/436/1/012078

Nuclear symmetry energy in charge and isotopic distributions

H Imal, N Buyukcizmeci and R Ogul

Department of Physics, Selcuk University, 42079 Kampus, Konya, Turkey

Abstract. Isotopic yields of nitrogen and boron fragments from the decay of the single nucleus with $A_0 = 186$, $Z_0 = 75$ were studied on the basis of the statistical multifragmentation model, and compared with data from the MSU experiments for central collisions of ^{124}Sn + ^{124}Sn at $E/A = 50$ MeV. We have reconfirmed that a significant reduction of the symmetry term coefficient is necessary to reproduce the experimental isotopic distributions.

1. Introduction

Nuclear multifragmentation has attracted great interest due to its similarity to the collapse and explosion of massive stars. By this reason, the importance of isotopic effects in nuclear multifragmentation are related to astrophysical processes such as supernova simulations and neutron star models. We have interpreted MSU data for central collisions of ^{124}Sn + ^{124}Sn at $E/A = 50$ MeV. In this presentation, we investigate the influence of the expected decrease of symmetry energy extracted from the isotopic distributions for N and B isotopes within the microcanonical Markov-chain version of the statistical multifragmentation model [1, 2]. For simulation we have considered the decay of a single source with $A_0 = 186$, $Z_0 = 75$, which is presumably formed during the central collisions of ^{124}Sn + ^{124}Sn. Then, we compare the results with MSU experimental data [3].

2. Theoretical calculations and results

For the calculations of charge and isotopic yields we consider the statistical multifragmentation model (SMM) [1]. In the SMM, it is assumed that a statistical equilibrium is reached at low density in the freeze-out region. It is also assumed that all breakup channels are composed of nucleons, and the laws of the conservation of energy E^*, momentum, angular momentum, mass number A and charge number Z are taken into account. Besides the breakup channels, the compound-nucleus channels are also included, and a competition among all channels is permitted. In this way, the SMM covers the conventional evaporation and fission processes occurring at low excitation energy as well as the transition region between the low- and high-energy de-excitation regimes. In the thermodynamic limit, SMM is consistent with the liquid-gas phase transition when the liquid phase is represented by an infinite nuclear cluster.

As was demonstrated earlier [4, 5], varying surface and symmetry terms can essentially influence the characterization of the fragments produced in multifragmentation. In Fig. 1, we show the results of a reasonably small variation of the surface energy term B_0 in between 16 and 20 MeV (panels a and b), and the results of the variation of the symmetry term (panels c and

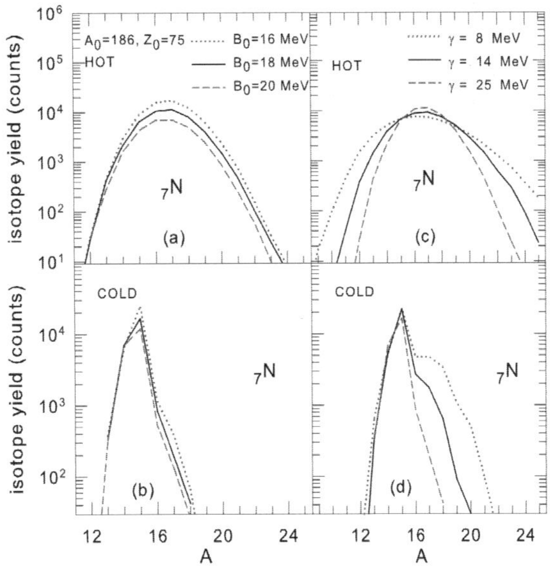

Figure 1. Predicted isotope distributions for nitrogen fragments from the multifragmentation of the single source $A_0 = 186$, $Z_0 = 75$, at various surface energy B_0 (panels a and b) and symmetry energy γ values (panels c and d).

d). It is seen that the surface energy term B_0 does not produce significant changes in the isotope distributions for nitrogen fragments, but the symmetry term causes significant changes. This was also confirmed in Ref. [4] within SMM-ensemble calculations performed for the interpretation of ALADIN experimental data. In Ref. [6], we showed that the symmetry energy has a small influence on fragment mass and charge distributions, however, its effect is more pronounced in

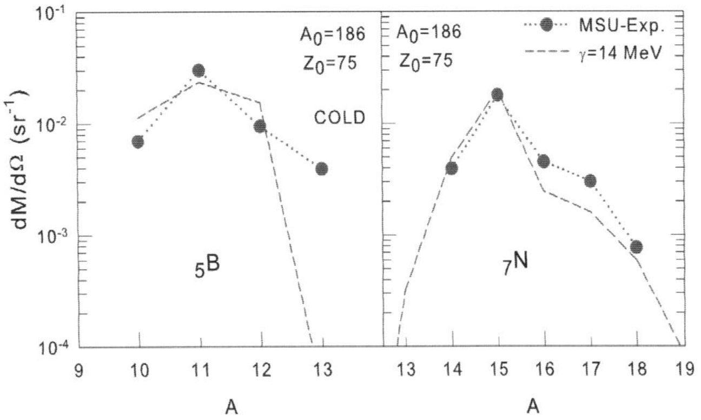

Figure 2. Predicted isotope distributions for boron and nitrogen fragments from the decay of the single nucleus with $A_0 = 186$, $Z_0 = 75$. The dashed lines are the present results for cold fragments at $\gamma = 14$ MeV, and the solid circles are the MSU data.

the isotopic distributions. By means of this sensitivity, we can estimate the symmetry energy by comparing the predicted distributions with experimental data.

In Fig. 2, we compare our results for isotopic yields with MSU data. One can conclude from this figure that our secondary cold fragment distributions compare well with the experimental data at the reduced gamma values around $\gamma = 14$ MeV, as was already obtained in our analyses of other observables [7, 8, 9].

3. Conclusions

It has been demonstrated that it is possible to investigate, in experiments, the modification of the liquid drop parameters of fragments, such as the symmetry energy coefficient, at subnuclear densities in the dense environment of the nucleus undergoing fragmentation. This is important for the equation of state of nuclear and stellar matter in terms of the density dependence of symmetry energy.

Acknowledgements

The authors thank A.S. Botvina and I.N. Mishustin for enlightening discussions.

References

[1] Bondorf J P, Botvina A S, Iljinov A S, Mishustin I N and Sneppen K 1995 *Phys. Rep.* **257** 133
[2] Botvina A S and Mishustin I N 2001 *Phys. Rev.* C **63** 061601(R)
[3] Liu T X *et al* 2004 *Phys. Rev.* C **69** 014603
[4] Ogul R, Botvina A S *et al* 2011 *Phys. Rev.* C **83** 024608
[5] Botvina A S *et al* 2006 *Phys. Rev.* C **74** 044609
[6] Buyukcizmeci N, Ogul R and Botvina A S 2005 *Eur. Phys. J.* A **25** 57
[7] Buyukcizmeci N *et al* 2011 *Acta Physica Polonica B* **42** 697
[8] Ogul R *et al* 2009 *J. Phys. G: Nucl. Part. Phys.* **36** 115106
[9] Buyukcizmeci N *et al* 2012 *J. Phys. G: Nucl. Part. Phys.* **39** 115102

Journal of Physics: Conference Series **436** (2013) 012079 doi:10.1088/1742-6596/436/1/012079

Pseudo-critical clusterization in nuclear multi-fragmentation

D Gruyer[1], J D Frankland[1], R Botet[2], M Płoszajczak[1], E Bonnet[1], A Chbihi[1] and P Marini[1] for the INDRA collaboration

[1]GANIL, CEA-DSM/CNRS-IN2P3, Bvd. Henri Becquerel, F-14076 Caen Cedex, France
[2]Laboratoire de Physique du Solide, Universit de Paris-Sud, F-91405 Orsay, France

E-mail: diego.gruyer@ganil.fr

Abstract. We show that the biggest fragment charge distribution in central collisions of ^{129}Xe+^{nat}Sn leading to multifragmentation is an admixture of two asymptotic distributions observed for the lowest and highest bombarding energies. The evolution of the relative weights of the two components with bombarding energy is shown to be analogous to that observed as a function of time for the largest cluster produced in irreversible aggregation for a finite system. We infer that the size distribution of the largest fragment in nuclear multifragmentation is also characteristic of the time scale of the process, which is largely determined by the onset of radial expansion in this energy range.

Phase transitions play a central role in many fields of physics. Indeed, they allow us to investigate the equation of state and phase diagram of the system under study. The case of nuclear multifragmentation, as observed in intermediate energy heavy-ion collisions [1], provides a unique opportunity to study not only thermodynamical properties of nuclear matter, but also phase transitions in *finite* systems.

In an infinite system, fluctuations are generally irrelevant and phase transitions can be characterized by the correlations between the control parameter and the order parameter (fig. 1(a)). With decreasing size of the system, fluctuations become more important and correlations are distorted (fig. 1(b)), such that for small system sizes (comparable to that of a nucleus), we are not able to identify the transition from this simple picture (fig. 1(c)). The universal character of order parameter fluctuations in finite systems [2] provides a good framework in which to address such questions. Whithin such a framework, it was shown that the size (atomic number) of the largest fragment produced in multifragmentation events, Z_{max}, behaves like an order parameter, *i.e.* the scaling properties of its fluctuations change with increasing energy [3]. In ref. [4] it was shown that a distinct asymptotic form of the Z_{max} distribution can be associated with each scaling regime: a quasi-Gaussian distribution at the lower bombarding energies ($E_{beam} \lesssim 30$ MeV/A), and a Gumbel distribution in the higher-energy disordered regime ($E_{beam} \gtrsim 40$ MeV/A).

Here we study in more detail the transition from one energy regime to the other, using new data on Z_{max} distributions for ^{129}Xe+^{nat}Sn central collisions measured with INDRA [4, 5], at eight beam energies, between 25 and 50 MeV/A. Within the hypothesis that at intermediate energies between the two regimes, Z_{max} distributions can be described by an admixture of the

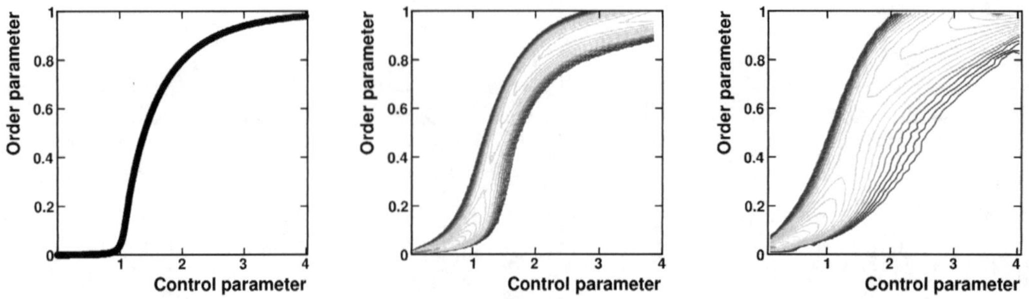

(a) $N \sim 10^4$ particles (b) $N \sim 200$ particles (c) $N \sim 50$ particles

Figure 1. (color online) Correlations between the order parameter and the control parameter for a critical phase transition occurring in systems of different sizes.

two asymptotic forms, we fit them with the following function :

$$f(\mathrm{Z_{max}}) = \eta f_{Ga}(\mathrm{Z_{max}}) + (1-\eta) f_{Gu}(\mathrm{Z_{max}}), \tag{1}$$

where, f_{Ga} and f_{Gu} are the Gaussian and the Gumbel component.

The results of the fits using eq. (1) are shown in fig. 2. It can be seen that, for all analyzed energies, experimental data are well reproduced by $f(\mathrm{Z_{max}})$. At the highest considered energy (50 MeV/A), $\mathrm{P(Z_{max})}$ is an almost pure Gumbel distribution (fig. 2(d)). For lower bombarding energies (fig. 2(a-c)), both Gaussian and Gumbel contributions are present, and the relative importance of the Gaussian component increases with decreasing energy. In fig. 2(e) we present the evolution of the relative weight between the two components as a function of the beam energy. We observe not only two asymptotic regimes, but a continuous and smooth evolution between them.

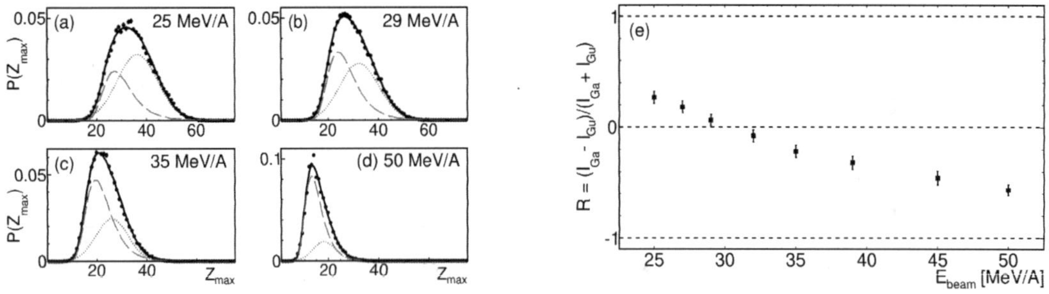

Figure 2. (Color online) Analysis of ^{129}Xe+^{nat}Sn experimental data: (a-d) largest fragment charge distributions, (Black solid curve) best fit to the data using eq. (1), (Red dashed curve) Gumbel component, (Blue dotted curve) Gaussian component; and (e) evolution of the relative weight between the two components, as a function of the beam energy.

To interpret such evolution, we consider the Smoluchowski irreversible aggregation model, which describes an out-of-equilibrium clusterization process. It exhibits a second-order phase transition after a critical time (t_c), whose order parameter is also the average size of the largest cluster, s_{max}. The order parameter distribution for irreversible aggregation is not known exactly.

For short time evolution ($t \ll t_c$), there are only few aggregation events. So $s_{\rm max}$ is the largest part of randomly distributed cluster sizes. $s_{\rm max}$ is then of extremal nature, and it results in its Gumbel distribution (fig. 3(a)). In this model, the average size of the largest cluster increases with time as more and more coalescence of smaller clusters takes place. At long times ($t \gg t_c$), the order parameter is then essentially of additive nature. From the central limit theorem, this results in an asymptotic Gaussian distribution (fig. 3(d)). In the critical domain, *finite size* fluctuations are so large that similarly prepared systems can exhibit one or the other behavior. In this domain, the $s_{\rm max}$ distribution is an admixture of the two asymptotic forms (fig. 3(b-c)), with a continuous evolution of their relative weight (fig. 3(e)).

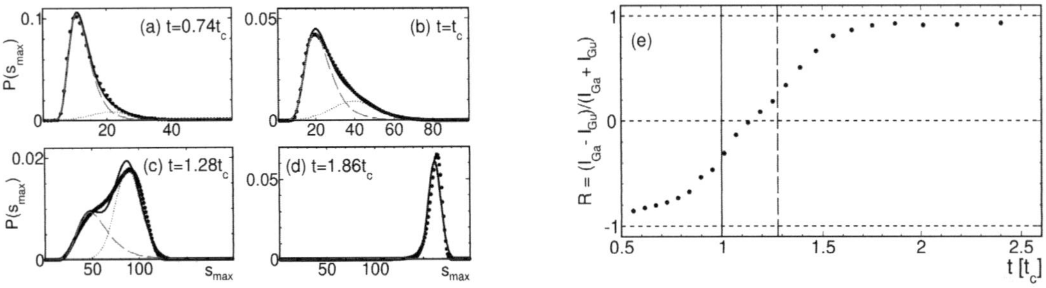

Figure 3. (Color online) Analysis of Smoluchowski calculations for $N = 216$ particles: (a-d) largest cluster size distributions, (Black solid curve) best fit to the data using eq. (2), (Red dashed curve) Gumbel component, (Blue dotted curve) Gaussian component; (e) time evolution of the relative weight between the two components (eq. (3)), (vertical dashed line) pseudo-critical time t_c^*.

A strong similarity can be seen between the results of the analysis of irreversible aggregation calculations (fig. 3) and experimental data on ^{129}Xe+^{nat}Sn central collisions (fig. 2). In the aggregation model the order parameter distribution depends on the time during which clusters can form. In central heavy-ion collisions, a determining factor for the time-scale of fragment formation is the radial expansion of the multifragmenting system, which increases with the bombarding energy [1]. Fragment sizes evolve as long as exchanges of nucleons take place between them, *i.e.* until the freeze-out condition is reached. It has been shown that, for central ^{129}Xe+^{nat}Sn reactions, the onset of significant radial expansion occurs for beam energies above 25 MeV/A [5]. Therefore, the similarity between figs. 3 and 2 can be interpreted in terms of fragment size distributions being determined on shorter and shorter time-scales due to increasingly rapid expansion.

We have shown that the largest fragment size distribution in multifragmentation events is an admixture of the two asymptotic forms. A similar decomposition is observed in critical aggregation, indicating that the critical domain lies around $E_{\rm beam} \approx$ 30MeV/A for the ^{129}Xe+^{nat}Sn system. We interpret such criticality as the onset of an 'explosive' multifragmentation regime.

References

[1] Borderie B and Rivet M-F 2008 *Progress in Particle and Nuclear Physics* **61** 551
[2] Botet R and Ploszajczak M 2000 *Physical Review* E **62** 1825
[3] Botet R, Ploszajczak M, Chbihi A, Borderie B, Durand D and Frankland J 2001 *Phys. Rev. Lett.* **86** 3514
[4] Frankland J *et al* 2005 *Phys. Rev.* C **71** 034607
[5] Bonnet E *et al* 2009 *Nucl. Phys.* A **816** 1

Journal of Physics: Conference Series **436** (2013) 012080 doi:10.1088/1742-6596/436/1/012080

Four- and five-body cluster structure of hypernuclei

E. Hiyama

Nishina Center for Accelerator-Based Science, RIKEN, Wako, Saitama, 351-0198, Japan

E-mail: hiyama@riken.jp

Abstract. We review interesting phenomena of hypernuclear structure from view points of clustering physics. The glue like role of Λ particle provide with size contraction of hypernuclei. For an example, we review the structure of ${}^{7}_{\Lambda}$Li, ${}^{13}_{\Lambda}$C. As an example, the energy gain by the addition of a Λ particle is dependent on the shell-like states or clustering states in core nuclei, we review the structure of ^{12}C and ${}^{13}_{\Lambda}$C. We show a very preliminary result of ${}^{6}_{\Lambda}$H within the framework of ${}^{5}\mathrm{H} + n + n + \Lambda$ four-body cluster model to demonstrate the stabilization of the state due to the attraction of ΛN interaction. Finally, we review the structure of double Λ hypernucleus, ${}^{11}_{\Lambda\Lambda}$Be with the $\alpha\alpha + \Lambda + \Lambda + n$ five-body cluster model.

1. Introduction

One of the purpose of hypernuclear physics is to study the new dynamical features induced by the Λ particle. For example, in light p-shell hypernuclei, the remarkable role of the Λ particle has been pointed out in Refs. [1]: the Λ participation gives rise to more bound states and the appreciable contraction of the system (this stabilization is called the 'gluelike' role of the Λ). In light nuclei near the neutron drip line, interesting phenomena concerning the neutron halo have been observed. If a Λ particle is added to a very weakly bound halo nucleus the resultant hypernucleus will become substantially stable against the neutron decay. Thanks to the gluelike role of the Λ particle, there is a new chance to produce a hypernuclear neutron (proton) halo state if the core nucleus has a weakly unbound (resonant) state with an appropriate energy above the particle decay threshold. In this paper, we show two examples such as ${}^{7}_{\Lambda}$He and ${}^{6}_{\Lambda}$H within the framework of $\alpha + \Lambda + N + N$ and $t + \Lambda + n + n$ four-body cluster model. As mentioned above, it is considered that the gluelike of Λ produce shrinkage effect of the core nuclei. Now, we have a question that all nuclei are compressed or not. In order to answer it, we review the structure of ${}^{13}_{\Lambda}$C within the framework of $3\alpha + \Lambda$ four-body model. Third, we have interesting property of the injection of Λ particle as follows: When a Λ particle is added to core nuclei with $A \geq 11$, the energy gain by addition of a Λ particle is dependent on the clustering states or shell-like states. For a example, we show the energy spectra of ${}^{13}_{\Lambda}$C. Another goal in hypernuclear physics is to understand the baryon-baryon interaction. The baryon-baryon interaction is fundamental and important for the study of nuclear physics. In order to understand the baryon-baryon interaction, two-body scattering experiments are the most useful tool. For this purpose, many NN scattering experiments have been done and the total number of NN data are more than 4,000. However, due to the difficulty of performing two-body hyperon(Y)-nucleon(N) and hyperon(Y)-hyperon(Y) scattering experiments, the total number of YN scattering data are very limited. Namely, the number of differential cross section are only about 40 and there is no YY scattering data. Then, YN and YY potential models so far proposed have large ambiguity. Therefore, as a substitute for the two-body limited YN and non-existent YY scattering data, the systematic investigation of light hypernuclear structure is essential. In fact, at J-PARC facility, they are planning to produce many double Λ hyprnuclei by emulsion experiment. However, it is difficult to determine the spin and parity and whether the observed state is the ground state or an excited state. Then, it is necessary to compare the data with the theoretical calculation for the identification of the state. For successful examples, we review the clustering structure of $^{10}_{\Lambda\Lambda}$Be and $^{11}_{\Lambda\Lambda}$Be.

2. Shrinkage effect in $^{7}_{\Lambda}$Li and $^{13}_{\Lambda}$C.

When a Λ particle is injected into the light nucleus, it is considered that nuclear size is reduced due to the ΛN interaction. Historically, Motoba *et al.* pointed out that the shrinkage effect was seen in the value of $B(E2)$ [1]. In Ref.[2], we suggested to experimentalist to measure $B(E2; 5/2^+ \rightarrow 1/2^+)$ in $^{7}_{\Lambda}$Li ,and to propose a prescription to derive the size of the ground state of $^{7}_{\Lambda}$Li from that $B(E2)$ value with the aid of the empirical $B(E2; 3^+ \rightarrow 1^+)$ and the size of the ground state of ^{6}Li. Within the framework of $^{5}_{\Lambda}$He $+ N + N$ three-body model, we predicted that core nucleus ^{6}Li will be shrunk by 25 %. The first observation of the hypernuclear $B(E2)$ strength was recently performed in the KEK-E419 experiment for $B(E2; 5/2^+ \rightarrow 1/2^+)$ in $^{7}_{\Lambda}$Li; the observed $B(E2)$ value was 3.6 ± 0.5 e^{2}fm^4 [3] and they reported the size of $\bar{R}_{\mathrm{core}-(np)}$ was shrunk by 19 ± 4 %, which is consistent with our prediction. This confirms, for the first time, the shrinkage of the nuclear size induced by the Λ particle.

One may ask the next question "Are all nuclei compress by the injection of a Λ particle?" Our answer is No. We predict that the nuclear density of the ground states in the stable nuclei with $A \leq 11$ is not compressed due to the addition of a Λ particle. On the other hand, the matter radius in some excited states are shrunk by as much as 30 % due to the injection of a Λ particle. This was precisely studied for ^{12}C and $^{13}_{\Lambda}$C within the framework of 3α and $3\alpha + \Lambda$ three- and four-body model [4].

It is well known that the 0^+_1 state in ^{12}C is shell-like ground state while the 0^+_2 state at $E_x = 7/65$ MeV is well-developed clustering state and that both states are simultaneously well described by the microscopic 3α cluster model. The calculated Λ separation energies of $1/2^+_1$ and $1/2^+_2$ states in $^{13}_{\Lambda}$C are 11.69 MeV and 8.59 MeV, respectively. The $1/2^+_1$ is dominantly composed of the $0^+_1(^{12}\mathrm{C})$ and $0s$ Λ particle and the $1/2^+_2$ is composed of the $0^+_2(^{12}\mathrm{C})$ and $0s$ Λ particle. We can expect to observe a very characteristic $E2$ transition from the second $1/2^+$ state to the spin-doublet $3/2^+$-$5/2^+$ states of the ground-state band. This transition corresponds, in the core nucleus ^{12}C, to the $E2$ transition which intervenes between the well-developed clustering state (0^+_2) and the shell-model dominating state (2^+_1). Calculation of the $B(E2)$ strength is in progress, but, referring to the size-shrinkage of the second $1/2^+$ state, we can speculate it to be about a half of that of the corresponding transition in ^{12}C.

We also calculated r.m.s. distance between two α clusters, $\bar{r}_{\alpha-\alpha}$ in the $^{13}_{\Lambda}$C and ^{12}C. The r.m.s. distance between two α clusters in the ground state of $^{13}_{\Lambda}$C is $\bar{r} = 2.87 - 2.90$fm, which represents little shrinkage from $\bar{r} = 3.02$ fm in the ground state of ^{12}C. This contraction is more vividly seen in Fig.1, which illustrates the probability density of finding two α at a distance $r_{\alpha-\alpha}$. The little change in $\bar{r}_{\alpha-\alpha}$ is also visualized in Fig. 1 by drawing two similar distributions of the probability to find two α clusters at a distance $\bar{r}_{\alpha-\alpha}$. This reasonably implies that shell-like states are not easily contracted when a Λ particle joins.

On the other hand, in the second $1/2^+$ state of $^{13}_{\Lambda}$C, we obtain $r_{\alpha-\alpha} = 4.46 - 4.58$ fm, and therefore a sizable contraction of the $\alpha - \alpha$ distance is seen as compared with $r_{\alpha-\alpha} = 6.25$ fm in the 0_2^+ state of ^{12}C. This significant contraction is seen in Fig.1. It is to be stressed that these theoretical results on the size-contraction in the excited $1/2^+$ state is owing to the four-body frame which is suited for describing the response of the 3α system to the addition of the Λ particle.

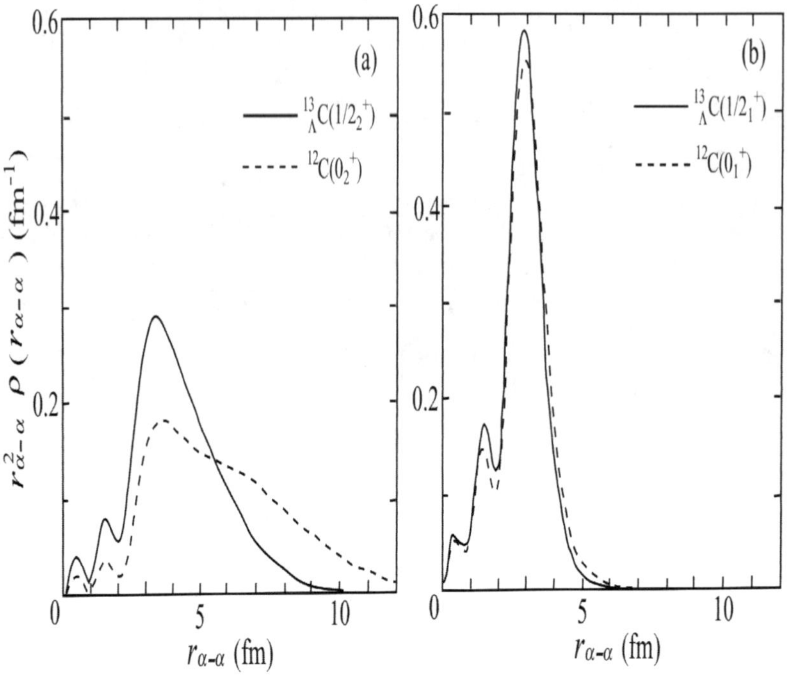

Figure 1. Probability densities of finding two α at a distance $r_{\alpha-\alpha}$ in the states (a) $^{12}C(0_2^+)$ and $^{13}_{\Lambda}C(1/2_2^+)$, and (b) $^{12}C(0_1^+)$ and $^{13}_{\Lambda}C(1/2_1^+)$. Shrinkage of the excited 0_2^+ in ^{12}C is drastic, but that in the ground state is little. This figure is taken from [4].

The characteristic feature of ^{12}C is the existence of cluster states, together with the ground band states $(0_1^+, 2_1^+, 4_1^+)$ with a shell-model-like nature. The 0_2^+ state at excitation energy, $E_x = 7.64$ MeV, called 'Hoyle state', has a loosely bound three α-cluster structure, for which a shell-model type of calculation would have great difficulties in reproducing its properties. In fact, the microscopic 3α cluster model succeeded in reproducing the Hoyle state, including the ground-band states and negative-parity states. Figure 1 shows the density distribution of the 0_1^+ and 0_2^+ states of ^{12}C calculated by the microscopic 3α cluster model. One can see a clear difference between the structures of the 0_1^+ and 0_2^+ states. On the other hand, the negative-parity states, 3^- and 1^-, are known to have intermediate characters between the shell-model compact structure and the loosely bound 3α structure. Therefore, it is a new issue in $^{13}_{\Lambda}$C *not* considered in $^{9}_{\Lambda}$Be what kinds of structures appear in addition of a Λ particle into ^{12}C with the three different types of structures, i.e. shell-model compact one, loosely bound 3α cluster one, and the intermediate one.

Journal of Physics: Conference Series **436** (2013) 012080 doi:10.1088/1742-6596/436/1/012080

3. Neutron-rich Λ hypernucleus, ${}^{6}_{\Lambda}$H

Recently, a neutron-rich Λ hypernucleus, ${}^{6}_{\Lambda}$H has been observed as a bound state by FINUDA experiment [5]. They reported the observed B_Λ to be 4.0 ± 1.1 MeV. Before measurement, some authors [6–8] studied the structure of this hypernucleus by shell model and G-matrix theory.

Now, we have an important issue: we can reproduce the observed data or not. Motivating the experiment, we calculated the binding energy of ${}^{6}_{\Lambda}$H within the framework of $t + n + n + \Lambda$ four-body model and here report a very preliminary result.

In order to discuss the binding energy of ${}^{6}_{\Lambda}$H, it is important to reproduce the observed energy of the core nucleus, ^{5}H, $E = 1.7 \pm 0.3, \Gamma = 1.9 \pm 0.4$ MeV.

To study the core nucleus, ^{5}H, R. De Diego *et al.*, calculated the energy and width within the framework of $t+n+n$ three-body model [9] We employ the $t - n$ potential employed by Ref.[9]. However, it should be noticed that they employed partial dependent $t-n$ potential. It is difficult to employ this potential directly. Then, here we employ parity dependent potential starting from $t - n$ potential by Ref.[9]. The two-body interaction does not reproduce the observed data of ^{5}H, then, we use a phenomenological $t - n - n$ three-body force introduce by Ref.[9]. Using this potential, the calculated energy of ^{5}H is 1.45 MeV with $\Gamma = 2.1$ MeV, which is in good agreement with the observed data within the error bar.

The employed $t - \Lambda$ potential reproduce the binding energies pf the 0^+ and 1^+ states of ${}^{4}_{\Lambda}$H. In this case, $\Lambda N - \Sigma N$ coupling is renormalized into ΛN interaction. And we employ AV8 NN realistic interaction. The ΛN potential reproduces the binding energies of ${}^{3}_{\Lambda}$H and ${}^{4}_{\Lambda}$H which is employed for the discussion on charge symmetry breaking effect of ${}^{4}_{\Lambda}$H and ${}^{4}_{\Lambda}$He.

Using all potential above, the preliminary result of ${}^{6}_{\Lambda}$H B_Λ is 2.41 MeV, which is unbound with respect to ${}^{4}_{\Lambda}\mathrm{H} + n + n$ breakup threshold.

The preliminary result concludes that we have no bound state as long as we take only Λ channel. When we take $\Lambda N - \Sigma N$ coupling effect, our conclusion might change. The further analysis is in progress.

4. $S = -2$ hypernuclei and YY interaction

It is interesting to investigate the structure of the multi-strangeness system when one or more Λ's are added to a $S = -1$ nucleus. It is conjectured that extreme limit, which includes many Λ's (and other hyperons) in nuclear matter is the core of a neutron star. In this sense, the sector of $S = -2$ nuclei, double Λ hypernuclei and Ξ hypernuclei, is just the entrance to the multi-strangeness world. However, we have hardly any knowledge of the YY interaction because there exists no YY scattering data. Then, in order to understand the YY interaction, it is crucial to study the structure of double Λ hypernuclei and Ξ hypernuclei. The equation of state with the strangeness degree of freedom is a crucial component in understanding neutron stars.

Recently, the epoch-making data has been reported by the KEK-E373 experiment. Namely, the double Λ hypernucleus ${}^{6}_{\Lambda\Lambda}$He was observed [10]. This observation was called NAGARA event. The formation of ${}^{6}_{\Lambda\Lambda}$He was uniquely identified by the observation of sequential weak decays, and the precise experimental value of the 2Λ binding (separation) energy, $B_{\Lambda\Lambda} = 7.25 \pm 0.19^{+0.18}_{-0.11}$ MeV, was obtained.

We studied double Λ hypernuclei with $A = 6 - 10$ [11]. Firstly, (1) we employed the $\Lambda\Lambda$ interaction of Nijmegen model D and performed an $\alpha + \Lambda + \Lambda$ three-body calculation for ${}^{6}_{\Lambda\Lambda}$He. (2) By comparing the theoretical result with the experimental data of the binding energy of ${}^{6}_{\Lambda\Lambda}$He, (3) we suggested reducing the strength of ${}^{1}S_0$ term of the $\Lambda\Lambda$ interaction by half to reproduce the data. Again, (2) using the improved potential, we predicted energy spectra of new double Λ hypernuclei with $A = 7 - 10$ [11], which is discussed below.

In fact, it is planned at J-PARC to produce many double Λ hypernuclei by emulsion experiment [12]. However, it will be difficult to determine spin-parities and to know whether the observed state is the ground state or an excited state. Therefore, it will be necessary to compare the data with any theoretical study for the identification of the state. The author's role is to contribute to the theoretical calculation using few-body techniques.

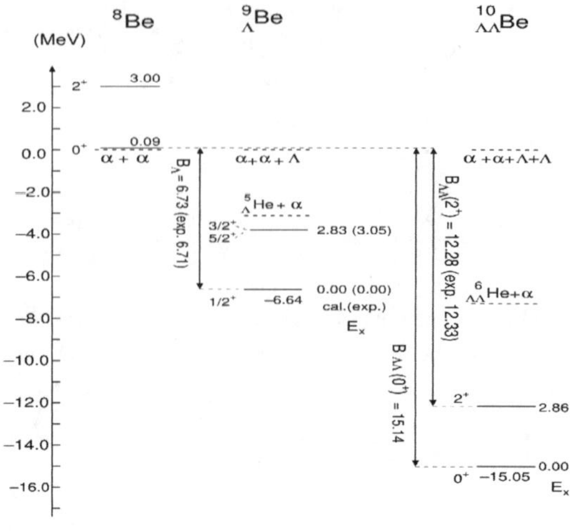

Figure 2. Calculated energy levels of ^{8}Be, $^9_\Lambda$Be and $^{10}_{\Lambda\Lambda}$Be on the basis of the $\alpha+\alpha$, $\alpha+\alpha+\Lambda$, and $\alpha+\alpha+\Lambda+\Lambda$ models, respectively. The level energies are measured from the particle breakup thresholds or are given by the excitation energies E_x. The calculated 2^+ state of $^{10}_{\Lambda\Lambda}$Be explains the Demachi-Yanagi event. This figures is taken from [11]

A successful example to determine spin-parity of double Λ hypernuclei is the case of $^{10}_{\Lambda\Lambda}$Be. There was one more event found in the E373 experiment named the 'Demachi-Yanagi' event [13]. The most probable interpretation of this event is the production of a bound state of $^{10}_{\Lambda\Lambda}$Be having $B^{\rm exp}_{\Lambda\Lambda} = 12.33^{+0.35}_{-0.21}$ MeV. But the experiment could not determine whether this state was the ground state or any excited state. In order to determine this, our calculation [11] mentioned above was useful. We studied $^{10}_{\Lambda\Lambda}$Be by employing an $\alpha+\alpha+\Lambda+\Lambda$ four-body model. The $\Lambda\Lambda$ interaction is the one improved from the Nijmegen Model D as mentioned above. The $\Lambda\Lambda$, $\alpha\Lambda$ and $\alpha\alpha$ interactions were chosen so as to reproduce the binding energies of all the subsystems, $^6_{\Lambda\Lambda}$He, $^5_\Lambda$He, ^{8}Be and $^9_\Lambda$Be. As shown in Fig.2, it is striking that our calculated value of $B_{\Lambda\Lambda}(^{10}_{\Lambda\Lambda}{\rm Be}(2^+))$ is 12.28 MeV that agrees with the experimental data. Therefore, the Demachi-Yanagi event can be interpreted most probably as the observation of the 2^+ excited state in $^{10}_{\Lambda\Lambda}$Be.

Furthermore, in the KEK-E373 experiments, they observed one more event, Hida event [13]. This observation is for $^{11}_{\Lambda\Lambda}$Be or $^{12}_{\Lambda\Lambda}$Be. The observed $B_{\Lambda\Lambda}$ for $^{11}_{\Lambda\Lambda}$Be is 20.49 ± 1.15 MeV and $B_{\Lambda\Lambda}$ for $^{12}_{\Lambda\Lambda}$Be is 22.06 ± 1.15 MeV [13]. The important issue is to interpret the Hida event as a ground state or excited state in $^{11}_{\Lambda\Lambda}$Be or $^{12}_{\Lambda\Lambda}$Be. We assume Hida event as $^{11}_{\Lambda\Lambda}$Be and calculate the $B_{\Lambda\Lambda}$ with $\alpha\alpha n\Lambda\Lambda$ five-body problem. This five-body calculation is numerically difficult since we have three kinds of particles such as α, Λ and neutron, and we have five different kinds of interactions such as $\Lambda\Lambda$, Λn, $\Lambda\alpha$, $n\alpha$ and $\alpha\alpha$, and we have Pauli principle between α and α, and between α and neutron. Recently, we succeeded in performing this calculation.

In the present $\alpha+\alpha+n+\Lambda+\Lambda$ five-body model for $^{11}_{\Lambda\Lambda}$Be, it is absolutely necessary that all sub-cluster systems composed of two α's, a neutron and two Λ's are described reasonably with the interactions among these units. In our previous work [11], our interactions, which include the $\alpha\alpha$, αn, $\alpha\Lambda$, Λn and $\Lambda\Lambda$ interactions, were determined so as to reproduce reasonably well the following observed quantities: (i) Energies of the low-lying states and scattering phase shifts in the $\alpha + n$ and $\alpha + \alpha$ systems, (ii) Λ-binding energies B_Λ in ${}^5_\Lambda$He $(= \alpha + \Lambda)$, ${}^6_\Lambda$He $(= \alpha + \Lambda + n)$ and ${}^9_\Lambda$Be $(= \alpha + \alpha + \Lambda)$, (iii) double-$\Lambda$ binding energies $B_{\Lambda\Lambda}$ in ${}^{\ 6}_{\Lambda\Lambda}$He $(= \alpha + \Lambda + \Lambda)$, the Nagara event. Then, as mentioned above, the Demachi-Yanagi event for ${}^{10}_{\Lambda\Lambda}$Be $(= \alpha + \alpha + \Lambda + \Lambda)$ was simultaneously reproduced with no additional adjustable parameter.

In the present work, we employ the same interactions of Ref.[11] so that those severe constraints are also successfully met in our two-, three- and four-body subsystems. But, as for the present core nucleus ^{9}Be $(= \alpha + \alpha + n)$, which does not belong to the subsystems studied previously, use of the interactions that explain well the property of the $\alpha\alpha$ and αn subsystems do not well reproduce the energies of the low-lying states of ^{9}Be measured from the $\alpha + \alpha + n$ threshold (the same property of the calculated result was reported in another microscopic $\alpha + \alpha + n$ cluster-model study [14]). Therefore, we additionally introduce a phenomenological $\alpha\alpha n$ three-body force with a Gaussian shape, $v_0 e^{-(r_{\alpha-\alpha}/r_0)^2 - (R_{\alpha\alpha-n}/R_0)^2}$, having $r_0 = 3.6$ fm, $R_0 = 2.0$ fm and $v_0 = -9.7$ MeV (+13.0 MeV) for the negative-parity (positive-parity) state; we thus reproduce well the observed energies of the $3/2^-_1, 5/2^-_1, 1/2^-_1$ and $1/2^+_1$ states of ^{9}Be. The energy level is illustrated in Fig. 3 together with that of ^{9}Be. Interestingly enough, the

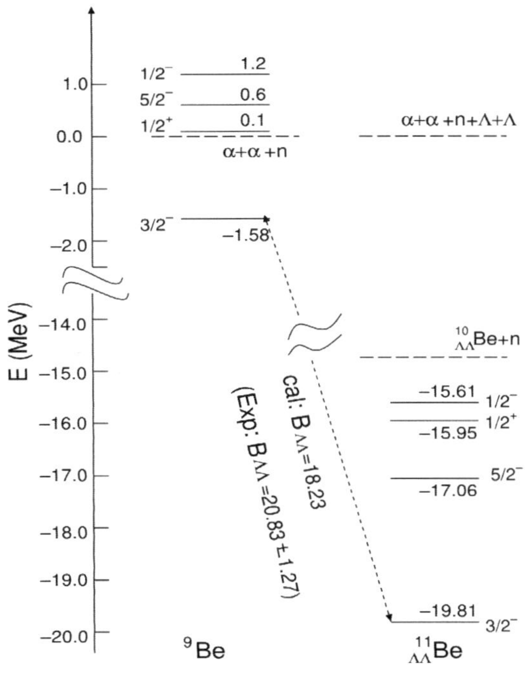

Figure 3. Calculated energy spectra of the low-lying states of ${}^{11}_{\Lambda\Lambda}$Be together with those of the core nucleus ^{9}Be.

order of the $1/2^+$ and $5/2^-$ states is reversed from ^{9}Be to ${}^{11}_{\Lambda\Lambda}$Be. This is because the energy gain due to the addition of the Λ-particle(s) is larger in the compactly coupled state $(5/2^-_1)$ than in the loosely coupled state $(1/2^+_1)$. Note that the same type of theoretical prediction was reported, in our early work [15] for ${}^{13}_{\Lambda}$C based on the $\alpha\alpha\alpha\Lambda$ model, for the Λ particle addition to the compactly coupled state (3^-_1) and to the loosely coupled state (0^+_2) in ^{12}C. As seen in Fig. 3 the calculated value of $B_{\Lambda\Lambda}({}^{11}_{\Lambda\Lambda}\mathrm{Be})$ is 18.23 MeV for the $3/2^-$ ground state, while for the excited states the $B_{\Lambda\Lambda}$ values are calculated to be less than 15.5 MeV. Therefore, the observed Hida event can be interpreted to be the ground state. When our calculated binding energy is compared with the experimental value of 20.83 MeV with a large uncertainty of σ=1.27 MeV, we can say at least that our result does not contradict the data within 2σ. Motivated by the recent observation of the Hida event for a new double Λ hypernucleus, we have succeeded in performing a five-body calculation of ${}^{11}_{\Lambda\Lambda}$Be using an $\alpha\alpha n\Lambda\Lambda$ cluster model. The calculated $\Lambda\Lambda$ binding energy does not contradict the interpretation that the Hida event is an observation of the ground state of ${}^{11}_{\Lambda\Lambda}$Be.

References

[1] Motoba T, Bandō H and Ikeda K 1983 *Prog. Theor. Phys.* **70** 189; Motoba T, Bandō, H Ikeda K and Yamada T 1985 *Prog. Theor. Phys. Suppl.* **81** 42
[2] Hiyama E, Kamimura M, Miyazaki K, Motoba T 1999 *Phys. Rev.* C **59** 2351
[3] Tanida K *et al* 2001 *Phys. Rev. Lett.* **86** 1982
[4] Hiyama E, Kamimura M, Motoba T, Yamada T and Yamamoto Y 1997 *Prog. Theor. Phys.* **97** 881
[5] Agnello M *et al* 2012 *Phys. Rev. Lett.* **108** 042501
[6] Dalitz R H and Levi-Stti R 1963 *Nuovo Cimento* **30** 489
[7] Majling L 1995 *Nucl. Phys.* A **585** 211c
[8] Akaishi Y and Yamazaki Y 1999 *Physics and Detectors for DAΨNE /* (Frascati Physics Series Vol. 16) (Rome: SIS INFN-LNF) p 59
[9] de Diego R *et al* 2007 *Nucl. Phys.* A **786** 71
[10] Takahashi H *et al* 2002 *Phys. Rev. Lett.* **87** 212502
[11] Hiyama E, Kamimura M, Motoba T, Yamada T and Yamamoto Y 2002 *Phys. Rev.* C **66** 024007
[12] Imai K, Nakazawa K, Tamura H *et al* 2006 *J-PARC proposal No.E07*
[13] Nakazawa K and Takahashi H 2010 *Prog. Theor. Phys. Suppl.* **185** pp 335-343
[14] Arai K, Descouvemont P, Baye D and Catford W N 2003 *Phys. Rev.* C **68** 014310
[15] Hiyama E, Kamimura M, Motoba T, Yamada T and Yamamoto Y 2000 *Phys. Rev. Lett.* **85** 270

Journal of Physics: Conference Series **436** (2013) 012081 doi:10.1088/1742-6596/436/1/012081

Few-nucleon systems interacting via non-local quark-model baryon-baryon interaction

Y Fujiwara[a] **and K Fukukawa**[b]

[a]Department of physics, Kyoto University, Kyoto 606-8502, Japan
[b]RIKEN Nishina Center, 2-1 Hirosawa, Wako, Saitama 351-0198, Japan

E-mail: `yfujiwar@scphys.kyoto-u.ac.jp`

Abstract. Low- and medium-energy nucleon-deuteron scattering less than the incident energy 65 MeV per nucleon is examined by using the quark-model nucleon-nucleon interaction fss2. The off-shell effect of this non-local interaction yields a part of the attractive effect given by the three-nucleon force in the standard description by meson-exchange potentials. The triton binding energy and the spin-doublet scattering length are well described by fss2 without introducing the three-nucleon force. The nucleon analyzing power is slightly improved from the results by the AV18 potential, but the so-called A_y puzzle in the low-energy region still remains. Similar reduction of the peak heights by the Coulomb force is also seen in the vector analyzing power of the deuteron, $iT_{11}(\theta)$. Other observables are rather similar to the predictions by meson-exchange potentials, including discrepancies of deuteron breakup differential cross sections from experiment for some particular kinematical configurations. The systematic KVI data of the $^1\mathrm{H}(\overrightarrow{d}, 2p)n$ scattering at E_d = 130 MeV are reasonably reproduced for breakup differential cross sections and deuteron analyzing powers.

1. Introduction

The QCD-inspired spin-flavor SU_6 quark model (QM) for the baryon-baryon interaction, developed by the Kyoto-Niigata group, has achieved accurate description of available nucleon-nucleon (NN) and hyperon-nucleon experimental data [1]. These QM baryon-baryon interactions are characterized by the nonlocality and the energy dependence inherent to the framework of the resonating-group method (RGM) for two-cluster systems composed of three-quark clusters. In the strangeness sector, the Pauli forbidden state exists as the result of the exact antisymmetrization of six quarks. The short-range repulsion of the baryon-baryon interaction is mainly described by the quark-exchange kernel, which gives quite different off-shell properties from standard meson-exchange potentials. The energy dependence of the interaction is eliminated by the standard off-shell transformation [2], utilizing the square-root inverse of the normalization kernel. This procedure yields an extra nonlocality, whose effect was examined in detail for the three-nucleon ($3N$) bound state and hypertriton [3]. We found a large triton binding energy by the QM NN interaction; namely, the deficiency of 350 keV, predicted by the model fss2, is almost half of the standard values 0.5 MeV - 1 MeV [4], given by the modern meson-exchange NN potentials. A similar situation is also found in the recent Faddeev-Yakubovsky calculation for the binding energy of the α-particle by fss2.

In this study, we apply our QM NN interaction fss2 to the neutron-deuteron (nd) and proton-deuteron (pd) scattering in the Faddeev formalism for systems of composite particles [5].

The Alt-Grassberger-Sandhas (AGS) equations [6] are solved in the momentum representation, using the off-shell RGM T-matrix obtained from the energy-independent renormalized RGM kernel [2]. The Gaussian nonlocal potential constructed from the fss2 is essentially used in the isospin basis [7]. The singularity of the NN T-matrix from the deuteron pole is handled in the Noyes-Kowalski method [8]. Another notorious moving singularity of the free three-body Green function is treated by the standard spline interpolation technique developed by the Böchum-Krakow group [9]. The Coulomb force in the pd scattering is approximately treated by the screened Coulomb force, by introducing a finite cutoff radius ρ [10]. We mainly use the channel-spin formalism with $S_c = 1 \times 1/2 = 1/2$ (doublet) + $3/2$ (quartet) channels, which is convenient to discuss the nucleon-deuteron (Nd) elastic scattering. The NN interaction up to $I_{\max} = 4$ is included in the present calculation, which leads to the nucleon incident energies discussed in this paper up to about $E_N \sim 65$ MeV.

2. Attractive features of the spin doublet Nd interaction in fss2

The spin-doublet ${}^2S_{1/2}$ state corresponds to the triton ground state and the Nd interaction in this channel is especially attractive because of the strong distortion effect of the deuteron. On the other hand, the spin-quartet ${}^4S_{3/2}$ Nd interaction is strongly repulsive owing to the Pauli principle. The $3N$ force is only influential in the $J^\pi = 1/2^+$ channel, and the Coulomb force is more important in the low-energy region. These features are clearly seen from the S-wave eigenphase shifts at $E_N = 3$ MeV, listed in Table 1. The fss2 predictions are very similar not to AV18 but to the AV18 plus Urbana (UR) $3N$ potentials (AV18+UR$3N$) [11] shown in the parentheses. Our nd and pd ${}^2S_{1/2}$ nuclear phase shifts are more attractive than AV18 by 3° - 5° at the energy range of $E_N = 1$ - 3 MeV. In the same energy range, the maximum difference from the AV18+UR$3N$ results is only 0.6°. This attractive behavior is correlated to the fact that fss2 predicts the almost correct triton binding energy. Table 1 also shows that our ${}^2S_{1/2}$ nuclear phase shift at E_p=3 MeV is not so attractive as to reproduce the result of the phase-shift analysis (PSA). This is apparently because the binding energy of ^{3}H is still insufficiently reproduced even by fss2.

In Table 2, the triton binding energy, spin doublet (${}^2a_{nd}$) and quartet (${}^4a_{nd}$) scattering lengths by fss2 are compared with some other extensive calculations using meson-exchange potentials and $3N$ forces [12]. Since the charge dependence is not included in our calculation, the effect is estimated form the Phillips line -0.686 fm/MeV for fss2, resulting in $\Delta({}^2a_{nd}) \sim 0.10 - 0.14$ fm. In the meson-exchange potentials, the experimental values of ${}^2a_{nd}$ and triton binding energy are reproduced only when the $3N$ force is included. The $3N$-force effect on the doublet scattering length ${}^2a_{nd}$ is more than 0.4 fm. The model fss2 almost reproduces the experimental values of the triton binding energy and ${}^2a_{nd}$ simultaneously without the $3N$ force.

Table 1. The nd and pd eigenphase shifts at $E_N = 3$ MeV, predicted by fss2. For the pd calculation, the cutoff Coulomb radius $\rho = 9$ fm is used. The corresponding parameters calculated by the Pisa group for the AV18 potential and AV18+UR$3N$ potentials (in the parentheses) are also listed for comparison [11].

Model	fss2 (nd)	AV18(+UR$3N$) (nd)	fss2 (pd)	AV18(+UR$3N$) (pd)	PSA
${}^2S_{1/2}$	149.2	144.7 (149.2)	152.8	147.8 (152.2)	155.15 ± 0.23
${}^4S_{3/2}$	-69.6	-69.9 (-69.7)	-62.5	-63.1 (-63.1)	-63.80 ± 0.11

Table 2. Comparison of the nd scattering lengths predicted by using fss2 ($I_{\rm max} = 4$) with other models [12] ($I_{\rm max} = 5$). In fss2, the charge dependence of the NN force is neglected. The heading NN denotes calculations using only the NN force, and NN+TM99 those including the Tucson-Melbourne 99 (TM99) 2π-exchange $3N$ force. The values of ${}^4a_{nd}$ are insensitive to the $3N$ force. The experimental values are taken from Ref. [13].

	$E_B(^3\mathrm{H})$ (MeV)		${}^2a_{nd}$ (fm)		${}^4a_{nd}$ (fm)
Model	NN	NN+TM99	NN	NN+TM99	NN(+TM99)
fss2	8.307	—	0.66	—	6.30
CD-Bonn 2000	8.005	8.482	0.925	0.569	6.347
AV18	7.628	8.482	1.248	0.587	6.346
Nijm I	7.742	8.485	1.158	0.594	6.342
exp.		8.482		0.65 ± 0.04	6.35 ± 0.02

3. *nd* and *pd* differential cross sections and polarization observables

Some examples for the elastic differential cross sections and polarization observables are shown in Fig. 1 for the nucleon incident energy $E_N = 5$ MeV. The solid curves represent the pd results with the Coulomb cutoff radius $\rho = 8$ fm, while the dashed curves nd without the Coulomb force.

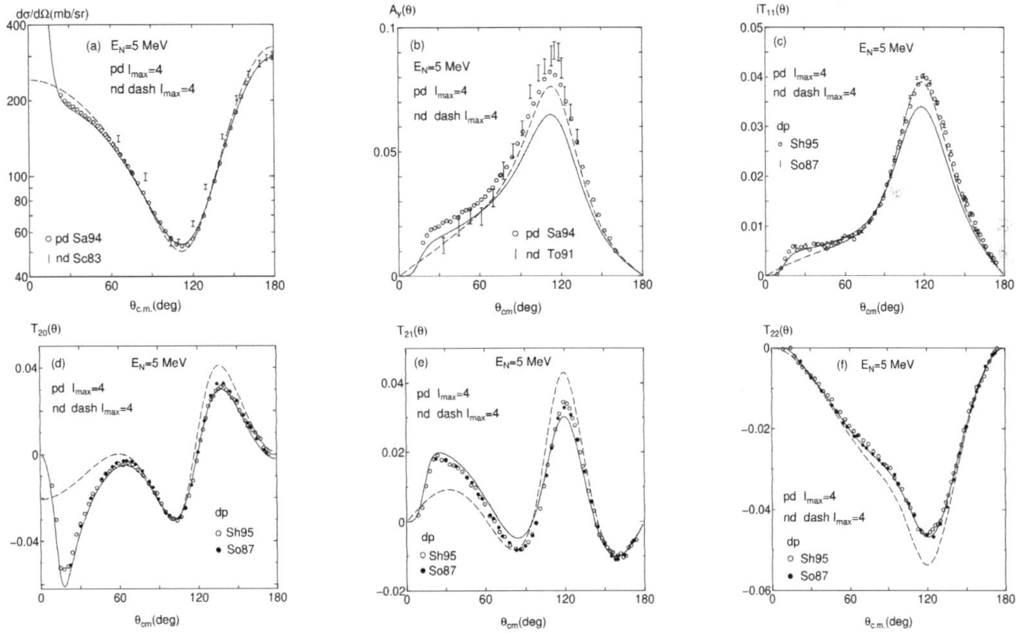

Figure 1. The elastic scattering observables at $E_N = 5$ MeV, compared with the nd and pd (or dp) experimental data; (a) the elastic differential cross section ($d\sigma/d\Omega$), (b) the nucleon vector analyzing power $A_y(\theta)$, (c) the deuteron vector analyzing power $iT_{11}(\theta)$, (d) the deuteron tensor analyzing power $T_{20}(\theta)$, (e) $T_{21}(\theta)$ and (f) $T_{22}(\theta)$. All the experimental data of the deuteron analyzing powers are for the dp scattering. See Ref. [14] for the experimental data.

Here, the Coulomb force is very important to reproduce spin observables in the whole angular region. When the energy becomes higher, the Coulomb effect is confined only to the forward angles. The experimental data are reasonably reproduced except for the vector analyzing power of the nucleon ($A_y(\theta)$) and that of the deuteron ($iT_{11}(\theta)$) in the low-energy region $E_N \leq 25$ MeV. These observables are very sensitive to the partial-wave truncation of the model space and to the Coulomb force, resulting in a large modification by a slight change of eigenphase shifts.

The long-standing A_y puzzle for the large discrepancy between the theory and experiment in the low-energy region $E_N \leq 25$ MeV still persists even in our calculations. In order to examine the discrepancy more quantitatively, we calculate the relative magnitude at the maximum point according to $\mathrm{RA}_{\mathrm{max}} = (A_y^{\mathrm{th}})_{\mathrm{max}}/(A_y^{\mathrm{exp}})_{\mathrm{max}}$, where $(A_y^{\mathrm{th}})_{\mathrm{max}}$ and $(A_y^{\mathrm{exp}})_{\mathrm{max}}$ are the maximum values of $A_y(\theta)$ in the theoretical predictions and the experimental data, respectively. In Fig. 2, we show $\mathrm{RA}_{\mathrm{max}}$ in the energy region $E_N \leq 19$ MeV, and compare them with the AV18 predictions in Ref. [15]. The deficiency of the $A_y(\theta)$ at the maximum point is about 15 – 20% in our case, which is smaller than 25 – 30% in the predictions by AV18. In our calculations, the A_y puzzle for the *pd* scattering is more serious than for the *nd* scattering. However, in the *pd* scattering, this discrepancy diminishes as the energy increases.

The energy dependence of the elastic differential cross sections at the minimum points (diffraction minima) is often discussed as Sagara discrepancy [16] and is very important to discuss the $3N$ force effect. In order to examine this quantitatively, we show in Fig. 3 the relative difference, which is defined by $\mathrm{RD}_{\mathrm{min}} = \left[(d\sigma/d\Omega)^{\mathrm{cal}}_{\mathrm{min}} - (d\sigma/d\Omega)^{\mathrm{exp}}_{\mathrm{min}}\right]/(d\sigma/d\Omega)^{\mathrm{exp}}_{\mathrm{min}}$ in the energy region $E_p < 23$ MeV. On the low-energy side $E_p < 18$ MeV, fss2 reproduces the minimum values of the *pd* differential cross sections at accuracy less than 2%. In Ref. [17], the Urbana IX (URIX) $3N$ potential lessens the cross sections at the diffraction minima, resulting in a good agreement with the experimental data. In our calculations, fss2 takes into account this attractive effect at the NN level. This result is in accordance with the good reproduction of the triton binding energy and *nd* doublet scattering length [18].

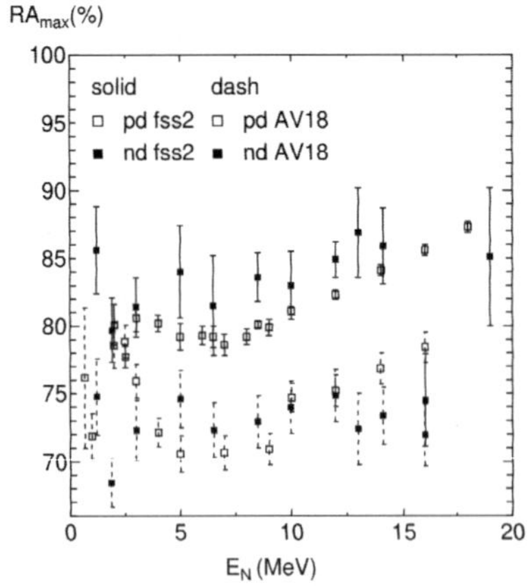

Figure 2. The theoretical to experimental ratio $\mathrm{RA}_{\mathrm{max}}$ for the maximum points of *pd* (open squares) and *nd* (filled squares) analyzing powers $A_y(\theta)$.

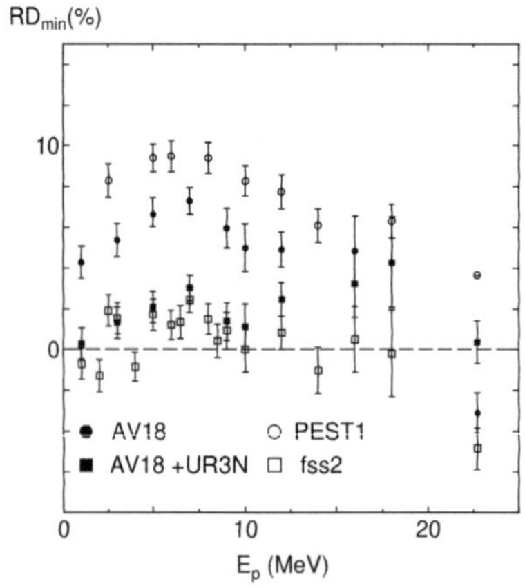

Figure 3. Deviation of the diffraction minimum from the experimental data measured by $\mathrm{RD}_{\mathrm{min}}$ (see the text.)

4. Deuteron breakup processes

Figure 4 shows the breakup differential cross sections for the reactions $d(p,2p)n$ and $d(n,2n)p$ at $E_N = 13$ MeV. The nd results are shown by the dashed curves and should be compared with the experimental data with error bars. We find that the Coulomb effect is rather small in these examples and our results are very similar to the predictions by meson-exchange potentials, given in Refs. [19, 20, 21, 22]. A slight overestimation of the peak in the quasi-free scattering (QFS) is reduced by the Coulomb effect and the agreement with the pd experimental data [23] or more recent one [24] is improved for $\rho = 16$ fm or 8 fm. The np final state interaction (FSI) peaks are well reproduced. In the collinear (COLL1, COLL2) and coplanar star (CST1) configurations, the nd data [25] agree well with theoretical predictions, but our CST1 result is at least 10 - 20% too small. The space star result is located just between the lower pd data and the higher nd data, which is the same feature as other predictions by meson-exchange potentials. This disagreement of breakup differential cross sections at $E_n = 13$ MeV is still an unsolved problem called the space star anomaly [25].

Figure 5 shows the systematic change of the deuteron breakup differential cross sections at $E_d = 130$ MeV for kinematical configurations with $\theta_1 = \theta_2 = 13°$ and $\phi_{12} = \phi_1 - \phi_2 = 20°$ - $180°$. For the pp final-state interaction with $\phi_{12} = 20°$, the choice $\rho = 8$ fm (dotted curve) is not good enough to reproduce the oscillatory structure, although a large peak in the no-Coulomb case (dashed curve) is strongly suppressed. The characteristic behavior of oscillation is only reproduced if ρ is taken to be sufficiently large like $\rho \geq 16$ fm. The analyzing power in the breakup reaction is even more sensitive to the treatment of the Coulomb force. (Not shown.) In most cases, the results with $\rho = 8$, 16 and 20 fm are rather similar to each other, but for some

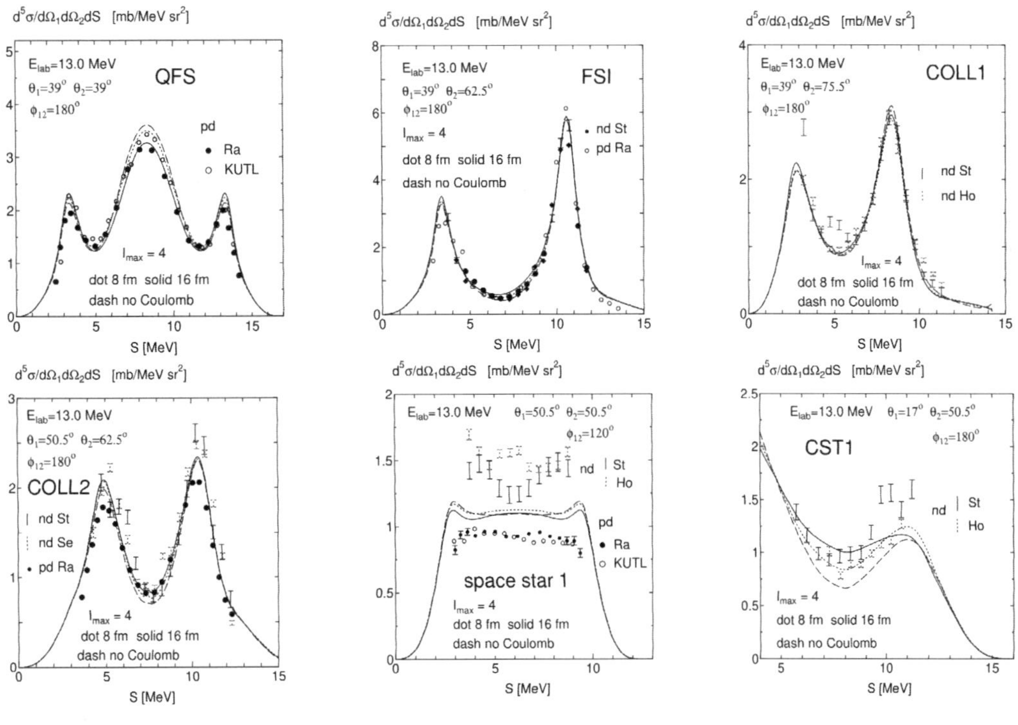

Figure 4. Breakup differential cross sections for the reactions $d(p,2p)n$ (solid curve) and $d(n,2n)p$ (dashed curve) at $E_N = 13$ MeV, compared with the experimental data [26].

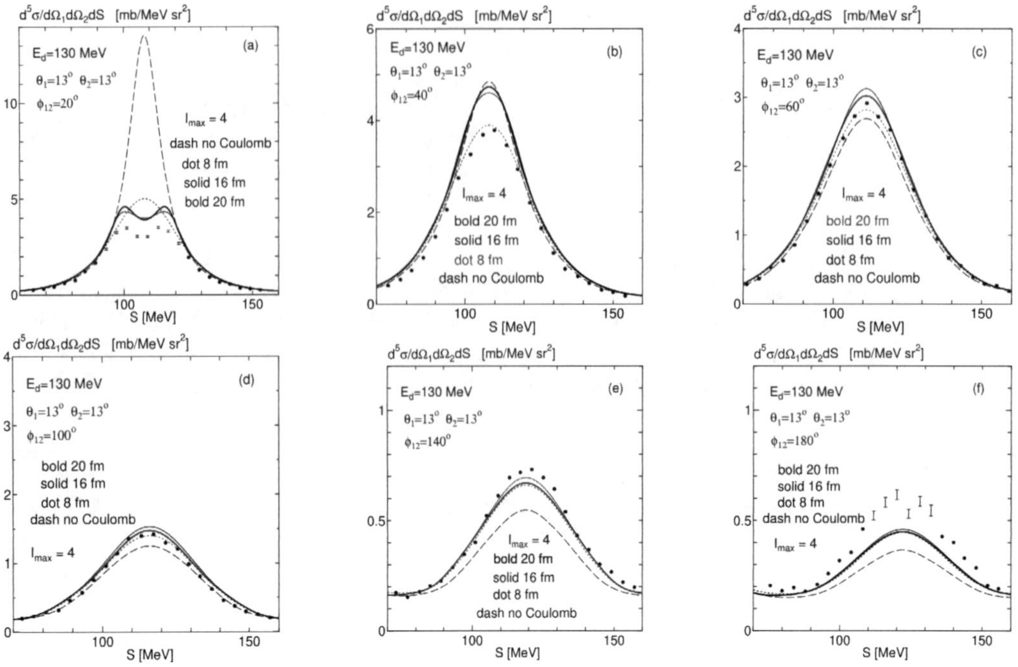

Figure 5. Breakup differential cross sections for the reaction $H(\overrightarrow{d},2p)n$, with the energy $E_d = 130$ MeV. The experimental data are taken from [27].

kinematical configurations with $\theta_1 \sim \theta_2$ and $\phi_{12} = 0°$ unpleasant oscillation appears for $\rho \geq 20$ fm. A larger ρ-value requires more partial waves and makes the solutions of AGS equations more singular. In spite of these limitations for the choice of large ρ, the results of $\rho = 8$ fm are rather stable and show that the Coulomb effect is not so large for $E_N = 65$ MeV except for the *pp* final-state interaction region.

5. Summary

In spite of some apparent disagreement between the theory and experiment in breakup observables, our QM NN interaction fss2 is still very successful in reproducing the essential features of almost all other experimental data in the three-nucleon system without reinforcing it with the three-body force. These include: 1) a nearly correct binding energy of the triton [3], 2) reproduction of the doublet and quartet S-wave scattering lengths, ${}^2a$ and ${}^4a$ [18], 3) low-energy differential cross sections of the *pd* elastic scattering up to $E_p \sim 35$ MeV at the diffraction minima [8], 4) improved maximum heights of the nucleon analyzing power $A_y(\theta)$ in the low-energy region $E_n \leq 25$ MeV, although still insufficient [14], 5) breakup differential cross sections with many kinematical configurations [26]. Many of these improvements are related to the sufficiently attractive nd interaction in the ${}^2S$ channel, in which the strong distortion effect of the deuteron is very sensitive to the treatment of the short-range repulsion of the NN interaction. In our QM NN interaction, this part is described by the quark exchange kernel of the color-magnetic quark-quark interaction. In the strangeness sector involving the ΛN and ΣN interactions, the effect of the Pauli repulsion on the quark level appears in some baryonic channels. It is therefore interesting to study $\Sigma^{\pm}$-deuteron scattering in the present framework to find the repulsive effect directly related to the quark degree of freedom.

Acknowledgments
This work was supported by a Grant-in-Aid for Scientific Research (C) from the Japan Society for the Promotion of Science (JSPS) (Grant No. 23540302), and by a Grant-in-Aid for the Global COE Program "The Next Generation of Physics, Spun from Universality and Emergence" from the Ministry of Education, Culture, Sports, Science and Technology (MEXT) of Japan. The numerical calculations were carried out on SR16000 at YITP in Kyoto University and on the high-performance computing system Intel Xeon X5680 at RCNP in Osaka University.

References
[1] Fujiwara Y, Suzuki Y and Nakamoto C 2007 *Prog. Part. Nucl. Phys.* **58** 439
[2] Suzuki Y, Matsumura H, Orabi M, Fujiwara Y, Descouvemont P, Theeten M and Baye D 2008 *Phys. Lett.* B **659** 160
[3] Fujiwara Y, Suzuki Y, Kohno M and Miyagawa K 2008 *Phys. Rev. C* **77** 027001
[4] Nogga A, Kamada H and Glöckle W 2000 *Phys. Rev. Lett.* **85** 944
[5] Fujiwara Y, Nemura H, Suzuki Y, Miyagawa K and Kohno M 2002 *Prog. Theor. Phys.* **107** 745
[6] Alt E O, Grassberger P and Sandhas W 1967 *Nucl. Phys.* B **2** 167
[7] Fukukawa K, Fujiwara Y and Suzuki Y 2009 *Mod. Phys. Lett.* A **24** 1035
[8] Fujiwara Y and Fukukawa K 2010 *Prog. Theor. Phys.* **124** 433
[9] Witala H, Cornelius T and Glöckle W 1988 *Few-Body Systems* **3** 123
[10] Fujiwara Y and Fukukawa K 2012 *Prog. Theor. Phys.* **128** 301
[11] Kievsky A, Rosati S, Tornow W and Viviani M 1996 *Nucl. Phys.* A **607** 402
[12] Witała H, Nogga A, Kamada H, Glöckle W, Golak J and Skibiński R 2003 *Phys. Rev. C* **68** 034002
[13] Dilg W, Koester L and Nistler W 1971 *Phys. Lett.* B **36** 208
[14] Fukukawa K and Fujiwara Y 2011 *Prog. Theor. Phys.* **125** 957
[15] Tornow W, Esterline J H and Weisel G J 2008 *J. of Phys.* G **35** 125104
[16] Sagara K, Oguri H, Shimizu S, Maeda K, Nakamura H, Nakashima T and Morinobu S 1994 *Phys. Rev. C* **50** 576
[17] Kievsky A, Viviani M and Rosati S 2001 *Phys. Rev. C* **64** 024002
[18] Fukukawa K and Fujiwara Y 2011 *Prog. Theor. Phys.* **125** 957
[19] Glöckle W, Witała H, Hüber D, Kamada H and Golak J 1996 *Phys. Rep.* **274** 107
[20] Deltuva A, Fonseca A C and Sauer P U 2005 *Phys. Rev. C* **72** 054004
[21] Ishikawa S 2009 *Phys. Rev. C* **80** 054002
[22] Kuroś-Żołnierczuk J, Witała H, Golak J, Kamada H, Nogga A, Skibinśki R and Gločkle W 2002 *Phys. Rev. C* **66** 024004
[23] Rauprich G, Lemaître S, Niessen P, Nyga K R, Reckenfelderbäumer R, Sydow L, gen Schieck H P, Witała H and Glöckle W 1991 *Nucl. Phys.* A **535** 313
[24] Kimura S, *et al.* 2012 *Few-Body Systems* Online First DOI 10.1007/s00601–012–0393–3
[25] Setze H R, *et al.* 1996 *Phys. Lett.* B **388** 229
[26] Fujiwara Y and Fukukawa K 2011 *Prog. Theor. Phys.* **125** 779
[27] Kistryn S, *et al.* 2006 *Phys. Lett.* B **641** 23

Journal of Physics: Conference Series **436** (2013) 012082 doi:10.1088/1742-6596/436/1/012082

Structure of Be hyper isotopes

M Isaka[1], **H Homma**[1] **and M Kimura**[1]

[1] Department of Physics, Hokkaido University, Sapporo 060-0810, Japan

E-mail: `isaka@nucl.sci.hokudai.ac.jp`

Abstract. The low-lying level structure of $^{10}_{\Lambda}$Be and $^{12}_{\Lambda}$Be was studied with the antisymmetrized molecular dynamics for hypernuclei (HyperAMD). It is found that the first excited state $1/2^+$ of ^{9}Be is shifted up by about 700 keV in $^{10}_{\Lambda}$Be. In $^{12}_{\Lambda}$Be, the parity-inverted ground state of ^{11}Be is reverted by adding the Λ hyperon. In this article, we discuss the reason for these structure changes and the deformation of $^{10}_{\Lambda}$Be and $^{12}_{\Lambda}$Be.

1. Introduction

We focus on the ground and low-lying sates of $^{10}_{\Lambda}$Be and $^{12}_{\Lambda}$Be, and discuss the changes of the level structure caused by the Λ hyperon. Since Be isotopes are composed of 2α cluster structure with surrounding neutrons, the ground and low-lying excited states have various deformation depending on the orbit of the extra neutron(s). For example, in ^{9}Be, the deformation of the first excited state $1/2^+$ is larger than that of the ground state $3/2^-$. In ^{11}Be, the ground state is $1/2^+$ which is inconsistent with ordinary shell-model-picture where the seven neutrons may have a $1/2^-$ state as the ground state. One of the reasons for the parity inversion is molecular-orbit structure of ^{11}Be [1]. In the $1/2^+$ state, the extra neutrons occupy the σ orbit around the 2α cluster structure and enhance the 2α clustering. The deformation of the $1/2^-$ state is smaller than the $1/2^-$ state, because the extra neutrons occupy the π orbit and reduce the clustering. By adding a Λ hyperon to such nuclei, it is expected that the binding energy of Λ is different depending on the deformation, and the level structure will be changed. In this article, we discuss the difference of the Λ binding energy depending on nuclear deformation and the changes of the level structure by Λ. To investigate it, we have applied the antisymmetrized molecular dynamics for hypernuclei (HyperAMD [2]) to $^{10}_{\Lambda}$Be and $^{12}_{\Lambda}$Be.

2. Theoretical Framework of HyperAMD

The Hamiltonian used in this study is given as,

$$\hat{H} = \hat{T}_N + \hat{T}_\Lambda - \hat{T}_g + \hat{V}_{NN} + \hat{V}_{\Lambda N} + \hat{V}_C \tag{1}$$

where $\hat{T}_N$, $\hat{T}_\Lambda$ and $\hat{T}_g$ are the kinetic energies of the nucleons, the Λ hyperon and the center-of-mass motion. We have used the Gogny D1S as an effective nucleon-nucleon interaction V_{NN}, that has been successfully applied to the stable and unstable nuclei. The YNG-NSC97f [3] and improved-YNG-NF [4] interactions have been employed as the Λ-nucleon interaction $V_{\Lambda N}$. Since the YNG interactions depend on the nuclear density through the Fermi momentum k_F, we adopted $k_F = 0.987$ fm^{-1} and $k_F = 0.973$ fm^{-1} for $^{10}_{\Lambda}$Be and $^{12}_{\Lambda}$Be, respectively. The Coulomb interaction is approximated by the sum of seven Gaussians.

The total wave function of the hypernuclear system is given as

$$|\Psi\rangle = \sum_n C^{(n)} |\phi_\Lambda^n(\vec{r}_\Lambda)\rangle \otimes \frac{1}{\sqrt{A!}} \det[\phi(\vec{r}_j)]. \tag{2}$$

Note that the single particle wave function of Λ is represented by a superposition of the Gaussian packets to describe various orbits of Λ. The configuration of the nucleons and the Λ hyperon is optimized through the frictional cooling method.

After the variational calculation, we perform the angular momentum projection and the generator coordinate method (GCM) calculation to obtain the excitation spectra.

3. Results and Discussions

3.1. Level structure of ${}^{10}_{\Lambda}$Be

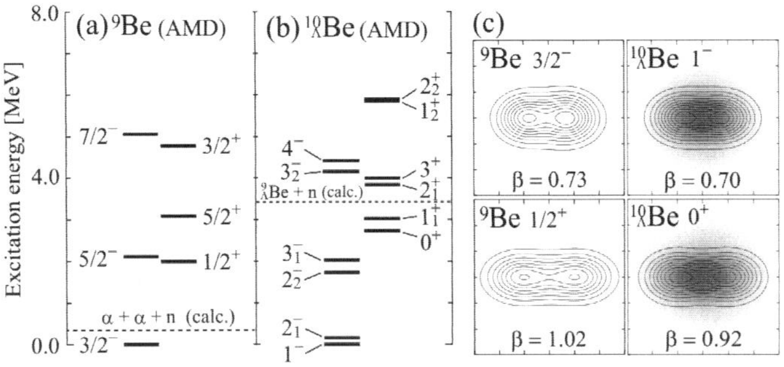

Figure 1. Calculated spectra of ^{9}Be (a) and ${}^{10}_{\Lambda}$Be (b). On panel (c), solid lines show the matter density distributions of the $3/2^-$ and $1/2^+$ states of ^{9}Be and the corresponding states in ${}^{10}_{\Lambda}$Be.

Figure 1(a) shows the excitation spectra of ^{9}Be obtained with AMD. The left panels of figure 1(c) show the intrinsic density distributions of the ground state $3/2^-$ and first excited state $1/2^+$. It shows that the deformation of each state is quite different. Namely, the deformation of the $1/2^+$ state is larger than that of the ground state. Indeed, the nuclear quadruple deformation parameters β listed in table 1 are quite different between the $3/2^-$ and $1/2^+$ states.

The calculated spectra of ${}^{10}_{\Lambda}$Be obtained with YNG-NSC97f [3] is presented in figure 1(b). We have obtained the negative and positive parity states of ${}^{10}_{\Lambda}$Be with Λ in s-orbit corresponding to them of ^{9}Be. Figure 1(a) and 1(b) show the shift up of the positive parity states by adding Λ. Indeed, the excitation energy of the $1/2^+$ state is shifted up by about 700 keV. This is because the Λ binding energies B_Λ are different between the $3/2^-$ and $1/2^+$ states. Here we define B_Λ as the sum of the expectation values of the Λ kinetic and ΛN potential energies. Table 1 shows that B_Λ of the $3/2^-$ state is larger than that of the $1/2^+$ state and this difference of B_Λ leads to the shift up of the $1/2^+$ state. This is because Λ coupled to the $3/2^-$ state with smaller deformation is more deeply bound than that coupled to the largely deformed $1/2^+$ state. This result is consistent with the predictions based on the four-body cluster model calculation [5, 6].

3.2. Parity reversion of the ${}^{12}_{\Lambda}$Be *ground state*

As shown in figure 2(a), the AMD model successfully describes the parity-inverted ground state of ^{11}Be. And the left panels of figure 2(c) present the nuclear density distribution of the $1/2^+$ and $1/2^-$ states. It shows that the deformation of the ground state $1/2^+$ is larger than that of the first excited state $1/2^-$. Since the deformation of the first excited state $1/2^-$ is smaller than that of the ground state $1/2^+$ and the excitation energy of the $1/2^-$ state is rather small, we can expect the parity reversion of the ${}^{12}_{\Lambda}$Be ground state.

Figure 2(b) shows the level structure of ${}^{12}_{\Lambda}$Be with Λ in s-orbit obtained with the improved YNG-NF [4]. It is found that the ground state parity of ${}^{12}_{\Lambda}$Be becomes negative, i.e. the parity

Table 1. Calculated binding and excitation energies B and E_x [MeV], and matter quadruple deformation β for the $3/2^-(1/2^-)$ and $1/2^+$ states of ^{9}Be (^{11}Be) and corresponding states $1_1^-(0^-)$ and 0^+ of $^{10}_{\Lambda}$Be ($^{12}_{\Lambda}$Be). Λ binding energy B_Λ [MeV], and expectation values of the Λ kinetic energy T_Λ [MeV] and ΛN potential energy $V_{\Lambda N}$ [MeV] are also shown in $^{10}_{\Lambda}$Be ($^{12}_{\Lambda}$Be).

	J^π	B	E_x	β	B_Λ	T_Λ	$V_{\Lambda N}$
^{9}Be	$3/2^-$	59.71	0.00	0.73			
^{9}Be	$1/2^+$	57.71	2.00	1.02			
$^{10}_{\Lambda}$Be	1^-	68.60	0.00	0.66	9.4	7.23	-16.63
$^{10}_{\Lambda}$Be	0^+	65.87	2.73	0.92	8.6	6.45	-15.04
^{11}Be	$1/2_1^-$	64.45	0.32	0.52			
^{11}Be	$1/2_1^+$	64.77	0.00	0.72			
$^{12}_{\Lambda}$Be	0_1^-	74.69	0.00	0.47	10.24	6.71	-16.93
$^{12}_{\Lambda}$Be	0_1^+	74.44	0.25	0.70	9.67	6.68	-16.42

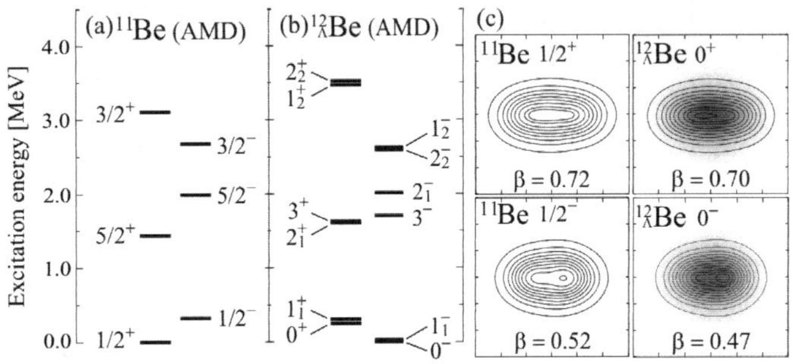

Figure 2. Same as figure 1 for ^{11}Be and $^{12}_{\Lambda}$Be.

reversion of the $^{12}_{\Lambda}$Be ground state will occur by adding Λ. As shown in table 1, B_Λ for the 0_1^- state is larger than the 0_1^+ state by about 500 keV. The difference of B_Λ mainly comes from in the ΛN potential energy $V_{\Lambda N}$, and it originates in the difference of the nuclear deformation.

4. Summary

In summary, we have applied an extended version of AMD for hypernuclei (HyperAMD) to $^{10}_{\Lambda}$Be and $^{12}_{\Lambda}$Be. The HyperAMD calculation predicts the positive and negative parity states of $^{10}_{\Lambda}$Be and $^{12}_{\Lambda}$Be with Λ in s-orbit. In $^{10}_{\Lambda}$Be, it is found that the first excited state $1/2^+$ of ^{9}Be is shifted up in $^{10}_{\Lambda}$Be. In $^{12}_{\Lambda}$Be, the parity-inverted ground state of ^{11}Be is reverted by Λ. These structure changes are mainly due to the difference of the Λ binding energy between the ground and first excited states, originated in the difference of their deformation. This is because Λ in s-orbit is more deeply bound with small deformation.

References

[1] Kanada-En'yo Y and Horiuchi H 2002 *Phys. Rev.* C **66** 024305
[2] Isaka M, Kimura M, Doté A and Ohnishi A 2011 *Phys. Rev.* C **83** 054304
[3] Yamamoto Y Private communication
[4] Hiyama E *et al* 1997 *Prog. Theor. Phys.* **97** 881
[5] Zhang Y Hiyama E and Yamamoto Y 2012 *Nucl. Phys.* A **881** 288
[6] Hiyama E and Yamamoto Y 2012 *Prog. Theor. Phys.* **128** 105

Some interesting and exotic applications of carbon-14 dating by accelerator mass spectrometry

A J T Jull[1,2]

[1] Department of Geosciences, University of Arizona, Tucson, Arizona 85721 USA
[2] Institute of Nuclear Research (ATOMKI), 4026 Debrecen, Hungary

Email: jull@email.arizona.edu

Abstract. There are many applications of ^{14}C dating and other measurements using accelerator mass spectrometry (AMS). In particular, applications to dating of archaeological samples and interesting artifacts are discussed. Other applications, such as to extraterrestrial materials such as lunar samples and meteorites show the broad range of topics that can be addressed with ^{14}C studies.

1. Introduction

Carbon-14 (or radiocarbon) is produced in the upper atmosphere by the action of secondary thermal neutrons on nitrogen, which has a relatively high cross section. It has a half-life of 5,730 years which means that amounts of ^{14}C produced cover the time scale of approximately 50,000 years, which is also the period of interest to archaeology and many other fields. Radiocarbon dating relies on the basic assumption that organic or inorganic materials in equilibrium with the production of ^{14}C in the atmosphere and its removal into the oceans, establish a consistent level of ^{14}C. This relies on the well-established radioactive decay equation, where the decay rate is equivalent to the number of atoms:

$$\frac{dN}{dt} = -\lambda N \quad (1)$$

Where N is the number of atoms, t is time and λ is the decay constant of the nuclide. When the animal or plant dies, it is removed from this equilibrium and so the level of ^{14}C decays according to equally recognizable equation:

$$\frac{N}{N_0} = e^{-\lambda t} \quad (2)$$

There are a vast number of applications of ^{14}C and we can touch on only a few in this paper. Originally, ^{14}C was counted by decay counting of the 156keV β- decay of the nuclide, however since 1977, this has been increasing supplanted by accelerator mass spectrometry (AMS), so that this now predominates. Indeed, AMS has become the method of choice for most measurements of longer-lived radionuclides, of which the most well-known is carbon-14 [1-4]. This method allowed much smaller samples of carbon to be measured than were previously possible using decay counting, because instead of waiting for decays to be counted, we can count the atoms directly. In practice, the measurement of samples of carbon of 0.05 to 0.5mg is now possible [4].

For historical reasons, radiocarbon ages are quoted in "years before present" (yr. BP), where "present" is 1950AD. In practice, the production rate of ^{14}C in the atmosphere has varied, so that it is important to calibrate raw radiocarbon ages derived from eqn. 2 to the "calendar age". This is achieved by using a calibration of radiocarbon ages against true age from tree-ring records up to 12,700 before present. Beyond that time, calibration is achieved by cross-referencing ^{14}C ages of corals to the U-Th ages of the same corals. This involves some assumptions about the constancy of the oceanic ^{14}C record, but we now have a calibration for the last 50,000 years, basically the entire range of radiocarbon dating [5]. An example of a calibration is shown in figure 1.

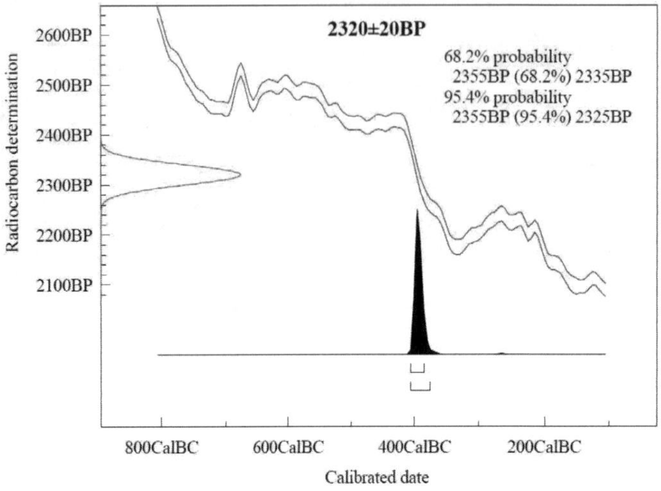

Figure 1: An example of a calibration of a radiocarbon age. The measured value is plotted on the vertical axis, with the 1σ error. The horizontal axis shows the combination of the analytical measurement with the calibration curve [5].

The example shown is very precise, since the curve is very steep at this point. However, an intersect with a flatter or varying part of the calibration curve can obviously give much wider error ranges for the resulting calibrated age.

Two other effects can change the ^{14}C in the atmosphere. Fossil-fuel burning has raised the level of CO_2 in the atmosphere from 280ppm in the 18th century to almost 400ppm today. This ^{14}C-free carbon added to the atmosphere dilutes the original signal. After 1950AD, we have an opposite effect on the radiocarbon curve. There is a large increase due to the atmospheric testing of nuclear weapons, which raised the atmospheric value in the northern hemisphere to 1.8 times the pre-bomb value. Since this testing mainly ceased after 1963, with some minor exceptions, the level in the atmosphere has now decreased to about 1.04 times the pre-bomb value. In this paper, we will review some of the basics of the method, and then give some examples of applications, that highlight the usefulness of these measurements to a wide variety of topics.

2. Basics of the AMS Method

1.1 Basic principles

The basic principles of AMS have been reviewed in many publications [1-4], so we will highlight only some brief examples here. For radiocarbon measurements, developments over the last 15 years have been reducing both the voltage and size requirements, so that nowadays, radiocarbon AMS may be conducted over a wide range of energies. However, the basic design of the machine remains the same.

Journal of Physics: Conference Series **436** (2013) 012083 doi:10.1088/1742-6596/436/1/012083

Figure 2: The Cs-sputter ion source in the Arizona 3MV AMS machine.

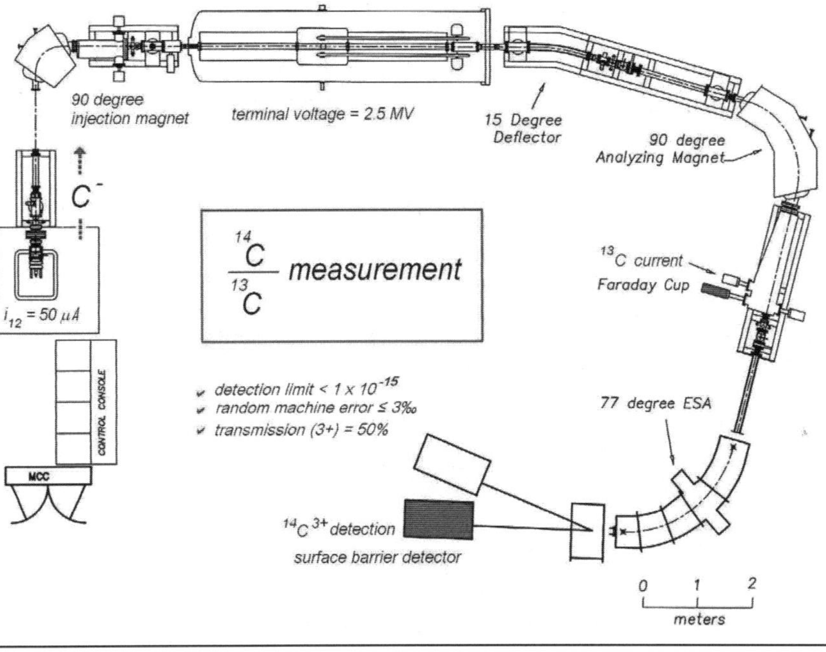

Figure 3: Layout of the 3MV National Electrostatics Corporation machine at the University of Arizona. The ion source is on the left-hand side, and the diagram shows the location of the various components, including the injection magnet, accelerator (terminal voltage at 2.5MV), and electrostatic and magnetic analysis at the high-energy end. In this case, the ions exiting the machine are mainly in the 3+ charge state.

The basic layout consists of the following components:

a. An ion source, that produces C^- ions by sputtering with Cs^+ ions onto a target, which is usually graphite, but can also be onto a Ta frit through which CO_2 gas is released. The gas is then effectively converted to graphite, as well as being ionized. This is often a complex mechanical device, which requires constant maintenance. The amount of material required varies but in general is between about 0.05 to 0.5mg C. A photograph of the ion source on the Arizona 3MV machine is shown in figure 2.

b. An injection magnet, which selects mass 14.
c. An electrostatic device (generally called the "bouncer") attached to the injection magnet chamber to allow selection of masses 12, 13 and 14 by applying a retarding potential to the magnet chamber.
d. An accelerator which accelerates the C^- ions, passes them through a gas (or sometimes a solid) stripper, which changes the charge to positive due to electronic collisions. Generally, for higher-voltage machines, the 3+ charge state is selected. For machines operating below about 500kV, C^+ is selected. If the singly-positive charged ions are used, care must be taken that molecular ions do not pass into the detector.
e. An analysing magnet at the high-energy end of the machine, that selects mass 14. Masses 12 and 13, if injected, are deflected more than mass 14 and hence are directed to Faraday cups, where the current can be measured.
f. An electrostatic analyser, which defines E/Q.
g. A detector, which can consist of as gas-filled cell, or a solid-state surface-barrier detector, where the ^{14}C ions are stopped and generate an electric pulse. These pulses are counted in the computer system and analysed.

An example of a typical system is shown in figure 3. Current technology has developed to the point where a machine can operate at 200kV or less [6]. Machines of this design are closer in size to a conventional mass spectrometer than older designs, which filled a room. A picture of such a small machine, the 200kV MICADAS system in Debrecen is shown in figure 4.

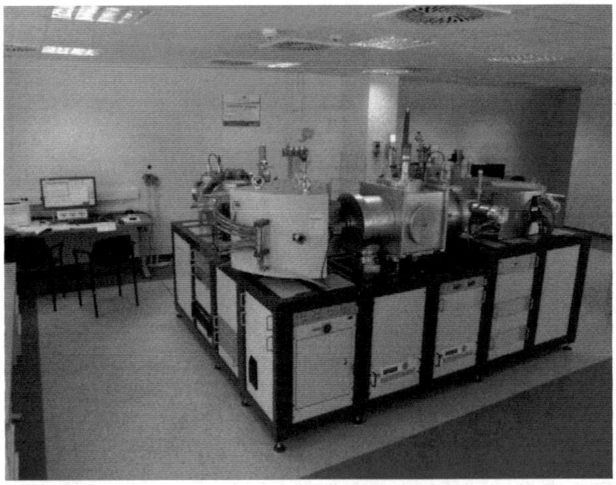

Figure 4: The 200kV MICADAS accelerator mass spectrometer installed at ATOMKI in Debrecen [7]. In contrast to figure 3, this machine is about 2m across.

1.2 Chemistry

As equal in importance as the AMS machine, is the sample-preparation chemistry. Samples such as wood and charcoal can be pretreated with a relative simple acid-alkali-acid treatment, but other samples such as bones and textiles require more complex pretreatments [8]. Bone samples are cleaned with acid, and then the collagen is extracted by a weak acid treatment. The remaining protein, or collagen, can then be either combusted directly or hydrolysed to its component amino acids. Textiles are often treated to a sequence of organic solvents, in order to remove any organic contaminants that might have been added to the sample, for example for conservation purposes. All samples are converted to CO_2 by combustion, or in the case of carbonates by acid hydrolysis. The CO_2 is converted to graphite over an Fe catalyst and pressed into a target holder for AMS analysis. In some laboratories, CO_2 can be injected directly into the ion source.

Journal of Physics: Conference Series **436** (2013) 012083 doi:10.1088/1742-6596/436/1/012083

2. Some examples of dating and other radiocarbon measurements.

It would be a vast task to try to summarize all the uses of AMS radiocarbon dating, so we will highlight just a few, to give some idea of the range of applications.

2.1 Unusual documents and textiles.

Some of the most interesting applications to radiocarbon dating involve sometimes controversial archaeological finds, ancient documents and textiles. At the Arizona laboratory, we have undertaken many such measurements, including dating of the Shroud of Turin [9], Dead Sea Scrolls [10,11], the Vinland Map [12], the Gospel of Judas and most recently, an unusual document called the Voynich manuscript [13]. An interesting feature of all these studies was the high interest of some sections of the public in these measurements. The Shroud of Turin, for example, clearly dates to the 14th century, with a calibrated radiocarbon age of 1260-1390AD [9], yet controversy still surrounds this object.

3.1.1 Dead Sea Scrolls

A less controversial example is the study of the Dead Sea Scrolls. These interesting documents, written on parchment or papyrus, contain detailed copies of books of the Old Testament, other religious commentaries on books of the bible of an esoteric nature, as well as more mundane business documents, such as financial transactions. These documents date from the mid-2nd century BC, for example the Book of Isaiah shown in figure 4, to the first centuries AD.

Figure 5: The Book of Isaiah
This parchment is on display in the Shrine of the Book in Jerusalem. It dates to 250-130BC according to radiocarbon dating, or 150-125BC from studies of the writing style [10,11].

3.1.2 Voynich Manuscript

The intriguing Voynich manuscript is currently in the possession of the Beineke Rare Books Library at Yale University in the USA. The document was known since about the 16th century, since it was at one time in the possession of the Holy Roman Emperor Rudolph II. It was acquired from the widow of Wilfred Voynich, who had apparently purchased the manuscript from the Jesuit College of Frascati, near Rome, about 1912. The document is enigmatic in that it is written in an indecipherable language, which is assumed to also be encrypted. Many have tried to decipher this code, but none have succeeded. The document consists of strange drawings of astronomical features, botanical drawings of unknown plants and ritual bathing. The purpose of the document remains unclear. Greg Hodgins at the University of Arizona worked with the Beineke Library to get the manuscript dated, which turned out to be in the 15th century, to 1404-1438AD [13].

2.2 Archaeology

The field of archaeology has relied heavily on radiocarbon dating since its inception in the early 1950's. Many radiocarbon dates are produced which find their way into archaeological reports of various kinds. It would be difficult to narrow down to a few examples. Recently, Ramsey et al. [14] have attempted to correlate radiocarbon dates on known materials from the tombs of Egyptian pharaohs, which gave good agreement for the Middle and New Kingdoms, but the dates for the Old Kingdom were generally older than archaeological estimates. Similarly, Friedrich et al. [15] dated an olive branch found in the destruction layer of the eruption of Santorini, generally assumed to be about

1550BC, but their radiocarbon dates came almost a century older. The reasons for these offsets are not yet established.

In the New World, much interest is focused on the arrival of early humans, which is considered to be sometime during the last Ice Age period, which ended about 11,000 years ago. A famous site close to Tucson, Arizona, known as the Murray Springs site [16] is well dated to 10,900±50 radiocarbon years before present (BP). At this location, the layer dated overlies the last evidence for mammoths in this region at a site where humans clearly killed the mammoths. This is clear evidence for humans, since arrow points are found here, and also the interaction of humans with the mammoths, that became extinct at the end of the last Ice Age. When we do the calibration [5], as discussed earlier, this corresponds to an absolute age range of 12930-12820 years before present. Recently, there is some evidence that humans arrived before this time [17], but this is still an on-going controversy. Jull and Burr [18] summarize some recent studies on this subject.

2.3 Meteorites

Another completely different approach is to use ^{14}C to study the times that meteorites have resided on the surface of the Earth. Most meteorites are stony-irons, they contain silicate and iron together. They are irradiated in space at approximately 1600 times the cosmic-ray exposure level at the surface of the Earth. The cosmic-ray exposure produces a cascade of secondary particles and ^{14}C and other radionuclides are produced by spallation reactions. This is a different production mechanism than in the atmosphere of the Earth. After the meteorite falls to the ground, it is shielded from most (but not all) cosmic radiation, and so the ^{14}C that has built up in space will decay with the characteristic half-life. We can then use this signal to estimate the residence time of the meteorites on the ground. Interestingly, the time-scale for the survival of meteorites in many desert locations is comparable to the radiocarbon time-scale [19]. This method has been used extensively to try to understand the infall rates of meteorites and their weathering in the terrestrial environment [19,20].

2.4 In situ ^{14}C production in rocks

A new method which uses AMS measurements is the use of cosmogenic radionuclides produced in situ at the surface of the Earth by interactions of cosmic radiation with the silicate in surface rocks [21]. Measurements of radionuclides produced *in situ* in the surfaces of rocks, soils can be used to estimate exposure age and erosion rates. This is particularly true for the nuclides ^{10}Be, ^{14}C, ^{26}Al and ^{36}Cl. These methods can be applied to landscape evolution variability, such as weathering, sediment transport and soil development, retreat and advance of glaciers, tectonics, volcanic flows, meteorite impacts and other phenomena. One recent example using ^{14}C is a study of rocks exposed by the recent retreat of the Rhone Glacier in Switzerland, that shows that these rocks were exposed for about 6,500 years within the last 10,000 years, even though they were covered by ice until recently [22]. This indicates that this glacier was small in the early Holocene period, which was warmer than current conditions.

2.5 ^{14}C from solar flares or supernovae?

A final example of an interesting application of ^{14}C is in tree rings. Although we use the tree ring chronology to establish a calibration curve [5], the fluctuations in the ^{14}C in tree rings of different age also show variations from climatic events, changes in the CO_2 in the atmosphere and changes in cosmic-ray flux. We only discuss one very interesting event here, which was shown recently by Japanese scientists [23]. In this study, an increase in ^{14}C in wood from AD 774-775 was discovered which does not have any explanation other than cosmic-ray effects. Hence, these authors proposed this signal is caused by intense solar flares or a nearby supernova. Further investigations from other sources of wood will determine if this is a global or local signal. In a completely different investigation, the effects of solar-cosmic rays on the surface of the Moon give us an excellent record of the integrated flux of solar radiation from integrating fluxes deduced from radionuclides of different half-life. From this record, it can be deduced that the solar flares could not have exceeded a total flux of ~$5x10^{13}$ p/cm^2 in the last 20,000 years. The tree-ring record over the last 7,000 years constrains this further to about $3x10^{11}$p/cm^2[24].

3. Conclusion

Carbon-14 is a very useful radionuclide which has extensive applications to dating of objects and as a tracer. This paper only presents a small aspect of the many diverse applications of this unique and versatile isotope.

Acknowledgements

I am grateful to the organizers of the Cluster Conference for their invitation to present something a little different in this meeting. I am also grateful to my hosts at ATOMKI for their gracious hospitality and to the Hungarian-American Fulbright Commission for partial support of my visit in Debrecen. The work in Tucson was partially supported by an equipment grant from the National Science Foundation.

References

[1] Jull A J T 2007. Radiocarbon Dating: AMS Method. In *Encyclopedia of Quaternary Science* ed Elias S (Amsterdam: Elsevier) pp. 2911-2918
[2] Kutschera W 2005 *Int. J. Mass Spectrom.* **242** 145
[3] Fifield L K 1999 *Reports on Progress in Physics* **62** 1223
[4] Tuniz C, Bird J R, Fink D and Herzog G F 1998 *Accelerator Mass Spectrometry: Ultasensitive Analysis for Global Science* (Boca Raton, FL: CRC Press).
[5] Reimer P, Baillie M G L, Bard E, Bayliss A, Beck J W, Blackwell P G, Bronk Ramsey C, Buck C E, Edwards R L, Friedrich M, Grootes P M, Guilderson T P, Hajdas I, Heaton T J, Hogg A G, Hughen K A, Kaiser K F, Kromer B, McCormac F G, Manning S W, Reimer R W, Richards D A, Southon J R, Talamo S, Turney C S M, van der Plicht J and Weyhenmeyer C 2009 *Radiocarbon* **51** 1111
[6] Wacker L, Bonani G, Friedrich M, Hajdas I, Kromer B, Nemec M, Ruff M, Suter M, Synal H-A and Vockenhuber C 2010 MICADAS: Routine and high-precision radiocarbon dating. *Radiocarbon* **52** 252
[7] Molnár M, Rinyu L, Veres M, Seiler M, Wacker L and Synal H-A 2013 *Radiocarbon* **55** in press.
[8] Jull A J T, Burr G S, Beck J W, Hodgins G W L, Biddulph D L, Gann J, Hatheway A L, Lange T E and Lifton N A 2006. Application of accelerator mass spectrometry to environmental and paleoclimate studies at the University of Arizona. In *Radionuclides in the Environment* ed P. Povinec and J. A. Sanchez-Cabrera (Amsterdam: Elsevier) pp. 3-23
[9] Damon P E, Donahue D J, Gore B H, Hatheway A L, Jull A J T, Linick T W, Sercel P J, Toolin L J, Bronk C R, Hall E T, Hedges R E M, Housley R, Law I A, Perry C, Bonani G, Trumbore S, Wölfli W, Bowman S G E, Leese M N and Tite M S 1989 *Nature* **337** 611
[10] Jull A J T, Donahue D J, Broshi M and Tov, E 1995 *Radiocarbon* **37** 11
[11] Bonani G, Ivy S, Wölfli W, Broshi M, Carmi I and Strugnell, J 1992 *Radiocarbon* **34** 843
[12] Donahue D J, Olin J S and Harbottle G 2002 *Radiocarbon* **44** 45
[13] Hodgins G W L 2011 Forensic investigation of the Voynich manuscript. Presented at Conference: Voynich 100. Frascati, Italy. May 2011
[14] Ramsey C B, Dee M W, Rowland J M, Higham T F G, Harris S A, Brock F, Quiles A, Wild E M, Marcus E S and Shortland A J 2010 *Science* **328** 1554
[15] Friedrich W L, Kromer B, Friedrich M, Heinemeier J, Pfeiffer T and Talamo S 2006 *Science* **312** 548
[16] Haynes C V 2007 *Murray Springs: A Clovis site with multiple activity areas in the San Pedro Valley, Arizona.* Anthropological Papers of the University of Arizona, no. 71. (Tucson: University of Arizona Press)
[17] Holliday V T 2009 Geoarchaeology and the search for the first Americans. *Catena* **78** 310

[18] Jull A J T and Burr G S 2013. Radiocarbon and Archaeology. In *Treatise of Geochemistry* ed K. K. Turekian and H. Holland (Amsterdam: Elsevier) in press.
[19] Jull A J T 2006 Terrestrial ages of meteorites. In *Meteorites and the Early Solar System II* ed D. Lauretta and H. Y. McSween Jr. (Tucson: University of Arizona Press) pp. 889-905
[20] Bland P A, Bevan A W R and Jull A J T 2000 Quarter. Res. **53** 131
[21] Gosse J and Phillips F M 2001 *Quarter. Sci. Rev.* **20** 1475
[22] Goehring B H, Schaefer J M, Schluechter C, Lifton N A, Jull A J T, Finkel R C, Akçar N and Alley R B 2011 *Geology* **39** 679
[23] Miyake F, Nagay K, Masuda K and Nakamura T 2012 *Nature* **486** 240
[24] Jull A J T, Cloudt S, Donahue D J, Sisterson J M, Reedy R C and Masarik J 1998 *Geochim. Cosmochim. Acta* **62** 3025.

Interaction of electromagnetic fields and atomic clusters

S Bubin, A G Russakoff and K Varga

Department of Physics and Astronomy, Vanderbilt University, TN-37240, Nashville, USA

E-mail: kalman.varga@vanderbilt.edu

Abstract. In the framework of real-time real-space time-dependent density functional theory complemented with Ehrenfest molecular dynamics we have studied the response of nanostructures to intense femtosecond laser pulses. Examples of applications include laser desorption of hydrogen from graphene and Coulomb explosion of hydrocarbon molecules.

1. Introduction

The basic idea of cluster models is to describe a complex physical system by using a simpler, physically motivated selection of the degrees of freedom. This approach has been successfully used not only in nuclear physics, but in many other areas of research including atomic and molecular physics and quantum chemistry. In nuclear physics, cluster models are used to build basis functions by separating the internal degrees of freedom of groups of nuclei and the wave functions of relative motions between clusters. Using this idea, we have developed a multidomain framework for electronic structure calculations of atomic clusters and nanostructures [1]. In this approach the basis functions of the system are generated from sets of cluster wave functions of groups of atoms. Many other ingredients of this approach, e.g. the use of Taylor time propagators of complex absorbing potentials is familiar and has been developed in nuclear physics as well.

In this contribution, we will present our study of electron and nuclear dynamics induced by strong laser pulses in the framework of the time-dependent density functional theory in real-time and real-space [2, 3, 4, 5, 6, 1]. With the advance of time-dependent probes [7] the investigation of the dynamical behaviour of matter at the nanoscale has become a very important research direction [8, 9, 10, 11, 12, 13, 14, 15, 16]. Comprehensive knowledge of the dynamic behaviour of electrons and ions in condensed-matter systems is pertinent to the development of many modern technologies, such as semiconductor and molecular electronics, optoelectronics, information processing and photovoltaics.

Density functional theory (DFT) is now well established as one of the most efficient methods for computing electronic ground state properties. In its original formulation, DFT only applies to the electronic ground state. Runge and Gross (RG) [17] generalized DFT to time-dependent (TD) systems. According to the RG theorem, for any given initial state of a many-electron system, a TD potential acting on it is uniquely determined by the subsequent time evolution of the one-electron density. Using this theorem, it is possible to formally establish a TD Kohn-Sham (KS) equation from which various one-particle properties of the system can be obtained as functions of time. The resulting theoretical framework is usually referred to as time-dependent density functional theory (TDDFT).

TDDFT has been successfully used in various time-dependent quantum mechanical simulations. The most important applications of the TDDFT approach are (i) non-perturbative calculations with systems in intense laser fields [18, 19, 20], (ii) calculations of optical response, dielectric functions and electronic transitions [21, 22, 23, 24, 25, 26, 27, 28, 29, 30], (iii) calculation of electronic excitations [31, 32, 33], and (iv) time-dependent transport calculations [34, 35, 36].

2. Method

The electron dynamics in our simulations is modelled within real-space real-time TDDFT using the adiabatic local density approximation (ALDA) with the parameterization of Perdew and Zunger [37]. Core electrons are represented by norm-conserving Troullier–Martins pseudopotentials [38]. At the first stage, the initial state of the system is prepared by performing the ground state DFT calculation. Next, the time-dependent Kohn–Sham orbitals, ψ_k, are determined by solving the equation

$$i\hbar\frac{\partial\psi_k(\mathbf{r},t)}{\partial t} = H\psi_k(\mathbf{r},t), \quad k = 1,\ldots,N_{\text{orb}} \tag{1}$$

with the Hamiltonian,

$$H = -\frac{\hbar^2}{2m}\nabla_{\mathbf{r}} + V_{\text{H}}[\rho](\mathbf{r},t) + V_{\text{XC}}[\rho](\mathbf{r},t) + V_{\text{ext}}(\mathbf{r},t), \tag{2}$$

where ρ is the electronic density,

$$\rho(\mathbf{r},t) = 2\sum_{k}^{N_{\text{orb}}} |\psi_k(\mathbf{r},t)|^2, \tag{3}$$

and V_{H} and V_{XC} are the Hartree and exchange-correlation potential, respectively. V_{ext} is the external potential, which includes the potential due to the ions, V_{ions}, and the explicitly time-dependent potential due to the laser electric field. In the dipole approximation, $V_{\text{laser}}(\mathbf{r},t) = \mathbf{r}\cdot\mathbf{E}(t)$. Assuming polarization along the x-axis, the electric field of the laser can be written as

$$\mathbf{E}(t) = \mathbf{e}_x E_{\text{max}} \exp\left[-\frac{(t-\tau)^2}{2a^2}\right]\sin(\omega t). \tag{4}$$

Here parameters a, τ, and E_{max} define the width, the position of the centre, and the maximum electric field amplitude of the Gaussian envelope respectively, while ω is the frequency of the laser.

The time-propagation of electronic orbitals and density is achieved through the action of the time-evolution operator,

$$U(0,t) = \mathcal{T}\exp\left[-\frac{i}{\hbar}\int H(\mathbf{r},t')dt'\right], \tag{5}$$

where $\mathcal{T}$ denotes time-ordering. In practice, the above expression for $U(0,t)$ is split into a product of multiple time-evolution operators, each corresponding to a short time step Δt,

$$U(0,t) = \prod_q U(t_q, t_q+\Delta t), \quad t_q = q\Delta t, \tag{6}$$

so that the Hamiltonian at time t_q remains nearly commutative with the Hamiltonian at time $t_q + \Delta t$. In all calculations performed in this work we used a time step of 0.72 attoseconds, which we found to provide an acceptable balance of speed, stability, and accuracy for the time-propagation. The actual value of the time step was chosen based on our experience with other hydrocarbons subjected to laser pulses of similar intensity and duration where we monitored the conservation of the total charge[1] and the convergence of the results with decreasing the time

[1] Then the imaginary absorbing potential was switched off. See later.

step. To represent $U(t_q, t_q + \Delta t)$ we employed the fourth order Taylor expansion,

$$\psi_k(\mathbf{r}, t_q + \Delta t) \approx \sum_{n=1}^{4} \frac{1}{n!} \left(-\frac{i\Delta t}{\hbar} H(\mathbf{r}, t_q) \right)^n \psi_k(\mathbf{r}, t_q). \tag{7}$$

In real-space DFT and TDDFT the Kohn–Sham orbitals are represented on discrete points in real space (usually organized in uniform rectangular grids). This brings certain advantages to calculations of sophisticated electron dynamics in the case when a strong laser pulse is applied. The dynamics may involve a significant variation of the electron density within a large volume as well as ionization and fragmentation of the system. It is generally not known in advance how this process occurs. The grid representation of the orbitals allows one to describe an essentially arbitrary time-evolution, as long as the simulation cell (i.e. the area of real space covered by the numerical grids) is sufficiently large. The accuracy of such a representation is easily controlled by a single parameter – the grid spacing. In our calculations we used uniform grids with x-, y-, and z-spacing of 0.25 Å.

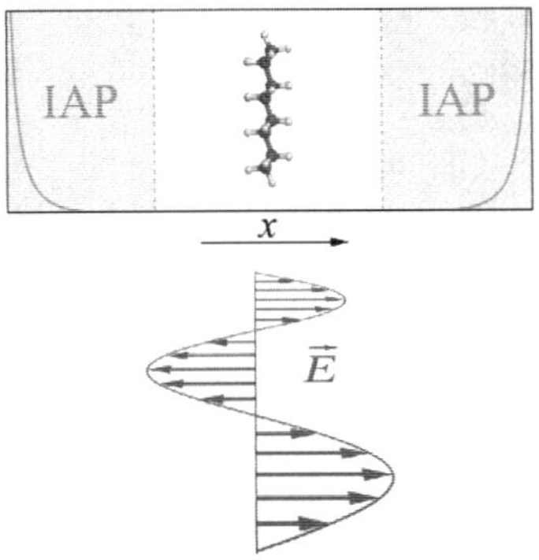

Figure 1. Schematic view of the simulation cell setup: the studied system is in the middle of the box, while the imaginary absorbing potential (IAP) acts on the left- and right-hand side and reaches its maximum at the walls. Laser polarization is along the x-axis.

To allow long time evolution in the calculations with real-space grids it is necessary to negate the effects of zero-boundary conditions, which causes an unphysical reflection of the wave function from the walls of the simulation cell. Given typical electron velocities and a length of the simulation cell that exceeds the system size by 5-10 Å in all directions, it only takes a few femtoseconds for the electrons to reach the boundaries. This problem is traditionally solved by using an imaginary absorbing potential (IAP) or a real mask-function (a concept closely related to IAP). In our calculations we employed an imaginary absorbing potential in the form proposed by Manolopoulos [39] with an additional softening parameter to suppress extremely large IAP magnitudes at the boundaries, which can cause numerical instabilities in grid-type approaches. Since the momentum of the electrons escaping from the system lies primarily in the x-direction, the IAP in our calculations was chosen such that it only had nonzero value in the region of space next to the x-boundaries. A schematic of this set up is shown in Fig. 1.

In the middle of the simulation cell, which contains the studied system, the IAP is zero by definition and does not influence the system in any way. While the ponderomotive amplitude for some of the laser intensities used in our calculations well exceeded the length of the IAP-free region, our previous experience with calculations having a similar computational setup and comparable laser pulses showed that one does not need to have an extremely large simulation cell to obtain converged results.

The ionic motion in this work is simulated by means of the Ehrenfest approach [40], in which the forces on ions are computed as the derivatives of the expectation value of the total electronic energy with respect to ionic positions, i.e.

$$M_i \frac{d^2\mathbf{R}_i}{dt^2} = \sum_{j \neq i}^{N_{\text{ions}}} \frac{Z_i Z_j (\mathbf{R}_i - \mathbf{R}_j)}{|\mathbf{R}_i - \mathbf{R}_j|^3} - \nabla_{\mathbf{R}_i} \int V_{\text{ion}}(\mathbf{r}, \{\mathbf{R}\}) \rho(\mathbf{r}, t) d\mathbf{r}. \tag{8}$$

In the above classical equation of motion M_i is the mass of the ith ion and Z_i is its valence charge (the charge of the atom minus the charge of the core used in the pseudopotential). The set of equations (8) for $i = 1 \ldots N_{\text{ions}}$ is integrated simultaneously with the set of time-dependent Kohn-Sham equations (1), although the integration time step for (8) may be chosen to be somewhat longer. To integrate equation (8) the well known Verlet algorithm is used.

3. Applications

In this section we present a few selected applications of the computational approach.

3.1. Graphene and graphane fragments in high-intensity laser pulses

An understanding of the interaction of intense, ultrashort laser pulses and graphene is a key prerequisite for both laser patterning and cleaning, as well as for applications of graphene in optical devices. Laser-induced bond breaking, defect formation, and damage mechanisms have been intensively investigated since the advent of high power laser pulses [41, 42]. Unlike their long-pulse counterparts, material processing with ultrashort-pulse lasers have negligible heat diffusion effects, minimal plasma absorption, and require smaller laser fluences for processing, making them ideal for achieving high spatial resolution [43, 44]. The effect of the laser pulse on the material strongly depends on the absorbed energy, which is determined by the material properties, as well as on the strength, intensity, shape, and duration of the pulse.

A general illustration of the hydrogen desorption process in a graphane fragment is given in Fig. 2. As the oscillation of the electric field grows, the amplitude of the electron density oscillations grows as well. Eventually, when it becomes large enough, the hydrogen ions on both sides of the surface start moving away (at that moment the entire system is heavily ionized). It is interesting to note that at the initial stage of the motion, when the hydrogen ions are still close (1-4 Å) to the carbon ions they were initially attached to, the charge density around the H ions changes dramatically depending on the instantaneous direction of the electric field. Essentially, the state of the hydrogen ions oscillates between a bare proton and a hydrogen atom. These oscillations decay as the hydrogen ions move away from the surface.

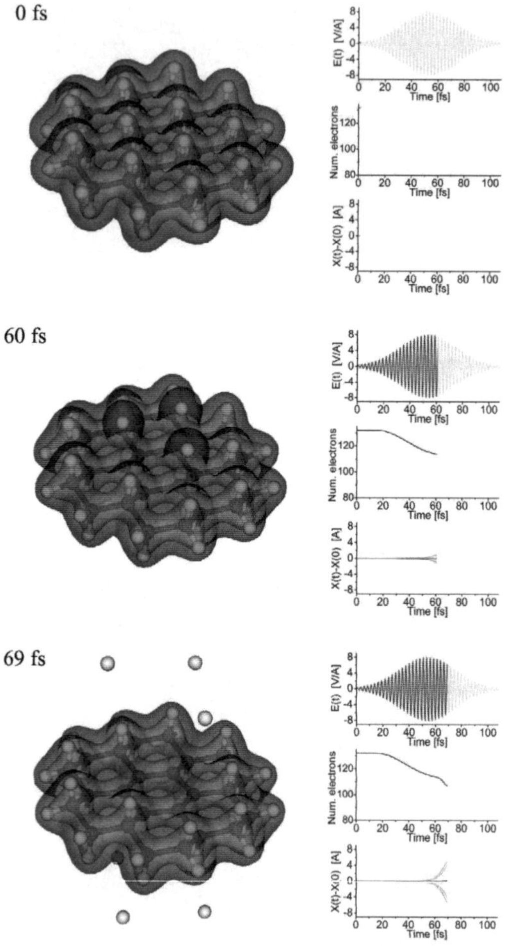

Figure 2. Time-evolution of the electron density and ionic positions in graphane fragment subjected to a 790nm, 35.4 fs laser pulse with the maximum electric field amplitude of 8 V/Å (enhanced online).

With each oscillation the laser continues to ionize the main target. Bare protons in close proximity (a few Å) to the main target, can recapture a small fraction of the escaped electrons.

However, when the direction of the electric field gets reversed, those electrons will not stay with the protons. At the time when the laser pulse is near its end the hydrogen ions are sufficiently far from the carbon surface and carry practically no electron charge with them.

Comparing graphene and graphane, the calculations show that graphene exhibits a considerably higher immediate damage threshold due to its strictly planar structure. Finite temperature simulations that distort the planar symmetry yield structural damage at lower intensities. The survival of irradiated graphene for a time that significantly exceeds tens of femtoseconds (after the pulse) is likely to depend on experimental conditions such as laser spot size and the presence of a substrate which determines its ability to quickly gain electrons from the bulk of the material to replace those lost during the laser pulse. The results obtained in the simulations of graphane demonstrate that by choosing the proper laser wavelength and intensity it is possible to achieve a desorption of hydrogen atoms without destroying the underlying layer of carbon atoms. Such a selective desorption achieved by tuning the two parameters is, in fact, a very basic realization of Quantum Optimal Control – a framework that is aimed to actively manipulate quantum dynamics using external time-dependent fields [45, 46, 47, 48, 49].

3.2. Coulomb explosion of hydrocarbon molecules

The strong-field ionization and fragmentation of hydrocarbon molecules is a prototypical example of the Coulomb explosion of polyatomic molecules, and it has been the subject of several experiments [50, 51, 52, 53, 54, 55]. An important quantity measured in these experiments is the kinetic energy distribution of the protons ejected during the fragmentation.

We have investigated the Coulomb explosion of methane, CH_4, and 1,3-butadiene, C_4H_6, in the framework of the time-dependent density functional theory and compared the results of the simulations to experimental data.

The dynamics of the ions, i.e. the displacements from their initial positions, $|\mathbf{R}_j(t) - \mathbf{R}_j(0)|$, and the evolution of kinetic energies, are illustrated in Figs. 3.2(c)-(f). As can be seen, the

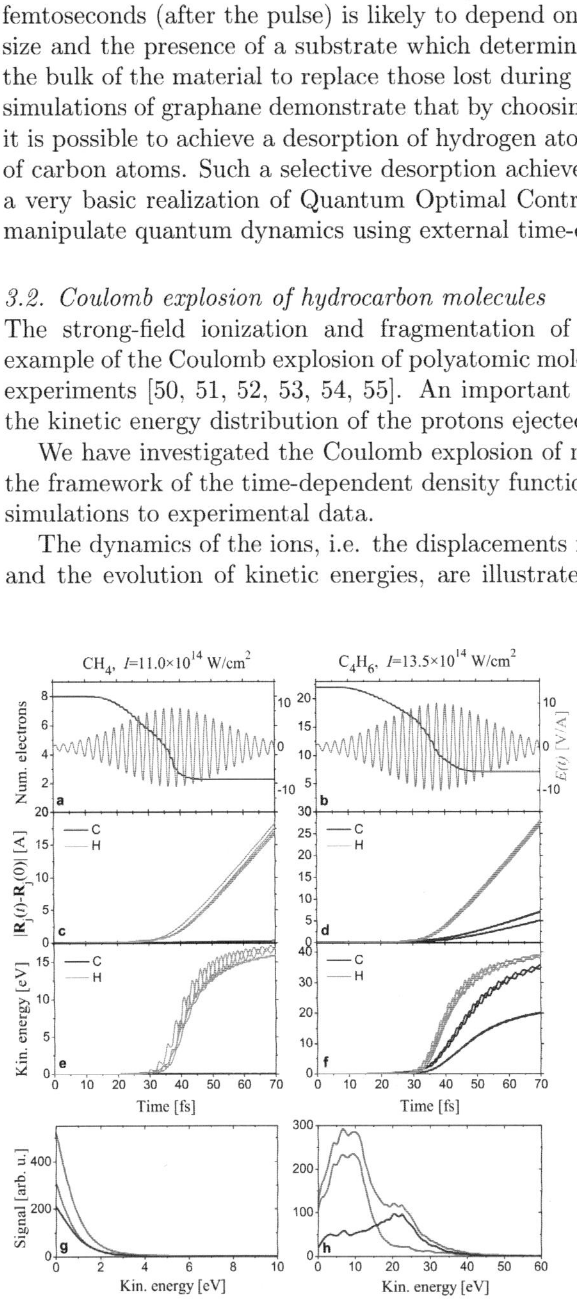

Figure 3. (a, b) Number of valence electrons remaining in the molecular system (blue lines) during ionization by a laser pulse (gray lines) for methane and 1,3-butadiene. (c, d) Displacements of individual protons (green lines) and carbon ions (dark red lines) from their initial positions. (e, f) Kinetic energies of individual protons (green lines) and carbon ions (dark red lines). While the data shown for the displacements and kinetic energies is for some randomly picked molecular spatial orientation, the same qualitative behaviour could be observed for the majority of other orientations. (g, h) Measured carbon energy spectrum (red line) decomposed into the contributions of singly and doubly charged carbons (gray lines).

molecules fragment completely and all protons get ejected simultaneously in a concerted process. The protons fly out in different directions as governed by the conservation of the total momentum but have very similar kinetic energies, see Figs. 3.2(e) and (f). The fact that the protons are ejected concertedly is also reflected by the near zero kinetic energy of the remaining carbon ion of CH_4 in both the simulated, Fig. 3.2(e), and measured data, Fig. 3.2(g). The measured carbon energy spectrum can be decomposed into the contributions of singly and doubly charged carbons (see the gray lines). Both of them peak at zero kinetic energy in accordance with a concerted, all-at-once emission of protons, which leaves the central carbon ion at rest. The similarity of the final kinetic energies of all ejected protons is most pronounced in the simulations with the highest laser intensity performed for the C_4H_6 molecule, Figs. 3.2(d) and (f). After the simultaneous departure of the protons the heavy carbon skeletal structure separates in two steps, with each of them involving the explosion of two carbon ions, leading to two distinctly different final carbon energies. These dynamics are nicely reflected in the measured bimodal carbon energy spectrum in Fig. 3.2(h) with peaks at approx. 8 and 22 eV. A decomposition of the measured spectrum into the contributions of the singly and doubly charged carbon ions, shown by gray lines, reveals that the two peaks in the spectrum are caused by doubly and singly charged carbon ions, respectively. Fig. 3.2(e) shows that for the slightly weaker laser peak intensity of 11.0×10^{14} W/cm^2 used for the CH_4 molecule, the high similarity of the proton energies observed for the high-intensity case of the 1,3-butadiene simulation becomes less pronounced. In the second simulation for CH_4 with the still smaller intensity of 6.7×10^{14} W/cm^2, the difference in the individual proton kinetic energies was quite noticeable and for some of the initial spatial orientations the molecule did not undergo a complete fragmentation during the simulation time.

3.3. Fragmentation of Uracil in Collisions with Neutral Potassium

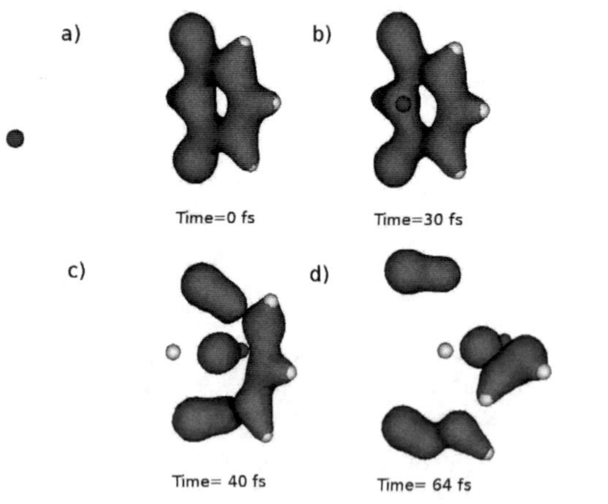

Figure 4. Time-evolution of the electron density and ionic positions of uracil during collision with a potassium atom with kinetic energy 200 eV, and initial separation of 10 Å.

An understanding of the interaction of moving particles with biological systems, RNA and DNA in particular, is an important prequisite for discerning the effects of radiation damage and for developing more efficient radiation therapies [56]. A complete description of the effects of particle collisions in DNA and RNA requires knowledge of both the direct damage from the particle, and the indirect damage caused by secondary particles, including low energy electrons ejected through ionization and particles ejected through molecular fragmentation [57]. Collisions involving neutral potassium are of particular interest since recent experimental work has shown that selective bond scission may be achieved by tuning the potassium atom's kinetic energy [58].

Figure 4 shows a simulation, within the framework of time dependent density functional theory, of the collision of a neutral potassium atom with the RNA nucleobase uracil, $C_4H_4N_2O_2$. The potassium atom's initial kinetic energy is 200 eV, and is separated from the uracil molecule by 10 Å. The total run time for the simulation is 64 fs. Prior to collision we see significant distortion of the uracil's structure, and subsequent fragmentation as the potassium atom passes through the uracil molecule.

Future work will include varying the potassium atom's kinetic energy and impact parameter, varying the orientation of the molecule, and a study of collisions with other nucleobases. Observables to be probed include the energy and charge transferred from the potassium to the molecule.

4. Summary

With the advent of powerful light sources [59, 60], the study of the interaction of strong short laser pulses with matter has become a centre of intense research interest. In this work, we have carried out first-principles time-dependent simulations of the interaction of intense short laser pulses with nanostructures using time-dependent density functional calculations complemented with Ehrenfest molecular dynamics. These first principles simulations are indispensable tools to gain insight into the physical mechanisms behind highly nonequilibrium and nonlinear processes, such as laser-induced structural deformation and Coulomb explosion. The presented TDDFT framework is a low-cost and high-accuracy approach that can be used to simulate the electron and nuclear dynamics in strong external fields, to explain experimental results and to design new experiments. Computational approaches developed in nuclear cluster theory have been utilized in this work, and with the continuous growth in computional power we expect that this cross fertilization can further advance other research fields as well.

References

[1] Goncharov V A and Varga K 2011 *Phys. Rev. B* **83** 035118
[2] Bubin S and Varga K 2012 *Phys. Rev. B* **85**(20) 205441
[3] Bubin S and Varga K 2011 *Appl. Phys. Lett.* **98** 154101
[4] Bubin S and Varga K 2011 *J. Appl. Phys.* **110** 064905
[5] Bubin S, Wang B, Pantelides S and Varga K 2012 *Phys. Rev. B* **85**(23) 235435
[6] Driscoll J A, Bubin S and Varga K 2011 *Phys. Rev. B* **83** 233405
[7] Krausz F and Ivanov M 2009 *Rev. Mod. Phys.* **81** 163–234
[8] Kruger M, Schenk M and Hommelhoff P 2011 *Nature* **475** 78–81 ISSN 0028-0836 URL `http://dx.doi.org/10.1038/nature10196`
[9] Fohlisch A, Feulner P, Hennies F, Fink A, Menzel D, Sanchez-Portal D, Echenique P M and Wurth W 2005 *Nature* **436** 373–376 ISSN 0028-0836 URL `http://dx.doi.org/10.1038/nature03833`
[10] Hamoudi H, Neppl S, Kao P, Schüpbach B, Feulner P, Terfort A, Allara D and Zharnikov M 2011 *Phys. Rev. Lett.* **107** 027801
[11] Young L, Kanter E P, Krassig B, Li Y, March A M, Pratt S T, Santra R, Southworth S H, Rohringer N, DiMauro L F, Doumy G, Roedig C A, Berrah N, Fang L, Hoener M, Bucksbaum P H, Cryan J P, Ghimire S, Glownia J M, Reis D A, Bozek J D, Bostedt C and Messerschmidt M 2010 *Nature* **466** 56–61 ISSN 0028-0836 URL `http://dx.doi.org/10.1038/nature09177`
[12] Ikeura-Sekiguchi H and Sekiguchi T 2007 *Phys. Rev. Lett.* **99** 228102
[13] Cavalieri A L, Muller N, Uphues T, Yakovlev V S, Baltuska A, Horvath B, Schmidt B, Blumel L, Holzwarth R, Hendel S, Drescher M, Kleineberg U, Echenique P M, Kienberger R, Krausz F and Heinzmann U 2007 *Nature* **449** 1029–1032 ISSN 0028-0836 URL `http://dx.doi.org/10.1038/nature06229`
[14] Haessler S, Caillat J, Boutu W, Giovanetti-Teixeira C, Ruchon T, Auguste T, Diveki Z, Breger P, Maquet A, Carre B, Taieb R and Salieres P 2010 *Nat Phys* **6** 200–206 ISSN 1745-2473 URL `http://dx.doi.org/10.1038/nphys1511`
[15] Nordlund D, Ogasawara H, Bluhm H, Takahashi O, Odelius M, Nagasono M, Pettersson L G M and Nilsson A 2007 *Phys. Rev. Lett.* **99** 217406
[16] Muino R D, Snchez-Portal D, Silkin V M, Chulkov E V and Echenique P M 2011 *Proceedings of the National Academy of Sciences* **108** 971–976 (*Preprint* `http://www.pnas.org/content/108/3/971.full.pdf+html`) URL `http://www.pnas.org/content/108/3/971.abstract`
[17] Runge E and Gross E K U 1984 *Phys. Rev. Lett.* **52** 997
[18] Yabana K and Bertsch G F 1996 *Phys. Rev. B* **54** 4484–4487
[19] Chu X and Chu S I 2001 *Phys. Rev. A* **64** 063404
[20] Kawashita Y, Nakatsukasa T and Yabana K 2009 *J. Phys. Condens. Matter* **21** 064222
[21] Iwata J I, Yabana K and Bertsch G F 2001 *J. Chem. Phys.* **115** 8773–8783
[22] Nakatsukasa T and Yabana K 2001 *J. Chem. Phys.* **114** 2550–2561

[23] Stott M J and Zaremba E 1980 *Phys. Rev. A* **21** 12–23
[24] Tsolakidis A, Sánchez-Portal D and Martin R M 2002 *Phys. Rev. B* **66** 235416
[25] Kootstra F, de Boeij P L and Snijders J G 2000 *J. Chem. Phys.* **112** 6517–6531
[26] Bertsch G F, Iwata J I, Rubio A and Yabana K 2000 *Phys. Rev. B* **62** 7998–8002
[27] van Gisbergen S J A, Snijders J G and Baerends E J 1998 *J. Chem. Phys.* **109** 10657–10668
[28] Frediani L, Rinkevicius Z and Ågren H 2005 *J. Chem. Phys.* **122** 244104
[29] Takimoto Y, Vila F D and Rehr J J 2007 *J. Chem. Phys.* **127** 154114
[30] Yabana K and Bertsch G F 1999 *Phys. Rev. A* **60** 1271–1279
[31] Görling A 1996 *Phys. Rev. A* **54** 3912–3915
[32] Doltsinis N L and Sprik M 2000 *Chem. Phys. Lett.* **330** 563 – 569
[33] Adamo C, Scuseria G E and Barone V 1999 *J. Chem. Phys.* **111** 2889–2899
[34] Kurth S, Stefanucci G, Almbladh C O, Rubio A and Gross E K U 2005 *Phys. Rev. B* **72** 035308
[35] Qian X, Li J, Lin X and Yip S 2006 *Phys. Rev. B* **73** 035408
[36] Marini A, Del Sole R and Rubio A 2003 *Phys. Rev. Lett.* **91** 256402
[37] Perdew J P and Zunger A 1981 *Phys. Rev. B* **23** 5048–5079
[38] Troullier N and Martins J L 1991 *Phys. Rev. B* **43** 1993–2006
[39] Manolopoulos D E 2002 *J. Chem. Phys.* **117** 9552
[40] Ehrenfest P 1927 *Z. Phys.* **45**(7) 455–457
[41] Carr C W, Radousky H B, Rubenchik A M, Feit M D and Demos S G 2004 *Phys. Rev. Lett.* **92**(8) 087401
[42] Lenzner M, Krüger J, Sartania S, Cheng Z, Spielmann C, Mourou G, Kautek W and Krausz F 1998 *Phys. Rev. Lett.* **80**(18) 4076–4079
[43] Fan C H and Longtin J P 2001 *Appl. Opt.* **40** 3124–3131
[44] Yao Y L, Chen H and Zhang W 2005 *Int. J. Adv. Manuf. Technol.* **26**(5) 598–608
[45] Assion A, Baumert T, Bergt M, Brixner T, Kiefer B, Seyfried V, Strehle M and Gerber G 1998 *Science* **282** 919–922
[46] Shapiro M and Brumer P 2003 *Principles of Quantum Control of Molecular Processes* (New York: Wiley)
[47] Werschnik J and Gross E K U 2007 *J. Phys. B* **40** R175
[48] Brif C, Chakrabarti R and Rabitz H 2010 *New J. Phys.* **12** 075008
[49] Krieger K, Castro A and Gross E K U 2011 *Chem. Phys.* **391** 50 – 61
[50] Cornaggia C, Normand D and Morellec J 1992 *J. Phys. B* **25** L415–L422
[51] Cornaggia C, Schmidt M and Normand D 1995 *Phys. Rev. A* **51** 1431–1437
[52] Markevitch A N, Romanov D A, Smith S M and Levis R J 2004 *Phys. Rev. Lett.* **92** 063001
[53] Shimizu S, Kou J, Kawato S, Shimizu K, Sakabe S and Nakashima N 2000 *Chem. Phys. Lett.* **317** 609–614
[54] Palaniyappan S, Mitchell R, Ekanayake N, Watts A, White S, Sauer R, Howard L, Videtto M, Mancuso C, Wells S, Stanev T, Wen B, Decamp M and Walker B 2010 *Phys. Rev. A* **82** 043433
[55] Roither S, Xie X, Kartashov D, Zhang L, Schöffler M, Xu H, Iwasaki A, Okino T, Yamanouchi K, Baltuska A and Kitzler M 2011 *Phys. Rev. Lett.* **106**(16) 163001
[56] Schulz-Ertner D, Nikoghosyan A, Thilmann C, Haberer T, Jkel O, Karger C, Kraft G, Wannenmacher M and Debus J 2004 *International Journal of Radiation Oncology*Biology*Physics* **58** 631 – 640 ISSN 0360-3016
[57] de Vries J, Hoekstra R, Morgenstern R and Schlathölter T 2003 *Phys. Rev. Lett.* **91**(5) 053401
[58] Almeida D, Ferreira da Silva F, Garcia G and Limão Vieira P 2013 *Phys. Rev. Lett.* **110**(2) 023201
[59] Protopapas M, Keitel C H and Knight P L 1997 *Rep. Prog. Phys.* **60** 389
[60] Mourou G A, Barry C P J and Perry M D 1998 *Phys. Today* **51** 22–28

Conference summary: an experimental perspective

W N Catford

Department of Physics, University of Surrey, Guildford GU2 7XH, UK

E-mail: w.catford@surrey.ac.uk

Abstract. The experimental results reported at this conference, the themes that emerged and the implications for future work are examined. Some example highlights, chosen from a personal perspective are discussed.

1. Introduction

A wide range of experimental work has been presented in this conference, including some new approaches and some new developments of older methods. Some common themes emerged during the week, and this summary aims to bring together those themes and to examine where they might lead us in the future. The aim is not to provide a complete summary of all contributions; all of the papers are here in the proceedings. Nor is the aim to provide an up-to-date overview of clustering studies; that is addressed more by the introductory article by Freer [1]. Rather, we summarise the snapshot of themes and directions offered by this conference, and discuss some illustrative examples.

Any non-exhaustive summary will inevitably represent a personal perspective. Therefore, it is perhaps acceptable to begin with a short introduction about my own history of studying clustering, partly to declare my personal position and prejudices. This conference, the tenth in the series of Cluster conferences, was the fourth of the series that I attended, after Chester, Strasbourg and Stratford-on-Avon. The very first topic in nuclear physics that I ever studied experimentally, in 1978, was the cluster structure of ^{16}O via inelastic scattering ^{12}C$(\alpha, \alpha_2)^{12}$C*, exploiting the selectivity imposed by requiring ^{12}C in the 0_2^+ 'Hoyle' state [2]. That work was based in turn upon a paper from ten years earlier, in 1967 [3]. We heard at this conference that essentially the same reaction is still being studied today, and with much the same objectives. My point is that there is something inherently difficult about these experiments — both technically and in the theoretical interpretation — or else we would surely have clear answers by now.

Clustering in nuclei is one of the most misunderstood aspects of nuclear structure. As I have remarked before [4], some of our colleagues are convinced that we *cluster people* are in a state of denial, because it has been evident since the earliest studies that α-particles cannot retain an existence as *real* α-particles when they are inside nuclei [5, 6]. Of course, we are not so misled, and the current theoretical models are extremely sophisticated. Nevertheless, the proper situation has not been transmitted to our non-clustering colleagues world-wide in an entirely effective fashion. One reason is experimental: what exactly is the experimental evidence that particular nuclides, or rather particular states within a nuclide, should be classified as *cluster states*? The values of partial widths for α-particle emission are driven by energetics and dynamics, as well as structure, and it can be challenging to disentangle the contributions. The presence,

Journal of Physics: Conference Series **436** (2013) 012085 doi:10.1088/1742-6596/436/1/012085

and mixing, of states with other structures (such as more-standard shell model structure) also complicates the interpretation of experiments. Nevertheless, better and better experiments are being performed and progress is being made both experimentally and theoretically. To return to a short review talk that I gave nearly fifteen years ago [4] I'd like to mention two points. Firstly, the comment was made with reference to ^{12}C: 'perhaps it is reasonable to think of this as a kind of Bose-Einstein condensate of α-particles'. This was a musing based on little real knowledge. In recent years, however, our theoretical colleagues have applied such ideas in a rigorous way in the THSR model [7], and this has given deep new insights into the structure of cluster nuclei. The BEC description is not applicable to all states, but provides an excellent description for particular states where the individual α-particle clusters are more distinct. The existence of such BEC states in a strong-force environment is an extraordinary feature of nuclear structure, as seen for example in the 7.65 MeV 'Hoyle' state in ^{12}C. It is worth commenting that the type of condensation seen, for example, in the ground state of ^{12}C is equally remarkable. In early models [8], and in more detailed microscopic models where the clustering arises naturally [9, 10, 11, 12], the individual α-like mass centres are evident even in the more-compact ^{12}C ground state. The cluster structure *persists*, even though the objects *overlap*, interact via the *strong* interaction and constantly *exchange* nucleons. As remarked by Kanada-En'yo [11], the Pauli principle inhibits the full rotational symmetry whilst still retaining strong structural symmetries. Perhaps this type of condensation is unique to nuclear physics. In that earlier review talk [4], a second remark concerned the predicted linear chain states of alpha-particles that were then popular in the literature [13] and asserted that they were 'not compellingly in evidence experimentally beyond ^{8}Be.' The same conclusion was reached in the comprehensive review published just before the previous cluster conference in Stratford [14]. In the present conference, we have seen tentative new evidence for pushing this experimental boundary up to ^{12}C [15], via a detailed analysis of inclusive and exclusive α-particle inelastic scattering. The analysis highlights the extreme difficulty in pinning down potential chain states experimentally, in regions of excitation energy where many broad underlying states coexist. Finally, from the earlier review talk [4], Figure 1 was used to summarise the different types of clustering discussed in the literature. All of these structures were discussed, in an experimental context, at the present meeting except for the extremely speculative *necklace* type of state [16] and the longer chain configurations. A decline in the emphasis on simple linear chains is a change over the last fifteen years, for reasons mentioned above. A newly emerging and related topic of study is the class of *covalently* bound states, whereby a chain is stabilized or a cluster-core interaction is modified by one or two covalently bound neutrons.

2. Experimental Themes

This conference was wide-ranging in its scope and impressive in the number of new results presented. Some recurring themes emerged, and we choose to limit the discussion here to some examples that fall under the following headings, representing four *physics* themes:

- ^{12}C: particular states, especially the 2^+ excitation of the Hoyle state, and 0^+ states
- BEC: the Bose-Einstein condensate in ^{12}C, fascinating theory for ^{16}O and a search in ^{24}Mg
- Detailed studies of certain light nuclei below ^{12}C
- Renewed interest in *sd*-shell cluster resonances,

and four *instrumentation/methodology* themes:

- Resonant scattering in inverse (or even normal) kinematics
- A resurgence in the use of magnetic spectrometers
- Renewed exploitation of gamma-ray decays to uncover cluster structure.
- Some insights and results relating to high energy knockout such as $(p, p'\alpha)$.

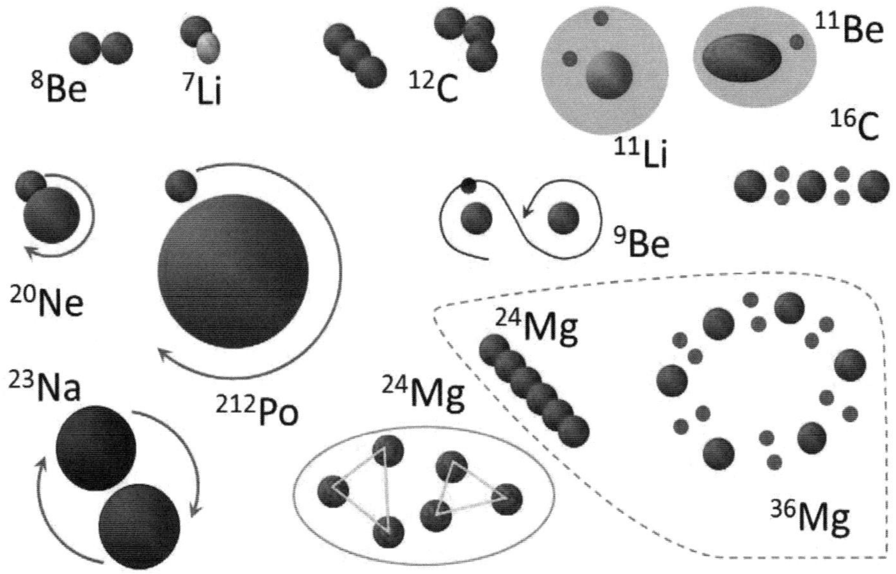

Figure 1. Different types of clustering in nuclei (adapted from Ref. [4]). Only those inside the dashed line were not discussed experimentally at this conference.

The fact that ^{12}C would, by itself, be a major theme of the conference was foreseen in the logo chosen by the organisers. The renewed interest in ^{12}C, one of the very first nuclei to be studied from a cluster perspective, was due to the exceptionally interesting questions that have become accessible experimentally and theoretically. Many of these were described in the introduction by Freer [1].

Exciting new results for ^{12}C were reported by Itoh [15]. In a previous publication [17] this group showed good experimental evidence for the 2^+ excitation of the Hoyle state, at 9.84 MeV. The analysis used the multipole decomposition (MD) technique whereby (α, α') angular distributions are decomposed, as a function of excitation energy, into a superposition of the expected distributions for each angular momentum transfer. Their analysis also showed evidence that the broad 0^+ strength near 10 MeV was split into two components centred at 9.04 and 10.56 MeV respectively. The analysis was admittedly complicated by the presence, right in the middle of this same region of energy, of the narrow, strongly excited 3^- state at 9.64 MeV. The new results from this conference [15] extend the experimental setup to measure coincident decay α-particles. They include the angular correlation functions (ACFs) for decays feeding the $^{8}\mathrm{Be}_{\mathrm{g.s.}}(0^+)$ and the $^{8}\mathrm{Be}(2^+)$ states, extracted for a series of narrow slices in the excitation energy of ^{12}C. The ACF and decay-branching analyses are consistent with the earlier MD conclusions and significantly strengthen the association of the two regions of 0^+ strength, respectively, with a higher nodal state of the Hoyle state (at 9.04 MeV) and the long sought-after 0^+ linear chain state in ^{12}C (at 10.56 MeV).

Equally exciting were the measurements reported by Gai [18] for the reaction $^{12}\mathrm{C}(\gamma, 3\alpha)$ studied by stepping the energy of an almost monoenergetic beam of gamma-rays through the excitation range 9.1 to 10.7 MeV in ^{12}C. The breakup of the ^{12}C into three α-particles was recorded optically in three dimensions in a time projection chamber that also served as the target. The angular correlation for the α+$^{8}\mathrm{Be}_{\mathrm{g.s.}}(0^+)$ channel was extracted at each incident beam energy. The excitation function was fitted, taking into account a 2^+ contribution (E2) and a 1^-

contribution (E1). Note that the excitation function was sensitive not only to the magnitudes and shapes of the E2 and E1 strengths, but also the phase between their contributions. An analysis including a phase difference that was consistent with theory gave precise parameters for the 2^+ state, with a resonance energy of 10.03 MeV and a peak cross section (owing to the energy-dependent Breit-Wigner resonance shape) of 9.8 MeV. Gai [18] discusses the properties of this 2^+ state in terms of an excitation of the Hoyle resonance. This state was also measured and discussed by Itoh [15] and Freer [1].

Kawabata [19] reported new results from searching in a heavier nucleus for a Hoyle-like state, described as an α-condensed state or a dilute-gas-like structure of α-particles. As in the work on ^{12}C [15], inelastic scattering of α-particles was studied at the Osaka cyclotron using the dispersion-matched Grand Raiden spectrometer, now with a ^{24}Mg target. Again, coincident particles from the decay of the excited nucleus were detected: in this case, proton, α and ^{8}Be decays of ^{24}Mg could be identified. The monopole strength from the MD analysis gave no candidate for a 6α condensate near the relevant energy threshold. Possible candidates were identified for α-condensed states around the cores ^{12}C and ^{16}O near the energies that had been predicted [20]. A state at 13.9 MeV that decays mostly by α-particle emission was considered to be the best candidate for a 2α-^{16}O condensed state. As a comment, naively the state seems remarkably narrow for a 0^+ $T = 0$ state at this energy, fitting within the 200 keV width of the bins in the MD analysis. A known isoscalar 0^+ state at 11.73 MeV already has a width of 10 keV and a 0^+ state at 13.04 MeV [21] has a width of 3 keV even though it is largely $T = 1$ in character (the 13.04 MeV $T = 1$ state is, as expected, seen only very weakly if at all, in the MD analysis). There is no doubt that the states reported by Kawabata are very interesting and merit further investigation.

The technique of inverse resonant scattering was clearly evident throughout the experimental sessions. In this technique a heavy beam is incident (in inverse kinematics) on light target nuclei comprising a thick target (which may be a gas). The beam will typically pass through various resonances as it slows down. Taking advantage of the large difference in energy loss between the beam and target, the depth of the reaction can be reconstructed for each event, assuming elastic scattering. Excitation functions can be measured relatively quickly over a wide range of energies. In one talk describing this technique, Norrby discussed a study of ^{32}S via the ^{28}Si+α reaction [22]. This revealed 30 new level assignments in a study spanning 132 resonances in total. This reaction has also been studied much more laboriously in normal kinematics using a ^{4}He beam, and a direct comparison of the two methods is shown in Figure 5 of Ref. [22]. The inverse technique has poorer energy resolution and a higher background, balanced against its speed advantages. Rogachev's talk [23] also reported work using the inverse scattering technique, including conclusive new results for the highly α-clustered state at 10.16 MeV in ^{10}Be, confirming the spin as 4^+. Another very exciting aspect of this talk was the description of the new device ANASEN [24] used in the experiment. This device comprises a cylindrical gas target volume surrounded by silicon strip detectors that are backed by CsI scintillators. Inside the radius of the silicon array, proportional counters within the gas volume allow tracking of the detected particles, which arise from reactions of the beam as it travels through the fill gas. This is truly the *Rolls Royce* of devices for this type of experiment, and heralds the arrival of a new generation of results exploiting this technique.

Inverse resonant scattering wasn't developed specifically for radioactive beam experiments, but is perfectly adapted to them. With a short-lived radioactive ion, the option of a normal kinematics experiment doesn't exist. The ANASEN detector mentioned above is in fact designed primarily for studies with radioactive beams. Such radioactive beam experiments at the low energy CRIB facility at RIKEN, including ^{7}Be+^{4}He, were described by Kubono [25]. Using a 90% pure beam of 3×10^8 pps, the reactions (α, α) and (α, p) were studied with an astrophysical motivation. The same resonant nucleus, ^{11}C, has also been studied using the inverse resonant scattering technique by Freer *et al.* with a weaker but pure beam of 5×10^7 pps at Louvain-la-Neuve [26], and this was extended to higher excitation energies using the much higher energy (pure ^{7}Be) beam of 5×10^4 pps available at the Oak Ridge tandem [26].

The resurgence of interest in the study of nucleon transfer and compound nuclear reactions using magnetic spectrometers, with their high energy resolution, is exemplified by the reports by Milin [27] and Wheldon [28]. An example discussed by Milin was the study [29] in which many states in ^{18}O were tentatively identified up to nearly 20 MeV excitation, and assigned to positive and negative parity rotational bands. This process is fraught with subjectivity because of the difficulty in obtaining conclusive spin assignments. This was recognised and addressed in Ref. [29] where the systematic dependence on spin of the differential cross section at a fixed angle was used to help validate the spin suggestions. The cross section should follow a $(2J + 1)$ dependence, which is statistical and independent of the reaction mechanism [29], but some scatter for specific states can, and does occur, making definitive assignments impossible. With tentative spins, the identification of rotational bands based largely on energy-spin systematics is dangerous, as discussed by Milin [27] with reference to ^{13}C (where subsequent measurements contradicted some previous suggestions).

Two talks, in particular, described a revival in the use of gamma-ray decays to understand possible cluster structure. Jenkins [30] and Courtin [31] concentrated on studies aimed at the molecular resonances seen in ^{12}C+^{12}C and ^{16}O+^{12}C, respectively. This work revisits pioneering studies by Sandorfi *et al.* [32, 33] in the 1980's, but with significant experimental improvements. Jenkins [34] reviewed a study of the region near 20 MeV excitation in ^{24}Mg which showed resonances with spins of 0 or 2 that decay via isovector M1 transitions through 1^+ $T = 1$ levels near 8–10 MeV. The decay of such resonances via high energy gamma-rays of order 10 MeV was predicted theoretically by Baye and Descouvemont [36] although their explanation didn't involve the $T = 1$ states identified here. Courtin [31] reported new results showing similar behaviour in ^{16}O+^{12}C. Resonances near 25 MeV excitation in ^{28}Si decayed via enhanced isovector M1 transitions through 1^+ and 2^+ $T = 1$ states. A ^{12}C+^{12}C resonance at 22 MeV in ^{24}Mg was studied in more detail using Gammasphere [35]. The decay was largely statistical but with some enhancement in the population of a particular prolate rotational band in ^{24}Mg. The idea that it might be possible to infer deformed cluster structure from the preferential population of prolate deformed bands was discussed earlier by Sandorfi and collaborators [33]. It also brings to mind other studies, for example that by Sanders *et al* [37] where the symmetric fission of a ^{56}Ni compound nucleus was seen to populate preferentially a prolate band in ^{28}Si. Courtin [31] also reminded us of an ambitious experiment by Haas *et al* to measure in-band gamma-ray transitions between high-lying rotational states in ^{24}Mg [38]. This experiment was also mentioned in the introduction by Freer [1]. The ^{24}Mg states had been seen previously to break up by fission into ^{12}C+^{12}C. If these states represent a very highly deformed rotational band, then an in-band gamma-ray transition could be sufficiently enhanced to compete with particle emission at a detectable level. In the experiment, ^{24}Mg was populated at an excitation energy of 30.5 MeV corresponding to a known 10^+ resonance, and then the missing kinetic energy in the ^{12}C+^{12}C fragments (Q-value) was compared with the observed gamma-ray energy. A measurement of $\Gamma_\gamma/\Gamma = (1.2 \pm 0.4) \times 10^{-5}$ was reported for the 10^+ to 8^+ transition ($E_\gamma = 6.9$ MeV) [38], which perhaps might best be interpreted as an upper limit given the extreme challenges of eliminating all background. In any case, this study has a potentially very important result and could usefully be revisited if the experimental setup can be improved.

The quasi-free knockout of preformed α-particles from nuclei via the 100 MeV/A reaction $(p, p'\alpha)$, was discussed by Cowley [39]. The results are extremely important for several high profile experiments planned around the world, such as the R3B experiment at FAIR [40], which include the possibility to study this reaction in inverse kinematics with radioactive beams. A recent result demonstrating quasi-free proton scattering from α-particles inside the ground state of ^{12}C was reviewed. With suitable kinematic cuts to satisfy the quasi-free conditions, the differential cross sections were close to identical in shape for scattering by free α-particles and scattering from those found inside the ^{12}C ground state. Further, the analysing power as a function of scattering angle was the same in the two cases. The important new result from this conference was the new analysis for a ^{40}Ca target, also for 100 MeV/A. Whereas a previous analysis had failed to show consistency with quasi-free scattering, an improved distorted wave treatment for α+^{36}Ar shows consistency with quasi-free scattering from α-particles inside ^{40}Ca. Another interesting knockout result was reported by Ye [41], who described a RIKEN experiment to study the knockout of a ^{6}He core via $(p, p'\ ^6\mathrm{He})$, performed in inverse kinematics with a ^{8}He beam at around 10 MeV/A.

Tamura reported on results from experiments at J-PARC to detect gamma-rays from the hypernuclei produced when Λ hyperons are introduced into ^{8}Be and ^{6}Li nuclei. Compared to ^{6}Li, which is well described by a cluster structure of α+d, the introduction of a Λ-particle (unhindered by the Pauli principle) makes $^7_\Lambda$Li and binds the two clusters closer together spatially, causing a shrinkage of order 20% [42]. The extension to results for heavier nuclei, and other future perspectives were also reviewed [43].

3. Outlook

This brief summary of highlights shows, very clearly, that we are at an exciting time in the experimental study of nuclear clustering. Some qualitatively new techniques have recently emerged, and some existing techniques have been adapted or extended to take advantage of technological advances. Of course, many other interesting results were reported, including those in the areas of: halo states, dripline structure, astrophysics, the trojan horse method, meson scattering, superheavy element studies, heavy cluster radioactivity and more.

However, even as the new results accrue, we are left with many puzzles and situations that look promising but are not resolved. More specifically, we find many experiments in which a plausible argument can be made, to link the observations with the theoretical expectations, but the measurements fall short of being conclusive. Maybe, with the newly emerging methods, we are in sight of being able to push past this limit, towards conclusive results. Certainly there is a need to achieve confident spin determinations and to measure absolute decay widths. The fact that an excited state decays by cluster emission is not sufficient, in itself, to demonstrate cluster structure - some sort of enhancement must be demonstrated. Some specific challenges spring to mind. In the case of excited molecular bands, to go beyond plausibility we need robust spin assignments, partial decay probabilities and probably B(E2) measurements. There is a lot of scope for α-particle transfer experiments to inform these studies, possibly using radioactive beams. In ^{16}O, what is the nature of the resonances seen by Chevallier *et al* [3] in the ^{12}C(α,^{8}Be) reaction? Are they related to a chain state, or a structure involving some degree of Bose-Einstein condensation, and can we decide for sure? In ^{20}Ne and ^{24}Mg, can the various expected Bose-Einstein condensates be located with confidence?

It is clear that leaps forward will be seen in the quality and extent of experimental results using resonant scattering in inverse kinematics. This is already happening with the emergence of devices such as ANASEN [23, 24]. It may be that through the study of hypernuclei, it will be possible to characterise some cluster configurations in light hypernuclei that cannot be studied in ordinary nuclei, taking advantage of the lowering in energy of the configuration due to the extra binding of the Λ particle. Finally, in my personal list, the next conference may bring forth results from proton induced knockout in inverse kinematics using radioactive beams. As well as $(p, p'p)$ or $(p, p'n)$ quasi-free knockout of nucleons, α-particle knockout $(p, p'\alpha)$ is due to be studied at facilities such as R3B at FAIR and similar setups in RIKEN, and can explore the theoretical predictions for the development of clustering towards the neutron dripline.

One last comment in this experimental summary, and one that was well demonstrated by this conference: clustering in nuclei is one of the very best parts of nuclear physics for nurturing a close and mutually beneficial relationship between experimenters and theorists. This close cooperation will surely continue, and will underpin the breakthroughs, insights and new results that we can anticipate at Napoli, when next we meet.

Acknowledgments

Thank you to the organisers for giving me the opportunity to summarise this conference. Mine was the final talk of the conference and, as I did in my talk, I would like to finish by congratulating the organisers on producing an excellent logo, an interesting programme and an inspiring week.

References

[1] Freer M 2012 contribution to the present proceedings
[2] Catford W N 1978 First Year PhD Research Report, University of Oxford, (unpublished), available at http://personal.ph.surrey.ac.uk/ phs1wc/report/WNC-01-78.pdf
[3] Chevallier P *et al* 1967 *Phys. Rev.* **160** 827
[4] Catford W N 1999 *Proc. Int. Conf. on The Nucleus: New Physics for the New Millenium, (Cape Town, 1999)* ed F D Smit *et al* (New York: Kluwer Academic/Plenum) p 127 (http://arxiv.org/abs/1302.3849)
[5] Wheeler J A 1941 *Phys. Rev.* **59** 16; *Phys. Rev.* **59** 27
[6] Wefelmeier W 1937 *Z. Phys.* **107** 332; *Naturwiss.* **25** 525
[7] Tohsaki A 2001 *Phys. Rev. Lett.* **87** 192501
[8] Brink D M 1966 *Proc. Int. School of Physics Enrico Fermi (Varenna, 1965)* Course **36** ed C Bloch (NewYork: Academic) p 247
[9] Kanada-En'yo Y and Horiuchi H 2001 *Prog. Theor. Phys. Suppl.* **142** 205
[10] Chernykh M *et al* 2007 *Phys. Rev. Lett.* **98** 032501
[11] Kanada-En'yo Y 2012 contribution to the present proceedings
[12] Neff T 2012 contribution to the present proceedings
[13] Merchant A C and Rae W D M 1992 *Nucl. Phys.* A **549** 431
[14] von Oertzen W, Freer M and Kanada-En'yo Y 2006 *Phys. Rep.* **432** 43
[15] Itoh M 2012 contribution to the present proceedings
[16] Wilkinson D H 1986 *Nucl. Phys.* A **452** 296
[17] Itoh M *et al* 2011 *Phys. Rev.* C **84** 054308
[18] Gai M 2012 contribution to the present proceedings
[19] Kawabata T 2012 contribution to the present proceedings
[20] Itagaki N, Kimura M, Kurokawa C, Ito M and von Oertzen W 2007 *Phys. Rev.* C **75** 037303
[21] Fifield L K *et al* 1978 *Nucl. Phys.* A **309** 77
[22] Lönnroth T *et al* 2010 *Eur. Phys. J.* **46** 5
[23] Rogachev G 2012 contribution to the present proceedings
[24] Matoš M *et al* 2010 *Proc. 11th Symposium on Nuclei in the Cosmos (Heidelberg, 2010)* p 226
[25] Kubono S 2012 contribution to the present proceedings
[26] Freer M *et al* 2012 *Phys. Rev.* C **85** 014304
[27] Milin M 2012 contribution to the present proceedings
[28] Wheldon C 2012 contribution to the present proceedings
[29] von Oertzen W *et al* 2010 *Eur. Phys. J.* **A43** 17
[30] Jenkins D G 2012 contribution to the present proceedings
[31] Courtin S 2012 contribution to the present proceedings
[32] Nathan A M *et al* 1981 *Phys. Rev.* C **24** 932
[33] Collins M T *et al* 1982 *Phys. Rev. Lett.* **49** 1553
[34] Jenkins D G *et al* 2007 *Phys. Rev.* C **76** 044310
[35] Jenkins D G *et al* 2011 *Phys. Rev.* C **84** 044332
[36] Baye D and Descouvemont P 1984 *Nucl. Phys.* A **419** 397
[37] Sanders S J *et al* 1994 *Phys. Rev.* C **49** 1016
[38] Haas F *et al* 1997 *Il Nuovo Cimento* A **110** 989
[39] Cowley A A 2012 contribution to the present proceedings
[40] Aumann T 2007 *Prog. in Part. Nucl. Phys.* **59** 3
[41] Ye Y 2012 contribution to the present proceedings
[42] Tanida K *et al* 2001 *Phys. Rev. Lett.* **86** 1982
[43] Tamura H *et al* 2012 *Nucl. Phys.* A **881** 310

Journal of Physics: Conference Series **436** (2013) 012086 doi:10.1088/1742-6596/436/1/012086

Summary talk – from a theoretical point of view

Y Suzuki[1,2]

[1] Physics Department, Niigata University, Niigata 950-2181, Japan

[2] RIKEN Nishina Center, Wako 351-0198, Japan

E-mail: suzuki@nt.sc.niigata-u.ac.jp

Abstract. This summary is based on the theoretical talks presented at the Tenth International Conference on Clustering Aspects of Nuclear Structure and Dynamics, Debrecen, Hungary, 24–28 September, 2012.

1. Introduction

The concept of clustering in nuclei is already fairly old but still frequently employed in studies on both nuclear structure and reactions. The basic idea of clustering is that a group of nucleons forms a spatially localized substructure and behaves as an entity rather than individual nucleons. The clustering assumption leads to simplicity of the description of complicated nuclear phenomena. Its physical foundation is primarily based on the saturation property of the nucleus though its microscopic foundation is a challenge even today.

Many of the plenary session titles of the Debrecen conference are the same as the keywords of the previous cluster conferences. They are more or less traditional subjects, but compared, e.g., to the Kyoto conference in 1988 [1], some new titles appear, such as *Few-body and ab initio approaches*, *Structure and reactions of exotic nuclei* and *Clustering in nuclear matter*. These obviously reflect a current trend in the nuclear physics community.

Some basic questions addressed include where do we expect cluster states, what about the coupling of cluster states to other degrees of freedom, how are cluster states formed in reactions, what role do cluster states play in reactions and fission, and what about a microcopic foundation of clustering with the use of realistic nuclear forces.

In his introductory talk on recent developments in cluster physics Horiuchi [2] has referred to four works, viz. on neutron-rich nuclei, cluster gas phase, coexistence of cluster and mean-field dynamics, and *ab initio* approaches.

2. Cluster structure in light nuclei

2.1. Alpha and exotic clustering

Alpha clustering plays a central role in cluster physics. As suggested by the Ikeda diagram [3], some cluster states may appear near the threshold for α breakup.

Using an orthogonality condition model of 4α for ^{16}O, Funaki has reported the existence of six 0^+ states [4], which are understood to arise from the excitation of ^{12}C and of the ^{12}C-α relative motion. In particular, the sixth state, which appears at 2 MeV above the 4α threshold is indicated to have a gas-like structure, which is a candidate for the observed 0^+ level at 15.1

MeV with a decay width of 170 keV. The small width has been explained by a small overlap of that special structure with the decaying-channel wave functions.

Motivated by the finding of resonances that preferentially decay to ^{6}He and ^{8}He, Ito has studied exotic cluster structure in Be isotopes using the two-centre cluster model of $\alpha+\alpha+xn$ [5]. Besides the two bound 0^+ states, two 0^+ resonances above the ^{6}He+^{8}He threshold were obtained in ^{14}Be by solving a scattering problem that takes into account the coupling to the continuum.

Since the α cluster is a composite particle consisting of two neutrons and two protons, it is subject to the Pauli principle in a microscopic cluster model. Its effect is especially important when the α cluster interacts with the other nucleons of the system at short distances. A calculation with a fully microscopic cluster model becomes increasingly hard when more than two clusters are employed though such calculations usually produce rewarding results; see, for example, the multi-cluster calculations for ^{9,10,11}Li [6]. Therefore, an approximate version has often been employed for many years.

In a Faddeev formalism Dufour has treated a 3α model for ^{12}C using local and non-local α-α potentials, and has shown that cluster-cluster potentials of the latter type appear to work better if the effect of the Pauli principle is taken into account [7]. It is worthwhile recalling that a non-local α-α potential that renormalizes the energy-dependent kernel of the resonating group method has already been constructed and found to lead to a fair description of both the ground and the Hoyle state of ^{12}C [8].

A semi-microscopic cluster model has been considered by Hess introducing σ and π bosons, and its geometry and phase transition have been discussed [9].

The techniques and lessons of cluster physics can be applied to hypernuclear physics as well. Since the hyperon-nucleon and hyperon-hyperon interactions are only poorly known, a theoretical analysis of light hypernuclei is expected to play an important role. A review talk on light Λ hypernuclei has been given by Hiyama [10]. Some examples of four- and five-body calculations have been presented and the glue-like nature of Λ has been stressed.

2.2. Molecular states and deformed states

The ^{12}C+^{12}C system is a well-known classical example that exhibits a number of narrow resonances near the Coulomb barrier [11]. The appearance of the narrow resonances is explained by a coupling to the inelastic channel involving the 2^+ state of ^{12}C as well as to other reaction channels, e.g., ^{8}Be+^{16}O and α+^{20}Ne.

It is interesting that the ^{24}Mg+^{24}Mg and ^{28}Si+^{28}Si systems present even narrower high spin resonances well above the Coulomb barrier. These examples indicate that a di-nuclear molecule can be formed provided the entrance channel satisfies a specific condition on the energy and the angular momentum. Uegaki has discussed a mechanism of the resonances in these two cases by analysing the normal modes of two deformed molecules (a prolate ^{24}Mg and an oblate ^{28}Si) around some equilibrium [12]. He has taken into account the wobbling motion perpendicular to the beam direction. The γ-ray yields from the resonance of the ^{28}Si+^{28}Si case at 55.8 MeV were successfully accounted for.

The possibility of producing strongly deformed (hyperdeformed) states near the Coulomb barrier together with the cooling by neutron emission has been discussed by Adamian [13] for several combinations of isotopes such as Ca+Sn, Ca+Xe, Ca+Ba etc.

The approaches mentioned above describe nuclear deformation in terms of geometrical shape parameters. In contrast, a microscopic approach has been undertaken by Taniguchi [14] to study possible largely deformed states in the $A \sim 40$ region. In particular, some excited states of ^{42}Ca have been described in deformed-basis Gaussian wave packets (antisymmetrized molecular dynamics) together with a mixing of α+^{38}Ar cluster configurations. Since various deformed states may appear more or less at the same energy, a generator coordinate method has been employed to take into account the coexistence of the rotational bands, which can be tested in a comparison to experimental data on electric quadrupole transitions. It is desirable that this type of microscopic approach may be further extended to an analysis of molecular resonance phenomena.

3. Coupling of cluster states to other degrees of freedom

The clusters are usually frozen to their model ground states, and only their relative motion needs to be determined carefully. Furthermore, the intrinsic wave functions of the constituent clusters are quite often assumed to be the simplest possible shell model wave functions. This is implicitly related to the fact that the nucleon-nucleon potential used in the cluster model is more or less phenomenological, and, for example, contains no tensor component. This appears to be reasonable because otherwise cluster model calculations become extremely hard if the number of nucleons is not very small. Under these conditions it is possible to study the effects of the coupling of the cluster states to structures of different types.

Core excitations in exotic nuclei have been studied by Descouvemont [15]. He showed that there are several states in ^{16}B that cannot be reproduced in a single-channel cluster model of ^{15}B+n but the inclusion of many excited states of ^{15}B can improve the description significantly. He has also shown that the core excitation has a crucially important role in explaining the $3p$ emission from ^{17}Na.

Kimura has discussed nuclear structure in the island of inversion with $N \approx 20$ in antisymmetrized molecular dynamics calculations with deformed Gaussian wave packets [16]. Using a variety of deformed configurations, he has stressed the importance of cluster correlation near the neutron drip line, particularly in the description of the halo property of ^{31}Ne and ^{37}Mg, which is indicated by recent experimental findings of the enhanced reaction cross sections [17].

An approach similar to the above has been employed by Kanada-En'yo [18] to discuss cluster correlation in lighter nuclei, such as Be and C isotopes. She has pointed out that the correlation can be different, depending on whether the system of interest has a normal or dilute density.

It is well-known that, owing to the Pauli principle, some harmonic oscillator shell-model configurations are expressible in terms of very few cluster-model wave functions. As one of the simplest examples, we recall that the closed-shell configuration of ^{16}O can be equivalently rewritten as an α+^{12}C cluster wave function. Based on this fact, most of the levels of ^{16}O were accounted for in the unified framework of the α+^{12}C cluster model [19]. Yamada [20] has discussed the isoscalar monopole excitation of ^{16}O in a 4α cluster model, and obtained considerable strength below 16 MeV excitation energy. The monopole operator can excite not only the cluster relative motion but also the cluster internal motion, so that the experimentally observed strength at higher excitation energy is attributed to the latter type.

The monopole operator belongs to the generators of the symplectic group $Sp(6, R)$ together with the quadrupole operator [21, 22]. Though the 4α model is more extensive than the α+^{12}C cluster model, Yamada's result appears to be consistent with the old cluster-symplectic mixed basis calculation [23].

An ambitious approach to the no-core shell model guided by the symplectic algebra (or more generally some suitable symmetry) has been reported by Draayer [24]. With an increasing number of $N_{\rm max}$ (the number of oscillator shells included), the energies of the second 0^+ and 2^+ states of ^{12}C come down dramatically, closely approaching the observed energies. It will be interesting to analyse a mechanism of this behaviour and a connection with the cluster model, especially that with localized clustering.

It is of great importance to realize spatially localized clustering in microscopic theories that have no *a priori* cluster assumption. In this respect the description of the first excited 0^+ states of ^{12}C and ^{16}O are still challenging even with effective nuclear forces, let alone with realistic forces. Some interesting attempts to describe clustering have been reported. One of such approaches is the description of ^{12}C by Fukuoka, who superposed a number of Slater determinants consisting of Gaussian wave packets chosen in a stochastic way [25]. Another work of this type is a five-body calculation of $p+p+n+n+{}^{12}$C presented by Horiuchi [26]. He uses correlated Gaussians whose parameters are chosen by a stochastic procedure and their coefficients are determined variationally. He has demonstrated that the experimental energies of both the ground and the first excited 0^+ state of ^{16}O can be reproduced in this model almost perfectly.

4. Clustering in reactions

Nuclei far from stability have been explored since the advent of research with the use of radioactive nuclear beams. Since they are studied through scattering and reactions, sound reaction theories are needed to extract structure information. Especially breakup effects are known to be quite important in scattering and reactions induced by weakly bound nuclei.

Capel has compared various reaction models like continuum-discretized coupled channels, time-dependent approach and eikonal approximations for a projectile of a core plus a nucleon [27]. He has found that the ratio of two angular distributions, breakup and elastic, is insensitive to the reaction mechanism, which may enable one to get structure information.

In a multi-cluster model Katō has presented several results including resonances of multi-particle systems, $\alpha+d$ scattering and Coulomb breakup reactions of ^{6}He, and, furthermore, 3α and 4α 0^+ states [28]. The relevant continuum states have been treated in a unified way with the use of the complex scaling method.

In nuclear reactions at intermediate energies hot nuclear matter is formed and subsequently decays into a number of fragment combinations. This reaction of nuclear multifragmentation has been discussed in two different approaches. Ono has described the fragmentation in the antisymmetrized molecular dynamics on Gaussian wave packets with the Skyrme interaction [29]. With the inclusion of the nucleon-nucleon collision, the multifragmentation data have been reproduced very well. The symmetry energy of the asymmetric nuclear matter has also been discussed. Raduţa has treated the multifragmentation in a statistical model with cluster degrees of freedom included, and discussed the finite-size effects on the liquid-gas phase transition [30].

5. Cluster decay in heavy nuclei and fission dynamics

It is an intriguing question what evidence we have for clustering in heavy nuclei. The enhancement of α-decay rate is a well-known fact. For example, the theoretical understanding of the α decay of ^{212}Po has a long history. A most probable explanation so far has been achieved by admixing an α-cluster configuration explicitly near the ^{208}Pb surface to the shell-model configurations [31]. The recent observation of enhanced $E1$ transitions in ^{212}Po may aslo indicate the importance of α clustering in heavy nuclei [32].

The α decay of ^{212}Po and ^{104}Te has been studied by Id Betan in the framework of a complex-energy shell model [33]. The idea was to take into account the continuum components to improve the tail behaviour of the α-formation amplitude.

In the region of the even larger masses of superheavy nuclei, α decay may compete with heavy-particle radioactivity. The current status of comprehensive and detailed calculations for the competition in the region of Z=104–124 has been reported by Poenaru. He treats radioactivity in a fission-type model [34]. He has stressed that further improved accurate atomic mass models are needed to make a reliable prediction for the half lives.

Nuclear fission has been known for many years but still offers intresting observations. The recent experiment that $^{180}_{80}$Hg shows an asymmetric mass distribution of fission frgaments rather than a symmetric distribution peaked at $^{90}_{40}$Zr has been explained by Antonenko [35] by comparing the potential energy surfaces calculated for different fragment pairs in a scission-point model.

A quantum mechanical description of fission, especially the scission point, has not been realized yet. Schunck talked about an attempt toward the understanding of the dynamics of fission in the density functional theory approach [36]. In particular, he has discussed the fission barrier and the temperature of the fission fragments for the neutron-induced fission of ^{240}Pu.

6. Clustering in nuclear matter

Schuck has first summarized the study of α condensation in light nuclei and then discussed the α condensation or quartetting in infinite nuclear matter as well as the critical temperature for α and deuteron condensation [37].

Usmani has discussed the binding energy per nucleon of nuclear matter in the limit of zero density and its connection with the clustering in the surface region [38].

7. Few-body and *ab initio* approaches

Few-body systems offer unique opportunities to test the interaction between the nucleons and theoretical methods of solving bound and continuum problems etc. Among others, *ab initio* calculations with realistic nuclear forces are appreciated very much because of their high predictive power. It should be recalled that some theoretical tools such as the resonating group method, the generator coordinate method and R-matrix method are tailored to nuclear cluster physics and are frequently employed as standard machinery.

As is well-known, realistic nuclear forces are characterized by strong repulsion at short distances and strong long-ranged tensor components. The latter mixes S and D waves, so that, for example, the α particle contains more than 10% D states [39]. It is particularly interesting and important to clarify whether or not we have deeper understanding of clustering phenomena starting from a realistic interaction.

Calculations with a direct use of the realistic potentials are obviously very hard. Arai has reported such type of calculations on d-induced reactions on d [40]. Compared to a pilot calculation using a tensor-force-free effective potential, he has convincingly shown that the tensor force plays an essential role in reproducing the astrophysical S-factors of reactions, $d(d,\gamma)^4$He, $d(d,p)^3$H and $d(d,n)^3$He, at $E \leq 0.1\,$MeV. This is attributed to the fact that two deuterons initially approaching each other in the S-wave go to the final state dominantly through the D-wave.

Employing a realistic low-momentum interaction generated from the unitary correlation operator method, Neff has reported calculations for cluster states of ^{12}C and for the ^{3}He$(\alpha,\gamma)^7$Be radiative capture reaction [41]. He superposed many basis states of Gaussian wave packets according to fermionic molecular dynamics, and has succeeded in obtaining results that agree very well with the relevant experiment. It is interesting to note, however, that his calculation on ^{3}H$(\alpha,\gamma)^7$Li gives slightly larger S-factor than experiment.

Fujiwara has presented n–d scattering cross sections calculated in a Faddeev formalism [42]. The potential used by him is generated from the quark cluster model. It is nonlocal and reproduces the nucleon-nucleon scattering phase shifts fairly well.

8. Conclusion

It is natural that I could not cover all the theory talks presented at this conference, mainly because some of them are beyond my expertise and partly because many of them were delivered in parallel sessions. I apologize even to those speakers whose talks I have covered for the possible incompleteness of my understanding, interpretation and appreciation of their work.

A number of talks have demonstrated the importance of clusters in both nuclear structure and reactions. I am especially impressed with the talks that have exhibited the vital role of clusters and a clear link between the cluster concept and the nuclear force.

It is true that cluster physics has come a long way since the first cluster conference in 1969. Having listened to the many talks delivered at the conference, however, I realize that there are still many intriguing and unsolved problems that will keep us busy.

If my observation is right, no talks from Debrecen or Hungary were presented at the conference despite the fact that a centre for nuclear science, Atomki, has been producing a number of works relevant to cluster physics. I missed some of them, but hope to remember the pleasant Debrecen atmosphere.

Acknowledgments

I thank the local organizing committee for sending me those manuscripts which were submitted to this Proceedings early enough. I am indebted to the bilateral collaboration project (No. 119), 2011.4–2013.3, between the Japan Society for the Promotion of Science and the Hungarian Academy of Sciences for its support.

References

[1] Ikeda K, Katori K and Suzuki Y (eds) 1989 Proc. Fifth Int. Conf. on Clustering Aspects in Nuclear and Subnuclear Systems, Kyoto, 1988 *J. Phys. Soc. Jpn. Suppl.* **58**
[2] Horiuchi H 2012 Debrecen conference
[3] Ikeda K, Takigawa N and Horiuchi H 1968 *Prog. Theor. Phys. Suppl.* **Extra Number** 464
[4] Funaki Y *et al* 2012 Debrecen conference
[5] Ito M 2012 Debrecen conference
[6] Varga K, Suzuki Y and Lovas R G 2002 *Phys. Rev.* **C 66** 041302(R)
[7] Dufour M 2012 Debrecen conference
[8] Suzuki Y, Matsumura H, Orabi M, Fujiwara Y, Descouvemont P, Theeten M and Baye D 2008 *Phys. Lett.* B **659** 160
[9] Hess P O 2012 Debrecen conference
[10] Hiyama E 2012 Debrecen conference
[11] Bromley D A, Kuehner J A and Almqvist E 1960 *Phys. Rev. Lett.* **4** 515
[12] Uegaki E 2012 Debrecen conference
[13] Adamian G G 2012 Debrecen conference
[14] Taniguchi Y 2012 Debrecen conference
[15] Descouvemont P 2012 Debrecen conference
[16] Kimura M 2012 Debrecen conference
[17] Takechi M *et al* 2010 *Mod. Phys. Lett.* **A 25** 1878
[18] Kanada-En'yo Y 2012 Debrecen conference
[19] Suzuki Y 1976 *Prog. Theor. Phys.* **55** 1751; *Prog. Theor. Phys.* **56** 111
[20] Yamada T 2012 Debrecen conference
[21] Rosensteel G and Rowe D J 1977 *Phys. Rev. Lett.* **38** 10
[22] Rowe D J 1985 *Rep. Prog. Phys.* **48** 1419
[23] Suzuki Y 1989 *J. Phys. Soc. Jpn. Suppl.* **58** 129; Suzuki Y and Hara S 1989 *Phys. Rev.* C **39** 658
[24] Draayer J P 2012 Debrecen conference
[25] Fukuoka Y 2012 Debrecen conference
[26] Horiuchi W 2012 Debrecen conference
[27] Capel P 2012 Debrecen conference
[28] Katō K 2012 Debrecen conference
[29] Ono A 2012 Debrecen conference
[30] Raduţa A R 2012 Debrecen conference
[31] Varga K, Lovas R G and Liotta R J 1992 *Nucl. Phys.* **A 550** 421
[32] Astier A, Petkov P, Porquet M G, Delin D S and Schuck P 2010 *Phys. Rev. Lett.* **104** 042701
[33] Id Betan R 2012 Debrecen conference
[34] Poenaru D N 2012 Debrecen conference
[35] Antonenko N V 2012 Debrecen conference
[36] Schunck N 2012 Debrecen conference
[37] Schuck P 2012 Debrecen conference
[38] Usmani Q N 2012 Debrecen conference
[39] Kamada H *et al* 2001 *Phys. Rev.* **C 64** 044001
[40] Arai K 2012 Debrecen conference
[41] Neff T 2012 Debrecen conference
[42] Fujiwara Y 2012 Debrecen conference

List of participants

Abe Yasuhisa <abey@rcnp.osaka-u.ac.jp>
Reseach Center for Nuclear Physics, Osaka University, Ibaraki, Osaka 567-0047, Japan

Adamian, Gurgen G <adamian@theor.jinr.ru>
Joint Institute for Nuclear Research, 141980, Dubna, Russia

Akimune Hidetoshi <akimune@konan-u.ac.jp>
Department of Physics, Konan University, Kobe 658-8501-1, Japan

Angeli István <angeli@tigris.unideb.hu>
Department of Experimental Physics, University of Debrecen, Debrecen, Bem tér 18/A, H-4026, Hungary

Antonenko, Nikolai V <antonenk@theor.jinr.ru>
Joint Institute for Nuclear Research, 141980, Dubna, Russia

Aoyama Shigeyoshi <aoyama@cc.niigata-u.ac.jp>
Center for Academic Information Service, Niigata University, Niigata 950-2181, Japan

Arai Koji <arai@nagaoka-ct.ac.jp>
Division of General Education, Nagaoka National College of Technology, 888 Nishikatakai, Nagaoka, Niigata, 940-8532, Japan

Baye, Daniel <dbaye@ulb.ac.be>
Physique Nucléaire Théorique et Physique Mathématique, C.P. 229, Université Libre de Bruxelles (ULB), B-1050 Brussels, Belgium

Beck, Christian <christian.beck@iphc.cnrs.fr>
Département de Recherches Subatomiques, Institut Pluridisciplinaire Hubert Curien, IN2P3-CNRS et Université de Strasbourg - 23, rue du Loess BP 28, F-67037 Strasbourg Cedex 2, France

Capel, Pierre <pierre.capel@ulb.ac.be>
Physique Nucléaire et Physique Quantique (CP 229), Université Libre de Bruxelles (ULB), 50 av. F. D. Roosevelt, B-1050 Brussels, Belgium

Catford, Wilton N. <w.catford@surrey.ac.uk>
Department of Physics, University of Surrey, Guildford, GU2 7XH, Surrey, 1

Courtin, Sandrine <sandrine.courtin@iphc.cnrs.fr>
Institut Pluridisciplinaire Hubert Curien, UMR 7178, Université de Strasbourg / CNRS-IN2P3. 23, rue du Loess, 67037 Strasbourg, France

Cowley, Anthony A <aac@sun.ac.za>
Department of Physics, Stellenbosch University, Private Bag X1, Matieland 7602, South Africa;
iThemba Laboratory for Accelerator Based Sciences, P O Box 722, Somerset West 7129, South Africa

Csatlós Margit <margo@atomki.hu>
MTA Institute of Nuclear Research, Debrecen, PO Box 51, H–4001, Hungary

Cseh József <cseh@atomki.hu>
MTA Institute of Nuclear Research, Debrecen, PO Box 51, H–4001, Hungary

Darai Judit <darai@atomki.hu>
Department of Experimental Physics, University of Debrecen, Debrecen, Bem tér 18/A, H-4026, Hungary

de la Mota, Virginia <delamota@subatech.in2p3.fr>
SUBATECH, Ecole des Mines, Université de Nantes, CNRS/IN2P3, Nantes, France

Descouvemont, Pierre <pdesc@ulb.ac.be>

Journal of Physics: Conference Series **436** (2013) 011005 doi:10.1088/1742-6596/436/1/011005

Physique Nucléaire Théorique et Physique Mathématique, C.P. 229, Université Libre de Bruxelles (ULB), B-1050 Brussels, Belgium

Dohet-Eraly, Jérémy <jdoheter@ulb.ac.be>
Physique Nucléaire et Physique Quantique, Université Libre de Bruxelles (ULB), CP229, B-1050 Brussels, Belgium

Dombrádi Zsolt <domb@atomki.hu>
MTA Institute of Nuclear Research, Debrecen, PO Box 51, H–4001, Hungary

Draayer, Jerry P <draayer@sura.org>
Department of Physics and Astronomy, Louisiana State University, Baton Rouge, Louisiana 70803, USA

Dufour, Marianne <Marianne.Dufour@IReS.in2p3.fr>
IPHC Bat27, IN2P3-CNRS / Université de Strasbourg BP28, F-67037 Strasbourg Cedex 2 - France

Düllmann, Christoph E <c.e.duellmann@gsi.de>
Johannes Gutenberg University Mainz, Germany;
GSI Helmholtz Center for Heavy Ion Research, Darmstadt, Germany;
Helmholtz Institute Mainz, Germany

Eren, Nuretdin <eren@selcuk.edu.tr>
Department of Physics, University of Selçuk, TR-42075 Konya, Turkey

Fényes Tibor
MTA Institute of Nuclear Research, Debrecen, PO Box 51, H–4001, Hungary

Freer, Martin <M.Freer@bham.ac.uk>
School of Physics and Astronomy, University of Birmingham, Birmingham, B15 2TT, UK

Fujiwara Yoshikazu <yfujiwar@scphys.kyoto-u.ac.jp>
Department of Physics, Kyoto University, Kyoto 606-8502, Japan

Fukuoka Yuta <fukuoka@nucl.ph.tsukuba.ac.jp>
Graduate School of Pure and Applied Sciences, University of Tsukuba, Tsukuba 305-8571, Japan

Fülöp Zsolt <fulop@atomki.hu>
MTA Institute of Nuclear Research, Debrecen, PO Box 51, H–4001, Hungary

Funaki Yasuro <funaki@riken.jp>
Nishina Center, RIKEN, Wako 351-0198, Japan

Gai, Moshe <moshe.gai@yale.edu>
LNS at Avery Point, University of Connecticut, Groton, CT 06340-6097, USA;
WNSL, Room 100, Yale University, New Haven, CT 06520-8124, USA

Georgieva, Ana I <anageorg@inrne.bas.bg>
Institute of Nuclear Research and Nuclear Energy, Bulgarian Academy of Sciences, 72 bul. Tzarigradsko Chaussee, Sofia 1784, Bulgaria

Giovinazzo, Jérôme <giovinaz@cenbg.in2p3.fr>
Centre d'Etudes Nucléaires de Bordeaux-Gradignan, CNRS/IN2P3-Univ. Bordeaux 1, Le Haut Vigneau BP 120, 33175 Gradignan, France

Gnesi, Ivan <gnesi@to.infn.it>
Dipartimento di Fisica Generale “A. Avogadro”, Università di Torino, via Pietro Giuria 1, 10125 Torino, Italy (present);
INFN, Sezione di Torino, Italy;
Centro Studi e Ricerche “Enrico Fermi”, Roma, Italy

Gruyer, Diego <diego.gruyer@ganil.fr>
GANIL, CEA-DSM/CNRS-IN2P3, Bvd. Henri Becquerel, F-14076 Caen Cedex, France

Gyürky György <gyurky@atomki.hu>
MTA Institute of Nuclear Research, Debrecen, PO Box 51, H–4001, Hungary

Halász Zoltán <fiser@atomki.hu>
MTA Institute of Nuclear Research, Debrecen, PO Box 51, H–4001, Hungary

Hess, Peter Otto <hess@nucleares.unam.mx>
Instituto de Ciencias Nucleares, UNAM, Circuito Exterior, C.U., A.P. 70-543, 04510 México, D.F., Mexico;
Frankfurt Institute for Advanced Studies (FIAS), J. W. von Goethe Universität, Ruth-Moufang-Str. 1,

60438 Frankfurt am Main, Germany;
GSI für Schwerionenforschung Helmholzzentrum GmbH, Max-Planck-Str. 1 64291 Darmstadt, Germany

Hiyama Emiko <hiyama@riken.jp>
Nishina Center for Accelerator-Based Science, The Institute of Physical and Chemical Research (RIKEN), Wako 351-0198, Saitama, Japan

Horiuchi Hisashi <horiuchi@rcnp.osaka-u.ac.jp>
Research Center for Nuclear Physics, Osaka University, Ibaraki, Osaka 567-0047;
International Institute for Advanced Studies, Kizugawa, Kyoto 619-0225, Japan

Horiuchi Wataru <whoriuchi@nucl.sci.hokudai.ac.jp>
Department of Physics, Hokkaido University, Sapporo 060-0810, Japan

Hornyák István <ihornyak@atomki.hu>
MTA Institute of Nuclear Research, Debrecen, PO Box 51, H–4001, Hungary

Hussein, Mahir S <hussein@if.usp.br>
Instituto de Estudos Avançados and Instituto de Física, Universidade de São Paulo, CP 66318, cep 05314-970, São Paulo, S.P., Brazil

Id Betan, Rodolfo <idbetan@ifir-conicet.gov.ar>
Department of Physics and Chemistry (FCEIA-UNR), Physics Institute of Rosario (CONICET), Av. Pellegrini 250, 2000 Rosario, Argentina;
Department of Physics and Astronomy, University of Tennessee, Knoxville, Tennessee 37996, USA;
Physics Division, Oak Ridge National Laboratory, P.O. Box 2008, Oak Ridge, Tennessee 37831, USA

Ikeda Kiyomi <k-ikeda@riken.jp>
Nishina Center for Accelerator-Based Sciences, The Institute of Physical and Chemical Research (RIKEN), Wako 351-0198, Japan

Isaka Masahiro <isaka@nucl.sci.hokudai.ac.jp>
Department of Physics, Hokkaido University, Sapporo 060-0810, Japan

Itoh Masatoshi <itoh@cyric.tohoku.ac.jp>
Cyclotron and Radioisotope Center (CYRIC), Tohoku University, Tohoku, Japan

Ito Makoto <itomk@kansai-u.ac.jp>
Department of Pure and Applied Physics, Kansai University, Yamate-cho 3-3-35, Suita, Osaka 564-8680, Japan;
Nishina Center for Accelerator-Based Sciences, RIKEN, Wako, 351-0198, Japan

Iwata Yoritaka <y.iwata@gsi.de>
GSI Helmholtzzentrum für Schwerionenforschung, Darmstadt, Germany

Jenkins, David <david.jenkins@york.ac.uk>
Department of Physics, University of York, York YO10 5DD, UK

Jull, Anthony John Timothy <jull@email.arizona.edu>
Department of Geosciences, University of Arizona, Tucson, Arizona 85721 USA;
MTA Institute of Nuclear Research and University of Debrecen, 4026 Debrecen, Hungary

Kanada-En'yo Yoshiko <yenyo@ruby.scphys.kyoto-u.ac.jp>
Department of Physics, Kyoto University, Kyoto 606-8502, Japan

Kasagi Jirohta <kasagi@lns.tohoku.ac.jp>
Research Center for Electron Photon Science, Tohoku University, Sendai, Japan

Katō Kiyoshi <kato@nucl.sci.hokudai.ac.jp>
Nuclear Reaction Data Centre, Faculty of Science, Hokkaido University, Sapporo 060-0810, Japan

Kawabata Takahiro <kawabata@scphys.kyoto-u.ac.jp>
Department of Physics, Kyoto University, Kyoto, Japan

Kikuchi Yuma <yuma@rcnp.osaka-u.ac.jp>
Research Center for Nuclear Physics (RCNP), Osaka University, Ibaraki 567-0047, Japan

Kimura Masaaki <masaaki@nucl.sci.hokudai.ac.jp>
Department of Physics, Hokkaido University, Sapporo, 011-0021, Japan

Kirsebom, Oliver Sølund <oliverk@triumf.ca>
TRIUMF, 4004 Wesbrook Mall, Vancouver BC, V6T 2A3, Canada

Kiss Gábor Gyula <ggkiss@atomki.hu>
MTA Institute of Nuclear Research, Debrecen, PO Box 51, H–4001, Hungary

Kobayashi Fumiharu <f-kobayashi@ruby.scphys.kyoto-u.ac.jp>
Kyoto University, Kyoto, 606-8502, Japan

Kobayashi Nobuyuki <kobayashi.n.aa@m.titech.ac.jp>
Department of Physics, Tokyo Institute of Technology, 2-12-1 O-Okayama, Meguro, Tokyo 152-8551, Japan

Krasznahorkay Attila <kraszna@atomki.hu>
MTA Institute of Nuclear Research, Debrecen, PO Box 51, H–4001, Hungary

Kruppa András Tibor <atk@atomki.hu>
MTA Institute of Nuclear Research, Debrecen, PO Box 51, H–4001, Hungary

Kubono Shigeru <kubono@riken.jp>
RIKEN Nishina Center, 2-1 Hirosawa, Wako, Saitama 351-0198, Japan;
Center for Nuclear Study, University of Tokyo, 2-1 Hirosawa, Wako, Saitama 351-0198, Japan;
Institute of Modern Physics, Chinese Academy of Sciences, Lanzhou, China

Kurihara Nozomi <kurihara@nucl.sci.hokudai.ac.jp>
Division of Physics, Graduate School of Science, Hokkaido University, Sapporo 060-0810, Japan

Lamia, Livio <llamia@lns.infn.it> Dipartimento di Fisica e Astronomia, Università degli Studi di Catania, Italy

Lépine-Szily Alinka <alinka@if.usp.br>
Instituto de Física, Universidade de São Paulo, CP 66318, 05389-970, São Paulo, Brazil

Lévai Géza <levai@atomki.hu>
MTA Institute of Nuclear Research, Debrecen, PO Box 51, H–4001, Hungary

Lombardo, Ivano <ivlombardo@na.infn.it>
Dipartimento di Scienze Fisiche, Università degli Studi di Napoli "Federico II" and INFN – Sezione di Napoli, Via Cinthia, 80126, Napoli, Italy

Lovas Rezső György <rgl@atomki.hu>
MTA Institute of Nuclear Research, Debrecen, PO Box 51, H–4001, Hungary

Masui Hiroshi <hgmasui@mail.kitami-it.ac.jp>
Information Processing Center, Kitami Institute of Technology, 090-8507 Kitami, Japan

Máté Zoltán <mate@atomki.hu>
MTA Institute of Nuclear Research, Debrecen, PO Box 51, H–4001, Hungary

Milin, Matko <matko.milin@phy.hr>
Physics Department, Faculty of Science, University of Zagreb, Zagreb, Croatia

Myo Takayuki <myo@ge.oit.ac.jp>
General Education, Faculty of Engineering, Osaka Institute of Technology, Osaka 535-8585, Japan

Neff, Thomas <t.neff@gsi.de>
GSI Helmholtzzentrum für Schwerionenforschung GmbH, Planckstraße 1, 64291 Darmstadt, Germany

Norrby, Markus <mnorrby@abo.fi>
Åbo Akademi University, Teacher Training School, Skolhusgatan 31, 65100 Vasa, Finland

Oğul, Rıza <rogul@selcuk.edu.tr>
Department of Physics, Selçuk University, TR-42079 Kampus, Konya, Turkey

Ono Akira <ono@nucl.phys.tohoku.ac.jp>
Department of Physics, Tohoku University, Sendai 980-8578, Japan

Pál Károly Ferenc <kfpal@atomki.hu>
MTA Institute of Nuclear Research, Debrecen, PO Box 51, H–4001, Hungary

Pinilla Beltràn, Edna Carolina <epinilla@ulb.ac.be>
Physique Nucléaire Théorique et Physique Mathématique, C.P. 229, Université Libre de Bruxelles (ULB), B 1050 Brussels, Belgium

Poenaru, Dorin N <D.Poenaru@fias.uni-frankfurt.de>
Horia Hulubei National Institute for Physics and Nuclear Engineering, PO Box MG-6, RO-077125 Bucharest-Măgurele, Romania;
Frankfurt Institute for Advanced Studies, Ruth-Moufang-Str. 1, 60438 Frankfurt-am-Main, Germany

Journal of Physics: Conference Series **436** (2013) 011005 doi:10.1088/1742-6596/436/1/011005

Prepolec, Lovro <lovro@lnr.irb.hr>
Ruđer Bošković Institute, Bijenička 54, HR-10000 Zagreb, Croatia

Raduţa, Adriana R <araduta@nipne.ro>
National Institute for Nuclear Physics and Engineering (IFIN-HH), RO-077125 Bucharest, Romania

Rogachev, Grigory <grogache@fsu.edu>
Department of Physics, Florida State University, Tallahassee, FL, USA;
National Superconducting Cyclotron Laboratory, Michigan State University, MI, USA

Röpke, Gerd <gerd.roepke@uni-rostock.de>
University of Rostock, Institut für Physik, 18051 Rostock, Germany

Salamon Péter <salamon@atomki.hu>
MTA Institute of Nuclear Research, Debrecen, PO Box 51, H–4001, Hungary

Schuck, Peter <schuck@ipno.in2p3.fr>
Institut de Physique Nucléaire, CNRS, UMR8608, Orsay, F-91406, France;
Université Paris-Sud, Orsay, F-91505, France;
Laboratoire de Physique et de Modélisation des Milieux Condensés, CNRS and Université Joseph Fourier, 25 Av. des Martyrs, BP 166, F-38042 Grenoble Cedex 9, France

Schunck, Nicolas <schunck1@llnl.gov>
Physics Division, Lawrence Livermore National Laboratory, Livermore, CA 94551, USA

Shimoyama Hirotaka <shimoyama@nt.sc.niigata-u.ac.jp>
Graduate School of Science and Technology, Niigata University, Niigata 950-2181, Japan

Simpson, Edward <e.simpson@surrey.ac.uk>
Department of Physics, Faculty of Engineering and Physical Sciences, University of Surrey, Guildford, Surrey, GU2 7XH, UK

Sohler Dóra <sohler@atomki.hu>
MTA Institute of Nuclear Research, Debrecen, PO Box 51, H–4001, Hungary

Somorjai Endre <somorjai@atomki.hu>
MTA Institute of Nuclear Research, Debrecen, PO Box 51, H–4001, Hungary

Suhara Tadahiro <suhara@nucl.ph.tsukuba.ac.jp>
Institute of Physics, University of Tsukuba, Tsukuba 305-8571, Japan

Suzuki Yasuyuki <suzuki@nt.sc.niigata-u.ac.jp>
Physics Department, Niigata University, Niigata 950-2181, Japan;
RIKEN Nishina Center, Wako 351-0198, Japan

Szűcs Tamás <tszucs@atomki.hu>
MTA Institute of Nuclear Research, Debrecen, PO Box 51, H–4001, Hungary

Tamura Hirokazu <tamura@lambda.phys.tohoku.ac.jp>
Department of Physics, Tohoku University, Sendai 908-8578, Japan

Taniguchi Yasutaka <yasutaka@ccs.tsukuba.ac.jp>
Center for Computational Sciences, University of Tsukuba, 1-1-1 Tennodai, Tsukuba-city, Ibaraki 305-8577, Japan

Tchuvil'sky, Yury M <tchuvl@nucl-th.sinp.msu.ru>
Skobeltsyn Institute of Nuclear Physics, Lomonosov Moscow State University, 119991, Vorob'evy Gory, Moscow, Russia

Tímár János <timar@atomki.hu>
MTA Institute of Nuclear Research, Debrecen, PO Box 51, H–4001, Hungary

Tohsaki Akihiro <tohsaki@rcnp.osaka-u.ac.jp>
Research Center for Nuclear Physics, Osaka University, Osaka 567-0047, Japan

Tokić, Vedrana <vtokic@irb.hr>
Ruđer Bošković Institute, Bijenička 54, HR-10000 Zagreb, Croatia

Tumino, Aurora <tumino@lns.infn.it>
Laboratori Nazionali del Sud - INFN, Catania, Italy;
Università degli Studi di Enna "Kore", Enna, Italy

Uegaki Eiji <uegaki@phys.akita-u.ac.jp>
Graduate School of Engineering and Resource Science, Akita University, Akita 010-8502, Japan

Uroić, Milivoj <milivoj.uroic@irb.hr>
Ruđer Bošković Institute, Bijenička 54, HR-10000 Zagreb, Croatia

Usmani, Qamar Nasir <qamar@unimap.edu.my>
Institute of Engineering Mathematics, University Malaysia Perlis, Malaysia;
School of Microelectronics Engineering, University Malaysia Perlis, Malaysia

Vardaci, Emanuele <Emanuele.Vardaci@unina.it>
Dipartimento di Scienze Fisiche, Università di Napoli Federico II and Istituto Nazionale di Fisica Nucleare, 80126 Napoli, Italy

Varga Kálmán <kalman.varga@vanderbilt.edu>
Department of Physics and Astronomy, Vanderbilt University, TN-37240, Nashville, USA

Vertse Tamás <vertse@atomki.hu>
MTA Institute of Nuclear Research, Debrecen, PO Box 51, H–4001, Hungary

Wheldon, Carl <c.wheldon@bham.ac.uk>
School of Physics and Astronomy, University of Birmingham, Edgbaston, Birmingham B15 2TT, UK

Wisniewski, Krzysztof <kwisnia@physi.uni-heidelberg.de>
University of Warsaw, Institute for Experimental Physics, L-00681 Warsaw, Poland

Xu Furong <frxu@pku.edu.cn>
State Key Laboratory of Nuclear Physics and Technology, School of Physics, Peking University, Beijing 100871, China

Yamada Taiichi <yamada@kanto-gakuin.ac.jp>
Laboratory of Physics, Kanto Gakuin University, Yokohama 236-8501, Japan

Ye Yanlin <yeyl@pku.edu.cn>
School of Physics and State Key Laboratory of Nuclear Physics and Technology, Peking University, Beijing, 100871 China

Yoshida Toru <yoshida@cns.s.u-tokyo.ac.jp>
Center for Nuclear Study, University of Tokyo, 7-3-1 Hongo, Bunkyo, Tokyo 113-0033, Japan

Zubov, Andrey S <azubov@theor.jinr.ru>
Joint Institute for Nuclear Research, 141980 Dubna, Russia